The Human Mitochondrial Genome

The Human Mitochondrial Genome

From Basic Biology to Disease

Edited by

Giuseppe Gasparre

Department of Medical and Surgical Sciences (DIMEC), Unit of Medical Genetics, University of Bologna, Bologna, Italy

Center for Applied Biomedical Research (CRBA), University of Bologna, Bologna, Italy

Anna Maria Porcelli

Department of Pharmacy and Biotechnology (FABIT), University of Bologna, Bologna, Italy

Interdepartmental Center for Industrial Research Life Sciences and Technologies for Health, University of Bologna, Bologna, Italy

Academic Press is an imprint of Elsevier
125 London Wall, London EC2Y 5AS, United Kingdom
525 B Street, Suite 1650, San Diego, CA 92101, United States
50 Hampshire Street, 5th Floor, Cambridge, MA 02139, United States
The Boulevard, Langford Lane, Kidlington, Oxford OX5 1GB, United Kingdom

British Library Cataloguing-in-Publication Data
A catalogue record for this book is available from the British Library

Library of Congress Cataloging-in-Publication Data
A catalog record for this book is available from the Library of Congress

ISBN: 978-0-12-819656-4

For Information on all Academic Press publications
visit our website at https://www.elsevier.com/books-and-journals

Publisher: Andre Gerhad Wolff
Acquisitions Editor: Peter B. Linsley
Editorial Project Manager: Billie Jean Fernandez
Production Project Manager: Sreejith Viswanathan
Cover Designer: Matthew Limbert

Typeset by MPS Limited, Chennai, India

Dedication

First of all, thanks to my friend and co-editor Anna Maria, for this umpteenth joint venture and for the many more to come.

I would like to dedicate this work to my two scientific mentors and life-coaches, Marcella Attimonelli and Giovanni Romeo, who first introduced me to the DNA molecule that has since decided the course of my research. Thank you for taking my hand and guiding me to be the professional I am today, you will always have my respect and deep affection.

Giuseppe Gasparre

I would like to dedicate this work to my family and in particular to my father and my husband. Thanks to my unique friend and colleague Giuseppe Gasparre for sharing with me these last 15 years of the mitochondrial life with the wish that this is only the beginning.

Anna Maria Porcelli

Contents

Dedication v
List of Contributors xvii
Editor's biographies xxi
Preface xxiii
Acknowledgments xxv

Part 1 Biology of human mtDNA 1

Chapter 1: mtDNA replication, maintenance, and nucleoid organization 3

Mara Doimo, Annika Pfeiffer, Paulina H. Wanrooij and Sjoerd Wanrooij

1.1 Human mitochondrial DNA 3
1.2 The process of mtDNA replication 5
1.3 The mitochondrial dNTP supply 12
1.4 Mitochondrial nucleoids 15
Research perspectives 21
Acknowledgments 22
References 22

Chapter 2: Human mitochondrial transcription and translation 35

Flavia Fontanesi, Marco Tigano, Yi Fu, Agnel Sfeir and Antoni Barrientos

2.1 Introduction 35
2.2 Coordination of mitochondrial DNA replication and transcription 36
2.3 Mitochondrial transcription and mitochondrial RNA transactions 40
2.4 Mitochondrial translation 50
2.5 Compartmentalization of gene expression 58
Research perspectives 61
References 62

Chapter 3: Epigenetic features of mitochondrial DNA 71

Takehiro Yasukawa, Shigeru Matsuda and Dongchon Kang

3.1 A brief overview of mitochondrial DNA 71
3.2 Does cytosine methylation occur in mtDNA? 72
3.3 Bisulfite sequencing analysis of mtDNA 76
3.4 Estimation of mtDNA methylation with McrBC endonuclease 78
3.5 Investigation of 5mC in mtDNA by nucleoside liquid chromatography/mass spectrometry 79
3.6 Epigenetic features of mammalian mtDNA 81
Research perspectives 83
Acknowledgments 83
References 83

Chapter 4: Heredity and segregation of mtDNA 87

Stephen P. Burr and Patrick F. Chinnery

4.1 Introduction 87
4.2 General principles of mtDNA segregation 87
4.3 Uniparental maternal inheritance of mitochondrial DNA 90
4.4 Paternal leakage during mtDNA inheritance 92
4.5 mtDNA mutations—homoplasmy versus heteroplasmy 93
4.6 Germline segregation of mtDNA mutations and the genetic bottleneck 94
4.7 Purifying selection against mtDNA mutations in the germline 97
4.8 Somatic mtDNA mutations and clonal expansion 100
4.9 Conclusions 102
Research perspectives 103
References 103

Part 2 mtDNA evolution and exploitation 109

Chapter 5: Haplogroups and the history of human evolution through mtDNA 111

Antonio Torroni, Alessandro Achilli, Anna Olivieri and Ornella Semino

5.1 Early restriction fragment length polymorphism studies 112
5.2 The advent of polymerase chain reaction in the mtDNA world 113
5.3 Haplogroup nomenclature of human mtDNA 114
5.4 The survey of entire mitogenomes 115
5.5 The “Out of Africa Exit” 118

5.6 The first peopling of the Americas ... 120
5.7 The peopling of an island in the Mediterranean Sea ... 121
Research perspectives ... 123
References ... 123

Chapter 6: Human nuclear mitochondrial sequences (NumtS) ... 131

Marcella Attimonelli and Francesco Maria Calabrese

6.1 NumtS definition and introduction ... 131
6.2 NumtS discovery ... 131
6.3 NumtS detection ... 132
6.4 Numtogenesis: Mechanisms of NumtS insertion ... 136
6.5 NumtS variability and polymorphisms ... 137
6.6 The role of NumtS in mtDNA sequencing and disease ... 140
6.7 NumtS annotation: Current and future roles of NumtS ... 140
Research perspectives ... 141
References ... 141

Chapter 7: mtDNA exploitation in forensics ... 145

Adriano Tagliabracci and Chiara Turchi

7.1 Introduction ... 145
7.2 mtDNA typing in historical forensic identification ... 146
7.3 mtDNA sequencing in forensic practice ... 149
7.4 Data analysis, alignment, and haplotype notation ... 153
7.5 Interpretation of mtDNA results ... 157
7.6 Mitochondrial DNA population databases used in forensics ... 161
7.7 Guidelines and recommendations ... 163
Research perspectives ... 165
References ... 165

Part 3 mtDNA mutations ... 171

Chapter 8: Human mitochondrial DNA repair ... 173

Elaine Ayres Sia and Alexis Stein

8.1 Base excision repair ... 175
8.2 Repair of bulky lesions ... 178
8.3 Double-strand break repair ... 180

8.4 Mismatch repair ... 184
8.5 Translesion synthesis ... 185
8.6 Concluding remarks ... 186
Research perspectives ... 187
References ... 188

Chapter 9: Mechanisms of onset and accumulation of mtDNA mutations ... 195

Ian James Holt and Antonella Spinazzola

9.1 Mitochondrial DNA abnormalities ... 195
9.2 Criteria to designate a primary mtDNA mutation as pathological ... 200
9.3 Clinical and biochemical correlates ... 201
9.4 Mitochondrial genetic rules ... 201
9.5 Selection and counterselection of deleterious mtDNA variants ... 202
9.6 Genetic drift ... 206
9.7 Mitochondrial DNA selection—more or less? ... 207
9.8 Stable heteroplasmy—the persistence of a fixed proportion of mutant and wild-type mtDNA ... 208
9.9 Mitochondrial DNA maintenance disorders ... 208
9.10 Ribonucleotide incorporation—a new mtDNA abnormality and a potential precursor or mitigator of mtDNA deletions and depletion ... 209
9.11 Overlaps between nuclear defects in the mtDNA maintenance system and primary mtDNA mutants ... 210
9.12 A mitochondrial DNA network and its implications for heteroplasmy ... 210
Research perspectives ... 213
Acknowledgments ... 213
References ... 214

Chapter 10: Mitochondrial DNA mutations and aging ... 221

Karolina Szczepanowska and Aleksandra Trifunovic

10.1 Introduction ... 221
10.2 Old and new mitochondrial theories of aging—how changes in mtDNA contribute to aging? ... 221
10.3 Mitochondrial genetics from the perspective of aging ... 223
10.4 mtDNA deletions and aging ... 224
10.5 mtDNA point mutations ... 227
10.6 How do somatic mtDNA mutations lead to aging? ... 233
10.7 mtDNA mutations and aging of stem cells ... 235

10.8 Conclusions 236
Acknowledgments 236
Research perspectives 237
References 237

Chapter 11: Methods for the identification of mitochondrial DNA variants 243

Claudia Calabrese, Aurora Gomez-Duran, Aurelio Reyes and Marcella Attimonelli

11.1 Introduction to human mtDNA variants detection 243
11.2 Techniques for detecting mitochondrial variants 244
11.3 Challenges in mitochondrial variant studies 257
11.4 Bioinformatics strategies to detect mitochondrial variants and heteroplasmy 259
Research perspectives 265
References 266

Chapter 12: Bioinformatics resources, databases, and tools for human mtDNA.... 277

Marcella Attimonelli, Roberto Preste, Ornella Vitale, Marie T. Lott, Vincent Procaccio, Zhang Shiping and Douglas C. Wallace

12.1 Introduction to human mtDNA variability 277
12.2 Human mtDNA genomes and variants 279
12.3 The Human MitoCompendium: HmtDB, HmtVar, and HmtPhenome 284
12.4 MSeqDR—Mitochondrial Disease Sequence Data Resource (MSeqDR) Consortium (Lott M) 291
12.5 Other specialized human mitochondrial databases 293
12.6 Tools for variant annotations 293
12.7 Nuclear encoded mitochondrial genes databases 295
Research perspectives 297
References 297
Further reading 303

Chapter 13: Methods and models for functional studies on mtDNA mutations 305

Luisa Iommarini, Anna Ghelli and Francisca Diaz

13.1 Introduction 305
13.2 Models for the study of mtDNA mutations: *in vitro* models 306
13.3 Animal models 311

13.4 Methods for assessment of functional defects induced by mtDNA alterations 321
Research perspectives 339
References 340

Part 4 mtDNA-determined diseases and therapies 351

Chapter 14: Mitochondrial DNA-related diseases associated with single large-scale deletions and point mutations 353

Robert D.S. Pitceathly and Shamima Rahman

14.1 Clinical syndromes of mitochondrial DNA-related diseases associated with single large-scale deletions and point mutations 353
14.2 Molecular genetics of mitochondrial DNA single large-scale deletions and point mutations 364
14.3 Diagnostic approach to mitochondrial DNA-related diseases associated with single large-scale deletions and point mutations 365
14.4 Management of mitochondrial DNA-related diseases associated with single large-scale deletions and point mutations 369
Research perspectives 370
Acknowledgments 371
References 371

Chapter 15: Nuclear genetic disorders of mitochondrial DNA gene expression 375

Ruth I.C. Glasgow, Albert Z. Lim, Thomas J. Nicholls, Robert McFarland, Robert W. Taylor and Monika Oláhová

15.1 Introduction 375
15.2 Mechanisms of mtDNA replication 376
15.3 Defects of mtDNA replication 379
15.4 Maintenance of dNTP pool 381
15.5 Defects of the dNTP salvage pathway and nucleotide metabolism 382
15.6 Mechanism of mitochondrial transcription 383
15.7 Defects of mitochondrial transcription 384
15.8 Transcript processing 385
15.9 Defects of maturation of pre mt-RNA 386
15.10 mt-mRNA maturation and turnover 387
15.11 Defects of mt-mRNA maturation and turnover 388
15.12 mt-tRNA maturation 389

15.13 Defects of mt-tRNA maturation and modification ... 389
15.14 mt-rRNA maturation ... 392
15.15 Defects of mt-rRNA maturation, modification, and stability ... 392
15.16 Mechanism of mitochondrial translation ... 393
15.17 Mutations in mitoribosomal proteins ... 394
15.18 Defects of translation initiation ... 395
15.19 Defects of translation elongation ... 396
15.20 Defects of translation termination and mitoribosome recycling ... 396
15.21 Defects of translational activation and coupling ... 397
15.22 IMM insertion of mtDNA-encoded OXPHOS proteins ... 397
Research perspectives ... 398
Acknowledgments ... 398
References ... 399

Chapter 16: mtDNA maintenance: disease and therapy ... 411

Corinne Quadalti and Caterina Garone

16.1 Introduction ... 411
16.2 Defects in mtDNA replisome ... 412
16.3 Defects in mitochondrial nucleotides pool balance ... 418
16.4 Defects in mitochondrial dynamics ... 420
16.5 Defect in nucleoid proteins ... 421
16.6 Experimental therapies ... 422
16.7 General pharmacological approaches ... 422
16.8 Disease-tailored therapies ... 427
Research perspectives ... 435
Acknowledgments ... 436
References ... 436

Chapter 17: mtDNA mutations in cancer ... 443

Giulia Girolimetti, Monica De Luise, Anna Maria Porcelli, Giuseppe Gasparre and Ivana Kurelac

17.1 The landscape of mtDNA mutations in cancer ... 443
17.2 Functional effects of mtDNA mutations in solid cancers ... 447
17.3 The fate of severely pathogenic mtDNA mutations in progressing solid tumors ... 452
17.4 Clinical potential of cancer-associated mtDNA mutations ... 457
17.5 Insights from next generation sequencing and bioinformatics approaches ... 465

Research perspectives ... 468
References ... 469

Chapter 18: MitoTALENs for mtDNA editing ... 481

Sandra R. Bacman and Carlos T. Moraes

18.1 Introduction ... 481
18.2 The use of specific endonucleases to target mtDNA ... 481
18.3 The use of mitoTALENs to target mtDNA ... 484
18.4 Structure of mitoTALENs ... 485
18.5 MitoTALENs targeting mutations in cybrids ... 487
18.6 MitoTALENs in a heteroplasmic mouse model carrying a $tRNA^{Ala}$ mutation ... 489
18.7 MitoTALENs and induced pluripotent stem cells ... 490
18.8 Other uses of mitoTALENs ... 490
18.9 Pros and cons of using mitoTALENs for gene therapy ... 491
Research perspectives ... 494
References ... 494

Chapter 19: Mitochondrially targeted zinc finger nucleases ... 499

Pedro Pinheiro, Payam A. Gammage and Michal Minczuk

19.1 Introduction ... 499
19.2 Zinc finger domain—structure and interaction with DNA ... 500
19.3 Designer zinc fingers ... 500
19.4 Chimeric zinc finger proteins—birth of zinc finger nuclease ... 502
19.5 Manipulation of the mammalian mitochondrial genome with mtZFNs ... 504
19.6 Concluding remarks ... 510
Research perspectives ... 511
Acknowledgments ... 511
References ... 511

Chapter 20: Mitochondrial movement between mammalian cells: an emerging physiological phenomenon ... 515

Michael V. Berridge, Patries M. Herst and Carole Grasso

20.1 Introduction ... 515
20.2 Cell-to-cell transfer of mitochondria with mtDNA: a brief overview ... 516
20.3 Translational benefits of mitochondrial transfer ... 517

20.4 Mechanisms of mitochondrial transfer 534
20.5 Mito-nuclear crosstalk: potential consequences of mitochondrial transfer/transplantation 534
20.6 Concluding statement 538
Research perspectives 538
Acknowledgments 538
References 538

Index* *547

List of Contributors

Alessandro Achilli Department of Biology and Biotechnology "Lazzaro Spallanzani", University of Pavia, Pavia, Italy

Marcella Attimonelli Department of Biosciences, Biotechnology and Biopharmaceutics, University "Aldo Moro", Bari, Italy

Sandra R. Bacman Department of Neurology, University of Miami Miller School of Medicine, Miami, FL, United States

Antoni Barrientos Department of Biochemistry and Molecular Biology, University of Miami, Miller School of Medicine, Miami, FL, United States; Department of Neurology, University of Miami, Miller School of Medicine, Miami, FL, United States

Michael V. Berridge Malaghan Institute of Medical Research, Wellington, New Zealand

Stephen P. Burr Department of Clinical Neurosciences and MRC Mitochondrial Biology Unit, University of Cambridge, Cambridge, United Kingdom

Claudia Calabrese Department of Clinical Neurosciences, University of Cambridge, Cambridge Biomedical Campus, Cambridge, United Kingdom; MRC Mitochondrial Biology Unit, University of Cambridge, Cambridge, United Kingdom

Francesco Maria Calabrese Department of Biosciences, Biotechnology and Biopharmaceutics, University "Aldo Moro", Bari, Italy; Department of Soil, Plant and Food Science, University of Bari Aldo Moro, Bari, Italy

Patrick F. Chinnery Department of Clinical Neurosciences and MRC Mitochondrial Biology Unit, University of Cambridge, Cambridge, United Kingdom

Monica De Luise Unit of Medical Genetics, Department of Medical and Surgical Sciences (DIMEC), University of Bologna, Bologna, Italy; Center for Applied Biomedical Research (CRBA), University of Bologna, Bologna, Italy

Francisca Diaz Department of Neurology, University of Miami, Miller School of Medicine, Miami, FL, United States

Mara Doimo Department of Medical Biochemistry and Biophysics, Umeå University, Umeå, Sweden

Flavia Fontanesi Department of Biochemistry and Molecular Biology, University of Miami, Miller School of Medicine, Miami, FL, United States

Yi Fu Skirball Institute of Biomolecular Medicine, Department of Cell Biology, NYU School of Medicine, New York, NY, United States

Payam A. Gammage CRUK Beatson Institute, Glasgow, United Kingdom; Institute of Cancer Sciences, University of Glasgow, Glasgow, United Kingdom

Caterina Garone Department of Medical and Surgical Sciences, Medical Genetics Unit, University of Bologna, Bologna, Italy; Center for Applied Biomedical Research, University of Bologna, Bologna, Italy

Giuseppe Gasparre Unit of Medical Genetics, Department of Medical and Surgical Sciences (DIMEC), University of Bologna, Bologna, Italy; Center for Applied Biomedical Research (CRBA), University of Bologna, Bologna, Italy

Anna Ghelli Department of Pharmacy and Biotechnology, University of Bologna, Bologna, Italy

Giulia Girolimetti Unit of Medical Genetics, Department of Medical and Surgical Sciences (DIMEC), University of Bologna, Bologna, Italy; Center for Applied Biomedical Research (CRBA), University of Bologna, Bologna, Italy

Ruth I.C. Glasgow Wellcome Centre for Mitochondrial Research, Newcastle University, Newcastle upon Tyne, United Kingdom; Newcastle University Translational and Clinical Research Institute, Newcastle University, Newcastle upon Tyne, United Kingdom

Aurora Gomez-Duran Department of Clinical Neurosciences, University of Cambridge, Cambridge Biomedical Campus, Cambridge, United Kingdom; MRC Mitochondrial Biology Unit, University of Cambridge, Cambridge, United Kingdom

Carole Grasso Malaghan Institute of Medical Research, Wellington, New Zealand

Patries M. Herst Malaghan Institute of Medical Research, Wellington, New Zealand; Department of Radiation Therapy, University of Otago, Wellington, New Zealand

Ian James Holt Biodonostia Health Research Institute, San Sebastián, Spain; IKERBASQUE, Basque Foundation for Science, Bilbao, Spain; CIBERNED (Center for Networked Biomedical Research on Neurodegenerative Diseases, Ministry of Economy and Competitiveness, Institute Carlos III), Madrid, Spain; Department of Clinical and Movement Neurosciences, UCL Queen Square Institute of Neurology, London, United Kingdom

Luisa Iommarini Department of Pharmacy and Biotechnology, University of Bologna, Bologna, Italy

Dongchon Kang Department of Clinical Chemistry and Laboratory Medicine, Graduate School of Medical Sciences, Kyushu University, Fukuoka, Japan

Ivana Kurelac Unit of Medical Genetics, Department of Medical and Surgical Sciences (DIMEC), University of Bologna, Bologna, Italy; Center for Applied Biomedical Research (CRBA), University of Bologna, Bologna, Italy

Albert Z. Lim Wellcome Centre for Mitochondrial Research, Newcastle University, Newcastle upon Tyne, United Kingdom; Newcastle University Translational and Clinical Research Institute, Newcastle University, Newcastle upon Tyne, United Kingdom

Marie T. Lott Center for Mitochondrial and Epigenomic Medicine, Children's Hospital of Philadelphia, Philadelphia, PA, United States

Shigeru Matsuda Department of Clinical Chemistry and Laboratory Medicine, Graduate School of Medical Sciences, Kyushu University, Fukuoka, Japan

Robert McFarland Wellcome Centre for Mitochondrial Research, Newcastle University, Newcastle upon Tyne, United Kingdom; Newcastle University Translational and Clinical Research Institute, Newcastle University, Newcastle upon Tyne, United Kingdom

Michal Minczuk MRC Mitochondrial Biology Unit, University of Cambridge, Cambridge, United Kingdom

Carlos T. Moraes Department of Neurology, University of Miami Miller School of Medicine, Miami, FL, United States

Thomas J. Nicholls Wellcome Centre for Mitochondrial Research, Newcastle University, Newcastle upon Tyne, United Kingdom; Newcastle University Biosciences Institute, Newcastle University, Newcastle upon Tyne, United Kingdom

Monika Oláhová Wellcome Centre for Mitochondrial Research, Newcastle University, Newcastle upon Tyne, United Kingdom; Newcastle University Biosciences Institute, Newcastle University, Newcastle upon Tyne, United Kingdom

Anna Olivieri Department of Biology and Biotechnology "Lazzaro Spallanzani", University of Pavia, Pavia, Italy

Annika Pfeiffer Department of Medical Biochemistry and Biophysics, Umeå University, Umeå, Sweden

Pedro Pinheiro MRC Mitochondrial Biology Unit, University of Cambridge, Cambridge, United Kingdom

Robert D.S. Pitceathly Department of Neuromuscular Diseases, UCL Queen Square Institute of Neurology and The National Hospital for Neurology and Neurosurgery, London, United Kingdom

Anna Maria Porcelli Department of Pharmacy and Biotechnology (FABIT), University of Bologna, Bologna, Italy; Interdepartmental Center for Industrial Research Life Sciences and Technologies for Health, University of Bologna, Ozzano dell'Emilia, Italy

Roberto Preste Department of Biosciences, Biotechnology and Biopharmaceutics, University "Aldo Moro", Bari, Italy

Vincent Procaccio Biochemistry and Genetics Department, MitoVasc Institute, UMR CNRS 6015 – INSERM U1083, CHU Angers, Angers, France

Corinne Quadalti Department of Medical and Surgical Sciences, Medical Genetics Unit, University of Bologna, Bologna, Italy; Center for Applied Biomedical Research, University of Bologna, Bologna, Italy

Shamima Rahman UCL Great Ormond Street Institute of Child Health, London, United Kingdom

Aurelio Reyes MRC Mitochondrial Biology Unit, University of Cambridge, Cambridge, United Kingdom

Ornella Semino Department of Biology and Biotechnology "Lazzaro Spallanzani", University of Pavia, Pavia, Italy

Agnel Sfeir Skirball Institute of Biomolecular Medicine, Department of Cell Biology, NYU School of Medicine, New York, NY, United States

Zhang Shiping Center for Mitochondrial and Epigenomic Medicine, Children's Hospital of Philadelphia, Philadelphia, PA, United States

Elaine Ayres Sia Department of Biology, University of Rochester, Rochester, NY, United States

Antonella Spinazzola Department of Clinical and Movement Neurosciences, UCL Queen Square Institute of Neurology, London, United Kingdom; Queen Square Centre for Neuromuscular Diseases, UCL Queen Square Institute of Neurology and National Hospital for Neurology and Neurosurgery, London, United Kingdom

Alexis Stein Department of Biology, University of Rochester, Rochester, NY, United States

Karolina Szczepanowska Cologne Excellence Cluster on Cellular Stress Responses in Ageing-Associated Diseases (CECAD) and Center for Molecular Medicine (CMMC), University of Cologne, Cologne, Germany; Institute for Mitochondrial Diseases and Ageing, Medical Faculty, University of Cologne, Cologne, Germany

Adriano Tagliabracci Section of Legal Medicine, Department of Excellence of Biomedical Sciences and Public Health, Polytechnic University of Marche, Ancona, Italy

Robert W. Taylor Wellcome Centre for Mitochondrial Research, Newcastle University, Newcastle upon Tyne, United Kingdom; Newcastle University Translational and Clinical Research Institute, Newcastle University, Newcastle upon Tyne, United Kingdom

Marco Tigano Skirball Institute of Biomolecular Medicine, Department of Cell Biology, NYU School of Medicine, New York, NY, United States

Antonio Torroni Department of Biology and Biotechnology "Lazzaro Spallanzani", University of Pavia, Pavia, Italy

Aleksandra Trifunovic Cologne Excellence Cluster on Cellular Stress Responses in Ageing-Associated Diseases (CECAD) and Center for Molecular Medicine (CMMC), University of Cologne, Cologne, Germany; Institute for Mitochondrial Diseases and Ageing, Medical Faculty, University of Cologne, Cologne, Germany

Chiara Turchi Section of Legal Medicine, Department of Excellence of Biomedical Sciences and Public Health, Polytechnic University of Marche, Ancona, Italy

Ornella Vitale Department of Biosciences, Biotechnology and Biopharmaceutics, University "Aldo Moro", Bari, Italy

Douglas C. Wallace Center for Mitochondrial and Epigenomic Medicine, Children's Hospital of Philadelphia, Philadelphia, PA, United States; Perelman School of Medicine, University of Pennsylvania, Philadelphia, PA, United States

Paulina H. Wanrooij Department of Medical Biochemistry and Biophysics, Umeå University, Umeå, Sweden

Sjoerd Wanrooij Department of Medical Biochemistry and Biophysics, Umeå University, Umeå, Sweden

Takehiro Yasukawa Department of Clinical Chemistry and Laboratory Medicine, Graduate School of Medical Sciences, Kyushu University, Fukuoka, Japan

Editor's biographies

Giuseppe Gasparre, born in 1979, is professor of medical genetics and head of the Center for Applied Biomedical Research at the University of Bologna and S. Orsola Hospital, Italy. He obtained a MS in pharmaceutical biotechnologies at the University of Bologna, Italy and a PhD in human genetics at the University of Turin, where he also served in the College of Mentors of the PhD program until 2018. His research focuses on mitochondrial genetics, with particular reference to mitochondrial metabolism in solid tumors, where he proposed the novel definition of oncojanus genes for those mitochondria-encoded genes whose role in tumor progression may be double-edged, according to the mutation type and load. He has long contributed to the curation of the Human Mitochondrial Database (HmtDB), and to the current workflows for the determination of pathogenicity of mtDNA mutations. He has also worked on the catalog of human NUMTS, and on the use of mtDNA pseudogenes as a phylogenetic marker. He has been a coordinator of the European project MEET—Mitochondrial European Educational Training (FP7 ITN-Marie Curie) and he is in the coordinating team and WP leader of the Horizon 2020 Marie Curie project TRANSMIT—Translating the Role of Mitochondria in Tumorigenesis. He teaches courses in cancer metabolism within the medicine program at the University of Bologna as well as within master programs in Italy. He coauthored 90 scientific publications on ISI-impacted journals and collaborates with several European and US research groups. His research is funded by the Italian Association for Cancer Research (AIRC), Fondazione Veronesi, the Italian Ministry of Health, the Italian Ministry of University, and the Worldwide Cancer Research, as well as by the EU.

Anna Maria Porcelli, born in 1972, is professor of biochemistry at the University of Bologna, Italy. She obtained a MS in biological sciences and a PhD in cellular biology and physiology at the University of Bologna, Italy, where she also serves in the College of Mentors of the PhD program. Her research focuses on bioenergetics and mitochondria biology with particular focus on the role of these organelles in the modulation of metabolic and hypoxic adaptation in solid tumors. Since 2010, she is a member of the Interdepartmental Center for Industrial Research of Emilia Romagna. She has been WP leader of the European project MEET—Mitochondrial European Educational Training

(FP7 ITN-Marie Curie) and she is a coordinator of Horizon 2020 Marie Curie project TRANSMIT—Translating the Role of Mitochondria in Tumorigenesis. She teaches courses in cellular biochemistry and molecular signaling transduction within the several biotechnology master degrees at the University of Bologna as well as within master programs in Italy. She coauthored 70 scientific publications on ISI-impacted journals and collaborates with several European and US research groups. She has been awarded the prestigious L'Oreal prize for Women in Science. Her research is funded by the Italian Association for Cancer Research (AIRC), Fondazione Veronesi, the Italian Ministry of University, and the EU.

Preface

Two adjectives have been recently attributed to the human mitochondrial genome (mtDNA), namely *neglected* [1] and *overlooked* [2]. Since 2001, the nuclear human genome is sequenced in the search for the determinants of variability and pathology, too often ignoring the smaller circular DNA harbored in the mitochondrial compartment. In 2012, indeed, the mtDNA was still being excluded as a contaminant from most deep sequencing practices, but at the beginning of the third decade of this century the role of this molecule in human physiopathology is far from being understood.

With the discovery of the first point mtDNA mutation as causative of Leber's Hereditary Optic Neuropathy [3] and of the first deletion responsible for a form of myopathy [4] in 1988, clinicians have started to consider that the etiology of a subset of human diseases resides in the mitochondrial, not the nuclear genome. Mitochondrial medicine is indeed nowadays a well-established branch of medicine facing the current challenge of developing efficient innovative therapies. Yet, this remains a niche within human pathology so that only very recently a common international effort has begun to harmonize the bulk of knowledge collected in the field of mtDNA research [5]. Very little is known on the role of mtDNA in human diseases other than canonical mitochondrial disorders, where it is likely that variants and/or mutations in mitochondria-encoded genes may act as modifiers or even causative, as it has been postulated in the oncology field.

As a growing number of researchers, we feel that it is becoming of paramount importance to uncover the still mysterious features of the mitochondrial genome, starting from the basic processes of replication, maintenance, and heredity, as well as how mtDNA evolves and mutates to give rise to human variability, adaptation, and disease. In the first place, this need stems from our personal experience as genetics and biochemistry teachers: both medicine and biology/biotechnology students face a cultural gap when it comes to understanding the significance of mtDNA in the advent and evolution of Life on Earth. Too many of them are not aware of how relevant is the role of mitochondria, with their own special DNA, in detoxifying the cellular environment from molecular oxygen, which only they are able to turn into an inert product such as water, while producing energy. They widen their eyes when they learn that multiple copies of the mtDNA reside embedded within the nuclear chromosomes, some of them being the ancestral remnants of a still ongoing transfer of DNA between cell compartments. They apprehend with awe that pregnancies with zygotes originating from three different parents, one of whom donating mtDNA, have been attempted with the aim to correct genetic defects. They find somewhat a vintage flavor in that the mtDNA does not obey the universal genetic code, making it impossible for mitochondria-encoded genes to be translated in the cytosol. They are amazed by the evidence that the mtDNA, and whole mitochondria, travel between cells in complex and fascinating tubular structures or vesicles, and may re-establish their function at the arrival site.

We address this book first of all to students and young scholars, with the wishful thinking that they may fall in love with a *Lord of the rings* that holds so much information in such a compact structure, on whose implications so much we have yet to learn. To this aim, the Research Perspectives boxes at the end of each chapter summarize the gaps to be filled, which will be hopefully picked up by the generations of scientists to come.

As senior researchers, we hope that our colleagues will benefit from this compendium, just as we have enjoyed reading all the valuable contributions from the most authoritative experts in the field. We trust this will be a chance to look at mtDNA from a different angle, spanning from the basic biology and physiology to clinics, touching evolution and frontier research aspects.

Giuseppe Gasparre
Department of Medical and Surgical Sciences, Unit of Medical Genetics, and Center for Applied Biomedical Research (CRBA), University of Bologna, Bologna, Italy

Anna Maria Porcelli
Department of Pharmacy and Biotechnology, Unit of Cell Biochemistry, University of Bologna, Bologna, Italy

References

[1] Pesole G, et al. The neglected genome. EMBO Rep 2012;13(6):473–4.
[2] Frezza C, Gammage P. Mitochondrial DNA: the overlooked oncogenome? BMC Biol 2019;17(1):53.
[3] Wallace D, et al. Mitochondrial DNA mutation associated with Leber's hereditary optic neuropathy. Science 1988;242(4884):1427–30.
[4] Holt IJ, et al. Deletions of muscle mitochondrial DNA in patients with mitochondrial myopathies. Nature 1988;331:717–19.
[5] Falk M, et al. Mitochondrial Disease Sequence Data Resource (MSeqDR): a global grass-roots consortium to facilitate deposition, curation, annotation, and integrated analysis of genomic data for the mitochondrial disease clinical and research communities. Mol Genet Metab 2015;114(3):388–96.

Acknowledgments

We would like to thank our creative postdoc fellows Monica De Luise, Stefano Miglietta, and Manuela Sollazzo for their contribution to the book cover design.

We are very grateful to all authors and the publisher for making this work possible with their enthusiastic and competent contribution.

PART 1

Biology of human mtDNA

CHAPTER 1

mtDNA replication, maintenance, and nucleoid organization

Mara Doimo, Annika Pfeiffer, Paulina H. Wanrooij and Sjoerd Wanrooij
Department of Medical Biochemistry and Biophysics, Umeå University, Umeå, Sweden

1.1 Human mitochondrial DNA

1.1.1 Characteristics of mitochondrial DNA

The evidence for mitochondrial deoxyribonucleic acid (mtDNA) dates back to the early 1960s when fibers with characteristics of DNA were first observed in the mitochondrial matrix of chick embryo mitochondria [1,2]. The human mitochondrial genome is a closed circular double-stranded DNA molecule with a length of about 5 μm [3,4]. MtDNA molecules mostly occur as monomers, but can also exist as catenated circles, which are most often dimers [4–6]. In accordance with its small size, mtDNA amounts to less than one percent of the total cellular DNA in animal cells [7]. Early estimations suggested that a single animal mitochondrion contains between two and more than ten mtDNA genomes [8], and—because there are multiple mitochondria per cell—human cells in general harbor from several hundred to over ten thousand mtDNA molecules per cell, depending on the cell type [9–12]. However, the mtDNA copy number of all cell types does not fit in this range, as, for example, human oocytes contain hundreds of thousands of mtDNA copies per cell [13–15].

Human mtDNA is maternally inherited [16]. A key characteristic of mtDNA that derives from its multicopy nature is that cells can contain more than one population of mtDNA genomes, a state referred to as heteroplasmy [17]. The level of heteroplasmy, that is, the frequency of a population of mutated mtDNA molecules, can determine the severity of symptoms in mitochondrial diseases that are caused by mutations in mtDNA [18]. Another peculiar feature that separates the mitochondrial genome from its nuclear counterpart is that mtDNA contains occasional embedded ribonucleotides, the building blocks of ribonucleic acid (RNA) [19,20].

The Human Mitochondrial Genome.
DOI: https://doi.org/10.1016/B978-0-12-819656-4.00001-2

1.1.2 Organization of the human mitochondrial genome

The human mitochondrial genome was the first fully sequenced mitochondrial genome [21]. It is 16.5 kilo base pairs (kbp) in length and encodes for 2 ribosomal RNAs (rRNAs), 22 transfer RNAs (tRNAs), and 13 protein-coding genes with very few or no noncoding bases between neighboring genes (Fig. 1.1A) [21]. All 13 protein-coding genes encode indispensable subunits of the ATP-producing oxidative phosphorylation (OXPHOS) system [21–26]. The two strands of mtDNA are designated as heavy (H) and light (L) due to differences in base composition (GT/CA ratio) that leads to their different buoyant densities on alkaline cesium chloride gradients [27]. Twelve of the protein-coding genes, the 12S and 16S rRNAs as well as fourteen tRNA genes are encoded on the H-strand, while the L-strand encodes one protein-coding gene and eight tRNA genes [21]. The H-strand is G-rich and thereby contains several sequence motifs with the potential to form specific secondary DNA

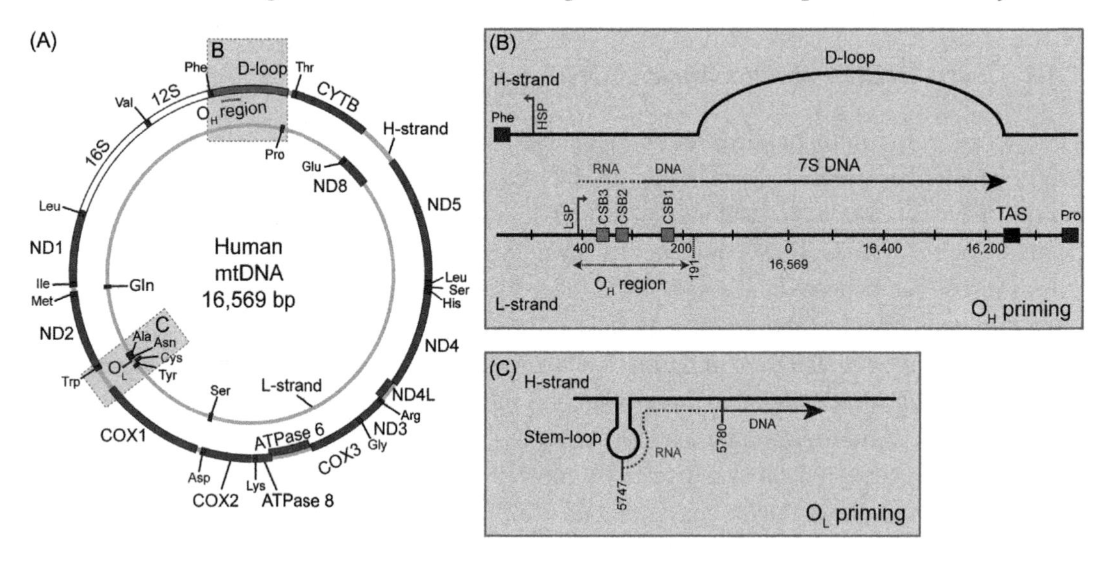

Figure 1.1: The human mitochondrial genome.
(A) The mitochondrial genome has a size of 16,568 base pairs in humans. The two strands are termed heavy (H-strand) and light strand (L-strand). The location of the displacement loop (D-loop) is indicated in the noncoding region between $tRNA^{Pro}$ and $tRNA^{Phe}$, and the position of the genes for complex I (NADH dehydrogenase; ND), complex III (cytochrome b; CYTB), complex IV (cytochrome c oxidase; COX), as well as complex V (ATP synthase; ATPase) subunits of the respiratory system are shown. Highlighted with gray boxes are the sites of replication priming in the O_H region (B) and at O_L (C). (B) Characteristic features of the noncoding region are the promoters HSP and LSP, the highly conserved sequence blocks (CSB 1–3), the single-termination sequence (TAS), and the DNA stretch (7S DNA) that forms the D-loop. Priming in the O_H region is initiated with a RNA primer and RNA-to-DNA transition sites were mapped to the CSB region. (C) Priming at O_L involves the formation of a DNA stem-loop structure and the synthesis of a short RNA primer by POLRMT. The RNA-to-DNA transition sites were mapped downstream of the O_L region.

structures called G-quadruplexes (G4s) [28,29]. These motifs often colocalize with mtDNA deletion break-points [29], suggesting that they may hamper the mtDNA replication process.

A circa 1100 bp long noncoding region (NCR), also referred to as the control region, is located between the $tRNA^{Pro}$ and $tRNA^{Phe}$ genes [21]. The NCR contains the origin of replication for H-strand synthesis (O_H) [21,30] as well as the H-strand promoter (HSP) and the L-strand promoter (LSP) for transcription of each mtDNA strand [31]. Although O_H has traditionally been annotated at nucleotide position 191 [21], H-strand replication actually initiates at LSP, approximately 200 nucleotides (nt) upstream of O_H [32]. We will therefore use the term "O_H region" to refer to the region containing both LSP and O_H [33]. The NCR also contains three highly conserved sequence blocks (CSB 1–3), which are considered to have a function in regulating replication [34], and a single termination–associated sequence (TAS) downstream of the three CSBs [35]. Early characterization of mtDNA in mouse [36–38] and later in human cells [39] revealed molecules with a short triple-stranded DNA region, the so-called displacement loop (D-loop), encompassing the O_H region. The D-loop is the result of synthesis of a ~600 nt long DNA stretch, termed 7S DNA, that is complementary to the L-strand and is formed by replication initiated in the O_H region followed by early termination at the TAS [35,40,41]. The second origin of replication on mtDNA, O_L, directs L-strand synthesis. O_L is located outside of the NCR, in a cluster of five tRNA genes two-thirds of the way around the genome from O_H [42].

1.2 The process of mtDNA replication

1.2.1 Replication mechanisms

In comparison to the nuclear DNA duplication process, human mitochondrial replication is distinct in terms of both the mechanism(s) and the proteins involved. The proteins of the mitochondrial replication fork duplicate the organelle's DNA by an asymmetric mechanism in which both strands are synthesized continuously [43]. This so-called strand-displacement model of mtDNA replication is based on early pulse and pulse-chase labeling experiments, electron microscopy and 5′-end characterization [7,44], as well as more recent biochemical reconstitution of mtDNA replication in vitro using recombinant proteins [45,46]. This mode of DNA replication is strikingly different from that of mammalian nuclear DNA, but it is not unique in biology since it is similar to replication of ColE1 plasmid DNA [47].

MtDNA replication starts with H-strand (leading strand) synthesis that initiates from the O_H region. As the mtDNA replication machinery synthesizes a nascent H-strand, the parental H-strand is displaced, giving rise to the triple-stranded D-loop structure (Fig. 1.2). For yet unknown reasons, mtDNA replication often terminates around the TAS region and only a minority of the nascent H-strands are extended past this point. In the cases where H-strand replication continues beyond the TAS region, it continues unidirectionally, displacing the

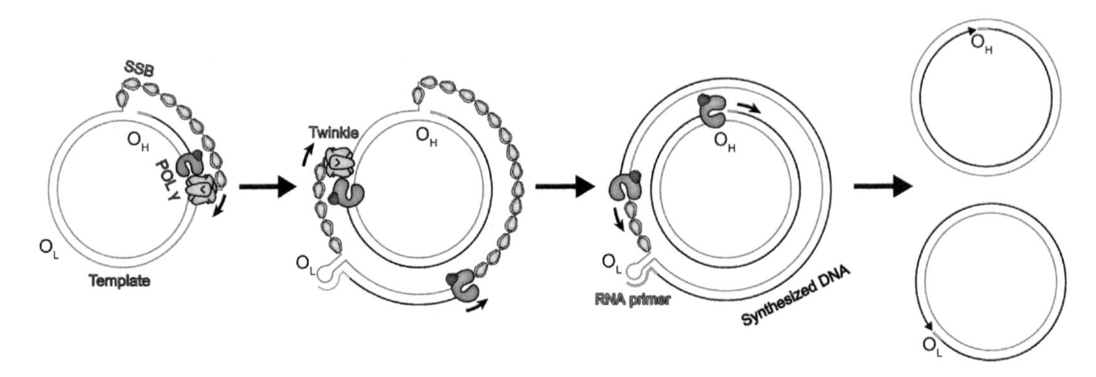

Figure 1.2: The model of mitochondrial DNA (mtDNA) replication.
Mitochondrial DNA is replicated by a unique enzymatic machinery that includes the heterotrimeric DNA polymerase POL γ (*pink*), the mitochondrial single-stranded DNA-binding protein mtSSB (*green*), and the DNA helicase Twinkle (*blue*). (1) After initiation at the leading strand origin (O_H), the replisome consisting of POL γ and Twinkle proceeds to unidirectionally replicate one new strand. The displaced, second strand is bound and stabilized by mtSSB (*green*). (2) When the replication machinery passes the second strand (lagging-strand) origin (O_L), a stem-loop structure is formed. A short primer is synthesized at the stem-loop, and is used to initiate the second strand (lagging strand) DNA synthesis. (3) Both the leading and lagging strand replication machineries proceed full-circle around the mtDNA molecule. (4) After completion of mtDNA strand synthesis, replication is terminated at either O_H or O_L, depending on where DNA synthesis was initiated.

parental H-strand until it reaches O_L. Unwinding of the DNA duplex at O_L allows the formation of a characteristic hairpin structure that leads to activation of this origin [48]. L-strand (lagging strand) replication thus initiates and continues unidirectionally from O_L in the opposite orientation to H-strand synthesis, and the synthesis of both strands proceeds until they have reached full-circle.

An alternative interpretation of the strand-displacement mode of replication has been made based on two-dimensional agarose gel electrophoresis experiments. However, the two models are very similar with the principal difference that in this so-called bootlace/RITOLS model [49,50], the displaced H-strand is coated with preformed RNA in contrast to the mitochondrial single-stranded DNA-binding protein (mtSSB) suggested in the strand-displacement model [51,52]. Both models agree on that mtDNA replication results in extensive ssDNA replication intermediates, which are indeed the dominant replication intermediates observed in dividing cells [53]. In nondividing cells, however, the majority of mtDNA replication intermediates are double-stranded in nature [53]. This could potentially be the consequence of enhanced promiscuous (O_L-independent) priming of L-strand replication [54], or, as proposed by others [55], the result of an alternative bidirectional mode of replication carried out by an as-yet unidentified set of proteins.

1.2.2 Priming

The first step of DNA replication is the synthesis of an RNA primer that can subsequently be elongated by the mtDNA polymerase γ (POL γ). A mitochondrial primase activity was isolated from human mitochondria as early as 1985 [56,57], but the identity of the protein was not elucidated at the time. Subsequent work has revealed that the primers at both origins of replication are in fact synthesized by the mitochondrial RNA polymerase (POLRMT) that is related to the RNA polymerases of the T-odd lineage of bacteriophages [58].

The primer for leading strand replication has a 5′ end at the LSP, the same site where L-strand transcripts initiate [32]. Hence, transcription by POLRMT from the LSP creates not only the near-genome-length transcripts that are processed to liberate tRNAs and mRNAs [59], but also the much shorter primer for leading strand synthesis (Fig. 1.1B). A G4 structure that forms at CSB2 causes premature termination of transcription about 120 nt downstream of LSP [60,61]. This prematurely terminated transcript remains stably hybridized to the DNA template as an RNA–DNA hybrid referred to as the mitochondrial R-loop [62,63] and is assumed to be used as the H-strand replication primer. In support of this model of H-strand priming, RNA-to-DNA transition sites have been mapped to the CSB region [32,60,64,65]. In an in vitro transcription assay based on purified recombinant proteins, the level of premature transcription termination—and thus the switch from transcription to replication—is modulated by the mitochondrial transcription elongation factor (TEFM) that facilitates the bypass of obstacles such as DNA secondary structures and oxidative lesions by POLRMT [66,67]. In accordance with TEFM's role in transcription elongation, it is essential in mice and heart-specific knockout causes a dramatic decrease in promoter-distal transcripts [68]. Because the majority of LSP-initiated transcripts fail to be elongated far enough to reach CSB2 in the absence of TEFM, de novo mtDNA replication is dramatically reduced in TEFM knockout hearts [68].

As described above, POLRMT synthesizes the H-strand primer using double-stranded mtDNA as a template. In contrast, L-strand replication is initiated when H-strand synthesis has reached two-thirds of the way around the genome and exposes O_L in single-stranded form (Fig. 1.2). Once single-stranded, O_L adopts a hairpin structure that directs POLRMT to initiate primer synthesis within a stretch of six Ts within the loop (nucleotide position 5747–5751) (Fig. 1.1C [48]). On this ssDNA template, POLRMT is nonprocessive and generates only short RNAs that can be elongated by POL γ [46,48]. In accordance with this model, in vivo RNA-to-DNA transition sites have been mapped just downstream of the O_L hairpin [48].

In contrast to the L-strand replication primer, the RNA primer generated in the O_H region cannot be directly elongated by POL γ. Instead, in vitro studies suggest that the H-strand primer may require processing by RNase H1 before elongation by POL γ is possible [65]. As discussed below, RNase H1 also plays a defined role in removal of the RNA primers at both mitochondrial origins [69]. Accordingly, mutations that decrease the activity of RNase H1 cause mtDNA depletion and deletions, manifesting as chronic progressive external ophthalmoplegia (CPEO) [70].

1.2.3 Elongation of mtDNA replication

After priming of mtDNA synthesis, elongation of the replication primer is catalyzed by the nuclear-encoded POL γ, which was the first polymerase isolated from mitochondria of human HeLa cells [71]. POL γ needs the help of the mtDNA helicase Twinkle and mtSSB to replicate mtDNA in human cells. Together, POL γ, Twinkle and mtSSB constitute the minimal mtDNA replisome (Fig. 1.3) and are essential for reconstitution of mitochondrial replication in vitro [72].

1.2.3.1 The mitochondrial DNA polymerase POL γ

POL γ was first described as an RNA-dependent DNA polymerase [73]. The holoenzyme consists of one 140-kDa POL γA catalytic subunit and two 55-kDa POL γB accessory subunits [74–76]. The catalytic subunit possesses DNA polymerase and 3′–5′ exonuclease activities in addition to 5′-deoxyribose phosphate lyase activity [77,78]. In general, the intrinsic exonuclease activity of POL γ functions as a proofreading mechanism during replication, which ensures fidelity of the polymerase through high nucleotide selectivity and slow extension of mismatches [79]. POL γ can switch between polymerase and exonuclease

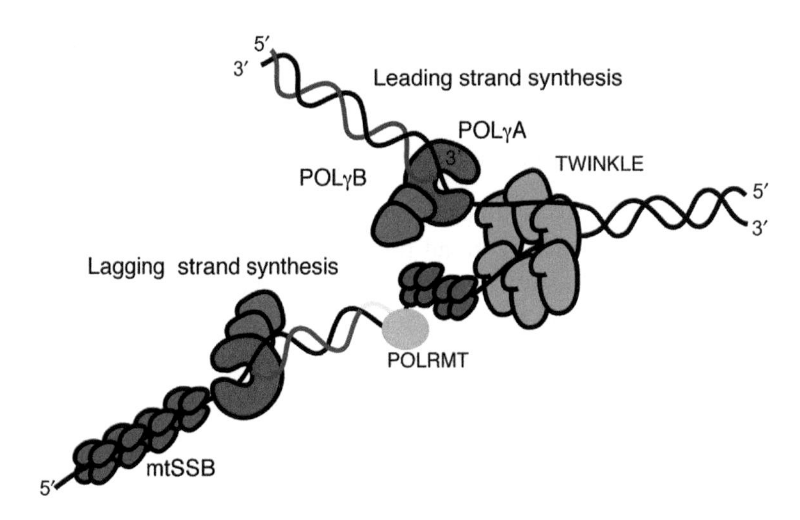

Figure 1.3: The core mitochondrial DNA (mtDNA) replication fork proteins.
The TWINKLE helicase (*blue*) moves in a 5′ to 3′ direction while unwinding dsDNA. The mtSSB protein (*dark green*) stabilizes the single-stranded conformation of DNA and stimulates DNA synthesis by POLγ (*red* (A) and *gray* (B)). POLRMT (*light green*) synthesizes the RNA primer (*yellow line*) needed for lagging strand DNA synthesis.

activities without dissociating from the DNA template because it can transfer the nascent DNA from the polymerase site to the exonuclease site and back through a mechanism of intramolecular strand transfer [80]. With an error rate of less than 1×10^{-6} per nucleotide, POL γ is one of the most accurate polymerases [79].

The accessory subunit POL γB is also known as the processivity factor because it increases the processivity of POL γA by increasing its DNA binding affinity. In addition, POL γB stimulates both the exonuclease and polymerase activities, and it can also improve nucleotide binding and incorporation, thus increasing the polymerization rate of the holoenzyme [81,82]. In solution, dimers of POL γB form a heterotrimer with the catalytic subunit POL γA through the tight binding of the POL γB dimer to the POL γA monomer [75]. Each POL γB unit in the holoenzyme has a specific function: the monomer proximal to POL γA in the holoenzyme increases the interaction with DNA, whereas the distal POL γB monomer is responsible for enhancing the reaction rate [83]. Modeling of POL γ binding to DNA suggests that POL γA binds approximately 10 bp of template DNA and that interaction with POL γB increases the footprint of POL γA on the DNA to 25 bp. Although POL γB has dsDNA-binding activity [84], it does not interact with the primer-template DNA [85]. While it is not essential for the stimulation of POL γA, the dsDNA-binding activity of POL γB is required for the function of the mtDNA replisome on a dsDNA template through coordination of POL γ and Twinkle at the replication fork [86].

1.2.3.2 The mitochondrial DNA helicase Twinkle

Twinkle (T7 gp4-like protein with intramitochondrial nucleoid localization) is the nuclear-encoded mitochondrial replicative helicase. It shares structural similarity with phage T7 primase/helicase and colocalizes with mtDNA in nucleoprotein complexes [87]. Twinkle is required for unwinding of the double-stranded DNA template ahead of POL γ that can only use ssDNA as a template [72]. It is an NTP-dependent DNA helicase that unwinds DNA in a $5' - 3'$ orientation and needs a fork-like structure with a single-stranded 5′-DNA loading site and a short 3′-tail to initiate unwinding [88]. The helicase activity of Twinkle is stimulated considerably by mtSSB [88], but it can only unwind longer dsDNA stretches in the presence of POL γ [72]. Earlier reports suggested that Twinkle is hexameric [87,89], while more recent analysis by electron microscopy revealed evidence of both hexameric and heptameric forms [90,91].

1.2.3.3 The mitochondrial single-stranded DNA-binding protein

Human mtSSB is a 15 kDa protein that shares sequence similarity with *Escherichia coli* SSB [92], and forms a tetramer that binds 59 nt of ssDNA [93]. Single-stranded DNA wraps once around the mtSSB tetramer [94,95]. MtSSB stimulates POL γ polymerase activity as well as Twinkle helicase activity [88,96,97]. The stimulatory effect on POL γ polymerase activity has been suggested to occur without any direct interaction between the proteins, through the ability of mtSSB to organize the ssDNA template and remove any secondary DNA structures [98]. Studies on *Drosophila* mtSSB showed that mtSSB increases *Drosophila* POL γ synthesis primarily by increasing primer recognition and binding and, as a result, increases the rate of initiation of DNA synthesis [99,100]. In addition to stimulating POL γ polymerase activity, *Drosophila* mtSSB can even stimulate the exonuclease activity of POL γ [101].

1.2.3.4 The mitochondrial DNA replisome

In biochemical assays, POL γ alone is unable to use dsDNA as a template, but the addition of Twinkle dramatically improves polymerization by POL γ on a primed mini-circle template and allows synthesis of extensive stretches of DNA in a rolling-circle replication mode [72]. The Twinkle-POL γ interaction also allows Twinkle to unwind longer stretches of dsDNA although the proteins do not seem to form a stable complex [72]. The addition of mtSSB further improves DNA synthesis allowing the formation of up to genome-length DNA products [72]. In conclusion, the mtDNA polymerase POL γ together with Twinkle and mtSSB constitute the core machinery necessary to replicate the mtDNA in human cells. The importance of these proteins for mtDNA maintenance in vivo is underscored by the fact that defects in these core components of the mtDNA replisome lead to mtDNA instability and mitochondrial disease [87,102–106].

1.2.4 Termination of mtDNA replication

Once the replisome has synthetized the mtDNA strands in their entirety, three sequential steps need to occur in order to give rise to the two daughter mtDNA molecules: (I) primer removal, (II) ligation, and (III) separation of the newly synthetized molecules.

1.2.4.1 Primer removal

As described previously, the process of mtDNA replication at both origins of replication starts with the synthesis of RNA primers by POLRMT. Long stretches of ribonucleotides are expected to impair mtDNA stability [107] and interfere with the replisome function [108]. Therefore these RNA primers need to be removed before the end of the replication process. The main enzyme implicated in removal of the primers at both O_H and O_L is RNase H1, as implicated by the finding that the loss of this enzyme causes the retention of the RNA primers at both origins [69]. In vitro, RNase H1 processes the RNA strand in RNA–DNA hybrid substrates and, although it does not exhibit any sequence specificity, it requires four consecutive ribonucleotides flanking the cleavage site [109]. Cleavage by RNase H1 leaves two ribonucleotides attached to the 5′-end of the nascent DNA strand [109]. Consequently, a second nuclease is required to remove these last ribonucleotides.

The removal of the primer at O_L has been reconstituted in vitro [110]. According to this model, RNase H1 processes the primer at O_L and leaves 1–3 ribonucleotides at the 5′-end of the nascent DNA that are subsequently removed by the flap-structure specific endonuclease 1 (FEN1). FEN1 is $5' - 3'$specific endonuclease with dual mitochondrial and nuclear localization [111]. It preferentially cuts short RNA or DNA 5′-flaps flanked by dsDNA. Among its other functions, FEN1 is implicated in the processing of nucleic acid intermediates that form during lagging strand replication in the nucleus [112]. However, its function inside the mitochondria is still debated because the variant of the protein that is imported into the mitochondrial matrix is truncated and, in vitro, cannot cleave 5′-ssDNA flaps [113]. Therefore another nuclease with Fen1-like activity might be involved in the removal of the primer at O_L in vivo.

The process of primer removal at O_H remains poorly understood. Initially, the mitochondrial genome maintenance exonuclease 1 (MGME1) was proposed to be involved [114]. This ssDNA-specific exonuclease, which localizes exclusively in the mitochondria, can process DNA flap substrates [115]. In vitro MGME1 works in combination with POL γ to remove DNA flaps that are originated by the strand-displacement activity of POL γ [116]. However, MGME1 cannot process short RNA flaps [116], indicating that an additional nuclease must be recruited to remove the ribonucleotides left by RNase H1. The involvement of MGME1 in mtDNA maintenance is nevertheless demonstrated by the fact that mutations in the *MGME1* gene are associated with a multisystemic mitochondrial disorder characterized by mtDNA depletion and the accumulation of multiple mtDNA deletions [115].

Recently, also the endonuclease/exonuclease G (EXOG) has been proposed to have a role in the removal of the ribonucleotides left behind by RNase H1 [117]. EXOG localizes to the mitochondria [118] and can process RNA–DNA hybrids in vitro [117]. Nonetheless, the in vivo involvement of EXOG in primer removal has not been clarified so far.

1.2.4.2 Ligation

After the removal of the primers, the nascent mtDNA strands are ligated by DNA ligase 3 (LIG3). Due to the presence of two in-frame ATGs in the coding sequence, the *LIG3* gene encodes both a nuclear and a mitochondrial variant of the enzyme [119]. Depletion of the protein in human cells results in reduced mtDNA content and in the accumulation of nicked mtDNA [120], indicating that LIG3 is essential for the maintenance of mtDNA.

1.2.4.3 Separation

At the end of the replication process, the two daughter molecules are still interlinked and need to be separated. Topoisomerase TOP3α was recently found to be the main enzyme involved in this process [121]. TOP3α is a type 1A topoisomerase that localizes both in the nucleus and in the mitochondria [122]. Enzymes belonging to this class of topoisomerases separate two catenated molecules of DNA by cleaving one of the two strands to allow the passage of the other strand [123]. In the mtDNA, at the end of replication, the newly synthetized molecules remain interlinked in a region close to O_H [121]. TOP3α is essential to separate these molecules and its depletion causes the accumulation of catenated mtDNA structures. The same abnormal mtDNA topology is detected also in the skeletal muscle of patients with CPEO caused by bi-allelic mutations in *TOP3A* gene [121].

1.2.5 Other proteins involved in mtDNA replication

Besides the above-described proteins, additional proteins whose precise role remains elusive have been implicated in mtDNA maintenance. TOP1MT is a mitochondria-specific topoisomerase [124] that is nonessential for mtDNA replication under basal conditions. However, TOP1MT binds to the NCR of mtDNA [125] and is required for proper mtDNA replication under specific stress conditions in mice [126]. Moreover, the chemical inhibition of yet another topoisomerase, TOP2β, was shown to reduce mtDNA replication initiation in human cells [127]. Although the mechanistic details of this defect were not addressed, it suggests that TOP2β has an important function in mtDNA replication.

The human nuclease/helicase DNA2 can be found inside the mitochondria and specifically localizes to nucleoids after induction of mtDNA replication stalling [128]. The implication of DNA2 in mtDNA maintenance is supported by the observation that mutations in the *DNA2* gene lead to mtDNA instability in patients with mitochondrial myopathies [129,130]. Despite valuable biochemical characterization of the purified human DNA2 [131], the precise mitochondrial import mechanism and function is not resolved. A second helicase that can be found in the nucleus and mitochondria of human cells and that has many alleged functions in DNA metabolism is PIF1 (petite frequency integration 1) [132]. Translation initiation at an alternative ATG of the *PIF1* gene produces a strictly mitochondrial isoform [133]. PIF1 knockout mice accumulate mtDNA alterations and develop mitochondrial myopathy [134], however, the exact role of mammalian PIF1 in mtDNA stability remains to be determined.

The DNA polymerase/primase PrimPol has roles in both nuclear and mtDNA maintenance [135]. PrimPol functionally interacts with mtSSB and Twinkle, but nothing is known about the mechanism by which it is targeted to the mitochondrion [136,137]. PrimPol was shown to be able to reinitiate stalled mtDNA replication by priming mtDNA replication at nonconventional replication origins [54]. However, many regulatory and mechanistic details of this mtDNA replication reinitiation still need to be disclosed.

1.3 The mitochondrial dNTP supply

The faithful maintenance of mtDNA requires not only a functional replication machinery, but also a sufficient and balanced supply of mitochondrial deoxyribonucleoside triphosphates (dNTPs), the building blocks of DNA. The importance of an appropriate dNTP supply is highlighted by the fact that defects in mitochondrial nucleotide metabolism give rise to mitochondrial diseases such as PEO and mtDNA depletion syndrome [138–142]. There are two principal ways to synthesize dNTPs: by producing them from ribonucleoside diphosphates (NDPs) through the de novo pathway, or by phosphorylating existing deoxyribonucleosides (dNs) by action of the salvage pathway. Both the de novo and the salvage pathways contribute to maintenance of mitochondrial dNTP pools (Fig. 1.4).

The main regulatory step of de novo dNTP synthesis is catalyzed by the enzyme ribonucleotide reductase (RNR) that is a heterotetramer consisting of two large RRM1 catalytic subunits and two small RRM2 accessory subunits [143]. RNR carries out the reduction of NDPs to dNDPs in the cytosol; the dNDPs can then be phosphorylated to dNTPs by nucleoside diphosphate kinases (NDPKs) that are present in both the cytosol and the mitochondria [144–146]. The synthesis of dTTP requires the contribution of additional enzymes. The activity of the de novo pathway follows the cell cycle, peaking in S phase when the demand for dNTPs is high due to the replication of nuclear DNA. This cell-cycle-phase-dependent regulation is mainly achieved by modulating the levels of the small RRM2 subunit of RNR, while the levels of the RRM1 subunit remain stable [143]. RRM2 is undetectable in nondividing cells [147] where RNR activity instead relies on the alternative small subunit RRM2B to provide the dNTPs required for DNA repair and mtDNA replication [148–151]. However, the levels of the RRM1/RRM2B complex in nondividing cells are clearly lower than those of the RRM1/RRM2 complex in S phase, resulting in ~18-fold lower dNTP pools in nondividing cells [152]. Despite the low activity of de novo dNTP synthesis in nondividing cells, its contribution is nonetheless critical for mtDNA maintenance as evidenced by the severe mtDNA depletion observed in patients with RRM2B defects [141].

Unlike de novo dNTP synthesis, mitochondrial dNTP salvage is considered to produce a low but constitutive level of mitochondrial dNTPs throughout the cell cycle [153]. The rate-limiting step of mitochondrial salvage is the phosphorylation of dNs to deoxyribonucleoside monophosphates (dNMPs), a step that is catalyzed by thymidine kinase 2 (TK2; phosphorylates dT, dU, and dC) and deoxyguanosine kinase (dGK; phosphorylates dG and dA) [153]. Accordingly, defects in either of these nucleoside kinases manifest as mtDNA depletion [138,139], and TK2 mutations can also result in multiple mtDNA deletions [142]. The dNMPs produced by TK2 and dGK are phosphorylated by mitochondrial nucleoside monophosphate kinases (NMPKs) to dNDPs and further by NDPKs to dNTPs (Fig. 1.4).

Both of the described pathways of mitochondrial dNTP synthesis involve import of either dNs or deoxyribonucleotides (dNMPs, dNDPs, or dNTPs) into mitochondria. In fact, mitochondrial dN and deoxyribonucleotide pools are believed to be in exchange with cytosolic pools [154–156] due to the presence of carrier proteins in the inner mitochondrial membrane. dNs, which constitute the substrates of mitochondrial salvage, can originate from the

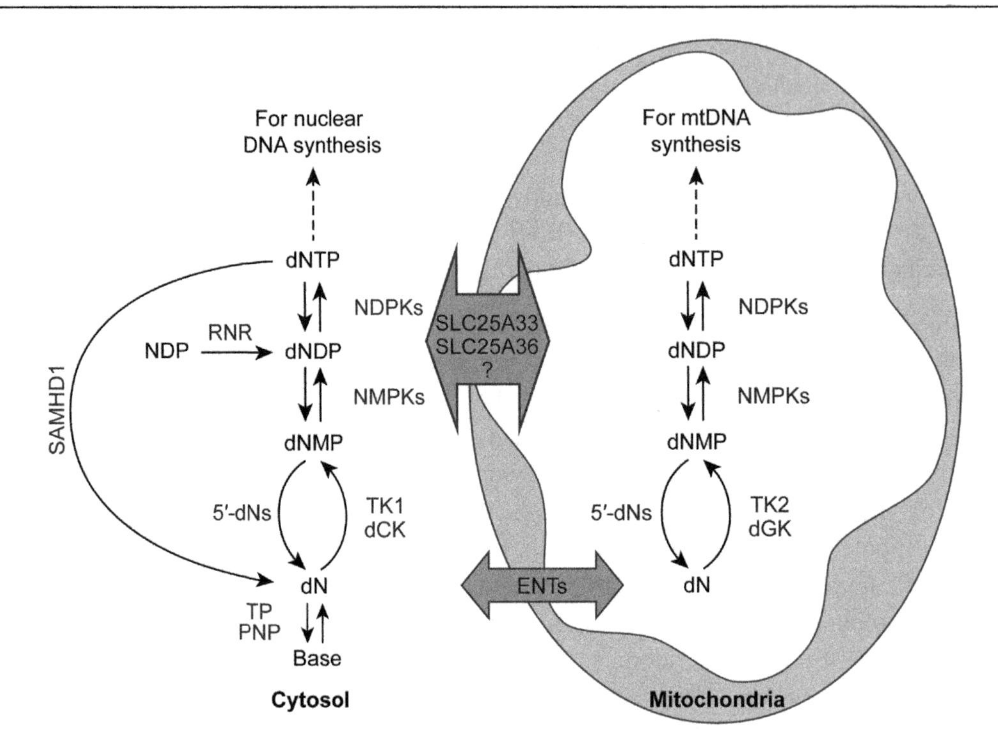

Figure 1.4: Nucleotide synthesis pathways that provide mitochondrial dNTPs.
Enzyme names are in *gray* type. The synthesis of dNTPs can occur through the de novo or the salvage pathways. In the de novo pathway, the cytosolic enzyme ribonucleotide reductase (RNR) catalyzes the reduction of NDPs to dNDPs that can subsequently be phosphorylated to dNTPs by nucleoside diphosphate kinases (NDPKs) that are present in both the cytosol (left-hand side) and the mitochondria (right-hand side). Alternatively, dNTPs can be produced through the salvage pathway that "recycles" bases or nucleosides (dNs) that are taken up from outside the cell or that derive from the breakdown of endogenous nucleotides or nucleic acids. The rate-limiting step of dNTP salvage synthesis is the phosphorylation of dNs to deoxyribonucleoside monophosphates (dNMPs). This step is carried out by thymidine kinase 1 (TK1) and deoxycytidine kinase (dCK) in the cytosol, and thymidine kinase 2 (TK2) and deoxyguanosine kinase (dGK) in the mitochondria. The dNMPs thus produced can be further phosphorylated to deoxyribonucleoside diphosphates (dNDPs) and then to dNTPs through the sequential action of nucleoside monophosphate kinases (NMPKs) and NDPKs. The synthesis of dNTPs is counteracted by catabolic enzymes such as 5′-deoxynucleotidases (5′-dNs) that dephosphorylate dNMPs to dNs in both the cytosol and the mitochondria, as well as nucleoside phosphorylases (thymidine phosphorylase [TP] and purine nucleoside phosphorylase [PNP]) that degrade dNs to bases in the cytosol. Also the dNTP phosphohydrolase SAMHD1 limits the levels of dNTPs in cells outside of S phase. Due to the presence of nucleotide carriers such as SLC25A33, SLC25A36 and as-yet unidentified carrier(s), the cytosolic and mitochondrial pools of deoxynucleotides (dNMPs, dNDPs, and/or dNTPs) are in rapid exchange. Similarly, equilibrative nucleoside transporters (ENTs) ensure the exchange of dNs between the two cellular compartments.

degradation of existing mitochondrial deoxyribonucleotides or be imported from the cytosol by equilibrative nucleoside transporters (ENTs) [157,158]. In contrast, the products of cytosolic de novo dNTP synthesis are imported into mitochondria in the form of dNMPs, dNDPs, and/or dNTPs. So far, a comprehensive understanding of mitochondrial nucleotide carriers and their individual contributions to maintenance of the intramitochondrial deoxyribonucleotide pool remains elusive. However, two solute carrier 25 family proteins, SLC25A33 and SLC25A36, are likely responsible for the exchange of mitochondrial and cytosolic pyrimidine nucleotides (U, T, C) [159–161].

The overall size of the cellular dNTP pool is determined not only by the anabolic pathways that synthesize dNTPs, but also by the activity of catabolic enzymes. These include 5′-deoxynucleotidases (5′-dNs) that dephosphorylate dNMPs to dNs in both the cytosol and the mitochondria as well as nucleoside phosphorylases that degrade dNs in the cytosol [162]. Defects in one such phosphorylase, thymidine phosphorylase (TP), imbalance cellular dNTP pools and manifest as mtDNA deletions and consequent mitochondrial disease [140,163]. Furthermore, the SAMHD1 (SAM domain and HD domain-containing protein 1) dNTP phosphohydrolase converts dNTPs to dNs in the cytosol and thereby limits cellular dNTP levels [164–166]. Due to the exchange of cytosolic and mitochondrial pools, the activity of SAMHD1 also reduces mitochondrial dNTP pools [167].

A peculiar feature of mtDNA is that it contains frequent ribonucleotides, the building blocks of RNA [19,20,168]. Owing to the great excess of free NTPs over dNTPs [169] in the cell, ribonucleotides are occasionally inserted in place of dNTPs during replication and repair of the nuclear and mitochondrial genomes [108,170–173]. However, while ribonucleotides incorporated in nuclear DNA are removed by a dedicated repair pathway [174–176], the ones in mtDNA are not effectively repaired and therefore persist [177,178]. Owing to the lack of repair, the main factor defining the frequency and identity of mtDNA ribonucleotides is the ratio of NTPs to dNTPs in the cell [177,179,180]. At this stage it is unclear whether the ribonucleotides in mtDNA serve a specific purpose or if they are merely tolerated. They are in any case neither essential for mtDNA maintenance nor exceedingly harmful because a dramatic reduction in ribonucleotide content has no striking effect on mtDNA stability during mouse lifespan [180].

1.4 Mitochondrial nucleoids

MtDNA localizes in specific areas of the mitochondrial matrix and is organized in nucleoprotein complexes called mitochondrial nucleoids. The concept of mitochondrial nucleoids dates back to the 1960s, when the term was used, in analogy with the bacterial nucleoids, to indicate specific, electron-transparent, DNA-containing small rod-like structures in the mitochondrial matrix [181]. In the following three decades, mtDNA-protein complexes were isolated from the slime mold *Physarum polycephalum* [182], HeLa cells [183], *Xenopus laevis* oocytes [184,185], yeast cells [186,187], and rat liver [188]. In more recent years, methods were developed to directly visualize nucleoids in cells using either fluorescent nucleic acid-binding dyes like 4′,6-diamidino-2-phenylindole (DAPI) [12,189] and PicoGreen [190,191] or antibodies against nucleoid-associated proteins [87,192].

The exact protein composition of the mitochondrial nucleoids is still unknown, but comprises the packaging factor TFAM, mtSSB, and other proteins involved in DNA replication, transcription, and mitochondrial metabolism. However, recent observations have pointed out that, with the exception of TFAM, the protein composition of nucleoids is not uniform and different subsets of nucleoids coexist within human cells [193], highlighting the extremely dynamic nature of these structures.

The organization of mtDNA into nucleoids might have a dual function: on one hand, the association with proteins protects the DNA and renders it more resistant to damaging agents such as reactive oxygen species produced by the respiratory chain complexes [194]. On the other hand, it provides the proper microenvironment for the different DNA transactions including mtDNA replication, transcription, and repair [195]. Therefore the mtDNA nucleoids are essential for the maintenance of mtDNA (Box 1.1).

BOX 1.1 Approaches to study nucleoids in cells and tissues

Nucleoids can be detected by staining the nucleic acid component with fluorescent dyes like DAPI [12] that binds to AT-rich regions of the genome and PicoGreen that selectively binds to dsDNA [191]. Conversely to DAPI, PicoGreen is permeable to cell membranes and can be used to visualize nucleoids in living cells [190].

Alternatively, nucleoids can be visualized by detecting the proteic-component with immunocyto- and histochemistry techniques employing specific antibodies [87,192].

Finally, the subpopulation of replicatively active nucleoids can be visualized either with antibodies against proteins of the mtDNA replisome (POL γ, Twinkle, or mtSSB) [196] or by treating the cells with thymidine analogs like BrdU (5-bromo-2′-deoxyuridine) or EdU (5-Ethynyl-2′-deoxyuridine) that will be incorporated during the replication process and can be detected with specific antibodies [106,192].

Because conventional fluorescence microscopy allows to reach resolution of 200–350 nm, and the size of a nucleoid is around 100 nm, high-resolution techniques must be used in order to visualize single nucleoids. These include STED (stimulation emission depletion) microscopy [197] and PALM (photoactivated localization) microscopy [198]. While based on different physical principles, both methods overcome the diffraction barrier and allow to resolve structures smaller than 50 nm [199].

1.4.1 Nucleoid composition

1.4.1.1 mtDNA

In relaxed form, the 16-kb-long human mtDNA molecule has a contour length of 5 μm [200]. In the nucleoids, the mtDNA is highly packaged into a condensed form that is only 100 nm in diameter [197]. Several groups have quantified the number of nucleoids and mtDNA copies in cells and found that in rapidly dividing cells, each nucleoid contains around 2–10 mtDNA molecules [201]. However, these observations were done using conventional fluorescence microscopy whose resolution does not allow resolving structures smaller than 200 nm. More recently, superresolution microscopy techniques have indicated an average of 1.1 mtDNA molecules per nucleoid [202]. Notably, two decades earlier, Satoh and Kuroiwa reached the same conclusion by calculating the fluorescence intensity of the ethidium bromide signal in the nucleoids [12]. A rise in the mtDNA copy number leads to an increased number of nucleoids rather than to a change in their morphology, further corroborating the ~1:1 ratio of mtDNA and nucleoids [202]. However, this ratio might not be applicable to all cell types and tissues, since mtDNA is not always present as single-monomeric molecule, but can also exist as catenated molecules containing multiple complete genomes [203,204].

1.4.1.2 Nucleoid-associated proteins

The first nucleoid-associated protein identified was the *Saccharomyces cerevisiae* ARS-binding factor 2 protein (Abf2p), originally named HM [205,206]. This 20-kDa protein contains two high mobility group (HMG) domains and is homologous to the DNA-binding HMG proteins located at the nuclear chromatin [207]. The protein is highly abundant in yeast and coats the entire mtDNA in a ratio of one molecule for every 15 bp [206]. The deletion of Abf2p causes the loss of mtDNA in strains grown on a fermentable carbon source (e.g., glucose), indicating the importance of this protein for the maintenance of the mtDNA. However, when *abf2*$^{-}$ cells are grown on nonfermentable carbon sources where the mitochondrial genome is not dispensable, they maintain normal levels of mtDNA [206], suggesting that also other factors are involved in yeast mtDNA maintenance. Nonetheless, *abf2*$^{-}$ cells grown under these conditions are more sensitive to damaging agents [208] and have altered nucleoid morphology and segregation [209]. The mtDNA instability due to loss of Abf2p can be rescued by expression of the 25-kDa human mitochondrial transcription factor A (TFAM) [210]. In mouse, *TFAM* is an essential gene, since its loss causes embryonic lethality and mtDNA depletion [211]. Like the yeast counterpart, it contains two HMG domains and it is sufficiently abundant to fully cover the mtDNA [212]. In vitro, the protein coordinates the packaging of DNA [213], and it colocalizes with mtDNA in vivo [192]. The crystal structure of the protein complexed with the LSP region of mtDNA revealed that TFAM, through its HMG domains, bends the DNA 180°, thus inducing a U-turn [214]. The same DNA distortion into a U-turn also occurs when TFAM binds to nonspecific DNA sequences [215]. A similar DNA bending-mode was detected also for Abf2p, indicating that it represents a common mechanism for mtDNA compaction by HMG proteins [216]. However, unlike Abf2p, TFAM plays an additional role in the initiation of mitochondrial transcription [217]. The two functions are ascribed to two different domains of the protein. Besides the HMG-box domains, TFAM contains an additional C-terminal tail that accounts for the specific binding to the LSP promoter and the role of TFAM in transcription initiation [218]. Accordingly, a chimeric Abf2p protein fused with the C-terminal tail of TFAM can activate transcription from the LSP promoter [219]. The mechanism by which TFAM coordinates both DNA packaging and transcription initiation is still unknown;[220] however, the amount of protein seems to be implicated in the switch between the two functions as high TFAM concentrations result in increased genome compaction and reduced mtDNA transcription and replication in vitro (Fig. 1.5B) [221].

Recently, it was proposed that TFAM alone can package single mtDNA molecules and form a nucleoid unit [202]. This is further corroborated by the findings that the components of the mtDNA replication machinery, such as Twinkle and mtSSB, colocalize in situ only with a subset of nucleoids, suggesting that this association is transient and that nucleoids do not have a uniform composition [193]. Besides the mtDNA replication factors, several other proteins have over the years been found to be associated with nucleoids. These factors were identified either by in situ colocalization with mtDNA or by purification or immuno-precipitation of nucleoids with antibodies against DNA, and known nucleoid-associated proteins, followed by mass-spectrometry analysis [222]. More recently, also proximity-dependent labeling methods have been used to find mtDNA-associated proteins [223]. These methods have identified several components of the mtDNA replication and transcription machineries as nucleoid proteins, but also proteins involved in RNA processing and translation, as well as other factors like the Lon protease and the ATPase family AAA-domain-containing 3A protein (ATAD3A) [224]. The Lon protease is an ATP-dependent protease that binds to mtDNA in living cells [225] and has role in the regulation of cellular levels of TFAM through degradation [226,227]. ATAD3A is a mitochondrial inner membrane (MIM) protein [228] implicated in mtDNA maintenance [229,230]. ATAD3A downregulation caused altered nucleoid distribution and the protein is proposed to be part of a protein platform that connects the nucleoids to the MIM, as discussed below [231].

1.4.2 Nucleoid topology

In human cells, mitochondrial nucleoids appear as spherical, ellipsoid structures with a diameter of around 100 nm [197,202]. When visualized with fluorescent dyes, nucleoids look like a punctate pattern inside the mitochondrial network (Fig. 1.5A). The total number of nucleoids per cell has been reported for several human and mouse cell types and ranges from 500 to a few thousand [232].

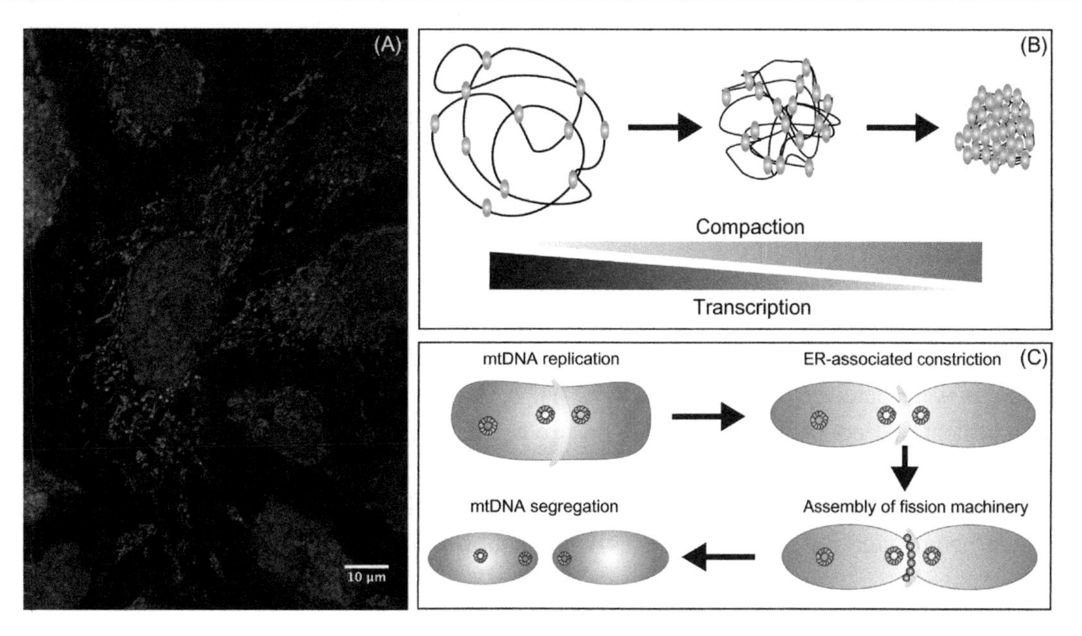

Figure 1.5: mtDNA nucleoids.
(A) In vivo imaging of HeLa cells stained with MitoTracker Red CMXRos (*red*), a fluorescent dye that accumulates into mitochondria depending on the membrane potential, and PICOGREEN (*green*), that specifically stains nucleic acids. In living cells, nucleoids appear like a punctate pattern inside the mitochondrial network. Cells were imaged with the Leica SP8 FALCON Confocal instrument equipped with a HC PL APO 63x/1.20 Water objective. (B) Schematic representation of the dual function of TFAM (*green dots*) in mtDNA metabolism. TFAM is implicated in the initiation of mtDNA transcription and in the compaction of mtDNA. Low protein abundance is linked in vitro to increased levels of mtDNA transcription, while higher TFAM concentrations induce the compaction of the mitochondrial genome. (C) Schematic representation of nucleoids segregation within the mitochondrial network. Actively replicating nucleoids (*green circles*), for example, containing POL γ (*purple*), are located in close proximity to a subset of ER-mitochondria contact site (*acid green*). Upon replication and generation of two novel mtDNA molecules, a constriction of the mitochondrial membranes occurs at these sites triggering the assembly of the fission machinery (*violet dots*). This results in the segregation of the two mtDNA molecules across the mitochondrial network.

The organization of the nucleoprotein complex is largely unknown and, as mentioned above, there is no consensus on what defines a nucleoid unit and which proteins are part of it. This uncertainty can be ascribed partly to the transient nature of the interactions of many mtDNA-associated factors with the mtDNA [193] and partly to the fact that the methods for nucleoid purification used by different groups differ in experimental conditions, and, as result, give different read-outs [222]. In 2008, Bogenhagen and coauthors [233] compared two different methods: a stringent purification that implied formaldehyde-cross-linking of the proteins to the mtDNA before nucleoid purification, and a "native" purification, in which nucleoids were immunoprecipitated using antibodies against TFAM or mtSSB after lysis of mitochondria with nonionic detergent. After comparing the cohort of proteins detected with the two methods, a layered model for mtDNA structure was proposed. According to this model, the proteins of the replication and transcription machineries are part of the central core of the nucleoids and are located in close proximity to the mtDNA, while components of the translation machinery as well as the RNA processing and respiratory chain complexes are located in the peripheral area of the nucleoids.

This model is further corroborated by the discovery that in mitochondria de novo mRNA is organized in mitochondrial RNA granules (MRGs), which are contiguous to the mitochondrial nucleoids [234,235]. Foci containing newly synthetized RNA a short distance from the nucleoids were detected already in 2004 [201]. A decade later, this novel mitochondrial subcompartment that contains the factors to ensure correct mtRNA processing [234,235] and degradation [236,237] was defined. More recently, it has been shown that the MRGs also contain the mtDNA replication-related proteins mtSSB and Twinkle, suggesting a cross-talk between these structures and the process of mtDNA replication [238].

1.4.3 Nucleoid localization

Nucleoids are not randomly distributed in the mitochondrial matrix, but at least a subpopulation interacts with the MIM [193]. The first evidence of this interaction dates back to the 1960s, when, with the use of electron microscopy techniques, the association of a large part of mtDNA molecules with mitochondrial membranes was reported [200]. This is reminiscent of the organization of the bacterial chromosome [239]. In HeLa cells, this association occurred through a region close to O_H [183]. The precise localization of nucleoids inside the mitochondrial matrix recently received renewed attention, when superresolution imaging allowed the detection of nucleoids in close proximity to the MIM, often close to mitochondrial cristae or wrapped around a cristae-like structure [198].

Only a subset of nucleoids contains the mtDNA replication factors mtSSB and Twinkle, and can thus be considered actively replicating. Moreover, Twinkle organizes in foci and localizes at the MIM even in cells devoid of mtDNA (ρ^0 cells). Together, these findings suggest that the subset of actively replicating nucleoids is associated with the membrane [193]. Interestingly, Twinkle is found in cholesterol-enriched areas of the MIM that are associated with endoplasmic reticulum (ER)-mitochondrial junctions [231]. Similarly, actively replicating nucleoids in *S. cerevisiae* are connected to the ER via the yeast-specific ERMES (ER-mitochondria encounter structure) complex [240,241]. The distribution of cholesterol in the MIM as well as that of membrane-bound nucleoids is regulated by ATAD3, supporting a role for this protein in nucleoid organization [231]. Finally, nucleoid organization might be linked to the organization of cristae structures, since the disruption of the MICOS (mitochondrial contact site and cristae organizing system) complex results in changes in nucleoid distribution and clustering, and reduces mtDNA transcription [242].

1.4.4 Nucleoid segregation

Mitochondria are dynamic organelles that constantly fuse and divide. These processes are coordinated and regulated by the fusion and fission machineries that work in concert with other cellular apparatus to guarantee mitochondrial function and ensure proper energy supply of the cell [243]. These machineries also play a role in the distribution of the nucleoids inside the mitochondrial network. Defects in the pro-fusion proteins Optic Atrophy 1 (OPA1) and the mitofusins MFN1 and MFN2 are associated with mtDNA depletion in patients and mouse models [244–246]. In contrast, the depletion of the pro-fission dynamin-related protein (DRP1) does not impact total mtDNA content, but alters nucleoid morphology. In Drp1 knock-down cells, nucleoids appear enlarged and aggregated in cristae-rich mitochondria, pointing toward a role of this protein in the distribution of nucleoids among the mitochondrial network [247].

More recently, it was shown that actively replicating nucleoids (e.g., containing POL γ) localized in close proximity of the ER-mitochondria contact sites that are engaged in mitochondrial division [196]. Upon replication and division of the mtDNA molecules, the fission machinery assembles at these specific contact sites resulting in the division of the mitochondrion into two organelles, each of them containing a nucleoid (Fig. 1.5C).

Research perspectives

Ever since the initial discovery of mtDNA c.60 years ago, we have learned a lot about its organization and maintenance. However, several key aspects remain unresolved. It is poorly understood how the mtDNA replication machinery adapts to different cellular and tissue-specific conditions. This adaptation could involve the use of alternative modes of replication, but could potentially also be achieved by merely adjusting a sole mode of replication. Elucidation of the precise role of proteins with as-yet undefined functions in mtDNA maintenance (e.g., PrimPol, Pif1, and DNA2) may shed light on the way mtDNA is duplicated in a variety of cellular circumstances. Future studies should also address the biological significance of single rNMPs in mtDNA and the regulation of mitochondrial nucleotide levels. Furthermore, novel approaches that can detect proteins of low abundance might provide a consensus regarding the exact protein composition and distribution of the mitochondrial nucleoids. Finally, recent data has shown that the processes of mtDNA replication and nucleoid segregation are intimately linked to the dynamics of the mitochondrial network. Therefore more focus will be needed in order to understand how the processes of mtDNA replication and nucleoid segregation integrate with mitochondrial and cellular functions.

Acknowledgments

We would like to acknowledge Dr. Paolo Lorenzon for helping in figure preparation.

References

[1] Nass MM, Nass S. Intramitochondrial fibers with DNA characteristics. I. Fixation and electron staining reactions. J Cell Biol 1963;19:593–611.

[2] Nass S, Nass MM. Intramitochondrial fibers with DNA characteristics. II. Enzymatic and other hydrolytic treatments. J Cell Biol 1963;19:613–29.

[3] Radloff R, Bauer W, Vinograd J. A dye-buoyant-density method for the detection and isolation of closed circular duplex DNA: the closed circular DNA in HeLa cells. Proc Natl Acad Sci U S A 1967;57 (5):1514–21.

[4] Hudson B, Vinograd J. Catenated circular DNA molecules in HeLa cell mitochondria. Nature. 1967;216 (5116):647–52.

[5] Clayton DA, Vinograd J. Circular dimer and catenate forms of mitochondrial DNA in human leukaemic leucocytes. Nature. 1967;216(5116):652–7.

[6] Clayton DA, Smith CA, Jordan JM, Teplitz M, Vinograd J. Occurrence of complex mitochondrial DNA in normal tissues. Nature 1968;220(5171):976–9.

[7] Clayton DA. Replication of animal mitochondrial DNA. Cell 1982;28(4):693–705.

[8] Borst P, Kroon AM. Mitochondrial DNA: physicochemical properties, replication, and genetic function. Int Rev Cytol 1969;26:107–90.

[9] Bogenhagen D, Clayton DA. The number of mitochondrial deoxyribonucleic acid genomes in mouse L and human HeLa cells. Quantitative isolation of mitochondrial deoxyribonucleic acid. J Biol Chem 1974;249(24):7991–5.

[10] King MP, Attardi G. Human cells lacking mtDNA: repopulation with exogenous mitochondria by complementation. Science. 1989;246(4929):500–3.

[11] D'Erchia AM, Atlante A, Gadaleta G, et al. Tissue-specific mtDNA abundance from exome data and its correlation with mitochondrial transcription, mass and respiratory activity. Mitochondrion. 2015;20:13–21.

[12] Satoh M, Kuroiwa T. Organization of multiple nucleoids and DNA molecules in mitochondria of a human cell. Exp Cell Res 1991;196(1):137–40.

[13] Chen X, Prosser R, Simonetti S, Sadlock J, Jagiello G, Schon EA. Rearranged mitochondrial genomes are present in human oocytes. Am J Hum Genet 1995;57(2):239–47.

[14] Reynier P, May-Panloup P, Chretien MF, et al. Mitochondrial DNA content affects the fertilizability of human oocytes. Mol Hum Reprod 2001;7(5):425–9.

[15] Steuerwald N, Barritt JA, Adler R, et al. Quantification of mtDNA in single oocytes, polar bodies and subcellular components by real-time rapid cycle fluorescence monitored PCR. Zygote. 2000;8(3):209–15.

[16] Giles RE, Blanc H, Cann HM, Wallace DC. Maternal inheritance of human mitochondrial DNA. Proc Natl Acad Sci U S A 1980;77(11):6715–19.

[17] Holt IJ, Harding AE, Morgan-Hughes JA. Deletions of muscle mitochondrial DNA in patients with mitochondrial myopathies. Nature. 1988;331(6158):717–19.

[18] Russell O, Turnbull D. Mitochondrial DNA disease-molecular insights and potential routes to a cure. Exp Cell Res 2014;325(1):38–43.

[19] Wong-Staal F, Mendelsohn J, Goulian M. Ribonucleotides in closed circular mitochondrial DNA from HeLa cells. Biochem Biophys Res Commun 1973;53(1):140–8.

[20] Grossman LI, Watson R, Vinograd J. The presence of ribonucleotides in mature closed-circular mitochondrial DNA. Proc Natl Acad Sci U S A 1973;70(12):3339–43.

[21] Anderson S, Bankier AT, Barrell BG, et al. Sequence and organization of the human mitochondrial genome. Nature. 1981;290(5806):457–65.

[22] Chomyn A, Cleeter MW, Ragan CI, Riley M, Doolittle RF, Attardi G. URF6, last unidentified reading frame of human mtDNA, codes for an NADH dehydrogenase subunit. Science. 1986;234 (4776):614–18.

[23] Chomyn A, Mariottini P, Cleeter MW, et al. Six unidentified reading frames of human mitochondrial DNA encode components of the respiratory-chain NADH dehydrogenase. Nature. 1985;314(6012):592–7.

[24] Chomyn A, Mariottini P, Gonzalez-Cadavid N, et al. Identification of the polypeptides encoded in the ATPase 6 gene and in the unassigned reading frames 1 and 3 of human mtDNA. Proc Natl Acad Sci U S A 1983;80(18):5535–9.

[25] Mariottini P, Chomyn A, Attardi G, Trovato D, Strong DD, Doolittle RF. Antibodies against synthetic peptides reveal that the unidentified reading frame A6L, overlapping the ATPase 6 gene, is expressed in human mitochondria. Cell. 1983;32(4):1269–77.

[26] Mariottini P, Chomyn A, Riley M, Cottrell B, Doolittle RF, Attardi G. Identification of the polypeptides encoded in the unassigned reading frames 2, 4, 4L, and 5 of human mitochondrial DNA. Proc Natl Acad Sci U S A 1986;83(6):1563–7.

[27] Corneo G, Zardi L, Polli E. Human mitochondrial DNA. J Mol Biol 1968;36(3):419–23.

[28] Bharti SK, Sommers JA, Zhou J, et al. DNA sequences proximal to human mitochondrial DNA deletion breakpoints prevalent in human disease form G-quadruplexes, a class of DNA structures inefficiently unwound by the mitochondrial replicative Twinkle helicase. J Biol Chem 2014;289(43):29975–93.

[29] Dong DW, Pereira F, Barrett SP, et al. Association of G-quadruplex forming sequences with human mtDNA deletion breakpoints. BMC Genomics. 2014;15:677.

[30] Crews S, Ojala D, Posakony J, Nishiguchi J, Attardi G. Nucleotide sequence of a region of human mitochondrial DNA containing the precisely identified origin of replication. Nature. 1979;277(5693):192–8.

[31] Chang DD, Clayton DA. Precise identification of individual promoters for transcription of each strand of human mitochondrial DNA. Cell. 1984;36(3):635–43.

[32] Chang DD, Clayton DA. Priming of human mitochondrial DNA replication occurs at the light-strand promoter. Proc Natl Acad Sci U S A 1985;82(2):351–5.

[33] Gustafsson CM, Falkenberg M, Larsson NG. Maintenance and expression of mammalian mitochondrial DNA. Annu Rev Biochem 2016;85:133–60.

[34] Walberg MW, Clayton DA. Sequence and properties of the human KB cell and mouse L cell D-loop regions of mitochondrial DNA. Nucleic Acids Res 1981;9(20):5411–21.

[35] Doda JN, Wright CT, Clayton DA. Elongation of displacement-loop strands in human and mouse mitochondrial DNA is arrested near specific template sequences. Proc Natl Acad Sci U S A 1981;78 (10):6116–20.

[36] Kasamatsu H, Robberson DL, Vinograd J. A novel closed-circular mitochondrial DNA with properties of a replicating intermediate. Proc Natl Acad Sci U S A 1971;68(9):2252–7.

[37] Robberson DL, Kasamatsu H, Vinograd J. Replication of mitochondrial DNA. Circular replicative intermediates in mouse L cells. Proc Natl Acad Sci U S A 1972;69(3):737–41.

[38] Robberson DL, Clayton DA, Morrow JF. Cleavage of replicating forms of mitochondrial DNA by EcoRI endonuclease. Proc Natl Acad Sci U S A 1974;71(11):4447–51.

[39] Brown WM, Vinograd J. Restriction endonuclease cleavage maps of animal mitochondrial DNAs. Proc Natl Acad Sci U S A 1974;71(11):4617–21.

[40] Brown WM, Shine J, Goodman HM. Human mitochondrial DNA: analysis of 7S DNA from the origin of replication. Proc Natl Acad Sci U S A 1978;75(2):735–9.

[41] Gillum AM, Clayton DA. Displacement-loop replication initiation sequence in animal mitochondrial DNA exists as a family of discrete lengths. Proc Natl Acad Sci U S A 1978;75(2):677–81.

[42] Tapper DP, Clayton DA. Mechanism of replication of human mitochondrial DNA. Localization of the 5′ ends of nascent daughter strands. J Biol Chem 1981;256(10):5109–15.

[43] Robberson DL, Davidson N. Covalent coupling of ribonucleic acid to agarose. Biochemistry. 1972;11 (4):533–7.

[44] Brown TA, Cecconi C, Tkachuk AN, Bustamante C, Clayton DA. Replication of mitochondrial DNA occurs by strand displacement with alternative light-strand origins, not via a strand-coupled mechanism. Genes Dev 2005;19(20):2466–76.
[45] Wanrooij S, Falkenberg M. The human mitochondrial replication fork in health and disease. Biochim Biophys Acta 2010;1797(8):1378–88.
[46] Wanrooij S, Fuste JM, Farge G, Shi Y, Gustafsson CM, Falkenberg M. Human mitochondrial RNA polymerase primes lagging-strand DNA synthesis in vitro. Proc Natl Acad Sci U S A 2008;105(32):11122–7.
[47] Masukata H, Tomizawa J. A mechanism of formation of a persistent hybrid between elongating RNA and template DNA. Cell. 1990;62(2):331–8.
[48] Fuste JM, Wanrooij S, Jemt E, et al. Mitochondrial RNA polymerase is needed for activation of the origin of light-strand DNA replication. Mol Cell. 2010;37(1):67–78.
[49] Yasukawa T, Reyes A, Cluett TJ, et al. Replication of vertebrate mitochondrial DNA entails transient ribonucleotide incorporation throughout the lagging strand. EMBO J. 2006;25(22):5358–71.
[50] Yasukawa T, Kang D. An overview of mammalian mitochondrial DNA replication mechanisms. J Biochem 2018;164(3):183–93.
[51] Wanrooij S, Miralles Fuste J, Stewart JB, et al. In vivo mutagenesis reveals that OriL is essential for mitochondrial DNA replication. EMBO Rep. 2012;13(12):1130–7.
[52] Pohjoismaki JLO, Forslund JME, Goffart S, Torregrosa-Munumer R, Wanrooij S. Known unknowns of mammalian mitochondrial DNA maintenance. Bioessays. 2018;40(9):e1800102.
[53] Pohjoismaki JL, Goffart S. Of circles, forks and humanity: topological organisation and replication of mammalian mitochondrial DNA. Bioessays. 2011;33(4):290–9.
[54] Torregrosa-Munumer R, Forslund JME, Goffart S, et al. PrimPol is required for replication reinitiation after mtDNA damage. Proc Natl Acad Sci U S A 2017;114(43):11398–403.
[55] Holt IJ, Lorimer HE, Jacobs HT. Coupled leading- and lagging-strand synthesis of mammalian mitochondrial DNA. Cell. 2000;100(5):515–24.
[56] Wong TW, Clayton DA. In vitro replication of human mitochondrial DNA: accurate initiation at the origin of light-strand synthesis. Cell. 1985;42(3):951–8.
[57] Wong TW, Clayton DA. Isolation and characterization of a DNA primase from human mitochondria. J Biol Chem 1985;260(21):11530–5.
[58] Tiranti V, Savoia A, Forti F, et al. Identification of the gene encoding the human mitochondrial RNA polymerase (h-mtRPOL) by cyberscreening of the Expressed Sequence Tags database. Hum Mol Genet 1997;6(4):615–25.
[59] D'Souza AR, Minczuk M. Mitochondrial transcription and translation: overview. Essays Biochem 2018;62(3):309–20.
[60] Pham XH, Farge G, Shi Y, Gaspari M, Gustafsson CM, Falkenberg M. Conserved sequence box II directs transcription termination and primer formation in mitochondria. J Biol Chem 2006;281(34):24647–52.
[61] Wanrooij PH, Uhler JP, Simonsson T, Falkenberg M, Gustafsson CM. G-quadruplex structures in RNA stimulate mitochondrial transcription termination and primer formation. Proc Natl Acad Sci U S A 2010;107(37):16072–7.
[62] Xu B, Clayton DA. RNA–DNA hybrid formation at the human mitochondrial heavy-strand origin ceases at replication start sites: an implication for RNA–DNA hybrids serving as primers. EMBO J. 1996;15 (12):3135–43.
[63] Wanrooij PH, Uhler JP, Shi Y, Westerlund F, Falkenberg M, Gustafsson CM. A hybrid G-quadruplex structure formed between RNA and DNA explains the extraordinary stability of the mitochondrial R-loop. Nucleic Acids Res 2012;40(20):10334–44.
[64] Kang D, Miyako K, Kai Y, Irie T, Takeshige K. In vivo determination of replication origins of human mitochondrial DNA by ligation-mediated polymerase chain reaction. J Biol Chem 1997;272(24):15275–9.
[65] Posse V, Al-Behadili A, Uhler JP, et al. RNase H1 directs origin-specific initiation of DNA replication in human mitochondria. PLoS Genet. 2019;15(1):e1007781.

[66] Posse V, Shahzad S, Falkenberg M, Hallberg BM, Gustafsson CM. TEFM is a potent stimulator of mitochondrial transcription elongation in vitro. Nucleic Acids Res 2015;43(5):2615–24.
[67] Agaronyan K, Morozov YI, Anikin M, Temiakov D. Mitochondrial biology. Replication-transcription switch in human mitochondria. Science 2015;347(6221):548–51.
[68] Jiang S, Koolmeister C, Misic J, et al. TEFM regulates both transcription elongation and RNA processing in mitochondria. EMBO Rep. 2019;20(6).
[69] Holmes JB, Akman G, Wood SR, et al. Primer retention owing to the absence of RNase H1 is catastrophic for mitochondrial DNA replication. Proc Natl Acad Sci U S A 2015;112(30):9334–9.
[70] Reyes A, Melchionda L, Nasca A, et al. RNASEH1 mutations impair mtDNA replication and cause adult-onset mitochondrial encephalomyopathy. Am J Hum Genet 2015;97(1):186–93.
[71] Bolden A, Noy GP, Weissbach A. DNA polymerase of mitochondria is a gamma-polymerase. J Biol Chem 1977;252(10):3351–6.
[72] Korhonen JA, Pham XH, Pellegrini M, Falkenberg M. Reconstitution of a minimal mtDNA replisome in vitro. EMBO J 2004;23(12):2423–9.
[73] Fridlender B, Fry M, Bolden A, Weissbach A. A new synthetic RNA-dependent DNA polymerase from human tissue culture cells (HeLa-fibroblast-synthetic oligonucleotides-template-purified enzymes). Proc Natl Acad Sci U S A 1972;69(2):452–5.
[74] Gray H, Wong TW. Purification and identification of subunit structure of the human mitochondrial DNA polymerase. J Biol Chem 1992;267(9):5835–41.
[75] Yakubovskaya E, Chen Z, Carrodeguas JA, Kisker C, Bogenhagen DF. Functional human mitochondrial DNA polymerase gamma forms a heterotrimer. J Biol Chem 2006;281(1):374–82.
[76] Carrodeguas JA, Theis K, Bogenhagen DF, Kisker C. Crystal structure and deletion analysis show that the accessory subunit of mammalian DNA polymerase gamma, Pol gamma B, functions as a homodimer. Mol Cell. 2001;7(1):43–54.
[77] Longley MJ, Prasad R, Srivastava DK, Wilson SH, Copeland WC. Identification of 5′-deoxyribose phosphate lyase activity in human DNA polymerase gamma and its role in mitochondrial base excision repair in vitro. Proc Natl Acad Sci U S A 1998;95(21):12244–8.
[78] Longley MJ, Ropp PA, Lim SE, Copeland WC. Characterization of the native and recombinant catalytic subunit of human DNA polymerase gamma: identification of residues critical for exonuclease activity and dideoxynucleotide sensitivity. Biochemistry. 1998;37(29):10529–39.
[79] Longley MJ, Nguyen D, Kunkel TA, Copeland WC. The fidelity of human DNA polymerase gamma with and without exonucleolytic proofreading and the p55 accessory subunit. J Biol Chem 2001;276(42):38555–62.
[80] Johnson AA, Johnson KA. Exonuclease proofreading by human mitochondrial DNA polymerase. J Biol Chem 2001;276(41):38097–107.
[81] Lim SE, Longley MJ, Copeland WC. The mitochondrial p55 accessory subunit of human DNA polymerase gamma enhances DNA binding, promotes processive DNA synthesis, and confers N-ethylmaleimide resistance. J Biol Chem 1999;274(53):38197–203.
[82] Johnson AA, Tsai Y, Graves SW, Johnson KA. Human mitochondrial DNA polymerase holoenzyme: reconstitution and characterization. Biochemistry. 2000;39(7):1702–8.
[83] Lee YS, Lee S, Demeler B, Molineux IJ, Johnson KA, Yin YW. Each monomer of the dimeric accessory protein for human mitochondrial DNA polymerase has a distinct role in conferring processivity. J Biol Chem 2010;285(2):1490–9.
[84] Carrodeguas JA, Pinz KG, Bogenhagen DF. DNA binding properties of human pol gammaB. J Biol Chem 2002;277(51):50008–14.
[85] Lee YS, Kennedy WD, Yin YW. Structural insight into processive human mitochondrial DNA synthesis and disease-related polymerase mutations. Cell. 2009;139(2):312–24.
[86] Farge G, Pham XH, Holmlund T, Khorostov I, Falkenberg M. The accessory subunit B of DNA polymerase gamma is required for mitochondrial replisome function. Nucleic Acids Res 2007;35(3):902–11.

[87] Spelbrink JN, Li FY, Tiranti V, et al. Human mitochondrial DNA deletions associated with mutations in the gene encoding Twinkle, a phage T7 gene 4-like protein localized in mitochondria. Nat Genet. 2001;28(3):223–31.

[88] Korhonen JA, Gaspari M, Falkenberg M. TWINKLE Has 5′ -> 3′ DNA helicase activity and is specifically stimulated by mitochondrial single-stranded DNA-binding protein. J Biol Chem 2003;278 (49):48627–32.

[89] Farge G, Holmlund T, Khvorostova J, Rofougaran R, Hofer A, Falkenberg M. The N-terminal domain of TWINKLE contributes to single-stranded DNA binding and DNA helicase activities. Nucleic Acids Res 2008;36(2):393–403.

[90] Ziebarth TD, Gonzalez-Soltero R, Makowska-Grzyska MM, Nunez-Ramirez R, Carazo JM, Kaguni LS. Dynamic effects of cofactors and DNA on the oligomeric state of human mitochondrial DNA helicase. J Biol Chem 2010;285(19):14639–47.

[91] Fernandez-Millan P, Lazaro M, Cansiz-Arda S, et al. The hexameric structure of the human mitochondrial replicative helicase Twinkle. Nucleic Acids Res 2015;43(8):4284–95.

[92] Tiranti V, Rocchi M, DiDonato S, Zeviani M. Cloning of human and rat cDNAs encoding the mitochondrial single-stranded DNA-binding protein (SSB). Gene. 1993;126(2):219–25.

[93] Curth U, Urbanke C, Greipel J, Gerberding H, Tiranti V, Zeviani M. Single-stranded-DNA-binding proteins from human mitochondria and *Escherichia coli* have analogous physicochemical properties. Eur J Biochem 1994;221(1):435–43.

[94] Yang C, Curth U, Urbanke C, Kang C. Crystal structure of human mitochondrial single-stranded DNA binding protein at 2.4 A resolution. Nat Struct Biol 1997;4(2):153–7.

[95] Kaur P, Longley MJ, Pan H, Wang H, Copeland WC. Single-molecule DREEM imaging reveals DNA wrapping around human mitochondrial single-stranded DNA binding protein. Nucleic Acids Res 2018;46 (21):11287–302.

[96] Oliveira MT, Kaguni LS. Functional roles of the N- and C-terminal regions of the human mitochondrial single-stranded DNA-binding protein. PLoS One. 2010;5(10):e15379.

[97] Oliveira MT, Kaguni LS. Reduced stimulation of recombinant DNA polymerase gamma and mitochondrial DNA (mtDNA) helicase by variants of mitochondrial single-stranded DNA-binding protein (mtSSB) correlates with defects in mtDNA replication in animal cells. J Biol Chem 2011;286 (47):40649–58.

[98] Ciesielski GL, Bermek O, Rosado-Ruiz FA, et al. Mitochondrial single-stranded DNA-binding proteins stimulate the activity of DNA polymerase gamma by organization of the template DNA. J Biol Chem 2015;290(48):28697–707.

[99] Williams AJ, Kaguni LS. Stimulation of Drosophila mitochondrial DNA polymerase by single-stranded DNA-binding protein. J Biol Chem 1995;270(2):860–5.

[100] Thommes P, Farr CL, Marton RF, Kaguni LS, Cotterill S. Mitochondrial single-stranded DNA-binding protein from Drosophila embryos. Physical and biochemical characterization. J Biol Chem 1995;270 (36):21137–43.

[101] Farr CL, Wang Y, Kaguni LS. Functional interactions of mitochondrial DNA polymerase and single-stranded DNA-binding protein. Template-primer DNA binding and initiation and elongation of DNA strand synthesis. J Biol Chem 1999;274(21):14779–85.

[102] Van Goethem G, Dermaut B, Lofgren A, Martin JJ, Van Broeckhoven C. Mutation of POLG is associated with progressive external ophthalmoplegia characterized by mtDNA deletions. Nat Genet. 2001;28 (3):211–12.

[103] Longley MJ, Clark S, Yu Wai Man C, et al. Mutant POLG2 disrupts DNA polymerase gamma subunits and causes progressive external ophthalmoplegia. Am J Hum Genet 2006;78(6):1026–34.

[104] Young MJ, Copeland WC. Human mitochondrial DNA replication machinery and disease. Curr Opin Genet Dev 2016;38:52–62.

[105] Del Dotto V, Ullah F, Di Meo I, et al. SSBP1 mutations cause mtDNA depletion underlying a complex optic atrophy disorder. J Clin Invest 2019.

[106] Piro-Megy C, Sarzi E, Tarres-Sole A, et al. Dominant mutations in mtDNA maintenance gene SSBP1 cause optic atrophy and foveopathy. J Clin Invest 2019.

[107] Wanrooij PH, Chabes A. Ribonucleotides in mitochondrial DNA. FEBS Lett. 2019.

[108] Kasiviswanathan R, Copeland WC. Ribonucleotide discrimination and reverse transcription by the human mitochondrial DNA polymerase. J Biol Chem 2011;286(36):31490–500.

[109] Lima WF, Rose JB, Nichols JG, et al. Human RNase H1 discriminates between subtle variations in the structure of the heteroduplex substrate. Mol Pharmacol. 2007;71(1):83–91.

[110] Al-Behadili A, Uhler JP, Berglund AK, et al. A two-nuclease pathway involving RNase H1 is required for primer removal at human mitochondrial OriL. Nucleic Acids Res 2018;46(18):9471–83.

[111] Liu P, Qian L, Sung JS, et al. Removal of oxidative DNA damage via FEN1-dependent long-patch base excision repair in human cell mitochondria. Mol Cell Biol 2008;28(16):4975–87.

[112] Stodola JL, Burgers PM. Mechanism of lagging-strand DNA replication in eukaryotes. Adv Exp Med Biol 2017;1042:117–33.

[113] Kazak L, Reyes A, He J, et al. A cryptic targeting signal creates a mitochondrial FEN1 isoform with tailed R-Loop binding properties. PLoS One. 2013;8(5):e62340.

[114] Uhler JP, Falkenberg M. Primer removal during mammalian mitochondrial DNA replication. DNA Repair (Amst) 2015;34:28–38.

[115] Kornblum C, Nicholls TJ, Haack TB, et al. Loss-of-function mutations in MGME1 impair mtDNA replication and cause multisystemic mitochondrial disease. Nat Genet. 2013;45(2):214–19.

[116] Uhler JP, Thorn C, Nicholls TJ, et al. MGME1 processes flaps into ligatable nicks in concert with DNA polymerase gamma during mtDNA replication. Nucleic Acids Res 2016;44(12):5861–71.

[117] Wu CC, Lin JLJ, Yang-Yen HF, Yuan HS. A unique exonuclease ExoG cleaves between RNA and DNA in mitochondrial DNA replication. Nucleic Acids Res 2019;47(10):5405–19.

[118] Cymerman IA, Chung I, Beckmann BM, Bujnicki JM, Meiss G. EXOG, a novel paralog of Endonuclease G in higher eukaryotes. Nucleic Acids Res 2008;36(4):1369–79.

[119] Lakshmipathy U, Campbell C. The human DNA ligase III gene encodes nuclear and mitochondrial proteins. Mol Cell Biol 1999;19(5):3869–76.

[120] Lakshmipathy U, Campbell C. Antisense-mediated decrease in DNA ligase III expression results in reduced mitochondrial DNA integrity. Nucleic Acids Res 2001;29(3):668–76.

[121] Nicholls TJ, Nadalutti CA, Motori E, et al. Topoisomerase 3alpha is required for decatenation and segregation of human mtDNA. Mol Cell. 2018;69(1):9–23 e26.

[122] Wang Y, Lyu YL, Wang JC. Dual localization of human DNA topoisomerase IIIalpha to mitochondria and nucleus. Proc Natl Acad Sci U S A 2002;99(19):12114–19.

[123] Pommier Y, Sun Y, Huang SN, Nitiss JL. Roles of eukaryotic topoisomerases in transcription, replication and genomic stability. Nat Rev Mol Cell Biol 2016;17(11):703–21.

[124] Zhang H, Barcelo JM, Lee B, et al. Human mitochondrial topoisomerase I. Proc Natl Acad Sci U S A 2001;98(19):10608–13.

[125] Dalla Rosa I, Huang SY, Agama K, Khiati S, Zhang H, Pommier Y. Mapping topoisomerase sites in mitochondrial DNA with a poisonous mitochondrial topoisomerase I (Top1mt). J Biol Chem 2014;289(26):18595–602.

[126] Khiati S, Baechler SA, Factor VM, et al. Lack of mitochondrial topoisomerase I (TOP1mt) impairs liver regeneration. Proc Natl Acad Sci U S A 2015;112(36):11282–7.

[127] Hangas A, Aasumets K, Kekalainen NJ, et al. Ciprofloxacin impairs mitochondrial DNA replication initiation through inhibition of Topoisomerase 2. Nucleic Acids Res 2018;46(18):9625–36.

[128] Duxin JP, Dao B, Martinsson P, et al. Human DNA2 is a nuclear and mitochondrial DNA maintenance protein. Mol Cell Biol 2009;29(15):4274–82.

[129] Ronchi D, Di Fonzo A, Lin W, et al. Mutations in DNA2 link progressive myopathy to mitochondrial DNA instability. Am J Hum Genet 2013;92(2):293–300.

[130] Phowthongkum P, Sun A. Novel truncating variant in DNA2-related congenital onset myopathy and ptosis suggests genotype-phenotype correlation. Neuromuscul Disord. 2017;27(7):616–18.

[131] Pinto C, Kasaciunaite K, Seidel R, Cejka P. Human DNA2 possesses a cryptic DNA unwinding activity that functionally integrates with BLM or WRN helicases. Elife 2016;5.
[132] Sabouri N. The functions of the multi-tasking Pfh1(Pif1) helicase. Curr Genet. 2017;63(4):621–6.
[133] Kazak L, Reyes A, Duncan AL, et al. Alternative translation initiation augments the human mitochondrial proteome. Nucleic Acids Res 2013;41(4):2354–69.
[134] Bannwarth S, Berg-Alonso L, Auge G, et al. Inactivation of Pif1 helicase causes a mitochondrial myopathy in mice. Mitochondrion 2016.
[135] Garcia-Gomez S, Reyes A, Martinez-Jimenez MI, et al. PrimPol, an archaic primase/polymerase operating in human cells. Mol Cell. 2013;52(4):541–53.
[136] Guilliam TA, Jozwiakowski SK, Ehlinger A, et al. Human PrimPol is a highly error-prone polymerase regulated by single-stranded DNA binding proteins. Nucleic Acids Res 2015;43(2):1056–68.
[137] Stojkovic G, Makarova AV, Wanrooij PH, Forslund J, Burgers PM, Wanrooij S. Oxidative DNA damage stalls the human mitochondrial replisome. Sci Rep. 2016;6:28942.
[138] Saada A, Shaag A, Mandel H, Nevo Y, Eriksson S, Elpeleg O. Mutant mitochondrial thymidine kinase in mitochondrial DNA depletion myopathy. Nat Genet. 2001;29(3):342–4.
[139] Mandel H, Szargel R, Labay V, et al. The deoxyguanosine kinase gene is mutated in individuals with depleted hepatocerebral mitochondrial DNA. Nat Genet. 2001;29(3):337–41.
[140] Nishino I, Spinazzola A, Hirano M. Thymidine phosphorylase gene mutations in MNGIE, a human mitochondrial disorder. Science. 1999;283(5402):689–92.
[141] Bourdon A, Minai L, Serre V, et al. Mutation of RRM2B, encoding p53-controlled ribonucleotide reductase (p53R2), causes severe mitochondrial DNA depletion. Nat Genet. 2007;39(6):776–80.
[142] Tyynismaa H, Sun R, Ahola-Erkkila S, et al. Thymidine kinase 2 mutations in autosomal recessive progressive external ophthalmoplegia with multiple mitochondrial DNA deletions. Hum Mol Genet 2012;21 (1):66–75.
[143] Nordlund P, Reichard P. Ribonucleotide reductases. Annu Rev Biochem 2006;75:681–706.
[144] Milon L, Meyer P, Chiadmi M, et al. The human nm23-H4 gene product is a mitochondrial nucleoside diphosphate kinase. J Biol Chem 2000;275(19):14264–72.
[145] Tsuiki H, Nitta M, Furuya A, et al. A novel human nucleoside diphosphate (NDP) kinase, Nm23-H6, localizes in mitochondria and affects cytokinesis. J Cell Biochem 1999;76(2):254–69.
[146] Chen CW, Wang HL, Huang CW, et al. Two separate functions of NME3 critical for cell survival underlie a neurodegenerative disorder. Proc Natl Acad Sci U S A 2019;116(2):566–74.
[147] Chabes A, Thelander L. Controlled protein degradation regulates ribonucleotide reductase activity in proliferating mammalian cells during the normal cell cycle and in response to DNA damage and replication blocks. J Biol Chem 2000;275(23):17747–53.
[148] Tanaka H, Arakawa H, Yamaguchi T, et al. A ribonucleotide reductase gene involved in a p53-dependent cell-cycle checkpoint for DNA damage. Nature. 2000;404(6773):42–9.
[149] Nakano K, Balint E, Ashcroft M, Vousden KH. A ribonucleotide reductase gene is a transcriptional target of p53 and p73. Oncogene. 2000;19(37):4283–9.
[150] Guittet O, Hakansson P, Voevodskaya N, et al. Mammalian p53R2 protein forms an active ribonucleotide reductase in vitro with the R1 protein, which is expressed both in resting cells in response to DNA damage and in proliferating cells. J Biol Chem 2001;276(44):40647–51.
[151] Pontarin G, Ferraro P, Hakansson P, Thelander L, Reichard P, Bianchi V. p53R2-dependent ribonucleotide reduction provides deoxyribonucleotides in quiescent human fibroblasts in the absence of induced DNA damage. J Biol Chem 2007;282(23):16820–8.
[152] Hakansson P, Hofer A, Thelander L. Regulation of mammalian ribonucleotide reduction and dNTP pools after DNA damage and in resting cells. J Biol Chem 2006;281(12):7834–41.
[153] Arner ES, Eriksson S. Mammalian deoxyribonucleoside kinases. Pharmacol Ther. 1995;67(2):155–86.
[154] Pontarin G, Gallinaro L, Ferraro P, Reichard P, Bianchi V. Origins of mitochondrial thymidine triphosphate: dynamic relations to cytosolic pools. Proc Natl Acad Sci U S A 2003;100(21):12159–64.

[155] Ferraro P, Nicolosi L, Bernardi P, Reichard P, Bianchi V. Mitochondrial deoxynucleotide pool sizes in mouse liver and evidence for a transport mechanism for thymidine monophosphate. Proc Natl Acad Sci U S A 2006;103(49):18586–91.
[156] Leanza L, Ferraro P, Reichard P, Bianchi V. Metabolic interrelations within guanine deoxynucleotide pools for mitochondrial and nuclear DNA maintenance. J Biol Chem 2008;283(24):16437–45.
[157] Lai Y, Tse CM, Unadkat JD. Mitochondrial expression of the human equilibrative nucleoside transporter 1 (hENT1) results in enhanced mitochondrial toxicity of antiviral drugs. J Biol Chem 2004;279(6):4490–7.
[158] Govindarajan R, Leung GP, Zhou M, Tse CM, Wang J, Unadkat JD. Facilitated mitochondrial import of antiviral and anticancer nucleoside drugs by human equilibrative nucleoside transporter-3. Am J Physiol Gastrointest Liver Physiol 2009;296(4):G910–922.
[159] Di Noia MA, Todisco S, Cirigliano A, et al. The human SLC25A33 and SLC25A36 genes of solute carrier family 25 encode two mitochondrial pyrimidine nucleotide transporters. J Biol Chem 2014;289 (48):33137–48.
[160] Floyd S, Favre C, Lasorsa FM, et al. The insulin-like growth factor-I-mTOR signaling pathway induces the mitochondrial pyrimidine nucleotide carrier to promote cell growth. Mol Biol Cell 2007;18 (9):3545–55.
[161] Franzolin E, Miazzi C, Frangini M, Palumbo E, Rampazzo C, Bianchi V. The pyrimidine nucleotide carrier PNC1 and mitochondrial trafficking of thymidine phosphates in cultured human cells. Exp Cell Res 2012;318(17):2226–36.
[162] Rampazzo C, Miazzi C, Franzolin E, et al. Regulation by degradation, a cellular defense against deoxyribonucleotide pool imbalances. Mutat Res. 2010;703(1):2–10.
[163] Song S, Wheeler LJ, Mathews CK. Deoxyribonucleotide pool imbalance stimulates deletions in HeLa cell mitochondrial DNA. J Biol Chem 2003;278(45):43893–6.
[164] Goldstone DC, Ennis-Adeniran V, Hedden JJ, et al. HIV-1 restriction factor SAMHD1 is a deoxynucleoside triphosphate triphosphohydrolase. Nature. 2011;480(7377):379–82.
[165] Powell RD, Holland PJ, Hollis T, Perrino FW. Aicardi-Goutieres syndrome gene and HIV-1 restriction factor SAMHD1 is a dGTP-regulated deoxynucleotide triphosphohydrolase. J Biol Chem 2011;286 (51):43596–600.
[166] Franzolin E, Pontarin G, Rampazzo C, et al. The deoxynucleotide triphosphohydrolase SAMHD1 is a major regulator of DNA precursor pools in mammalian cells. Proc Natl Acad Sci U S A 2013;110 (35):14272–7.
[167] Franzolin E, Salata C, Bianchi V, Rampazzo C. The deoxynucleoside triphosphate triphosphohydrolase activity of SAMHD1 protein contributes to the mitochondrial DNA depletion associated with genetic deficiency of deoxyguanosine kinase. J Biol Chem 2015;290(43):25986–96.
[168] Miyaki M, Koide K, Ono T. RNase and alkali sensitivity of closed circular mitochondrial DNA of rat ascites hepatoma cells. Biochem Biophys Res Commun 1973;50(2):252–8.
[169] Kong Z, Jia S, Chabes AL, et al. Simultaneous determination of ribonucleoside and deoxyribonucleoside triphosphates in biological samples by hydrophilic interaction liquid chromatography coupled with tandem mass spectrometry. Nucleic Acids Res 2018;46(11):e66.
[170] Nick McElhinny SA, Watts BE, Kumar D, et al. Abundant ribonucleotide incorporation into DNA by yeast replicative polymerases. Proc Natl Acad Sci U S A 2010;107(11):4949–54.
[171] Clausen AR, Zhang S, Burgers PM, Lee MY, Kunkel TA. Ribonucleotide incorporation, proofreading and bypass by human DNA polymerase delta. DNA Repair (Amst) 2013;12(2):121–7.
[172] Goksenin AY, Zahurancik W, LeCompte KG, Taggart DJ, Suo Z, Pursell ZF. Human DNA polymerase epsilon is able to efficiently extend from multiple consecutive ribonucleotides. J Biol Chem 2012;287 (51):42675–84.
[173] Forslund JME, Pfeiffer A, Stojkovic G, Wanrooij PH, Wanrooij S. The presence of rNTPs decreases the speed of mitochondrial DNA replication. PLoS Genet. 2018;14(3):e1007315.
[174] Sparks JL, Chon H, Cerritelli SM, et al. RNase H2-initiated ribonucleotide excision repair. Mol Cell. 2012;47(6):980–6.

[175] Reijns MA, Rabe B, Rigby RE, et al. Enzymatic removal of ribonucleotides from DNA is essential for mammalian genome integrity and development. Cell. 2012;149(5):1008–22.

[176] Hiller B, Achleitner M, Glage S, Naumann R, Behrendt R, Roers A. Mammalian RNase H2 removes ribonucleotides from DNA to maintain genome integrity. J Exp Med 2012;209(8):1419–26.

[177] Berglund AK, Navarrete C, Engqvist MK, et al. Nucleotide pools dictate the identity and frequency of ribonucleotide incorporation in mitochondrial DNA. PLoS Genet. 2017;13(2):e1006628.

[178] Wanrooij PH, Engqvist MKM, Forslund JME, et al. Ribonucleotides incorporated by the yeast mitochondrial DNA polymerase are not repaired. Proc Natl Acad Sci U S A 2017;114 (47):12466–71.

[179] Moss CF, Dalla Rosa I, Hunt LE, et al. Aberrant ribonucleotide incorporation and multiple deletions in mitochondrial DNA of the murine MPV17 disease model. Nucleic Acids Res 2017;45 (22):12808–15.

[180] Wanrooij PH, Tran P, Thompson LJ, et al. Age-dependent loss of mitochondrial DNA integrity in mammalian muscle. bioRxiv 2019;746719.

[181] Nass MM. Mitochondrial DNA: advances, problems, and goals. Science. 1969;165(3888):25–35.

[182] Kuroiwa T, Kawano S, Hizume M. A method of isolation of mitochondrial nucleoid of *Physarum polycephalum* and evidence for the presence of a basic protein. Exp Cell Res 1976;97(2):435–40.

[183] Albring M, Griffith J, Attardi G. Association of a protein structure of probable membrane derivation with HeLa cell mitochondrial DNA near its origin of replication. Proc Natl Acad Sci U S A 1977;74 (4):1348–52.

[184] Barat M, Rickwood D, Dufresne C, Mounolou JC. Characterization of DNA-protein complexes from the mitochondria of *Xenopus laevis* oocytes. Exp Cell Res 1985;157(1):207–17.

[185] Pinon H, Barat M, Tourte M, Dufresne C, Mounolou JC. Evidence for a mitochondrial chromosome in *Xenopus laevis* oocytes. Chromosoma. 1978;65(4):383–9.

[186] Miyakawa I, Sando N, Kawano S, Nakamura S, Kuroiwa T. Isolation of morphologically intact mitochondrial nucleoids from the yeast, *Saccharomyces cerevisiae*. J Cell Sci 1987;88(Pt 4):431–9.

[187] Rickwood D, Chambers JA, Barat M. Isolation and preliminary characterisation of DNA-protein complexes from the mitochondria of *Saccharomyces cerevisiae*. Exp Cell Res 1981;133(1):1–13.

[188] Van Tuyle GC, McPherson ML. A compact form of rat liver mitochondrial DNA stabilized by bound proteins. J Biol Chem 1979;254(13):6044–53.

[189] Sando N, Miyakawa I, Nishibayashi S, Kuroiwa T. Arrangement of mitochondrial nucleoids during life-cycle of *Saccharomyces cerevisiae*. J Gen Appl Microbiol 1981;27(6):511–16.

[190] Ashley N, Harris D, Poulton J. Detection of mitochondrial DNA depletion in living human cells using PicoGreen staining. Exp Cell Res 2005;303(2):432–46.

[191] Bereiter-Hahn J, Vöth M. Distribution and dynamics of mitochondrial nucleoids in animal cells in culture. EBO — Experimental biology online annual, vol. 1996/1997. Berlin, Heidelberg: Springer; 1998.

[192] Garrido N, Griparic L, Jokitalo E, Wartiovaara J, van der Bliek AM, Spelbrink JN. Composition and dynamics of human mitochondrial nucleoids. Mol Biol Cell 2003;14(4):1583–96.

[193] Rajala N, Gerhold JM, Martinsson P, Klymov A, Spelbrink JN. Replication factors transiently associate with mtDNA at the mitochondrial inner membrane to facilitate replication. Nucleic Acids Res 2014;42 (2):952–67.

[194] Miyakawa I. Organization and dynamics of yeast mitochondrial nucleoids. Proc Jpn Acad Ser B Phys Biol Sci 2017;93(5):339–59.

[195] Spelbrink JN. Functional organization of mammalian mitochondrial DNA in nucleoids: history, recent developments, and future challenges. IUBMB Life. 2010;62(1):19–32.

[196] Lewis SC, Uchiyama LF, Nunnari J. ER-mitochondria contacts couple mtDNA synthesis with mitochondrial division in human cells. Science. 2016;353(6296):aaf5549.

[197] Kukat C, Wurm CA, Spahr H, Falkenberg M, Larsson NG, Jakobs S. Super-resolution microscopy reveals that mammalian mitochondrial nucleoids have a uniform size and frequently contain a single copy of mtDNA. Proc Natl Acad Sci U S A 2011;108(33):13534–9.

[198] Brown TA, Tkachuk AN, Shtengel G, et al. Superresolution fluorescence imaging of mitochondrial nucleoids reveals their spatial range, limits, and membrane interaction. Mol Cell Biol 2011;31 (24):4994–5010.

[199] Jakobs S, Wurm CA. Super-resolution microscopy of mitochondria. Curr Opin Chem Biol 2014;20:9–15.

[200] Nass MM. Mitochondrial DNA. I. Intramitochondrial distribution and structural relations of single- and double-length circular DNA. J Mol Biol 1969;42(3):521–8.

[201] Iborra FJ, Kimura H, Cook PR. The functional organization of mitochondrial genomes in human cells. BMC Biol. 2004;2:9.

[202] Kukat C, Davies KM, Wurm CA, et al. Cross-strand binding of TFAM to a single mtDNA molecule forms the mitochondrial nucleoid. Proc Natl Acad Sci U S A 2015;112(36):11288–93.

[203] Piko L, Matsumoto L. Complex forms and replicative intermediates of mitochondrial DNA in tissues from adult and senescent mice. Nucleic Acids Res 1977;4(5):1301–14.

[204] Pohjoismaki JL, Goffart S, Tyynismaa H, et al. Human heart mitochondrial DNA is organized in complex catenated networks containing abundant four-way junctions and replication forks. J Biol Chem 2009;284(32):21446–57.

[205] Caron F, Jacq C, Rouviere-Yaniv J. Characterization of a histone-like protein extracted from yeast mitochondria. Proc Natl Acad Sci U S A 1979;76(9):4265–9.

[206] Diffley JF, Stillman B. A close relative of the nuclear, chromosomal high-mobility group protein HMG1 in yeast mitochondria. Proc Natl Acad Sci U S A 1991;88(17):7864–8.

[207] Landsman D, Bustin M. A signature for the HMG-1 box DNA-binding proteins. Bioessays. 1993;15 (8):539–46.

[208] Chen XJ, Wang X, Kaufman BA, Butow RA. Aconitase couples metabolic regulation to mitochondrial DNA maintenance. Science. 2005;307(5710):714–17.

[209] Miyakawa I, Kanayama M, Fujita Y, Sato H. Morphology and protein composition of the mitochondrial nucleoids in yeast cells lacking Abf2p, a high mobility group protein. J Gen Appl Microbiol 2010;56 (6):455–64.

[210] Parisi MA, Xu B, Clayton DA. A human mitochondrial transcriptional activator can functionally replace a yeast mitochondrial HMG-box protein both in vivo and in vitro. Mol Cell Biol 1993;13(3):1951–61.

[211] Larsson NG, Wang J, Wilhelmsson H, et al. Mitochondrial transcription factor A is necessary for mtDNA maintenance and embryogenesis in mice. Nat Genet. 1998;18(3):231–6.

[212] Alam TI, Kanki T, Muta T, et al. Human mitochondrial DNA is packaged with TFAM. Nucleic Acids Res 2003;31(6):1640–5.

[213] Kaufman BA, Durisic N, Mativetsky JM, et al. The mitochondrial transcription factor TFAM coordinates the assembly of multiple DNA molecules into nucleoid-like structures. Mol Biol Cell 2007;18 (9):3225–36.

[214] Rubio-Cosials A, Sidow JF, Jimenez-Menendez N, et al. Human mitochondrial transcription factor A induces a U-turn structure in the light strand promoter. Nat Struct Mol Biol 2011;18(11):1281–9.

[215] Ngo HB, Lovely GA, Phillips R, Chan DC. Distinct structural features of TFAM drive mitochondrial DNA packaging versus transcriptional activation. Nat Commun. 2014;5:3077.

[216] Chakraborty A, Lyonnais S, Battistini F, et al. DNA structure directs positioning of the mitochondrial genome packaging protein Abf2p. Nucleic Acids Res 2017;45(2):951–67.

[217] Shi Y, Dierckx A, Wanrooij PH, et al. Mammalian transcription factor A is a core component of the mitochondrial transcription machinery. Proc Natl Acad Sci U S A 2012;109(41):16510–15.

[218] Malarkey CS, Bestwick M, Kuhlwilm JE, Shadel GS, Churchill ME. Transcriptional activation by mitochondrial transcription factor A involves preferential distortion of promoter DNA. Nucleic Acids Res 2012;40(2):614–24.

[219] Dairaghi DJ, Shadel GS, Clayton DA. Addition of a 29 residue carboxyl-terminal tail converts a simple HMG box-containing protein into a transcriptional activator. J Mol Biol 1995;249(1):11–28.

[220] Farge G, Falkenberg M. Organization of DNA in mammalian mitochondria. Int J Mol Sci 2019;20.

[221] Farge G, Mehmedovic M, Baclayon M, et al. In vitro-reconstituted nucleoids can block mitochondrial DNA replication and transcription. Cell Rep. 2014;8(1):66–74.

[222] Hensen F, Cansiz S, Gerhold JM, Spelbrink JN. To be or not to be a nucleoid protein: a comparison of mass-spectrometry based approaches in the identification of potential mtDNA-nucleoid associated proteins. Biochimie. 2014;100:219–26.

[223] Han S, Udeshi ND, Deerinck TJ, et al. Proximity biotinylation as a method for mapping proteins associated with mtDNA in living cells. Cell Chem Biol 2017;24(3):404–14.

[224] Gilkerson R, Bravo L, Garcia I, et al. The mitochondrial nucleoid: integrating mitochondrial DNA into cellular homeostasis. Cold Spring Harb Perspect Biol 2013;5(5):a011080.

[225] Lu B, Yadav S, Shah PG, et al. Roles for the human ATP-dependent Lon protease in mitochondrial DNA maintenance. J Biol Chem 2007;282(24):17363–74.

[226] Lu B, Lee J, Nie X, et al. Phosphorylation of human TFAM in mitochondria impairs DNA binding and promotes degradation by the AAA + Lon protease. Mol Cell. 2013;49(1):121–32.

[227] Matsushima Y, Goto Y, Kaguni LS. Mitochondrial Lon protease regulates mitochondrial DNA copy number and transcription by selective degradation of mitochondrial transcription factor A (TFAM). Proc Natl Acad Sci U S A 2010;107(43):18410–15.

[228] Hubstenberger A, Merle N, Charton R, Brandolin G, Rousseau D. Topological analysis of ATAD3A insertion in purified human mitochondria. J Bioenerg Biomembr 2010;42(2):143–50.

[229] Cooper HM, Yang Y, Ylikallio E, et al. ATPase-deficient mitochondrial inner membrane protein ATAD3A disturbs mitochondrial dynamics in dominant hereditary spastic paraplegia. Hum Mol Genet 2017;26(8):1432–43.

[230] Desai R, Frazier AE, Durigon R, et al. ATAD3 gene cluster deletions cause cerebellar dysfunction associated with altered mitochondrial DNA and cholesterol metabolism. Brain. 2017;140(6):1595–610.

[231] Gerhold JM, Cansiz-Arda S, Lohmus M, et al. Human mitochondrial DNA-protein complexes attach to a cholesterol-rich membrane structure. Sci Rep. 2015;5:15292.

[232] Bogenhagen DF. Mitochondrial DNA nucleoid structure. Biochim Biophys Acta 2012;1819 (9–10):914–20.

[233] Bogenhagen DF, Rousseau D, Burke S. The layered structure of human mitochondrial DNA nucleoids. J Biol Chem 2008;283(6):3665–75.

[234] Jourdain AA, Koppen M, Wydro M, et al. GRSF1 regulates RNA processing in mitochondrial RNA granules. Cell Metab. 2013;17(3):399–410.

[235] Antonicka H, Sasarman F, Nishimura T, Paupe V, Shoubridge EA. The mitochondrial RNA-binding protein GRSF1 localizes to RNA granules and is required for posttranscriptional mitochondrial gene expression. Cell Metab. 2013;17(3):386–98.

[236] Borowski LS, Dziembowski A, Hejnowicz MS, Stepien PP, Szczesny RJ. Human mitochondrial RNA decay mediated by PNPase-hSuv3 complex takes place in distinct foci. Nucleic Acids Res 2013;41 (2):1223–40.

[237] Pietras Z, Wojcik MA, Borowski LS, et al. Dedicated surveillance mechanism controls G-quadruplex forming non-coding RNAs in human mitochondria. Nat Commun. 2018;9(1):2558.

[238] Hensen F, Potter A, van Esveld SL, et al. Mitochondrial RNA granules are critically dependent on mtDNA replication factors Twinkle and mtSSB. Nucleic Acids Res 2019;47(7):3680–98.

[239] Leibowitz PJ, Schaechter M. The attachment of the bacterial chromosome to the cell membrane. Int Rev Cytol 1975;41:1–28.

[240] Hobbs AE, Srinivasan M, McCaffery JM, Jensen RE. Mmm1p, a mitochondrial outer membrane protein, is connected to mitochondrial DNA (mtDNA) nucleoids and required for mtDNA stability. J Cell Biol 2001;152(2):401–10.

[241] Meeusen S, Nunnari J. Evidence for a two membrane-spanning autonomous mitochondrial DNA replisome. J Cell Biol 2003;163(3):503–10.

[242] Li H, Ruan Y, Zhang K, et al. Mic60/Mitofilin determines MICOS assembly essential for mitochondrial dynamics and mtDNA nucleoid organization. Cell Death Differ 2016;23(3):380–92.

[243] Westermann B. Molecular machinery of mitochondrial fusion and fission. J Biol Chem 2008;283 (20):13501–5.
[244] Amati-Bonneau P, Valentino ML, Reynier P, et al. OPA1 mutations induce mitochondrial DNA instability and optic atrophy 'plus' phenotypes. Brain. 2008;131(Pt 2):338–51.
[245] Chen H, Vermulst M, Wang YE, et al. Mitochondrial fusion is required for mtDNA stability in skeletal muscle and tolerance of mtDNA mutations. Cell. 2010;141(2):280–9.
[246] Silva Ramos E, Motori E, Bruser C, et al. Mitochondrial fusion is required for regulation of mitochondrial DNA replication. PLoS Genet 2019;15(6):e1008085.
[247] Ban-Ishihara R, Ishihara T, Sasaki N, Mihara K, Ishihara N. Dynamics of nucleoid structure regulated by mitochondrial fission contributes to cristae reformation and release of cytochrome c. Proc Natl Acad Sci U S A 2013;110(29):11863–8.

CHAPTER 2

Human mitochondrial transcription and translation

Flavia Fontanesi[1], Marco Tigano[2], Yi Fu[2], Agnel Sfeir[2] and Antoni Barrientos[1,3]

[1]Department of Biochemistry and Molecular Biology, University of Miami, Miller School of Medicine, Miami, FL, United States, [2]Skirball Institute of Biomolecular Medicine, Department of Cell Biology, NYU School of Medicine, New York, NY, United States, [3]Department of Neurology, University of Miami, Miller School of Medicine, Miami, FL, United States

2.1 Introduction

Mitochondria are semiautonomous eukaryotic organelles of endosymbiotic bacterial origin that have retained a vestige of their ancestral genome: the mitochondrial DNA (mtDNA). Along with evolution, most genes required to sustain the transcription of the mtDNA, as well as the translation of the resulting mRNAs, have been transferred to the nucleus. Therefore mitochondrial gene expression requires the cooperation and coordination of two physically separated genomes.

The human mtDNA is a 16.6 kbp circular DNA that is present in 100–1000 copies per cell and is critical for proper cellular function. Purification of mtDNA using high sucrose density gradients revealed two strands, heavy and light (H and L), with different sedimentation properties owing to a slight skew in purine versus pyrimidine composition. The human mtDNA encodes for a total of 37 genes (13 proteins, 22 transfer RNAs, and 2 ribosomal RNAs) and is transcribed using two major promoters, the H-strand (HSP) and L-strand (LSP) promoters. Transcription yields large precursor polycistronic RNA molecules [1,2] that are later processed into discrete transcripts by specific enzymes [3]. The HSP and LSP promoters are located in a noncoding region, which also contains a displacement loop (D-loop) that is critical for mtDNA replication and coordinates the switch between replication and transcription. The 13 polypeptides expressed from the mtDNA are all essential components of the oxidative phosphorylation (OXPHOS) system enzymes. These proteins are synthesized in mitochondrial ribosomes, which differ from their bacterial and cytoplasmic counterparts, and are specialized in the synthesis of hydrophobic proteins. Mitoribosomes are anchored to the matrix face of the mitochondrial inner membrane to facilitate the

The Human Mitochondrial Genome.
DOI: https://doi.org/10.1016/B978-0-12-819656-4.00002-4

cotranslational membrane insertion of the newly synthesized polypeptides and coordinate their assembly with nucleus-encoded protein partners to form the OXPHOS enzymatic complexes (reviewed in Refs. [4,5]). With a proteome of ~1500 factors that are predominantly encoded by the nucleus, proper mitochondrial gene expression and function is largely dependent on the nuclear genome.

In this chapter, we will describe and discuss the fundamental processes of mitochondrial gene expression: transcription and translation.

2.2 Coordination of mitochondrial DNA replication and transcription

2.2.1 Overview of mitochondrial DNA replication

The human mtDNA is replicated by DNA polymerase γ (Polγ) in concert with critical components of the replication machinery, as extensively discussed in Chapter 1, MtDNA replication, maintenance, and nucleoid organization. Briefly, the minimal mitochondrial replisome has been reconstituted *in vitro*. It is composed of the mitochondrial replicative DNA polymerase Polγ subunit A (PolγA, coded by the gene *POLG*) complexed with the dimeric Polγ subunit B (PolγB, coded by the gene *POLG2*), the hexameric helicase Twinkle (coded by the gene *TWNK* or *PEO1*) and mtSSB (coded by the gene *SSBP1*). In addition to the minimal replisome components, several proteins play instrumental roles in mtDNA replication and maintenance *in vivo*. These include the mitochondrial transcription factor A (TFAM), mitochondrial genome maintenance exonuclease 1 (MGME1), DNA ligase 3 (LIG3), Ribonuclease H1 (RNASEH1), two topoisomerases (TOP1mt and TOP3A), and the mitochondrial RNA polymerase (POLRMT). POLRMT is a single-subunit DNA-directed RNA polymerase. It plays central roles in both mitochondrial transcription (see Section 2.3) and mtDNA replication. During replication, POLRMT serves as a primase for mtDNA replication on both strands. POLRMT initiates H-strand transcription ~200 nt upstream of O_H, and the transcript is used by POLG as a primer for H-strand synthesis [6]. L-strand replication starts when leading-strand synthesis copies past the O_L. *In vitro* assays allowed identifying a stem-loop structure within the O_L that allows POLRMT binding to synthesize primer for lagging strand replication [7,8].

Given its dual function in transcription and replication, the regulation of POLRMT is key to allow the coordination of the two processes. The switch between transcription and replication was recently characterized in depth and shown to be mediated by the transcription elongation factor (TEFM) [9]. In the absence of TEFM, POLRMT commits to replication by generating primers for H-strand and then L-strand replication. On the other hand, the interaction with TEFM boosts the processivity of POLRMT to produce full-length transcripts from HSP and LSP.

2.2.2 The mitochondrial DNA control region

The noncoding portion of mtDNA (NCR) is a 1.1 kbp region spanning from $tRNA^{Phe}$ to $tRNA^{Pro}$. Being the only noncoding portion of mtDNA, it has been reported to accumulate the highest number of base substitutions, both inter and intraspecies [10,11]. Despite being prone to mutations, the NCR harbors constrained sequences that are important for mtDNA replication and transcription, including the LSP and HSP promoters and the origin of replication of the H-strand—O_H. Extended sequence analysis of the NCR across different organisms identified three conserved regions—CSBs 1–3—located between LSP and O_H [12]. CSB1 (25 bp) is present in all organisms, CSB2 (17 bp) is partially conserved, and CSB3 (18 bp) is absent in certain branches of evolution [13]. The NCR often exists in D-loop configuration, whereby a fragment of DNA that is 650 nt in length and termed 7S DNA displaces the L- and H-strands [14]. The proportion of mtDNA molecules in this conformation can vary from 10% to 90% in a cell-type and species-specific fashion (reviewed in Ref. [10]).

It is well-established that mitochondrial transcription and replication are mutually exclusive, a step that is necessary to avoid replication fork collapses due to collision with the transcription machinery. Although, the initial steps of both processes are highly overlapping, and both processes share a common RNA primer that can either prime transcription or replication (see below). Accordingly, the mtDNA NCR is a key regulator that dictates the choice between the two DNA transactions.

2.2.2.1 The mitochondrial displacement loop

The 7S DNA is an abortive replication event that initiated from an RNA primer generated at the LSP by POLRMT [7,15]. This RNA primer is about 200 nt long [16], and while the transition to DNA has not been mapped to a single nucleotide level, it can take place between CSB1–CSB2 and the O_H [6]. This suggests that the RNA primer is subject to significant processing by RNAseH1 or MGME1 before replication. The total levels of 7S DNA are controlled by Polγ and other replisome proteins. They include PolγB and mtSSB, as well as mitochondrial transcriptional factors, such as TFAM [17,18]. It is estimated that 95% of the replication events fired from RNA primers are used to generate 7S DNA molecules instead of full-size mtDNA, reinforcing the critical role of this structure in mtDNA metabolism.

The 7S DNA extends until a termination region known as TAS, which is in close proximity to $tRNA^{Pro}$ [19]. The process of replication termination at TAS remains uncharacterized [20], but might be influenced by the levels of Polγ and Twinkle. Specifically, under physiological conditions, Polγ occupancy within the TAS region is high while Twinkle binding is low, leading to a stalled/paused replication fork. However, when mtDNA copy number decreases and synthesis is needed, the levels of Twinkle at TAS increase to promote the restart of replication and allowing the replisome to bypass the 7S DNA [21]. Accordingly, the regulation of mtDNA replication appears to be dictated by pretermination as opposed to replication initiation, and this is in strike contrast to the mechanisms that govern the control of replication firing in the nuclear genome.

Transcriptional regulators can also influence replication. For example, MTERF1, a mitochondrial transcription terminator, has been suggested to influence the pausing of replication at different sites throughout the mitochondrial genome [22]. Whether MTERF1 or another unknown factor regulates the pausing at TAS is yet to be determined. Notably, footprint analysis of bovine TAS showed a protected region compatible with a molecularly unknown protein of 48 kDa [23]. Lastly, a new idea has emerged, suggesting that the regulation of mtDNA replication could also be governed spatially by contact sites between mitochondria and the ER [24].

2.2.2.2 The switch between replication and transcription

As aforementioned, mtDNA molecules undergo replication or transcription in a mutually exclusive manner, and the switch between the two processes is influenced by the sequence of the NCR in the CSB region (Fig. 2.1). The initiation of transcription is common to both LSP and HSP promoters and requires TFAM [25]. TFAM binds upstream of promoter sequences prompting the bending of DNA in a typical U-shape that is then able to recruit POLRMT [26]. The preinitiation complex transiently recruits TFB2M to assist in promoter melting. In contrast to T7 RNA polymerase, POLRMT does not make contact with downstream DNA and therefore proceeds very slowly, often leading to pausing [27,28]. The recruitment of the elongation factor TEFM stimulates the processivity of POLRMT [9,29]. Following initiation, transcription proceeds from the LSP and HSP promoters to generate genome size polycistronic RNAs that are processed in RNA granules (see sections below). In the case of LSP, the sequence composition of CSB2 causes POLRMT to stall. Specifically, the polymerase stalls within a G-rich stretch of 16 nt that have the propensity to generate a bulky secondary G-quadruplex (G4) structure that can terminate transcription [30]. The presence of TEFM destabilizes the G-quadruplex and thus acts as a transcriptional antiterminator that allows POLRMT to bypass CSB2 and transcribe the entire strand. In contrast, when TEFM is not loaded, transcription by POLRMT terminates at CSB2, generating a replicative primer for Polγ [9,31,32].

The commitment to either replication or transcription can be envisioned as a multistep process that is subject to tight regulation by several factors, including TFAM, POLRMT, and TEFM. Moreover, the switch is influenced by the sequence of the control region (Fig. 2.1). Additional *in vivo* and *in vitro* experiments are required to define better the molecular steps that regulate this critical mtDNA metabolic choice. Nevertheless, the existence of a tight mechanism that regulates the balance between transcription and replication might have necessitated high conservation of the underlying D-loop region.

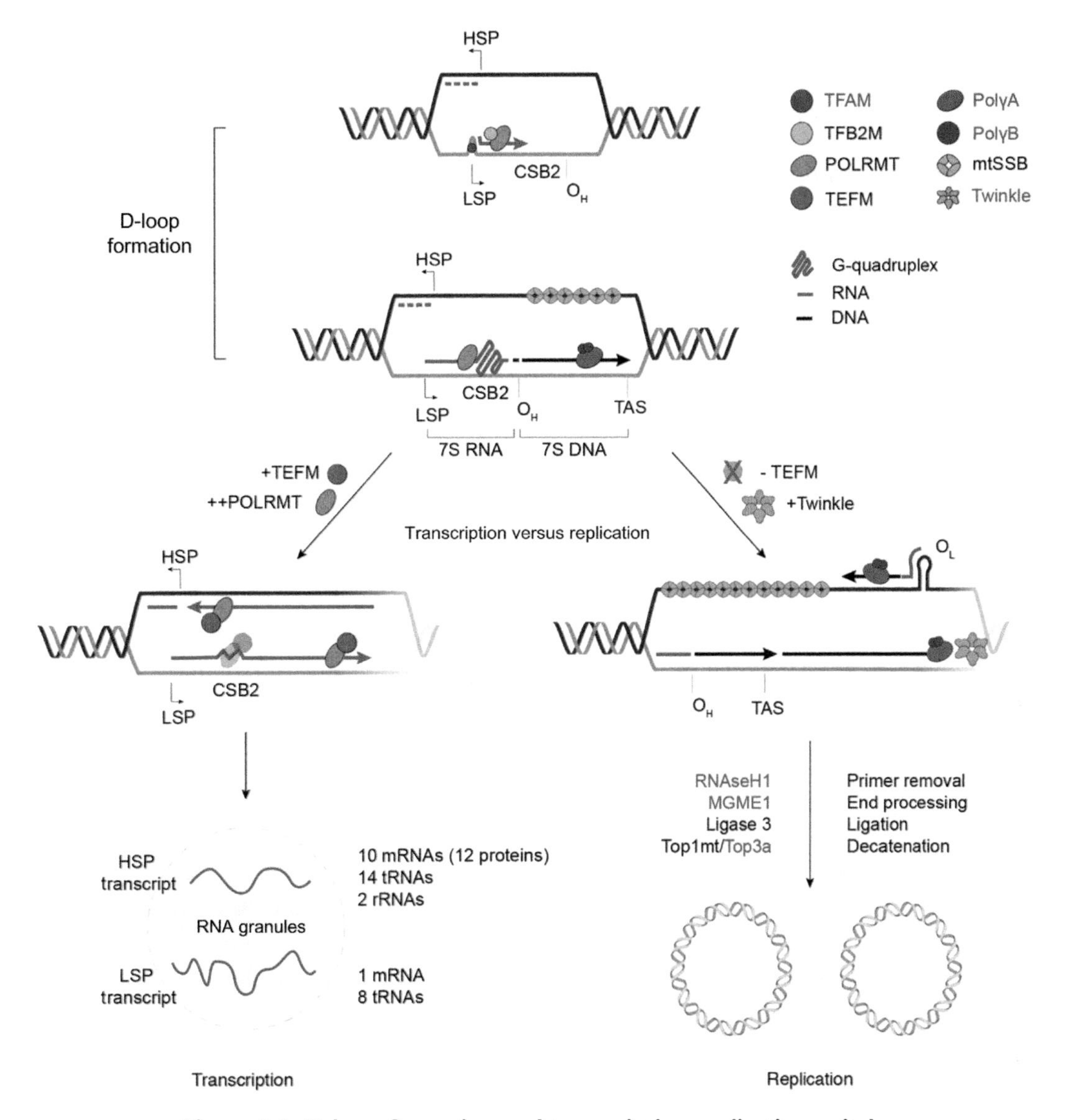

Figure 2.1: D-loop formation and transcription-replication switch.
(Top) The D-loop is generated from a transcriptional event promoted by TFAM at the light strand promoter LSP. POLRMT and TFB2M are recruited to LSP and transcribe an RNA molecule of 200 nt until a G-quadruplex sequence blocks the process. This primer can be used by Polγ to generate the 7S DNA that displaces the hybridization between heavy- and light-strand while mtSSB stabilizes the single stranded DNA. (Lower Left) Higher levels of POLRMT and recruitment of TEFM allow to bypass the G-quadruplex block and start genome size transcriptions from both LSP and HSP. The two large polycistronic RNA molecules are further processed in the RNA granules. (Lower Right) Lack of TEFM and recruitment of Twinkle at the TAS allow a fully functional replisome to be assembled and replicate the mtDNA accordingly to the strand-displacement model. Additional factors will complete the process by removing RNA primers, processing and ligating the ends and decatenating the two newly synthetized mtDNA molecules. Mutations in proteins indicated in *red* (black in print version) are associated with human mitochondrial pathologies.

2.3 Mitochondrial transcription and mitochondrial RNA transactions

2.3.1 Mitochondrial DNA transcription

Mitochondrial transcription starts from strand-specific promoters located in the D-loop and is carried out by the single-subunit POLRMT that is related to the RNA polymerase of T7 bacteriophage, both at the level of sequence homology and structurally. The X-ray structure of human POLRMT revealed that the homology is especially evident for both the C-terminal catalytic domain and the N-terminal promoter-binding domain [33]. The POLRMT catalytic domain resembles a hand composed of palm, fingers, and thumb domains [33]. Within the palm domain localizes the active site of the enzyme that catalyzes nucleic acid synthesis by a canonical two-metal-dependent reaction. Promoter recognition and binding are mediated by the AT-rich loop located in the N-terminal domain. A second element of the N-terminus domain, the intercalating hairpin, and the specificity loop in the C-terminus domain are involved in the formation of the transcription bubble at the promoter site [33]. However, despite POLRMT being able to bind to the promoter region in a sequence-specific manner, it is incapable of initiating transcription on its own. The presence of unique domains, including the flexible N-terminal extension and the pentatricopeptide repeat (PPR) domain, could potentially explain the requirement for transcription factors during transcription initiation.

Mitochondrial transcription consists of three steps: initiation, elongation, and termination. As for mtDNA replication, mitochondrial transcription entirely depends on nucleus-encoded transcription factors. In the last decade, numerous structural and biochemical studies have shed light onto this process, although still uncharacterized aspects remain as well as controversies mainly regarding the role of transcriptional factors [34]. According to the prevailing model (Fig. 2.2), the transcription factor A (TFAM) binds to the promoter upstream of the transcription start site and recruits POLRMT. Additional binding of the transcription factor B2 (TFB2M) to the initiation complex favors promoter melting and initiation of RNA synthesis. A transcription factor exchange leads to the incorporation of the elongation factor (TEFM), which allows POLRMT to synthesize complete transcripts. Transcription termination is mediated by the termination factor 1 (MTERF1) and, most likely additional factors yet to be identified.

Transcription of the L-strand starts at the LSP in the D-loop and generates a large polycistronic transcript encompassing eight tRNAs, and the *MTND6* mRNA. Additionally, transcription from LSP is responsible for the synthesis of the H-strand replication primer. Contrary to the well-defined LSP, transcription of the H-strand has been proposed to involve two H-strand promoters, HSP1 and HSP2, located 19 bp upstream of the $tRNA^{Phe}$ gene, and 2 bp upstream of the *12S-rRNA* gene, respectively [1]. Whereas HSP1 drives the synthesis of a primary transcript encompassing the rRNAs, $tRNA^{Phe}$ and $tRNA^{Val}$,

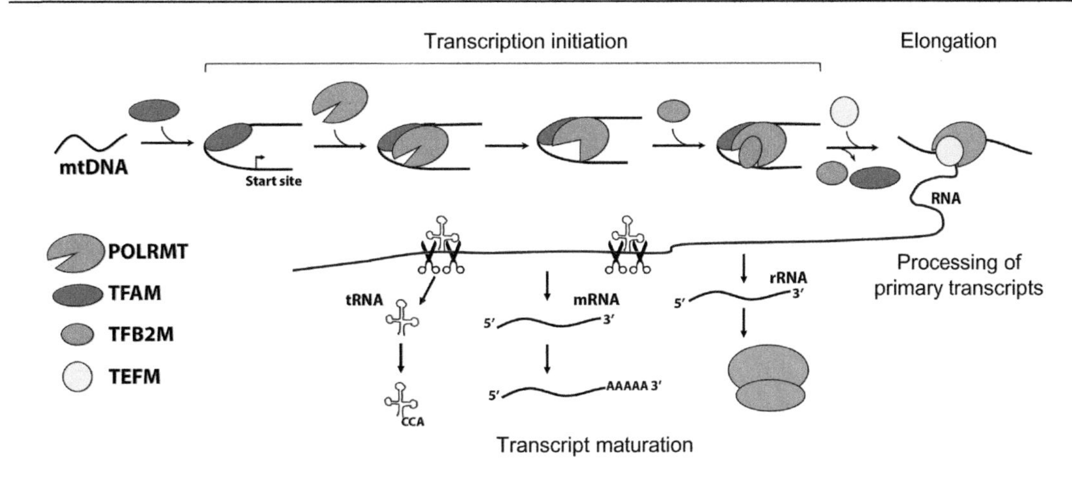

Figure 2.2: Mitochondrial transcription and transcript maturation.
(Top) Mitochondrial transcription initiates with the binding of TFAM to the promoter region upstream of the transcription start site. POLRMT is recruited by TFAM to the promoter, where it binds to a specific DNA sequence. POLRMT undergoes a conformational change accompanied by the binding of TFB2M to form the initiation complex. Exchange of initiation factors TFAM and TFB2M with the elongation factor TEFM drives the transition from transcription initiation to elongation. (Bottom) Mitochondrial transcription generates long polycistronic transcripts in which mRNA and rRNA sequences are intercalated by tRNAs. Processing of the primary transcripts at the tRNA sites liberates the messenger and ribosomal RNAs that are further modified to generate mature RNAs.

transcription from HSP2 generates a genome sized polycistronic transcript. The existence of two H-strand promoters was initially proposed to explain the higher abundance in rRNAs and how differential regulation in mRNA and rRNA expression could be achieved [35]. However, the functionality of HSP2 has been recently questioned by the observation that in *MTERF1* knockout mice no changes in RNA relative abundances were observed [36], which would be expected if MTERF1 stimulates transcription from the HSP1, as it does *in vitro*, through DNA looping between the promoter and the $tRNA^{Leu}$ LSP/HSP1 common termination site [37].

Some experimental approaches commonly used to study mitochondrial transcription are listed in Box 2.1.

2.3.1.1 Transcription initiation

TFAM belongs to the high mobility group (HMG-box) domain protein family and can bind, unwind, and bend mtDNA without sequence specificity. Based on these properties, TFAM plays a crucial role in mtDNA maintenance and organization into nucleoids [38–40] (see Chapter 1: MtDNA replication, maintenance, and nucleoid organization). Additionally, TFAM is necessary for mitochondrial transcription initiation [41], an activity that requires

BOX 2.1 RNA-seq-based approaches for the study of the mitochondrial transcriptome

In *RNA-seq or RNA deep-sequencing* analysis, transcripts are isolated from biological samples (cultured cells or tissues) and converted into cDNA by reverse transcription. cDNAs are then shredded into fragments, which are ligated to sequencing adapters for next-generation sequencing (NGS). NGS generates a sequencing library, which is composed of a large number of reads. These reads are aligned with the genome to identify novel transcripts and quantified to determine differential gene expression. Modifications of this approach, such as *Parallel Analysis of RNA ends (PARE)*, have been developed to sequence the 3′ and 5′ transcript ends and map RNA processing sites. Moreover, increased sequencing error can be used as a marker for nucleotide modification. RNA-seq-based approaches are also used to study RNA–protein interactions. In the case of *RNA footprinting*, all transcript regions protected from ribonuclease digestion are detected, although this approach does not identify the polypeptide bound to them. Protein-specific binding sites can be characterized by RNA-seq upon protein-RNA *UV-cross-linking and immunoprecipitation (CLIP)* of the polypeptide of interest.

its C-terminal tail [42] and binds with high affinity and in a sequence-specific manner to the D-loop region, 10–35 nt upstream of the transcription initiation site. TFAM binding to the promoter induces a stable DNA U-turn, and it recruits POLRMT to form the preinitiation or closed initiation complex [26,27,32,43,44], which has an identical topology at both L- and H-strand promoters [25]. The POLRMT N-terminal extension domain mediates its interaction with TFAM and positions the enzyme active site near to the point of transcription initiation. In particular, the POLRMT intercalating hairpin is located 4 nt upstream of the initiation site, where it separates the DNA strands [45]. Recruitment of TFB2M to POLRMT stabilizes this melting-competent conformation to form the open initiation complex [46]. TFB2M and its paralog TFB1M were initially identified based on their sequence homology with the yeast mitochondrial transcriptional activator Mtf1. The three proteins have structural similarities with rRNA methyltransferase domains [47]. However, whereas TFB1M has retained an rRNA methyltransferase function, TFB2M is an *in bona fide* transcriptional activator that facilitates promoter melting and is also required for the formation of the first RNA phosphodiester bond [48]. Recent *in vitro* and *in vivo* studies have suggested that under certain conditions, which favor DNA strand spontaneous separation, POLRMT and TFB2M could initiate transcription in the absence of TFAM [49,50]. However, the biological relevance of these observations remains to be fully elucidated.

Certainly, TFAM plays complex and diverse roles in mitochondrial biogenesis, which include regulation of mitochondrial gene expression by TFAM levels, as explained in the replication section. A high intra-mitochondrial TFAM concentration inhibits mitochondrial transcription and replication, thus favoring nucleoids packaging. On the contrary, a low concentration of this factor stimulates transcription differentially for the different promoters, with HSP2 as the most active [51]. Thus, whereas in the presence of low TFAM amounts, expression of OXPHOS subunits is favored, at intermediate concentrations, transcription of rRNA and primers for mtDNA replication is preferred.

Of the several genes involved in mitochondrial transcription, to date, only *TFAM* has been associated with human diseases. *TFAM* pathogenic mutations have been reported in patients with progressive, fatal liver failure accompanied by mtDNA depletion [52] (see Chapter 14: Mitochondrial DNA-related diseases associated with single large-scale deletions and point mutations). Variations in TFAM steady-state levels and mtDNA content have also been reported in several neurodegenerative diseases [53].

2.3.1.2 Transcription elongation

The initiation-to-elongation transition involves the exchange of the transcription factors interacting with POLRMT. After promoter escape by POLRMT, the initiation factors TFAM and TFB2M are released, and the elongation factor TEFM is recruited to form the elongation or antitermination complex [54]. TEFM forms a homodimer through the interaction of its C-terminal domains and occupies a POLRMT-binding site overlapping with TFB2M, which explains the necessity of TFB2M release from the initiation complex for TEFM binding. The N-terminus portion of the protein contains a resolvase-like domain, homolog to bacterial Holliday-junction resolvases, which has been repurposed for mitochondrial transcription [32]. Recent structural analysis has shown that TEFM stabilizes the intercalating hairpin domain of POLRMT, separating the nascent RNA strand from the template DNA and contributes to the formation of a narrow RNA exit channel [32]. In this way, TEFM strongly promotes POLRMT processivity and allows the synthesis of near-genome-length transcripts by preventing the formation of secondary structure in the nascent RNA [29,32]. Ultimately, TEFM helps POLRMT bypassing RNA regions prone to form complex secondary structures as well as oxidative lesions, such as 8-oxo-guanosine, which may cause premature transcription termination [29]. In the absence of TEFM, transcription from the LSP is often terminated around the strong G-quadruplex-forming region CSB2, which is believed to favor mtDNA replication [9,30,55], as discussed above.

2.3.1.3 Transcription termination

At the end of each transcription cycle, POLRMT dissociates from the mtDNA. The termination site for transcription initiated at the LSP is a 22-nucleotide sequence located within the $tRNA^{Leu}$ gene and recognized by the termination factor MTERF1 [56]. Indeed, the L-strand does not encode any genes beyond this point. The crystal structure of MTERF1 revealed repeated two-α-helices motifs resembling domains present in other nucleic acid-binding proteins [57,58]. The helical fold allows MTFER1 an extensive surface interaction with the major groove of the substrate DNA duplex. MTERF1 binding induces a DNA 25° bend, duplex melting, and flipping of three nucleotides [57,58]. Base-flipping is essential for binding stability and transcription termination. It has been proposed that MTERF1 acts as a roadblock interfering with the transcription elongation complex [57,58]. This would imply that MTERF1 could arrest transcription bi-directionally [59], as it does in vitro at a close to 1:1 protein—DNA molar ratio [60]. However, a distinct polarity has been observed for MTERF1 activity *in vivo*: whereas MTERF1 is required to terminate L-strand transcription, it only partially terminates transcription of the H-strand [36,60]. In contrast to the well-characterized process of L-strand transcription termination, insights into the mechanism of H-strand transcription termination are currently limited.

From a physiological perspective, a role of MTERF1 in the termination of transcripts initiated at the HSP1 was proposed to explain the higher steady-state levels of rRNA compared to mRNAs. Alternatively, MTERF1 could have an indirect effect on rRNA levels since the termination of LSP transcription at the $tRNA^{Leu}$ site may prevent the synthesis of antisense rRNA transcripts that could interfere with mitoribosome biogenesis [36,61]. Interestingly, several mitochondrial diseases associated mutations in the $tRNA^{Leu}$ gene occur in the MTERF1 binding site. Among them, the most common mtDNA mutation, the A3243G transition, which accounts for about 80% of MELAS cases [62], and it has been shown to affect transcription *in vitro* [63]. Although the pathogenic effects of tRNA mutations are typically attributed to defective mitochondrial translation, it cannot be ruled out the possibility that alterations in transcriptional termination could contribute to the disease phenotype.

Lastly, three human MTERF1 paralogs have been identified, MTERF2–4 [64]. Their structure is similar to MTERF1, suggesting a role in nucleic acid binding [58,65–67]. However, none of them seems to play a role in transcription termination. Recent studies in mouse models indicate a role for MTERF3 and 4 in mitoribosome biogenesis [68,69], although their functions remain to be fully characterized.

2.3.2 The mitochondrial transcriptome

Although human mitochondria have evolved a small and compact genome, the corresponding transcriptome is remarkably complex. Indeed, despite their common polycistronic origin, the documented wide variation in the abundance of mature mitochondrial transcripts [70] suggests the existence of extensive posttranscriptional regulatory processes. It has been estimated that the rate of synthesis of mitochondrial mRNAs in HeLa cells is 50–100 times lower to the rate of rRNA transcription leading to a disparity in their steady-state levels ranging from 50- to 300-fold [71]. However, more recent studies based on RNA deep-sequencing approaches (see Box 2.1) in 143B cells reported an rRNA to mRNA ratio of 10:1 and substantial differences among individual mRNA abundance, with the lower levels detected for *MTND6* mRNA [70]. The observed differences may be attributed to the technology used for the analysis. However, cell line- or tissue-specificity may also contribute to the apparent discrepancy.

Mitochondrial transcription produces three primary transcripts, which are processed into the mature coding and noncoding RNAs that constitute the mitochondrial transcriptome. Mitochondrial mRNAs lack conventional 5′- and 3′-untranslated regions and introns. Exceptions are the *MTCO1*, *MTND1*, and *MTATP8* mRNAs, which have 3 nt, 2 nt, and 1 nt, respectively, preceding the start codon, and the bicistronic transcripts *MTATP8/ATP6*, and *MTND4L/ND4*, in which the ORFs in the second position have a significant 5′- leader sequence. Mitochondrial noncoding RNAs include the ribosomal *12S* and *16S*-rRNA and a complete set of 22 transfer RNAs (tRNAs). Recently, long and small noncoding RNAs (lncRNAs and sRNAs) have been also identified [70]. Three lncRNAs (*lncND5*, *lncND6*, and *lncCYTB*) have been detected in mammalian mitochondria with multiple approaches [70,72]. In cell cultures, their abundance was reported to be comparable to their complementary mRNA. However, different ratios of lncRNA/mRNA were observed in human tissues, indicating that the expression of complementary coding and noncoding RNAs can be regulated independently. Moreover, lncRNAs are resistant to cleavage by single strand-specific ribo-endonucleases suggesting that they form duplexes with their complementary mRNAs and that they could play a regulatory function in mammalian mitochondrial gene expression [73]. Regarding the mitochondrial sRNAs, whereas two main classes of 21 and 26 nt respectively, mainly derived from tRNA genes, have been identified, their physiological roles remain unclear [70].

2.3.3 Mitochondrial RNA-binding proteins and RNA biology

Posttranscriptional regulation of mitochondrial gene expression is mediated by an array of RNA-binding proteins (RBPs), which are involved in all aspects of RNA biology, including transcript processing, maturation, stability, and degradation. Global mitochondrial RNA–protein interactions have been mapped by UV-cross-linking and immunoprecipitation (CLIP). An array of polypeptides, albeit unidentified, was detected, ranging from 15 to 120 kDa [74]. More recent studies based on RNase footprinting identified 88 distinct protein-binding sites, 33 of which were within mRNAs, 8 in rRNAs, 7 in tRNAs, and 40 in transcription regulatory sites and noncoding transcripts [75]. Furthermore, the RNA interactome is known to be highly dynamic and change under different physiological conditions. For example, inhibition of mitochondrial translation with chloramphenicol revealed 270 RNA sites protected from endonuclease cleavage [75]. From these, 124 protein footprints located within mitochondrial mRNAs, only 22 of which were recognized as mitochondrial ribosome stalling sites [75].

The known mitochondrial RBPs represent a diverse group of proteins, which includes members of the PPR and Fas-activated serine/threonine kinase (FASTK) families and other unique RBPs. PPR proteins were first identified in plants, where they represent a large family of sequence-specific RBPs involved in all aspects of organellar gene expression [76]. They are defined by the presence of degenerated 35-amino acid sequences, the PPR motifs, repeated in tandem 2–30 times [77,78]. Each PPR motif folds in two antiparallel α-helices and contacts one RNA base. In tandem, multiple PPR motifs are predicted to form a solenoid structure, where the central groove is involved in the RNA binding [33,79]. To date, seven PPR proteins were identified in mammalian mitochondria: POLRMT, the leucine-rich PPR cassette (LRPPRC), the PPR domain-containing proteins 1 and 2 (PTCD1 and 2), the mitoribosome SSU proteins mS27 and mS39, and the mitochondrial RNase P protein 3 (MRPP3) [80]. Additionally, the mitochondrial FASTK family comprises six members, FASTK and its homologs FASTKD1–5 [81]. All the members of the family contain three conserved C-terminal domains, namely FAST1, FAST2, and RAP. While RAP is believed to bind RNA [82], the precise function of the FAST domains remains unknown. However, the lack of conservation in key enzymatic residues has questioned the initially proposed FASTK kinase activity [83]. Interestingly, structural modeling has suggested that FASTK proteins share a repeated α-helix motif architecture with PPR proteins [81].

Mitochondrial RBPs play fundamental roles in the regulation of mitochondrial gene expression, which can occur at the transcriptional and posttranscriptional levels, including translational and posttranslational regulation, by an array of regulatory mechanisms whose relative contributions remain to be fully elucidated.

2.3.3.1 Mitochondrial RNA processing

Transcription of the H- and L-strands generates long polycistronic transcripts that require extensive processing to produce mature RNAs. Processing involves cleavage of primary transcripts, polyadenylation, and base modification (Fig. 2.2).

In the mitochondrial genome, the rRNA and most coding sequences are separated by tRNAs. According to the tRNA punctuation model, endonucleolytic excision of the tRNAs releases ribosomal and messenger RNAs [84]. Processing of tRNAs from the primary transcripts is performed by RNaseP and the RNaseZ enzyme ELAC2 at the 5′- and 3′-end, respectively. The mitochondrial RNaseP is an entirely proteinaceous endonuclease formed by three subunits MRPP1–3. It lacks the catalytic RNA component typical of bacterial RNaseP enzymes. MRPP3 is the catalytic subunit of the complex and contains three PPR motifs and a putative metallonuclease domain. MRPP1 and 2 subunits form a subcomplex that remains bound to the RNA after MRPP3-mediate processing and could play a role in promoting 3′-end maturation by ELAC2 and/or tRNA stability. This is in line with recent findings indicating that primary transcript processing follows a hierarchical order according to which 5′-end cleavage of tRNAs precedes the 3′-end cleavage [85–87]. The existence of an order of processing also suggests that the evolutionary conserved ELAC2 enzyme could only act on smaller RNA substrates, whose complex secondary structure has been resolved by the mitochondria-specific

RNaseP complex [88]. Lastly, ELAC2 interacts with the PPR protein PTDC1, which is also involved in tRNA processing and can specifically act as a *tRNA*^{Leu} negative regulator, although its mechanism of action remains to be elucidated [3].

The tRNA punctuation model does not explain all primary transcript cleavage events, as four sites exist in which mRNAs are not flanked by tRNAs. These are the 3′-end of *MTND6* mRNA, the 5′-ends of *MTCO1* and *MTCYB* mRNAs, and the junction between *MTATP6* and *MTCO3* mRNAs. RNaseP performs the cleavage *MTCO1* mRNA 5′-end [3,89]. However, neither RNaseP nor ELAC2 are involved in the processing of the three remaining non-tRNA containing junctions. A role in the cleavage of the *MTCYB-ND5* precursor mRNA has been proposed for the PPR protein PTCD2 [90] and FASTK proteins D4 and D5 [91,92]. Moreover, several FASTK proteins, including FASTK and FASTKD1–5 participate in RNA processing at noncanonical sites and affect mRNA steady-state levels [81], although their exact mechanism of action remains to be fully understood.

Defects in mitochondrial primary transcript processing due to mutations in MRPP1 and 2, ELAC2, and FASTKD2 have been identified in patients with infantile multisystemic mitochondrial disorders (reviewed in Ref. [93]) (see Chapter 14: Mitochondrial DNA-related diseases associated with single large-scale deletions and point mutations).

2.3.3.2 Mitochondrial RNA maturation

After processing of the primary transcripts, all mRNAs, except *MTND6*, are polyadenylated by the poly(A) polymerase MTPAP, which adds an average of 45 nt at the 3′-end of the transcripts [94]. Seven mitochondrial mRNAs do not contain a stop codon, and polyadenylation is required to complete their coding sequence correctly. For these mRNAs, polyadenylation has been shown to increase their stability [95]. However, the opposite effect has been observed for some other mRNAs as a consequence of attenuated polyadenylation. Whereas this has suggested that the role of polyadenylation is mRNA-dependent, distinct pools of adenylated and nonadenylated transcripts exist in mitochondria [96,97]. MTPAP forms a homodimer with a high affinity for ATP and UTP [98,99]. However, it lacks a canonical RNA-binding domain, and it does not require RNA-binding cofactors. Thus, it is unclear how MTPAP recognizes its substrates. Spurious MTPAP polyadenylation activity on rRNA and mature tRNAs has been reported, and it is believed to be mitigated by the exonuclease PDE12 [100]. Polyadenylation is essential for mitochondrial gene expression, and mutations in MTPAP have been associated with progressive spastic ataxia and optic atrophy [101] (see Chapter 14: Mitochondrial DNA-related diseases associated with single large-scale deletions and point mutations).

Differently from mRNA maturation, mitochondrial noncoding RNAs undergo chemical nucleotide modification. Base modifications of *12S* and *16S*-rRNAs include 2′-*O*-ribose-methylation and pseudouridylation. A single pseudouridylated site has been identified in position 1397 of human *16S*-rRNA. This modification is most likely catalyzed by the RPUSD4 Ψ synthase and is essential for mitoribosome assembly and stability [102,103]. More numerous are the known sites of rRNA methylation. Two adenines (human A936 and A937) in the *12S*-rRNA are methylated by the transcription factor TFB2M paralog, TFB1M that has retained its methyltransferase activity [104,105]. Additionally, methylation of the *12S*-rRNA cysteine 841 has been shown to be catalyzed by the NSUN4 enzyme [69]. Regarding the *16S*-rRNA, three nucleotides located in the peptidyl transferase center (PTC) are methylated by a group of closely related methyltransferase enzymes, MRM1–3 [106,107]. Lastly, the recently reported methylation of the *16S*-rRNA adenine 947 is catalyzed by the tRNA methyltransferase TRMT61B [108]. Although the exact role of rRNA methylation in mitoribosome assembly, stability, or function has not been dissected in all cases, the rRNA base modifications are ultimately necessary for efficient and accurate mitochondrial translation.

Even more extensive modifications occur during tRNA maturation. The first step in mitochondrial tRNA maturation is the addition of a CCA sequence at their 3′-end, which is not encoded in the mtDNA but is posttranscriptionally synthesized by the tRNA nucleotidyltransferase 1 enzyme TRNT1 [109]. CCA addition is required for aminoacyl-tRNA synthetase binding and amino acid loading, as well as for tRNA interaction with the translational elongation factor Tu [110]. CCA addition is followed by tRNA base modifications at numerous specific positions, which are also necessary for accurate tRNA folding and function (reviewed in Ref. [111]). Several mitochondrion-specific enzymes responsible for tRNA modification have been identified in humans and shown to be frequently associated with severe infantile mitochondrial disorders (reviewed in Refs. [93,111]). The most common base modification is pseudouridylation, which provide structural rigidity to the tRNA molecules [112]. Of particular functional relevance are the modifications of the first position of the tRNA anticodon, which generate unusual bases, also referred to as wobble bases, known to facilitate non-Watson–Crick base pairing and then the recognition of multiple codons by the same tRNA [113]. This gain in anticodon pairing ability plays a key role in mitochondrial protein synthesis, which relies on the minimal number of tRNAs for translation (the 22 tRNAs encoded by the mtDNA).

2.3.3.3 Mitochondrial RNA chaperones and mRNA stability

The stability of the mitochondrial mRNAs generated by H-strand transcription is regulated by LRPPRC, a leucine-rich PPR protein. LRPPRC interacts with the RBP SLIRP to form a complex [114], which extensively binds mRNAs throughout the transcriptome. The LRPPRC–SLIRP complex acts as an RNA chaperone [115]. By resolving mRNA secondary structures, LRPPRC–SLIRP enables transcript polyadenylation and facilitates translation [116]. Indeed, LRPPRC amounts in mitochondria correlate with mRNA polyadenylation and steady-state levels [117,118]. Moreover, LRPPRC knockout in mice leads to a severe mitochondrial translation defect [116].

Whereas the mRNA stabilization function of LPPRC is associated with polyadenylation, the FASTK protein mediates the processing and stability of the only nonpolyadenylated transcript, the *MTND6* mRNA [91]. Both LPPRC and FASTK functions are closely associated with the RNA degradosome, which mediates RNA decay in mitochondria. The human RNA degradosome is a heterodimer formed by the $3' - 5'$ exoribonuclease PNPase and the NTP-dependent helicase SUV3 [119]. Lack of either degradosome complex component leads to the accumulation of RNA degradation intermediates and aberrant and polyadenylated RNA species [119,120]. Moreover, the dinucleotidase REXO2 has been shown in human mitochondria to degrade short oligonucleotides with a preference for RNA substrates [121,122]. Notably, the accumulation of dinucleotides in REXO2 mutant mice favors aberrant translation initiation from noncanonical sites [122].

A homozygous founder mutation in *LRPPRC* gene was identified in the Quebec region as one of the first nuclear gene mutations associated with mitochondrial disease. LRPPRC mutations result in the French-Canadian variant of Leigh syndrome associated with respiratory chain complex IV (or cytochrome *c* oxidase) deficiency [123] (see Chapter 14: Mitochondrial DNA-related diseases associated with single large-scale deletions and point mutations).

2.3.3.4 Mitochondrial RNA translation activators

The characteristics of mitochondrial mRNAs condition their recognition by mitoribosomes for translation initiation, a process that remains poorly understood (see section 2.4). Human mitochondrial mRNAs lack Shine/Dalgarno sequences, canonical 5′-untranslated region (5′-UTR), and a 5′-methylguanosine cap [80]. Thus mammalian mitochondrial ribosomes do not recognize the start codon using the Shine/Dalgarno interaction between the mRNA and the *16S*-rRNA as in prokaryotes. Further, this system does not use a cap-binding and scanning mechanism as in the eukaryotic cytoplasm. In yeast, mRNAs do not contain Shine–Dalgarno sequences either, but they have long 5′-UTRs where mRNA-specific translational activators bind to facilitate translation initiation [124]. The existence of translation factors in human mitochondria has been a lasting object of debate. A single translational activator, TACO1 has been so far identified [125]. TACO1 has an N-terminus positively charged domain that binds to *MTCO1* mRNA specifically to promote its association with the mitochondrial small ribosomal subunit and subsequent translation [125,126]. In humans, mutations in *TACO1* have been associated with complex IV deficiency and late-onset Leigh syndrome [125,127] (see Chapter 14: Mitochondrial DNA-related diseases associated with single large-scale deletions and point mutations). The identification of TACO1 suggests that mitochondrial RBPs can act on specific mRNAs to selectively regulate their translation. In this line, FASTK3 is also required for efficient *MTCO1* mRNA translation without affecting mRNA steady-state levels [128]. Beyond the potential existence of ternary factors, a general role in mRNA recognition has been proposed for the mitoribosome-specific protein mS39, which locates in close proximity to the mtSSU mRNA entry channel and contains nine PPR domains [129]. In a recent structural analysis of the translating mitoribosome by cryoelectron microscopy (cryo-EM), mS39 has been shown to directly interact with *MTCO3* mRNA and proposed to mediate the initial interaction between mRNAs and the mitoribosome initiation complex [130].

2.4 Mitochondrial translation

Protein synthesis is universally catalyzed by the ribosome. Within mitochondria, the mitoribosomes synthesize a small set of proteins encoded in mtDNA. The assembly and function of the mitochondrial translation machinery require the participation of both the nuclear and mitochondrial genetic systems. While the two mitochondrial ribosomal RNAs (*12S* and *16S*-rRNAs) and a full set of 22 mitochondrial tRNAs are transcribed from mtDNA, all ribosomal proteins (MRPs), mitoribosome assembly factors as well as aminoacyl-tRNA synthetases and translation factors are synthesized in the cytosol and imported into mitochondria.

The mitochondrial translation system evolved from that of the bacterial ancestor of mitochondria. As a consequence, the catalytic properties of mitochondrial and bacterial ribosomes are similar. This is evident from the fact that the proteins and RNA domains of mitochondria and bacteria that contribute to decoding and peptide bond formation share a high degree of similarity. Also, whereas mitoribosomes are not sensitive to some inhibitors of cytoplasmic ribosomes, such as cycloheximide, they are sensitive to antibiotics such as chloramphenicol. Translation factors are conserved, and several mitochondrial factors can functionally replace their homologs in bacteria [131]. Nevertheless, evolution of the mitochondrial system resulted in (1) deviations in the genetic code [132,133], (2) significant differences in the actual process of translation [134], and (3) the formation of mitochondrial ribosomes that differ significantly in structure and composition, not only in comparison to their bacterial relatives but also among different species [135,136]. A recent study has examined the composition and evolutionary history of mitoribosomes across the phylogenetic tree by combining three-dimensional structural information with a comparative analysis of the secondary structures of rRNAs and published proteomic data. This has allowed for the generation of a map of the acquisition of structural variation and reconstruction of the stages that shaped the evolution of at least the mtLSU and led to this diversity, allowing the authors to conclude that structural patching has fostered mitoribosome divergence [136].

The mitochondrial translation machinery is biomedically relevant because mitoribosomes share a sensitivity to antibiotics similar to common infectious bacteria. Also, they are emerging as new targets for cancer therapeutics (reviewed in Ref. [137]). Additionally, mutations in mitoribosome proteins, rRNAs, and translation factors are responsible for a heterogeneous group of human multisystemic OXPHOS disorders, frequently involving sensorineural hearing loss, encephalomyopathy, and hypertrophic cardiomyopathy (reviewed in Ref. [5]).

In this section, we will briefly summarize the elements and functions of the mitochondrial translation machinery. Some experimental approaches commonly used to study mitochondrial translation are listed in Box 2.2.

2.4.1 The mitochondrial translation machinery

2.4.1.1 Mitoribosome structure

The mammalian mitoribosomes were first isolated in the late 1960s/early 1970s [138–140]. They differ from bacterial (70S) and cytoplasmic ribosomes (80S) in their lower RNA:protein ratio, where significant amounts of RNA have been replaced by mitochondrion-specific proteins. The mammalian mitoribosome is a 55S RNA–protein complex, formed by a 39S large subunit (mtLSU) composed of 52 mitoribosome proteins (MRPs), a *16S*-rRNA and a structural tRNA (Val in human cells), and a 28S small subunit (mtSSU) formed by 30 MRPs and a *12S*-rRNA [141,142]. Approximately half of the mitochondrial ribosome proteins have bacterial homologs, although they frequently contain N- or C-terminal extensions of unclear function [135]. However, cryoelectron microscopic analyses of bovine mitochondrial ribosomes have shown that the catalytic region at the interface of both subunits is largely conserved [141,142]. The mitochondrion-specific subunits physically compensate for the loss of structural RNA, which is evident in the human and porcine mitoribosome [143–147]. These mitochondrion-specific proteins are mainly peripherally distributed over the solvent-accessible surface, forming clusters at the central protuberance, the L7/L12 stalk, and adjacent to the polypeptide exit site. Similar to the protein extensions, these proteins accommodate novel positions rather than compensate for the missing rRNA, and could have been recruited to stabilize the general structure of the ribosome [148]. Additionally, some mitochondria-specific proteins appear to be important for establishing intersubunit bridges [149], a feature that is different from cytoplasmic ribosomes that typically contain RNA–RNA intersubunit connections. A specific characteristic of the mitoribosome is the absence of a *5S*-rRNA, as it had long been speculated, but the presence in the mtLSU central protuberance of $tRNA^{Val}$, which plays an integral structural role. In the mtDNA, the $tRNA^{Val}$ gene is located in the middle of the genes encoding for the *12S* and *16S*-rRNAs. These three RNAs are transcribed together as a polycistronic transcript similar to that of the bacterial rRNA operon.

BOX 2.2 Experimental approaches for the study of mitochondrial translation

Despite multiple attempts to develop cell-free approaches to analyze mitochondrial protein synthesis, an *in vitro* mitochondrial translation system remains to be established. Methodologies currently used to study mitochondrial translation in isolated mitochondria (*in organello*) and in whole cultured cells include:

Metabolic labeling. This approach takes advantage of the different sensitivity of cytoplasmic and mitochondrial ribosomes to antibiotics to semiquantitatively study translation rates in isolated mitochondria (which always have cytoplasmic ribosomes bound to the outer membrane) or whole cells. Following inhibition of exclusively cytoplasmic protein synthesis with cycloheximide or emetine, a radiolabeled precursor, usually ^{35}S-methionine, is added and its incorporation into newly synthesized mitochondrial proteins can be assessed in a time-dependent manner.

Ribosome profiling. Originally developed for the analysis of cytosolic protein synthesis, ribosome profiling has been recently adapted for mitochondrial translation studies. The method is based on deep sequencing of mRNA fragments protected from nuclease digestion by the mitochondrial ribosomes during their translation. It provides a snapshot of all transcripts being actively translated in mitochondria at a specific time point. Additionally, it allows identifying ribosome pause sites within the transcriptome.

Cryogenic-electron microscopy (cryo-EM). For cryo-EM analysis, an aqueous solution of the molecule of interest is applied to a grid, snap-frozen, and imaged by electron microscopy. Recent advancements in EM and computer algorithms for the analysis of EM single-particle images have allowed the determination of the mitochondrial ribosome structure at near-atomic resolution. These studies have provided information regarding mitoribosome components and their arrangement as well as insights into the molecular mechanisms of mitochondrial translation.

Stable Isotope Labeling with Amino Acids in Cell Culture (SILAC). SILAC is a metabolic labeling method that uses *in vivo* metabolic incorporation of "heavy" 13C- or 15N-labeled amino acids into proteins followed by mass spectrometry (MS) analysis for comprehensive identification, characterization, and quantitation of proteins. The SILAC approach has been used to study mitochondrial ribosome biogenesis. For this purpose, cells are initially cultured in the presence of "light" (regular) 12C- or 14N-containing amino acids, followed by transfer in media containing "heavy" amino acids for increasing time periods. Quantitative isotope incorporation into a polypeptide is detected by MS and provides a time-course of protein assembly into multimeric macro-molecules, such as the mitoribosome.

The recent reconstructions of the 55S human [141,150,151] and porcine [142,152,153] mitoribosome at high resolution by cryo-EM (see Box 2.2), have allowed distinguishing important features in the mammalian mtSSU. Similar to the mtLSU, several peripheral rRNA helices present in bacteria are either truncated or missing in the *12S*-rRNA, but the tertiary structure of the *12S*-rRNA core and the overall positions of the MRPs with bacteria homologs are preserved in the structure. A significant structural remodeling is observed at the mRNA entrance of the mtSSU compared to the bacterial SSU. This remodeling serves to accommodate mammalian mitochondrial mRNAs, which have either null or very short 5′-UTRs [154], as mentioned earlier.

An additional major remodeling is observed at the A (aminoacyl) and P (peptidyl) tRNA binding sites, where some subunits present in bacterial ribosomes (e.g., uL5 at the P-site, bL25 and the A-site) have been lost in order to accommodate human tRNAs, which contain highly variable loops at the tRNA elbow. However, the P-site finger, unique to the mammalian ribosome, compensates for these missing interactions. A distinctive property of mitoribosomes is the acquisition of an intrinsic GTPase activity through the GTP-binding protein, mS29 of the 28S mtSSU subunit [141,142,155]. The GTPase activity is probably linked to subunit association since mS29 locates at the subunit interface and is involved in coordinating two mitochondria-specific bridges.

The polypeptide exit tunnel is adapted to the transit of hydrophobic nascent peptides [151,153]. The tunnel exit site consists of conserved proteins from bacteria, namely bL23, bL29, bL22, bL24, and bL17, which create a ring around the exit site. Moreover, this conserved core is surrounded by a second layer of protein density, consisting of bL33 and mL45, which promote anchoring of the mtLSU to the inner mitochondrial membrane [151,153]. The structure of the human mitoribosome with nascent polypeptide shows its extensive interactions with specific hydrophobic residues of the tunnel wall [141]. Membrane anchoring aligns the polypeptide exist site with the OXA1L translocon to facilitate cotranslational membrane insertion of newly synthesized proteins. The overall tunnel path is similar to that of bacterial and cytoplasmic ribosomes but different from yeast mitoribosomes.

Mitoribosomes are bound to the inner membrane, presumably as a result of their specialization on the synthesis of hydrophobic membrane proteins, which are cotranslationally inserted into the inner membrane [146,156]. Several inner membrane proteins mediate membrane binding of ribosomes such as the OXA1L machinery, which facilitates the insertion of nascent polypeptides into the inner membrane [157–163].

2.4.1.2 Mitoribosome biogenesis

The process of mitoribosome assembly requires the assistance of a growing number of nonribosomal proteins. These proteins include RNA modification enzymes, guanosine triphosphatases (GTPases), DEAD-box RNA helicases, and kinases [164,165]. They act as assembly factors to guide the processing and modification of mitoribosomal components and their temporal association to form preribosomal particles during the assembly of individual subunits, and formation of the monosome. Mitoribosome biogenesis follows a maturation pathway that is just starting to emerge and involves the cooperative assembly of protein sets, forming structural clusters and preassembled modules [166,167]. SILAC (Stable Isotope Labeling with Amino Acids in Cell Culture)-based metabolic labeling followed by proteomics studies (see Box 2.2) have shown that for each subunit, the protein components are synthesized in excess and imported into mitochondria, where their stoichiometric accumulation is regulated by degradation of the nonassembled free protein fractions [167]. The biogenesis of the two mitoribosome subunits is coordinated. It starts cotranscriptionally with mtLSU proteins forming a subcomplex on an unprocessed RNA containing the *16S rRNA*, whose formation is required for precursor RNA processing and liberation of the *12S-rRNA* as a condition for mtSSU protein incorporation [85]. Recent investigations have revealed quality control mechanisms that are in place to ensure that only the mature mtSSU and mtLSU are assembled into functional monosomes [69,165,168,169]. Although the number of mitoribosome assembly factors identified is increasing steadily, their specific functions and the molecular details of mitoribosomal biogenesis remain, in most cases, to be fully understood. A recent manuscript has reported the structures of two late-stage assembly intermediates of the human mtLSU isolated from a native pool within a human cell line and solved by cryo-EM. A comparison of the structures allowed for the identification of new assembly factors and to reveal insights into the timing of rRNA folding and protein incorporation during the final steps of mtLSU maturation [150].

2.4.2 Mitochondrial protein synthesis

2.4.2.1 Mitochondrial translation initiation

The overall steps of protein synthesis in mammalian mitochondrial ribosomes are similar to those in bacteria. However, mitochondrial translation differs substantially from bacterial or cytosolic translation systems, particularly at the initiation step. Key differences are seen in the interaction of the mitoribosome with the mRNAs, the tRNAs, and the mitochondrial translational factors. An important difference involves the lack of 5′ leader sequences by most mitochondrial mRNAs to promote their binding to the ribosome through the potential action of translation activators as it has been shown in yeast mitochondria [170]. Moreover, mitochondria contain a single $tRNA^{Met}$ that fulfills the dual role of the initiator and elongator $tRNA^{Met}$. A fraction of met-$tRNA^{Met}$ is formylated by mitochondrial methionyl-tRNA formyltransferase (MTFMT) to generate *N*-formylmethionine-$tRNA^{Met}$ (fMet-$tRNA^{Met}$), which is used for translation initiation [171,172]. The differences are also contributed by the fact that mitochondrial translation initiation lacks initiation factor 1, which is essential in all other translation systems [173]. Whereas translation initiation in bacteria involves three essential canonical initiation factors IF1, IF2, and IF3, mammalian mitochondria have homologs of the latter two, namely mtIF2 and mtIF3 [174,175].

Mitochondrial translation starts on an mtIF3-bound mtSSU that would then recruit the mt-mRNA (Fig. 2.3). Mammalian mtIF3 retains only ~25% sequence homology with respect to its bacterial counterparts [175], although its basic domain organization, comprising N- and C-terminal domains that are connected through a flexible linker, is similar to bacterial IF3. However, mtIF3 has ~30-amino-acid-long N- and C-terminal extensions (NTE and CTE), which are important for the destabilization of incorrect initiation complexes (the CTE) or to control the affinity of IF3mt for the 39S subunit (the NTE) [175–178]. A cryo-EM structural characterization of the mtSSU–mtIF3 complex has shown that the N-terminal domain of mtIF3 interacts with more mtSSU elements than its bacterial counterpart, thus enhancing its affinity for the mtSSU [178]. The mtIF3-C-terminal domain sits on the mtSSU in the vicinity of the P-site, and it interacts with several *12S*-rRNA helices. Its position would be expected to prevent the joining of the 39S subunit with the 28S subunit by directly interfering with the formation of two of the conserved intersubunit bridges [178]. It has been proposed that in this way, mtIF3 prevents the premature joining of the mtLSU until a proper preinitiation complex composed of IF2mt, mRNA, and initiator tRNA has been formed [178], a mechanism conserved in the bacterial system. How mtIF3 detects the presence of a bound mRNA and prevents the accommodation of the initiator tRNA in the absence of the mRNA remains to be fully understood. However, superimposition of the mtSSU–mtIF3 structure with the structure of the 55S-fMet-$tRNA^{Met}$-mtIF2 complex [130] have allowed to interpret the role of the C-terminal domain and its mitochondrion-specific

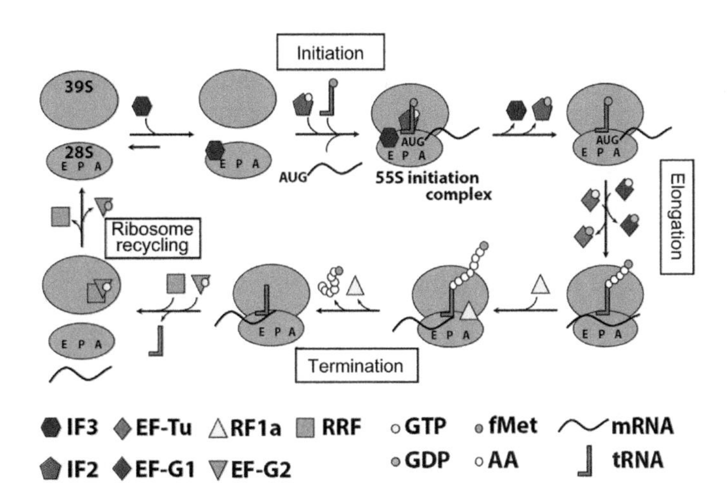

Figure 2.3: Mitochondrial translation in mitoribosomes.
Schematic representation of the four stages of protein synthesis, and the basic translation factors involved. Initiation involves mtIF2 and mtIF3; elongation is catalyzed by mtEF-Tu, mtEF-Ts, and mtEF-G1; termination is facilitated by the release factor mtRF1a, and ribosome recycling requires mtRRF and mtEF-G2. See a detailed explanation in the text.

CTE in destabilizing the initiator tRNA in the absence of mRNA, and to highlight the potential role of several lysine residues on the mtIF3-C-terminal domain [178].

Following the recruitment of the mt-mRNA to the mtIF3-bound mtSSU, GTP-bound IF2mt stimulates the binding of fMet-$tRNA^{Met}$ to the mtSSU [174]. Although charged $tRNA^{Met}$ and mtIF2 can bind the mtSSU in the absence of the mt-mRNA; the association is weak [172]. When a positive codon:anticodon interaction occurs, a stable complex is formed and triggers the recruitment of the mtLSU. A recent structure determination by cryo-EM of the complete translation initiation complex has revealed key features of mammalian mitochondrial translation initiation [130]. The study identified unique features of mtIF2 that are required for specific recognition of fMet-$tRNA^{Met}$ and regulation of its GTPase activity. Mitochondrial IF2 contains a mitochondrion-specific 37-amino-acid-long domain insertion that has been proposed to mimic the function IF1 plays in bacteria by sterically blocking the binding of initiator tRNA to the ribosomal A site [130]. Although mtIF2 does not form specific interactions with mRNA, by closing the decoding center, it may help to stabilize the binding of leaderless mRNAs. The cryo-EM structure showed *MTCO3* mRNA to be engaged with the PPR protein mS39, which is found at the entrance of the mt-mRNA channel. As mentioned earlier, this interaction may facilitate subsequent threading of the mRNA into the mRNA channel for start codon–anticodon interaction. The mRNA-mS39 contacts are not expected to be sequence- or even structure-specific, but it has been noted that starting at codon #7 the mt-mRNAs frequently show U-rich sequences that may be the determinant for PPR association and promote initial binding of the mitochondrial mRNAs to the initiation complex [130]. The mRNA channel is lined by a positively charged mitochondrion-specific extension of uS5m positioned between the entrance and the A site and has been proposed to guide the mRNA toward the P-site, where the codon–anticodon interaction fixes the initiation codon and stabilizes the mRNA binding in frame [130].

Recruitment of the mtLSU triggers hydrolysis of the mtIF2-bound GTP to GDP concomitantly with the release of both mtIF2 and mtIF3 from the mtSSU. If f-Met-$tRNA^{Met}$ is not available or if the start codon is not present in the P-site, the inspection by the initiation factors fails, the monosome is not formed, and the mRNA is released.

2.4.2.2 Mitochondrial translation elongation

Once the monosome is formed, elongation of the nascent chain proceeds by cycles of aminoacyl-tRNAs binding, peptide bond formation, and displacement of deacylated tRNAs (Fig. 2.3). The main factor involved in this step is the mitochondrial elongation factor Tu (mtEF-Tu), a highly conserved GTPase, which, in its GTP-bound activated form, delivers the aminoacyl-tRNAs to the mitoribosome A site through the formation of a ternary complex. This process uses the energy released from GTP hydrolysis, which yields an inactive GDP-bound mtEF-Tu that is then released from the ribosome and is regenerated by the elongation factor Ts (mtEF-Ts), a nucleotide exchange factor that promotes the exchange of GDP with GTP [179].

The release of mtEF-Tu leads to the formation of the peptide bond, catalyzed at the mtLSU PTC. Once the bond is formed, the mitoribosome P-site is occupied by a deacylated mt-tRNA, and the dipeptidyl-tRNA is found in the A-site. Mitochondria possess two elongation factor Gs, one of which, mtEF-G1 acts during elongation by catalyzing both the translocation of dipeptidyl-tRNA from the A site to the P-site and the movement of the mRNA to expose the next codon in the A site. During translocation, the rate-limiting step for tRNA-mRNA movement is a rearrangement of the ribosome (so-called unlocking), which is induced by the binding of GTP-bound mtEF-G1 to the ribosome and accelerated by GTP hydrolysis and consequent release of GDP-bound mtEF-G1 from the mitoribosome [180]. The cryo-EM structures of the human and porcine mitoribosomes have recently confirmed the presence of the E-site [141,142], and despite its differences from the characteristic bacterial structure, the deacylated mt-tRNA moves to the E-site before exiting the mitoribosome. The translation elongation process is repeated by multiple cycles until a stop codon is positioned in the A-site.

2.4.2.3 Mitochondrial translation termination and mitoribosome recycling

When the translation machinery encounters a stop codon, protein synthesis is complete, and the termination codon is recognized by a ribosome release factor (mtRF1a) that in the presence of GTP induces the release of the newly formed polypeptide from the mtLSU [181] through a conserved mechanism [182] (Fig. 2.3). Although the mtRF1a is believed to be sufficient to terminate all 13 ORFs, this factor is a class I RF, which specifically recognizes the codons UAA and UAG. In human mitochondria, UAA and UAG are used as stop codons to terminate nine monocistronic and two bicistronic ORFs, respectively. However, although the triplets following the coding sequence of ORFs *MTCO1* and *MTND6* are AGA and AGG, respectively, the termination codons of transcripts positioned at the A-site of the human mitoribosome in intact cells were found to terminate with the classical UAG codon. The mechanism involves a −1 frameshift, potentially driven by structured RNA immediately downstream of the termination codons within the transcripts [154,183]. This mechanism is plausible for humans, but would not work in all vertebrates, which has suggested that other release factors might be involved in terminating the translation of these two mt-mRNAs, perhaps in other mammalian species [184,185].

Following the action of the release factor, two ribosomal recycling factors, mtRRF1 [186], and the second mitochondrial member of the elongation factor G family, mtEF-G2 [187], promote the disruption of the mitoribosome posttermination complex by enabling the dissociation of the ribosomal subunits and the release of mt-mRNA and deacylated mt-tRNA (Fig. 2.3). The binding of GTP-bound mtEF-G2 to the mtRRF1-bound mitoribosome post-termination complex dissociates the ribosome into its two subunits, and at least the release of mtEF-G2 from the mtLSU requires the hydrolysis of GTP [187]. A recent cryo-EM structure of the human 55S mitoribosome–mtRRF1 complex has revealed that a mitochondrion-specific N-terminal extension of the recycling factor makes multiple mitochondrion-specific interactions with functionally critical regions of the mitoribosome, including rRNA segments that form the PTC and those that connect PTC with the mtLSU GTPase-associated center [188]. The interactions of mtRRF1 with the mitoribosome have been proposed to ensure complete inaccessibility to tRNAs and other ligands to both PTC and the entrance of the polypeptide exit tunnel of the mitoribosome [188].

Once the mtRRF1 and mtEF-G2 are finally released from the mtLSU, the translation cycle can reinitiate.

2.4.2.4 Cotranslational membrane insertion of newly synthesized polypeptides

Mitoribosomes are bound to the inner membrane, presumably as a result of their specialization on the synthesis of hydrophobic membrane proteins, which are cotranslationally inserted into the inner membrane. Visualization of human mitoribosomes studied *in situ* by cryoelectron tomography (cryo-ET) has revealed a single major contact site of the mtLSU with the inner mitochondrial membrane, mediated by the mitochondria-specific protein mL45 [189]. Further cryo-EM studies showed that mL45 sits at the exit of the polypeptide exit tunnel at the interface of the mitoribosome and the mitochondrial inner membrane [130,151,153]. A central domain in mL45 shows homology to the carboxy-terminal domain, membrane-binding segment, of TIM44, a subunit of the inner membrane translocase essential for protein import [190]. Because TIM44 is known to establish interactions with cardiolipin, the signature lipid of the mitochondrial inner membrane, it is possible that mL45 could also mediate cardiolipin-mediated tethering of the mitoribosome to the inner membrane in an orientation that facilitates cotranslational membrane insertion of newly synthesized polypeptides as they exit the tunnel. Although the precise mechanism involved remains to be understood, cryo-EM analyses have shown that the N-terminal tail of mL45 (amino acids K38–N64) inserts into the exit tunnel and occludes it during translation initiation [130]. Considering that the N-terminal extension of mL45 must be displaced from the tunnel during translation elongation, it has been hypothesized that the mL45 N-tail could facilitate membrane insertion of nascent chains by recruiting the OXA1L translocon machinery.

2.5 Compartmentalization of gene expression

An important aspect of mitochondrial gene expression relates to the compartmentalization of the processes involved. At least three distinct types of foci relevant to mtDNA expression have been identified and visualized within the mitochondrial matrix of human cells. Those are the mitochondrial nucleoids, RNA granules, and the RNA degradosome [119].

2.5.1 Mitochondrial DNA nucleoids

As explained in Chapter 1, MtDNA replication, maintenance, and nucleoid organization, eukaryotic cells typically contain hundreds to thousands of mtDNA genomes assembled into hundreds of dense proteinaceous structures known as nucleoids [191,192]. This organization in nucleoids provides stability to the mitochondrial genomes and is essential for their inheritance and segregation [193].

Early quantitative analyses of the size and mtDNA content of nucleoids in cultured mammalian cells suggested that an average nucleoid could contain 5–7 mtDNA genomes packed in a space with a diameter of only 70 nm [194]. However, recent superresolution microscopy data have revealed that mammalian mitochondrial nucleoids have a uniform size of ~100 nm and contain only 1–2 copies of mtDNA [195]. The packaging of mammalian mtDNA in nucleoids is accomplished with the assistance of DNA-binding proteins, including two particularly abundant proteins, essential for mtDNA replication: TFAM and mtSSB [38,192]. Studies in *Xenopus* oocytes have shown that the ratio of TFAM per mtDNA is developmentally regulated and inversely correlates with the number of mtDNA molecules that are engaged in replication or transcription [196]. Nucleoid purification approaches followed by mass spectrometry to analyze their protein composition, revealed a large variety of proteins involved in all aspects of mitochondrial gene expression [192,197,198]. The use of formaldehyde cross-linking allowed determining the nucleoid proteins that are in close contact with the mtDNA, and proposing a model for a layered structure of mtDNA nucleoids. In this model, replication and transcription occur in the central core, whereas RNA processing, translation, and OXPHOS complex assembly may occur in the peripheral region [192], which transitions to a different compartment known as the RNA granule.

2.5.2 Mitochondrial RNA granules

As a breakthrough in the compartmentalization of mtDNA expression field, analysis of newly transcribed RNA in human mitochondria from cultured cells by bromouridine (BrU) labeling identified distinct BrU-positive RNA foci in close proximity to the nucleoids [194]. These foci, termed RNA granules, contain the protein GRSF1 (G-rich sequence binding factor 1) and RNase P and accumulate mt-RNAs to regulate their processing, storage, sorting, or translation [199,200]. These foci also contain ribosomal proteins [200] and ribosomal RNA modifying enzymes [201]. Furthermore, GRSF1 depletion induces a defect in the assembly of the mitoribosomal SSU and a marked attenuation in mitochondrial protein synthesis [199,200].

This has raised the possibility that ribosome biogenesis could occur near or within the RNA granules. Affinity purification of tagged GRSF1 identified ribosomal proteins [200] and the *16S*-rRNA methyltransferases MRM1, MRM2, and MRM3/RMTL1 [201]. Affinity purification of tagged MRM3 coeluted ribosomal proteins and ribosome assembly factors that colocalize to the RNA granules, such as GRSF1 and the DEAD-box helicase DDX28 [201,202]. Recent comprehensive affinity purification studies of GRSF1 and DDX28, followed by mass spectrometry analysis by two independent groups, have yielded similar results [92,203]. In the study performed by our group, the DDX28 interactome was analyzed in extracts prepared in the presence of increasing salt concentrations to assess the strength of the interactions. DDX28 was found associated with five major groups of relevant proteins [165,203]. (1) Mitoribosome proteins; (2) mitoribosome assembly factors; (3) mitochondrial translational factors; (4) RNA metabolism factors, including transcription factors, RNA processing and modifying enzymes, and RNA stability and degradosome proteins; and (5) proteins previously identified as mtDNA nucleoid components, such as the helicase DHX30 that acts as a transcriptional regulator.

The overlapping pool of proteins associated with either mitochondrial nucleoids or RNA granules reflects their proximity and the lack of membranous boundaries sealing these compartments [119]. Their functional and physical dynamics to attend the necessities of newly transcribed RNAs is highlighted by the observation by fluorescence microscopy that some 10%–15% of nucleoids and RNA granules do actually overlap in most studies, probably reflecting sites of active transcription [119,194,199,200]. A quantitative assessment of these dynamics showed that after an initial short pulse of 20 minutes, most BrU-positive RNA foci colocalize within 200 nm of mtDNA, but after more extended periods (or a pulse-chase) the Br-positive RNA foci become distributed randomly [200]. Thus mitoribosome assembly could initiate in granule-nucleoid overlapping foci [167,204], possibly in a cotranscriptional manner as it occurs in bacteria [205], to end within the granule environment [92,165,203,206], or even outside the granule, as it has been recently suggested [207].

As yet another connection between mtDNA nucleoids and RNA granules, two mtDNA replication factors, Twinkle and mtSSB, have been found to participate in granule formation and mtRNA processing/degradation in an interplay with GRSF1 [207], which highlights the role of at least these proteins in multiple aspects of mtDNA gene expression.

2.5.3 Mitochondrial RNA degradosome

The RNA granules occasionally colocalize with RNA breakdown complexes termed RNA degradosomes (the PNPase-SUV3 helicase complex) formed only in specific foci (the D-foci) [119].

RNA granules and D-foci are intimately connected not only physically, but also functionally. For example, a translational variant of FASTK colocalizes with mitochondrial RNA granules and is required to regulate specifically the expression of the

MTND6 mRNA, the only protein-coding sequence located on the L-strand transcript. Mechanistically, FASTK binds the L-strand precursor RNA at multiple sites both within and downstream of the *ND6*-coding sequence and, together with the mitochondrial degradosome, participates in generating the mature form of the *MTND6* mRNA [91]. In another example, GRSF1 has been shown to cooperate with the mitochondrial degradosome in the degradation of mitochondrial long noncoding lncRNAs that contain G-quadruplexes (G4s), which are prone to be formed due to the GC skew in the mtDNA. GRSF1 promotes G4 melting to facilitate degradosome-mediated decay [208]. Furthermore, whereas the lncRNAs are most prominently derived from the L-strand surrounding the *ND6* gene, it has been shown that the mt-RNA patterns for a double mtSSB/GRSF1 knockdown resemble those of degradosome-deficient mitochondria [207]. These observations have raised the hypothesis that mtSSB could participate in the degradation of G4-containing RNAs by binding unfolded RNAs following GRSF1 action and prevent them from reforming G-quadruplexes [207].

Research perspectives

Over the last 50 years, dramatic progress has been made in the understanding of mitochondrial replication, transcription, and translation systems, and the mechanisms and factors involved in each process. The recent development of high-resolution cryo-EM methods, in combination with novel proteomics and classical genetic and biochemical procedures, and mouse and human cell gene-editing approaches is speeding the understanding of mitochondrial gene expression at the molecular, cellular, and organismal level.

Several essential questions that are still open are anticipated to be the focus of future investigations over the next few years. Potential open questions, scattered through the text, and future aims for the field include: disclosing the determinants of mtDNA copy number control across different cell types; devising approaches to edit the mtDNA; fully understanding the switch between replication and transcription; dissect the mechanism of transcription termination; disclosing how are processed all the non-tRNA containing junctions in polycistronic transcripts; developing an *in vitro* mitochondrial translation system; performing mitoribosome structural studies capturing the simultaneous presence of mRNA and the two translation initiation factors to understand better the mechanism of mitochondrial translation initiation, or capturing the presence of the two factors involved in recycling to disclose their interplay in resetting the translation machinery; identifying all mitoribosome assembly factors and their specific roles; or better characterizing the functional overlaps among the components of the nucleoid the RNA granule and the degradosome.

References

[1] Montoya J, et al. Identification of initiation sites for heavy-strand and light-strand transcription in human mitochondrial DNA. Proc Natl Acad Sci USA 1982;79(23):7195–9.

[2] Chang DD, Clayton DA. Precise identification of individual promoters for transcription of each strand of human mitochondrial DNA. Cell 1984;36(3):635–43.

[3] Sanchez MI, et al. RNA processing in human mitochondria. Cell Cycle 2011;10(17):2904–16.

[4] Mick DU, Fox TD, Rehling P. Inventory control: cytochrome c oxidase assembly regulates mitochondrial translation. Nat Rev Mol Cell Biol 2011;12(1):14–20.

[5] De Silva D, et al. Mitochondrial ribosome assembly in health and disease. Cell Cycle 2015;14 (14):2226–50.

[6] Pham XH, et al. Conserved sequence box II directs transcription termination and primer formation in mitochondria. J Biol Chem 2006;281(34):24647–52.

[7] Fuste JM, et al. Mitochondrial RNA polymerase is needed for activation of the origin of light-strand DNA replication. Mol Cell 2010;37(1):67–78.

[8] Wanrooij S, et al. Human mitochondrial RNA polymerase primes lagging-strand DNA synthesis in vitro. Proc Natl Acad Sci U S A 2008;105(32):11122–7.

[9] Agaronyan K, et al. Mitochondrial biology. Replication-transcription switch in human mitochondria. Science 2015;347(6221):548–51.

[10] Nicholls TJ, Minczuk M. In D-loop: 40 years of mitochondrial 7S DNA. Exp Gerontol 2014;56:175–81.

[11] Wei W, et al. Frequency and signature of somatic variants in 1461 human brain exomes. Genet Med 2019;21(4):904–12.

[12] Bibb MJ, et al. Sequence and gene organization of mouse mitochondrial DNA. Cell 1981;26(2 Pt 2):167–80.

[13] Sbisa E, et al. Mammalian mitochondrial D-loop region structural analysis: identification of new conserved sequences and their functional and evolutionary implications. Gene 1997;205(1–2):125–40.

[14] Doda JN, Wright CT, Clayton DA. Elongation of displacement-loop strands in human and mouse mitochondrial DNA is arrested near specific template sequences. Proc Natl Acad Sci U S A 1981;78 (10):6116–20.

[15] Kuhl I, et al. POLRMT regulates the switch between replication primer formation and gene expression of mammalian mtDNA. Sci Adv 2016;2(8):e1600963.

[16] Ojala D, et al. A small polyadenylated RNA (7S RNA), containing a putative ribosome attachment site, maps near the origin of human mitochondrial DNA replication. J Mol Biol 1981;150(2):303–14.

[17] Ruhanen H, et al. Mitochondrial single-stranded DNA binding protein is required for maintenance of mitochondrial DNA and 7S DNA but is not required for mitochondrial nucleoid organisation. Biochim Biophys Acta 2010;1803(8):931–9.

[18] Milenkovic D, et al. TWINKLE is an essential mitochondrial helicase required for synthesis of nascent D-loop strands and complete mtDNA replication. Hum Mol Genet 2013;22(10):1983–93.

[19] Fish J, Raule N, Attardi G. Discovery of a major D-loop replication origin reveals two modes of human mtDNA synthesis. Science 2004;306(5704):2098–101.

[20] Pereira F, et al. Evidence for variable selective pressures at a large secondary structure of the human mitochondrial DNA control region. Mol Biol Evol 2008;25(12):2759–70.

[21] Jemt E, et al. Regulation of DNA replication at the end of the mitochondrial D-loop involves the helicase TWINKLE and a conserved sequence element. Nucleic Acids Res 2015;43(19):9262–75.

[22] Hyvarinen AK, et al. The mitochondrial transcription termination factor mTERF modulates replication pausing in human mitochondrial DNA. Nucleic Acids Res 2007;35(19):6458–74.

[23] Ghivizzani SC, et al. In organello footprint analysis of human mitochondrial DNA: human mitochondrial transcription factor A interactions at the origin of replication. Mol Cell Biol 1994;14(12):7717–30.

[24] Lewis SC, Uchiyama LF, Nunnari J. ER-mitochondria contacts couple mtDNA synthesis with mitochondrial division in human cells. Science 2016;353(6296):aaf5549.

[25] Morozov YI, Temiakov D. Human mitochondrial transcription initiation complexes have similar topology on the light and heavy strand promoters. J Biol Chem 2016;291(26):13432–5.
[26] Ngo HB, Kaiser JT, Chan DC. The mitochondrial transcription and packaging factor Tfam imposes a U-turn on mitochondrial DNA. Nat Struct Mol Biol 2011;18(11):1290–6.
[27] Yakubovskaya E, et al. Organization of the human mitochondrial transcription initiation complex. Nucleic Acids Res 2014;42(6):4100–12.
[28] Gaspari M, et al. The mitochondrial RNA polymerase contributes critically to promoter specificity in mammalian cells. EMBO J 2004;23(23):4606–14.
[29] Posse V, et al. TEFM is a potent stimulator of mitochondrial transcription elongation in vitro. Nucleic Acids Res 2015;43(5):2615–24.
[30] Wanrooij PH, et al. G-quadruplex structures in RNA stimulate mitochondrial transcription termination and primer formation. Proc Natl Acad Sci USA 2010;107(37):16072–7.
[31] Hillen HS, et al. Mechanism of transcription anti-termination in human mitochondria. Cell 2017;171 (5):1082–1093.e13.
[32] Hillen HS, et al. Structural basis of mitochondrial transcription initiation. Cell 2017;171(5):1072–1081. e10.
[33] Ringel R, et al. Structure of human mitochondrial RNA polymerase. Nature 2011;478(7368):269–73.
[34] Shokolenko IN, Alexeyev MF. Mitochondrial transcription in mammalian cells. Front Biosci (Landmark Ed) 2017;22:835–53.
[35] Montoya J, Gaines GL, Attardi G. The pattern of transcription of the human mitochondrial rRNA genes reveals two overlapping transcription units. Cell 1983;34(1):151–9.
[36] Terzioglu M, et al. MTERF1 binds mtDNA to prevent transcriptional interference at the light-strand promoter but is dispensable for rRNA gene transcription regulation. Cell Metab 2013;17(4):618–26.
[37] Martin M, et al. Termination factor-mediated DNA loop between termination and initiation sites drives mitochondrial rRNA synthesis. Cell 2005;123(7):1227–40.
[38] Kaufman BA, et al. The mitochondrial transcription factor TFAM coordinates the assembly of multiple DNA molecules into nucleoid-like structures. Mol Biol Cell 2007;18(9):3225–36.
[39] Kanki T, et al. Architectural role of mitochondrial transcription factor A in maintenance of human mitochondrial DNA. Mol Cell Biol 2004;24(22):9823–34.
[40] Alam TI, et al. Human mitochondrial DNA is packaged with TFAM. Nucleic Acids Res 2003;31 (6):1640–5.
[41] Fisher RP, Clayton DA. A transcription factor required for promoter recognition by human mitochondrial RNA polymerase. Accurate initiation at the heavy- and light-strand promoters dissected and reconstituted in vitro. J Biol Chem 1985;260(20):11330–8.
[42] Dairaghi DJ, Shadel GS, Clayton DA. Addition of a 29 residue carboxyl-terminal tail converts a simple HMG box-containing protein into a transcriptional activator. J Mol Biol 1995;249(1):11–28.
[43] Morozov YI, et al. A novel intermediate in transcription initiation by human mitochondrial RNA polymerase. Nucleic Acids Res 2014;42(6):3884–93.
[44] Rubio-Cosials A, et al. Human mitochondrial transcription factor A induces a U-turn structure in the light strand promoter. Nat Struct Mol Biol 2011;18(11):1281–9.
[45] Morozov YI, et al. A model for transcription initiation in human mitochondria. Nucleic Acids Res 2015;43(7):3726–35.
[46] Posse V, Gustafsson CM. Human mitochondrial transcription factor B2 is required for promoter melting during initiation of transcription. J Biol Chem 2017;292(7):2637–45.
[47] Schubot FD, et al. Crystal structure of the transcription factor sc-mtTFB offers insights into mitochondrial transcription. Protein Sci 2001;10(10):1980–8.
[48] Lodeiro MF, et al. Identification of multiple rate-limiting steps during the human mitochondrial transcription cycle in vitro. J Biol Chem 2010;285(21):16387–402.
[49] Shutt TE, et al. Core human mitochondrial transcription apparatus is a regulated two-component system in vitro. Proc Natl Acad Sci USA 2010;107(27):12133–8.

[50] Wang J, et al. Dilated cardiomyopathy and atrioventricular conduction blocks induced by heart-specific inactivation of mitochondrial DNA gene expression. Nat Genet 1999;21(1):133–7.
[51] Lodeiro MF, et al. Transcription from the second heavy-strand promoter of human mtDNA is repressed by transcription factor A in vitro. Proc Natl Acad Sci USA 2012;109(17):6513–18.
[52] Stiles AR, et al. Mutations in TFAM, encoding mitochondrial transcription factor A, cause neonatal liver failure associated with mtDNA depletion. Mol Genet Metab 2016;119(1–2):91–9.
[53] Kang I, Chu CT, Kaufman BA. The mitochondrial transcription factor TFAM in neurodegeneration: emerging evidence and mechanisms. FEBS Lett 2018;592(5):793–811.
[54] Minczuk M, et al. TEFM (c17orf42) is necessary for transcription of human mtDNA. Nucleic Acids Res 2011;39(10):4284–99.
[55] Yu H, et al. TEFM enhances transcription elongation by modifying mtRNAP pausing dynamics. Biophys J 2018;115(12):2295–300.
[56] Kruse B, Narasimhan N, Attardi G. Termination of transcription in human mitochondria: identification and purification of a DNA binding protein factor that promotes termination. Cell 1989;58(2):391–7.
[57] Yakubovskaya E, et al. Helix unwinding and base flipping enable human MTERF1 to terminate mitochondrial transcription. Cell 2010;141(6):982–93.
[58] Jimenez-Menendez N, et al. Human mitochondrial mTERF wraps around DNA through a left-handed superhelical tandem repeat. Nat Struct Mol Biol 2010;17(7):891–3.
[59] Shang J, Clayton DA. Human mitochondrial transcription termination exhibits RNA polymerase independence and biased bipolarity in vitro. J Biol Chem 1994;269(46):29112–20.
[60] Asin-Cayuela J, et al. The human mitochondrial transcription termination factor (mTERF) is fully active in vitro in the non-phosphorylated form. J Biol Chem 2005;280(27):25499–505.
[61] Hyvarinen AK, et al. Effects on mitochondrial transcription of manipulating mTERF protein levels in cultured human HEK293 cells. BMC Mol Biol 2010;11:72.
[62] Manwaring N, et al. Population prevalence of the MELAS A3243G mutation. Mitochondrion 2007;7 (3):230–3.
[63] Hess JF, et al. Impairment of mitochondrial transcription termination by a point mutation associated with the MELAS subgroup of mitochondrial encephalomyopathies. Nature 1991;351(6323):236–9.
[64] Linder T, et al. A family of putative transcription termination factors shared amongst metazoans and plants. Curr Genet 2005;48(4):265–9.
[65] Spahr H, et al. Structure of mitochondrial transcription termination factor 3 reveals a novel nucleic acid-binding domain. Biochem Biophys Res Commun 2010;397(3):386–90.
[66] Spahr H, et al. Structure of the human MTERF4-NSUN4 protein complex that regulates mitochondrial ribosome biogenesis. Proc Natl Acad Sci USA 2012;109(38):15253–8.
[67] Yakubovskaya E, et al. Structure of the essential MTERF4:NSUN4 protein complex reveals how an MTERF protein collaborates to facilitate rRNA modification. Structure 2012;20(11):1940–7. Available from: https://doi.org/10.1016/j.str.2012.08.027 Epub 2012 Sep 27.
[68] Wredenberg A, et al. MTERF3 regulates mitochondrial ribosome biogenesis in invertebrates and mammals. PLoS Genet 2013;9(1):e1003178.
[69] Metodiev MD, et al. NSUN4 is a dual function mitochondrial protein required for both methylation of 12S rRNA and coordination of mitoribosomal assembly. PLoS Genet 2014;10(2):e1004110.
[70] Mercer TR, et al. The human mitochondrial transcriptome. Cell 2011;146(4):645–58. Available from: https://doi.org/10.1016/j.cell.2011.06.051.
[71] Gelfand R, Attardi G. Synthesis and turnover of mitochondrial ribonucleic acid in HeLa cells: the mature ribosomal and messenger ribonucleic acid species are metabolically unstable. Mol Cell Biol 1981;1 (6):497–511.
[72] Lung B, et al. Identification of small non-coding RNAs from mitochondria and chloroplasts. Nucleic Acids Res 2006;34(14):3842–52.
[73] Rackham O, et al. Long noncoding RNAs are generated from the mitochondrial genome and regulated by nuclear-encoded proteins. RNA 2011;17(12):2085–93.

[74] Koc EC, Spremulli LL. RNA-binding proteins of mammalian mitochondria. Mitochondrion 2003;2(4):277–91.
[75] Liu G, et al. Mapping of mitochondrial RNA–protein interactions by digital RNase footprinting. Cell Rep 2013;5(3):839–48.
[76] Nakamura T, Yagi Y, Kobayashi K. Mechanistic insight into pentatricopeptide repeat proteins as sequence-specific RNA-binding proteins for organellar RNAs in plants. Plant Cell Physiol 2012;53(7):1171–9.
[77] Lurin C, et al. Genome-wide analysis of Arabidopsis pentatricopeptide repeat proteins reveals their essential role in organelle biogenesis. Plant Cell 2004;16(8):2089–103.
[78] Small ID, Peeters N. The PPR motif—a TPR-related motif prevalent in plant organellar proteins. Trends Biochem Sci 2000;25(2):46–7.
[79] Williams-Carrier R, Kroeger T, Barkan A. Sequence-specific binding of a chloroplast pentatricopeptide repeat protein to its native group II intron ligand. RNA 2008;14(9):1930–41.
[80] Rackham O, Mercer TR, Filipovska A. The human mitochondrial transcriptome and the RNA-binding proteins that regulate its expression. Wiley Interdiscip Rev RNA 2012;3(5):675–95.
[81] Jourdain AA, et al. The FASTK family of proteins: emerging regulators of mitochondrial RNA biology. Nucleic Acids Res 2017;45(19):10941–7.
[82] Lee I, Hong W. RAP—a putative RNA-binding domain. Trends Biochem Sci 2004;29(11):567–70.
[83] Simarro M, et al. Fast kinase domain-containing protein 3 is a mitochondrial protein essential for cellular respiration. Biochem Biophys Res Commun 2010;401(3):440–6.
[84] Ojala D, Montoya J, Attardi G. tRNA punctuation model of RNA processing in human mitochondria. Nature 1981;290(5806):470–4.
[85] Rackham O, et al. Hierarchical RNA processing is required for mitochondrial ribosome assembly. Cell Rep 2016;16(7):1874–90.
[86] Manam S, Van Tuyle GC. Separation and characterization of 5′- and 3′-tRNA processing nucleases from rat liver mitochondria. J Biol Chem 1987;262(21):10272–9.
[87] Kuznetsova I, et al. Simultaneous processing and degradation of mitochondrial RNAs revealed by circularized RNA sequencing. Nucleic Acids Res 2017;45(9):5487–500.
[88] Lee RG, et al. Is mitochondrial gene expression coordinated or stochastic? Biochem Soc Trans 2018;46(5):1239–46.
[89] Brzezniak LK, et al. Involvement of human ELAC2 gene product in 3′ end processing of mitochondrial tRNAs. RNA Biol 2011;8(4):616–26.
[90] Xu F, et al. Disruption of a mitochondrial RNA-binding protein gene results in decreased cytochrome b expression and a marked reduction in ubiquinol-cytochrome c reductase activity in mouse heart mitochondria. Biochem J 2008;416(1):15–26.
[91] Jourdain AA, et al. A mitochondria-specific isoform of FASTK is present in mitochondrial RNA granules and regulates gene expression and function. Cell Rep 2015;10(7):1110–21.
[92] Antonicka H, Shoubridge EA. Mitochondrial RNA granules are centers for posttranscriptional RNA processing and ribosome biogenesis. Cell Rep 2015;10(6):920–32.
[93] Boczonadi V, Ricci G, Horvath R. Mitochondrial DNA transcription and translation: clinical syndromes. Essays Biochem 2018;62(3):321–40.
[94] Tomecki R, et al. Identification of a novel human nuclear-encoded mitochondrial poly(A) polymerase. Nucleic Acids Res 2004;32(20):6001–14.
[95] Nagaike T, et al. Human mitochondrial mRNAs are stabilized with polyadenylation regulated by mitochondria-specific poly(A) polymerase and polynucleotide phosphorylase. J Biol Chem 2005;280(20):19721–7.
[96] Slomovic S, et al. Polyadenylation and degradation of human mitochondrial RNA: the prokaryotic past leaves its mark. Mol Cell Biol 2005;25(15):6427–35.
[97] Wydro M, et al. Targeting of the cytosolic poly(A) binding protein PABPC1 to mitochondria causes mitochondrial translation inhibition. Nucleic Acids Res 2010;38(11):3732–42.

[98] Bai Y, et al. Structural basis for dimerization and activity of human PAPD1, a noncanonical poly(A) polymerase. Mol Cell 2011;41(3):311–20.
[99] Lapkouski M, Hallberg BM. Structure of mitochondrial poly(A) RNA polymerase reveals the structural basis for dimerization, ATP selectivity and the SPAX4 disease phenotype. Nucleic Acids Res 2015;43 (18):9065–75.
[100] Pearce SF, et al. Maturation of selected human mitochondrial tRNAs requires deadenylation. Elife 2017;6.
[101] Crosby AH, et al. Defective mitochondrial mRNA maturation is associated with spastic ataxia. Am J Hum Genet 2010;87(5):655–60.
[102] Antonicka H, et al. A pseudouridine synthase module is essential for mitochondrial protein synthesis and cell viability. EMBO Rep 2017;18(1):28–38.
[103] Zaganelli S, et al. The pseudouridine synthase RPUSD4 is an essential component of mitochondrial RNA granules. J Biol Chem 2017;292(11):4519–32.
[104] Metodiev MD, et al. Methylation of 12S rRNA is necessary for in vivo stability of the small subunit of the mammalian mitochondrial ribosome. Cell Metab 2009;9(4):386–97.
[105] Seidel-Rogol BL, McCulloch V, Shadel GS. Human mitochondrial transcription factor B1 methylates ribosomal RNA at a conserved stem-loop. Nat Genet 2003;33(1):23–4.
[106] Lee KW, Bogenhagen DF. Assignment of 2′-O-methyltransferases to modification sites on the mammalian mitochondrial large subunit 16S rRNA. J Biol Chem 2014;289(36):24936–42.
[107] Rorbach J, et al. MRM2 and MRM3 are involved in biogenesis of the large subunit of the mitochondrial ribosome. Mol Biol Cell 2014;25(17):2542–55.
[108] Bar-Yaacov D, et al. Mitochondrial 16S rRNA is methylated by tRNA methyltransferase TRMT61B in all vertebrates. PLoS Biol 2016;14(9):e1002557.
[109] Nagaike T, et al. Identification and characterization of mammalian mitochondrial tRNA nucleotidyltransferases. J Biol Chem 2001;276(43):40041–9.
[110] Levinger L, Morl M, Florentz C. Mitochondrial tRNA 3′ end metabolism and human disease. Nucleic Acids Res 2004;32(18):5430–41.
[111] D'Souza AR, Minczuk M. Mitochondrial transcription and translation: overview. Essays Biochem 2018;62(3):309–20.
[112] Patton JR, et al. Mitochondrial myopathy and sideroblastic anemia (MLASA): missense mutation in the pseudouridine synthase 1 (PUS1) gene is associated with the loss of tRNA pseudouridylation. J Biol Chem 2005;280(20):19823–8.
[113] Agris PF, Vendeix FA, Graham WD. tRNA's wobble decoding of the genome: 40 years of modification. J Mol Biol 2007;366(1):1–13.
[114] Sasarman F, et al. LRPPRC and SLIRP interact in a ribonucleoprotein complex that regulates posttranscriptional gene expression in mitochondria. Mol Biol Cell 2010;21(8):1315–23.
[115] Siira SJ, et al. LRPPRC-mediated folding of the mitochondrial transcriptome. Nat Commun 2017;8 (1):1532.
[116] Ruzzenente B, et al. LRPPRC is necessary for polyadenylation and coordination of translation of mitochondrial mRNAs. EMBO J 2012;31(2):443–56.
[117] Wilson DN. Ribosome-targeting antibiotics and mechanisms of bacterial resistance. Nat Rev Microbiol 2014;12(1):35–48.
[118] Chujo T, et al. LRPPRC/SLIRP suppresses PNPase-mediated mRNA decay and promotes polyadenylation in human mitochondria. Nucleic Acids Res 2012;40(16):8033–47.
[119] Borowski LS, et al. Human mitochondrial RNA decay mediated by PNPase-hSuv3 complex takes place in distinct foci. Nucleic Acids Res 2013;41(2):1223–40.
[120] Szczesny RJ, et al. Human mitochondrial RNA turnover caught in flagranti: involvement of hSuv3p helicase in RNA surveillance. Nucleic Acids Res 2010;38(1):279–98.
[121] Bruni F, et al. REXO2 is an oligoribonuclease active in human mitochondria. PLoS One 2013;8(5): e64670.

[122] Nicholls TJ, et al. Dinucleotide degradation by REXO2 maintains promoter specificity in mammalian mitochondria. Mol Cell 2019;76(5):784–796.e6.
[123] Mootha VK, et al. Identification of a gene causing human cytochrome c oxidase deficiency by integrative genomics. Proc Natl Acad Sci U S A 2003;100(2):605–10.
[124] Fontanesi F. Mechanisms of mitochondrial translational regulation. IUBMB Life 2013;65 (5):397–408.
[125] Weraarpachai W, et al. Mutation in TACO1, encoding a translational activator of COX I, results in cytochrome c oxidase deficiency and late-onset Leigh syndrome. Nat Genet 2009;41(7):833–7.
[126] Richman TR, et al. Loss of the RNA-binding protein TACO1 causes late-onset mitochondrial dysfunction in mice. Nat Commun 2016;7:11884.
[127] Seeger J, et al. Clinical and neuropathological findings in patients with TACO1 mutations. Neuromuscul Disord 2010;20(11):720–4.
[128] Boehm E, et al. Role of FAST kinase domains 3 (FASTKD3) in post-transcriptional regulation of mitochondrial gene expression. J Biol Chem 2016;291(50):25877–87.
[129] Davies SM, et al. Pentatricopeptide repeat domain protein 3 associates with the mitochondrial small ribosomal subunit and regulates translation. FEBS Lett 2009;583(12):1853–8.
[130] Kummer E, et al. Unique features of mammalian mitochondrial translation initiation revealed by cryo-EM. Nature 2018;560(7717):263–7.
[131] Spremulli LL, et al. Initiation and elongation factors in mammalian mitochondrial protein biosynthesis. Prog Nucleic Acid Res Mol Biol 2004;77:211–61.
[132] Osawa S, et al. Evolution of the mitochondrial genetic code. II. Reassignment of codon AUA from isoleucine to methionine. J Mol Evol 1989;29(5):373–80.
[133] Jukes TH, Osawa S. The genetic code in mitochondria and chloroplasts. Experientia 1990;46 (11–12):1117–26.
[134] Mai N, Chrzanowska-Lightowlers ZM, Lightowlers RN. The process of mammalian mitochondrial protein synthesis. Cell Tissue Res 2017;367(1):5–20.
[135] Smits P, et al. Reconstructing the evolution of the mitochondrial ribosomal proteome. Nucleic Acids Res 2007;35(14):4686–703.
[136] Petrov AS, et al. Structural patching fosters divergence of mitochondrial ribosomes. Mol Biol Evol 2019;36(2):207–19.
[137] Kim HJ, Maiti P, Barrientos A. Mitochondrial ribosomes in cancer. Semin Cancer Biol 2017;47:67–81.
[138] O'Brien TW. The general occurrence of 55S ribosomes in mammalian liver mitochondria. J Biol Chem 1971;246(10):3409–17.
[139] O'Brien TW, Kalf GF. Ribosomes from rat liver mitochondria. I. Isolation procedure and contamination studies. J Biol Chem 1967;242(9):2172–9.
[140] Grivell LA, Reijnders L, Borst P. Isolation of yeast mitochondrial ribosomes highly active in protein synthesis. Biochim Biophys Acta 1971;247(1):91–103.
[141] Amunts A, et al. The structure of the human mitochondrial ribosome. Science 2015;348(6230):95–8.
[142] Greber BJ, et al. Ribosome. The complete structure of the 55S mammalian mitochondrial ribosome. Science 2015;348(6232):303–8.
[143] Koc EC, et al. Identification and characterization of CHCHD1, AURKAIP1, and CRIF1 as new members of the mammalian mitochondrial ribosome. Front Physiol 2013;4:183.
[144] Suzuki T, et al. Proteomic analysis of the mammalian mitochondrial ribosome. Identification of protein components in the 28S small subunit. J Biol Chem 2001;276(35):33181–95.
[145] Koc EC, et al. The large subunit of the mammalian mitochondrial ribosome. Analysis of the complement of ribosomal proteins present. J Biol Chem 2001;276(47):43958–69.
[146] Greber BJ, et al. Architecture of the large subunit of the mammalian mitochondrial ribosome. Nature 2014;505(7484):515–19.
[147] Kaushal PS, et al. Cryo-EM structure of the small subunit of the mammalian mitochondrial ribosome. Proc Natl Acad Sci USA 2014;111(20):7284–9.

[148] Mears JA, et al. A structural model for the large subunit of the mammalian mitochondrial ribosome. J Mol Biol 2006;358(1):193–212.
[149] Sharma MR, et al. Structure of the mammalian mitochondrial ribosome reveals an expanded functional role for its component proteins. Cell 2003;115(1):97–108.
[150] Brown A, et al. Structures of the human mitochondrial ribosome in native states of assembly. Nat Struct Mol Biol 2017;24(10):866–9.
[151] Brown A, et al. Structure of the large ribosomal subunit from human mitochondria. Science 2014;46 (6210):718–22.
[152] Greber BJ, Ban N. Structure and function of the mitochondrial ribosome. Annu Rev Biochem 2016;85:103–32.
[153] Greber BJ, et al. The complete structure of the large subunit of the mammalian mitochondrial ribosome. Nature 2014;515(7526):283–6.
[154] Temperley RJ, et al. Human mitochondrial mRNAs—like members of all families, similar but different. Biochim Biophys Acta 2010;1797(6–7):1081–5.
[155] Denslow ND, Anders JC, O'Brien TW. Bovine mitochondrial ribosomes possess a high affinity binding site for guanine nucleotides. J Biol Chem 1991;266(15):9586–90.
[156] Amunts A, et al. Structure of the yeast mitochondrial large ribosomal subunit. Science 2014;343:1485–9.
[157] Jia L, et al. Yeast Oxa1 interacts with mitochondrial ribosomes: the importance of the C-terminal region of Oxa1. EMBO J 2003;22(24):6438–47.
[158] Jia L, Kaur J, Stuart RA. Mapping of the *Saccharomyces cerevisiae* Oxa1-mitochondrial ribosome interface and identification of MrpL40, a ribosomal protein in close proximity to Oxa1 and critical for oxidative phosphorylation complex assembly. Eukaryot Cell 2009;8(11):1792–802.
[159] Szyrach G, et al. Ribosome binding to the Oxa1 complex facilitates co-translational protein insertion in mitochondria. EMBO J 2003;22(24):6448–57.
[160] Kohler R, et al. YidC and Oxa1 form dimeric insertion pores on the translating ribosome. Mol Cell 2009;34(3):344–53.
[161] Bauerschmitt H, et al. Ribosome-binding proteins Mdm38 and Mba1 display overlapping functions for regulation of mitochondrial translation. Mol Biol Cell 2010;21(12):1937–44.
[162] Gruschke S, et al. Proteins at the polypeptide tunnel exit of the yeast mitochondrial ribosome. J Biol Chem 2010;285(25):19022–8.
[163] Ott M, et al. Mba1, a membrane-associated ribosome receptor in mitochondria. EMBO J 2006;25 (8):1603–10.
[164] De Silva D, Fontanesi F, Barrientos A. The DEAD-Box protein Mrh4 functions in the assembly of the mitochondrial large ribosomal subunit. Cell Metab 2013;18:712–25.
[165] Maiti P, et al. Human GTPBP10 is required for mitoribosome maturation. Nucleic Acids Res 2018;13.
[166] Zeng R, Smith E, Barrientos A. Yeast mitoribosome large subunit assembly proceeds by hierarchical incorporation of protein clusters and modules on the inner membrane. Cell Metab 2018;27(3):645–56.
[167] Bogenhagen DF, et al. Kinetics and mechanism of mammalian mitochondrial ribosome assembly. Cell Rep 2018;22(7):1935–44.
[168] Kim H-J, Barrientos A. MTG1 couples mitoribosome large subunit assembly and intersubunit bridge formation. Nucleic Acid Res 2018;.
[169] Lavdovskaia E, et al. The human Obg protein GTPBP10 is involved in mitoribosomal biogenesis. Nucleic Acids Res 2018;2(5063820).
[170] Poutre CG, Fox TD. PET111, a *Saccharomyces cerevisiae* nuclear gene required for translation of the mitochondrial mRNA encoding cytochrome c oxidase subunit II. Genetics 1987;115(4):637–47.
[171] Kuzmenko A, et al. Mitochondrial translation initiation machinery: conservation and diversification. Biochimie 2014;100:132–40.
[172] Tucker EJ, et al. Mutations in MTFMT underlie a human disorder of formylation causing impaired mitochondrial translation. Cell Metab 2011;14(3):428–34.

[173] Atkinson GC, et al. Evolutionary and genetic analyses of mitochondrial translation initiation factors identify the missing mitochondrial IF3 in *S. cerevisiae*. Nucleic Acids Res 2012;40(13):6122–34.

[174] Spencer AC, Spremulli LL. The interaction of mitochondrial translational initiation factor 2 with the small ribosomal subunit. Biochim Biophys Acta 2005;1750(1):69–81.

[175] Koc EC, Spremulli LL. Identification of mammalian mitochondrial translational initiation factor 3 and examination of its role in initiation complex formation with natural mRNAs. J Biol Chem 2002;277(38):35541–9.

[176] Bhargava K, Spremulli LL. Role of the N- and C-terminal extensions on the activity of mammalian mitochondrial translational initiation factor 3. Nucleic Acids Res 2005;33(22):7011–18.

[177] Haque ME, et al. Contacts between mammalian mitochondrial translational initiation factor 3 and ribosomal proteins in the small subunit. Biochim Biophys Acta 2011;1814(12):1779–84.

[178] Koripella RK, et al. Structure of human mitochondrial translation initiation factor 3 bound to the small ribosomal subunit. iScience 2019;12:76–86.

[179] Cai YC, et al. Interaction of mammalian mitochondrial elongation factor EF-Tu with guanine nucleotides. Protein Sci 2000;9(9):1791–800.

[180] Bhargava K, Templeton P, Spremulli LL. Expression and characterization of isoform 1 of human mitochondrial elongation factor G. Protein Expr Purif 2004;37(2):368–76.

[181] Soleimanpour-Lichaei HR, et al. mtRF1a is a human mitochondrial translation release factor decoding the major termination codons UAA and UAG. Mol Cell 2007;27(5):745–57.

[182] Schmeing TM, et al. An induced-fit mechanism to promote peptide bond formation and exclude hydrolysis of peptidyl-tRNA. Nature 2005;438(7067):520–4.

[183] Temperley R, et al. Hungry codons promote frameshifting in human mitochondrial ribosomes. Science 2010;327(5963):301.

[184] Chrzanowska-Lightowlers ZM, Pajak A, Lightowlers RN. Termination of protein synthesis in mammalian mitochondria. J Biol Chem 2011;286(40):34479–85.

[185] Young DJ, et al. Bioinformatic, structural, and functional analyses support release factor-like MTRF1 as a protein able to decode nonstandard stop codons beginning with adenine in vertebrate mitochondria. RNA 2010;16(6):1146–55.

[186] Rorbach J, et al. The human mitochondrial ribosome recycling factor is essential for cell viability. Nucleic Acids Res 2008;36(18):5787–99.

[187] Tsuboi M, et al. EF-G2mt is an exclusive recycling factor in mammalian mitochondrial protein synthesis. Mol Cell 2009;35(4):502–10.

[188] Koripella RK, et al. Structural insights into unique features of the human mitochondrial ribosome recycling. Proc Natl Acad Sci USA 2019;116(17):8283–8.

[189] Englmeier R, Pfeffer S, Forster F. Structure of the human mitochondrial ribosome studied in situ by cryoelectron tomography. Structure 2017;25(10):1574–81.

[190] Schneider HC, et al. Mitochondrial Hsp70/MIM44 complex facilitates protein import. Nature 1994;371(6500):768–74.

[191] Kucej M, Butow RA. Evolutionary tinkering with mitochondrial nucleoids. Trends Cell Biol 2007;17(12):586–92.

[192] Bogenhagen DF, Rousseau D, Burke S. The layered structure of human mitochondrial DNA nucleoids. J Biol Chem 2008;283(6):3665–75. Available from: https://doi.org/10.1074/jbc.M708444200 Epub 2007 Dec 6.

[193] Cao L, et al. The mitochondrial bottleneck occurs without reduction of mtDNA content in female mouse germ cells. Nat Genet 2007;39(3):386–90.

[194] Iborra FJ, Kimura H, Cook PR. The functional organization of mitochondrial genomes in human cells. BMC Biol 2004;2:9.

[195] Kukat C, et al. Super-resolution microscopy reveals that mammalian mitochondrial nucleoids have a uniform size and frequently contain a single copy of mtDNA. Proc Natl Acad Sci USA 2011;108(33):13534–9.

[196] Shen EL, Bogenhagen DF. Developmentally-regulated packaging of mitochondrial DNA by the HMG-box protein mtTFA during Xenopus oogenesis. Nucleic Acids Res 2001;29(13):2822–8.
[197] Wang Y, Bogenhagen DF. Human mitochondrial DNA nucleoids are linked to protein folding machinery and metabolic enzymes at the mitochondrial inner membrane. J Biol Chem 2006;281(35):25791–802.
[198] He J, et al. Human C4orf14 interacts with the mitochondrial nucleoid and is involved in the biogenesis of the small mitochondrial ribosomal subunit. Nucleic Acids Res 2012;40:6097–108.
[199] Antonicka H, et al. The mitochondrial RNA-binding protein GRSF1 Localizes to RNA granules and Is required for posttranscriptional mitochondrial gene expression. Cell Metab 2013;17(3):386–98.
[200] Jourdain AA, et al. GRSF1 regulates RNA processing in mitochondrial RNA granules. Cell Metab 2013;17(3):399–410.
[201] Lee KW, et al. Mitochondrial rRNA methyltransferase family members are positioned to modify nascent rRNA in foci near the mtDNA nucleoid. J Biol Chem 2013;288(43):31386–99.
[202] Hess KC, et al. A mitochondrial CO_2-adenylyl cyclase-cAMP signalosome controls yeast normoxic cytochrome c oxidase activity. FASEB J 2014;28(10):4369–80.
[203] Tu YT, Barrientos A. The human mitochondrial DEAD-box protein DDX28 resides in RNA granules and functions in mitoribosome assembly. Cell Rep 2015;10(6):854–64.
[204] Bogenhagen DF, Martin DW, Koller A. Initial steps in RNA processing and ribosome assembly occur at mitochondrial DNA nucleoids. Cell Metab 2014;19(4):618–29. Available from: https://doi.org/10.1016/j.cmet.2014.03.013.
[205] Shajani Z, Sykes MT, Williamson JR. Assembly of bacterial ribosomes. Annu Rev Biochem 2011;80:501–26.
[206] Barrientos A. Mitochondriolus: assembling mitoribosomes. Oncotarget 2015;6(19):16800–1.
[207] Hensen F, et al. Mitochondrial RNA granules are critically dependent on mtDNA replication factors Twinkle and mtSSB. Nucleic Acids Res 2019;47(7):3680–98.
[208] Pietras Z, et al. Dedicated surveillance mechanism controls G-quadruplex forming non-coding RNAs in human mitochondria. Nat Commun 2018;9(1):2558.

CHAPTER 3

Epigenetic features of mitochondrial DNA

Takehiro Yasukawa, Shigeru Matsuda and Dongchon Kang

Department of Clinical Chemistry and Laboratory Medicine, Graduate School of Medical Sciences, Kyushu University, Fukuoka, Japan

3.1 A brief overview of mitochondrial DNA

The maintenance of mitochondrial DNA (mtDNA) is fundamental to the oxidative phosphorylation (OXPHOS) system that produces ATP and therefore to life. Mammalian mtDNA is a circular molecule of approximately 16 kb [1,2], comprising two strands designated as heavy (H) and light (L) strands according to the biased nucleotide composition, with the H and L strands being abundant in G and C, respectively. Despite the small genome size, mRNAs transcribed from mtDNA constitute a high proportion of the total cellular mRNA content in humans—nearly 30% in the heart and 5%–25% in other tissues [3]. Unlike nuclear DNA, mtDNA is maternally inherited [4] and replicates not only in proliferating cells but also in differentiated cells, such as neuronal and cardiac cells [5,6]. mtDNA mutations and deletions and copy number reduction are associated with various human diseases, including neurological and muscular diseases [e.g., mitochondrial encephalopathy, lactic acidosis, stroke-like episodes (MELAS), mtDNA depletion syndrome, and Leigh syndrome] [7–9]. Further, the proper control of mtDNA copy number was proposed as being important for early development and for iPS cell reprogramming process [10].

All of these facts suggest the importance of the proper maintenance and expression of mtDNA, which depends on key processes such as replication (Chapter 1: MtDNA replication, maintenance and nucleoid organization) [11], transcription (Chapter 2: Human mitochondrial transcription and translation), repair (Chapter 8: Human mitochondrial DNA repair), and segregation (Chapter 4: Heredity and Segregation of mtDNA). If mitochondria possess DNA methylation system, then it can be added to the process list. Considering that mtDNA is maternally inherited, there should be no need for genome imprinting. Moreover, since mtDNA encodes essential OXPHOS subunits and functional RNAs, which are transcribed polycistronically (Chapter 2: Human mitochondrial transcription and translation),

The Human Mitochondrial Genome.
DOI: https://doi.org/10.1016/B978-0-12-819656-4.00003-6

it is unlikely that the expression of one or two specific genes is suppressed in a tissue-specific manner. Therefore, should mtDNA methylation really occur, it must play a distinct role from the roles in the nucleus, regulating mtDNA in a unique way.

3.2 Does cytosine methylation occur in mtDNA?

In mammalian *nuclear* DNA, methylation is predominantly found in cytosines of the dinucleotide sequence CG in a palindromic manner (Fig. 3.1). It occurs in cytosine ring carbon C5 and is a crucial epigenetic process contributing to the regulation of gene expression, genomic imprinting, X chromosome inactivation, retroviral silencing, and cellular reprogramming [12,13]. In human somatic cells, 5-methylcytosine (5mC), designated as the "fifth base" of the human genome, accounts for about 4%–5% of cytosine and is present in about 80% of the CG sequence [14,15]. Aberrant nuclear DNA methylation has been implicated in human diseases, such as cancer [16], imprinting disorders [17], and neurodegeneration [18,19]. Mammalian 5mC modification is catalyzed by three active DNA (cytosine-5)-methyltransferases (DNMTs): DNMT1, DNMT3A, and DNMT3B. Classically, it was considered that while DNMT3A and DNMT3B perform the *de novo* methylation, DNMT1 maintains the methylation patterns by copying the existing patterns upon DNA replication. However, as evidence accumulated that indicates that DNMT1 cooperates with DNMT3 enzymes in the *de novo* methylation and that DNMT3A and DNMT3B play the role in the maintenance of methylation, more sophisticated action of DNMTs is now recognized [20].

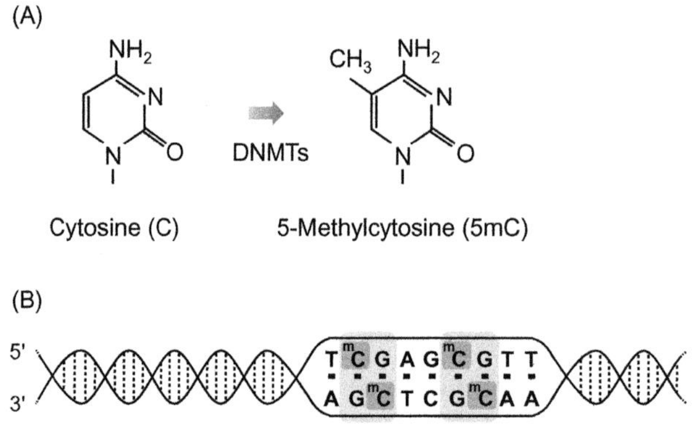

Figure 3.1: Schematic diagram of DNA methylation.
(A) Cytosine (C) is modified to 5-methylcytosine (5mC) by DNA (cytosine-5)-methyltransferases (DNMTs). DNMTs transfer the methyl group at the fifth position of the carbon of cytosine residues in DNA using *S*-adenosyl methionine (not shown in the diagram) as a donor. (B) Schematic representation of nuclear DNA methylation. Most 5mCs are found in cytosine positions in the CG-dinucleotide sequence in a palindromic manner.

Because of the significant roles and clinical relevance of nuclear DNA methylation, whether mtDNA is methylated or not attracts much attention in recent years following the publication of three papers postulating the occurrence of methylation in mtDNA in 2011. This chapter does not intend to review publications investigating mtDNA methylation one by one. Rather, we introduce several papers that we consider triggered a recent wave of mtDNA methylation research. For an overview of the comprehensive list of publications including pioneering works on this subject [21–26], please see other articles, such as Refs. [27–29]. In 2011, Infantino et al. [30] performed nucleoside mass spectrometric analysis of mtDNA from Epstein–Barr virus-immortalized lymphoblastoid cells of Down's syndrome and control subjects and reported that 25% and 13% of cytosines in the control- and patient-derived cells, respectively, were modified to 5mC. The proportion of 5mC in mtDNA proposed in the report is substantially higher than that in nuclear DNA as ~5% of cytosines are methylated in nuclear DNA. Shock et al. [31] found that the DNMT1 gene contains a putative mitochondrial targeting signal (MTS) sequence at the 5′-end in humans and other mammals. In cultured cells, green fluorescent proteins (GFPs) fused with the predicted MTS peptides appear to be localized to mitochondria, supporting the possibility that native DNMT1 proteins are imported into the mitochondrial matrix. Further, the presence of 5mC and 5-hydroxymethylcytosine (5hmC) [15] in mtDNA from HCT116 cells was proposed by analyzing five target regions [in 12S rRNA, 16S rRNA, COXII, and ATP6 genes and in the major noncoding region (NCR)] using methylated DNA immunoprecipitation (meDIP), followed by quantitative-PCR amplification of the immunoprecipitates [31]. Chestnut et al. [32] showed a strong western-blotting band of DNMT3A in mitochondrial fractions of human motor cortex homogenates, with a comment that mitochondrial localization of this protein is tissue-specific. In the same study, the presence of 5mC in mitochondria of mouse motor neuron was suggested by immunostaining with an anti-5mC antibody.

Following the publication of those papers, a number of studies investigated mtDNA methylation, often through the use of bisulfite sequencing [33,34], a frequently used gold standard method to examine the DNA methylation status. Bisulfite sequencing can be applied either to a limited region by conventional cycle sequencing or to a whole genome by next-generation sequencing at base resolution. It comprises two processes, namely, bisulfite conversion and DNA sequencing (Fig. 3.2), of which the former is the crucial step to detect 5mC. The reaction converts unmodified cytosine to uracil, and 5mC remains unchanged. Following the conversion step, the primer extension and/or DNA amplification steps introduce thymine to uracil positions and cytosine to 5mC positions. Sequencing the resulting DNA provides information on the extent of methylation at each cytosine position. In the case of targeted analysis with cycle sequencing, after bisulfite treatment, the region of interest is amplified by PCR and cloned into a vector, after which the resulting plasmids are transformed into *Escherichia coli* (*E. coli*) competent cells and numerous clones are sequenced. In the case of deep sequencing, millions of reads are obtained from DNA

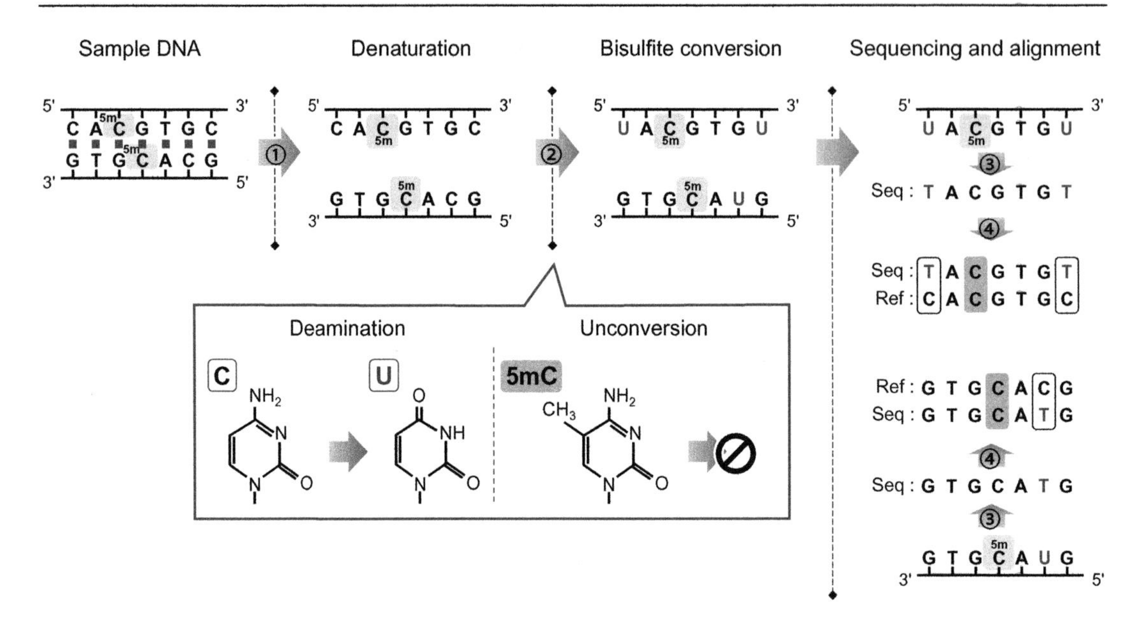

Figure 3.2: Principle of bisulfite sequencing.

Prior to bisulfite conversion, sample DNA must be denatured (①). Bisulfite conversion is then performed, leading to conversion of unmodified cytosine (C) into uracil (U) (deamination), and 5-methylcytosine (5mC), which is resistant to the treatment, remains unchanged (②). Treated DNA sample is then processed for sequencing, during which 5mC and U are changed to C and T, respectively, and the resulting DNA strands are sequenced by cyclesequencing or deepsequencing (③). Obtained sequence (Seq) is aligned to the reference sequence (Ref). A site of sequenced DNA containing C at a cytosine site in the reference sequence is judged to contain 5mC in the original sample DNA. On the other hand, if the sequenced DNA contains T at a cytosine site in the reference sequence, it is judged to be unmodified C in sample DNA. In sample DNA, the extent of methylation at a given cytosine site is estimated by assembling the cytosine status information of multiple reads (methylated or unmethylated in each read).

libraries constructed with bisulfite-treated samples and reads are mapped to the reference sequence. Both methods provide the frequency of cytosine and thymine occurrence at each cytosine position, from which the extent of methylation can then be calculated. Bellizzi et al. [35] used bisulfite sequencing to investigate the NCR of mtDNA in the blood and cultured cells of humans and mice. They proposed that numerous cytosines are completely methylated and that methylation was detected exclusively in the L strand both at the CG and non-CG sites. Their results were intriguing, especially because the NCR contains transcription and replication initiation sites [11]. However, it may be pointed out that both the CG and non-CG sites had comparable methylation frequencies and that the patterns of 5mC-accumulated sites in the NCR of human mtDNA appear to be different from those of mouse mtDNA (see Figures 1 and 2 in Ref. [35]). On the other hand, Bianchessi et al. [36] used bisulfite sequencing but reported a distinct methylation feature in the NCR in human endothelial cells. They also proposed a substantial occurrence of methylation both at CG and non-CG sites, and both in H and L strands. Whether methylation occurs in both strands or only in L stand is a fundamental difference, unlikely to be reconciled by the divergence in cell types used in different studies. In the same year as Bellizzi et al., Hong et al. [37] made an even more controversial proposal: the absence of mtDNA methylation. Using bisulfite sequencing, they examined the four HCT116 mtDNA regions that were proposed by Shock et al. [31] to contain 5mC and 5hmC and detected effectively no methylated cytosine from the regions. Although bisulfite sequencing cannot discriminate between 5mC and 5hmC, it can detect both bases [38]. To resolve the discrepancy of the two reports, Hong et al. carefully examined the work of Shock et al. and proposed that the meDIP used by them is not reliable when the number of 5mC in DNA fragments is low. Further, Hong et al. subjected HCT 116 cells to next-generation bisulfite sequencing and analyzed several published data of genome-wide bisulfite sequencing, from which they concluded that mtDNA effectively lacks cytosine methylation [37].

Along with the above publications which we consider led many researchers to wonder about mtDNA methylation, a number of papers were published since 2011 ([39–48] and references in the reviews [27–29]), indicating a high attention to this subject. However, it appears that no study has unequivocally demonstrated the occurrence of methylation specifically at CG sequence nor at any other sequence with a rational rule through investigating the whole mtDNA. Moreover, if mtDNA methylation occurs, then there must be an active DNMT(s) in the mitochondrial matrix. Several publications have addressed this critical issue. As described above, Shock et al. [31] showed the localization of DNMT1 MTS-GFP fusion proteins to mitochondria by fluorescent microscopy and observed native DNMT1 in mitochondrial fractions prepared from HCT116 cells and mouse embryonic fibroblasts by western blotting, and DNMT3A and DNMT3B were absent in the same fractions. Bellizzi et al. [35] also performed cell fractionation with HeLa and 3T3-L1 cells and suggested the presence of DNMT1 and DNMT3B and the absence of DNMT3A in the mitochondrial preparations. On the other hand, Chestnut et al. [32] reported clear DNMT3A detection in mitochondrial fractions of human motor cortex homogenates. The same group expanded their study using nycodenz step-gradient centrifugation for mitochondria preparation and concluded that DNMT3A is present in the mitochondria of mouse skeletal muscle, brain, spinal cord, heart, and testes, while its levels are very low or undetectable in the liver, spleen, kidney, and lung mitochondria [40]. However, in the same study, analysis of CG sequence cytosines in mtDNA from the brain, testes, and liver suggested the presence of 5mC at comparable levels, which does not appear to match the tissue specificity of the mitochondrial localization of DNMT3A. Additionally, they reported the absence of DNMT1 in the mitochondria of skeletal muscle, brain, and spinal cord and the absence of DNMT3B in the skeletal muscle mitochondria [40].

The above-described inconsistencies regarding the mitochondrial localization of DNMT enzymes, mtDNA methylation patterns, and even the occurrence of methylation do not allow us to draw a clear picture of epigenetic features of mtDNA. Therefore, we performed a thorough investigation of mtDNA methylation using three different methods to detect 5mC in DNA: bisulfite sequencing, McrBC methylcytosine-sensitive endonuclease cleavage assay, and nucleoside liquid chromatography/mass spectrometry. After cautious data interpretation, we concluded that if ever mtDNA undergoes 5mC modification, the frequency is very low and it has no positional specificity, and therefore 5mC is unlikely to have a universal role in mtDNA gene expression and metabolism [49]. In the following sections, we will discuss how we reached this conclusion with consideration of why variable results were reported from different groups. We hope that this chapter will contribute for a more accurate understanding of the mitochondrial genome on cytosine methylation.

3.3 Bisulfite sequencing analysis of mtDNA

As described above, this method enables researchers to reveal the methylation status of each cytosine in the target DNA. Unmodified cytosine and 5mC can be discriminated through their different reactivity to bisulfite. Although unmodified cytosine is deaminated by bisulfite and converted to uridine, under the experimental conditions, 5mC is resistant to the chemical and remains unaffected. This is a powerful method to determine the extent of methylation at base resolution (Fig. 3.2). However, if unmodified cytosine is not converted for some reason, it will be false-positively counted as 5mC, which is the critical point of this method. Bisulfite conversion is highly single-strand specific [50,51]. Therefore, DNA denaturation prior to bisulfite conversion is crucial, and thus linearization should be important for efficient denaturation when analyzing circular molecules, such as mammalian mtDNA. In fact, Liu et al. [45] and Mechta et al. [47] compared the bisulfite conversion efficiency between untreated human mtDNA and that treated with restriction enzyme digestion and reported a higher conversion efficiency in linearized mtDNA. In addition to linearization, it is important to design control experiments carefully. A portion of mtDNA amplified by PCR can be used as a 5mC-free control. It looks appropriate especially if the amplified portion is the target region of native mtDNA, since the absence of unconverted cytosines in the PCR products could be taken as a ground of efficient bisulfite conversion of native mtDNA. However, it is possible that although cytosines in PCR-generated short fragments are completely converted, the 16.3-kb-long native mtDNA is not converted to the same extent as the control fragments because of the substantial length difference between control fragments and native mtDNA.

Considering the above-described points, we digested mouse mtDNA (16,299 bp) with BglII restriction enzyme which cuts it once. Then, we generated "synthetic mtDNA" sized 16,291 bp by PCR using a pair of primers which anneal both ends of BglII-digested native mtDNA. Thus, synthetic mtDNA presents an identical sequence and virtually the same structure and length as BglII-digested native mtDNA. Using the synthetic mtDNA as the 5mC-free control, we performed bisulfite-sequencing analysis on mtDNA from mouse tissues and cultured embryonic stem cells (ESCs) using a next-generation sequencer. Surprisingly, in all analyzed samples, including synthetic DNA, roughly 30% of L strand-mapped reads were not converted, viz., none of the cytosines in such reads were converted. Mapping the reads with $\geq$90% unconversion levels to the reference mtDNA sequence provided similar patterns in native and synthetic mtDNA (see Supplementary Figure S4 in Matsuda et al. [49]), indicating that such reads do not stem from the presence of 5mC but are attributable to certain characteristics of the L strand sequence, in conjunction with our experimental conditions. This interpretation was only possible by analyzing the synthetic mtDNA in parallel with the native mtDNA. This highlights that including control DNA identical to native mtDNA in sequence, length, and linearization is crucial. To our knowledge, no other studies have performed robust control experiments as were done in our study. Additionally, we would like to point out that the experimental conditions and mtDNA status (e.g., bisulfite conversion conditions, mtDNA preparation methods, and mtDNA digestion prior to denaturation) may influence mtDNA resistance to bisulfite conversion. We therefore considered that such unconverted cytosines were false-positive and discarded reads with $\geq$90% unconversion level. Mapping of the remaining reads and comparison of the mapping results between mtDNA samples suggested that 5mC is not present at any specific position(s) in mtDNA with reliable frequencies, which is in accordance with Hong et al. [37]. It should be noted that after excluding reads with $\geq$90% unconversion level, a small fraction of unconverted cytosines was still observed in native and synthetic mtDNA and in 5mC-free λDNA that was spiked in mtDNA samples as an internal control. Comparison between the native and synthetic mtDNA suggested that the unconverted cytosines in native mtDNA likely escaped from bisulfite conversion, as were the cases with synthetic mtDNA and 5mC-free λDNA. This also indicates the importance of parallel examination of synthetic mtDNA to interpret the data of native mtDNA. Under these considerations, we proposed two possibilities: mitochondria are devoid of a cytosine methylation mechanism, or mtDNA contains 5mC with no positional specificity and its frequency at each cytosine site is below the reliable detection sensitivity of our examination [49].

When analyzing mtDNA using bisulfite sequencing, an accurate interpretation of the results relies on the following points:

1. Prior to bisulfite conversion, the circular structure of mtDNA should be resolved.
2. Control DNA identical to native mtDNA in sequence, length, and end positions (restriction enzyme-digested positions) should be analyzed in parallel with sample mtDNA.
3. Upon analysis of next-generation bisulfite sequencing data, setting "exclusion conditions" will influence the output of mtDNA methylation/unconversion landscape. In our

own bisulfite sequencing experiments, reads with ≥90% levels of unconverted cytosines were excluded.

4. If an mtDNA methylation signature has no sequence rule (i.e., CG dinucleotide sequence) or functionally relevant positional specificity (i.e., the NCR), then the nature of that signature should be cautiously interpreted as it may have been generated by imperfect conversion.

3.4 Estimation of mtDNA methylation with McrBC endonuclease

As described above, bisulfite sequencing did not suggest the occurrence of mtDNA methylation, which is inconsistent with many other studies published during and after the year 2011. It is therefore important to examine mtDNA using methods that are based on different 5mC detection principles. Since bisulfite sequencing relies on chemical conversion of bases, we selected an enzymatic detection method. McrBC endonuclease is a heterodimeric enzyme expressed in *E. coli* K-12, which plays a role in the self-defense system of the organism. It recognizes 5mC in DNA and digests the DNA in the presence of GTP. Importantly, McrBC has a rather weak sequence specificity; the required recognition elements are two $R^{m}C$ sequences (R = G or A; ^{m}C = methylated cytosine), which can be optimally separated at 40–80 bp, but as far as 3 kb [52–54]. This property is suitable in investigating the presence of 5mC at a relatively low frequency. We performed quantitative estimation by establishing an McrBC cleavage assay protocol for mtDNA using nucleic acids from purified mitochondria [49]. Since ribonucleotides are embedded in mtDNA strands [55–57], RNase treatment was excluded. Mitochondrial nucleic acids were first incubated with a restriction enzyme that cuts mtDNA at a single site. The resulting sample was then treated with McrBC in the presence or absence of GTP, and the linearized full-length mtDNA band was quantified using Southern hybridization after agarose gel electrophoresis. Incubation without GTP serves as the negative control. If mtDNA molecules are cleaved by McrBC at least once, then the band intensity of the full-length mtDNA will be reduced. In theory, if two 5mCs preceded by R are present in mtDNA molecules within 3 kb distance, the molecules would be digested (Fig. 3.3). Under these conditions, mtDNAs from mouse liver, brain, and cultured ESCs were subjected to McrBC cleavage assays, and no significant reduction in the intensity of mtDNA bands was observed, suggesting that mtDNA does not contain appreciable amounts of 5mC [49]. This is consistent with the results of our bisulfite sequencing.

Our McrBC cleavage assay examines the presence of 5mC in mtDNA by detecting whether or not McrBC reduces the intensity of mtDNA band. This assay can be the first choice in investigating mtDNA methylation, since it is quicker and simpler than analyses involving bisulfite sequencing and mass spectrometry (see below). If positive results

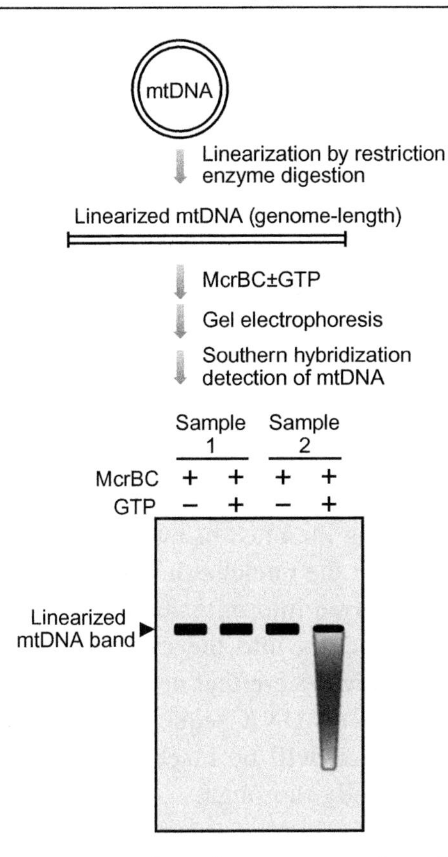

Figure 3.3: Schematic presentation of an McrBC endonuclease cleavage assay.
A scheme of mtDNA methylation analysis with McrBC is described. Suppose two samples are analyzed. While mtDNA in sample 1 is not digested by McrBC, mtDNA in sample 2 is digested. The result suggests that mtDNA in sample 1 contains no or very few 5mCs and mtDNA in sample 2 contains substantial amount of 5mC.

are obtained, it implies that mtDNA may contain substantial amounts of 5mC with reasonable sequence rule and positional specificity, which should then be detected by other methods.

3.5 Investigation of 5mC in mtDNA by nucleoside liquid chromatography/mass spectrometry

In addition to bisulfite sequencing and McrBC cleavage assays, we examined the presence of 5mC in mtDNA using liquid chromatography/mass spectrometry (LC/MS). Advantages of this method are high sensitivity and quantitative nature, whereas the challenging feature when investigating mtDNA is that unlike the other two methods, it cannot depend on mtDNA sequence information as it requires digestion of samples into mononucleosides. Although mtDNA is present in hundreds to thousands of copies per cell, the mass of mtDNA is much smaller than that of nuclear DNA. In humans, for example, mtDNA mass is about 1% of nuclear DNA mass (if there are 4000 copies of mtDNA per cell, then [mtDNA mass (16,569 bp $\times$ 4000 copies)/nuclear DNA mass (3.2×10^9 bp $\times$ 2 copies)] $\times$ 100 = 1%). About 5% of cytosines in nuclear DNA are methylated. Therefore, purity of mtDNA is crucial for LC/MS analysis, since contaminations of nuclear DNA in mtDNA preparations cannot be distinguished in this method, leading to inaccurate 5mC estimations in mtDNA. The same criterion is applied to an enzyme-linked immunosorbent assay (ELISA), which detects the 5mC present in any DNA in the sample solution through anti-5mC antibodies, providing no information on the origin of 5mC. Similarly, the purity of mitochondrial preparation critically influences the western-blotting examination of mitochondrial localization of proteins. For example, considering the mass of nuclear DNA and mtDNA, in proliferating cells, the levels of DNMT1 in the nucleus should be 100 times higher than those in mitochondria if it is indeed imported into mitochondria. Under this assumption, a trace contamination of the nuclear fraction into the mitochondrial preparation would spoil the analysis. It should be emphasized here that detection of 5mC by LC/MS, ELISA, or any other method unable to rely on DNA sequence information and western-blotting examination of DNMT localization will be inaccurate in case mtDNA and mitochondria preparations are not sufficiently thorough.

Considering the above-described requirement of LC/MS analysis seriously, we performed multiple-step purification to prepare mtDNA at very high purity (Fig. 3.4). After disruption of mouse liver tissue, differential centrifugation was performed to obtain mitochondrial fraction, which is sometimes called crude mitochondrial fraction. It was then subjected to ultracentrifugation with a sucrose density-gradient to increase mitochondrial purity, after which the mitochondrial fraction was treated with nuclease and proteinase to degrade residually contaminating nuclear DNA in the fraction, followed by removal of the nuclear DNA by extensive wash of mitochondria. The mitochondrial nucleic acids were then isolated from the resulting mitochondrial fraction. Furthermore, mitochondrial nucleic acids were treated with EcoRV restriction enzyme to generate two mouse mtDNA fragments and gel-electrophoresed to fractionate the mtDNA fragments. (In addition, mitochondrial RNA was degraded by RNase T1 treatment before electrophoresis.) mtDNA bands were then purified from gels, which provided an additional opportunity to remove nuclear DNA contaminations that might still have remained in mtDNA preparations. Finally, the purified mtDNA was enzymatically digested into mononucleosides for LC/MS analysis [49]. In addition to careful mtDNA purification, it is important to consider that ionization efficiency differs between nucleosides in LC/MS analysis. It is therefore necessary to generate standard curves with authentic chemicals to compare the

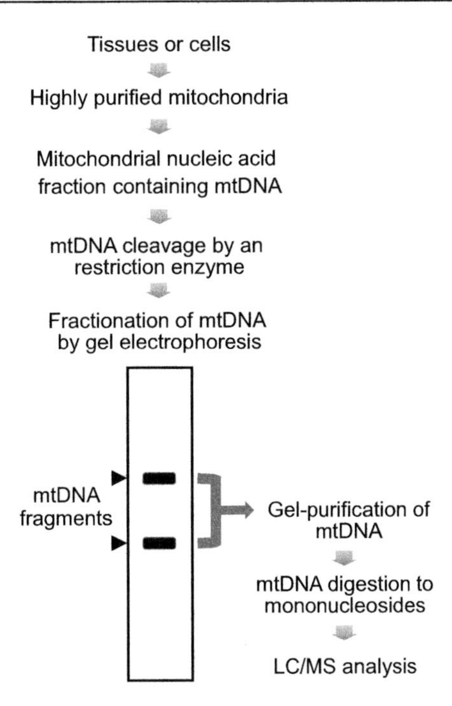

Figure 3.4: Experimental flow of mtDNA preparation for LC/MS analysis.
Multiple-step mtDNA purification procedure is briefly summarized. The schematic image represents that two mtDNA fragments generated by restriction enzyme digestion are visualized after gel electrophoresis. mtDNA bands are excised from the gel and purified mtDNA was degraded into mononucleosides for LC/MS analysis. Details can be found in Ref. [49].

amount of different nucleosides in a given DNA sample. With the standard curves of 5-methyldeoxycytidine (m^5dC) and deoxycytidine (dC), we estimated that the relative amount of m^5dC against dC is about 0.4%, which is equivalent to 24 5mC residues per molecule of mtDNA [49]. This ratio is significantly lower than that seen in nuclear DNA.

3.6 Epigenetic features of mammalian mtDNA

Although LC/MS detected 5mC in mtDNA preparations, the estimated levels were very low, and neither bisulfite sequencing nor McrBC cleavage assays suggested specific rules or patterns for 5mC modification in mtDNA. Therefore, it may be concluded that mammalian mitochondria lack a *mechanism* of cytosine methylation with substantial biological function. Thus we proposed that 5mC is not present at any specific position(s) in mtDNA and levels of the methylated cytosine are extremely low, provided the modification occurs [49]. If mtDNA underwent functional methylation, then 5mC should be present at a specific region of mtDNA with a certain sequence specificity, likely to CG dinucleotide sequence as a DNMT must be involved in the methylation process.

We infer the following possible sources for the 5mC detected by our mass spectrometric analysis: it could result from trace contamination of nuclear DNA in mtDNA preparations or from the genuine presence of 5mC in mtDNA. These hypotheses are not mutually exclusive. Since the mass of mtDNA is less than 1% of that of nuclear DNA in the mouse liver [58], even after extensive mtDNA purification, it might have been difficult to absolutely eliminate nuclear DNA from mtDNA preparations. We have to consider that even if 99.9% of nuclear DNA had been removed in the course of mtDNA preparation, a relative 5mC amount of 0.5% against unmodified cytosine would be given to mtDNA (in this calculation, the mass of mtDNA is supposed to be 1% of that of nuclear DNA whose relative content of 5mC against unmodified cytosine is ~5%). Three hypotheses can be speculated for the latter alternative. A fraction of DNMT1, DNMT3A, or DNMT3B could translocate to the mitochondria and modify cytosines at a frequency too low to be identified by bisulfite sequencing and McrBC cleavage assays. It should be pointed out, however, that the mitochondrial localization of DNMTs is not consensual, as discussed in this chapter and in Ref. [28]. 5-Methyl-dCTP could be incorporated into mtDNA during normal mtDNA replication. 5mC is one of the modified bases in mitochondrial RNAs [59], and mtDNA and mitochondrial RNAs are present in the same compartment, the mitochondrial matrix. Therefore, it is possible that 5mC is somehow converted into 5-methyl-dCTP in mitochondrial nucleotide metabolism pathways and taken up to mtDNA strands. An RNA methyltransferase residing in the matrix might modify mtDNA nonspecifically.

Additionally, we would like to comment on the use of total cellular nucleic acid preparation for mtDNA methylation analysis. Although it depends on the sample preparation procedure, it is possible that DNMTs which are localized in the nucleus *in vivo* encounter some mtDNA molecules upon cell disruption, adding methyl groups to mtDNA.

Our investigation suggests that 5mC is unlikely to play a universal role in the regulation of mtDNA. However, establishing the absolute absence of mtDNA cytosine methylation with biological function is *probatio diabolica*. Our work is limited to mouse liver, brain, and cultured ESCs [49] and cannot refute the possibility that functional mtDNA methylation occurs in certain specific tissues or life stages. Under this supposition, it can be speculated that the 5mC detected in mouse liver mtDNA by LC/MS may be partly, if not entirely, due to trace residual activity of methylation, which is programmed to cease in somatic tissues. The methylation activity might be generated by DNMT1 as this gene is predicted to contain an MTS sequence at the N terminus [31].

Research perspectives

Since mtDNA methylation is likely to be generally absent, upon detection of a positive 5mC signature, a thorough investigation of the entire mtDNA region is crucial. Moreover, it is useful to investigate mtDNA methylation in various materials, especially human materials, using robust controls and strict sample preparations as discussed in this article. Such efforts will help in establishing the landscape of epigenetic features of mtDNA. Although the absence of (functional) 5mC modification in human mitochondria may be the final answer, this will be a very important conclusion for biology and for mitochondrial medicine.

Acknowledgments

We thank our collaborators on the investigation of mtDNA methylation; Prof. Tsutomu Suzuki (The University of Tokyo), Prof. Hiroyuki Sasaki (Kyushu University), Prof. Kenji Ichiyanagi (Nagoya University), Dr. Yuriko Sakaguchi (The University of Tokyo), Dr. Motoko Unoki (Kyushu University), Dr. Kei Fukuda (RIKEN), and Dr. Kazuhito Gotoh (Kyushu University). This work was supported by Grants-in-Aid for Scientific Research from the Japan Society for the Promotion of Science (17K07504).

References

[1] Anderson S, Bankier AT, Barrell BG, de Bruijn MH, Coulson AR, Drouin J, et al. Sequence and organization of the human mitochondrial genome. Nature 1981;290:457–65.

[2] Andrews RM, Kubacka I, Chinnery PF, Lightowlers RN, Turnbull DM, Howell N. Reanalysis and revision of the Cambridge reference sequence for human mitochondrial DNA. Nat Genet 1999;23:147.

[3] Mercer TR, Neph S, Dinger ME, Crawford J, Smith MA, Shearwood AM, et al. The human mitochondrial transcriptome. Cell 2011;146:645–58.

[4] Giles RE, Blanc H, Cann HM, Wallace DC. Maternal inheritance of human mitochondrial DNA. Proc Natl Acad Sci U S A 1980;77:6715–19.

[5] Magnusson J, Orth M, Lestienne P, Taanman JW. Replication of mitochondrial DNA occurs throughout the mitochondria of cultured human cells. Exp Cell Res 2003;289:133–42.

[6] Ylikallio E, Tyynismaa H, Tsutsui H, Ide T, Suomalainen A. High mitochondrial DNA copy number has detrimental effects in mice. Hum Mol Genet 2010;19:2695–705.

[7] Greaves LC, Reeve AK, Taylor RW, Turnbull DM. Mitochondrial DNA and disease. J Pathol 2012;226:274–86.

[8] Schon EA, DiMauro S, Hirano M. Human mitochondrial DNA: roles of inherited and somatic mutations. Nat Rev Genet 2012;13:878–90.

[9] Ylikallio E, Suomalainen A. Mechanisms of mitochondrial diseases. Ann Med 2012;44:41–59.

[10] Sun X, St John JC. The role of the mtDNA set point in differentiation, development and tumorigenesis. Biochem J 2016;473:2955–71.

[11] Yasukawa T, Kang D. An overview of mammalian mitochondrial DNA replication mechanisms. J Biochem 2018;164:183–93.

[12] Bird A. DNA methylation patterns and epigenetic memory. Genes Dev 2002;16:6–21.

[13] Jaenisch R, Bird A. Epigenetic regulation of gene expression: how the genome integrates intrinsic and environmental signals. Nat Genet 2003;33 Suppl:245–54.

[14] Laurent L, Wong E, Li G, Huynh T, Tsirigos A, Ong CT, et al. Dynamic changes in the human methylome during differentiation. Genome Res 2010;20:320–31.

[15] Breiling A, Lyko F. Epigenetic regulatory functions of DNA modifications: 5-methylcytosine and beyond. Epigenetics Chromatin 2015;8:24.
[16] Klutstein M, Nejman D, Greenfield R, Cedar H. DNA methylation in cancer and aging. Cancer Res 2016;76:3446–50.
[17] Elhamamsy AR. Role of DNA methylation in imprinting disorders: an updated review. J Assist Reprod Genet 2017;34:549–62.
[18] Qureshi IA, Mehler MF. Epigenetic mechanisms governing the process of neurodegeneration. Mol Asp Med 2013;34:875–82.
[19] Hwang JY, Aromolaran KA, Zukin RS. The emerging field of epigenetics in neurodegeneration and neuroprotection. Nat Rev Neurosci 2017;18:347–61.
[20] Jeltsch A, Jurkowska RZ. New concepts in DNA methylation. Trends Biochem Sci 2014;39:310–18.
[21] Nass MM. Differential methylation of mitochondrial and nuclear DNA in cultured mouse, hamster and virus-transformed hamster cells. In vivo and in vitro methylation. J Mol Biol 1973;80:155–75.
[22] Dawid IB. 5-methylcytidylic acid: absence from mitochondrial DNA of frogs and HeLa cells. Science 1974;184:80–1.
[23] Vanyushin BF, Kirnos MD. The nucleotide composition and pyrimidine clusters in DNA from beef heart mitochondria. FEBS Lett 1974;39:195–9.
[24] Groot GS, Kroon AM. Mitochondrial DNA from various organisms does not contain internally methylated cytosine in -CCGG- sequences. Biochim Biophys Acta 1979;564:355–7.
[25] Shmookler Reis RJ, Goldstein S. Mitochondrial DNA in mortal and immortal human cells. Genome number, integrity, and methylation. J Biol Chem 1983;258:9078–85.
[26] Pollack Y, Kasir J, Shemer R, Metzger S, Szyf M. Methylation pattern of mouse mitochondrial DNA. Nucleic Acids Res 1984;12:4811–24.
[27] Iacobazzi V, Castegna A, Infantino V, Andria G. Mitochondrial DNA methylation as a next-generation biomarker and diagnostic tool. Mol Genet Metab 2013;110:25–34.
[28] Maresca A, Zaffagnini M, Caporali L, Carelli V, Zanna C. DNA methyltransferase 1 mutations and mitochondrial pathology: is mtDNA methylated? Front Genet 2015;6:90.
[29] Mposhi A, Van der Wijst MG, Faber KN, Rots MG. Regulation of mitochondrial gene expression, the epigenetic enigma. Front Biosci (Landmark Ed) 2017;22:1099–113.
[30] Infantino V, Castegna A, Iacobazzi F, Spera I, Scala I, Andria G, et al. Impairment of methyl cycle affects mitochondrial methyl availability and glutathione level in Down's syndrome. Mol Genet Metab 2011;102:378–82.
[31] Shock LS, Thakkar PV, Peterson EJ, Moran RG, Taylor SM. DNA methyltransferase 1, cytosine methylation, and cytosine hydroxymethylation in mammalian mitochondria. Proc Natl Acad Sci U S A 2011;108:3630–5.
[32] Chestnut BA, Chang Q, Price A, Lesuisse C, Wong M, Martin LJ. Epigenetic regulation of motor neuron cell death through DNA methylation. J Neurosci 2011;31:16619–36.
[33] Hayatsu H, Wataya Y, Kai K, Iida S. Reaction of sodium bisulfite with uracil, cytosine, and their derivatives. Biochemistry 1970;9:2858–65.
[34] Frommer M, McDonald LE, Millar DS, Collis CM, Watt F, Grigg GW, et al. A genomic sequencing protocol that yields a positive display of 5-methylcytosine residues in individual DNA strands. Proc Natl Acad Sci U S A 1992;89:1827–31.
[35] Bellizzi D, D'Aquila P, Scafone T, Giordano M, Riso V, Riccio A, et al. The control region of mitochondrial DNA shows an unusual CpG and non-CpG methylation pattern. DNA Res 2013;20:537–47.
[36] Bianchessi V, Vinci MC, Nigro P, Rizzi V, Farina F, Capogrossi MC, et al. Methylation profiling by bisulfite sequencing analysis of the mtDNA non-coding region in replicative and senescent endothelial cells. Mitochondrion 2016;27:40–7.
[37] Hong EE, Okitsu CY, Smith AD, Hsieh CL. Regionally specific and genome-wide analyses conclusively demonstrate the absence of CpG methylation in human mitochondrial DNA. Mol Cell Biol 2013;33:2683–90.

[38] Huang Y, Pastor WA, Shen Y, Tahiliani M, Liu DR, Rao A. The behaviour of 5-hydroxymethylcytosine in bisulfite sequencing. PLoS One 2010;5:e8888.
[39] Chen H, Dzitoyeva S, Manev H. Effect of valproic acid on mitochondrial epigenetics. Eur J Pharmacol 2012;690:51–9.
[40] Wong M, Gertz B, Chestnut BA, Martin LJ. Mitochondrial DNMT3A and DNA methylation in skeletal muscle and CNS of transgenic mouse models of ALS. Front Cell Neurosci 2013;7:279.
[41] Byun HM, Panni T, Motta V, Hou L, Nordio F, Apostoli P, et al. Effects of airborne pollutants on mitochondrial DNA methylation. Part Fibre Toxicol 2013;10:18.
[42] Sun Z, Terragni J, Borgaro JG, Liu Y, Yu L, Guan S, et al. High-resolution enzymatic mapping of genomic 5-hydroxymethylcytosine in mouse embryonic stem cells. Cell Rep 2013;3:567–76.
[43] Mishra M, Kowluru RA. Epigenetic modification of mitochondrial DNA in the development of diabetic retinopathy. Invest Ophthalmol Vis Sci 2015;56:5133–42.
[44] Jia L, Li J, He B, Jia Y, Niu Y, Wang C, et al. Abnormally activated one-carbon metabolic pathway is associated with mtDNA hypermethylation and mitochondrial malfunction in the oocytes of polycystic gilt ovaries. Sci Rep 2016;6:19436.
[45] Liu B, Du Q, Chen L, Fu G, Li S, Fu L, et al. CpG methylation patterns of human mitochondrial DNA. Sci Rep 2016;6:23421.
[46] Saini SK, Mangalhara KC, Prakasam G, Bamezai RNK. DNA Methyltransferase1 (DNMT1) Isoform3 methylates mitochondrial genome and modulates its biology. Sci Rep 2017;7:1525.
[47] Mechta M, Ingerslev LR, Fabre O, Picard M, Barres R. Evidence suggesting absence of mitochondrial DNA methylation. Front Genet 2017;8:166.
[48] Owa C, Poulin M, Yan L, Shioda T. Technical adequacy of bisulfite sequencing and pyrosequencing for detection of mitochondrial DNA methylation: Sources and avoidance of false-positive detection. PLoS One 2018;13:e0192722.
[49] Matsuda S, Yasukawa T, Sakaguchi Y, Ichiyanagi K, Unoki M, Gotoh K, et al. Accurate estimation of 5-methylcytosine in mammalian mitochondrial DNA. Sci Rep 2018;8:5801.
[50] Clark SJ, Harrison J, Paul CL, Frommer M. High sensitivity mapping of methylated cytosines. Nucleic Acids Res 1994;22:2990–7.
[51] Warnecke PM, Stirzaker C, Song J, Grunau C, Melki JR, Clark SJ. Identification and resolution of artifacts in bisulfite sequencing. Methods 2002;27:101–7.
[52] Raleigh EA, Wilson G. *Escherichia coli* K-12 restricts DNA containing 5-methylcytosine. Proc Natl Acad Sci U S A 1986;83:9070–4.
[53] Sutherland E, Coe L, Raleigh EA. McrBC: a multisubunit GTP-dependent restriction endonuclease. J Mol Biol 1992;225:327–48.
[54] Panne D, Raleigh EA, Bickle TA. The McrBC endonuclease translocates DNA in a reaction dependent on GTP hydrolysis. J Mol Biol 1999;290:49–60.
[55] Yang MY, Bowmaker M, Reyes A, Vergani L, Angeli P, Gringeri E, et al. Biased incorporation of ribonucleotides on the mitochondrial L-strand accounts for apparent strand-asymmetric DNA replication. Cell 2002;111:495–505.
[56] Berglund AK, Navarrete C, Engqvist MK, Hoberg E, Szilagyi Z, Taylor RW, et al. Nucleotide pools dictate the identity and frequency of ribonucleotide incorporation in mitochondrial DNA. PLoS Genet 2017;13:e1006628.
[57] Moss CF, Dalla Rosa I, Hunt LE, Yasukawa T, Young R, Jones AWE, et al. Aberrant ribonucleotide incorporation and multiple deletions in mitochondrial DNA of the murine MPV17 disease model. Nucleic Acids Res 2017;45:12808–15.
[58] Malik AN, Czajka A, Cunningham P. Accurate quantification of mouse mitochondrial DNA without co-amplification of nuclear mitochondrial insertion sequences. Mitochondrion 2016;29:59–64.
[59] Bohnsack MT, Sloan KE. The mitochondrial epitranscriptome: the roles of RNA modifications in mitochondrial translation and human disease. Cell Mol Life Sci 2018;75:241–60.

CHAPTER 4

Heredity and segregation of mtDNA

Stephen P. Burr and Patrick F. Chinnery

Department of Clinical Neurosciences and MRC Mitochondrial Biology Unit, University of Cambridge, Cambridge, United Kingdom

4.1 Introduction

The nuclear (n)DNA, which accounts for more than 99.9% of the DNA in a cell, contains 20,000–25,000 genes that encode the genetic information required by the cell for its maintenance and function. The heredity of these genes, or "units of inheritance" was first studied scientifically by Gregor Mendel in the 19th century, and his Mendelian laws of inheritance still form the foundation for our understanding of nuclear gene transmission from one generation to the next.

One of the central tenets of modern human genetics is the biparental transmission of genes to offspring by sexual reproduction, with one allele of each gene inherited from the mother and the other from the father. Combined with genetic recombination in the germline, this promotes genetic diversity, hindering the accumulation of DNA mutations and allowing for natural selection and adaptation over many generations.

Transmission of the mitochondrial genome, however, is not governed by these same rules, meaning that the genetics of mitochondrial (mt)DNA inheritance is entirely different to that found in the nucleus (Table 4.1). In this chapter we will explore the unique aspects of mtDNA inheritance and how this impacts the transmission and segregation of mtDNA variants, both in the germline and in somatic tissues.

4.2 General principles of mtDNA segregation

Unlike the nuclear genome, which is diploid in somatic cells and haploid in the germline, each cell in the body (apart from mature erythrocytes) contains hundreds to thousands of copies of the mitochondrial genome (mtDNA). This means that copies of mtDNA containing sequence variants, such as SNPs or deletions, can coexist with wild-type genomes in the same cell—a phenomenon termed heteroplasmy (Fig. 4.1A). If the level of heteroplasmy in

The Human Mitochondrial Genome.
DOI: https://doi.org/10.1016/B978-0-12-819656-4.00004-8

Table 4.1: Characteristics of the human nuclear and mitochondrial genomes.

	Nuclear DNA	Mitochondrial DNA
Number of genes	20,000–25,000	37
Structure	Linear	Circular
Inheritance	Biparental	Uniparental (maternal)
Recombination	Yes	No
Ploidy	Diploid/Haploid	Polyploid
Mutation rate	Low	High
Segregation pattern	Mendelian	Homoplasmy/Heteroplasmy

a cell changes in favor of the wild-type sequence then the variant may disappear, but conversely, a change in favor of the variant sequence may allow it to overtake the wild-type genome and become fixed—a phenomenon termed homoplasmy (Fig. 4.1A). This differential segregation of coexisting mtDNA genomes within cells is a defining feature of many mitochondrial diseases [1].

Two main processes are thought to contribute to changing heteroplasmy levels in cells: vegetative segregation and relaxed replication (Fig. 4.1B). Vegetative selection occurs in dividing mitotic cells, where the mtDNA present in the parent cell is split between the two daughter cells upon each cell division. If a heteroplasmic variant is present, then one daughter cell, by chance, may receive a slightly higher proportion of mutant mtDNA genomes, resulting in a small difference in the heteroplasmy level between the two cells (Fig. 4.1B) [1], thus the level of heteroplasmy "drifts" to a new level. Over many cell divisions these repeated small drifts can lead to significant changes in heteroplasmy level and, if the variant is pathogenic, may result in some cells developing a biochemical defect affecting oxidative phosphorylation (OXPHOS), affecting their ability to replicate further and effectively selecting against the mutation [1]. This results in negative selection, acting at the cellular level, and is thought to be responsible for the gradual decrease of certain heteroplasmic mtDNA mutations in blood cells seen during human life [3].

While vegetative segregation is restricted to dividing cells, heteroplasmy levels can also change in nondividing cells, since mtDNA is constantly being replicated and destroyed independently of the cell-cycle. Unlike the nuclear DNA, mtDNA replication is not directly linked to the cell-cycle, a situation referred to as "relaxed replication" (Fig. 4.1B) [1]. Under relaxed replication, it is possible for a heteroplasmic variant to be replicated more frequently than the wild-type genome, leading to a shift in the mean heteroplasmy level in a cell. Again, these changes can result in "random drift" of heteroplasmy levels. In silico modeling suggests that, over time, these shifts can lead to a significant change in heteroplasmy level and may explain how low-level heteroplasmies can clonally expand during life and eventually cause mitochondrial disease, even in postmitotic (nondividing) tissues such as neurons and muscle [4].

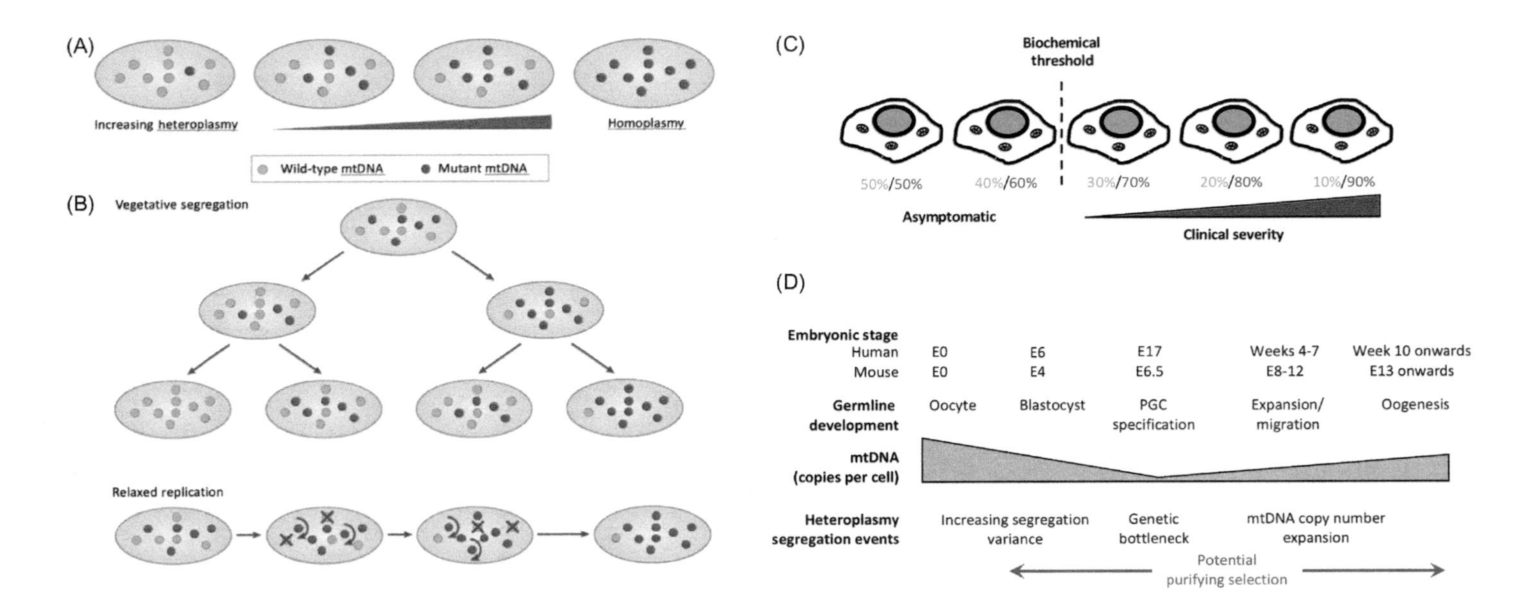

Figure 4.1: Mitochondrial DNA heteroplasmy and segregation.

(A) Due to the multicopy nature of mtDNA, a mixture of wild-type and mutant mitochondrial genomes can exist in cells; a situation termed heteroplasmy. If the proportion of mutant genomes increases to completely replace the wild-type mtDNA then the mutation becomes homoplasmic and is subsequently fixed in the cell. (B) When heteroplasmic cells divide, the proportion of wild-type and mutant genomes passed to each daughter cell can vary slightly, leading to changing heteroplasmy levels in successive generations through vegetative segregation. Heteroplasmy levels can also change in nondividing cells, as the mtDNA is constantly replicated and destroyed independent of the cell-cycle in a process termed relaxed replication. If one genome is replicated more frequently than the other then heteroplasmy levels will change over time. (*Arrows* represent mtDNA replication, *crosses* represent mtDNA destruction). (C) When the heteroplasmy level of a pathogenic mtDNA mutation (*red figures; black in print version*) is low compared to the wild-type (*green figures; gray in print version*), the cell is able to function normally. Beyond a certain mutant heteroplasmy threshold, the remaining wild-type mtDNA can no longer compensate for the presence of the mutation and the cell develops a respiratory chain deficit. As the mutation level rises further past this biochemical threshold, the frequency and severity of clinical mitochondrial disease tends to increase. (D) A schematic representation of the stages of female germline development in humans and mice, with associated events thought to contribute to mtDNA heteroplasmy segregation through the germline genetic bottleneck. Source: *Adapted from Stewart JB, Chinnery PF. The dynamics of mitochondrial DNA heteroplasmy: implications for human health and disease. Nat Rev Genet 2015;16(9):530–42 [1]; Burr SP, Pezet M, Chinnery PF. Mitochondrial DNA heteroplasmy and purifying selection in the mammalian female germ line. Dev Growth Differ 2018;60(1):21–32 [2].*

Vegetative segregation and relaxed replication are not necessarily mutually exclusive, and both likely combine to influence the dynamics of heteroplasmy segregation in vivo. For example, the compartmentalization of mtDNA into nucleoids, or within the mitochondrial network, is also likely to influence segregation of mtDNA heteroplasmy. In addition, selection may act at the molecular, organellar, and cellular level to influence segregation. These different processes could come into play at different times, in different tissues, and vary from mutation to mutation. In this chapter we consider two different situations where heteroplasmy segregation occurs: in the germline, and in somatic tissues [5].

4.3 Uniparental maternal inheritance of mitochondrial DNA

Unlike the nuclear genome, in animals the mitochondrial genome is predominantly transmitted uniparentally, and usually down the maternal line. Both male and female offspring receive mtDNA only from their mother, reflecting the asexual reproduction of the mitochondrion's ancient α-protobacterial ancestor. Evidence for maternal inheritance of mtDNA in eukaryotes was first observed in the fungus *Neurosporacrassa* [6], with subsequent work confirming a similar inheritance pattern in amphibians [7], mammals [8], and humans [9]. While there are some animal species where inheritance of paternal mtDNA is known to occur, such as the saltwater mussel genus Mytilus [10], multiple studies in humans and other mammalian species have shown that strict maternal inheritance of the mtDNA is the prevailing mode of transmission [11,12], and is therefore a defining characteristic of mtDNA genetics.

To maintain strict maternal inheritance of the mitochondrial genome, it is necessary to ensure that any paternal mtDNA present in the sperm cell is either excluded or removed from the zygote following fertilization. The mechanisms employed to achieve paternal mitochondrial elimination (PME) appear to be species-specific, and have been well defined in the invertebrate model organisms *Caenorhabditis elegans* (nematode worm) and *Drosophila melanogaster* (fruit fly).

In *C. elegans* around 60 paternal mitochondria are delivered into the oocyte at fertilization, but are quickly ubiquitinated and completely degraded via the autophagic pathway by the 16-cell stage [13,14]. Blocking autophagy in the early embryo results in persistence of paternal mitochondria through to the first larval stage [14], confirming that this pathway is crucial for maintaining uniparental inheritance in this species. Concurrently, the paternal mtDNA is specifically degraded by mitochondrial endonuclease G, accelerating PME [15].

In *D. melanogaster*, endonuclease G is also involved in PME, but, rather than being eliminated after fertilization, the paternal mtDNA is removed from the sperm mitochondria in the final stages of maturation via a mechanism involving both endonuclease G and the mitochondrial DNA polymerase gamma [16,17]. The resulting mature sperm still contains mitochondria, but these lack detectable paternal mtDNA and are degraded by autophagy soon after fertilization [18].

Despite these in-depth studies in invertebrate model organisms, it is not clear whether these mechanisms have been conserved through evolution, and the fate of the paternal mtDNA following fertilization in mammalian species is currently an area of active debate. In some species, such as the Chinese hamster (*Cricetulus griseus*), it appears that paternal mitochondria do not enter the ovum upon fertilization, with the sperm midpiece and tail being excluded from entry [19]. However, in the majority of mammals, including mice [20] and humans [21], the entire sperm enters the ovum at fertilization and therefore embryos need to eliminate any paternal mtDNA contained in the sperm to maintain uniparental inheritance. The processes that orchestrate PME in early mammalian embryos are currently not well understood and there is some uncertainty in the literature regarding the exact mechanisms involved.

The two main models currently proposed to account for the fate of paternal mtDNA in mammalian embryos are the "Passive Dilution" model and the "Active Elimination" model [22] (Fig. 4.2). The "Passive Dilution" model has arisen following the observation that in mouse embryos paternal mitochondria appear to persist until at least the morula stage, 3–4 days after fertilization, leading to the suggestion that PME in mice may be a passive process, by which paternal mitochondria are gradually diluted out over subsequent cell divisions [23]. According to this model, uniparental maternal inheritance is maintained in the majority of cases due to very low mtDNA copy number in mature sperm, estimated by some to be as low as 1–2 copies per cell in humans [24] and mice [23], due to mtDNA depletion during the maturation process [25]. Upon fertilization, the few surviving copies of paternal mtDNA would be negligible when compared to the >100,000 copies present in the oocyte [26] and would be effectively eliminated from the majority of embryonic cells as the mtDNA segregates through the initial cell divisions [23]. However, these very low mtDNA copy number estimates are not universally supported, with other studies suggesting that there are anywhere from 50 to several hundred copies present in mature sperm cells [27,28].

The "Active Elimination" model proposes that any paternal mitochondria and mtDNA present in the mature sperm are actively removed from the embryo following fertilization, thus maintaining uniparental maternal inheritance. It has long been established that mammalian sperm mitochondria are ubiquitinated during maturation in the male reproductive tract [29], with prohibitin being the likely target protein [30], suggesting that the paternal mitochondria arrive in the oocyte "pretagged" for rapid degradation. Subsequent studies in a number of mammalian species have strengthened the case for active elimination of paternal mitochondria by the autophagic pathway [13,31,32] and in mice this appears to be via a PARKIN/MUL1-dependent pathway [31]. Despite growing evidence of active PME in mammalian

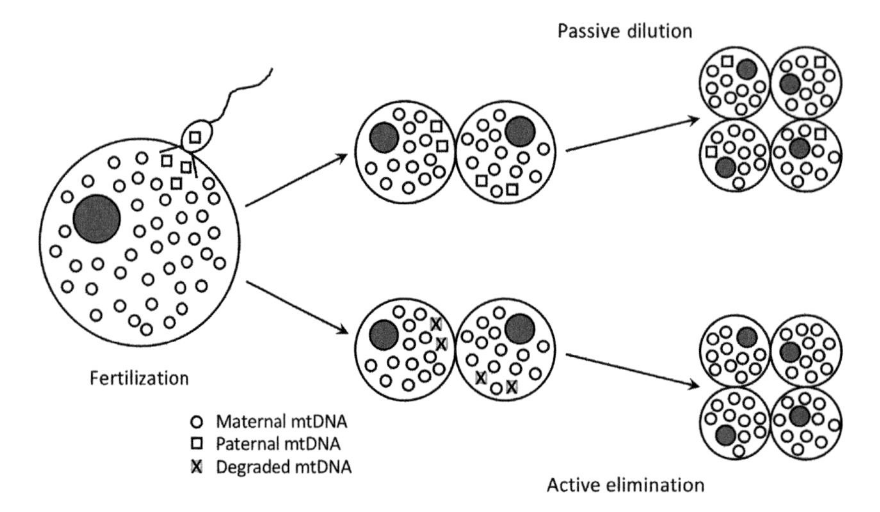

Figure 4.2: Models of paternal mtDNA elimination.
Two models have been proposed to explain how paternal mtDNA is eliminated from the early embryo to maintain strict maternal inheritance. The "Passive Dilution" model proposes that the relatively small number of paternal mtDNA genomes introduced to the zygote at fertilization is gradually diluted out to negligible levels by the maternal genome as the embryo grows. The "Active Elimination" model proposes that the paternal mitochondria are actively targeted for degradation soon after fertilization, possibly via the autophagic pathway, ensuring that no paternal mtDNA genomes remain in the developing embryo. It is currently unclear which model applies in human embryos.

embryos, there is no conclusive data to prove that paternal mtDNA is also degraded during this process. Coupled with the absence of similar studies human embryos (due in a large part to the ethical concerns regarding such experiments), the exact fate of paternal mtDNA following fertilization in humans remains elusive.

4.4 Paternal leakage during mtDNA inheritance

Although the evidence for strict maternal inheritance of mtDNA in mammals is compelling in the vast majority of cases, there have been occasional reports of paternal mtDNA being successfully transmitted to offspring. This raises the possibility that sperm mtDNA may occasionally be able to survive following fertilization and persist in the developing embryo, a phenomenon termed paternal leakage.

If the "Active Elimination" model of uniparental inheritance is correct, then paternal leakage requires sperm mitochondria to either evade degradation, or survive due to a defect in the zygote's ability to eliminate them. On the other hand, if the "Passive Dilution" model is true, then paternal leakage merely requires fertilization by a sperm cell containing sufficient paternal mtDNA molecules to make a detectable contribution to the embryo's mtDNA pool [23].

Several of the studies reporting evidence of paternal leakage utilized interspecific crosses, in which the parental mtDNA sequences are sufficiently different to allow easy differentiation of maternal and paternal genomes. In mice, paternal mtDNA has been detected in offspring from interspecific crosses of *Mus musculus* and *Mus spretus* [33–35], and has also been reported in crosses between the domestic cow and the Asian wild guar [29]. However, paternal leakage is likely to be more prevalent in interspecific crosses compared to intraspecific crosses, as paternal mitochondria from a different species may be able to evade species-specific degradation pathways in the zygote. Indeed, in mice it has been shown that intraspecific crossing of congenic *M. musculus* strains carrying *M. spretus* mtDNA with wild-type *M. musculus* does not result in paternal leakage, suggesting that this phenomenon is likely to be unusual in nature, where interspecific breeding is rare.

Apparently genuine cases of naturally occurring paternal mtDNA inheritance have been reported in sheep [36] and in humans [37,38], although multiple studies in independent patient cohorts have found no evidence of widespread paternal leakage in the human population [39–42] and some have even questioned the validity of the most recently reported case of paternal inheritance [43].

Although controversial, the possibility of paternal leakage is still an important consideration in human reproductive medicine, since biparental transmission of mtDNA is potentially deleterious. Mice inheriting a mixture of maternal and paternal mtDNAs display physiological, behavioral, and cognitive defects when compared to wild-types [44], and one reported case of paternal mtDNA inheritance in a human patient resulted in the development of mitochondrial myopathy [37]. This is of particular importance in human assisted reproductive therapies, since paternal mtDNA is known to persist after introduction to somatic cells [45] and in abnormal embryos [46], raising the possibility that artificial manipulation of fertilization and early embryonic development may increase the chance of paternal leakage occurring, potentially leading to pathology in the offspring.

4.5 mtDNA mutations—homoplasmy versus heteroplasmy

Due to the strict uniparental maternal inheritance of mtDNA, most individuals possess a single dominant mtDNA genome, matching that of their mother, present in multiple copies in each cell but all identical in sequence, or homoplasmic. However, due to the relatively high nucleotide substitution rate (estimated to be as much as 20 times higher than nDNA in mammals [47]) and continuous turnover of mtDNA, de novo genetic variants arise frequently, resulting in a heteroplasmic mixture of mutant and nonmutant genomes

(Fig. 4.1A). In the case of potentially pathogenic mutations, heteroplasmy can be present at low levels without obvious clinical symptoms or detectable mitochondrial dysfunction. Indeed, deep sequencing of mtDNA from healthy volunteers suggests that very low-level heteroplasmy (i.e., mutation levels <1%) may be present in the majority of the population [48]. As the proportion of pathogenic mutant mtDNA molecules increases, the ability of the remaining wild-type mtDNA to maintain normal OXPHOS function is reduced until the so-called "Biochemical Threshold" is reached (Fig. 4.1C). Beyond this threshold, cells are no longer able to function normally and, if the heteroplasmy is widespread, clinical symptoms begin to manifest as mitochondrial disease, most frequently in tissues with high ATP requirements such as skeletal muscle and the nervous system.

4.6 Germline segregation of mtDNA mutations and the genetic bottleneck

Heteroplasmic mtDNA mutations that are already present in the female germline are the major cause of mitochondrial disease, since the mutation present in the oocyte will ultimately propagate to all of the tissues in the offspring during embryonic development. The frequency of inherited pathogenic mtDNA mutations has been estimated at around 1 in 200 live births [49], and the presentation and severity of disease symptoms is largely dictated by the level of heteroplasmy initially present in the oocyte at fertilization, with oocytes carrying high levels of mutation being more likely to produce offspring with overt clinical symptoms.

Early studies investigating the germline transmission of a heteroplasmic mtDNA polymorphism in a Holstein cow lineage identified significant shifts in the level of heteroplasmy between mothers and their progeny over just a few generations [50], suggesting some mechanism during germline and/or embryonic development that resulted in differential segregation of heteroplasmy across individual offspring from the same mother. Similar shifts were subsequently seen in mice carrying a heteroplasmic mixture of NZB and BALBc mtDNA genotypes [51]. Examination of heteroplasmy segregation during germline development of NZB/BALBc embryos suggested that the heteroplasmy shifts occur during the early stages of germ cell development (Box 4.1) and are already established in mature oocytes (Fig. 4.1D) [51]. The authors hypothesized that the rapid shifts in heteroplasmy could be explained by random genetic drift following a reduction in mtDNA copy number to around 200 copies per cell [51]. This copy number reduction would introduce a strong sampling effect, or "bottleneck" at each cell division, increasing the likelihood that daughter cells, by chance, inherit significantly different proportions of mutant and wild-type mtDNA genotypes (Fig. 4.3A).

BOX 4.1 Development of the human female germline

Oocytes are the female germ cell, residing in the ovaries, with each ovum having the potential to pass on its genome, including the maternal mtDNA, to the next generation if successfully fertilized by a sperm after ovulation. By the time a female child is born oogenesis is largely complete; all of her oocytes are arrested in the first stage of meiosis and remain suspended in this state until their final maturation in adulthood, immediately prior to ovulation. This means that the majority of germline development occurs during embryogenesis, and it is in these very early stages of life that mtDNA heteroplasmy segregation is thought to occur. Development of the germline in both males and females begins around 17 days after conception in humans (Fig. 4.1D) [52]. At this point a small population of around 30–40 cells in the proximal epiblast begin to express the transcriptional regulators *Blimp1* and *SOX17* in response to BMP4 signaling from the extra-embryonic mesoderm [53]. By 21 days postconception these *Blimp1*-positive cells have specified to become primordial germ cells (PGCs), the precursor cells of the germline, and begin expressing a range of germ cell marker genes including *Dazl* and *TNAP* [52]. Beginning at around week four onwards, the PGCs begin to proliferate and migrate from their original position in the proximal epiblast, through the hindgut to reach the genital ridges by around week seven [54] (Fig. 4.1D). From week ten onwards the PGCs have settled in the gonads and, in females, continue to develop into primary oocytes (Fig. 4.1D). Thus the discussion of germline segregation of mtDNA heteroplasmy in this chapter must be considered in the context of these critical developmental processes occurring in the PGCs within the early embryo.

This so-called "Germline Genetic Bottleneck" theory (reviewed in [55]) has been supported by in silico modeling and quantitative mtDNA copy number measurements made at various stages of germ cell development, both in mice [56] and humans [57], and there is also evidence for an embryonic germline genetic bottleneck in several other vertebrate species, including zebra fish [58] and sheep [59]. While it is generally accepted that a germline bottleneck is responsible for the differential segregation of heteroplasmy between an affected mother's offspring, there is still significant debate over the timing and exact mechanism of the bottleneck. Work by Cree et al. hypothesized an early bottleneck, with absolute mtDNA copy number dropping to approximately 200 copies soon after specification of the primordial germ cells (PGCs) [56] (Box 4.1 and Fig. 4.1D), in very close agreement with the value predicted by Jenuth et al. [51], and subsequently backed up by further experimental data from mice [60] and humans [57]. In contrast, Cao et al. reported seeing no significant reduction of mtDNA levels in early stage PGCs, with copy number remaining >1500 at all stages of germ cell development. In this case a bottleneck caused by packaging of mtDNA molecules into a small number of homoplasmic clusters, or "segregation units," rather than a small absolute mtDNA copy number, was proposed [61] (Fig. 4.3B). A third possibility was proposed by Wai et al. who found no evidence of differential heteroplasmy segregation in PGCs, despite seeing a reduction in mtDNA copy number around the time of PGC

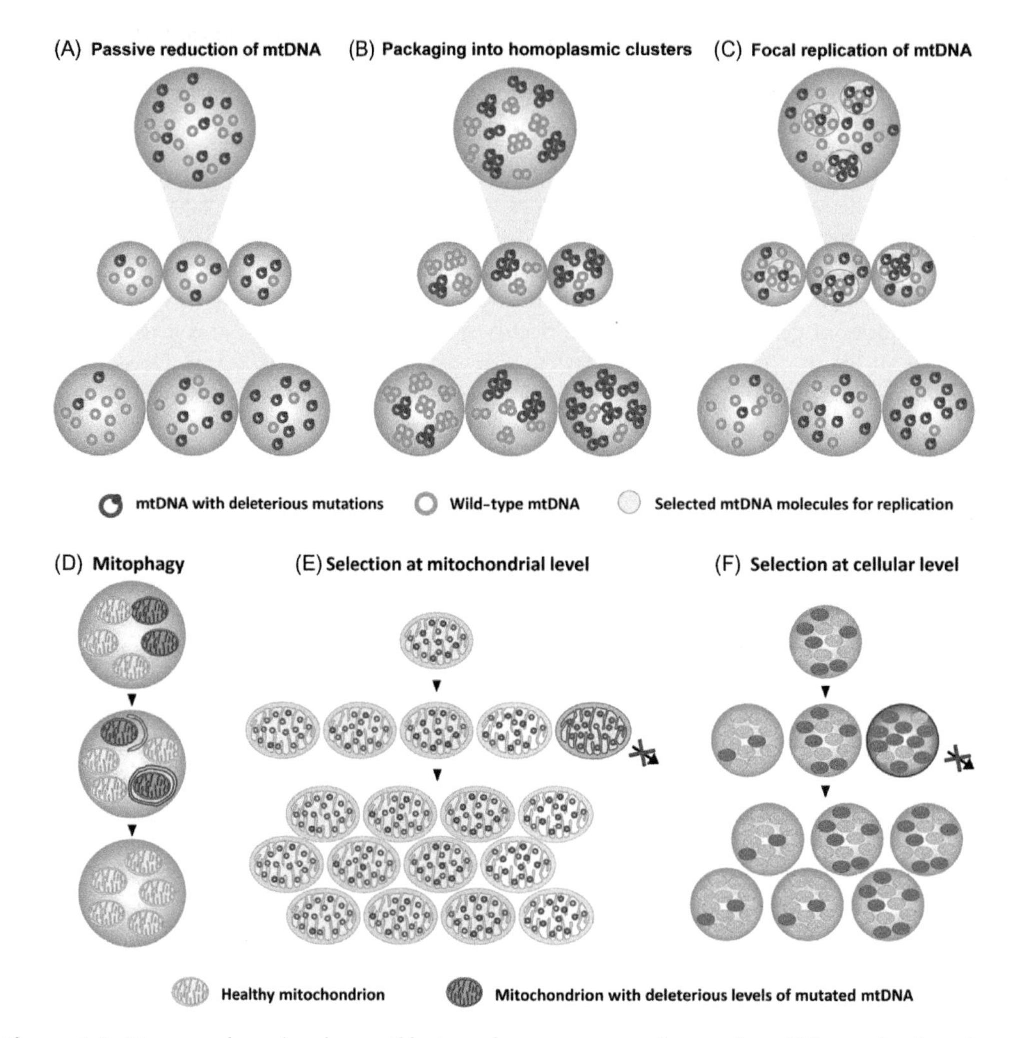

Figure 4.3: Proposed mechanisms of heteroplasmy segregation and purifying selection through the germline genetic bottleneck.

(A) Vegetative segregation in proliferating PGCs, combined with passive reduction in mtDNA copy number during PGC specification results in significant stochastic shifts in heteroplasmy levels. (B) Packaging of wild-type and mutant genomes into homoplasmic replicating clusters reduces the number of effective segregating units, creating a bottleneck effect without an overall reduction in mtDNA copy number. (C) Selective focal replication of a subpopulation of mtDNAs introduces a sampling effect that increases heteroplasmy segregation, this phenomenon could be temporally separate from the decrease in mtDNA copy number seen in early embryogenesis, resulting in later developmental bottlenecks. (D) Reduced ATP production secondary to OXPHOS deficiency leads to targeting of mitochondria containing high levels of mutant mtDNA for degradation via the autophagic pathway, therefore preferentially removing mutant genomes from the cell. (E) Selection at the level of the organelle may also occur if mitochondria carrying high levels of mutation are unable to replicate themselves (or their mtDNA) as effectively as healthy mitochondria. (F) Selection at the cellular level occurs if the number of defective organelles in a cell is sufficient to prevent cell division or cause cell death. Source: *From Zhang H, Burr SP, Chinnery PF. The mitochondrial DNA genetic bottleneck: inheritance and beyond. Essays Biochem 2018;62(3):225–34 [55].*

specification, similar to that reported by Cree et al. [56]. Instead, they suggested a late bottleneck, with shifts in heteroplasmy occurring during postnatal folliculogenesis via preferential replication of a subpopulation of mtDNA genomes in each oocyte [62] (Fig. 4.3C).

These contrasting theories, each backed up by experimental evidence, highlight the current uncertainty regarding the exact mechanisms dictating the dynamics of heteroplasmy transmission from mother to child. While the different findings may in part be due to the technical difficulties of identifying and isolating embryonic germ cells at various stages of development, followed by accurate measurement single-cell mtDNA copy number, it is also likely that genuine interstrain and interspecific biological variation. Regardless, this question remains a key focus of ongoing research efforts, since the provision of accurate prognostic advice to prospective mothers who carry a known pathogenic mtDNA mutation is currently very challenging [63].

4.7 Purifying selection against mtDNA mutations in the germline

If left unchecked, heteroplasmic mutations are either lost, or can eventually fix in the population (i.e., become homoplasmic). Given the high mutation rate of mtDNA and its asexual mode of transmission, the progressive accumulation of fixed homoplasmic mutations is predicted to result in an inevitable mutational meltdown—a process known as Muller's Ratchet [64]. One proposed mechanism to avoid such a meltdown is the presence of purifying selection against deleterious heteroplasmies during transmission of the mtDNA in the female germline. A key aspect of germline mtDNA segregation that is not yet fully understood is whether or not the genetic bottleneck results in some level of purifying selection against potentially pathogenic heteroplasmies. If such selection exists, then a better understanding of underlying mechanisms may enable the development of new therapeutic interventions aimed at reducing the incidence of inherited mitochondrial disease.

Early studies on heteroplasmy segregation suggested that random genetic drift was the major factor influencing heteroplasmy shifts during germline transmission [50,51]. However, since no pathology was reported in the animals used in these experiments, it is probable that the heteroplasmic variants present were functionally neutral, and therefore less likely to be subject to selection.

In recent years, a number of mouse models carrying pathogenic mtDNA heteroplasmies have been developed, and have provided compelling evidence to support the hypothesis that selective mechanisms are active against deleterious mutations during germline and embryonic development. Fan et al. generated a mouse line carrying two pathogenic heteroplasmic mtDNA mutations, one in the ND6 gene causing severe pathology, and another in the COXI gene causing much milder features [65]. When the germline transmission of these heteroplasmies was tracked, the ND6 mutation was rapidly eliminated over just four generations, while the COXI mutation persisted as a stable heteroplasmy in the line, despite causing gross pathology in the muscle and heart [65]. This suggests that purifying selection may act more strongly against severely pathogenic mutations, a theory supported in a contemporary study by Stewart et al. that analyzed the transmission of random mtDNA mutations arising in the *PolgA exo*$^-$ mtDNA mutator mouse model [66]. In this context, nonsynonymous mutations, which change the amino acid sequence of the affected protein, and are therefore possibly pathogenic, were much more likely to be rapidly eliminated from the germline compared to synonymous mutations, which are unlikely to cause pathology [66]. Subsequent mouse studies have further strengthened the case for purifying selection against deleterious heteroplasmy in the germline [44,60,67], although the point at which the selection occurs during development is currently not well defined, with two of the aforementioned studies reporting that selection appears to occur after birth [60,67], rather than during germline development as might be expected if the genetic bottleneck is a key step when purifying selection occurs. It is also unclear whether selection occurs at the molecular, organellar, or cellular level (Fig. 4.3D–F). These uncertainties highlight the current lack of data available regarding the mechanisms of selection against mtDNA heteroplasmy, and significant further work is required to fully understand this important process.

Although mouse models provide an invaluable tool for investigating the transmission of heteroplasmic mtDNA mutations from mother to offspring, a good understanding of how closely this mirrors the situation in humans will ultimately be required to facilitate future advances in treatment and/or prevention of inherited mitochondrial disease. Unfortunately, studying the dynamics of heteroplasmy segregation during human embryonic development is incredibly challenging, partly due to the ethical and logistical challenges of obtaining the tissues required for such work, but also due to the inherent bias introduced when pedigrees are obtained based on identification of an initial affected proband [68]. Despite these difficulties, there have been a small number of studies that have attempted to look for evidence of purifying selection against heteroplasmy in families carrying known pathogenic mtDNA heteroplasmies. Several studies in small patient cohorts carrying common pathogenic heteroplasmies have suggested that segregation is largely governed by random genetic drift, with little evidence of selection against the mutations [69–71]. However, data obtained from 39 healthy mother–child pairs carrying mtDNA mutations at $>1\%$ heteroplasmy showed that nonsynonymous (i.e., potentially pathogenic) mutations are transmitted to offspring at a lower rate than synonymous mutations, suggesting that purifying selection against deleterious heteroplasmies does occur in humans [72]. This was further supported by an analysis of mtDNA heteroplasmy transmission dynamics in the Genomes of the Netherlands (GoNL) dataset, comprising complete mtDNA genome sequences from 246 families, with strong evidence seen of negative selection against novel, potentially deleterious heteroplasmies [73]. To date, very little in vivo data is available to suggest possible timings and mechanisms of purifying selection in the human germline. One small-scale study in oocytes from nine healthy female donors by De Fanti et al. suggested that selection may occur during oocyte maturation, between the expulsion of the first and second polar bodies [74]. Contrastingly, Floros et al. reported evidence that purifying selection against nonsynonymous mtDNA mutations occurs during the early embryonic stages as the PGCs are proliferating and migrating to the gonads [57], suggesting that an early embryonic genetic bottleneck may indeed play a role in purging potentially deleterious mtDNA mutations from the germline.

While the studies discussed here have begun to give us some initial insight into the complex dynamics of mtDNA heteroplasmy transmission and segregation through the germline, we are still far from having a complete understanding of this critical process; a fact highlighted by the many contentious results obtained from studies in both murine models and human patients. It is quite possible that segregation of, and selection against, pathogenic mtDNA mutations occurs at multiple levels, at multiple time points and via a number of different methods, with all of these factors potentially varying on a mutation-by-mutation basis. As such, this remains an area of active research. While these studies provide strong experimental evidence that purifying selection against deleterious mtDNA mutations occurs in the mammalian germline, the mechanisms that underpin this selection are currently poorly understood [2]. However, recent work in the model organism Drosophila, which also exhibits purifying selection against pathogenic mtDNA mutations in the germline [75,76], has begun to give some insight into the molecular mechanism of purifying selection. Lieber et al. found that the critical period for purifying selection in Drosophila germline development corresponds with reduced expression of the pro-mitochondrial fusion protein Mitofusin, suggesting that the level of fragmentation in the mitochondrial network may be an important factor [77]. Further investigation showed that mitochondrial fragmentation is necessary for purifying selection to occur, with knockdown of Mitofusin (i.e., increased fragmentation) enhancing selection against mutant mtDNA in the germline [77]. The authors confirmed that there is reduced ATP synthesis in fragmented mitochondria containing mutant mtDNA, resulting in removal of these mitochondria by autophagy, therefore preferentially clearing mutant DNA from the germ cells [77]. This study marks an important step forward, but since germline development is significantly different in flies compared to mammals, it remains to be seen whether this is a fundamental mechanism that is conserved in evolution.

Another intriguing aspect of mtDNA heteroplasmy transmission to emerge in recent years is the possibility of positive selection acting on certain detrimental mtDNA mutations in the germline. This so-called "selfish drive" was first described in *Drosophila*, where introduction of a competing mtDNA genome carrying a detrimental mutation into *D. melanogaster* cells resulted in "selfish" transmission of the mutant mtDNA and death of the colony [78]. In-depth analysis of 1526 human mother-offspring pairs by Wei et al. has since provided evidence that positive selection may also be active in the human germline [12]. The authors observed selection both for and against heteroplasmic mtDNA variants, with known variants more likely to be transmitted to offspring at higher heteroplasmy levels than previously unidentified ones [12]. The genomic location of the variant also appears to influence the likelihood of positive versus negative selection, with D-loop variants the most likely to be transmitted to offspring at a higher heteroplasmy and rRNA variants the least likely [12]. Intriguingly, Wei et al. found that the nuclear genetic background also influences the dynamics of germline heteroplasmy transmission and selection in humans. In cases of mtDNA/nuclear "mismatch" (e.g., an individual with an Asian mtDNA haplogroup and a European nuclear ancestry), analysis showed that the heteroplasmies present on the mtDNA genome were more likely to match the nuclear genetic ancestry than the ancestry of the mtDNA on which the heteroplasmy occurred [12], suggesting that heteroplasmic mtDNA variants are selected to match the nuclear genetic background when such a mismatch is present, potentially helping to maximize mtDNA/nuclear compatibility.

Although much progress has been made in recent years, our overall understanding of germline transmission and selection for and against mtDNA mutations is far from complete, with a number of conflicting and seemingly incompatible theories proposed. As already discussed, the process of segregation and selection of pathogenic mtDNA mutations is likely to be multifactorial, and Burr et al. recently provided an in-depth review of the current tools available to study these mechanisms. Burr et al. recently provided an in-depth review of the current tools available for future studies [2].

4.8 Somatic mtDNA mutations and clonal expansion

While the majority of mitochondrial disease is caused by mutations inherited through the female germline, de novo mutations in somatic cells arising postfertilization may also result in the development of disease. In addition, somatic mtDNA mutations occur as part of the ageing process, and may contribute to late-onset multifactorial diseases including neurodegenerative disorders, such as Parkinson's disease [79] and cancer [80]. The number of different tissues affected by de novo mutations depends on the developmental stage at which the initial mutation occurs. De novo pathogenic mtDNA mutations that occur during early embryonic development in pluripotent stem cells may end up propagating to multiple tissues as the cell carrying the heteroplasmy proliferates and differentiates to a range of cell lineages. Although rarer than inherited mutations (around 1 in 1000 live births compared to 1 in 200 for inherited heteroplasmy [49]), these early embryonic somatic mutations have the potential to cause overt clinical symptoms, since they affect numerous organ systems. Similar widespread accumulation of de novo somatic mtDNA mutations may also occur secondary to inherited mutations in nuclear genes that are responsible for normal replication and maintenance of the mitochondrial genome [81]. In both cases, if mutant heteroplasmy levels increase sufficiently to reach the biochemical threshold (a process termed "clonal expansion") then tissue function can be affected, resulting in mitochondrial disease.

In healthy adults, the accumulation of clonally expanded somatic mtDNA mutations is a hallmark of the ageing process [82,83] (Fig. 4.4) and clonal expansion of pathogenic heteroplasmies undoubtedly contributes to age-related deterioration of tissue function, both in mouse models such as the *PolgA exo*$^-$ mtDNA mutator mouse, which displays a progeroid phenotype [84], and in humans, where it has been shown to result in age-related pathology in a number of tissue types, both mitotic, for example, colonic crypt epithelial cells [85], and nonmitotic, including skeletal muscle [86] and neurons [87]. It is important to note that in many cases it is difficult to distinguish whether clonal expansion occurs following genuine de novo mutations or from extremely low-level inherited heteroplasmy [48], although in both cases clonal expansion of the mutant genome must take place to reach the biochemical threshold and cause disease.

Although the role of somatic mtDNA mutations in ageing is well documented, the mechanisms that allow preferential expansion of mutant mtDNA genomes are currently not clear and a number of different theories have been proposed to explain clonal expansion. The simplest hypothesis suggests that, as the mtDNA is continuously turned over, random intracellular drift could be sufficient to explain the clonal expansion seen in older individuals

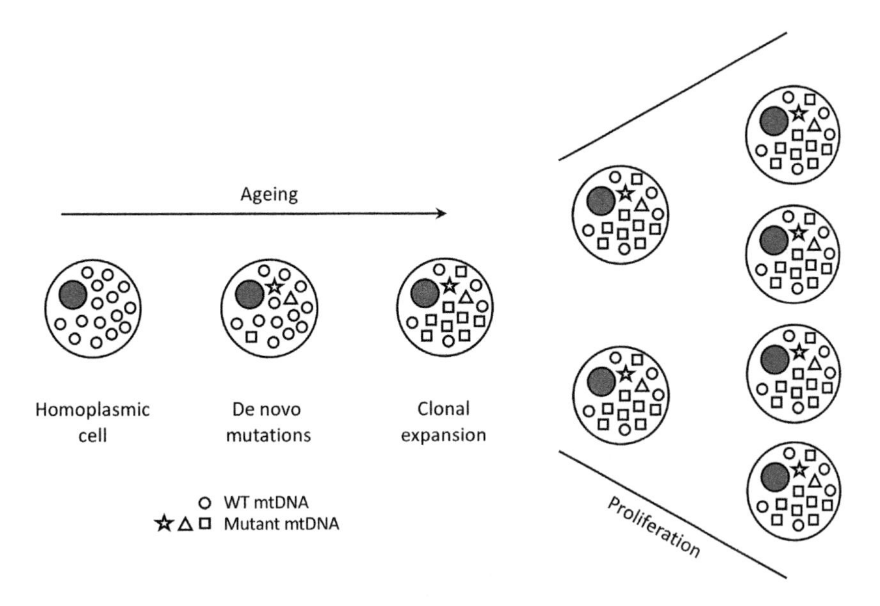

Figure 4.4: De novo mtDNA mutations and clonal expansion in ageing tissues. As cells age they accumulate de novo heteroplasmic mtDNA mutations, which can then undergo clonal expansion to become the dominant genome. Proliferating mitotic cells that accumulate clonally expanded pathogenic mutations can transmit these mutations to daughter cells, potentially leading to widespread tissue OXPHOS deficiencies.

[88]. However, while random drift models appear to work in long-lived species such as humans, the same models fail to correctly predict experimental observations of clonal expansion in shorter-lived species, leading some to question the validity of this theory [89].

Alternative theories have focused on identifying a selective advantage that allows mutant mtDNA molecules to be preferentially replicated in favor of wild-type molecules. Since the majority of age-related mtDNA mutations are deletions, rather than point mutations [90], one early hypothesis was that, after acquiring the deletion, the smaller mutant mtDNA genome could be replicated more quickly than, and would eventually expand to outnumber, the wild-type genome [91]. This is now considered unlikely to be true, as the time taken to replicate the wild-type genome is much shorter than the average time between replications and is therefore not a rate limiting step [92,93].

It has also been suggested that mitochondrial reactive oxygen species may play a role in clonal expansion via so-called "reactive biogenesis" [94], where increased ROS production in the mitochondria containing mutant mtDNA acts as a signal for mitochondrial biogenesis, thus increasing replication of the mutant genome. However, the importance of ROS in the ageing process is contentious [95], and further experimental data would be required to back this theory up.

More recently, Kowald and Kirkwood have theorized that wild-type genomes are subject to a feedback mechanism that downregulates mtDNA transcription rates when mtDNA-encoded protein levels are sufficient, but that this feedback mechanism is lost in some deleted mtDNA genomes [96]. Since initiation of mtDNA replication requires transcription of the mitochondrial mRNA [92], this would mean that the replication rate of wild-type mtDNA genomes is subject to the same feedback mechanism, while mutant mtDNA genomes can replicate unrestricted [96]. As with all of the previous hypotheses, there is currently no published in vivo or in vitro experimental evidence to support this theory.

As with inherited mtDNA mutations, it is clear that while we have begun to understand some of the key aspects of clonal expansion, and the role of somatic mtDNA mutations in ageing and the development of mitochondrial disease, we are still a long way from gaining a complete understanding of these critical processes and much work remains to be done before significant headway can be made with new therapeutic approaches in this field.

4.9 Conclusions

Over the last decade it has become clear that, contrary to the previous view, mtDNA homoplasmy is probably not the norm. The high mutation rate of mtDNA ensures that mixed populations of mtDNA are common, accumulating throughout life in our somatic tissues, and being transmitted down the female germline. Heteroplasmy seems all pervasive, but despite this, we are only just beginning to understand the subcellular mechanisms that can influence this. These mechanisms are important, because they determine whether a cell or organ accumulates a sufficient mutation burden to cause disease; and if transmitted down the germline, they potentially lead to severe mitochondrial disorders. Several laboratories have developed new approaches to prevent the transmission of pathogenic mtDNA mutations in humans [97,98], which may benefit families known to harbor pathogenic mtDNA mutations; and more sophisticated approaches are being developed that may limit the mutational burden in somatic tissues [99]—but de novo mutations will occur, and some tissues will remain inaccessible during life. Thus the mechanism of mtDNA segregation is not only fascinating biology, but has translational implications. Fortunately, new tools are emerging that will advance our understanding at a greater pace over the next five years, allowing us to harness this new knowledge for therapeutic benefit.

Research perspectives

Applying new high-resolution microscopic techniques will cast light on the subcellular mechanics of mtDNA segregation in the germline, and particularly the segregation of mtDNA nucleoids and their relationship to mitochondrial and cell partitioning. Although not possible at present, accurately measuring heteroplasmy in living cells would prove to be invaluable. New high through-put single cell transcriptomic approaches have the potential to reveal the mechanisms of selection—both for and against specific genotypes. The role of the cell nucleus and the cellular environmental context are likely to be important, and new cell and animal models provide the opportunity to advance understanding in both areas. Ultimately, this will open up new opportunities to develop treatments aimed at modulating the transmission of mtDNA heteroplasmy, and thus prevent the transmission of pathogenic mtDNA mutations.

References

[1] Stewart JB, Chinnery PF. The dynamics of mitochondrial DNA heteroplasmy: implications for human health and disease. Nat Rev Genet 2015;16(9):530–42.

[2] Burr SP, Pezet M, Chinnery PF. Mitochondrial DNA heteroplasmy and purifying selection in the mammalian female germ line. Dev Growth Differ 2018;60(1):21–32.

[3] Rajasimha HK, Chinnery PF, Samuels DC. Selection against pathogenic mtDNA mutations in a stem cell population leads to the loss of the 3243A-- > G mutation in blood. Am J Hum Genet 2008;82(2):333–43.

[4] Chinnery PF, Samuels DC. Relaxed replication of mtDNA: a model with implications for the expression of disease. Am J Hum Genet 1999;64(4):1158–65.

[5] Aryaman J, et al. Mitochondrial network state scales mtDNA genetic dynamics. Genetics 2019;212 (4):1429–43.

[6] Reich E, Luck DJ. Replication and inheritance of mitochondrial DNA. Proc Natl Acad Sci U S A 1966;55 (6):1600–8.

[7] Dawid IB, Blackler AW. Maternal and cytoplasmic inheritance of mitochondrial DNA in Xenopus. Dev Biol 1972;29(2):152–61.

[8] Hutchison 3rd CA, et al. Maternal inheritance of mammalian mitochondrial DNA. Nature 1974;251 (5475):536–8.

[9] Giles RE, et al. Maternal inheritance of human mitochondrial DNA. Proc Natl Acad Sci U S A 1980;77 (11):6715–19.

[10] Zouros E, et al. Direct evidence for extensive paternal mitochondrial DNA inheritance in the marine mussel Mytilus. Nature 1992;359(6394):412–14.

[11] Elson JL, et al. Analysis of European mtDNAs for recombination. Am J Hum Genet 2001;68(1):145–53.

[12] Wei W, et al. Germline selection shapes human mitochondrial DNA diversity. Science 2019;364(6442).

[13] Al Rawi S, et al. Postfertilization autophagy of sperm organelles prevents paternal mitochondrial DNA transmission. Science 2011;334(6059):1144–7.

[14] Sato M, Sato K. Degradation of paternal mitochondria by fertilization-triggered autophagy in *C. elegans* embryos. Science 2011;334(6059):1141–4.

[15] Zhou Q, et al. Mitochondrial endonuclease G mediates breakdown of paternal mitochondria upon fertilization. Science 2016;353(6297):394–9.

[16] DeLuca SZ, O'Farrell PH. Barriers to male transmission of mitochondrial DNA in sperm development. Dev Cell 2012;22(3):660–8.

[17] Yu Z, et al. The mitochondrial DNA polymerase promotes elimination of paternal mitochondrial genomes. Curr Biol 2017;27(7):1033–9.

[18] Politi Y, et al. Paternal mitochondrial destruction after fertilization is mediated by a common endocytic and autophagic pathway in Drosophila. Dev Cell 2014;29(3):305–20.

[19] Yanagimachi R, et al. Gametes and fertilization in the Chinese hamster. Gamete Res 1983;8(2):97–117.

[20] Simerly CR, et al. Tracing the incorporation of the sperm tail in the mouse zygote and early embryo using an anti-testicular alpha-tubulin antibody. Dev Biol 1993;158(2):536–48.

[21] Sathananthan AH, et al. Human sperm–egg interaction in vitro. Gamete Res 1986;15(4):317–26.

[22] Carelli V. Keeping in shape the dogma of mitochondrial DNA maternal inheritance. PLoS Genet 2015;11 (5):e1005179.

[23] Luo SM, et al. Unique insights into maternal mitochondrial inheritance in mice. Proc Natl Acad Sci U S A 2013;110(32):13038–43.

[24] May-Panloup P, et al. Increased sperm mitochondrial DNA content in male infertility. Hum Reprod 2003;18(3):550–6.

[25] Rantanen A, et al. Downregulation of Tfam and mtDNA copy number during mammalian spermatogenesis. Mamm Genome 2001;12(10):787–92.

[26] Wai T, et al. The role of mitochondrial DNA copy number in mammalian fertility. Biol Reprod 2010;83 (1):52–62.

[27] Orsztynowicz M, et al. Mitochondrial DNA copy number in spermatozoa of fertile stallions. Reprod Domest Anim 2016;51(3):378–85.

[28] Diez-Sanchez C, et al. Mitochondrial DNA content of human spermatozoa. Biol Reprod 2003;68 (1):180–5.

[29] Sutovsky P, et al. Ubiquitin tag for sperm mitochondria. Nature 1999;402(6760):371–2.

[30] Thompson WE, Ramalho-Santos J, Sutovsky P. Ubiquitination of prohibitin in mammalian sperm mitochondria: possible roles in the regulation of mitochondrial inheritance and sperm quality control. Biol Reprod 2003;69(1):254–60.

[31] Rojansky R, Cha MY, Chan DC. Elimination of paternal mitochondria in mouse embryos occurs through autophagic degradation dependent on PARKIN and MUL1. Elife 2016;5:e17896.

[32] Song WH, et al. Autophagy and ubiquitin–proteasome system contribute to sperm mitophagy after mammalian fertilization. Proc Natl Acad Sci U S A 2016;113(36):E5261–70.

[33] Gyllensten U, et al. Paternal inheritance of mitochondrial DNA in mice. Nature 1991;352(6332):255–7.

[34] Kaneda H, et al. Elimination of paternal mitochondrial DNA in intraspecific crosses during early mouse embryogenesis. Proc Natl Acad Sci U S A 1995;92(10):4542–6.

[35] Shitara H, et al. Maternal inheritance of mouse mtDNA in interspecific hybrids: segregation of the leaked paternal mtDNA followed by the prevention of subsequent paternal leakage. Genetics 1998;148 (2):851–7.

[36] Zhao X, et al. Further evidence for paternal inheritance of mitochondrial DNA in the sheep (*Ovis aries*). Heredity (Edinb) 2004;93(4):399–403.
[37] Schwartz M, Vissing J. Paternal inheritance of mitochondrial DNA. N Engl J Med 2002;347(8):576–80.
[38] Luo S, et al. Biparental inheritance of mitochondrial DNA in humans. Proc Natl Acad Sci U S A 2018;115(51):13039–44.
[39] Schwartz M, Vissing J. No evidence for paternal inheritance of mtDNA in patients with sporadic mtDNA mutations. J Neurol Sci 2004;218(1-2):99–101.
[40] Taylor RW, et al. Genotypes from patients indicate no paternal mitochondrial DNA contribution. Ann Neurol 2003;54(4):521–4.
[41] Pyle A, et al. Extreme-depth re-sequencing of mitochondrial DNA finds no evidence of paternal transmission in humans. PLoS Genet 2015;11(5):e1005040.
[42] Rius R, et al. Biparental inheritance of mitochondrial DNA in humans is not a common phenomenon. Genet Med 2019;21:2823–6.
[43] Lutz-Bonengel S, Parson W. No further evidence for paternal leakage of mitochondrial DNA in humans yet. Proc Natl Acad Sci U S A 2019;116(6):1821–2.
[44] Sharpley MS, et al. Heteroplasmy of mouse mtDNA is genetically unstable and results in altered behavior and cognition. Cell 2012;151(2):333–43.
[45] Manfredi G, et al. The fate of human sperm-derived mtDNA in somatic cells. Am J Hum Genet 1997;61(4):953–60.
[46] St John J, et al. Failure of elimination of paternal mitochondrial DNA in abnormal embryos. Lancet 2000;355(9199):200.
[47] Allio R, et al. Large variation in the ratio of mitochondrial to nuclear mutation rate across animals: implications for genetic diversity and the use of mitochondrial DNA as a molecular marker. Mol Biol Evol 2017;34(11):2762–72.
[48] Payne BA, et al. Universal heteroplasmy of human mitochondrial DNA. Hum Mol Genet 2013;22(2):384–90.
[49] Elliott HR, et al. Pathogenic mitochondrial DNA mutations are common in the general population. Am J Hum Genet 2008;83(2):254–60.
[50] Hauswirth WW, Laipis PJ. Mitochondrial DNA polymorphism in a maternal lineage of Holstein cows. Proc Natl Acad Sci U S A 1982;79(15):4686–90.
[51] Jenuth JP, et al. Random genetic drift in the female germline explains the rapid segregation of mammalian mitochondrial DNA. Nat Genet 1996;14(2):146–51.
[52] Leitch HG, Tang WW, Surani MA. Primordial germ-cell development and epigenetic reprogramming in mammals. Curr Top Dev Biol 2013;104:149–87.
[53] Sybirna A, Wong FCK, Surani MA. Genetic basis for primordial germ cells specification in mouse and human: conserved and divergent roles of PRDM and SOX transcription factors. Curr Top Dev Biol 2019;135:35–89.
[54] Richardson BE, Lehmann R. Mechanisms guiding primordial germ cell migration: strategies from different organisms. Nat Rev Mol Cell Biol 2010;11(1):37–49.
[55] Zhang H, Burr SP, Chinnery PF. The mitochondrial DNA genetic bottleneck: inheritance and beyond. Essays Biochem 2018;62(3):225–34.
[56] Cree LM, et al. A reduction of mitochondrial DNA molecules during embryogenesis explains the rapid segregation of genotypes. Nat Genet 2008;40(2):249–54.
[57] Floros VI, et al. Segregation of mitochondrial DNA heteroplasmy through a developmental genetic bottleneck in human embryos. Nat Cell Biol 2018;20(2):144–51.
[58] Otten AB, et al. Replication errors made during oogenesis lead to detectable de novo mtDNA mutations in zebrafish oocytes with a low mtDNA copy number. Genetics 2016;204(4):1423–31.
[59] Cotterill M, et al. The activity and copy number of mitochondrial DNA in ovine oocytes throughout oogenesis in vivo and during oocyte maturation in vitro. Mol Hum Reprod 2013;19(7):444–50.

[60] Freyer C, et al. Variation in germline mtDNA heteroplasmy is determined prenatally but modified during subsequent transmission. Nat Genet 2012;44(11):1282–5.
[61] Cao L, et al. The mitochondrial bottleneck occurs without reduction of mtDNA content in female mouse germ cells. Nat Genet 2007;39(3):386–90.
[62] Wai T, Teoli D, Shoubridge EA. The mitochondrial DNA genetic bottleneck results from replication of a subpopulation of genomes. Nat Genet 2008;40(12):1484–8.
[63] Chinnery PF, et al. The challenges of mitochondrial replacement. PLoS Genet 2014;10(4):e1004315.
[64] Muller HJ. The relation of recombination to mutational advance. Mutat Res 1964;106:2–9.
[65] Fan W, et al. A mouse model of mitochondrial disease reveals germline selection against severe mtDNA mutations. Science 2008;319(5865):958–62.
[66] Stewart JB, et al. Strong purifying selection in transmission of mammalian mitochondrial DNA. PLoS Biol 2008;6(1):e10.
[67] Kauppila JHK, et al. A phenotype-driven approach to generate mouse models with pathogenic mtDNA mutations causing mitochondrial disease. Cell Rep 2016;16(11):2980–90.
[68] Wilson IJ, et al. Mitochondrial DNA sequence characteristics modulate the size of the genetic bottleneck. Hum Mol Genet 2016;25(5):1031–41.
[69] Monnot S, et al. Segregation of mtDNA throughout human embryofetal development: m.3243A > G as a model system. Hum Mutat 2011;32(1):116–25.
[70] Brown DT, et al. Random genetic drift determines the level of mutant mtDNA in human primary oocytes. Am J Hum Genet 2001;68(2):533–6.
[71] Steffann J, et al. Analysis of mtDNA variant segregation during early human embryonic development: a tool for successful NARP preimplantation diagnosis. J Med Genet 2006;43(3):244–7.
[72] Rebolledo-Jaramillo B, et al. Maternal age effect and severe germ-line bottleneck in the inheritance of human mitochondrial DNA. Proc Natl Acad Sci U S A 2014;111(43):15474–9.
[73] Li M, et al. Transmission of human mtDNA heteroplasmy in the Genome of the Netherlands families: support for a variable-size bottleneck. Genome Res 2016;26(4):417–26.
[74] De Fanti S, et al. Intra-individual purifying selection on mitochondrial DNA variants during human oogenesis. Hum Reprod 2017;32(5):1100–7.
[75] Hill JH, Chen Z, Xu H. Selective propagation of functional mitochondrial DNA during oogenesis restricts the transmission of a deleterious mitochondrial variant. Nat Genet 2014;46(4):389–92.
[76] Ma H, Xu H, O'Farrell PH. Transmission of mitochondrial mutations and action of purifying selection in *Drosophila melanogaster*. Nat Genet 2014;46(4):393–7.
[77] Lieber T, et al. Mitochondrial fragmentation drives selective removal of deleterious mtDNA in the germline. Nature 2019;570(7761):380–4.
[78] Ma H, O'Farrell PH. Selfish drive can trump function when animal mitochondrial genomes compete. Nat Genet 2016;48(7):798–802.
[79] Muller-Nedebock AC, et al. The unresolved role of mitochondrial DNA in Parkinson's disease: an overview of published studies, their limitations, and future prospects. Neurochem Int 2019;129: 104495.
[80] Gammage PA, Frezza C. Mitochondrial DNA: the overlooked oncogenome? BMC Biol 2019;17(1):53.
[81] Viscomi C, Zeviani M. MtDNA-maintenance defects: syndromes and genes. J Inherit Metab Dis 2017;40 (4):587–99.
[82] Su T, Turnbull DM, Greaves LC. Roles of mitochondrial DNA mutations in stem cell ageing. Genes (Basel) 2018;9(4):E182.
[83] Brierley EJ, et al. Role of mitochondrial DNA mutations in human aging: implications for the central nervous system and muscle. Ann Neurol 1998;43(2):217–23.
[84] Trifunovic A, et al. Somatic mtDNA mutations cause aging phenotypes without affecting reactive oxygen species production. Proc Natl Acad Sci U S A 2005;102(50):17993–8.
[85] Taylor RW, et al. Mitochondrial DNA mutations in human colonic crypt stem cells. J Clin Invest 2003;112(9):1351–60.

[86] Bua E, et al. Mitochondrial DNA-deletion mutations accumulate intracellularly to detrimental levels in aged human skeletal muscle fibers. Am J Hum Genet 2006;79(3):469–80.
[87] Keogh MJ, Chinnery PF. Mitochondrial DNA mutations in neurodegeneration. Biochim Biophys Acta 2015;1847(11):1401–11.
[88] Elson JL, et al. Random intracellular drift explains the clonal expansion of mitochondrial DNA mutations with age. Am J Hum Genet 2001;68(3):802–6.
[89] Kowald A, Kirkwood TB. Mitochondrial mutations and aging: random drift is insufficient to explain the accumulation of mitochondrial deletion mutants in short-lived animals. Aging Cell 2013;12(4):728–31.
[90] Trifunov S, et al. Clonal expansion of mtDNA deletions: different disease models assessed by digital droplet PCR in single muscle cells. Sci Rep 2018;8(1):11682.
[91] Wallace DC. Mitochondrial genetics: a paradigm for aging and degenerative diseases? Science 1992;256 (5057):628–32.
[92] Shadel GS, Clayton DA. Mitochondrial DNA maintenance in vertebrates. Annu Rev Biochem 1997;66:409–35.
[93] Kowald A, Dawson M, Kirkwood TB. Mitochondrial mutations and ageing: can mitochondrial deletion mutants accumulate via a size based replication advantage? J Theor Biol 2014;340:111–18.
[94] Lane N. Mitonuclear match: optimizing fitness and fertility over generations drives ageing within generations. Bioessays 2011;33(11):860–9.
[95] Speakman JR, Selman C. The free-radical damage theory: accumulating evidence against a simple link of oxidative stress to ageing and lifespan. Bioessays 2011;33(4):255–9.
[96] Kowald A, Kirkwood TBL. Resolving the enigma of the clonal expansion of mtDNA deletions. Genes (Basel) 2018;9(3):126.
[97] Ma H, et al. Correction of a pathogenic gene mutation in human embryos. Nature 2017;548 (7668):413–19.
[98] Hyslop LA, et al. Towards clinical application of pronuclear transfer to prevent mitochondrial DNA disease. Nature 2016;534(7607):383–6.
[99] Wu TH, et al. Mitochondrial transfer by photothermal nanoblade restores metabolite profile in mammalian cells. Cell Metab 2016;23(5):921–9.

PART 2

mtDNA evolution and exploitation

CHAPTER 5

Haplogroups and the history of human evolution through mtDNA

Antonio Torroni, Alessandro Achilli, Anna Olivieri and Ornella Semino
Department of Biology and Biotechnology "Lazzaro Spallanzani", University of Pavia, Pavia, Italy

What is a haplogroup? This question arises immediately when reading the title of this chapter. This term was first used in two companion papers, published in *The American Journal of Human Genetics* in 1993. These studies analyzed mitochondrial DNA (mtDNA) variation in Native Americans and Aboriginal Siberians, respectively [1,2]. Its meaning is rather straightforward: it is the contraction of "haplotype group" into a single word. Haplogroup members share one or more distinguishing mutations that derive by descent from the same ancestral mtDNA molecule. From a phylogenetic perspective, a haplogroup corresponds with a well-defined branch (or clade) in the tree and its ancestral (or founder) mutational motif is the one situated at the branch node. The term was coined in the setting of human studies, but then came to be commonly used in evolutionary and phylogeographic studies of many other species [3]. From mtDNA, its use later spread to other uniparentally transmitted genetic systems, particularly the Y chromosome [4] as well as chloroplast DNA [5,6]. It is sometimes even employed when assessing variation of nuclear genes [7,8] and viral genomes [9].

It is important to underscore that for more than two decades, prior to the technological revolution based on next generation sequencing and the subsequent advent of nuclear comparative genomics, mtDNA was probably the most employed and successful genetic tool for investigating the evolution, origins, and migration patterns of our species. Indeed, starting from the early 1980s, despite its small size (only 16,569 base pairs), it has provided a disproportionate amount of valuable information.

This success is due to a number of distinguishing features that allowed mtDNA data to be much more readily acquired and assessed relative to its nuclear counterpart. First, mtDNA is maternally transmitted [10]. This means that its sequence variation is not rearranged by recombination and is only due to the sequential accumulation on preexisting haplotypes of new mutations, along radiating maternal lineages. Second, it evolves much faster than nuclear genes [11] and, as a consequence of these random mutations, modern mitogenomes

The Human Mitochondrial Genome.
DOI: https://doi.org/10.1016/B978-0-12-819656-4.00005-X

harbor extensive sequence variation even though living humans share a rather recent common female ancestor. Third, the entire human mtDNA molecule was completely sequenced in 1981 [12,13], so a reference sequence was available much earlier than for the nuclear genome. Fourth, mtDNA is characterized by a high copy number per cell; and, since it is a small circular DNA molecule with a cytoplasmic location, it could be purified from the rest of cellular DNA with approaches (e.g., cesium chloride density gradients) that were in common use during the 1980s. This latter feature provided a large boost to the initial mtDNA variation studies already in pre-polymerase chain reaction (PCR) times when only surveys with restriction enzymes, Southern blots, ^{32}P labeled probes, and autoradiography could be performed.

5.1 Early restriction fragment length polymorphism studies

Early restriction fragment length polymorphism (RFLP) studies, in which each mtDNA was digested with several or many endonucleases, were extremely laborious and time consuming, especially when assessed through the eyes of a modern PhD student in genetics or molecular anthropology, but they were the setting in which the word "haplogroup" was later coined and they provided striking results for the times. Not only did they reveal that mtDNAs of higher primates, including apes and humans, harbored very distinctive cleavage patterns [11,14], they also detected quite a lot of unexpected interindividual sequence variation among humans. Moreover, many mtDNA "morphs" (cleavage patterns resulting from the use of a single endonuclease) and mtDNA "types" (cleavage patterns resulting from the use of several or many endonucleases) were found to characterize only certain geographic areas or major ethnic groups [15–17]. By assuming that morphs were due to point mutations that either generated or eliminated specific restriction sites, and by mapping the novel site gains with double digestions and using the complete mtDNA sequence published in 1981, the "types" identified by these early RFLP studies could be related to each other allowing the construction of the first mtDNA phylogenies [18,19]. It is worth noting that the term "haplotype," initially developed in the context of HLA studies [20] and now so commonly used, replaced the term "type" and entered human mtDNA genetics only in 1988 [21].

However, the scientific community only came to fully appreciate the power of mtDNA studies with the 1987 *Nature* article by Allan C. Wilson's group in Berkeley [22]. By digesting 147 mtDNAs of different continental origins—almost all purified from placentas—with 12 restriction enzymes (4–6 bp cutters), Rebecca Cann and collaborators were able to identify an extremely large number (133) of mtDNA types. These types formed a network with two main branches when they were related to each other according to the presence or absence of restriction sites and using maximum parsimony. Moreover, by rooting the network at the midpoint of the mutational path connecting the two branches, they were able to identify the phylogenetic location and the RFLP motif of the ancestral mtDNA from whom all 147 modern mtDNAs were derived.

The ancestral woman carrying this mtDNA, who quickly became dubbed "Mitochondrial Eve," had two striking features: "anthropologically speaking" she lived in the recent past, only about 140,000–290,000 years ago using the molecular clock available at that time—about 180,000 years ago according to more recent calibrations [23–25]—and she was African as indicated by the fact that one of the two main derived branches of the rooted tree was present only in Africans and the second one also encompassed Africans in addition to all of the other continental groups.

Another important result of the study was the detection of numerous "clusters of region-specific types." These type clusters formed ethnically or geographically restricted branches (clades) in the tree. For the non-African clusters, the geographic-specificity suggested that their defining mutation(s) had occurred at different times and locations after what we now call the "Out of Africa Exit" (see below), while modern humans were spreading all over the world.

In the meantime, RFLP surveys of mtDNAs were also carried out in relatively large samples from single human populations, either by digesting mtDNAs purified from placentas, for instance in Japanese [26] and Papuans [27], or by digesting total DNA, usually from blood cells, and then using Southern blots and purified ^{32}P labeled mtDNA as a probe. Examples include the Pima of Arizona [28], Arabs and Jews [29], the Tharu of Nepal [30], Italians [31], and Senegalese [32].

5.2 The advent of polymerase chain reaction in the mtDNA world

PCR technique has divided biological research into two epochs: pre-PCR and post-PCR. This division obviously also applies to mtDNA genetics. Initially it was employed only in a few specialized labs [33], but by 1987 both *Taq* polymerase and the first PCR machine became commercially available, and the two leading research groups in the field, that of the already mentioned Allan Wilson and the one led by Douglas C. Wallace at Emory University, were among the first to amplify human mtDNA fragments [34–36].

With the diffusion of PCR in the world of human mtDNA variation, two main research paths were initially followed. One began to PCR amplify and then to sequence the mtDNA control region with the Sanger method. This region, also called the D-loop, evolves much faster than the rest of the mitogenome, so in spite of the fact that usually only the first hypervariable segment I (HVSI) of the control region was sequenced and that HVSI is only 341 bp long (MITOMAP; https://www.mitomap.org/MITOMAP) [37], this approach allowed the detection of extensive sequence variation [38–41]. However, the advantage was sometimes also a disadvantage because of the rather high levels of recurrent mutations that could partially blur the structure of the tree.

The other path blended PCR and RFLP surveys together. MtDNA amplicons were first obtained in large amounts and they were then digested with restriction enzymes [42]. This latter approach probably reached the highest level of sophistication when each human mtDNA began to be amplified in 9 overlapping PCR fragments encompassing the entire molecule, and each fragment was then independently digested with 14 individual restriction endonucleases. High-resolution RFLP haplotypes obtained in this way allowed an approximate 20% survey of each mitogenome, thus mainly targeting the more stable mtDNA coding regions; they were first published in *Genetics* by Douglas C. Wallace's laboratory in three companion papers [43–45].

A major outcome of most of the studies mentioned above was the production of mtDNA phylogenetic trees. However, the comparison of these initial trees was rather complex. Not only were the mtDNAs from different populations, each with some ethnic-specific type clusters, some trees were based on control-region sequences while others were based on RFLP haplotypes, often obtained with different and only partially overlapping endonuclease subsets. This issue was addressed by labeling the branches of the tree and by linking the name of each identified branch with its distinguishing mutation(s).

5.3 Haplogroup nomenclature of human mtDNA

The current mtDNA haplogroup nomenclature is illustrated by PhyloTree (https://www.phylotree.org/) [46,47]. This site provides a comprehensive worldwide phylogeny of human mtDNA variation. The current version (mtDNA tree Build 17; February 18, 2016), despite it is not fully updated, comprises over 5400 haplogroups and subhaplogroups, each with its own name. If you look closely at the main branches of the phylogeny, you will immediately notice that the haplogroup names are sometimes awkward, but there is a good explanation for this. The current nomenclature stems from one of the three aforementioned companion papers published in *Genetics*, a study in which high-resolution RFLP analysis was used to unravel the origins of Native Americans [45], that is, the people that live in the continental area most recently colonized by humans and harbor mtDNA branches that are relatively recent compared to many later identified in Old World populations. Current mtDNA haplogroup nomenclature would probably have been more straightforward if that initial paper had surveyed sub-Saharan Africans instead.

The tree of Native American mtDNA showed that almost all identified haplotypes (48 out of 50) clustered into only four distinct basal branches, each defined by some restriction site variants. These four "haplotype groups" were simply named alphabetically as A, B, C, and D [45,48]. The following year, in the two studies in which the word "haplogroup" was first used [1,2], mtDNA samples began to be analyzed in parallel for both RFLPs and HVSI sequence variation, thus the initial very short list of haplogroup-specific mutations for A, B, C, and D became enriched with some additional HVSI markers (Fig. 5.1A). The survey of Siberian and Asian samples made it clear that haplogroups A, B, C, and D were also present in Asia, thus supporting the Asian ancestral origin of Native Americans.

After Native Americans, the same approach was employed to survey other human populations and to identify additional "continent-specific haplogroups." Thus the Asian-specific haplogroups E, F, and G were identified in Tibetans [49], haplogroups H, I, J, and K in North Americans of Europeans descent [50], haplogroups L, L1, and L2, the most ancient of all continent-specific haplogroups, were found in sub-Saharan Africans [51], the Asian-specific haplogroup M was named according to Ballinger et al. [44] and Torroni et al. [49], haplogroups T, U, V, W, and X were identified in Europeans [52], and so forth. This means that some nested and more recent branches (subhaplogroups) were discovered earlier and the alphabetical letter used as a name, thus preceded the name of the superbranch to which they belonged; an example is haplogroup K that is a subbranch of haplogroup U. We should be grateful to Martin Richards and collaborators for defining simple and clear cladistic rules for hierarchically naming and ordering the growing number of subhaplogroups [53], but at that juncture the initial names of the main branches of the mtDNA tree were already well established and thus remained as initially proposed with only limited changes [54,55]. The same rules were also later applied to the Y-chromosome phylogeny when it became clear that it was going to face the same nomenclature problems as mtDNA [56].

5.4 The survey of entire mitogenomes

The human mtDNA tree finally came of age when entire mtDNA sequences began to be published in adequate numbers [57–60]. At that point, trees began to reach the highest possible levels of molecular and phylogenetic resolution, with the "phylogeography approach," as initially envisioned by John Avise and collaborators [61], reaching apotheosis. This approach requires the combined assessment of three elements: the phylogenetic tree, the geographic distribution of the haplogroups and subhaplogroups in the tree, and the dating of the tree nodes from which haplogroups stem. Knowledge of the three elements was rapidly progressing, including the estimation of haplogroup ages. Indeed, molecular clocks were refined [62,63] compared with earlier studies, even though some further calibration adjustments were later applied when concerns regarding purifying selection were taken into account [23,25] and ancient mitogenomes with reliable radiocarbon dating became available [64].

Fig. 5.2 schematically illustrates how to classify newly sequenced mitogenomes into haplogroups and how to include them into a phylogenetic tree. Three recently published

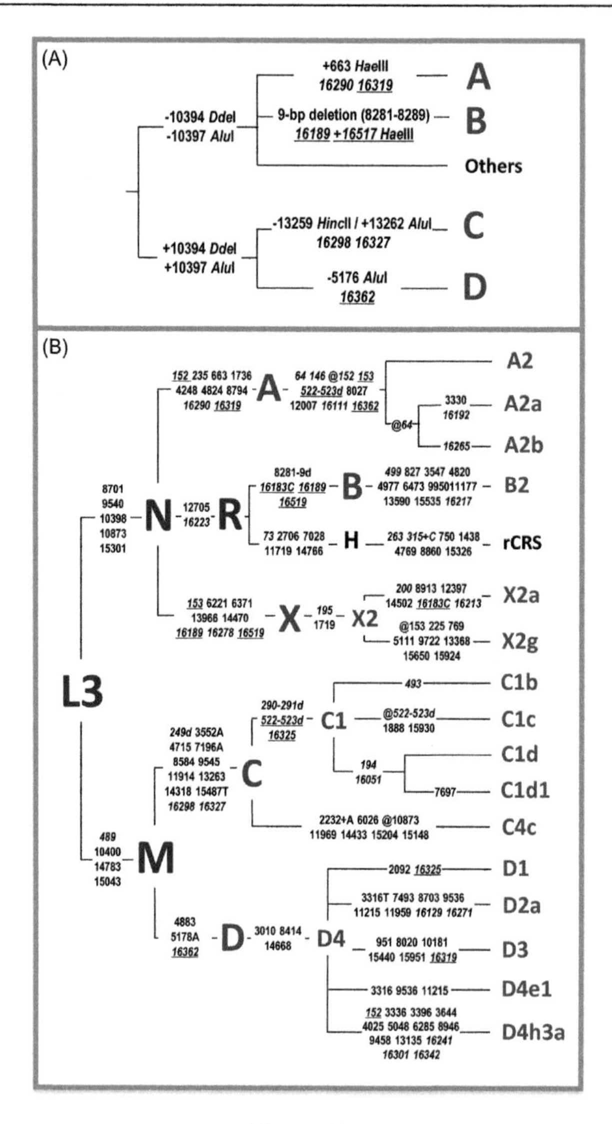

Figure 5.1

The diagnostic mutational motifs of Native American haplogroups. (A) The distinguishing RFLP and HVSI marker mutations of haplogroups A, B, C, and D according to the initial study [1]. (B) The diagnostic mutations of the 16 currently known Native American founder mtDNAs. Mutational motifs are shown on the branches. "Others" refers to a small number of mtDNAs found in North American natives that did not belong to either A, B, C, or D. Most of these were later identified as members of haplogroup X2a. The position of the revised Cambridge reference sequence (rCRS) [13], a member of the western Eurasian haplogroup H, is indicated for reading off-sequence motifs. Mutations are transitions unless a base is explicitly indicated; those in the control region are in italics. The prefix @ designates reversions, while suffixes indicate transversions (to A, G, C, or T) or indels (+, d). Recurrent mutations within the tree are *underlined*. L3 is the African haplogroup involved in the Out of Africa Exit. M, N, and R, which encompass all Eurasian-, Native American-, and Oceanian-specific haplogroups, arose in southwestern Asia at an early stage of the dispersal process.

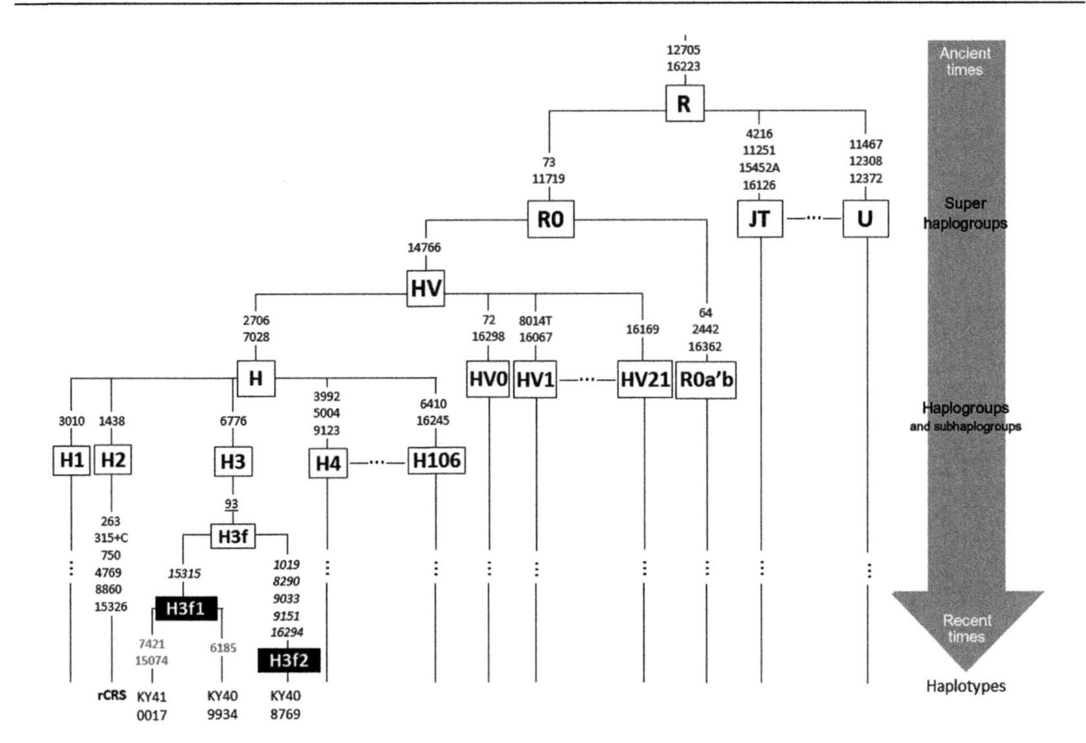

Figure 5.2

How to classify mtDNAs into haplogroups and how to include them into a phylogenetic tree: the case of three Sardinian mitogenomes. Haplogroup-diagnostic mutations, relative to the reference sequence (rCRS) [13], are shown at the root of major branches within macro-haplogroup R, which arose soon after the Out of Africa Exit. R includes, among others, superhaplogroups R0, JT, and U, which encompass most European mtDNAs. The haplotypes of three recently published mitogenomes (KY410017, KY409934, and KY408769) belonging to the Sardinian-specific haplogroups (SSHs), H3f1 and H3f2 [65], are used as examples and inserted in the tree. The transition at nucleotide position (np) 6776 is diagnostic of haplogroup H3, while the downstream mutation at np 93 defines its subclade H3f. Diagnostic mutations of H3f1 and H3f2 (in *black field*) are in italics. Private mutations, if present, are reported in *light gray*. Thus by reading off the mutations from the tip of rCRS, which is a member of haplogroup H2, to the tip of the newly analyzed mitogenome, its complete haplotype can be reconstructed. For example, KY410017 harbors 12 mutations relative to rCRS and bears the following haplotype: 93 263 315 + C 750 1438 4769 6776 7421 8860 15315 15074 15326. To guide the reader through available mtDNA tools: complete and/or partial mtDNA sequences can be classified into haplogroups and inserted in a tree by using the bioinformatics tool Haplogrep 2.0 (https://haplogrep.i-med.ac.at/) [66,67], which is based on PhyloTree (http://phylotree.org/) [46].

mitogenomes belonging to two Sardinian-specific haplogroups (SSHs), H3f1 and H3f2 [65], are used as examples. SSHs will be further discussed later in this chapter.

Initially, sequencing studies of entire mitogenomes rarely focused on populations, but rather on haplogroups, especially on those that appeared to be restricted to specific geographic areas [68–73]. Indeed, such geographic-specificity suggested that they had arisen while women were spreading throughout the world.

It should be emphasized that mtDNA as compared to nuclear genes does not reflect the full complexity of past demographic processes; it is a single maternally inherited locus and is particularly prone to genetic drift because of its reduced effective population size. However, the sequence variation of large datasets of entire mitogenomes allows extremely detailed reconstruction, potentially mutation by mutation and node by node, of the nesting relationships on each branch of the phylogeny, a feature that can be extremely informative for dating migration events.

Thus the history of human migrations and the peopling of continents, specific regions, islands, and so forth, began to be reconstructed through the molecular and phylogenetic dissection of mtDNA haplogroups. As a result, world maps with haplogroup names linked to arrows marking the paths of ancient human migrations came into common use [74].

Published mtDNA studies are now so numerous that even a simple overview is far beyond the scope of this chapter. However, as examples of the level of information that can be gleaned from mtDNA, the results of studies that have focused on three prehistoric migration events are summarized: (1) the "Out of Africa Exit," (2) the first peopling of a (double) continent—the Americas, and (3) the human settlement of an island in the Mediterranean Sea—Sardinia. These events are chronologically and geographically distant from each other and have impacted areas of very different sizes.

5.5 The "Out of Africa Exit"

Some early studies detected the highest mtDNA diversity in Africa [22,51,75], thus supporting an "Out of Africa" scenario. An important piece of the puzzle was resolved with the finding that all Eurasian mtDNAs derive from only one (L3) of the identified African super-haplogroups [76] and that haplogroups M and N, twin non-African derived subbranches of L3 (Fig. 5.1B), are the founders of all Eurasian mtDNAs [62,77].

Once the sequencing of entire mitogenomes became routine, a model suggesting a single dispersal from the Horn of Africa around 60–70,000 years ago was outlined. It was proposed that rather than the route through North Africa and the Levant, the East African founders carrying L3 mtDNAs would have spread following a southern coastal path. They and their descendants first reached the Arabian Peninsula (or the now-submerged Gulf region) and then spread both eastwards and northwards. Mutation and genetic drift apparently played major roles in shaping their mtDNA pool: haplogroups M, N, and R (a direct derivative of N) arose soon after the initial dispersal, while the ancestral L3 haplotype was lost [78,79]. It is important to note that no "pre-L3" mitogenomes have been detected so far in Eurasians. This implies that the age of L3, currently estimated at about 70,000 years,

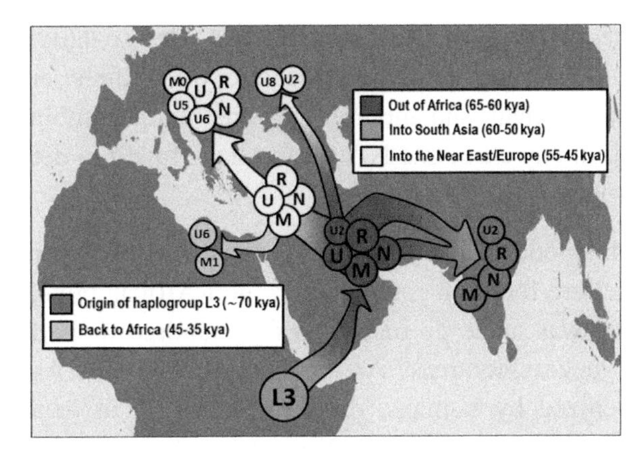

Figure 5.3

The early spread of mtDNAs at the crossroads between Africa, Asia, and Europe. Haplogroups M and N are derivatives of L3, haplogroup R derives from N, and haplogroup U from R. The illustrated scenario also takes into account the proto-European origin of U5 [69,81,82] that has been confirmed by its common detection in early Europeans [83]; the back migration to Africa from the Levant of [84]; and the discoveries of U2, U6, U8 mitogenomes in early European hunter-gatherers [83,85–87] as well as three mitogenomes that depart directly from the root of haplogroup M [64] and form the European-specific subhaplogroup M0 [79]. *kya*, thousand years ago.

represents the upper time limit for the Out of Africa dispersal [80], or rather the upper time limit for the dispersal of the founder mtDNAs from which all modern nonAfrican mitogenomes derive. A schematic mtDNA perspective of some of the events that occurred following the dispersal from Africa at the crossroads between Africa, Asia, and Europe is shown in Fig. 5.3.

Beyond the many successes of mtDNA, there are also some "failures." One of these is best portrayed by the July 11, 1997 cover story in *Cell*. Because of its abundance in the cell, mtDNA obviously led the way to ancient DNA studies and archeogenomics, a research avenue that is now progressing at an incredible speed [88]. The title of that cover story was "*Neanderthal were not our ancestors*" and synthesized the major outcomes of the first successful recovery and sequencing of Neanderthal mtDNA [89]. The HVSI sequence of the specimen discovered in the Neander Valley in 1856 was in fact found to form a sister branch relative to the one including all modern humans, a finding now confirmed by the sequencing of numerous entire Neanderthal mitogenomes [90,91]. This observation was not a surprise. It was already clear since 1987 that the age of the "Mitochondrial Eve" was not compatible with mtDNA inputs from archaic humans [22]. However, we now know that Neanderthals as well as Denisovans contributed a nonnegligible portion (1%–4%) of the nuclear genome in non-African populations [92–95], thus we should ponder the reasons for the apparent complete absence of Neanderthal (and Denisovan) mtDNAs (as well as Y-chromosomes) in the same populations. Genetic drift, directional mating, selection, incompatibility between nuclear and mtDNA genomes or a combination of these are all plausible answers [96]. In any case, mtDNA studies have failed to detect the contribution of archaic humans to our gene pool.

Apparently, only the L3 nodal haplotype was involved in the dispersal from Africa, or at least only its derivatives are found in contemporary non-African populations. This suggested that the founding group was made up of only a few hundred individuals [78]. What about the peopling of other geographic areas? As mentioned above, the Americas, the last continental area to be colonized by humans, drew the attention of geneticists and molecular anthropologists starting with the earliest mtDNA studies.

5.6 The first peopling of the Americas

The early studies of the 1990s showed that Native Americans exhibit a low variability when compared to other continental contexts and that only four haplogroups (A, B, C, and D), later re-labeled as A2, B2, C1, and D1 [55], encompass the vast majority of mtDNAs in the entire double continent. Forster and collaborators [55] also identified a fifth more rare haplogroup in the Americas. It was termed haplogroup X because that was the name that had been attributed to the same branch in a study that concomitantly was surveying mtDNAs in Europe [52]. All five Native American haplogroups are derivatives of the two superhaplogroups that arose shortly after the dispersal from Africa; haplogroups C and D arose from M, while A, B, and X are offshoots of N (Fig. 5.1B).

The presence of X mtDNAs in both Europe and North America, together with its almost complete absence in Siberia and Eastern Asia, raised the possibility of a direct genetic link between Native Americans and Europeans [97]. However, such a scenario was not supported by the studies later carried out at the level of entire mitogenomes. These showed that X mitogenomes from North America and Europe clustered into distinct subhaplogroups of X and all those from America (except one single X2g, see below) were members of X2a, a North American-specific branch [73,98]; but, it still took a few years before that the controversial Solutrean hypothesis of an Atlantic glacial entry route into North America [99] could be definitively dismissed [100].

The detection of some additional uncommon Native American haplogroups proceeded in parallel with the phylogenetic dissection of A, B, C, and D, both in Asia [101] and the Americas [102–104]. Overall, 16 maternal founding lineages of Beringian or Asian ancestry have been identified so far in Native Americans (Fig. 5.1B). Thus the number of female settlers was probably larger than the one involved in the Out of Africa Exit. Eight of these haplogroups (A2, B2, C1b, C1c, C1d, C1d1, D1, and D4h3a) are defined as "pan-American," as they are found throughout the double continent. All were probably present among the initial Paleo-Indian settlers who rapidly began to spread from eastern Beringia to South America along the Pacific coastal route as early as 16,000 years ago [105–107]. Others such as X2g [107] or D4e1 [108] are extremely rare. For instance, no additional X2g mtDNA has yet been reported after the initial discovery of this haplogroup in a single Ojibwa from Manitoulin Island in Lake Huron.

The remaining haplogroups are restricted to the populations of the North American arctic and subarctic regions and most (A2a, A2b, D2a, and D3) are the result of rather recent arrivals [109–111]. However, two have extensively contributed to the debate concerning Native American origins. Indeed, it has been proposed that the already mentioned X2a as well as haplogroup C4c (and possibly also the enigmatic X2g) might have marked the spread of a second group of Paleo-Indians from eastern Beringia [100,107,112] which, at the same time or shortly afterward, followed an alternative entry route—the ice free-corridor of western Canada. A scenario that, despite novel archeological and paleoecological evidence remains controversial, but that archeogenomics will hopefully clarify soon [88,113].

5.7 The peopling of an island in the Mediterranean Sea

As mentioned above, phylogeographic studies of entire mitogenomes, when carried out on very large population samples, might allow an extremely accurate reconstruction of the nesting relationships within a haplogroup, potentially identifying all of its nodes, mutation after mutation from the most ancient splits to the most recent ones. As a result, they also allow the identification of subbranches that have arisen very recently and characterize a single local community or even only the maternally related members of a single extended family. These high-resolution phylogeographic mitogenomes surveys are now coming of age.

The full power of this type of analysis has been recently tested in Sardinia [65], an island that remained unconnected with the mainland even when the sea level was at its lowest during the Last Glacial Maximum. It was probably also the last of the large Mediterranean islands to be colonized by modern humans. This insular context is of particular interest for European prehistory since modern Sardinians are "outliers" in the European genetic landscape and, according to archeogenomic nuclear data, the closest to early European Neolithic farmers [114]. This observation has led to the view that, in modern Europe, Sardinians may have best preserved the gene pool of early farmers, because after their arrival, the island remained relatively isolated from later Neolithic and Bronze Age expansions that instead changed the genetic landscape of continental Europe [115]. Still, the earliest human remains from the island date to the Upper Paleolithic, with an approximately 20,000 years old phalanx discovered at Corbeddu Cave. This early presence might have been very limited, even though several later Mesolithic settlements have also been discovered [116], thus the earliest settlers of Sardinia arrived prior to the Neolithic and some of their genes may have been retained in the contemporary population of the island.

Olivieri and collaborators tried to assess this issue by sequencing 3491 modern mitogenomes from the island as well as 21 mitogenomes from prehistoric subjects. These turned out to belong to the major haplogroup branches that characterize Europeans: HV, JT, and U. This however is not unexpected [117]; rather, the novelty is that almost 80% of modern mtDNAs cluster into a very large number (89) of SSHs [65]. In this study, a haplogroup was defined as "Sardinian-specific" (SSH) when three requirements were fulfilled: (1) inclusion of only mtDNAs of Sardinian origins, (2) encompassment of at least three mitogenomes and a minimum of two haplotypes, and (3) presence of at least one stable mutation (i.e., not recurrent in the tree) in the mutational motif at its root. The assessment of whether a haplogroup was Sardinian-specific was carried out by comparing its diagnostic motif with those reported in all available worldwide mitogenome datasets. This approach not only allowed investigators to calculate the coalescence age of each SSH node, but also to identify and date its closest upstream node (in the phylogeny) from which non-Sardinian mitogenomes radiate. In this way, minimum and maximum time estimates for the presence of each SSH on the island were estimated.

We should not overlook the possibility that the mutational motifs of some of the postulated SSHs actually arose outside the island but, after their arrival in Sardinia, were lost in the ancestral homeland, or they were simply not yet sampled in surrounding modern Mediterranean populations. Future studies of these populations and ancient DNA studies will help to clarify the issue. Still, the finding that such a large proportion of modern Sardinians mtDNAs are members of haplogroups that probably arose in situ in the island is somehow unexpected for a modern European population; it is rather the outcome that you would expect from the analysis of a rather isolated tribal population.

In brief, almost all SSHs were found to coalesce in the post-Nuragic, Nuragic and Neolithic-Copper Age periods in agreement with archeological evidence. However, at least two (K1a2d and U5b1i1) that together comprise about 3% of modern Sardinians coalesce prior to 7800 years ago, the postulated archeologically based starting time of the Neolithic in Sardinia. K1a2d is of Late Paleolithic Near Eastern ancestry, whereas U5b1i1 harbors deep ancestral roots in Paleolithic Western Europe. Such a dual ancestral origin highlights the complexity of past demographic events and raises additional novel and possibly controversial issues, but at least answers the initial question: Mesolithic inhabitants of Sardinia have indeed left a small but still detectable genetic trace in the mtDNA pool of modern islanders.

To conclude, it is clear that mtDNA has played a leading role in the reconstruction of the origins and migration patterns of our species in the last 40 years. One might wonder about the future of mtDNA studies now that we are in the population genomics era. Well, as mentioned above, the mitogenome is small and is essentially a single locus; therefore it obviously does not possess the informative power of the whole nuclear genome. However, nuclear genomic studies still face difficulties in making full use of the information present in haplotypes (and haplogroups), something that instead large-scale surveys of entire mitogenomes do extremely well. Moreover, whole genome studies lack, at least for the moment, the genealogical resolution provided by nonrecombining mtDNA. Therefore mtDNA data should remain a major complement to nuclear genome data in the genomic era; not to mention that the mitogenome is still sometimes the only one that is successfully retrieved in ancient samples, especially the oldest ones and those poorly preserved.

Research perspectives

Despite decades of studies, the often postulated role of mitogenome sequence variation in environmental adaptation and predisposition to diseases and, in general, the impact of selection (*vs* genetic drift) on the evolution of regional branches of the human mtDNA tree, are still rather poorly understood [74,118]. This is a fascinating and promising area of research, especially when considering that the interplay of the mitogenome with the nuclear genome is also largely unexplored [119].

Obviously, the rapid and continuous technological revolution [120] in the recovery of ancient DNA also affects mtDNA. To date, the oldest (almost) complete hominin mitogenome has been obtained from remains (Sima de los Huesos) dating to ~400 kya [121,122]. The retrieval of other Middle Pleistocene mitogenomes, or even older ones, is thus a very likely and exciting scenario.

References

[1] Torroni A, Schurr TG, Cabell MF, Brown MD, Neel JV, Larsen M, et al. Asian affinities and continental radiation of the four founding Native American mtDNAs. Am J Hum Genet 1993;53(3):563–90.

[2] Torroni A, Sukernik RI, Schurr TG, Starikorskaya YB, Cabell MF, Crawford MH, et al. mtDNA variation of aboriginal Siberians reveals distinct genetic affinities with Native Americans. Am J Hum Genet 1993;53(3):591–608.

[3] Troy CS, MacHugh DE, Bailey JF, Magee DA, Loftus RT, Cunningham P, et al. Genetic evidence for Near-Eastern origins of European cattle. Nature 2001;410(6832):1088–91.

[4] Underhill PA, Jin L, Lin AA, Mehdi SQ, Jenkins T, Vollrath D, et al. Detection of numerous Y chromosome biallelic polymorphisms by denaturing high-performance liquid chromatography. Genome Res 1997;7(10):996–1005.

[5] Wang L, Wu ZQ, Bystriakova N, Ansell SW, Xiang QP, Heinrichs J, et al. Phylogeography of the Sino-Himalayan fern *Lepisorus clathratus* on "the roof of the world". PLoS One 2011;6(9):e25896.

[6] Morris GP, Grabowski PP, Borevitz JO. Genomic diversity in switchgrass (*Panicum virgatum*): from the continental scale to a dune landscape. Mol Ecol 2011;20(23):4938–52.

[7] Warby SC, Montpetit A, Hayden AR, Carroll JB, Butland SL, Visscher H, et al. CAG expansion in the Huntington disease gene is associated with a specific and targetable predisposing haplogroup. Am J Hum Genet 2009;84(3):351–66.
[8] Ennis S, Murray A, Morton NE. Haplotypic determinants of instability in the FRAX region: concatenated mutation or founder effect? Hum Mutat 2001;18(1):61–9.
[9] Nguyen L, Li M, Chaowanachan T, Hu DJ, Vanichseni S, Mock PA, et al. CCR5 promoter human haplogroups associated with HIV-1 disease progression in Thai injection drug users. AIDS 2004;18(9):1327–33.
[10] Giles RE, Blanc H, Cann HM, Wallace DC. Maternal inheritance of human mitochondrial DNA. Proc Natl Acad Sci USA 1980;77(11):6715–19.
[11] Brown WM, George Jr M, Wilson AC. Rapid evolution of animal mitochondrial DNA. Proc Natl Acad Sci USA 1979;76(4):1967–71.
[12] Anderson S, Bankier AT, Barrell BG, de Bruijn MH, Coulson AR, Drouin J, et al. Sequence and organization of the human mitochondrial genome. Nature 1981;290(5806):457–65.
[13] Andrews RM, Kubacka I, Chinnery PF, Lightowlers RN, Turnbull DM, Howell N. Reanalysis and revision of the Cambridge reference sequence for human mitochondrial DNA. Nat Genet 1999;23(2):147.
[14] Ferris SD, Brown WM, Davidson WS, Wilson AC. Extensive polymorphism in the mitochondrial DNA of apes. Proc Natl Acad Sci USA 1981;78(10):6319–23.
[15] Brown WM. Polymorphism in mitochondrial DNA of humans as revealed by restriction endonuclease analysis. Proc Natl Acad Sci USA 1980;77(6):3605–9.
[16] Denaro M, Blanc H, Johnson MJ, Chen KH, Wilmsen E, Cavalli-Sforza LL, et al. Ethnic variation in Hpa 1 endonuclease cleavage patterns of human mitochondrial DNA. Proc Natl Acad Sci USA 1981;78(9):5768–72.
[17] Johnson MJ, Wallace DC, Ferris SD, Rattazzi MC, Cavalli-Sforza LL. Radiation of human mitochondria DNA types analyzed by restriction endonuclease cleavage patterns. J Mol Evol 1983;19(3-4):255–71.
[18] Cann RL, Brown WM, Wilson AC. Polymorphic sites and the mechanism of evolution in human mitochondrial DNA. Genetics 1984;106(3):479–99.
[19] Ferris SD, Wilson AC, Brown WM. Evolutionary tree for apes and humans based on cleavage maps of mitochondrial DNA. Proc Natl Acad Sci USA 1981;78(4):2432–6.
[20] Piazza A, Mattiuz PL, Ceppellini R. Combination of haplotypes of the HL-A system as a possible mechanism for gametic or zygotic selection. Haematologica 1969;54(10):703–20.
[21] Santachiara-Benerecetti AS, Scozzari R, Semino O, Torroni A, Brega A, Wallace DC. Mitochondrial DNA polymorphisms in Italy. II. Molecular analysis of new and rare morphs from Sardinia and Rome. Ann Hum Genet 1988;52(1):39–56.
[22] Cann RL, Stoneking M, Wilson AC. Mitochondrial DNA and human evolution. Nature 1987;325(6099):31–6.
[23] Behar DM, van Oven M, Rosset S, Metspalu M, Loogväli EL, Silva NM, et al. "Copernican" reassessment of the human mitochondrial DNA tree from its root. Am J Hum Genet 2012;90(4):675–84.
[24] Kivisild T, Shen P, Wall DP, Do B, Sung R, Davis K, et al. The role of selection in the evolution of human mitochondrial genomes. Genetics 2006;172(1):373–87.
[25] Soares P, Ermini L, Thomson N, Mormina M, Rito T, Röhl A, et al. Correcting for purifying selection: an improved human mitochondrial molecular clock. Am J Hum Genet 2009;84(6):740–59.
[26] Horai S, Gojobori T, Matsunaga E. Mitochondrial DNA polymorphism in Japanese. I. Analysis with restriction enzymes of six base pair recognition. Hum Genet 1984;68(4):324–32.
[27] Stoneking M, Bhatia K, Wilson AC. Rate of sequence divergence estimated from restriction maps of mitochondrial DNAs from Papua New Guinea. Cold Spring Harb Symp Quant Biol 1986;51(Pt 1):433–9.
[28] Wallace DC, Garrison K, Knowler WC. Dramatic founder effects in Amerindian mitochondrial DNAs. Am J Phys Anthropol 1985;68(2):149–55.
[29] Bonné-Tamir B, Johnson MJ, Natali A, Wallace DC, Cavalli-Sforza LL. Human mitochondrial DNA types in two Israeli populations—a comparative study at the DNA level. Am J Hum Genet 1986;38(3):341–51.

[30] Brega A, Gardella R, Semino O, Morpurgo G, Astaldi Ricotti GB, Wallace DC, et al. Genetic studies on the Tharu population of Nepal: restriction endonuclease polymorphisms of mitochondrial DNA. Am J Hum Genet 1986;39(4):502–12.
[31] Brega A, Scozzari R, Maccioni L, Iodice C, Wallace DC, Bianco I, et al. Mitochondrial DNA polymorphisms in Italy. I. Population data from Sardinia and Rome. Ann Hum Genet 1986;50(4):327–38.
[32] Scozzari R, Torroni A, Semino O, Sirugo G, Brega A, Santachiara-Benerecetti AS. Genetic studies on the Senegal population. I. Mitochondrial DNA polymorphisms. Am J Hum Genet 1988;43(4):534–44.
[33] Saiki RK, Scharf S, Faloona F, Mullis KB, Horn GT, Erlich HA, et al. Enzymatic amplification of beta-globin genomic sequences and restriction site analysis for diagnosis of sickle cell anemia. Science 1985;230(4732):1350–4.
[34] Vigilant L, Stoneking M, Wilson AC. Conformational mutation in human mtDNA detected by direct sequencing of enzymatically amplified DNA. Nucleic Acids Res 1988;16(13):5945–55.
[35] Wallace DC, Singh G, Lott MT, Hodge JA, Schurr TG, Lezza AM, et al. Mitochondrial DNA mutation associated with Leber's hereditary optic neuropathy. Science 1988;242(4884):1427–30.
[36] Wrischnik LA, Higuchi RG, Stoneking M, Erlich HA, Arnheim N, Wilson AC. Length mutations in human mitochondrial DNA: direct sequencing of enzymatically amplified DNA. Nucleic Acids Res 1987;15(2):529–42.
[37] MITOMAP <https://www.mitomap.org/MITOMAP>.
[38] Vigilant L, Pennington R, Harpending H, Kocher TD, Wilson AC. Mitochondrial DNA sequences in single hairs from a southern African population. Proc Natl Acad Sci USA 1989;86(23):9350–4.
[39] Ward RH, Frazier BL, Dew-Jager K, Pääbo S. Extensive mitochondrial diversity within a single Amerindian tribe. Proc Natl Acad Sci USA 1991;88(19):8720–4.
[40] Di Rienzo A, Wilson AC. Branching pattern in the evolutionary tree for human mitochondrial DNA. Proc Natl Acad Sci USA 1991;88(5):1597–601.
[41] Richards M, Côrte-Real H, Forster P, Macaulay V, Wilkinson-Herbots H, Demaine A, et al. Paleolithic and neolithic lineages in the European mitochondrial gene pool. Am J Hum Genet 1996;59(1):185–203.
[42] Schurr TG, Ballinger SW, Gan YY, Hodge JA, Merriwether DA, Lawrence DN, et al. Amerindian mitochondrial DNAs have rare Asian mutations at high frequencies, suggesting they derived from four primary maternal lineages. Am J Hum Genet 1990;46(3):613–23.
[43] Brown MD, Voljavec AS, Lott MT, Torroni A, Yang CC, Wallace DC. Mitochondrial DNA complex I and III mutations associated with Leber's hereditary optic neuropathy. Genetics 1992;130(1):163–73.
[44] Ballinger SW, Schurr TG, Torroni A, Gan YY, Hodge JA, Hassan K, et al. Southeast Asian mitochondrial DNA analysis reveals genetic continuity of ancient mongoloid migrations. Genetics 1992;130(1):139–52.
[45] Torroni A, Schurr TG, Yang CC, Szathmary EJ, Williams RC, Schanfield MS, et al. Native American mitochondrial DNA analysis indicates that the Amerind and the Nadene populations were founded by two independent migrations. Genetics 1992;130(1):153–62.
[46] PhyloTree <https://www.phylotree.org/>.
[47] van Oven M, Kayser M. Updated comprehensive phylogenetic tree of global human mitochondrial DNA variation. Hum Mutat 2009;30(2):E386–94.
[48] Wallace DC, Torroni A. American Indian prehistory as written in the mitochondrial DNA: a review. Hum Biol 1992;64(3):403–16.
[49] Torroni A, Miller JA, Moore LG, Zamudio S, Zhuang J, Droma T, et al. Mitochondrial DNA analysis in Tibet: implications for the origin of the Tibetan population and its adaptation to high altitude. Am J Phys Anthropol 1994;93(2):189–99.
[50] Torroni A, Lott MT, Cabell MF, Chen YS, Lavergne L, Wallace DC. mtDNA and the origin of Caucasians: identification of ancient Caucasian-specific haplogroups, one of which is prone to a recurrent somatic duplication in the D-loop region. Am J Hum Genet 1994;55(4):760–76.
[51] Chen YS, Torroni A, Excoffier L, Santachiara-Benerecetti AS, Wallace DC. Analysis of mtDNA variation in African populations reveals the most ancient of all human continent-specific haplogroups. Am J Hum Genet 1995;57(1):133–49.

[52] Torroni A, Huoponen K, Francalacci P, Petrozzi M, Morelli L, Scozzari R, et al. Classification of European mtDNAs from an analysis of three European populations. Genetics 1996;144(4):1835–50.
[53] Richards MB, Macaulay VA, Bandelt HJ, Sykes BC. Phylogeography of mitochondrial DNA in western Europe. Ann Hum Genet 1998;62(Pt 3):241–60.
[54] Torroni A, Achilli A, Macaulay V, Richards M, Bandelt HJ. Harvesting the fruit of the human mtDNA tree. Trends Genet 2006;22(6):339–45.
[55] Forster P, Harding R, Torroni A, Bandelt HJ. Origin and evolution of Native American mtDNA variation: a reappraisal. Am J Hum Genet 1996;59(4):935–45.
[56] YCC. A nomenclature system for the tree of human Y-chromosomal binary haplogroups. Genome Res 2002;12:339–48.
[57] Torroni A, Rengo C, Guida V, Cruciani F, Sellitto D, Coppa A, et al. Do the four clades of the mtDNA haplogroup L2 evolve at different rates? Am J Hum Genet 2001;69(6):1348–56.
[58] Finnilä S, Lehtonen MS, Majamaa K. Phylogenetic network for European mtDNA. Am J Hum Genet 2001;68(6):1475–84.
[59] Ingman M, Kaessmann H, Pääbo S, Gyllensten U. Mitochondrial genome variation and the origin of modern humans. Nature 2000;408(6813):708–13.
[60] Richards M, Macaulay V. The mitochondrial gene tree comes of age. Am J Hum Genet 2001;68 (6):1315–20.
[61] Avise JC, Arnold J, Ball RM, Bermingham E, Lamb T, Neigel JE, et al. Intraspecific phylogeography: the molecular bridge between population genetics and systematics. Ann Rev Ecol Syst 1987;18:489–522.
[62] Forster P, Torroni A, Renfrew C, Röhl A. Phylogenetic star contraction applied to Asian and Papuan mtDNA evolution. Mol Biol Evol 2001;18(10):1864–81.
[63] Mishmar D, Ruiz-Pesini E, Golik P, Macaulay V, Clark AG, Hosseini S, et al. Natural selection shaped regional mtDNA variation in humans. Proc Natl Acad Sci USA 2003;100(1):171–6.
[64] Posth C, Renaud G, Mittnik A, Drucker DG, Rougier H, Cupillard C, et al. Pleistocene mitochondrial genomes suggest a single major dispersal of non-Africans and a late glacial population turnover in Europe. Curr Biol 2016;26(6):827–33.
[65] Olivieri A, Sidore C, Achilli A, Angius A, Posth C, Furtwängler A, et al. Mitogenome diversity in Sardinians: a genetic window onto an island's past. Mol Biol Evol 2017;34(5):1230–9.
[66] Haplogrep 2.0. https://haplogrep.i-med.ac.at/
[67] Weissensteiner H, Pacher D, Kloss-Brandstätter A, Forer L, Specht G, Bandelt HJ, et al. HaploGrep 2: mitochondrial haplogroup classification in the era of high-throughput sequencing. Nucleic Acids Res 2016;44((W1):W58–63.
[68] Achilli A, Rengo C, Magri C, Battaglia V, Olivieri A, Scozzari R, et al. The molecular dissection of mtDNA haplogroup H confirms that the Franco-Cantabrian glacial refuge was a major source for the European gene pool. Am J Hum Genet 2004;75(5):910–18.
[69] Achilli A, Rengo C, Battaglia V, Pala M, Olivieri A, Fornarino S, et al. Saami and Berbers—an unexpected mitochondrial DNA link. Am J Hum Genet 2005;76(5):883–6.
[70] Friedlaender J, Schurr T, Gentz F, Koki G, Friedlaender F, Horvat G, et al. Expanding Southwest Pacific mitochondrial haplogroups P and Q. Mol Biol Evol 2005;22(6):1506–17 Erratum in: Mol Biol Evol 2005;22(11):2313.
[71] Kivisild T, Reidla M, Metspalu E, Rosa A, Brehm A, Pennarun E, et al. Ethiopian mitochondrial DNA heritage: tracking gene flow across and around the Gate of Tears. Am J Hum Genet 2004;75(5):752–70 Epub 2004 Sep 27. Erratum in: Am J Hum Genet 2006;78(6):1097.
[72] Palanichamy Mg, Sun C, Agrawal S, Bandelt HJ, Kong QP, Khan F, et al. Phylogeny of mitochondrial DNA macrohaplogroup N in India, based on complete sequencing: implications for the peopling of South Asia. Am J Hum Genet 2004;75(6):966–78.
[73] Reidla M, Kivisild T, Metspalu E, Kaldma K, Tambets K, Tolk HV, et al. Origin and diffusion of mtDNA haplogroup X. Am J Hum Genet 2003;73(5):1178–90.
[74] Wallace DC. Mitochondrial DNA variation in human radiation and disease. Cell 2015;163(1):33–8.

[75] Vigilant L, Stoneking M, Harpending H, Hawkes K, Wilson AC. African populations and the evolution of human mitochondrial DNA. Science 1991;253(5027):1503–7.
[76] Watson E, Forster P, Richards M, Bandelt HJ. Mitochondrial footprints of human expansions in Africa. Am J Hum Genet 1997;61(3):691–704.
[77] Quintana-Murci L, Semino O, Bandelt HJ, Passarino G, McElreavey K, Santachiara-Benerecetti AS. Genetic evidence of an early exit of *Homo sapiens sapiens* from Africa through eastern Africa. Nat Genet 1999;23(4):437–41.
[78] Macaulay V, Hill C, Achilli A, Rengo C, Clarke D, Meehan W, et al. Single, rapid coastal settlement of Asia revealed by analysis of complete mitochondrial genomes. Science 2005;308(5724):1034–6.
[79] Richards MB, Soares P, Torroni A. Palaeogenomics: mitogenomes and migrations in Europe's past. Curr Biol 2016;26(6):R243–6.
[80] Soares P, Alshamali F, Pereira JB, Fernandes V, Silva NM, Afonso C, et al. The expansion of mtDNA haplogroup L3 within and out of Africa. Mol Biol Evol 2012;29(3):915–27.
[81] Malyarchuk B, Derenko M, Grzybowski T, Perkova M, Rogalla U, Vanecek T, et al. The peopling of Europe from the mitochondrial haplogroup U5 perspective. PLoS One 2010;5(4):e10285.
[82] Richards M, Macaulay V, Hickey E, Vega E, Sykes B, Guida V, et al. Tracing European founder lineages in the Near Eastern mtDNA pool. Am J Hum Genet 2000;67(5):1251–76.
[83] Fu Q, Posth C, Hajdinjak M, Petr M, Mallick S, Fernandes D, et al. The genetic history of Ice Age Europe. Nature 2016;534(7606):200–5.
[84] Olivieri A, Achilli A, Pala M, Battaglia V, Fornarino S, Al-Zahery N, et al. The mtDNA legacy of the Levantine early Upper Palaeolithic in Africa. Science 2006;314(5806):1767–70.
[85] Hervella M, Svensson EM, Alberdi A, Günther T, Izagirre N, Munters AR, et al. The mitogenome of a 35,000-year-old *Homo sapiens* from Europe supports a Palaeolithic back-migration to Africa. Sci Rep 2016;6:25501.
[86] Seguin-Orlando A, Korneliussen TS, Sikora M, Malaspinas AS, Manica A, Moltke I, et al. Paleogenomics. Genomic structure in Europeans dating back at least 36,200 years. Science 2014;346 (6213):1113–18.
[87] Sikora M, Seguin-Orlando A, Sousa VC, Albrechtsen A, Korneliussen T, Ko A, et al. Ancient genomes show social and reproductive behavior of early Upper Paleolithic foragers. Science 2017;358 ((6363):659–62.
[88] Achilli A, Olivieri A, Semino O, Torroni A. Ancient human genomes—keys to understanding our past. Science 2018;360(6392):964–5.
[89] Krings M, Stone A, Schmitz RW, Krainitzki H, Stoneking M, Pääbo S. Neandertal DNA sequences and the origin of modern humans. Cell 1997;90(1):19–30.
[90] Bokelmann L, Hajdinjak M, Peyrégne S, Brace S, Essel E, de Filippo C, et al. A genetic analysis of the Gibraltar Neanderthals. Proc Natl Acad Sci USA 2019;116(31):15610–15.
[91] Posth C, Wißing C, Kitagawa K, Pagani L, van Holstein L, Racimo F, et al. Deeply divergent archaic mitochondrial genome provides lower time boundary for African gene flow into Neanderthals. Nat Commun 2017;8:16046.
[92] Browning SR, Browning BL, Zhou Y, Tucci S, Akey JM. Analysis of human sequence data reveals two Pulses of archaic Denisovan admixture. Cell 2018;173(1) 53-61.e9.
[93] Green RE, Krause J, Briggs AW, Maricic T, Stenzel U, Kircher M, et al. A draft sequence of the Neandertal genome. Science 2010;328(5979):710–22.
[94] Reich D, Green RE, Kircher M, Krause J, Patterson N, Durand EY, et al. Genetic history of an archaic hominin group from Denisova Cave in Siberia. Nature 2010;468(7327):1053–60.
[95] Dannemann M, Racimo F. Something old, something borrowed: admixture and adaptation in human evolution. Curr Opin Genet Dev 2018;53:1–8.
[96] Sharbrough J, Havird JC, Noe GR, Warren JM, Sloan DB. The mitonuclear dimension of Neanderthal and Denisovan ancestry in modern human genomes. Genome Biol Evol 2017;9(6):1567–81.

[97] Brown MD, Hosseini SH, Torroni A, Bandelt HJ, Allen JC, Schurr TG, et al. mtDNA haplogroup X: an ancient link between Europe/Western Asia and North America? Am J Hum Genet 1998;63(6):1852–61.
[98] Bandelt HJ, Herrnstadt C, Yao YG, Kong QP, Kivisild T, Rengo C, et al. Identification of Native American founder mtDNAs through the analysis of complete mtDNA sequences: some caveats. Ann Hum Genet 2003;67(Pt 6):512–24.
[99] Straus LG, David JM, Goebel T. Ice Age Atlantis? Exploring the Solutrean-Clovis 'connection'. World Archaeol 2005;37:507–32.
[100] Hooshiar Kashani B, Perego UA, Olivieri A, Angerhofer N, Gandini F, Carossa V, et al. Mitochondrial haplogroup C4c: a rare lineage entering America through the ice-free corridor? Am J Phys Anthropol 2012;147(1):35–9.
[101] Kong QP, Bandelt HJ, Sun C, Yao YG, Salas A, Achilli A, et al. Updating the East Asian mtDNA phylogeny: a prerequisite for the identification of pathogenic mutations. Hum Mol Genet 2006;15 (13):2076–86.
[102] Achilli A, Perego UA, Bravi CM, Coble MD, Kong QP, Woodward SR, et al. The phylogeny of the four pan-American MtDNA haplogroups: implications for evolutionary and disease studies. PLoS One 2008;3 (3):e1764.
[103] Perego UA, Angerhofer N, Pala M, Olivieri A, Lancioni H, Hooshiar Kashani B, et al. The initial peopling of the Americas: a growing number of founding mitochondrial genomes from Beringia. Genome Res 2010;20(9):1174–9.
[104] Tamm E, Kivisild T, Reidla M, Metspalu M, Smith DG, Mulligan CJ, et al. Beringian standstill and spread of Native American founders. PLoS One 2007;2(9):e829.
[105] Bodner M, Perego UA, Huber G, Fendt L, Röck AW, Zimmermann B, et al. Rapid coastal spread of First Americans: novel insights from South America's Southern Cone mitochondrial genomes. Genome Res 2012;22(5):811–20.
[106] de Saint Pierre M, Gandini F, Perego UA, Bodner M, Gómez-Carballa A, Corach D, et al. Arrival of Paleo-Indians to the southern cone of South America: new clues from mitogenomes. PLoS One 2012;7 (12):e51311.
[107] Perego UA, Achilli A, Angerhofer N, Accetturo M, Pala M, Olivieri A, et al. Distinctive Paleo-Indian migration routes from Beringia marked by two rare mtDNA haplogroups. Curr Biol 2009;19(1):1–8.
[108] Kumar S, Bellis C, Zlojutro M, Melton PE, Blangero J, Curran JE. Large scale mitochondrial sequencing in Mexican Americans suggests a reappraisal of Native American origins. BMC Evol Biol 2011;11:293.
[109] Gilbert MT, Kivisild T, Grønnow B, Andersen PK, Metspalu E, Reidla M, et al. Paleo-Eskimo mtDNA genome reveals matrilineal discontinuity in Greenland. Science 2008;320(5884):1787–9.
[110] Achilli A, Perego UA, Lancioni H, Olivieri A, Gandini F, Hooshiar Kashani B, et al. Reconciling migration models to the Americas with the variation of North American native mitogenomes. Proc Natl Acad Sci USA 2013;110(35):14308–13.
[111] Raghavan M, DeGiorgio M, Albrechtsen A, Moltke I, Skoglund P, Korneliussen TS, et al. The genetic prehistory of the New World Arctic. Science 2014;345(6200):1255832.
[112] O'Rourke DH, Raff JA. The human genetic history of the Americas: the final frontier. Curr Biol 2010;20(4):R202–7.
[113] Potter BA, Baichtal JF, Beaudoin AB, Fehren-Schmitz L, Haynes CV, Holliday VT, et al. Current evidence allows multiple models for the peopling of the Americas. Sci Adv 2018;4(8):eaat5473.
[114] Lazaridis I, Patterson N, Mittnik A, Renaud G, Mallick S, Kirsanow K, et al. Ancient human genomes suggest three ancestral populations for present-day Europeans. Nature 2014;513(7518):409–13.
[115] Haak W, Lazaridis I, Patterson N, Rohland N, Mallick S, Llamas B, et al. Massive migration from the steppe was a source for Indo-European languages in Europe. Nature 2015;522(7555):207–11.
[116] Francalacci P, Morelli L, Angius A, Berutti R, Reinier F, Atzeni R, et al. Low-pass DNA sequencing of 1200 Sardinians reconstructs European Y-chromosome phylogeny. Science 2013;341(6145):565–9.
[117] Morelli L, Grosso MG, Vona G, Varesi L, Torroni A, Francalacci P. Frequency distribution of mitochondrial DNA haplogroups in Corsica and Sardinia. Hum Biol 2000;72(4):585–95.

[118] Wallace DC. Genetics: mitochondrial DNA in evolution and disease. Nature 2016;535(7613):498–500.
[119] Wei W, Tuna S, Keogh MJ, Smith KR, Aitman TJ, Beales PL, et al. Germline selection shapes human mitochondrial DNA diversity. Science 2019;364(6442).
[120] van Dijk EL, Jaszczyszyn Y, Naquin D, Thermes C. The third revolution in sequencing technology. Trends Genet 2018;34(9):666–81.
[121] Meyer M, Fu Q, Aximu-Petri A, Glocke I, Nickel B, Arsuaga JL, et al. A mitochondrial genome sequence of a hominin from Sima de los Huesos. Nature 2014;505(7483):403–6.
[122] Meyer M, Arsuaga JL, de Filippo C, Nagel S, Aximu-Petri A, Nickel B, et al. Nuclear DNA sequences from the Middle Pleistocene Sima de los Huesos hominins. Nature 2016;531(7595):504–7.

CHAPTER 6

Human nuclear mitochondrial sequences (NumtS)

Marcella Attimonelli[1] and Francesco Maria Calabrese[1,2]

[1]Department of Biosciences, Biotechnology and Biopharmaceutics, University "Aldo Moro", Bari, Italy, [2]Department of Soil, Plant and Food Science, University of Bari Aldo Moro, Bari, Italy

6.1 NumtS definition and introduction

Browsing nuclear DNA in eukaryotic cells reveals the presence of a great quantity of mitochondrial DNA fragments. After ascertaining the mitochondrial fragments origin, the "NumtS" acronym, which stands for nuclear mitochondrial sequences, was created. To clarify, this acronym is meaningful for mitochondrial DNA fragments that escape the mitochondrion and then integrate into the nuclear DNA. The study and inspection of "alien" contaminants populating the nuclear DNA are not trivial. There are many open questions dealing with NumtS evolution and their insertion. This chapter is hence aimed at describing and clarifying (1) when and how NumtS were discovered (Section 6.2); (2) the technical procedures related to their recognition, also including the detection of personal intraspecies NumtS sets (Section 6.3); and (3) the principal mechanisms implicated in NumtS generation and the possibility to date insertion events (Section 6.4). In addition, the effect of NumtS on intra- and interspecies genomic variability (Section 6.5), their involvement in the diagnosis of mitochondrial diseases (Section 6.6) and, finally, the role they exert in genome plasticity (Section 6.7) are described.

6.2 NumtS discovery

The discovery of NumtS dates back to the 1980s, when the first complete mitochondrial genomes and fragments of nuclear eukaryotic genomes were both sequenced. In 1983, Tsuzuky et al. observed two lambda phage clones containing human DNA fragments similar to the mitochondrial 16S rRNAs gene but flanked by specific nuclear DNA. In 1986, Jacobs and Grimes focused their studies on Sea Urchin nuclear genomic regions that exhibited high

The Human Mitochondrial Genome.
DOI: https://doi.org/10.1016/B978-0-12-819656-4.00006-1

similarity with CO1 and 16S rRNA mitochondrial genes and called them "mitochondrial pseudogenes." These observations irrefutably proved mitochondrial DNA translocation and insertion in the nuclear DNA.

Later, the advent of sequencing technologies, the availability of entire eukaryotic genome sequences for many species and the application of in silico approaches for NumtS detection (Section 6.3) confirmed all the preliminary evidence as described in Section 6.3.1. To date, more than 700 human NumtS have been identified in silico by blasting the reference human mtDNA–rCRS [1] against the reference human nuclear genome [2–5]. Moreover, several studies browsing NumtS in various lineages have been published. Recently, 23 eukaryotic species also covering ancient lineages have been blasted and the NumtS compilations have been produced [6]. Furthermore, a recent paper based on the analysis of 64 different bird species reveals a "burst of homoplasmy-free NumtS," that is, insertion of NumtS not derived from the same ancestor and hence not homologous [7]. However, in line with the topic of this book, we will only describe features and challenges centered on human NumtS. Before going into detail, it is worth mentioning the concept introduced by Parr that it is more appropriate to refer to NumtS as the mitochondrial pseudo-genome [8], since within the human NumtS dataset, NumtS covering the entire human mitochondrial genome have been mapped as shown in Fig. 6.1 [4] and demonstrated in Hazkani-Covo et al. [9].

6.3 NumtS detection

The first compilations of species-specific NumtS were generated by scanning the entire nuclear reference genome (Section 6.3.1) which, as in the case of the human genome, represents a consensus derived from the sequencing of several samples. In this context, one aspect that should be highlighted, even if not strictly connected to human mitochondrial DNA, concerns the fact that in the procedure to detect NumtS on the basis of reference genomes, it could be possible to underestimate the resulting set of NumtS because these may not have been included in the assembly process.

Furthermore, considering the polymorphic nature of NumtS, the compilations produced are not exhaustive to contribute to the recognition of NumtS (1) as population markers; (2) as a cause of artifacts of mtDNA variants; and (3) in the reconstruction of the evolutionary process of NumtS. For this purpose, it is necessary to sequence the entire DNA (nuclear and mitochondrial) of each sample considered in any specific study (Section 6.3.2).

6.3.1 In silico human NumtS detection based on reference genomes

The in silico identification of NumtS based on reference genomes involves the application of BLAST, one of the most exploited and fruitful software packages that offers the highest number of adaptable options [10]. In the similarity search for NumtS detection, the main actors of the local alignment the blast performs, keeping the matches that work the best, are the mitochondrial genome used as the query and the reference nuclear genome used as target. In the process of generation of the human NumtS compilation, the nuclear reference genome sequences represent the consensus derived from the pooling and sequencing of six different samples as described in the first publication of the human genome in February 2001 [11]. The pioneering studies aimed at the detection of human NumtS were based on the blasting of the revised Cambridge Reference Sequence, rCRS (GenBank AC-number J01415.2 [1]) against the hg15, hg16, and hg17 human genome reference release. The definition of a revised human NumtS compilation was later based on the hg18 release, improved in its quality also with respect to previous assembling. Based on these data, the RHNumtS compilation was published in 2008 [2]. The revised compilation published in 2011 better describes the massive colonization within the human nuclear chromosomes of more than 700 hits of mtDNA sequences of different lengths [4]. Going into the practical steps required in one of the applied protocols, the human chromosome sequences in FASTA format were used as input for the tools "makeblastbd," provided within the blast suite, in order to build up a single blast genomic database. The incomparable advantage of having the total human nuclear genome as a unique database relies on the faster detection of tossed out NumtS hits and on the possibility of avoiding multiple runs of the software on each chromosome. The Blast algorithm involves the usage of several parameters that strongly influence the resulting hits, better reported in Blast as HSPs (high scoring pairs), thus influencing the number and the length of detectable NumtS. The Blastn parameters implemented by Simone and collaborators [4] were set as: 2 for match reward, −3 for mismatch penalty, −5 for gap opening, −2 for gap extension, and an e-value, that is, the probability of false results, equal to 0.001. Another no less important parameter is the word size set by default at 11 in the Blastn. The usage of different parameters may result in obtaining different alignment lengths and numbers of HSPs.

Upon collecting the entire set of Blastn human hits, some criteria were used to concatenate NumtS and consequently to ascribe some of them to the same insertion event from the mitochondria. The main criteria were the location on the same strand (orientation) and the distance between two blast hits, which was fixed at 2 kb. In order to trace NumtS hotspot region distribution and relate their presence to other genomic elements, the annotation in hg18 and hg19 human released onto the UCSC human genome browser (https://genome.ucsc.edu/index.html) was a milestone.

Nuclear and mitochondrial genomic NumtS contexts can be compared using the "ad hoc" publicly available "NumtS" and "NumtS on mitochondrion" UCSC tracks available both within the hg18 and hg19 releases.

These two tracks are interchangeably connected through an external HTML link, which allows the genomic context to be shifted from mtDNA to its counterpart on the nuclear

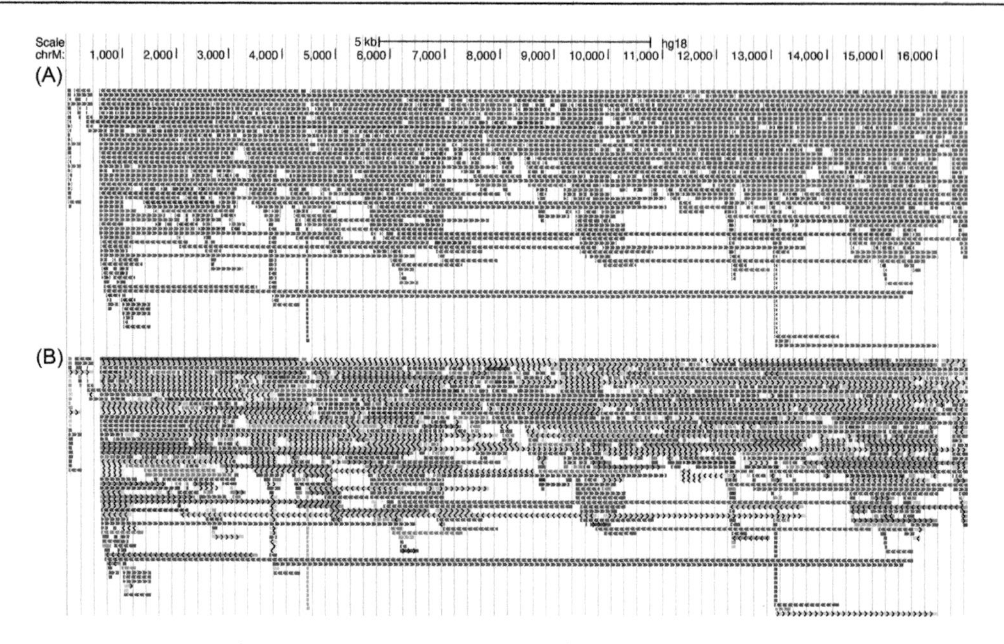

Figure 6.1: UCSC NumtS on mitochondrion track.
(A) NumtS mapped on the mitochondrial DNA region. (b) NumtS mapped on the mitochondrial DNA region and colored based on UCSC chromosome colors.

chromosome. Moreover, another useful NumtS track "NumtS on mitochondrion with chromosome placement" available on the hg18 human release, displays the mitochondrial NumtS content using assigned colors per nuclear chromosome as defined within the UCSC browser (Fig. 6.1). Finally 353 out of the 700 human NumtS sequences of the RHNumtS compilation as published in [4] and in part validated [12] are available on the NCBI Nucleotide database and can be downloaded searching for "NumtS AND Calabrese FM."

6.3.2 Detection of sample-specific NumtS

Once the presence of more than 700 NumtS on the human nuclear genome had been assessed, a new question arose: are all these mapped NumtS the result of a de novo insertion or may they also be derived from recombination events that have caused duplication and/or inversions of genomic regions containing NumtS, thus generating polymorphic and species-specific NumtS? This point implies the setting of methods allowing for the recognition of personal human datasets with the aim of avoiding mtDNA variants artifacts, recognizing NumtS markers of population, and recognizing NumtS useful for phylogenetic analysis.

Indeed, as reported in Ref. [13], the recent advancement of high throughput sequencing has enabled greater exploration of the diversity of polymorphic human NumtS, previously described by Lang et al. [12] and further inspected in Ref. [14]. The next-generation sequencing techniques grant really huge yields and are cheaper than in the past. Nowadays, complete human genome sequencing is thus a realistic effort for many research groups. The technical advance in sequencing required the set up of "ad hoc" designed methods aimed at increasing the accuracy of the detection. The abundance of Whole Genome Sequencing (WGS) data and the usage of new methods not based on comparison against a reference genome make it possible to find out polymorphic NumtS. Thanks to these technological improvements, in addition to the previously mapped NumtS annotated through the UCSC genome browser tracks [4], 141 further new sites of NumtS insertion in 20 populations [13] from the 1000 Genomes Project [15] were isolated. Among the 141 new NumtS, Dayama et al. [13] recognized nearly full-length mitochondrial genome insertions. Thus the presence/absence of NumtS in human populations and the comparison with closely related primate genomes ensured the identification of a subset of NumtS that can be advisedly considered as human specific and may vary between human populations. In the light of the assessed polymorphic status of NumtS, the availability of rapid and efficient protocols able to identify personal human NumtS became an important step. In this regard, the MToolBox package described elsewhere in this book [16] greatly contributes to reconstructing human mitochondrial DNA, avoiding the risk of variant artifacts and at the same time producing a sample-specific NumtS dataset.

6.3.3 In vitro NumtS identification

In order to validate human in silico predicted NumtS, PCR amplification and sequencing can be easily performed, just keeping in mind some tips on the design of primers in order to avoid mtDNA coamplification [2,4]. In more detail, to ensure amplification even in the case of an absence of NumtS, the oligos must amplify a certain margin of flanking regions too. This approach has been applied to a subcategory of NumtS polymorphic in humans [12]. Specifically, by applying a comparative genomic in silico approach, which relies on checking the NumtS presence in the phylogenetic closest species to humans (i.e., chimpanzee, orangutan, macaque, and marmoset), a set of 53 human-specific NumtS were identified and annotated in the NCBI nucleotide database searching for "NumtS" and "human-specific." The in silico predictions were confirmed by PCR amplification and sequencing of the same NumtS in chimpanzee. However, considering that above all the risk of attributing false variants of mtDNA due to coamplification of NumtS is high, an accurate screening of the primer datasets is required. A great improvement in the design of primers to discriminate NumtS from mtDNA sequences is reported in Ref. [17], where mitochondrial locations of all subsequences that are common or similar (one mismatch allowed) between nuclear DNA and mitochondrial DNA have been recognized.

Further approaches aimed at recognizing personal NumtS may be based on hybridization approaches that avoid the most expensive and time-consuming approach based on WGS. An attempt in this direction could be performed by trying sequence hybridization between the nuclear and the mitochondrial genome through the application of the FISH (fluorescence in situ hybridization) technique [18]. Recently, the implementation of the FIBER FISH technique was used to detect NumtS [19].

As reported by Koo et al. [19], "FIBER-FISH is a technique that permits high resolution mapping of genes and chromosomal regions on DNA fibers, allowing physical location of DNA probes down to a resolution of 1000 base pairs."

6.4 Numtogenesis: Mechanisms of NumtS insertion

The most established and validated theories concerning creation of NumtS report that they are inserted into the nuclear genome via two different mechanisms: (1) the nonhomologous end-joining (NHEJ) machinery [20,21] during repair of DNA double-strand breaks (DSBs) or (2) through homologous recombination. The two different mechanisms may be associated with different times and modes as described below. NumtS insertions arise due to the action of endogenous and exogenous agents such as ionizing radiation and aging (the most accredited cause) and are strictly dependent on the rate of DSBs in nuclear DNA [22]. As reported by Jensen-Seaman et al. [23] in a comparative study of NumtS inserted after the divergence between human and chimpanzee, most of the studied integrations were accompanied by micro homology and short indels of the kind typically observed in the nonhomologous end-joining pathway of DNA DSB repair (Fig. 6.2).

To support this feature, a model has been proposed by Hazkany-Covo and Covo [24] according to which NumtS can seal complex DSBs, thus reducing the risk of deleterious repair.

This concept invokes the creation of new genomic elements through genomic tinkering, that is, random combinations of preexisting elements [25]. In other words, the pool of mitochondrial fragments that escapes the mitochondria is randomly used for nuclear needs such as DSB repair but also to contribute to the settings of new functions, when and if the genome requires them. According to these mechanisms, the generation of NumtS can be ascribed to two different evolutionary time lapses, the first of which refers to the batch of mitochondrial fragment migrations that occurred contextually with the symbiont nuclear genome size reduction. The second time lapse accounted for all the consecutive speciation events occurring after the endosymbiosis (Fig. 6.3).

Through speciation, the fixed NumtS core (the one relative to the last common ancestor) underwent events of duplication, loss, and mutation. Moreover, new events of migration from mitochondrial to nuclear DNA occurred and still occur [24,26] with major evidence in

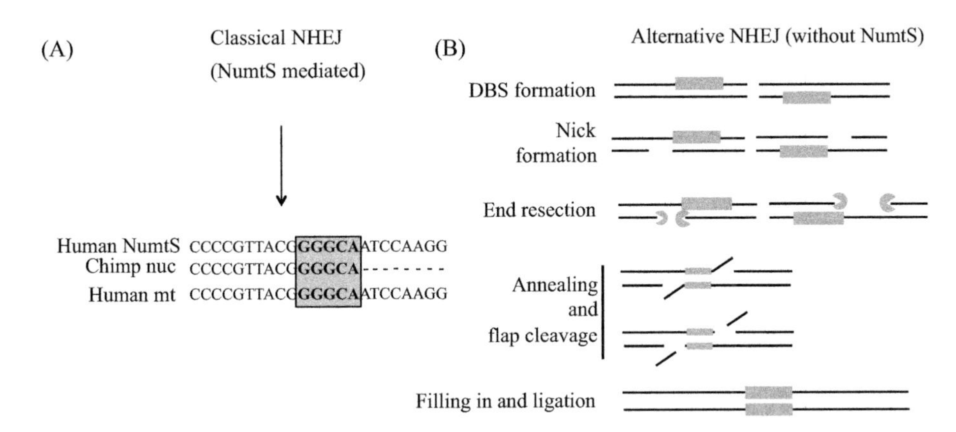

Figure 6.2: NHEJ mechanisms without or with NumtS.
Double-strand breaks (DSBRs) can be repaired following the accredited mechanisms of nonhomologous end-joining recombination (NHEJ) with (A) or without (B) the usage of NumtS as fillers. In both cases, microhomology-mediated end-joining (MMEJ) machinery repairs DNA breaks by using microhomology and results in deletions. When NumtS are used to seal DSBRs (NumtS mediated NHEJ) the risk of large deletions is reduced. The multialignment reported in part A is relative to one human-specific NumtS and allows the presence of one microhomology region used in the annealing step (when the mtDNA is used as filler) to be highlighted. Specifically, the chimpanzee (Chimp) nuclear genome does not contain the NumtS but a short microhomology region can be detected, which proves to be shared between the human mitochondrial region and its nuclear counterpart. Source: *Adapted from Hazkani-Covo E, Covo S. Numt-mediated double-strand break repair mitigates deletions during primate genome evolution. PLoS Genet 2008;4:e1000237. https://doi.org/10.1371/journal.pgen.1000237.*

pathogenetic processes defined by Singh et al. [27] as "numtogenesis." Hence, generation of NumtS sets can be hypothetically split into two subparts. The first part gathers all the insertions that can be ascribed to the endosymbiotic event. The second refers to all the mitochondrial fragments that reached the nucleus after endosymbiosis, were inserted into ancestral species, and then fixed there, following the dynamics of the genomic context according to the selective pressures acting differently on nuclear chromosomes. Hence, after a speciation event, the set of species-specific NumtS may be modified: some NumtS may be lost or may be subjected to accumulation of mutations based on the genomic context in which they are inserted, whereas others may be duplicated.

6.5 NumtS variability and polymorphisms

NumtS play an important role in genomic variability and random genetic drift may be considered as the driving force for NumtS acquisition. However, although the debate about their origin is still open, thanks to their intraspecies variability, in terms of sequence, homo/heterozygote status, and their presence/absence at a specific locus, NumtS may be

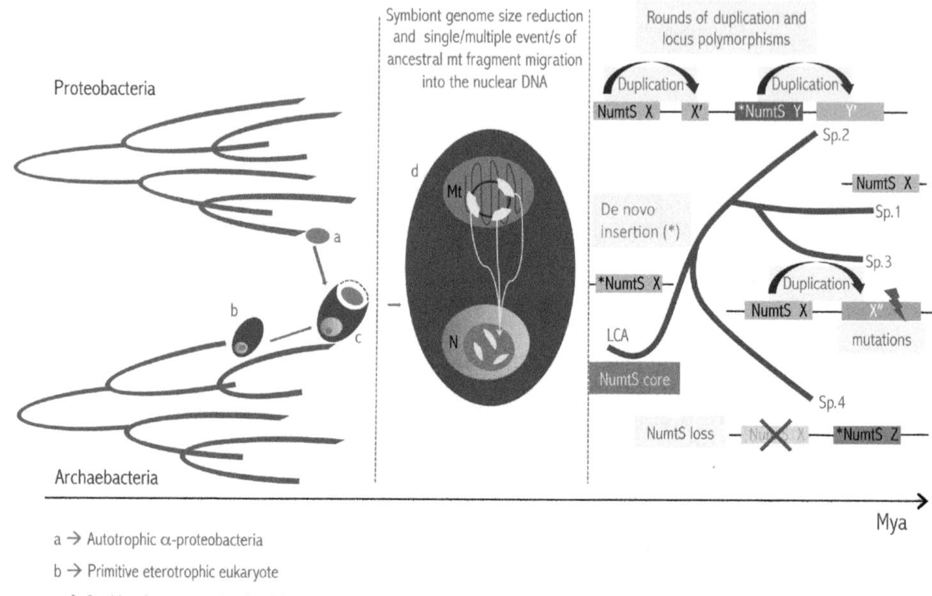

Figure 6.3: NumtS origin and evolution.
The primitive heterotrophic eukaryote (b) engulfed the autotrophic alpha proteobacteria (a) actively via phagocytosis; the symbiont (c) started its transformation to generate the mitochondrion by undergoing genome size reduction during which a first massive quantity of mitochondrial fragments reached the nuclear genome (d) of the heterotrophic eukaryote thus creating a NumtS pool that led to its first nuclear colonization. Through speciation, the fixed NumtS core (the one relative to the last common ancestor, LCA) underwent duplication events, NumtS loss and mutation. Possible events impacting on NumtS locus evolution are described on the right. Specifically, the NumtS X de novo insertion, which happened after the LCA, is maintained within all the subsequent species except for species 4 in which its deletion occurred. All the de novo insertion events are flagged with asterisks. Species 1 maintained the same LCA NumtS set, while in species 2 a de novo insertion plus a NumtS duplication occurred. In species 3 the NumtS X was duplicated and a mutation (*red flash; black flash in print version*) occurred in NumtS X, independently duplicated from the X locus.

considered population markers while the interspecies analysis reveals a conundrum of hypotheses that suggests different times of insertion during evolution as mentioned above [28]. During evolution, NumtS accumulate mutations depending on the genomic context in which they are inserted. The number of mutations results in sequence similarity; the more ancient the NumtS insertion time into the nuclear chromosome is, the lower the similarity to the mitochondrial genome of that species will be and the higher its phylogenetic distance will be. On the contrary, recent NumtS insertions will display a high similarity with their mitochondrial counterpart, thus causing a high risk of artifacts as frequently occurred in disease studies (Section 6.6). The polymorphic character of NumtS related to presence/absence, to homo/heterozygote status or to single nucleotide variants also allows recognition of putative polymorphic NumtS that could be used as population markers and hence adopted in Forensics as reported in Ref. [29], where a novel approach based on phylogenetic principles finalized to parse true variation (i.e., signal) from noise is described in its application to a dataset of 41 human mitochondrial genomes from a population in Rio de Janeiro. In that case, 451 putative NumtS were reconstructed and of these 147 were ancestral already noted as NumtS, while 122 represented different haplotypes for a single nucleotide, and none of them exactly coincided with the mtDNA of the considered subjects. Once again, the NumtS track can be used to understand the number of mismatched NumtS sequences when compared to the mitochondrial region which originated them. On the hg19 human release, by clicking on the NumtS code, the resulting alignment between the NumtS and the reference mtDNA is loaded. This schematic alignment allows inspection of mismatches occurring in both mitochondrial and nuclear sequences (Fig. 6.4).

Alignment of HSA_NumtS_227_b2 to chrM:577-1539:

```
001 CAGTTTATGTAGCTTAATTATTAAAAGCAAGACACTGAAAATGTCTAGAC 050
>>> ||||||||||||||||  |  | ||||||| ||||||||||||| ||||| >>>
577 cagtttatgtagcttacctcctcaaagcaatacactgaaaatgtttagac 626

051 GGACTTA..TTACCCCATAAACAGATAGGTTTGGTTCTGGCCTTTCTGTT 098
>>> || || |  | |||||||||||| ||||||||||| || |||||||| || >>>
627 gggctcacatcaccccataaacaaataggtttggtcctagcctttctatt 676

099 AACTCTTAGTAAGATTACACATGCAAGTATCACCATCCTAGTGAAAATAC 148
>>> | ||||||||||||||||||||||||| ||| || | | |||||    || >>>
677 agctcttagtaagattacacatgcaagcatccccgttccagtgagttcac 726

149 CCTCTAAATCATTATGATCAAAAGGAGTAAGAATCAAGCACAGACAAATG 198
>>> |||||||||||  | ||||||||||   ||| |||||||||  |  |||| >>>
727 cctctaaatcaccacgatcaaaagggacaagcatcaagcacgcagcaatg 776

199 CAGCTCAAAACACTTTGCTCGGCCACACCCCCACAGGAAGCAGCAGTGAT 248
>>> ||||||||||| ||| ||   ||||||||||||| |||| |||||||||| >>>
777 cagctcaaaacgcttagcctagccacacccccacgggaaacagcagtgat 826
```

Figure 6.4: Sequence difference between NumtS and its mtDNA counterpart.
Alignment between the mtDNA region (from 577 to 1539) and the corresponding NumtS obtained produced by browsing UCSC NumtS tracks.

6.6 The role of NumtS in mtDNA sequencing and disease

The presence of pathogenic variants, both heteroplasmic and homoplasmic, the increase in the number of mtDNA copies, extensive mtDNA depletions, and mutations of nuclear genes involved in mitochondrial processes are all causes and effects of degenerative processes with mitochondrial involvement. A search in PubMed of reviews associated with the terms "mitochondria and disease" produces more than sixty thousand articles, a result that highlights the strong role of mitochondria not only in classic rare mitochondrial diseases but also in chronic diseases as well as a great deal in the processes of tumor progression. With the improvement in sequencing technologies and the very accurate ability to read the mitochondrial DNA of even a single cell, the accuracy of the diagnosis through the recognition of causes and effects described above could certainly be optimized, but the presence of NumtS represents a considerable risk that must not be neglected so as to avoid false results. Many artifacts of mitochondrial pathogenic variants were reported in the literature at a time when NumtS were either unknown or ignored [30]. The knowledge of NumtS and the availability of the human NumtS compilations through UCSC surely may decrease the risk of artifacts by guiding the design of primers toward regions with the lowest similarity between mtDNA and nuclear DNA. However, considering that the entire mitochondrial DNA is present as NumtS on the nuclear genome and that the number of copies of the same fragment is reported many times although with different similarity (Fig. 6.1), the risk of coamplification survives also when the increase in the number of copies of mtDNA, due for example, to cancerous processes, could lead us to suppose that the estimated heteroplasmic fraction is with high probability exclusively ascribed to mtDNA. Last but not least, the risk of false pathogenicity assignments to human mtDNA variants is great when the samples considered include personal NumtS. This scenario could be surely explored through the phylogenetic approach mentioned in Section 6.5 [29] and again, based on the availability of NGS readings, it is possible to discriminate readings derived from mtDNA and those derived from NumtS thanks to the different rates of variation between nuclear and mtDNA [31]. More details regarding the wet and dry approaches contributing to reduce the risk for variant artifacts are detailed elsewhere in this book.

6.7 NumtS annotation: Current and future roles of NumtS

The human genome browser UCSC allows the inspection of genomic regions of interest for NumtS content and offers the possibility to visualize all the collected information for other genomic elements located in the same spot. As an example of related data onto the browser, the phylogenetic conservation of NumtS can be studied at a glance by using an interspecies approach based on UCSC tracks. By just connecting the NumtS tracks and the Chain/Net UCSC tracks annotating data regarding syntenic regions, users can visually evaluate species-specific or interspecies shared NumtS. Moreover, the crossing between the NumtS tracks and GenCode track [32] allows us to test and eventually recognize functions that NumtS have assumed over time as well as alert researcher about any potential functional alteration due to the presence of NumtS in intronic or, worse still, exonic regions. The most assessed data regards the functional NumtS named "humanin" which, as reported by Bodzioch et al. [33], are "neuroprotective and anti-apoptotic peptides derived from a portion of the mitochondrial MT-RNR2 gene for which bioinformatics and expression data suggest the existence of 13 MT-RNR2-like nuclear loci predicted to maintain the open reading frames of 15 distinct full-length HN-like peptides. At least ten of these nuclear genes are expressed in human tissues, and respond to staurosporine (STS) and beta-carotene." Moreover, the merging between the NumtS tracks and the microRNA track highlights examples of regions containing expressed NumtS, while the merger with the GenBank mRNA tracks reports several cases of patented NumtS in relationship with microRNA expression. Furthermore, the hypothesis of microRNA expressed by mtDNA has recently been reported by Pozzi and Dowling [34], but no wet proofs are available in this regard. Lastly, evidence of an entire mitochondrion inside a nucleus in cancerous samples is reported by Singh et al. [27].

Research perspectives

This chapter deals with some of the topics related to nuclear fragments of mitochondrial origin. The inspection of these fragments within the human genome and the broadening to other species phylogenetically closer to *homo sapiens* allows the in-depth characterization of the NumtS scenario.

The reader who wishes to pursue a research involving one of the treated aspects can find here the basic notions related to artifacts, due to NumtS/mtDNA coamplification, which are one of the main causes preventing easy personal NumtS set detection. Therefore future efforts should drive the ongoing design and implementation of more accurate dry and wet methods contributing to recognize new NumtS.

References

[1] Andrews RM, Kubacka I, Chinnery PF, Lightowlers RN, Turnbull DM, Howell N. Reanalysis and revision of the Cambridge reference sequence for human mitochondrial DNA. Nat Genet 1999;23:147. Available from: https://doi.org/10.1038/13779.

[2] Lascaro D, Castellana S, Gasparre G, Romeo G, Saccone C, Attimonelli M. The RHNumtS compilation: features and bioinformatics approaches to locate and quantify human NumtS. BMC Genomics 2008;9:267. Available from: https://doi.org/10.1186/1471-2164-9-267.

[3] Ramos A, Barbena E, Mateiu L, del Mar González M, Mairal Q, Lima M, et al. Nuclear insertions of mitochondrial origin: database updating and usefulness in cancer studies. Mitochondrion 2011;11:946–53. Available from: https://doi.org/10.1016/j.mito.2011.08.009.

[4] Simone D, Calabrese FM, Lang M, Gasparre G, Attimonelli M. The reference human nuclear mitochondrial sequences compilation validated and implemented on the UCSC genome browser. BMC Genomics 2011;12:517. Available from: https://doi.org/10.1186/1471-2164-12-517.

[5] Tsuji J, Frith MC, Tomii K, Horton P. Mammalian NUMT insertion is non-random. Nucleic Acids Res 2012;40:9073–88. Available from: https://doi.org/10.1093/nar/gks424.

[6] Calabrese FM, Balacco DL, Preste R, Diroma MA, Forino R, Ventura M, et al. NumtS colonization in mammalian genomes. Sci Rep 2017;7:16357. Available from: https://doi.org/10.1038/s41598-017-16750-2.

[7] Liang B, Wang N, Li N, Kimball RT, Braun EL. Comparative genomics reveals a burst of homoplasy-free numt insertions. Mol Biol Evol 2018;35:2060–4. Available from: https://doi.org/10.1093/molbev/msy112.

[8] Parr RL, Maki J, Reguly B, Dakubo GD, Aguirre A, Wittock R, et al. The pseudo-mitochondrial genome influences mistakes in heteroplasmy interpretation. BMC Genomics 2006;7:185. Available from: https://doi.org/10.1186/1471-2164-7-185.

[9] Hazkani-Covo E, Sorek R, Graur D. Evolutionary dynamics of large numts in the human genome: rarity of independent insertions and abundance of post-insertion duplications. J Mol Evol 2003;56:169–74. Available from: https://doi.org/10.1007/s00239-002-2390-5.

[10] Camacho C, Coulouris G, Avagyan V, Ma N, Papadopoulos J, Bealer K, et al. BLAST+: architecture and applications. BMC Bioinforma 2009;10:421. Available from: https://doi.org/10.1186/1471-2105-10-421.

[11] Lander ES, Linton LM, Birren B, et al. International Human Genome Sequencing Consortium. Initial sequencing and analysis of the human gemome. Nature 2001;409:860–921. Available from: https://doi.org/10.1038/35057062.

[12] Lang M, Sazzini M, Calabrese FM, Simone D, Boattini A, Romeo G, et al. Polymorphic NumtS trace human population relationships. Hum Genet 2012;131:757–71. Available from: https://doi.org/10.1007/s00439-011-1125-3.

[13] Dayama G, Emery SB, Kidd JM, Mills RE. The genomic landscape of polymorphic human nuclear mitochondrial insertions. Nucleic Acids Res 2014;42:12640–9. Available from: https://doi.org/10.1093/nar/gku1038.

[14] Hazkani-Covo E, Martin WF. Quantifying the number of independent organelle DNA insertions in genome evolution and human health. Genome Biol Evol 2017;9:1190–203. Available from: https://doi.org/10.1093/gbe/evx078.

[15] 1000 Genomes Project Consortium, Auton A, Brooks LD, Durbin RM, Garrison EP, Kang HM, et al. A global reference for human genetic variation. Nature 2015;526:68–74. Available from: https://doi.org/10.1038/nature15393.

[16] Calabrese C, Simone D, Diroma MA, Santorsola M, Guttà C, Gasparre G, et al. MToolBox: a highly automated pipeline for heteroplasmy annotation and prioritization analysis of human mitochondrial variants in high-throughput sequencing. Bioinforma Oxf Engl 2014;30:3115–17. Available from: https://doi.org/10.1093/bioinformatics/btu483.

[17] Albayrak L, Khanipov K, Pimenova M, Golovko G, Rojas M, Pavlidis I, et al. The ability of human nuclear DNA to cause false positive low-abundance heteroplasmy calls varies across the mitochondrial genome. BMC Genomics 2016;17:1017. Available from: https://doi.org/10.1186/s12864-016-3375-x.

[18] Pinkel D, Landegent J, Collins C, Fuscoe J, Segraves R, Lucas J, et al. Fluorescence in situ hybridization with human chromosome-specific libraries: detection of trisomy 21 and translocations of chromosome 4. Proc Natl Acad Sci U S A 1988;85:9138–42. Available from: https://doi.org/10.1073/pnas.85.23.9138.

[19] Koo D-H, Singh B, Jiang J, Friebe B, Gill BS, Chastain PD, et al. Single molecule mtDNA fiber FISH for analyzing numtogenesis. Anal Biochem 2018;552:45–9. Available from: https://doi.org/10.1016/j.ab.2017.03.015.

[20] Ricchetti M, Tekaia F, Dujon B. Continued colonization of the human genome by mitochondrial DNA. PLoS Biol 2004;2:E273. Available from: https://doi.org/10.1371/journal.pbio.0020273.

[21] Blanchard JL, Schmidt GW. Mitochondrial DNA migration events in yeast and humans: integration by a common end-joining mechanism and alternative perspectives on nucleotide substitution patterns. Mol Biol Evol 1996;13:537–48. Available from: https://doi.org/10.1093/oxfordjournals.molbev.a025614.
[22] Gaziev AI, Shaĭkhaev GO. Nuclear mitochondrial pseudogenes. Mol Biol (Mosk) 2010;44:405–17.
[23] Jensen-Seaman MI, Wildschutte JH, Soto-Calderón ID, Anthony NM. A comparative approach shows differences in patterns of numt insertion during hominoid evolution. J Mol Evol 2009;68:688–99. Available from: https://doi.org/10.1007/s00239-009-9243-4.
[24] Hazkani-Covo E, Covo S. Numt-mediated double-strand break repair mitigates deletions during primate genome evolution. PLoS Genet 2008;4:e1000237. Available from: https://doi.org/10.1371/journal.pgen.1000237.
[25] Jacob F. Evolution and tinkering. Science 1977;196:1161–6. Available from: https://doi.org/10.1126/science.860134.
[26] Schiavo G, Hoffmann OI, Ribani A, Utzeri VJ, Ghionda MC, Bertolini F, et al. A genomic landscape of mitochondrial DNA insertions in the pig nuclear genome provides evolutionary signatures of interspecies admixture. DNA Res Int J Rapid Publ Rep Genes Genomes 2017;24:487–98. Available from: https://doi.org/10.1093/dnares/dsx019.
[27] Singh KK, Choudhury AR, Tiwari HK. Numtogenesis as a mechanism for development of cancer. Semin Cancer Biol 2017;47:101–9. Available from: https://doi.org/10.1016/j.semcancer.2017.05.003.
[28] Gherman A, Chen PE, Teslovich TM, Stankiewicz P, Withers M, Kashuk CS, et al. Population bottlenecks as a potential major shaping force of human genome architecture. PLoS Genet 2007;3:e119. Available from: https://doi.org/10.1371/journal.pgen.0030119.
[29] Smart U, Budowle B, Ambers A, SoaresMoura-Neto R, Silva R, Woerner AE. A novel phylogenetic approach for de novo discovery of putative nuclear mitochondrial (pNumt) haplotypes. Forensic Sci Int Genet 2019;43:102146. Available from: https://doi.org/10.1016/j.fsigen.2019.102146.
[30] Wallace DC, Stugard C, Murdock D, Schurr T, Brown MD. Ancient mtDNA sequences in the human nuclear genome: a potential source of errors in identifying pathogenic mutations. Proc Natl Acad Sci U S A 1997;94:14900–5. Available from: https://doi.org/10.1073/pnas.94.26.14900.
[31] Petruzzella V, Carrozzo R, Calabrese C, Dell'Aglio R, Trentadue R, Piredda R, et al. Deep sequencing unearths nuclear mitochondrial sequences under Leber's hereditary optic neuropathy-associated false heteroplasmic mitochondrial DNA variants. Hum Mol Genet 2012;21:3753–64. Available from: https://doi.org/10.1093/hmg/dds182.
[32] Harrow J, Frankish A, Gonzalez JM, Tapanari E, Diekhans M, Kokocinski F, et al. GENCODE: the reference human genome annotation for The ENCODE Project. Genome Res 2012;22:1760–74. Available from: https://doi.org/10.1101/gr.135350.111.
[33] Bodzioch M, Lapicka-Bodzioch K, Zapala B, Kamysz W, Kiec-Wilk B, Dembinska-Kiec A. Evidence for potential functionality of nuclearly encoded human in isoforms. Genomics 2009;94(4):247–56. Available from: https://doi.org/10.1016/j.ygeno.2009.05.006.
[34] Pozzi A, Dowling DK. The genomic origins of small mitochondrial RNAs: are they transcribed by the mitochondrial DNA or by mitochondrial pseudogenes within the nucleus (NUMTs)? Genome Biol Evol 2019;11:1883–96. Available from: https://doi.org/10.1093/gbe/evz132.

CHAPTER 7

mtDNA exploitation in forensics

Adriano Tagliabracci and Chiara Turchi

Section of Legal Medicine, Department of Excellence of Biomedical Sciences and Public Health, Polytechnic University of Marche, Ancona, Italy

7.1 Introduction

Forensic genetic investigations result in obtaining a genetic profile aimed at individual identification. It is widely established by the forensic scientific community that genetic profiles suitable for forensic identification purposes are only those obtained from a specific set of autosomal short tandem repeat (STR) systems. However, dealing with degraded DNA samples or ancient specimens, the conventional forensic STR markers fail to yield successful profiles. Mitochondrial DNA (mtDNA) analysis can provide useful results in forensic samples with damaged DNA. The typical forensic specimens typed with mtDNA analysis are bone remains, teeth, hair, and hair shafts.

The main reason, which allows recovery of mtDNA from degraded samples, lies in the fact that in a human cell, mtDNA is present in a much higher copy number than nuclear DNA. Because there are hundreds to thousands of copies of the mitochondrial genome in each cell, in samples containing very low DNA, ancient remains or degraded specimens, mtDNA molecules are more likely to survive than autosomal DNA. In addition, the mitochondrial molecule is more protected from enzymatic degradation with respect to genomic DNA, as the circular shape makes it less susceptible to exonucleases activity. Therefore the probability of obtaining an mtDNA profile is higher than the probability of obtaining a useful profile with nuclear STR typing systems.

The human mitochondrial genome is maternally inherited, as mitochondria are passed from mothers to their offspring without any male influence. For this reason, mtDNA is also defined as "uniparentally inherited" in the same way as the Y chromosome for the paternal counterpart. Since maternal mtDNA does not come into contact with a paternal mitochondrial genome, it is not subject to the phenomenon of genetic recombination. Excluding the occurrence of mutational events during DNA transmission, this peculiar feature allows the mitochondrial molecules to be passed without modification to the mother's children.

The Human Mitochondrial Genome.
DOI: https://doi.org/10.1016/B978-0-12-819656-4.00007-3

Therefore maternal relatives share an identical mtDNA genome, or in other words, maternal relatives belong to the same mitochondrial lineage. As a consequence, one mtDNA genome is shared by numerous people that are close maternal relatives or that share the same maternal ancestor in the past (distant maternal relatives). For these reasons, mtDNA sequence polymorphisms cannot be used for the identification of individuals in the same way of nuclear DNA markers. Although autosomal DNA markers are much more informative and valuable for identification purposes, an mtDNA result is better than no result at all.

The main exploitation of mtDNA in forensic casework is related to the reconstruction of genetic relationships in case of missing person's identification or mass disaster investigations. The main use of mtDNA in these applications is the fact that all related maternal relatives (also distant relatives) may be used for comparison and to confirm the identity of a missing person.

7.2 mtDNA typing in historical forensic identification

mtDNA typing has been used in several interesting forensic identifications starting from the early 1990s. The first relevant report of the forensic application of mtDNA polymorphism was published in 1991 by Stoneking and collaborators [1]. The report described a method for detecting sequence variation of hypervariable segments of the mtDNA control region, using hybridization of sequence-specific oligonucleotide (SSO) probes to DNA target regions that have been amplified with PCR. Moreover, in order to demonstrate the usefulness of mtDNA typing for forensic identification, the authors described the results of a case analysis, related to the identification of bone remains found in 1986. The mtDNA sequence of the remains was compared to those of the parents of a 3-years-old child, who disappeared from her home two years earlier. Bone remains and the mother's sample shared the same mtDNA type, corroborating the hypothesis that the skeletal remains belonged to the missing child.

In a different scenario, mtDNA analyses were used in the identification of war victims, for the identification of remains of the Vietnam War [2], the Second World War [3–5], and the Korean war [6]. mtDNA analysis was also performed to identify victims of the "eccidio delle fosse Ardeatine," which occurred near Rome in 1944 during the Second World War, when 335 Italians were massacred by the occupying forces of Nazi Germany [7].

mtDNA analysis has been proven to be a powerful tool in several interesting historical identifications in the past years. One of the most famous and extremely intriguing cases is the identification of the remains of Romanov family, the last Russian Royal family (Fig. 7.1). In July 1918, Tsar Nicholas II, his wife Tsarina Alexandra, and their five children were executed in Ekaterinburg. In the late 1970s, a local geologist found the mass grave containing the remains of nine human bodies, supposed to belong to five of the seven members of the

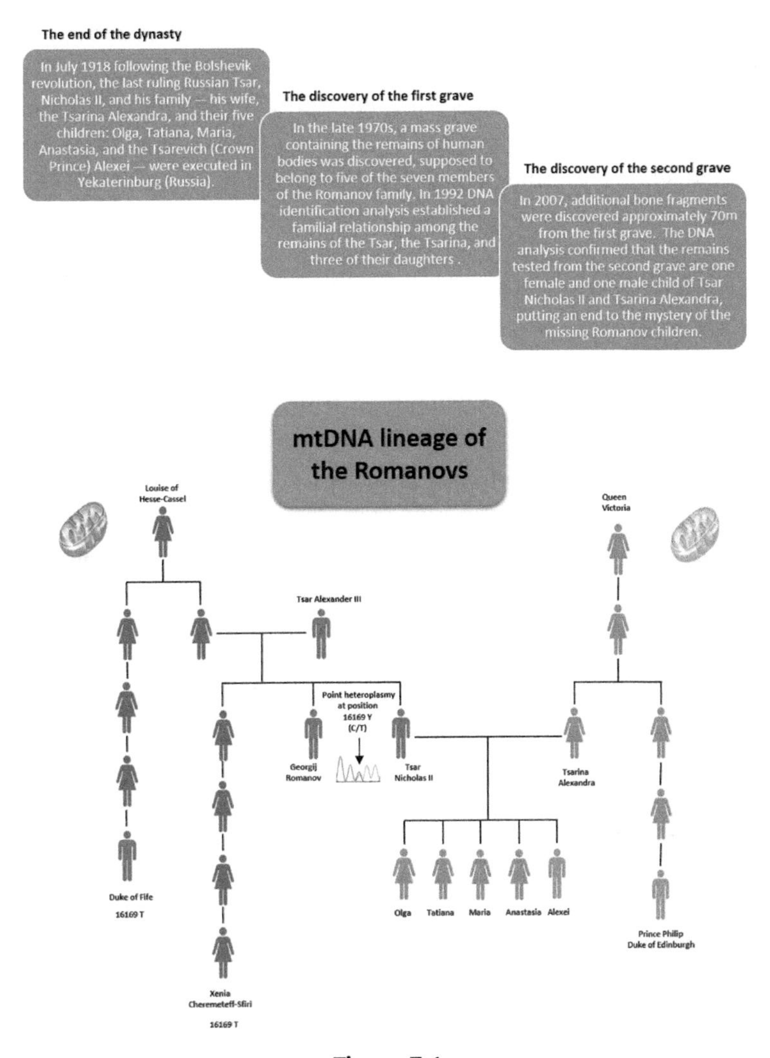

Figure 7.1

mtDNA lineage of Tsar Nicholas II: the Tsar's haplotype was compared with two maternal relatives, the Duke of Fife and Xenia Cheremeteff-Sfiri. The mtDNA types were identical, except for a single-point heteroplasmy at position 16169 (Y = C/T) observed in the Tsar, and absent in his maternal relatives which were 16169T. Subsequent analysis was performed to compare the mtDNA haplotype from the remains of Grand Duke Georgij Romanov, brother of Tsar Nicholas II. Both samples shared the same heteroplasmic status at 16169, confirming the authenticity of the heteroplasmy and the maternal relationship. mtDNA lineage of Tsarina Alexandra: the comparison of the Tsarina's haplotype to her distant cousin, HRH Prince Philip, was an exact match, confirming the maternal relationship. Moreover the mtDNA tests confirmed the maternal relationship with her daughters and son recovered from the grave.

Romanov family and four members of their staff. In 1991, bone remains were disinterred from the mass grave and in 1992, DNA identification analysis was performed by Peter Gill and Pavel Ivanov [8]. The autosomal STR profiles were used to sort, sex and show familial relationships among the Romanov family members, while mtDNA testing was used to link the tsar and tsarina to living maternal relatives. The mtDNA testing and comparison with maternal-related descendants of Tsarina Alexandra allowed the identification of the Tsarina and confirmed the maternal relationship among the Tsarina and three of her daughters recovered from the grave. Another comparison was made to confirm the identity of the Tsar, by comparing the mtDNA sequences of bone remains with two maternal relatives of Nicholas II. The mtDNA types of maternal relatives matched mtDNA type of remains of the Tsar, except for a single-point heteroplasmy (heteroplasmy condition is the presence of more than one mtDNA type in an individual) at position 16169 (C/T) observed in the mtDNA sequence of the Tsar, and absent in his maternal relatives. Subsequent analysis was performed by comparing the mtDNA haplotype from the remains of Grand Duke Georgij Romanov, brother of Tsar Nicholas II [9]. Both samples shared the same heteroplasmy at 16169, although in differing proportions, as the Tsar was mostly C while his brother was mostly T. Despite the forensic genetic evidence, the two children missing from the mass grave still raised doubts about the authenticity of the remains. In 2007, additional bone fragments were discovered approximately 70 m from the first grave and two independent DNA testing of the remains from the second grave were performed in the same year [10]. The DNA analysis confirmed that the remains tested from the second grave were one female and one male child of Tsar Nicholas II and Tsarina Alexandra, putting an end to the mystery of the missing Romanov children. The relevance of this forensic case lies not only in the historical reconstruction of the end of the Romanov dynasty, but also because it raised the issue of heteroplasmy in mtDNA forensic analysis. This issue will be discussed in more detail later in the chapter.

Other examples of forensic historical identifications with mtDNA typing concern the molecular genetic investigations on Austria's patron saint Leopold III [11], the mtDNA analysis on remains of a putative son [12] and daughter [13] of Louis XVI, King of France and Marie-Antoinette, and the identification of the remains believed to belong to King Richard III [14].

A more recent interesting identification case involving mtDNA analysis is related to the fascinating "mystery of Nanga Parbat" [15]. Günther Messner, brother of the world-famous mountaineer Reinhold Messner, did not return from an expedition to the world's ninth highest mountain Nanga Parbat (8125 m) in which he participated with his brother Reinhold in the summer of 1970 and the circumstances of his disappearance have given rise to controversy.

Reinhold claimed that he and Günther climbed the summit via the difficult east-oriented Rupal face and then they descended through the easier, west-oriented Diamir face together, when an avalanche swept away his brother. Other members of expedition doubted this version and accused Reinhold of having abandoned his weakened brother during the difficult climb. Reinhold Messner refuted these claims and returned to Nanga Parbat several times to search for the remains of his brother. In 2000, climbers discovered a human fibula at the Diamir face and, in order to evaluate potential kinship with the Messner family, the bone was subjected to forensic DNA analyses. Five years later, other remains were found nearby the earlier location, and a proximal phalanx protected in a mountain boot was subjected to forensic DNA typing. Autosomal STR marker profiles obtained from the fibula and from the two living brothers, Reinhold and Hubert Messner, indicated strong evidence for a kinship relationship. mtDNA sequencing analysis of the control region showed that the fibula and the two reference samples shared the same mtDNA haplotype (A16233G C16256T T16311C A16343G C16355TA73G C150TA263G 309.1C 315.1C A523del C524del). This haplotype, belonging to haplogroup U3, had not been observed before in more than 30,000 records, thus further corroborating the probability of maternal kinship. In addition, the proximal phalanx analyzed in 2005 strongly confirmed the earlier findings on the fibula and brotherhood was further supported by the analysis of 17 Y-STR loci, which resulted in a full match between the three samples.

In conclusion, the forensic DNA profiling provided no reasonable doubt that the analyzed bone remains indeed belonged to the brother of Reinhold Messner. With the discovery of Günther Messner's remains on the Diamir face of Nanga Parbat, one of Himalayan mountaineering's greatest controversies appears to be resolved.

7.3 mtDNA sequencing in forensic practice

The mtDNA sequencing analysis in forensic casework involves different steps that primarily consist of mtDNA extraction, PCR amplification of target regions, sequencing reactions of both DNA strands, comparison to reference sequences, reviewing, and editing of sequenced data.

The high copy number per cell makes mtDNA testing extremely sensitive to contamination. Isolation of the mitochondrial genome from forensic specimens needs to be performed following measures that mitigate risk of contamination. Best practice to reduce or minimize contamination often used by forensic laboratories performing mtDNA testing includes appropriate laboratory conditions as well as dedicated spaces with physically separate pre- and postamplification areas, dedicated instruments, equipment, and reagents. The use of protective lab wear, cleaning procedures with bleach and UV irradiation of lab bench surfaces should also be established. The use of controls as reagent blanks, negative controls in PCR and sequencing steps, as well as positive controls should be also carried out to monitor levels of exogenous DNA in reagents, in the laboratory environment or on instruments. Moreover, it is recommended to process the reference samples after the evidence samples have been completely analyzed.

7.3.1 Extraction

Forensic mtDNA analysis is typically performed on specimens where poor DNA is present and/or where the DNA templates are often damaged. Typical materials often used for mtDNA analysis in forensic casework consist of not only bones, especially ribs, the femur, and humerus, but also teeth and hair. Concerning bone remains, the mtDNA can be extracted using different methods [16,17] and then purified to remove PCR inhibitors that could be coextracted. A demineralization extraction protocol has improved success rates with mtDNA analysis [18]. It has been observed that the best sequencing results are obtained from ribs and femurs.

Hair shafts do not contain sufficient nuclear DNA for STR typing; therefore the unique genetic material that can be investigated in this kind of forensic sample is mtDNA. All the protocols provide that the outer surface of 1–2 cm of the hair shaft is carefully cleaned, before starting extraction. The extraction methods are based on methods that break down the keratin structure of the hair. A tissue grinder can be used to break down the structure of the hair and to release the mtDNA molecules [19]. Other methods rely on the hair digestion protocol to release nuclear DNA and mtDNA for analysis. Compared with head, pubic, and axillary, the highest success rate was obtained with head hair shafts. It was observed that the addition of BSA (bovine serum albumin) helps in reducing the PCR inhibitory effects of melanin.

7.3.2 mtDNA quantification by real-time PCR

Measuring the quantity of DNA isolated from a forensic sample is an important step to choose the optimal conditions to perform subsequent PCR and sequencing assays. Moreover, with particular relevance to highly compromised specimens, a PCR may fail not only due to the insufficient DNA quantity, but also due to the presence of coextracted inhibitors and highly degraded DNA. The method of choice for DNA quantification in forensics is the real-time PCR assay or qPCR due to its sensitivity and the low quantity of input DNA required, but especially because it can accurately reflect both the quality and the quantity of the DNA template. Several mtDNA quantification methods by qPCR were published, concerning the quantification of mtDNA alone or the simultaneous quantification of nuclear DNA (nDNA) and mtDNA [20–23], but only one assay gives information about degradation of mtDNA [24]. Recently, a new assay that combines one nDNA target and two different-sized mtDNA targets to provide information on the degradation state of the quantified mtDNA was published [25]. Moreover, this assay includes an internal positive control (IPC) to monitor potential PCR inhibition. Briefly, the developed real-time PCR assay included four different Taqman probes capable of detecting and quantifying one nDNA target (70 bp), two different-sized mtDNA targets (143 and 69 bp), and IPC (70 bp). This last probe is complementary to an artificial oligonucleotide that is always included in the reaction mix and therefore always detected by the assay, unless inhibitors were present in the sample that would limit PCR efficiency or even prevent PCR amplification both of the IPC and the genomic target. The two different-sized mtDNA targets were used for DNA degradation estimation, as with increasing degradation, longer amplicon targets tend to decrease disproportionately relative to shorter amplicon targets. As a result, the ratio of quantification results between the small target and large target may indicate potential DNA degradation of samples.

7.3.3 Targeted region and PCR amplification

The mitochondrial genome is featured by polymorphic sites spread throughout the molecule, but the majority of quickly evolving sites relevant for mitochondrial types (or haplotypes) discrimination in forensics are found within the control region (CR), or displacement loop (D-loop). Conventionally, CR is about 1100 nucleotides long and it spans between the 16024 nucleotide position and the 576 position. Several protocols have been developed to amplify by PCR the hypervariable regions within the D-loop known as hypervariable region I (HV1 or HVS-I, 16024–16365), hypervariable region II (HV2 or HVS-II, 73–340), and hypervariable region III (HV3 or HVS-III, 340–576). The first two regions, HVS-I and HVS-II, are the most frequently analyzed in forensic casework.

The FBI laboratory for mtDNA sequencing commonly uses the PCR primers reported in Ref. [19]. The nomenclature of primers is based on the strand corresponding to the primers (L for light and H for heavy) and the 3′ nucleotide position. Thus the primer named as L15997 corresponds to the light strand of the rCRS and ends at position 15997. The Armed Forces DNA Identification Laboratory (AFDIL) uses different primers and a different nomenclature that permits an easier determination of the overall PCR product size: strand designation in this case is by forward (F) and reverse (R) and a 5′ nucleotide position is noted. It is worth noting that the two nomenclature system could result in different primer names even though their nucleotide sequences are identical.

All these protocols used independent amplification and different combinations of primers, but despite the simplicity, the restricted approach proposed suffers from a number of implications, especially in population studies when several samples were analyzed at the same time. The independent amplification of hypervariable regions, combined with the manual processing of multiple samples at one time, has been shown to increase the risk of artificial recombination, that generates the so-called chimeric haplotypes or artificial recombinants that are nonexisting mtDNA types caused by the inadvertent mix-up of mtDNA regions from different individuals [26,27].

To avoid the risk of artificial recombination, laboratory protocols have been developed to amplify the entire CR in a single amplicon [27,28] in samples used for population databases. In forensic casework analysis, the occurrence of artificial recombination is restricted, as usually only one sample is processed at a time, and when possible the test is replicated. In any case, because the sequence information from the entire CR generally increases, the discriminatory power of mtDNA testing, the recovery of the entire CR sequencing data, instead of only HVS-I and HVS-II regions, is always desirable also in forensic casework.

However, forensic specimens submitted to mtDNA typing often contain highly degraded DNA, where the molecules are fragmented to small sizes. In these kinds of samples, the amplification of the entire CR in a single amplicon (about 1100 bp) is very difficult, if not impossible. As a result, protocols based on the amplification of small overlapping fragments spanning the entire control region have been developed to improve the recovery of more information from highly degraded DNA [29–31].

7.3.4 mtDNA sequencing

The gold standard technology widely used for mtDNA sequencing in forensic genetics is the Sanger method combined with capillary electrophoresis separation.

Laboratory protocols for mtDNA sequencing usually include the following steps: (1) PCR amplification of the entire control region in a single amplicon or in shorter overlapping fragments; (2) purification of PCR products in order to remove remaining dNTPs and primers using filter devices or through enzymatic digestion with Exo-Sap (exonuclease I and shrimp alkaline phosphatase); (3) determination of PCR product quantity; (4) DNA sequencing reactions of both forward and reverse strands; (5) removal of unincorporated fluorescent dye terminators and salts from the completed sequencing reaction usually through specific solutions that eliminate unincorporated dye terminators and free salts from the postsequencing reaction or spin column filtration; (6) separation of sequenced fragment through a capillary electrophoresis instrument; and (7) data analysis, alignment, and interpretation.

mtDNA sequencing is usually performed in both the forward and reverse directions to ensure at least double coverage of every nucleotide in order to allow more accurate base calling and perform a quality control check. In some circumstances, as around the C-stretch, where it is not possible to get readable sequences from both strands, the interpretable strand can be sequenced twice in separate reactions.

7.3.5 Rapid screening assay for mtDNA type

The mtDNA control region is characterized by polymorphic sites spread throughout its length; however, there are hotspots or hypervariable sites where most of the variation is clustered. Because sequence information of the entire CR requires effort both in terms of time and work, the availability of screening approaches and rapid low-resolution typing assays alternative to CR Sanger sequencing may be used to analyze samples that can be easily excluded from each other. Several assays for rapidly screening mtDNA polymorphisms at hypervariable hotspots have been developed and even validated for use in screening forensic casework. These methods are based on SSO probes [1], mini-sequencing [32], denaturing gradient gel electrophoresis [33], restriction digest assay [34], and a reverse dot blot or linear array assay approach [35].

7.3.6 Massive parallel sequencing of full mitochondrial genome

In the last years, it has been largely demonstrated that massive parallel sequencing (MPS or next-generation sequencing, NGS) technologies offer new possibilities for forensic genetics, both for the increased information that may be obtained in a single experiment from a unique sample by analyzing combinations of markers and for the analytical cost-effectiveness. The full mitochondrial genome was one of the first forensic genetic markers evaluated with MPS [36–39] and it has been demonstrated that MPS provides higher throughput and sensitivity than Sanger sequencing. Massively parallel sequencing of entire mtDNA increases the discrimination power in samples sharing a common mtDNA control region haplotype, as well as providing a high phylogenetic resolution, with respect to sequence information recovered from the solely CR. Moreover, it allows for a more comprehensive heteroplasmy detection, as minor alleles can be detected at lower levels [40,41]. New developments in MPS technologies have also demonstrated that a full mtDNA sequence can be obtained even from degraded forensic samples [38,42,43].

The studies performed until now are concordant about the great effectiveness of the mtDNA sequencing using MPS in a forensic context, even though more comprehensive studies to provide data to support validation protocols and to assess heteroplasmy in relation to the average read depth per nucleotide position, should be performed before this new technology should become a routine application in forensic casework.

7.4 Data analysis, alignment, and haplotype notation

7.4.1 Alignment

The mtDNA sequence for each sample is aligned and compared to the revised version of the first human mtDNA sequence (rCRS) [44,45].

Sequence data analysis is aided by computer programs, such as SeqScape (ThermoFisher) or Sequencer (GeneCodes) that perform a sequence editing process by aligning the multiple sequencing electropherograms, in both forward and reverse directions, generated over a region of the same sample. Then, the same software makes a comparison relative to the rCRS, highlighting the nucleotide positions in which discrepancies between sample and reference sequences are observed. Nevertheless, despite the improvement of sequence chemistries and instruments that lead to more even peaks, better sensitivity, and less noise, no software can still reliably evaluate mtDNA sequences. As a result, the process of mtDNA data editing has not yet been completely automated and the manual intervention of an expert analyst is still required, in order to review and potentially edit the base calls for each nucleotide. In this perspective, it is strongly recommended that two forensic analysts must independently examine, interpret, and edit sequence matching results as a final quality assurance measure.

7.4.2 Notation for forensics purposes

When the alignment and editing processes are completed, the mtDNA haplotype for each sample is reported by noting only the differences relative to the rCRS. The base on the cytosine-rich light strand (L-strand) is always quoted. Moreover, the interpretation range (excluding primer sequence information) must be reported to allow unbiased sample comparison or database searches. Differences relative to the rCRS are noted with nucleotide position and the altered base [46–50]. For example, a C to T transition observed at nucleotide position 16069 should be noted as C16069T, where C is the base present in the rCRS and T is the base observed in the questioned sample. It should be noted that also the notation 16069T, without the C preceding the nucleotide position, is widely accepted. In this format, all other nucleotides are assumed to be identical to the rCRS. Base deletions should be indicated by "DEL," "del," or "-" rather than "D" or "d," as the latter could be misleading with the International Union of Pure and Applied Chemistry (IUPAC) code, where "D" codes for a mixture of G, A, and T. Insertions should be noted with the nucleotide position immediately 5′ to the insertion followed by the designation ".1" (for the first base insertion), a ".2" if there is a second insertion, and so on, and by the nucleotide that is inserted. For example, the notation 44.1C means that a C insertion has occurred between base positions 44 and 45. In cases of insertions at homopolymeric tracts (i.e., C-stretches), the exact location of the insertion is indeterminable. The standard generally adopted is to assume that the insertion has occurred at the end of the strand and therefore the last C at 3′ is considered as the reference for annotation of insertion. One of the C-stretches most affected by insertion events consists of five Cs in position 311 through 315 (referring to rCRS) and if a C insertion has occurred inside it, it should be designated as 315.1C. Heteroplasmic positions should be reported using the IUPAC nomenclature system (i.e., a mixture of C and T should be noted as "Y," while a mixture of A and G should be noted as "R"). If all four bases are present at a single position or if no base call can be made at a given position, the site is represented by an "N." A more exhaustive explanation of mtDNA haplotype notation is reported in Ref. [49].

The system of reporting mtDNA sequences relative to a reference could result in different notations, if multiple alignments of the same sequencing product are feasible. Therefore conventions are required to avoid that identical haplotypes are reported in different ways and standardization in designating of mtDNA sequences is desirable to generate comparable data that can be shared between laboratories. Different mtDNA nomenclature approaches were proposed, which employed either a set of formal rules based on maximum parsimony relative to the rCRS [51,52] or a phylogenetic approach that bases the alignment on the established mutation patterns of the mitochondrial phylogeny [53].

Although no nomenclature system can easily address the full complexity of mtDNA diversity, the phylogenetic approach seems to be preferred, as it provides a biological basis for the representation. A tool to assist with the notation of mtDNA sequences is available at http://empop.org/, that has been adopted by the Scientific Working Group on DNA Methods (SWGDAM) in the United States (2019) and the International Society for Forensic Genetics (ISFG) [49].

It is desirable that an alternative approach that relies on string searches that utilize the full mtDNA sequence is developed, in order to avoid ambiguity and potential mismatches that can occur when reporting haplotypes as differences relative to a reference sequence using either hierarchical rules or a phylogenetic approach.

7.4.3 Heteroplasmy

The issue of heteroplasmy is very complex and it has multiple implications in the medical field for mtDNA disorders, but in this context, we have focused only on the forensic genetic aspect of heteroplasmic conditions.

A heteroplasmy condition can be observed at different levels: multiple mtDNA populations may occur within single mitochondrion, single cell or between cells, and it is believed that all individuals present some level of heteroplasmy, even if below the limits of detection of DNA sequence analysis [54–56]. Different tissues display different rate of heteroplasmy, and it was observed that it occurs more often in tissue with a high metabolic activity [57] and when mtDNA molecules are subjected to narrow bottleneck during development such as hairs. It should be also considered that individuals may have multiple mtDNA molecules in a single tissue, or they may exhibit one mtDNA type in one tissue and a different type in another tissue, or may be heteroplasmic in one tissue sample and homoplasmic in another tissue sample [58].

In forensics, we usually distinguish between two types of heteroplasmy: point (or sequence) heteroplasmy (PHP) and length heteroplasmy (LHP), which differ in electrophoretic pattern, frequency, and cause.

Sequence or point heteroplasmy is typically featured by the presence of two nucleotides at a single site, which shows up as overlapping peaks in a sequence electropherogram. The observation of PHP at single nucleotide position across the mtDNA control region is quite frequent, while it is rare to find two or three heteroplasmic sites in a single sample [59]. Therefore the presence of more than two PHP positions may raise suspicion about possible contamination of the sample. PHP ratio is affected by the dye terminator sequencing chemistry, depending on the nucleotide position and primer used. Therefore it is impossible to determine numerical threshold values for the detection of PHP and it should be evaluated and called considering the overall sequence quality and the corroboration of different sequencing strands in both forward and reverse directions. To better describe the nucleotide composition of the mixture, PHP should be noted with the appropriate IUPAC codes. The occurrence of PHP varies in relation to different tissue types. One of the major challenges of PHP is that the ratio of peak bases may not be the same across different tissues, such as blood and hair or between multiple hairs. Nucleotide sites most affected by PHP have been identified in mtDNA control region [56,59,60] and very often these affected positions coincided with evolutionary hot-spot mutations.

LHP often occurs beyond the homopolymeric tracts, commonly referred to as "C-stretches," when more than eight identical nucleotides compound the stretch. LHP shows up as a series of overlapping peaks until the end of the sequencing electropherogram. As a result, the sequenced region downstream the LHP position is generally uninterpretable.

Different regions through the control region are affected by LHP. In HV1 between positions 16184 and 16193, a T to C transition at position 16189 leads to a stretch of more than 10 Cs, resulting in a downstream extensive LHP (Fig. 7.2). In HV2 between positions 302 and 310, both single and multiple C insertions or a T to C transition at position 310 produce an uninterrupted long C-stretches, with a resulting PHP. Finally, in HV3, LHP is observed when a C insertion occurs between positions 568 and 573.

Similarly to PHP, detection of LHP depends on the sequencing chemistry and technology used and its occurrence varies within and between tissues of an individual. As a result, LHP variations are disregarded in direct forensic comparisons and database searches. On the other hand, for population genetic databasing useful in forensics, it is recommended to report the dominant LHP variant (major molecule) in the data [49,61].

Once acknowledged that heteroplasmy happens, careful attention should be paid during raw data analysis, haplotype reporting, and interpretation (i.e., handling differences between known and questioned forensic samples). The most recent ISFG guidelines for mtDNA typing [49] suggest that forensic laboratories "must establish their own interpretation and reporting guidelines for observed length and point heteroplasmy." Heteroplasmy evaluation

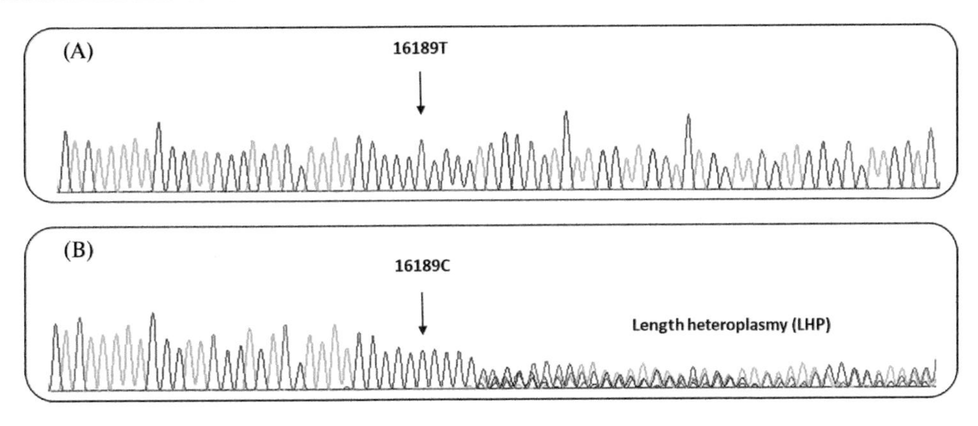

Figure 7.2
Different sequence electropherograms in samples without (A) and with (B) length heteroplasmic (LHP) condition. In (B) a T to C transition at position 16189 leads to a stretch of more than 10 Cs. The occurrence of two or more C-stretch length variants (i.e., LHP) creates a situation where the sequence products are out of phase with one another, resulting in poor quality and an uninterpretable sequence downstream the string of cytosines.

depends on the limitations of the technology and the quality of the sequencing reactions as well as the "experience of the laboratory." Moreover, the same guidelines point out that "differences in both PHP and LHP do not constitute evidence for excluding two otherwise identical haplotypes as deriving from the same source or same maternal linage." Although occurrence of heteroplasmy can sometimes complicate the interpretation of mtDNA results, the presence of the same sequence heteroplasmy can improve the strength of the evidence, such as seen in the Romanov family identification case.

7.5 Interpretation of mtDNA results

7.5.1 Sequence comparison

One of the main purposes of a forensic genetic analysis involving the mtDNA typing is to determine if an individual can or cannot be excluded as a possible source of the mtDNA evidence in casework. Generally, after completion of mtDNA sequencing analysis and the achievement of mtDNA haplotypes from a questioned specimen, comparisons are made between the questioned and reference samples. Following the comparison, mtDNA sequence results can be grouped into three categories: exclusion, inconclusive, or failure to exclude. The SWGDAM guidelines for mtDNA interpretation [46] make the following recommendations for comparisons of control region data:

- *Exclusion*: if samples differ at two or more nucleotide positions (excluding LHP), they can be excluded as coming from the same source or maternal lineage.
- *Inconclusive*: the comparison should be reported as inconclusive if samples differ at a single position only (whether or not they share a common length variant between positions 302–310).
- *Cannot exclude*: if samples have the same sequence, or are concordant (sharing a common DNA base at every nucleotide position), they cannot be excluded as coming from the same source or maternal lineage.

Because mutation events have been observed between mother and children, a single nucleotide difference between two sequences cannot be used to support an interpretation of exclusion and therefore the comparison result should be reported as "inconclusive." In forensic casework, if a maternal relative is used as a reference sample, the possibility of a single nucleotide difference may exist between two samples that are in fact maternally related.

A common base is defined as a shared base in the case of ambiguity (e.g., heteroplasmy) in the sequence [47]. If one haplotype displays point heteroplasmy at a site and another does not, they cannot be excluded as coming from the same source or maternal lineage, as heteroplasmy occurrence may vary within and between tissues of an individual or between samples belonging to the same maternal lineage. For the same reason, a LHP alone, especially in the C-stretch between positions 302 and 310, cannot be used to support an interpretation of exclusion (Table 7.1). As already mentioned (Section 7.4.3), laboratories should develop guidelines for the evaluation of forensic cases that involve heteroplasmy.

7.5.2 Statistical evaluation: weight of evidence

When the mtDNA profile of a questioned evidence sample and a reference sample cannot be excluded as coming from the same source or maternal lineage, a statistical estimate of significance of a match should be assessed in order to provide a statistical weight to support the conclusion drawn. This is achieved by estimating the frequency of the mtDNA type in an appropriate and relevant population database.

The maternal inheritance without recombination means that the nucleotide variations that characterize an mtDNA haplotypes must be treated as a single locus. The common practice to determine the rarity of an mtDNA type among individuals and consists in counting the number of times a particular haplotype is observed in a database [62], a method referred to as the "counting method." As a result, the choice of population database is an extremely important factor and it should be: representative of the appropriate population, be of an

Table 7.1: Examples of mtDNA sequence comparisons between a questioned (Q) and known (K) samples in forensic casework.

Sequence results	Observation	Interpretation
Q: TTACTGCCAGCCACCATGAATATTGTACGGTACCATAA K: TTACTGCCAGCCACCATGAATATTGTACGGTACCATAA	Sequences are fully concordant with common bases at every position	Cannot exclude
Q: GAATATTGTACAGTACCATAAATACTTGACTACCTGTAGT K: GAATATTGTACGGTACCATAAATACTTGACCACCTGTAGT	Sequences differ at two positions	Exclusion
Q: AATTAATGCTTGTAGGACATARTAATAACAATTGAATGTC K: AATTAATGCTTGTAGGACATAATAATAACAATTGAATGTC	Point heteroplasmy at position in one sample that is not present in the other; common base at every position (A in both samples)	Cannot exclude
Q: GTCTTTGATTCCTGCCTCATCCCATTATTTATCGCACCTAC K: GTCTTTGATTCCTGCCTCATCCTATTATTTATCGCACCTAC	Sequences identical at every position except one; no indication of heteroplasmy	Inconclusive
Q: AAATTTCCACCAAACCCCCCCCTCCCCCCGCTTCTGGCCACA K: AAATTTCCACCAAACCCCCCCCTCCCCCCGCTTCTGGCCACA	Length heteroplasmy in one sample that is not present in the other; common base at every position	Cannot exclude

(Continued)

Table 7.1: (Continued)

Sequence results	Observation	Interpretation
AATTAATGCTTGTAGGACATARTAATAACAATTGAATGTC Q AATTAATGCTTGTAGGACATARTAATAACAATTGAATGTC K	Point heteroplasmy at the same position in both samples; common base at every position	Cannot exclude

Source: (Adapted from J. M. Butler (2011), Advanced Topics in Forensic DNA Typing: Methodology (Ch.14), Academic Press (Ed.) Wyman Street, Waltham, MA 02451, USA [48])

adequate size and employ quality measures to assess the data entered into the database. In cases where the choice of database is questionable, it can be appropriate to report the haplotype frequencies in different databases.

In cases where the database size is small, and it does not represent all the potential contributors of the mtDNA type and the statistics for a random match, the frequency estimate could be misrepresented. The application of a confidence interval accounts for database size and sampling variations. Confidence intervals can be used to estimate the upper and lower bounds of a frequency calculation [50,63] and can offer assurance that all data obtained should include the true value of the parameter the proportion of time set by the confidence level.

In cases where an mtDNA haplotype is observed, a particular number of times (X) in a population database containing N profiles, its frequency (p) can be calculated as follows:

$$p = \frac{X}{N}$$

Alternatively, point estimation for probabilistic approaches is calculated as follows [64,65], where the uncertainty is due to sampling errors, it is approximately achieved by adding the case profiles to the database:

$$p = \frac{(X+1)}{(N+1)}$$

or

$$p = \frac{(X+2)}{(N+2)}$$

When there are one or more samples in the database, a 95% upper bound confidence interval can be placed on the profile's frequency using:

$$p + 1.96\sqrt{\frac{p(1-p)}{N}}$$

An alternative approach, the Clopper–Pearson method [66], may also be used to provide a more conservative estimate for the upper 95% confidence interval, when very low counts are observed from a haplotype database.

In cases where the profile has not been observed in a database, the 95% upper bound on the confidence interval is

$$1 - \alpha^{1/N}$$

where $\alpha = 0.05$ is the confidence coefficient and N is the number of individuals in the database. The rarity of an mtDNA profile is calculated as follows:

$$p = 1 - (0.05)^{1/N}$$

This confidence interval has been widely used, but it is known to be problematic in situations with small sample sizes or very few observations. An interesting comparison of methods for assessing the significance of mtDNA matching, with different accepted approaches, is reported in Ref. [49].

Likelihood ratios (LRs) for mtDNA are estimated by 1/(match probability). In forensic cases involving both mtDNA typing and autosomal STRs, it may be useful to combine the LR results. Indeed, LRs may be combined by multiplication, provided that the databases for each typed system adequately represent the same population and that the stated hypotheses are the same, and it is taken into consideration that maternal relatives cannot be distinguished by mtDNA.

7.6 Mitochondrial DNA population databases used in forensics

Population databases are an extremely important factor in estimating the expected frequency of mtDNA haplotypes in order to assess a statistical estimation of the significance of a match. The use of high-quality mtDNA sequence databases is strongly recommended, in order to make a reliable estimate of the frequency for a random match.

A European DNA Profiling Group mitochondrial DNA population database project (EMPOP) has collected thousands of mtDNA sequences and constructed a high-quality mtDNA database that can be accessed at http://www.emop.org. The EMPOP was developed by the Institute of Legal Medicine (GMI), Medical University of Innsbruck and the Institute of Mathematics,

University of Innsbruck. The EMPOP database aims to collect, high quality control and searchable presentation of mtDNA haplotypes from all over the world and one of its most important features is that the available primary sequence lane data are permanently linked to the database entries. The current version of the database is a result of international teamwork and continues to evolve by fruitful collaboration [28,67–71]. Moreover, EMPOP uses SAM, a string-based search algorithm that converts query and database sequences to position-free nucleotide strings and thus eliminates the possibility that identical sequences will be missed in a database query. SAM 2, an updated and optimized version of the software, is currently used to assist statistical evaluation of the evidence in forensic practice [72].

It has been observed that several errors have been detected in a number of mtDNA population genetic studies [73]. These errors can be segregated into different types such as mistakes in the course of transcription of the results and clerical errors, sample mix-up (e.g., artificial recombination), and contamination and use of different nomenclatures. The use of best laboratory practice (e.g., independent double evaluation of the data, electronic data reporting, sequencing both forward and reverse strands and in overlapping segments) can reduce the occurrence of these errors. Moreover, additional quality control measures should be applied. Quality control of population data can be assessed by using a statistical analysis clustering approach or phylogenetic analysis that offers a posteriori tool for detecting errors. Using phylogenetic analysis, multiple and closely related DNA sequences can be compared systematically in order to identify samples that are extremely different. Unusual differences may be an indication that the sample was contaminated or the sequence data were incorrectly recorded. The "Network" tool based on quasimedian network analysis has been developed [74,75] and it is available on the EMPOP website to examine the quality of an mtDNA dataset.

An additional tool for the evaluation of mtDNA datasets is haplogroup assignment, which provides useful information about the composition of the population studied and about the distribution of individual haplotypes on a global scale. In the last years, haplogrouping has been simplified by the establishment of Phylotree (www.phylotree.org), a comprehensive phylogenetic tree of worldwide human mtDNA variations. The EMMA tool for haplogroup assignment was developed and is available on the EMPOP website. This tool uses both Phylotree and a selection of 20,000 haplotypes to assign haplogroups.

Another mtDNA variation database is the FBI mtDNA Population Database, also known as CODIS^{mt}. It has a forensic component that contains almost 5000 mtDNA HV1 and HV2 profiles from 14 populations, while about 6000 additional published profiles were taken from literature data.

mtDNAmanager is an mtDNA population database developed by a research group from Yonsei University, Seoul [76]. Since August 2019, this database contains 9294 mtDNA control region sequences and they are grouped in the following five subsets: African (1496),

West Eurasian (3673), East Asian (2326), Oceanian (114), and Admixed (1685). All these mtDNA control region sequences are shown with the estimated haplogroup affiliations (both expected and estimated haplogroups) using the bioinformatics resources of the mtDNAmanager (http://mtmanager.yonsei.ac.kr/).

7.7 Guidelines and recommendations

Recommendations have been issued on the use of mtDNA typing in forensic cases and address the need of appropriate laboratory practice, alignment and nomenclature guidelines for sequencing variations and heteroplasmy notation, quality control issues, as well as guidance on interpretation, reporting and statistics both in mtDNA casework and mtDNA reference population databasing applications.

The DNA Commission of the International Society for Forensic Genetics (ISFG) has issued two guidelines concerning the use of mtDNA typing in forensics [49,58] and a list of the recommendations published in 2014 is reported in Table 7.2.

The Scientific Working Group on DNA Analysis Methods, better known by its acronym SWGDAM, has issued the SWGDAM Interpretation Guidelines for Mitochondrial DNA Analysis by Forensic DNA Testing Laboratories approved on April 23, 2019, in order to update the previous version (2013) with next-generation sequencing technology topics [50].

Table 7.2: List of ISFG recommendations published in 2014 on mitochondrial DNA typing.

DNA commission of the International Society for Forensic Genetics: revised and extended guidelines for mitochondrial DNA typing [49].	
Generation of mtDNA data, good laboratory practice	
Recommendation #1	Good laboratory practice and specific protocols for work with mtDNA must be followed in accordance with previous guidelines [58].
Recommendation #2	Negative and positive controls as well as extraction reagent blanks must be carried through the entire laboratory process.
Recommendation #3	Reported consensus sequences must be based on redundant sequence information, using forward and reverse sequencing reactions whenever practical.
Recommendation #4	Manual transcription of data should be avoided and independent confirmation of consensus haplotypes by two scientists must be performed.
Recommendation #5	Laboratories using mtDNA typing in forensic casework shall participate regularly in suitable proficiency testing programs.
Recommendation #6	In population genetic studies for forensic databasing purposes, the entire mitochondrial DNA control region should be sequenced.

(*Continued*)

Table 7.2: (Continued)

DNA commission of the International Society for Forensic Genetics: revised and extended guidelines for mitochondrial DNA typing [49].	
Data analysis, alignment, and interpretation	
Recommendation #7	mtDNA sequences should be aligned and reported relative to the revised Cambridge reference sequence (rCRS, NC001807), and should include the interpretation range (excluding primer sequence information).
Recommendation #8	IUPAC conventions using capital letters shall be used to describe differences to the rCRS and (point heteroplasmic) mixtures. Lower case letters should be used to indicate mixtures between deleted and nondeleted (inserted and noninserted) bases. N-designations should only be used when all four bases are observed at a single position (or if no base call can be made at a given position). For the representation of deletions, "DEL," "del," or "-" shall be used.
Recommendation #9	The alignment and notation of mtDNA sequences should be performed in agreement with the mitochondrial phylogeny (established patterns of mutations). Tools to assist with the notation of mtDNA sequences are available at http://empop.org/.
Recommendation #10	In forensic casework, laboratories must establish their own interpretation and reporting guidelines for observed length and point heteroplasmy. The evaluation of heteroplasmy depends on the limitations of the technology and the quality of the sequencing reactions as well as the experience of the laboratory. Differences in both PHP and LHP do not constitute evidence for excluding two otherwise identical haplotypes as deriving from the same source or same maternal lineage.
Recommendation #11	For population database samples, length heteroplasmy in homopolymeric sequence stretches should be interpreted by calling the dominant variant, which can be determined by identifying the position with the highest representation of a nonrepetitive peak downstream of the affected stretch.
Quality control of population data	
Recommendation #12	mtDNA population data should be subjected to analytical software tools that facilitate phylogenetic checks for data quality control. A comprehensive suite of QC tools is provided by EMPOP.
Databases and database searches	
Recommendation #13	The entire database of available sequences should be searched with respect to the sequencing (interpretation) range to avoid biased query results.
Recommendation #14	Laboratories must be able to justify the choice of database(s) and statistical approach used in reporting.
Recommendation #15	Laboratories must establish statistical guidelines for use in reporting an mtDNA match between two samples.
Recommendation #16	Highly variable positions such as length variants in homopolymeric stretches should be disregarded from searches for determining frequency estimates. Heteroplasmic calls should be queried in a manner that does not exclude any of the heteroplasmic variants.

Research perspectives

New technologies are regularly introduced and validated in forensic genetic laboratories. As a result, forensic DNA protocols can be expected to become more rapid, sensitive, and informative and to provide stronger conclusions, mostly with regard to challenging samples. This framework also includes the future of forensic mtDNA, which will mainly concern the strengthening of the full mitochondrial genome by NGS technologies and the collection of larger population databases to improve haplotype frequency estimates. Complete mitogenome sequencing produces the highest resolution possible from forensically relevant samples, including degraded DNA and it improves detection of heteroplasmy, which is desirable in the forensic context. Nevertheless, NGS technologies have not yet become routine applications for forensic casework analysis. The move toward NGS-mtDNA typing in casework application will come with the harmonization of testing protocols within the community, as well as with an improvement of reliable software solutions. Challenging points will concern degraded samples, amplification of numt sequences, point and LHP interpretation. These issues have been encountered previously and should be addressed in future studies aiming at a better understanding. Moreover, an improvement in knowledge of the variability between software, laboratories, and instruments may be helpful to issue recommendations and guidelines for the interpretation of mtDNA sequence data generated by NGS in the forensic genetic context.

References

[1] Stoneking M, Hedgecock D, Higuchi RG, Vigilant L, Erlich HA. Population variation of human mtDNA control region sequences detected by enzymatic amplification and sequence-specific oligonucleotide probes. Am J Hum Genet 1991;48(2):370–82.

[2] Holland MM, Fisher DL, Mitchell LG, Rodriquez WC, Canik JJ, Merril CR, et al. Mitochondrial DNA sequence analysis of human skeletal remains: identification of remains from the Vietnam War. J Forensic Sci 1993;38(3):542–53.

[3] Dudas E, Susa E, Pamjav H, Szabolcsi Z. Identification of World War II bone remains found in Ukraine using classical anthropological and mitochondrial DNA results. Int J Leg Med 2019;134(2):487–9.

[4] Ossowski A, Diepenbroek M, Kupiec T, Bykowska-Witowska M, Zielinska G, Dembinska T, et al. Genetic identification of communist crimes' victims (1944–1956) based on the analysis of one of many mass graves discovered on the Powazki Military Cemetery in Warsaw, Poland. J Forensic Sci 2016;61(6):1450–5.

[5] Palo JU, Hedman M, Soderholm N, Sajantila A. Repatriation and identification of the Finnish World War II soldiers. Croat Med J 2007;48(4):528–35.

[6] Lee HY, Kim NY, Park MJ, Sim JE, Yang WI, Shin KJ. DNA typing for the identification of old skeletal remains from Korean War victims. J Forensic Sci 2010;55(6):1422–9.

[7] Pilli E, Boccone S, Agostino A, Virgili A, D'Errico G, Lari M, et al. From unknown to known: identification of the remains at the mausoleum of fosse Ardeatine. Sci Justice 2018;58(6):469–78.

[8] Gill P, Ivanov PL, Kimpton C, Piercy R, Benson N, Tully G, et al. Identification of the remains of the Romanov family by DNA analysis. Nat Genet 1994;6(2):130–5.

[9] Ivanov PL, Wadhams MJ, Roby RK, Holland MM, Weedn VW, Parsons TJ. Mitochondrial DNA sequence heteroplasmy in the Grand Duke of Russia Georgij Romanov establishes the authenticity of the remains of Tsar Nicholas II. Nat Genet 1996;12(4):417–20.

[10] Coble MD, Loreille OM, Wadhams MJ, Edson SM, Maynard K, Meyer CE, et al. Mystery solved: the identification of the two missing Romanov children using DNA analysis. PLoS One 2009;4(3):e4838.

[11] Bauer CM, Bodner M, Niederstatter H, Niederwieser D, Huber G, Hatzer-Grubwieser P, et al. Molecular genetic investigations on Austria's patron saint Leopold III. Forensic Sci Int Genet 2013;7(2):313–15.

[12] Jehaes E, Decorte R, Peneau A, Petrie JH, Boiry PA, Gilissen A, et al. Mitochondrial DNA analysis on remains of a putative son of Louis XVI, King of France and Marie-Antoinette. Eur J Hum Genet 1998;6 (4):383–95.

[13] Parson W, Berger C, Sanger T, Lutz-Bonengel S. Molecular genetic analysis on the remains of the Dark Countess: revisiting the French Royal family. Forensic Sci Int Genet 2015;19:252–4.

[14] King TE, Fortes GG, Balaresque P, Thomas MG, Balding D, Maisano Delser P, et al. Identification of the remains of King Richard III. Nat Commun 2014;5:5631.

[15] Parson W, Brandstatter A, Niederstatter H, Grubwieser P, Scheithauer R. Unravelling the mystery of Nanga Parbat. Int J Leg Med 2007;121(4):309–10.

[16] Hochmeister MN, Budowle B, Borer UV, Eggmann U, Comey CT, Dirnhofer R. Typing of deoxyribonucleic acid (DNA) extracted from compact bone from human remains. J Forensic Sci 1991;36(6):1649–61.

[17] Ye J, Ji A, Parra EJ, Zheng X, Jiang C, Zhao X, et al. A simple and efficient method for extracting DNA from old and burned bone. J Forensic Sci 2004;49(4):754–9.

[18] Loreille OM, Diegoli TM, Irwin JA, Coble MD, Parsons TJ. High efficiency DNA extraction from bone by total demineralization. Forensic Sci Int Genet 2007;1(2):191–5.

[19] Wilson MR, DiZinno JA, Polanskey D, Replogle J, Budowle B. Validation of mitochondrial DNA sequencing for forensic casework analysis. Int J Leg Med 1995;108(2):68–74.

[20] Alonso A, Martin P, Albarran C, Garcia P, Garcia O, de Simon LF, et al. Real-time PCR designs to estimate nuclear and mitochondrial DNA copy number in forensic and ancient DNA studies. Forensic Sci Int 2004;139(2-3):141–9.

[21] Andreasson H, Gyllensten U, Allen M. Real-time DNA quantification of nuclear and mitochondrial DNA in forensic analysis. Biotechniques 2002;33(2):402–4 407-11.

[22] Goodwin C, Higgins D, Tobe SS, Austin J, Wotherspoon A, Gahan ME, et al. Singleplex quantitative real-time PCR for the assessment of human mitochondrial DNA quantity and quality. Forensic Sci Med Pathol 2018;14(1):70–5.

[23] Niederstatter H, Kochl S, Grubwieser P, Pavlic M, Steinlechner M, Parson W. A modular real-time PCR concept for determining the quantity and quality of human nuclear and mitochondrial DNA. Forensic Sci Int Genet 2007;1(1):29–34.

[24] Kavlick MF. Development of a triplex mtDNA qPCR assay to assess quantification, degradation, inhibition, and amplification target copy numbers. Mitochondrion 2019;46:41–50.

[25] Xavier C, Eduardoff M, Strobl C, Parson W. SD quants—sensitive detection tetraplex-system for nuclear and mitochondrial DNA quantification and degradation inference. Forensic Sci Int Genet 2019;42:39–44.

[26] Bandelt HJ, Salas A, Lutz-Bonengel S. Artificial recombination in forensic mtDNA population databases. Int J Leg Med 2004;118(5):267–73.

[27] Parson W, Bandelt HJ. Extended guidelines for mtDNA typing of population data in forensic science. Forensic Sci Int Genet 2007;1(1):13–19. Available from: https://doi.org/10.1016/j.fsigen.2006.11.003.

[28] Brandstatter A, Peterson CT, Irwin JA, Mpoke S, Koech DK, Parson W, et al. Mitochondrial DNA control region sequences from Nairobi (Kenya): inferring phylogenetic parameters for the establishment of a forensic database. Int J Leg Med 2004;118(5):294–306.

[29] Berger C, Parson W. Mini-midi-mito: adapting the amplification and sequencing strategy of mtDNA to the degradation state of crime scene samples. Forensic Sci Int Genet 2009;3(3):149–53.

[30] Eichmann C, Parson W. 'Mitominis': multiplex PCR analysis of reduced size amplicons for compound sequence analysis of the entire mtDNA control region in highly degraded samples. Int J Leg Med 2008;122(5):385–8.

[31] Gabriel MN, Huffine EF, Ryan JH, Holland MM, Parsons TJ. Improved mtDNA sequence analysis of forensic remains using a "mini-primer set" amplification strategy. J Forensic Sci 2001;46(2):247–53.

[32] Tully G, Sullivan KM, Nixon P, Stones RE, Gill P. Rapid detection of mitochondrial sequence polymorphisms using multiplex solid-phase fluorescent minisequencing. Genomics 1996;34(1):107–13.
[33] Steighner RJ, Tully LA, Karjala JD, Coble MD, Holland MM. Comparative identity and homogeneity testing of the mtDNA HV1 region using denaturing gradient gel electrophoresis. J Forensic Sci 1999;44(6):1186–98.
[34] Butler JM, Wilson MR, Reeder DJ. Rapid mitochondrial DNA typing using restriction enzyme digestion of polymerase chain reaction amplicons followed by capillary electrophoresis separation with laser-induced fluorescence detection. Electrophoresis 1998;19(1):119–24.
[35] Gabriel MN, Calloway CD, Reynolds RL, Primorac D. Identification of human remains by immobilized sequence-specific oligonucleotide probe analysis of mtDNA hypervariable regions I and II. Croat Med J 2003;44(3):293–8.
[36] Holland MM, McQuillan MR, O'Hanlon KA. Second generation sequencing allows for mtDNA mixture deconvolution and high resolution detection of heteroplasmy. Croat Med J 2011;52(3):299–313.
[37] Li M, Stoneking M. A new approach for detecting low-level mutations in next-generation sequence data. Genome Biol 2012;13(I):R34.
[38] Parson W, Strobl C, Huber G, Zimmermann B, Gomes SM, Souto L, et al. Reprint of: Evaluation of next generation mtGenome sequencing using the Ion Torrent Personal Genome Machine (PGM). Forensic Sci Int Genet 2013;7(6):632–9. Available from: https://doi.org/10.1016/j.fsigen.2013.09.007.
[39] Seo SB, King JL, Warshauer DH, Davis CP, Ge J, Budowle B. Single nucleotide polymorphism typing with massively parallel sequencing for human identification. Int J Leg Med 2013;127(6):1079–86.
[40] Gallimore JM, McElhoe JA, Holland MM. Assessing heteroplasmic variant drift in the mtDNA control region of human hairs using an MPS approach. Forensic Sci Int Genet 2018;32:7–17. Available from: https://doi.org/10.1016/j.fsigen.2017.09.013.
[41] Just RS, Irwin JA, Parson W. Mitochondrial DNA heteroplasmy in the emerging field of massively parallel sequencing. Forensic Sci Int Genet 2015;18:131–9.
[42] Eduardoff M, Xavier C, Strobl C, Casas-Vargas A, Parson W. Optimized mtDNA control region primer extension capture analysis for forensically relevant samples and highly compromised mtDNA of different age and origin. Genes (Basel) 2017;8(10):237. Available from: https://doi.org/10.3390/genes8100237.
[43] Strobl C, Eduardoff M, Bus MM, Allen M, Parson W. Evaluation of the precision ID whole mtDNA genome panel for forensic analyses. Forensic Sci Int Genet 2018;35:21–5.
[44] Anderson S, Bankier AT, Barrell BG, de Bruijn MH, Coulson AR, Drouin J, et al. Sequence and organization of the human mitochondrial genome. Nature 1981;290(5806):457–65.
[45] Andrews RM, Kubacka I, Chinnery PF, Lightowlers RN, Turnbull DM, Howell N. Reanalysis and revision of the Cambridge reference sequence for human mitochondrial DNA. Nat Genet 1999;23(2):147. Available from: https://doi.org/10.1038/13779.
[46] The Scientific Working Group on DNA Analysis Methods (SWGDAM), Interpretation Guidelines for Mitochondrial DNA Analysis by Forensic DNA Testing Laboratories, approved on April 23, 2019.
[47] Isenberg AR. Forensic mitochondrial DNA analysis. In: Saferstein R, editor. Forensic science handbook, Vol. II. Upper Saddle River, New Jersey: Pearson Prentice Hall; 2004.
[48] Butler JM. Advanced topics in forensic DNA typing: methodology (Ch.14), Academic Press (Ed.) Wyman Street, Waltham, MA 02451, USA. 2011.
[49] Parson W, Gusmao L, Hares DR, Irwin JA, Mayr WR, Morling N, et al. DNA commission of the International Society for Forensic Genetics: revised and extended guidelines for mitochondrial DNA typing. Forensic Sci Int Genet 2014;13:134–42. Available from: https://doi.org/10.1016/j.fsigen.2014.07.010.
[50] Tully G, Bar W, Brinkmann B, Carracedo A, Gill P, Morling N, et al. Considerations by the European DNA profiling (EDNAP) group on the working practices, nomenclature and interpretation of mitochondrial DNA profiles. Forensic Sci Int 2001;124(1):83–91.
[51] Budowle B, Polanskey D, Fisher CL, Den Hartog BK, Kepler RB, Elling JW. Automated alignment and nomenclature for consistent treatment of polymorphisms in the human mitochondrial DNA control region. J Forensic Sci 2010;55(5):1190–5.

[52] Wilson MR, Allard MW, Monson K, Miller KW, Budowle B. Recommendations for consistent treatment of length variants in the human mitochondrial DNA control region. Forensic Sci Int 2002;129(1):35–42.
[53] Bandelt HJ, Parson W. Consistent treatment of length variants in the human mtDNA control region: a reappraisal. Int J Leg Med 2008;122(1):11–21. Available from: https://doi.org/10.1007/s00414-006-0151-5.
[54] Bendall KE, Macaulay VA, Baker JR, Sykes BC. Heteroplasmic point mutations in the human mtDNA control region. Am J Hum Genet 1996;59(6):1276–87.
[55] Comas D, Paabo S, Bertranpetit J. Heteroplasmy in the control region of human mitochondrial DNA. Genome Res 1995;5(1):89–90.
[56] Tully LA, Parsons TJ, Steighner RJ, Holland MM, Marino MA, Prenger VL. A sensitive denaturing gradient-gel electrophoresis assay reveals a high frequency of heteroplasmy in hypervariable region 1 of the human mtDNA control region. Am J Hum Genet 2000;67(2):432–43.
[57] Calloway CD, Reynolds RL, Herrin Jr. GL, Anderson WW. The frequency of heteroplasmy in the HVII region of mtDNA differs across tissue types and increases with age. Am J Hum Genet 2000;66(4):1384–97.
[58] Carracedo A, Bar W, Lincoln P, Mayr W, Morling N, Olaisen B, et al. DNA commission of the international society for forensic genetics: guidelines for mitochondrial DNA typing. Forensic Sci Int 2000;110(2):79–85.
[59] Irwin JA, Saunier JL, Niederstatter H, Strouss KM, Sturk KA, Diegoli TM, et al. Investigation of heteroplasmy in the human mitochondrial DNA control region: a synthesis of observations from more than 5000 global population samples. J Mol Evol 2009;68(5):516–27. Available from: https://doi.org/10.1007/s00239-009-9227-4.
[60] Brandstatter A, Parson W. Mitochondrial DNA heteroplasmy or artefacts—a matter of the amplification strategy? Int J Leg Med 2003;117(3):180–4.
[61] Berger C, Hatzer-Grubwieser P, Hohoff C, Parson W. Evaluating sequence-derived mtDNA length heteroplasmy by amplicon size analysis. Forensic Sci Int Genet 2011;5(2):142–5.
[62] Budowle B, Wilson MR, DiZinno JA, Stauffer C, Fasano MA, Holland MM, et al. Mitochondrial DNA regions HVI and HVII population data. Forensic Sci Int 1999;103(1):23–35.
[63] Holland MM, Parsons TJ. Mitochondrial DNA sequence analysis—validation and use for forensic casework. Forensic Sci Rev 1999;11(1):21–50.
[64] Balding DJ, Nichols RA. DNA profile match probability calculation: how to allow for population stratification, relatedness, database selection and single bands. Forensic Sci Int 1994;64(2-3):125–40.
[65] Egeland T, Salas A. Estimating haplotype frequency and coverage of databases. PLoS One 2008;3(12):e3988.
[66] Clopper CJ, Pearson ES. The use of confidence or fiducial limits illustrated in the case of the binomial. Biometrika 1934;26:404–13.
[67] Parson W, Brandstatter A, Alonso A, Brandt N, Brinkmann B, Carracedo A, et al. The EDNAP mitochondrial DNA population database (EMPOP) collaborative exercises: organisation, results and perspectives. Forensic Sci Int 2004;139(2-3):215–26.
[68] Prieto L, Zimmermann B, Goios A, Rodriguez-Monge A, Paneto GG, Alves C, et al. The GHEP-EMPOP collaboration on mtDNA population data—a new resource for forensic casework. Forensic Sci Int Genet 2011;5(2):146–51.
[69] Turchi C, Buscemi L, Previdere C, Grignani P, Brandstatter A, Achilli A, et al. Italian mitochondrial DNA database: results of a collaborative exercise and proficiency testing. Int J Leg Med 2008;122(3):199–204. Available from: https://doi.org/10.1007/s00414-007-0207-1.
[70] Turchi C, Stanciu F, Paselli G, Buscemi L, Parson W, Tagliabracci A. The mitochondrial DNA makeup of Romanians: a forensic mtDNA control region database and phylogenetic characterization. Forensic Sci Int Genet 2016;24:136–42. Available from: https://doi.org/10.1016/j.fsigen.2016.06.013.
[71] Zimmermann B, Brandstatter A, Duftner N, Niederwieser D, Spiroski M, Arsov T, et al. Mitochondrial DNA control region population data from Macedonia. Forensic Sci Int Genet 2007;1(3-4):e4–9. Available from: https://doi.org/10.1016/j.fsigen.2007.03.002.

[72] Huber N, Parson W, Dur A. Next generation database search algorithm for forensic mitogenome analyses. Forensic Sci Int Genet 2018;37:204–14.

[73] Salas A, Carracedo A, Macaulay V, Richards M, Bandelt HJ. A practical guide to mitochondrial DNA error prevention in clinical, forensic, and population genetics. Biochem Biophys Res Commun 2005;335 (3):891–9.

[74] Bandelt HJ, Dur A. Translating DNA data tables into quasi-median networks for parsimony analysis and error detection. Mol Phylogenet Evol 2007;42(1):256–71.

[75] Parson W, Dur A. EMPOP—a forensic mtDNA database. Forensic Sci Int Genet 2007;1(2):88–92.

[76] Lee HY, Song I, Ha E, Cho SB, Yang WI, Shin KJ. mtDNAmanager: a web-based tool for the management and quality analysis of mitochondrial DNA control-region sequences. BMC Bioinforma 2008;9:483.

PART 3

mtDNA mutations

CHAPTER 8

Human mitochondrial DNA repair

Elaine Ayres Sia and Alexis Stein

Department of Biology, University of Rochester, Rochester, NY, United States

A few decades ago, genetic human diseases were shown to result from mitochondrial DNA mutations. Typically, these diseases result from inheritance of a specific mutation in a subset of an affected individual's mitochondrial genomes [1]. More recently, researchers have examined the phenotypic consequences of increasing the rate at which mutations occur in mammalian mitochondrial DNA. Such studies have demonstrated the importance of maintaining the mitochondrial genome with high fidelity in mammalian cells. The most extensive set of experiments of this type have been performed by researchers examining "mitochondrial mutator mice." These mice carry a mutation that abrogates proofreading by the replicative mitochondrial polymerase, DNA Pol γ, and were shown to have phenotypes consistent with premature aging [2]. Study of cardiac cells from these mitochondrial mutator mice shows reduced flux through ETC, suggestive of depletion of mitochondrially encoded components of the inner mitochondrial membrane complexes with increased electron leak and reactive oxygen species (ROS) production [3], which is predicted to be a source of additional damage.

ROS are unstable oxygen containing molecules, including hydrogen peroxide (H_2O_2), superoxide anions (O_2^-), and hydroxyl radicals ($^{\bullet}OH$). It has been known since the 1960s that mitochondria are an endogenous source of ROS [4]. The primary site of ROS production in mitochondria are complexes I and III of the electron transport chain [5–9]; however, there are other potential mitochondrial sources of ROS including matrix proteins, such as α-ketoglutarate dehydrogenase [10,11]. The damage within the cell that results from ROS generated by any particular source is difficult to trace and accurately measure in vivo, in part due to our incomplete understanding of mitochondrial function [8,12,13]. While the precise levels of ROS in vivo are currently unknown, we can detect oxidative damage to the mitochondrial genome, by the measuring elevated levels of oxidized nucleotides in mitochondrial DNA [14].

The reactive nature of ROS means these molecules can damage important cellular components, including lipids, proteins, and DNA. Oxidative damage to DNA primarily leads to base lesions, in which the structure of the nitrogenous base in the nucleotide has been altered by

The Human Mitochondrial Genome.
DOI: https://doi.org/10.1016/B978-0-12-819656-4.00008-5

interaction with ROS. There have been more than 20 base lesions identified, with the focus of most research on 7,8-dihydro-8-oxo-2′-deoxyguanosine (8-oxodG) (Fig. 8.1) [15]. Oxidative alterations to the bases can lead to a deviation from the standard Watson–Crick base-pairing scheme by replicative DNA polymerases, which may lead to a mutation if the lesions are not properly repaired prior to replication [16]. Unrepaired oxidized bases may also become substrates for translesion polymerases that have reduced replication fidelity, and are an important source of point mutations [17,18]. DNA oxidative damage is not limited to the base but can also occur at the sugar residue in a nucleotide, leading to both single and double-strand breaks (DSBs) that must be repaired [19,20].

In addition to fixed mitochondrial DNA mutations, base lesions may contribute to the synthesis of incorrect protein products via errors in transcription. Although many oxidative base lesions will not directly block transcription machinery, their presence can result in mutant proteins because RNA polymerases, like DNA polymerases are prone to misincorporation opposite damaged sites [21].

Oxidative damage can also generate intermolecular crosslinks that can form between DNA and the proteins associated with it [22,23], as well as interstrand crosslinks between the two strands of DNA. Both protein-DNA and interstrand crosslinks will lead to cell death if not repaired. Considering the negative impact ROS can have on DNA, it is not scqurprising that the cell has evolved a multifaceted approach to mitigate the effect of ROS. This includes enzymatic (e.g., superoxide dismutase, catalase, and glutathione peroxidase) and nonenzymatic (e.g., vitamins C, E, and β-carotene) antioxidants to reduce the amount of ROS present in the cell and extensive DNA repair pathways including nucleotide excision repair (NER), base excision repair (BER), single-strand break (SSB), and double-strand break repair (DSBR) to process the resulting DNA damage [24].

DNA repair pathways are highly conserved from bacteria to humans. As a result, the early identification of DNA repair mechanisms in *Escherichia coli* laid the foundation for

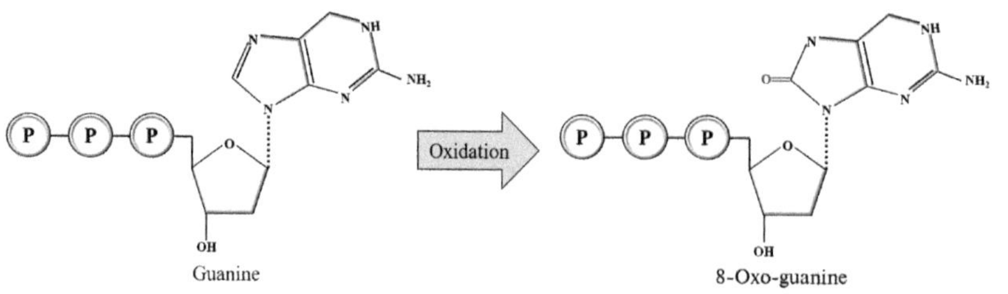

Figure 8.1

The most thoroughly studied oxidized nucleotide is 8-oxodG. The structural differences between guanine and 8-oxodG are small, yet significant, as they can lead to changes in base pairing from G-C to G-A. The *dotted line* represents the glycosidic covalent bond between the nitrogenous base and the deoxyribose sugar.

discovery of repair pathways in eukaryotic cells. Many of the known mechanisms for eukaryotic nuclear repair have been found to operate in mitochondria, often using many of the same proteins. For some repair mechanisms, a subset of the protein components in the nuclear pathway has been found in the mitochondria, while others have not, suggesting that mitochondrial DNA repair may consist of both shared and unique pathway components, or that mitochondria utilize known repair proteins in unique ways. Mitochondrial DNA repair pathways, as we currently understand them, are described in the following sections.

8.1 Base excision repair

BER pathways are the earliest repair mechanisms defined for mtDNA [25]. Although BER is often described as a "pathway," it consists of several partially redundant mechanisms to remove and replace damaged bases in DNA in both the nuclear and mitochondrial genomes (Fig. 8.2). Specialized for the repair of base damage and abasic sites, these pathways are predicted to be the primary mechanism for repair of ROS damage to DNA.

The basic steps of BER consist of N-glycosylase enzyme recognition of the damaged base and cleavage of the N-glycosidic bond that attaches it to the deoxyribose sugar, leaving an abasic site. Next, the abasic sugar must be removed. Two reactions are required to generate a 5′ phosphate and 3′ hydroxyl on either side of the abasic site to fully remove the residue from the DNA. The phosphodiester bond is cleaved on the 5′ side of the abasic site by AP endonuclease, generating a free 3′ end that can be extended by a polymerase. The remaining 5′ deoxyribose may be removed 3′ of the abasic site to leave a single nucleotide gap by lyase activity provided by polymerase β or as a second activity of some of the N-glycosylase enzymes in the case of "short-patch" BER (SP-BER). Alternatively, in "long-patch" BER, strand displacement synthesis occurs at the site of the 5′ nick, and the resulting 5′ flap is cleaved by a flap endonuclease (LP-BER). In either case, the free 3′ end is extended to fill the gap, and the nick in the newly synthesized duplex DNA is sealed by ligase to complete repair.

The DNA N-glycosylase enzymes are specialized for recognition of particular types of base damage, with some redundancy. Seven of 11 glycosylases in mammalian cells have, thus far, been found in mitochondrial fractions, representing a broad spectrum of recognition of base modifications. These consist of both monofunctional enzymes that lack lyase activity (UNG1, AAG1, and MYH) and bifunctional enzymes, with both glycosylase and lyase activities (NTHL1, OGG1, NEIL1, and NEIL2) [26]. It is optimal for these enzymes to act in the context of double-stranded DNA sequence because effective repair requires templating information on the opposite strand. In addition, glycosylase cleavage of ssDNA would lead to DSBs during replication. For this reason, several of these enzymes have been shown to interact with the mitochondrial single-stranded binding protein, which inhibits their activity [27–29].

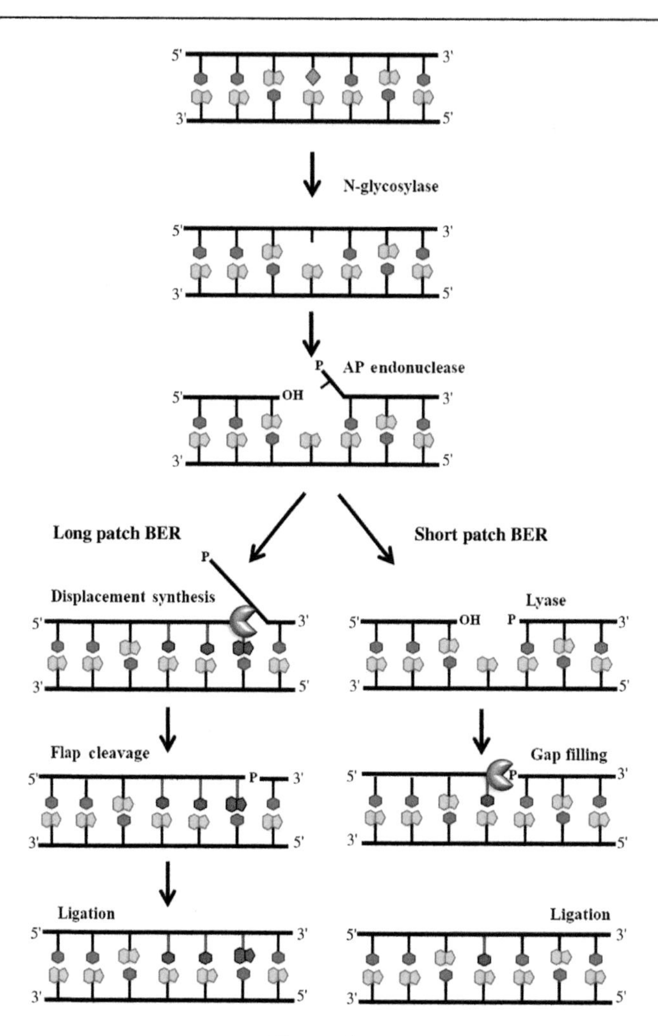

Figure 8.2

Mechanisms of base excision repair. In both short-patch and long-patch repair pathways, the damaged base is removed by an N-glycosylase. An AP endonuclease will then cleave the sugar phosphate backbone 5′ of the abasic site. The repair of abasic sites that may result from spontaneous depurination will begin at this step. In short-patch repair, lyase activity will remove the small deoxyribose flap. A polymerase will fill the single nucleotide gap and ligase will seal the nick. In long-patch repair, displacement synthesis by a DNA polymerase will extend the 3′ end generated by AP endonuclease cleavage, and displace the damaged 5′ end to generate a short flap. This flap is cleaved by a flap endonuclease, and the resulting nick is sealed by ligase. Both pathways result in the removal of the damaged base.

Two homologs of the *E. coli* AP endonuclease have been identified in mammalian cells, APE1 and APE2 [30,31]. Loss of APE1 (also called Ref-1) is lethal in mice [32]. It is widely believed to be the primary source of AP endonuclease activity in mammalian BER, since APE2 does not complement yeast mutants lacking AP endonuclease activity [33], and recombinant APE2 has been reported to lack AP endonuclease activity in biochemical assays [34]. While both APE1 and APE2 are found in mitochondrial fractions [35,36], biochemical studies suggest that APE1 is also the primary AP endonuclease in mitochondrial BER as well [37].

As in nuclear DNA, it seems likely that at least two alternative polymerases may be used in mitochondrial BER. DNA Pol γ was identified in rat mitochondria in 1978 [38], and quickly determined to be a replicative mitochondrial polymerase [39]. It was much later that other DNA polymerases were demonstrated to localize to mitochondria, and thus our models for mitochondrial repair have been revised to include other DNA polymerase enzymes, including DNA Pol β, a polymerase with lower processivity, no proofreading exonuclease, and associated lyase activity [40,41].

Human DNA Pol γ has been shown to fill single-strand gaps inefficiently. In addition, Pol γ can perform strand displacement synthesis in vitro, but this activity is suppressed by the proofreading exonuclease activity. It has been proposed that this suppression may be modulated in vivo to allow Pol γ to generate flaps during synthesis in mitochondrial LP-BER [42]. Lyase activity is also associated with the DNA Pol γ [43]; however, in vitro studies with mitochondrial extracts suggest that DNA Pol β may be responsible for the majority of mitochondrial SP-BER [41].

With the exception of DNA Pol γ, all of the enzymes shown so far to be important for mitochondrial BER are also found in the nucleus. Studies of localization of some of these well-studied nuclear BER enzymes to the mitochondria reveal complexity in the regulation of subcellular localization of repair proteins. The uracil N-glycosylase (UNG1) in mitochondria is encoded by the same gene, *UNG*, that encodes the nuclear isoform, but it results from the use of alternative transcriptional starts and alternative splicing [44,45]. The UNG1 isoform contains both an N-terminal mitochondrial targeting sequence, and residues important for nuclear localization, but it appears to localize preferentially to mitochondria [45].

Localization of the AP endonuclease, APE1 is particularly complex. Distribution to mitochondria relies on sequences near the C-terminus of the protein [46]; however, the protein appears to accumulate in the intermembrane space prior to matrix localization [47]. The amount of mitochondrial APE1 is reduced following perturbation of the MIA pathway, which supports the proper folding and disulfide bond formation of proteins in the mitochondrial intermembrane space. Unfolded proteins are transported via the TOM complex in the mitochondrial outer membrane, where they interact with Mia40 in the intermembrane space. This interaction is critical for proper folding and disulfide bond formation in substrate proteins, which typically remain in the intermembrane space [48]. However, in yeast, the mitochondrial ribosomal protein, Mrp10, is a substrate for the MIA pathway, prior to import into the mitochondrial matrix (Fig. 8.3). In the absence of Mia40 interaction, the incorrectly

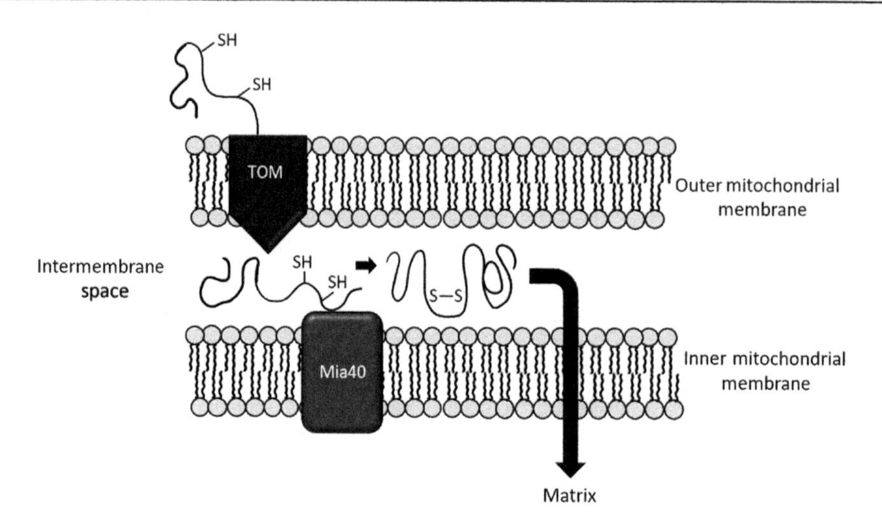

Figure 8.3
Unfolded proteins, such as APE1, enter the intermembrane space through the TOM complex where they interact with Mia40. Mia40 promotes proper disulfide bond formation and folding which increases protein stability. The properly folded protein may, in some cases, be imported into the matrix of mitochondria via an unknown mechanism.

processed protein is rapidly degraded [49]. It is unclear whether human APE1 interaction with the MIA pathway serves a similar purpose, and the mechanism of transport across the inner mitochondrial membrane is still unknown.

A number of the other BER pathway components are predicted to have N-terminal mitochondrial targeting sequences [26]; however, Pol β localization to the nucleus requires an N-terminal NLS sequence, which precludes a canonical N-terminal mitochondrial targeting sequence [50]. Pol β localization to mitochondria appears to require an internal, as yet unidentified, targeting sequence, or posttranslational modification [40]. The diverse mechanisms of mitochondrial localization for the proteins involved in BER suggest that the localization of repair proteins in the pathways described below will be equally complex.

8.2 Repair of bulky lesions

In the eukaryotic nucleus, pyrimidine dimers and other bulky lesions are primarily repaired by the NER pathway. While NER is perhaps best known for repair of UV-induced damage, multiple studies have demonstrated an additional role in the repair of oxidative damage, both directly, via recognition of bulky lesions, but also in the promotion of repair by BER [51].

In NER, endonucleases cleave the damaged strand on both sides of the damaged nucleotides, and a short single-stranded patch of DNA containing the damage is removed. This gap is filled by DNA polymerase, and then the DNA is ligated to seal the nick. There are two modes of damage recognition in NER, defining two related subpathways, global genome nucleotide excision repair (GG-NER), and transcription coupled repair (TCR) [52]. Because some of the proteins important for these subpathways are found in mitochondria, they will be described briefly below (Fig. 8.4).

In GG-NER, bulky lesions are recognized by an XPC/RAD23b complex, or in the case of cyclopyrimidine dimers, first by UV-DDB then by XPC/RAD23b. Bound XPC/RAD23b then recruits TFIIH, which includes XPB and XPD. The helicase activity of XPD unwinds the duplex surrounding the lesion to allow incision by ERCC1-XPF and XPG [53].

TCR results in a faster repair of the template strand in transcribed regions. Bulky lesions not only block the progress of replicative DNA polymerases but also cause stalling of RNA polymerase II. These stalled RNA Pols are recognized by CSB, which is important for recruitment of additional TCR factors, including the proteins CSA and UVSSA (UV-stimulated scaffold protein A). Subsequently, TFIIH is recruited and repair continues much like GG-NER [54].

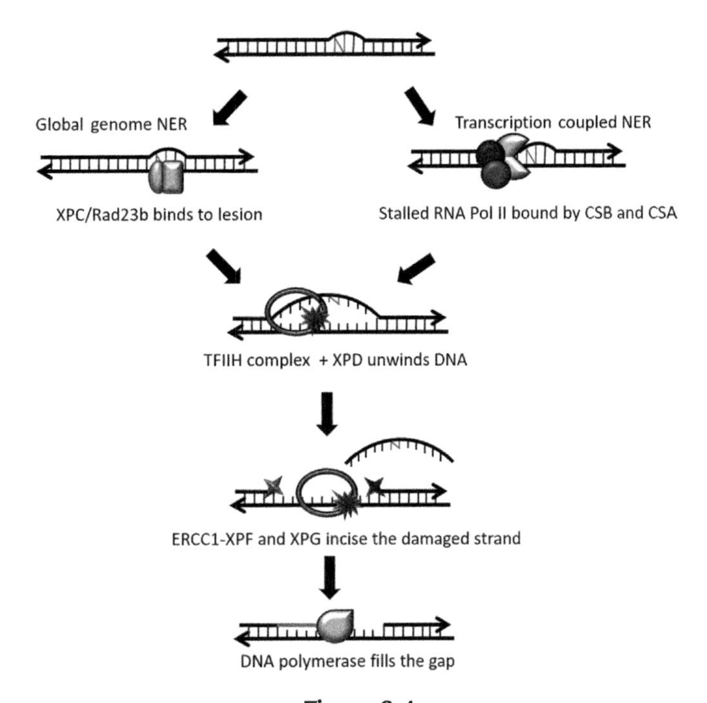

Figure 8.4

Bulky lesions can be repaired with GG-NER or TCR depending on how they are detected. In GG-NER, the lesion is detected by XPC/Rad23b complex. In TCR, a stalled RNA Pol II is recognized by the CSA and CSB. Both types of NER then recruit the TFIIH complex to the lesion. XPD, a helicase, is part of this complex and unwinds the helix at the site of the lesion. ERCC1-XPF and XPG will then nick the strand containing the lesion on either side. The resulting gap is filled in by DNA polymerase and sealed with ligase.

Early studies showed no evidence for repair of UV-induced damage in mitochondria, suggesting that this organelle lacks functional NER to repair pyrimidine photoproducts [55]. However, subsequent studies have demonstrated the localization of a number of NER protein components to mitochondrial fractions, and their role in mitochondrial DNA repair remains an open question.

The TCR proteins, CSA and CSB, were the first of the NER proteins found in mitochondrial fractions purified from mammalian cells [56,57]. These proteins are named for Cockayne Syndrome (CS), a genetic disorder which manifests in infancy or early childhood, and has complex neurological symptoms. CS patients have symptoms suggestive of accelerated aging, and their cells show significant mitochondrial dysfunction. In fact, based on the collection of symptoms, CS has been categorized as a mitochondrial disease by some researchers [58]. The mitochondrial levels of both CSA and CSB increase in response to oxidative stress, and mitochondrial oxidative damage and mitochondrial mutations increase in CSA- and CSB-deficient mammalian cells [56,57]. A direct role for CSB in mitochondrial BER has been proposed, as CSB is found in a complex with mitochondrial OGG1 glycosylase and mitochondrial DNA [56].

More recently, both XPD and RAD23A were also found in mitochondrial fractions, and like CSA and CSB reduce mitochondrial oxidative damage and mitochondrial mutagenesis [59,60]. A specific mechanism for the action of these proteins in mitochondria has not yet been defined, and may involve additional interacting proteins, including MMS19. In the yeast model system, MMS19 was identified many years ago as a NER gene [61]. The human ortholog was shown to interact with XPD and to impact NER in the nucleus. MMS19 was identified as a component of the cytosolic iron–sulfur cluster assembly complex (CIA) [62], where it has been proposed to aid in the assembly of the Fe–S cluster in XPD, which in turn is required for the assembly of XPD into the TFIIH complex [63]. Thus it may be proposed that the function of MMS19 in NER is indirect, via the assembly of TFIIH complex, as has been proposed for the budding yeast homolog [64]. However, human MMS19 is also found in mitochondria, and knockdown of MMS19 results in an increase in the appearance of a large deletion following hydrogen peroxide exposure [65]. This is an unexpected result, if the role of MMS19 in repair is due solely to its function in cytoplasmic iron–sulfur protein maturation, as the CIA complex is not required for the assembly of Fe–S clusters in mitochondrial proteins [66]. Additional studies will be required to determine whether the mitochondrial function of MMS19 is dependent on XPD, or other mitochondrially localized repair proteins, and to elucidate the mechanism by which each of the NER components reduce mitochondrial oxidative damage and damage-induced mutations.

8.3 Double-strand break repair

DSBs are a particularly toxic type of DNA damage because the failure to repair these lesions can result in significant loss of genomic sequences. Even when repair occurs, several mechanisms are error-prone, resulting in mutations. DSBs can be induced by a variety of exogenous and endogenous sources. Ionizing radiation can directly break the sugar phosphate backbone generating DSBs [67]. DSBs can also spontaneously arise as a result of endogenous DNA metabolic activities. For example, if BER has been initiated, but is not completed, replication-blocking lesions or SSBs will persist. Stalled forks and single-strand DNA breaks generated by BER enzymes result in the formation of DSBs and have been shown to stimulate HR in mammalian cells [68].

Mechanisms in place to accommodate the topological problems associated with opening a helical molecule can also be a source of DNA strand breaks. During replication and transcription, topoisomerases are required to maintain proper DNA topology [69,70]. Type II topoisomerases achieve this by introducing a DSB that reduces the tension generated during transcription and DNA replication and to promote the decatenation of circular genomes, like the mitochondrial genome, after replication [71]. During this reaction, the enzyme is covalently bound to the 5′ end of the DNA strands. If these reactions are incomplete or the topoisomerase becomes trapped in DNA-Top II complexes, persistent DSBs will result [72].

All of these mechanisms of DSB formation are likely to occur more frequently in the mitochondria where the proximity to elevated ROS contributes to DNA and protein oxidation. Due to our current incomplete understanding of both mitochondrial DNA metabolism and ROS production, we cannot state with any certainty which source of breaks is of primary concern.

In general, DSBs are repaired by two main categories of DNA repair, homologous recombination (HR) and nonhomologous end joining (NHEJ). Classical HR will use a homologous DNA molecule to serve as a template in the synthesis of DNA on the molecule with the DSB (Fig. 8.3) [73]. When repair is completed in this fashion, it is error free and the integrity of the genome is maintained. The multicopy nature of the mitochondrial genome provides accessible homologous templates for this type of repair under all stages of the cell cycle.

While homologous recombination is often described as an error-free pathway of DSB repair, there are subcategories of HR, such as single-strand annealing (SSA), that are mutagenic. In SSA, during the initial resection of the DSB, repetitive DNA sequences (indicated by the boxes in the Fig. 8.5) are revealed. These sequences then anneal, the resulting flaps are cleaved, gaps are filled, and the nicks sealed. This will result in the deletion of one of the repeat sequences and all the intervening DNA [74].

Classical NHEJ is considered mutagenic and normally results in small deletions (0–25 bp) or insertions at the site of the break due to the processing of the ends of the break followed directly by ligation [75]. During NHEJ, the ends of the DSB are initially bound by Ku70/80 heterodimer that recruits nucleases that process the ends of the break, followed by the recruitment of polymerases that can add nucleotides in a template dependent or independent

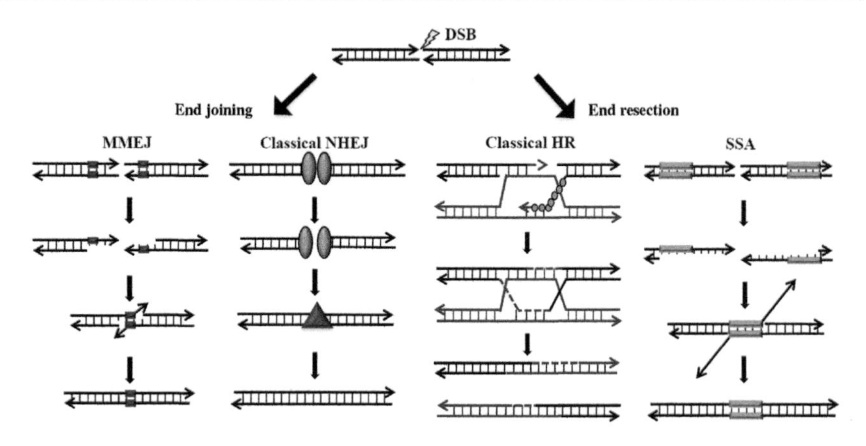

Figure 8.5

Models for double-strand break repair. Double-strand break repair pathways are broadly categorized as end-joining pathways and homologous recombination pathways. Classical NHEJ requires minimal processing and resection of the ends of the break, followed by ligation. The resulting repair product often exhibits small deletions or insertions. Microhomology-mediated end joining (MMEJ) involves more extensive resection and the presence of small direct repeats (5–25 bps) that can anneal. The resulting repair product often results in deletions larger than those seen in NHEJ. Classical HR uses a homologous sequence as a template to repair the double-strand break. When completed, the repair product is error free. SSA utilizes many HR proteins, but during the resection process, repetitive DNA sequences are revealed. The subsequent annealing of these sequences, and cleavage of the nonhomologous 3′ ssDNA tails by flap endonuclease enzymes, will lead to repair products that have undergone significant deletions.

manner. Finally, the nick is sealed by ligase IV in eukaryotes [76]. An alternative form of NHEJ, termed microhomology-mediated end joining (MMEJ), depends on small repeat homologies (5–25 bps) that are exposed during more extensive processing of the DSB, is capable of generating large deletions between these microhomologies and does not require the same proteins as classical NHEJ (Fig. 8.5) [77,78].

Purified human, mouse, and other vertebrate mitochondrial extracts exhibit robust HR activity in vitro [79,80]. Several key HR proteins have been shown to localize to the human mitochondria including Rad51, Rad50, MRE11, and NIBRIN [79,81,82]. Immunodepletion of the HR proteins Rad51, MRE11, and NIBRIN from mitochondrial extracts prior to in vitro HR assays led to significant decrease in detected HR events, suggesting that these proteins function in mitochondrial HR pathways, just as they do in the nucleus [79].

Elucidating the impact of HR in vivo has been significantly more challenging, due to the uniparental inheritance of mitochondrial DNA. Since essentially all mitochondrial genomes are identical within a cell, products of recombination are difficult, or impossible to detect. In a patient with both maternal and paternal mitochondrial DNA, recombinant molecules containing maternal and paternal DNA sequences were detected, indicating that spontaneous

HR does occur in vivo [83]. Further support for HR in vivo comes from the detection of recombination intermediates, consisting of four-way junctions, that have been observed in the heart, brain, and other highly oxidative tissues. The presence of these DNA species in cell types with a high energetic demand suggest a link between the increase of HR intermediates and an increase in the levels of ROS in these tissues [84–86].

To examine DSBR directly, in vivo studies have been performed using mouse and tissue culture models. Mitochondrial DSBs are induced via mitochondrially targeted restriction endonucleases. In these studies, little recombinant product could be detected, and the majority of the mitochondrial DNA appeared to be degraded upon DSB induction instead of repaired [87]. These endonucleases introduced multiple, chronic DSBs into the mitochondrial genome and it may be that this level of damage overwhelmed the DSB repair pathways.

While these results may argue against robust mitochondrial HR in DSBR, these findings posed an interesting treatment strategy for individuals harboring disease-causing mitochondrial DNA mutations. If DSBs could be introduced to only the mitochondrial DNA molecules with mutations, these mitochondrial DNA molecules may be targeted for degradation, allowing the healthy genomes to replicate and replace them, thus reducing the severity of symptoms or eliminating the disease [88–93]. While these strategies were effective in tissue culture, a gene therapy for introducing these enzymes into humans has yet to be developed.

In 2019, it was reported that human mitochondrial DNA could be edited using a Mito-CRISPR/Cas 9 system [94]. The CRISPR/Cas9 system works by introducing a DSB into a specific DNA sequence by the Cas 9 endonuclease and allowing the cell to repair the DSB using an exogenous DNA template via HR. The successful editing of human mitochondrial DNA in vivo using this system provides evidence that mitochondrial DSBs can be repaired via a HR pathway in vivo without complete degradation of the mitochondrial genome.

Currently, there is no biochemical evidence for classical NHEJ activity in mammalian mitochondria and the localization of several key nuclear NHEJ proteins to mitochondria remains to be conclusively demonstrated, with the exception of truncated isoform of Ku80 [95,96]. There has been clear in vitro evidence for MMEJ using rat mitochondrial extracts. The activity of MMEJ was dependent on CtIP, FEN1, MRE11, and PARP1 proteins [95]. The identification of a robust MMEJ pathway in mitochondria is interesting considering that 85% of all deletions identified in human mitochondrial DNA are flanked by short microhomologies, including the 4977 bp common deletion that occurs during aging and appears in many cancer types [97]. This would suggest that this error-prone pathway may play a significant role in the generation of these deletions.

Recently, an alternative model of deletion formation has been proposed, termed replication-mediated repair of SSBs or DSBs [98]. This model proposes that frequent stalling of the replication fork allows for the mispairing of the microhomologies flanking the common deletion. This mispairing leaves a large loop of single-stranded DNA exposed to potential damage, which is then lost during break repair. This pathway is unique from both SSA and MMEJ, in that it is dependent on the mitochondrial Pol γ, Twinkle (helicase), and MGME1 (mitochondrial nuclease) [98]. In support of this model, patients with a pathogenic mutation in Twinkle exhibit increased frequency of stalled replication forks and an increased incidence of deletions in their mitochondrial DNA [98,99].

8.4 Mismatch repair

During DNA replication, incorrectly paired nucleotides may be removed immediately following insertion by the replicative polymerase's 3′ to 5′ "proofreading" exonuclease activity. This is true in the mitochondria as well because DNA Pol γ possesses proofreading activity. However, once a misincorporated nucleotide is extended, it will no longer be removed by the polymerase [100]. In the immediate aftermath of DNA replication, mismatched bases can be removed via postreplication mismatch repair. Mismatch repair is a highly conserved pathway and was studied first in *E. coli*. MutS homodimers recognize and bind mismatched residues in the DNA [101]. Following mismatch recognition, the DNA-bound MutS dimers interact with MutL [102]. In *E. coli*, the DNA is transiently hemi-methylated after new DNA synthesis, with methylation only found on the template strand. MutH recognizes hemi-methylated GATC sequences that will subsequently be modified by the *dam* methylase. Interaction between MutS/MutL and MutH in the context of a mismatched DNA sequence activates the endonuclease activity of MutH, which will cleave the unmethylated strand [103]. A helicase loads at the nick and unwinds the newly synthesized strand, allowing exonucleolytic digestion. In this way, the newly synthesized strand can be recognized and removed from the nearest hemi-methylated sequence to the mismatch. The DNA is then resynthesized across this single-stranded gap by extending the 3′ end [104]. In the absence of MutH, strand bias of repair can be directed by a single-strand nick [105].

Eukaryotic nuclear mismatch repair proteins were identified by virtue of their homology to the bacterial MutS and MutL proteins. In the eukaryotic nucleus, there are multiple homologs of both MutS (MSH) and MutL (MLH/PMS), which form heterodimers [106–111]. No MutH homolog was found, indicating that the strand discrimination signal differs in prokaryotes and eukaryotes. The eukaryotic MMR proteins interact with the replicative sliding clamp, and evidence suggests that this interaction with replication complexes and discontinuities in the nascent strands provide the signal for strand discrimination [112].

In 2003, Mason et al. found that lysates from rat liver mitochondria were capable of repairing mismatched duplexes; however, they observe neither a strand bias in repair nor an MSH2 protein localized to mitochondria [113]. The presence of mismatch repair activity in purified mammalian mitochondria was subsequently confirmed using mitochondria from human cells, and YB-1 protein was purified by virtue of its binding to mismatched DNA.

Previously YB-1 was identified in screens for proteins that bind to Y-box transcription factor binding sites [114], and independently due to its ability to bind depurinated DNA [115]. Consistent with the diverse DNA substrates YB-1 interacts with, it has been shown to be a multifunctional protein with roles in transcriptional regulation of a number of genes, mRNA splicing, translation, and DNA repair [116].

Even in the nucleus, the function of YB-1 in DNA repair is unclear. Purified YB-1 protein has DNA melting activity, which is stimulated in the presence of cisplatin adducts, or a base mismatch [117]. YB-1 interacts with DNA repair proteins from a number of different DNA repair pathways including MSH2 (MMR), Ku80 and WRN (DSB repair) [117], XPC-HR23B (NER) [118], and APE1, NEIL1, NEIL2 and DNA Pol β (BER) [119,120]. Some of these interactions have been shown to result in stimulation of repair activity [118,121]; however, in one study, YB-1 was shown to compete with the MutS homologs for binding to mismatched DNA, and as a result inhibit mismatch repair [122].

The presence of YB-1 in mitochondria was verified, and depletion of YB-1 from mitochondria resulted in an increase in mitochondrial DNA mutations [123]. It is still not known whether the binding of YB-1 to mismatches reflects a direct role in the mismatch repair activity seen in mitochondrial extracts, or what other proteins may act in the pathway.

8.5 Translesion synthesis

The active site of high-fidelity replicative polymerases is restricted to ensure correct nucleotide selection, and these enzymes do not easily accommodate bulky lesions. In addition, the proofreading activity of these polymerases inhibits synthesis opposite bulky lesions and abasic sites [124–126]. Such lesions would result in a failure to replicate the genome if not repaired prior to DNA synthesis; however, cells have a number of lower fidelity, translesion polymerases that are capable of bypass of lesions that would otherwise be lethal [127]. In organisms from bacteria to humans, replication-blocking lesions, such as thymidine dimers and abasic sites, are mutagenic due to the low fidelity synthesis opposite damaged sites by translesion polymerases.

As in bacterial [128], eukaryotic nuclear DNA [129], and yeast mitochondrial DNA [130], exposure to UV light results in an increase in UV-induced point mutations in the human mitochondrial genome [131], suggesting error-prone bypass. DNA Pol γ replicates the mitochondrial genome with high fidelity, and contains an associated 3′–5′ proofreading exonuclease activity that removes erroneously inserted nucleotides during synthesis [132,133]. Like most high-fidelity enzymes, DNA Pol γ has a limited ability to bypass many DNA lesions, and when it occurs, translesion synthesis is generally error-prone [124,134–136]; however, several other polymerases have been found in mitochondria and these enzymes may contribute to lesion bypass.

In 2013, a new human DNA polymerase was identified that has both primase and DNA polymerase activities. This enzyme, named PrimPol, is found in both the nucleus and mitochondria [137]. This enzyme does not efficiently bypass abasic sites, cyclobutene dimers, or thymine glycol, but can synthesize across from 8-oxo-G and 6-4 UV photoproducts [137–139].

Mammalian cells contain an array of lower fidelity DNA polymerases that contribute to the tolerance of different types of replication-blocking DNA damage [127]. One of these polymerases, translesion polymerase, Pol ζ is conserved among eukaryotes, and is responsible for much of the nuclear mutagenesis in response to damage by a number of mutagens. In addition, Pol ζ has been shown to play an important role in nuclear NHEJ [140]. Unlike other translesion polymerases, Pol ζ is essential in mammals, resulting in embryonic lethality early in development [141–143]. It is unclear why loss of Pol ζ is lethal in mammalian cells, when it is not essential in yeast, and it has not been demonstrated that this lethality is a direct result of loss of translesion synthesis. Human Pol ζ is comprised of four subunits, Pol31, Pol32, Rev7, and the catalytic subunit, Rev3 [144]. Research has demonstrated that Rev3$^{-/-}$ cells show an accumulation of DSBs and translocations [145]; thus the lethality may reflect a primary defect in DSBR.

An isoform of Rev3 has been found in the mitochondria of human cells, and Rev3$^{-/-}$ cells are reported to have decreased OXPHOS complex IV activity and increased glucose consumption [146]. At this time, it is not known whether other subunits of DNA Pol ζ are found in mitochondria, or the precise role of Rev3 in maintaining normal mitochondrial function.

More recently, DNA polymerase θ has been shown to be concentrated in mitochondria of human cells, following a screen for DNA repair proteins that are required to tolerate mitochondrial DNA damaging agents. As with Pol ζ, cells that have lost Pol θ show decreased OXPHOS activity. In addition, loss of Pol θ results in a reduction in mitochondrial DNA mutagenesis, suggesting a role for this enzyme in mutagenic DNA synthesis, potentially during lesion bypass [59]; however, Pol θ also promotes error-prone NHEJ in the nucleus, and this activity could be the source of mitochondrial mutations [147].

8.6 Concluding remarks

For many reasons, the study of mitochondrial DNA repair has lagged behind the study of similar pathways in the nucleus. These delays in part result from technical difficulties, such as challenges in manipulating the mammalian mitochondrial genome, the essential function of mitochondrial DNA in the cells of multicellular eukaryotes, and the complication of working with a multicopy genome that is uniparentally inherited. However, other delays resulted from an early assumption that the mitochondrial genome was not subject to repair. A picture of mitochondria is now emerging in which some components of repair pathways are shared with the nucleus, but these components interact with proteins that are unique to the mitochondria. Thus the mitochondrial DNA repair pathways are still being defined by ongoing research.

It can be particularly difficult to assess the repercussions of loss of mitochondrial DNA repair when many of the components are shared between mitochondrial and nuclear repair, as defects in nuclear repair will be observed more readily. In addition, since nuclear genes encode proteins that are important for mitochondrial function, mutations in these nuclear genes can have indirect effects on the mitochondrial function or mutagenesis. For example, Pol $\beta^{-/-}$ mice die shortly after birth [148], but mitochondrial phenotypes can be assessed in homozygous mutant cells in culture. Dramatic changes to mitochondrial morphology have been observed in these cells; however, at this time, it is not possible to determine which phenotypes result from direct effects on mitochondrial DNA [40,41].

There is considerable redundancy in nuclear repair pathways, and extensive crosstalk between repair pathways is important for maintaining genome integrity. It seems likely that repair pathways interact in mitochondria as well. How interactions between repair proteins, for example, those proteins found in BER and nuclear NER, may contribute to the reduction of oxidative damage to the mitochondrial genome will require continued analysis.

Research perspectives

It is now clear that many repair proteins are shared between the nucleus and mitochondria, but we understand very little about how the subcellular localization of repair proteins is managed. For some proteins, little beyond an observation of their presence in mitochondrial fractions is known. Even for proteins with clear mitochondrial targeting signals, the dynamic localization of the protein in different cells, and under different intracellular conditions, has not been extensively studied. Does the import of the same protein to the nucleus or mitochondria depend or respond to damage in either compartment? If so, what are the signals that direct localization to mitochondria in response to damage?

In humans, we know that the structure of mitochondria and the copy number of the mitochondrial genomes vary depending on tissue type [149,150]. There is intriguing new evidence that the mode of mitochondrial DNA replication may also be tissue specific, and raises the possibility that cell type-specific differences may also impact repair pathways [84]. A comprehensive picture of mitochondrial DNA repair will not be possible until we have an integrated understanding of mitochondrial bioenergetics, ROS production, morphology, tissue-specific differences, and how various disease states such as cancer influence mitochondrial DNA metabolism.

References

[1] Carelli V, La Morgia C. Clinical syndromes associated with mtDNA mutations: where we stand after 30 years. Essays Biochem 2018;62:235–54.

[2] Trifunovic A, et al. Premature ageing in mice expressing defective mitochondrial DNA polymerase. Nature 2004;429:417–21.

[3] McLaughlin KL, McClung JM, Fisher-Wellman KH. Bioenergetic consequences of compromised mitochondrial DNA repair in the mouse heart. Biochem Biophys Res Comm 2018;504:742–8.

[4] Jensen PK. Antimycin-insensitive oxidation of succinate and reduced nicotinamide-adenine dinucleotide in electron-transport particles. II. Steroid effects. Biochim Biophys Acta 1966;122:167–74. Available from: https://doi.org/10.1016/0926-6593(66)90058-0.

[5] Cadenas E, Boveris A, Ragan CI, Stoppani AO. Production of superoxide radicals and hydrogen peroxide by NADH-ubiquinone reductase and ubiquinol-cytochrome c reductase from beef-heart mitochondria. Arch Biochem Biophys 1977;180:248–57. Available from: https://doi.org/10.1016/0003-9861(77)90035-2.

[6] Hinkle PC, Butow RA, Racker E, Chance B. Partial resolution of the enzymes catalyzing oxidative phosphorylation. XV. Reverse electron transfer in the flavin-cytochrome beta region of the respiratory chain of beef heart submitochondrial particles. J Biol Chem 1967;242:5169–73.

[7] Hirst J, King MS, Pryde KR. The production of reactive oxygen species by complex I. Biochem Soc Trans 2008;36:976–80. Available from: https://doi.org/10.1042/BST0360976.

[8] Murphy MP. How mitochondria produce reactive oxygen species. Biochem J 2009;417:1–13. Available from: https://doi.org/10.1042/BJ20081386.

[9] Turrens JF, Alexandre A, Lehninger AL. Ubisemiquinone is the electron donor for superoxide formation by complex III of heart mitochondria. Arch Biochem Biophys 1985;237:408–14. Available from: https://doi.org/10.1016/0003-9861(85)90293-0.

[10] Starkov AA, et al. Mitochondrial alpha-ketoglutarate dehydrogenase complex generates reactive oxygen species. J Neurosci 2004;24:7779–88. Available from: https://doi.org/10.1523/JNEUROSCI.1899-04.2004.

[11] Tretter L, Adam-Vizi V. Generation of reactive oxygen species in the reaction catalyzed by alpha-ketoglutarate dehydrogenase. J Neurosci 2004;24:7771–8. Available from: https://doi.org/10.1523/JNEUROSCI.1842-04.2004.

[12] Griendling KK, et al. Measurement of reactive oxygen species, reactive nitrogen species, and redox-dependent signaling in the cardiovascular system: a scientific statement from the American Heart Association. Circ Res 2016;119:e39–75. Available from: https://doi.org/10.1161/RES.0000000000000110.

[13] Kowaltowski AJ. Strategies to detect mitochondrial oxidants. Redox Biol 2019;21:101065. Available from: https://doi.org/10.1016/j.redox.2018.101065.

[14] Richter C, Park JW, Ames BN. Normal oxidative damage to mitochondrial and nuclear DNA is extensive. Proc Natl Acad Sci 1988;85:6465–7. Available from: https://doi.org/10.1073/pnas.85.17.6465.

[15] Cooke MS, Evans MD, Dizdaroglu M, Lunec J. Oxidative DNA damage: mechanisms, mutation, and disease. FASEB J 2003;17:1195–214. Available from: https://doi.org/10.1096/fj.02-0752rev.

[16] Shibutani S, Takeshita M, Grollman AP. Insertion of specific bases during DNA synthesis past the oxidation-damaged base 8-oxodG. Nature 1991;349:431–4. Available from: https://doi.org/10.1038/349431a0.

[17] Kunkel TA. DNA replication fidelity. J Biol Chem 2004;279:16895–8. Available from: https://doi.org/10.1074/jbc.r400006200.

[18] Waters LS, et al. Eukaryotic translesion polymerases and their roles and regulation in DNA damage tolerance. Microbiol Mol Biol Rev 2009;73:134–54. Available from: https://doi.org/10.1128/mmbr.00034-08.

[19] Regulus P, et al. Oxidation of the sugar moiety of DNA by ionizing radiation or bleomycin could induce the formation of a cluster DNA lesion. Proc Natl Acad Sci 2007;104:14032–7. Available from: https://doi.org/10.1073/pnas.0706044104.

[20] Rokita SE, Romero-Fredes L. The ensemble reactions of hydroxyl radical exhibit no specificity for primary or secondary structure of DNA. Nucleic Acids Res 1992;20:3069–72. Available from: https://doi.org/10.1093/nar/20.12.3069.

[21] Dutta A, Yang C, Sengupta S, Mitra S, Hegde ML. New paradigms in the repair of oxidative damage in human genome: mechanisms ensuring repair of mutagenic base lesions during replication and involvement of accessory proteins. Cell Mol Life Sci 2015;72:1679–98. Available from: https://doi.org/10.1007/s00018-014-1820-z.

[22] Groehler A, et al. Oxidative cross-linking of proteins to DNA following ischemia-reperfusion injury. Free Radic Biol Med 2018;120:89–101. Available from: https://doi.org/10.1016/j.freeradbiomed.2018.03.010.

[23] Perrier S, et al. Characterization of lysine–guanine cross-links upon one-electron oxidation of a guanine-containing oligonucleotide in the presence of a trilysine peptide. J Am Chem Soc 2006;128:5703–10. Available from: https://doi.org/10.1021/ja057656i.

[24] Birben E, Sahiner UM, Sackesen C, Erzurum S, Kalayci O. Oxidative stress and antioxidant defense. World Allergy Organ J 2012;5:9–19. Available from: https://doi.org/10.1097/WOX.0b013e3182439613.

[25] Mandavilli BS, Santos JH, van Houten B. Mitochondrial DNA repair and aging. Mutat Res 2002;509:127–51.

[26] Prakash A, Doublie S. Base excision repair in the mitochondria. J Cell Biochem 2015;116:1490–9.

[27] Sharma N, Chakravarthy S, Longley MJ, Copeland WC, Prakash A. The C-terminal tail of the NEIL1 DNA glycosylase interacts with the human mitochondrial single-stranded DNA binding protein. DNA Repair 2018;65:11–19.

[28] Wollen Steen K, et al. mtSSB may sequester UNG1 at mitochondrial ssDNA and delay uracil processing until the dsDNA conformation is restored. DNA Repair 2012;11:82–91.

[29] van Loon B, Samson L. Alkyladenine DNA glycosylase (AAG) localizes to mitochondria and interacts with mitochondrial single-stranded binding protein (mtSSB). DNA Repair 2013;12:177–87.

[30] Hadi MZ, Wison DMI. Second human protein with homology to the *Escherichia coli* abasic endonuclease exonuclease III. Env Mol Mutagen 2000;36:312–24.

[31] Demple B, Herman T, Chen DS. Cloning and expression of APE, the cDNA encoding the major human apurinic endonuclease: definition of a family of DNA repair enzymes. Proc Natl Acad Sci U S A 1991;88:11450–4.

[32] Xanthoudakis S, Smeyne RJ, Wallace JD, Curran T. The redox/DNA repair protein, Ref-1, essential for early embryonic development in mice. Proc Natl Acad Sci U S A 1996;93:8919–23.

[33] Ribar B, Izumi T, Mitra S. The major role of human AP-endonuclease homolog Apn2 in repair of abasic sites in *Schizosaccharomyces pombe*. Nucleic Acids Res 2004;32:115–26.

[34] Wiederhold L, et al. AP endonuclease-independent DNA base excision repair in human cells. Mol Cell 2004;15:209–20.

[35] Chattopadhyay R, et al. Identification and characterization of mitochondrial abasic (AP)-endonuclease in mammalian cells. Nucleic Acids Res 2006;34:2067–76.

[36] Tsuchimoto D, et al. Human APE2 protein is mostly localized in the nuclei and to some extent in the mitochondria, while nuclear APE2 is partly associated with proliferating cell nuclear antigen. Nucleic Acids Res 2001;29:2349–60.

[37] Akbari M, Otterlei M, Pena-Diaz J, Krokan HE. Different organization of base excision repair of uracil in DNA in nuclei and mitochondria and selective upregulation of mitochondrial uracil-DNA glycosylase after oxidative stress. Neuroscience 2007;145:1201–12.

[38] Tanaka M, Koike M. DNA polymerase-γ is localized in mitochondria. Biochem Biophys Res Comm 1978;81:791–7.

[39] Hubscher U, Kuenzle CC, Spadari S. Functional roles of DNA polymerases beta and gamma. Proc Natl Acad Sci U S A 1979;76:2316–20.

[40] Sykora P, et al. DNA polymerase beta participates in mitochondrial DNA repair. Mol Cell Biol 2015;37 e00237-00217.

[41] Prasad R, et al. DNA polymerase b: a missing link of the base excision repair machinery in mammalian mitochondria. DNA Repair 2017;60:77–88.

[42] He Q, Shumate CK, White MA, Molineux IJ, Yin YW. Exonuclease of human DNA polymerase gamma disengages its strand displacement function. Mitochondrion 2013;13:592–601.
[43] Longley MJ, Prasad R, Srivastava DK, Wilson SH, Copeland WC. Identification of 5′-deoxyribose phosphate lyase activity in human DNA polymerase γ and its role in mitochondrial base excision repair *in vitro*. Proc Natl Acad Sci U S A 1998;95:12244–8.
[44] Nilsen H, et al. Nuclear and mitochondrial uracil-DNA glycosylases are generated by alternative splicing and transcription from different positions in the UNG gene. Nucleic Acids Res 1997;25:750–5.
[45] Otterlei M, et al. Nuclear and mitochondrial splice forms of human uracil-DNA glycosylase contain a complex nuclear localisation signal and a strong classical mitochondrial localisation signal, respectively. Nucleic Acids Res 1998;26:4611–17.
[46] Li M, et al. Identification and characterization of mitochondrial targeting sequence of human apurinic/apyrimidinic endonuclease 1. J Biol Chem 2010;285:14871–81.
[47] Vascotto C, et al. Knock-in reconstitution studies reveal an unexpected role of Cys-65 in regulating APE1/Ref-1 subcellular trafficking and function. Mol Biol Cell 2011;22:3887–901.
[48] Mordas A, Tokatlidis K. The MIA pathway: a key regulator of mitochondrial oxidative protein folding and biogenesis. Acc Chem Res 2015;48:2191–9.
[49] Longen S, Woellhaf MW, Petrungaro C, Riemer J, Herrmann JM. The disulfide relay of the intermembrane space oxidizes the ribosomal subunit Mrp10 on its transit into the mitochondrial matrix. Dev Cell 2014;28:30–42.
[50] Kirby TW, et al. DNA polymerase b contains a functional nuclear localization signal at its N-terminus. Nucleic Acids Res 2016;45:1958–70.
[51] Melis JPM, Van Steeg H, Luijten M. Oxidative DNA damage and nucleotide excision repair. Antioxid Redox Signal 2013;18:2409–19.
[52] Spivak G. Nucleotide excision repair in humans. DNA Repair 2015;36:13–18.
[53] Mu H, Geacintov NE, Broyde S, Yeo J-E, Scharer OD. Molecular basis for damage recognition and verification by XPC-RAD23B and TFIIH in nucleotide excision repair. DNA Repair 2018;71:33–42.
[54] Geijer ME, Marteijn JA. What happens at the lesion does not stay at the lesion: transcription-coupled nucleotide excision repair and the effects of DNA damage on transcription *in cis* and *trans*. DNA Repair 2018;71:56–68.
[55] Clayton DA, Doda JN, Friedberg EC. The absence of a pyrimidine dimer repair mechanism in mammalian mitochondria. Proc Natl Acad Sci U S A 1974;71:2777–81.
[56] Kamenisch Y, et al. Proteins of nucleotide and base excision repair pathways interact in mitochondria to protect from loss of subcutaneous fat, a hallmark of aging. J Exp Med 2010; epub ahead of print.
[57] Aamann MD, et al. Cockayne syndrome group B protein promotes mitochondrial DNA stability by supporting the DNA repair association with the mitochondrial membrane. FASEB J 2010;24:2334–46.
[58] Karikkineth AC, Scheibye-Knudsen M, Fivenson E, Croteau DL, Bohr VA. Cockayne syndrome: clinical features, model systems and pathways. Ageing Res Rev 2017;33:3–17.
[59] Wisnovsky S, Jean SR, Liyanage S, Schimmer A, Kelley SO. Mitochondrial DNA repair and replication proteins revealed by targeted chemical probes. Nat Chem Biol 2016;12:567–73.
[60] Liu J, et al. XPD localizes in mitochondria and protects the mitochondrial genome from oxidative DNA damage. Nucleic Acids Res 2015;43:5476–88.
[61] Prakash L, Prakash S. Three additional genes involved in pyrimidine dimer removal in *Saccharomyces cerevisiae*: *RAD7, RAD14*, and *MMS19*. Mol Gen Genet 1979;176:351–9.
[62] Gari K, et al. MMS19 links cytoplasmic iron-sulfur cluster assembly to DNA metabolism. Science 2012;337:243–6.
[63] Vashisht AA, Yu CC, Sharma T, Ro K, Wohlschlegel JA. The association of the Xeroderma pigmentosum group D DNA helicase (XPD) with transcription factor IIH is regulated by the cytosolic iron-sulfur cluster assembly pathway. J Biol Chem 2015;290:14218–25.
[64] Kou H, Zhou Y, Gorospe RMC, Wang Z. Mms19 protein functions in nucleotide excision repair by sustaining an adequate cellular concentration of the TFIIH component Rad3. Proc Natl Acad Sci U S A 2008;105:15714–19.

[65] Wu R, et al. MMS19 localizes to mitochondria and protects the mitochondrial genome from oxidative damage. Biochem Cell Biol 2018;96:44–9.
[66] Stehling O, Lill R. The role of mitochondria in cellular iron-sulfur protein biogenesis: mechanisms, connected processes, and diseases. Cold Spring Harb Perspect Biol 2013;5:a011312.
[67] Roots R, Kraft G, Gosschalk E. The formation of radiation-induced DNA breaks: the ratio of double-strand breaks to single-strand breaks. Int J Radiat Oncol Biol Phys 1985;11:259–65. Available from: https://doi.org/10.1016/0360-3016(85)90147-6.
[68] Kiraly O, et al. DNA glycosylase activity and cell proliferation are key factors in modulating homologous recombination *in vivo*. Carcinogenesis 2014;35:2495–502.
[69] Schoeffler AJ, Berger JM. DNA topoisomerases: harnessing and constraining energy to govern chromosome topology. Q Rev Biophys 2008;41:41–101. Available from: https://doi.org/10.1017/S003358350800468X.
[70] Wang JC. Cellular roles of DNA topoisomerases: a molecular perspective. Nat Rev Mol Cell Biol 2002;3:430–40. Available from: https://doi.org/10.1038/nrm831.
[71] Nitiss JL. DNA topoisomerase II and its growing repertoire of biological functions. Nat Rev Cancer 2009;9:327–37. Available from: https://doi.org/10.1038/nrc2608.
[72] Lu HR, et al. Reactive oxygen species elicit apoptosis by concurrently disrupting topoisomerase II and DNA-dependent protein kinase. Mol Pharmacol 2005;68:983–94. Available from: https://doi.org/10.1124/mol.105.011544.
[73] San Filippo J, Sung P, Klein H. Mechanism of eukaryotic homologous recombination. Annu Rev Biochem 2008;77:229–57. Available from: https://doi.org/10.1146/annurev.biochem.77.061306.125255.
[74] Morrical SW. DNA-pairing and annealing processes in homologous recombination and homology-directed repair. Cold Spring Harb Perspect Biol 2015;7:a016444. Available from: https://doi.org/10.1101/cshperspect.a016444.
[75] Lieber MR. The mechanism of double-strand DNA break repair by the nonhomologous DNA end-joining pathway. Annu Rev Biochem 2010;79:181–211. Available from: https://doi.org/10.1146/annurev.biochem.052308.093131.
[76] Chang HHY, Pannunzio NR, Adachi N, Lieber MR. Non-homologous DNA end joining and alternative pathways to double-strand break repair. Nat Rev Mol Cell Biol 2017;18:495–506. Available from: https://doi.org/10.1038/nrm.2017.48.
[77] Riballo E, et al. A pathway of double-strand break rejoining dependent upon ATM, Artemis, and proteins locating to gamma-H2AX foci. Mol Cell 2004;16:715–24. Available from: https://doi.org/10.1016/j.molcel.2004.10.029.
[78] Wang M, et al. PARP-1 and Ku compete for repair of DNA double strand breaks by distinct NHEJ pathways. Nucleic Acids Res 2006;34:6170–82. Available from: https://doi.org/10.1093/nar/gkl840.
[79] Dahal S, Dubey S, Raghavan SC. Homologous recombination-mediated repair of DNA double-strand breaks operates in mammalian mitochondria. Cell Mol Life Sci 2018;75:1641–55. Available from: https://doi.org/10.1007/s00018-017-2702-y.
[80] Thyagarajan B, Padua RA, Campbell C. Mammalian mitochondria possess homologous DNA recombination activity. J Biol Chem 1996;271:27536–43. Available from: https://doi.org/10.1074/jbc.271.44.27536.
[81] Dmitrieva NI, Malide D, Burg MB. Mre11 is expressed in mammalian mitochondria where it binds to mitochondrial DNA. Am J Physiol Regul Integr Comp Physiol 2011;301:R632–40. Available from: https://doi.org/10.1152/ajpregu.00853.2010.
[82] Sage JM, Gildemeister OS, Knight KL. Discovery of a novel function for human Rad51: maintenance of the mitochondrial genome. J Biol Chem 2010;285:18984–90. Available from: https://doi.org/10.1074/jbc.M109.099846.
[83] Kraytsberg Y. Recombination of human mitochondrial DNA. Science 2004;304:981. Available from: https://doi.org/10.1126/science.1096342.
[84] Herbers E, Kekäläinen NJ, Hangas A, Pohjoismäki JL, Goffart S. Tissue specific differences in mitochondrial DNA maintenance and expression. Mitochondrion 2019;44:85–92. Available from: https://doi.org/10.1016/j.mito.2018.01.004.

[85] Pohjoismaki JL, et al. Human heart mitochondrial DNA is organized in complex catenated networks containing abundant four-way junctions and replication forks. J Biol Chem 2009;284:21446–57. Available from: https://doi.org/10.1074/jbc.M109.016600.
[86] Pohjoismaki JL, et al. Overexpression of Twinkle-helicase protects cardiomyocytes from genotoxic stress caused by reactive oxygen species. Proc Natl Acad Sci U S A 2013;110:19408–13. Available from: https://doi.org/10.1073/pnas.1303046110.
[87] Srivastava S, Moraes CT. Double-strand breaks of mouse muscle mtDNA promote large deletions similar to multiple mtDNA deletions in humans. Hum Mol Genet 2005;14:893–902. Available from: https://doi.org/10.1093/hmg/ddi082.
[88] Bacman SR, Williams SL, Duan D, Moraes CT. Manipulation of mtDNA heteroplasmy in all striated muscles of newborn mice by AAV9-mediated delivery of a mitochondria-targeted restriction endonuclease. Gene Ther 2012;19:1101–6. Available from: https://doi.org/10.1038/gt.2011.196.
[89] Bacman SR, Williams SL, Garcia S, Moraes CT. Organ-specific shifts in mtDNA heteroplasmy following systemic delivery of a mitochondria-targeted restriction endonuclease. Gene Ther 2010;17:713–20. Available from: https://doi.org/10.1038/gt.2010.25.
[90] Bacman SR, Williams SL, Pinto M, Peralta S, Moraes CT. Specific elimination of mutant mitochondrial genomes in patient-derived cells by mitoTALENs. Nat Med 2013;19:1111–13. Available from: https://doi.org/10.1038/nm.3261.
[91] Gammage PA, Rorbach J, Vincent AI, Rebar EJ, Minczuk M. Mitochondrially targeted ZFNs for selective degradation of pathogenic mitochondrial genomes bearing large-scale deletions or point mutations. EMBO Mol Med 2014;6:458–66. Available from: https://doi.org/10.1002/emmm.201303672.
[92] Srivastava S. Manipulating mitochondrial DNA heteroplasmy by a mitochondrially targeted restriction endonuclease. Hum Mol Genet 2001;10:3093–9. Available from: https://doi.org/10.1093/hmg/10.26.3093.
[93] Tanaka M, et al. Gene therapy for mitochondrial disease by delivering restriction endonuclease SmaI into mitochondria. J Biomed Sci 2002;9:534–41. Available from: https://doi.org/10.1159/000064726.
[94] Bian W-P, et al. Knock-in strategy for editing human and zebrafish mitochondrial DNA using mito-CRISPR/Cas9 system. ACS Synth Biol 2019;8:621–32. Available from: https://doi.org/10.1021/acssynbio.8b00411.
[95] Tadi SK, et al. Microhomology-mediated end joining is the principal mediator of double-strand break repair during mitochondrial DNA lesions. Mol Biol Cell 2016;27:223–35. Available from: https://doi.org/10.1091/mbc.E15-05-0260.
[96] Coffey G. An alternate form of Ku80 is required for DNA end-binding activity in mammalian mitochondria. Nucleic Acids Res 2000;28:3793–800. Available from: https://doi.org/10.1093/nar/28.19.3793.
[97] Yusoff AAM, Abdullah WSW, Khair SZNM, Radzak SMA. A comprehensive overview of mitochondrial DNA 4977-bp deletion in cancer studies. Oncol Rev 2019;13:409. Available from: https://doi.org/10.4081/oncol.2019.409.
[98] Phillips AF, et al. Single-molecule analysis of mtDNA replication uncovers the basis of the common deletion. Mol Cell 2017;65:527–538.e526. Available from: https://doi.org/10.1016/j.molcel.2016.12.014.
[99] Martin-Negrier ML, et al. TWINKLE gene mutation: report of a French family with an autosomal dominant progressive external ophthalmoplegia and literature review. Eur J Neurol 2011;18:436–41. Available from: https://doi.org/10.1111/j.1468-1331.2010.03171.x.
[100] Kunkel TA. Evolving views of DNA replication (in)fidelity. Cold Spring Harb Symp Quant Biol 2009;74:91–101.
[101] Su SS, Modrich P. *Escherichia coli mutS*-encoded protein binds to mismatched DNA base pairs. Proc Natl Acad Sci U S A 1986;83:5057–61.
[102] Grilley M, Welsh KM, Su SS, Modrich P. Isolation and characterization of the *Escherichia coli mutL* gene product. J Biol Chem 1989;264:1000–4.
[103] Au KG, Welsh K, Modrich P. Initiation of methyl-directed mismatch repair. J Biol Chem 1992;267:12142–8.
[104] Cooper DL, Lahue RS, Modrich P. Methyl-directed mismatch repair is bidirectional. J Biol Chem 1993;268:11823–9.

[105] Lahue RS, Su SS, Modrich P. Requirement for d(GATC) sequences in *Escherichia coli* mutHLS mismatch correction. Proc Natl Acad Sci U S A 1987;84:1482–6.
[106] Reenan RA, Kolodner RD. Isolation and characterization of two *Saccharomyces cerevisiae* genes encoding homologs of the bacterial HexA and MutS mismatch repair proteins. Genetics 1992;132:963–73.
[107] Iaccarino I, et al. MSH6, a *Saccharomyces cerevisiae* protein that binds to mismatches as a heterodimer with MSH2. Curr Biol 1996;6:484–6.
[108] Palombo F, et al. hMutSbega, a heterodimer of hMSH2 and hMSH3, binds to insertion/deletion loops in DNA. Curr Biol 1996;6:1181–4.
[109] Habraken Y, Sung P, Prakash L, Prakash S. Binding of insertion/deletion DNA mismatches by the heterodimer of yeast mismatch repair proteins MSH2 and MSH3. Curr Biol 1996;6:1185–7.
[110] Kramer W, Kramer B, Williamson MS, Fogel S. Cloning and nucleotide sequence of DNA mismatch repair gene PMS1 from *Saccharomyces cerevisiae*: homology of PMS1 to procaryotic MutL and HexB. J Bacteriol 1989;171:5339–46.
[111] Prolla TA, Christie DM, Liskay RM. Dual requirement in yeast DNA mismatch repair for MLH1 and PMS1, two homologs of the bacterial *mutL* gene. Mol Cell Biol 1994;14:407–15.
[112] Pavlov YI, Mian IM, Kunkel TA. Evidence for preferential mismatch repair of lagging strand DNA replication errors in yeast. Curr Biol 2003;13:744–8.
[113] Mason PA, Matheson EC, Hall AG, Lightowlers RN. Mismatch repair activity in mammalian mitochondria. Nucleic Acids Res 2003;31:1052–8.
[114] Didier DK, Schiffenbauer J, Woulfe SL, Zacheis M, Schwartz BD. Characterization of the cDNA encoding a protein binding to the major histocompatibility complex class II Y box. Proc Natl Acad Sci U S A 1988;85:7322–6.
[115] Hasegawa SL, et al. DNA binding properties of YB-1 and dbpA: binding to double-stranded, single-stranded, and abasic site containing DNAs. Nucleic Acids Res 1991;19:4915–20.
[116] Lyabin DN, Eliseeva IA, Ovchinnikov LP. YB-1 protein: functions and regulation. Wiley Interdiscip Rev: RNA 2013;5:95–110.
[117] Gaudreault I, Guay D, Lebel M. YB-1 promotes strand separation *in vitro* of duplex DNA containing either mispaired bases or cisplatin modifications, exhibits endonucleolytic activities and binds several DNA repair proteins. Nucleic Acids Res 2004;32:316–27.
[118] Fomina EE, et al. Y-Box binding protein 1 (YB-1) promotes detection of DNA bulky lesions by XPC-HR23B factor. Biochem (Mosc) 2015;80:219–27.
[119] Alemasova EE, et al. Y-box-binding protein 1 as a non-canonical factor of base excision repair. Biochem Biophys Acta 2016;1864:1631–40.
[120] Das S, et al. Stimulation of NEIL2-mediated oxidized base excision repair via YB-1 interaction during oxidative stress. J Biol Chem 2007;282:28474–84.
[121] Alemasova EE, Naumenko KN, Moor NA, Lavrik OI. Y-box-binding protein 1 stimulates abasic site cleavage. Biochem (Mosc) 2017;82:1521–8.
[122] Chang Y-W, et al. YB-1 disrupts mismatch repair complex formation, interferes with MutSa recruitment on mismatch and inhibits mismatch repair through interacting with PCNA. Oncogene 2014;33:5065–77.
[123] de Souza-Pinto NC, et al. Novel DNA mismatch repair activity involving YB-1 in human mitochondria. DNA Repair 2009;8:704–19.
[124] Pinz KG, Shibutani S, Bogenhagen DF. Action of mitochondrial DNA polymerase γ at sites of base loss or oxidative damage. J Biol Chem 1995;270:9202–6.
[125] Schmitt MW, Matsumoto Y, Loeb LA. High fidelity and lesion bypass capability of human DNA polymerase delta. Biochimie 2009;91:1163–72.
[126] Schmitt MW, et al. Active site mutations in mammalian DNA polymerase delta alter accuracy and replication fork progression. J Biol Chem 2010;285:32264–72.
[127] Quinet A, Lerner LK, Martins DJ, Menck CFM. Filling gaps in translesion DNA synthesis in human cells. Mutat Res Gen Tox En 2018;836:127–42.

[128] Lawrence CW, Gibbs PE, Borden A, Horsfall MJ, Kilbey BJ. Mutagenesis induced by single UV photoproducts in *E. coli* and yeast. Mutat Res 1993;299:157–63.
[129] Ikehata H, Ono T. Mechanisms of UV mutagenesis. J Radiat Res 2011;52:115–25.
[130] Ejchart A, Putrament A. Mitochondrial mutagenesis in *Saccharomyces cerevisiae* I. Ultraviolet radiation. Mutat Res 1979;60:173–80.
[131] Pascucci B, et al. DNA repair of UV photoproducts and mutagenesis in human mitochondrial DNA. J Mol Biol 1997;273:417–27.
[132] Longley MJ, Nguyen D, Kunkel TA, Copeland WC. The fidelity of human DNA polymerase gamma with and without exonucleolytic proofreading and the p55 accessory subunit. J Biol Chem 2001;276:38555–62.
[133] Kunkel TA, Mosbaugh DW. Exonucleolytic proofreading by a mammalian DNA polymerase. Biochemistry 1989;28:988–95.
[134] Graziewicz MA, Sayer JM, Jerina DM, Copeland WC. Nucleotide incorporation by human DNA polymerase γ opposite benzo[a] pyrene and benzo[c] phenanthrene diol epoxide adducts of deoxyguanosine and deoxyadenosine. Nucleic Acids Res 2004;32:397–405.
[135] Kasiviswanathan R, Minko I, Lloyd RS, Copeland WC. Translesion synthesis past acrolein-derived DNA adductions by human mitochondrial DNA polymerase gamma. J Biol Chem 2013;288:12247–4255.
[136] Kasiviswanathan R, Gustafson MA, Copeland WC, Meyer JN. Human mitochondrial DNA polymerase gamma exhibits potential for bypass and mutagenesis at UV-induced cyclobutane thymine dimers. J Biol Chem 2012;287:9222–9.
[137] Garcia-Gomez S, et al. PrimPol, an archaic primase/polymerase operating in human cells. Mol Cell 2013;52:541–53.
[138] Bianchi J, et al. PrimPol bypasses UV photoproducts during eukaryotic chromosomal DNA replication. Mol Cell 2013;52:566–73.
[139] Zafar MK, Ketkar A, Lodeiro MF, Cameron CE, Eoff RL. Kinetic analysis of human PrimPol DNA polymerase activity reveals a generally error-prone enzyme capable of accurately bypassing 7,8-dihydro-8-oxo-2′-deoxyguanosine. Biochemistry 2014;53:6584–94.
[140] Schenten D, et al. Pol zeta ablation in B cells impairs the germinal center reaction, class switch recombination, DNA break repair, and genome stability. J Exp Med 2009;206:477–90.
[141] Esposito G, et al. Disruption of the Rev3l-encoded catalytic subunit of polymerase zeta in mice results in early embryonic lethality. Curr Biol 2000;10:1221–4.
[142] Wittschieben J, et al. Disruption of the developmentally regulated *Rev3l* gene causes embryonic lethality. Curr Biol 2000;10:1217–20.
[143] Bemark M, Khamlichi AA, Davies SL, Neuberger MS. Disruption of mouse polymerase zeta (*Rev3*) leads to embryonic lethality and impairs blastocyst development *in vitro*. Curr Biol 2000;10:1213–16.
[144] Makarova AV, Burgers PM. Eukaryotic DNA polymerase zeta. DNA Repair 2015;29:47–55.
[145] Van Sloun PP, et al. Involvement of mouse Rev3 in tolerance of endogenous and exogenous DNA damage. Mol Cell Biol 2002;22:2159–69.
[146] Singh B, et al. Human REV3 DNA polymerase zeta localizes to mitochondria and protects the mitochondrial genome. PLoS One 2015;10:e140409.
[147] Mateos-Gomez PA. Mammalian polymerase theta promotes alternative NHEJ and suppresses recombination. Nature 2015;518:254–7.
[148] Sugo N, Aratani Y, Nagshima Y, Kubota Y, Koyama H. Neonatal lethality with abnormal neurogenesis in mice deficient in DNA polymerase β. EMBO J 2000;19:1397–404.
[149] Sun X, St. John JC. The role of the mtDNA set point in differentiation, development and tumorigenesis. Biochemical J 2016;473:2955–71.
[150] Scheffler IE. Structure and morphology: integration into the cell. Mitochondria. Wiley-Liss; 1999 [chapter 3].

CHAPTER 9

Mechanisms of onset and accumulation of mtDNA mutations

Ian James Holt[1,2,3,4] and Antonella Spinazzola[4,5]

[1]Biodonostia Health Research Institute, San Sebastián, Spain, [2]IKERBASQUE, Basque Foundation for Science, Bilbao, Spain, [3]CIBERNED (Center for Networked Biomedical Research on Neurodegenerative Diseases, Ministry of Economy and Competitiveness, Institute Carlos III), Madrid, Spain, [4]Department of Clinical and Movement Neurosciences, UCL Queen Square Institute of Neurology, London, United Kingdom, [5]Queen Square Centre for Neuromuscular Diseases, UCL Queen Square Institute of Neurology and National Hospital for Neurology and Neurosurgery, London, United Kingdom

9.1 Mitochondrial DNA abnormalities

9.1.1 Primary mitochondrial DNA mutants

We define as primary mtDNA disorders those arising from a specific variant of this genome. Because mtDNA is transmitted exclusively from mother to offspring with no paternal contribution, primary pathological mutants are exclusively maternally transmitted. However, as detailed below, the more egregious the mutant, the less likely it is to be transmitted to the offspring, at least in mammals.

9.1.1.1 Rearrangements: deletions and duplications

The first pathological mutations discovered in mtDNA were unequivocal, and provided several important insights into mtDNA disorders [1]. Many molecules lacked one or more kilobases of mtDNA, including at least 1 tRNA gene (Fig. 9.1A and B), and therefore in isolation, they could not yield the mitochondrial proteins required for oxidative phosphorylation (OXPHOS). However, the (partially) deleted molecules coexisted with full-length mtDNA, and so the biochemical and clinical phenotypes depend on the extent to which the wild-type mtDNAs are able to complement the mutant molecules.

Although the size and position of the deleted portion of the mitochondrial genome differed among patients, each individual lacked only one specific region. Strikingly, the mtDNA deletions were evident in muscle but not blood, suggesting that in proliferating cells, the

The Human Mitochondrial Genome.
DOI: https://doi.org/10.1016/B978-0-12-819656-4.00009-7

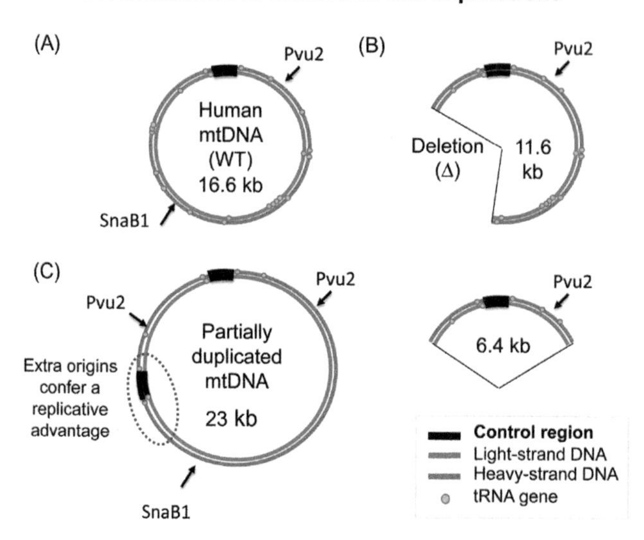

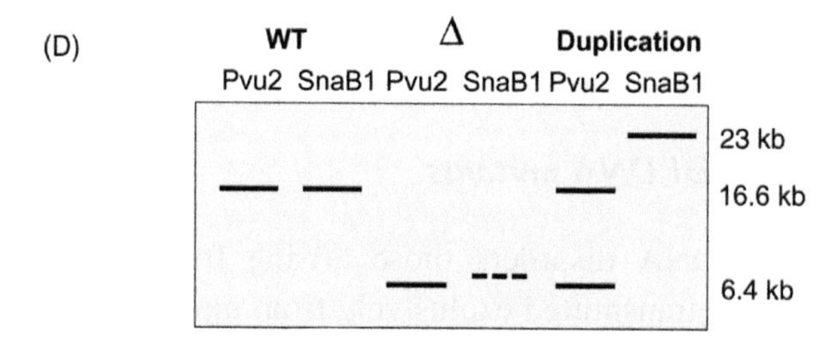

Figure 9.1: Pathological mtDNA variants and biased segregation.
(A) Wild type of 16.5 kb of duplex DNA. (B) A partially deleted molecule lacking 5 kb of mtDNA note the molecule will be a closed circle in situ. (C) A partially duplicated mtDNA molecule with additional origins of replication within and adjacent to the control region that can confer a replicative advantage, with the theoretical deleted equivalent to the right (see main text and Ref. [2]). (D) Illustrates how Southern hybridization can be used to distinguish duplications from deletions. Restriction enzymes Pvu2 and SnaB1 both cleave human mtDNA at a single site, yielding a linear band of 16.6 kb from normal control subjects (wild type, WT). A deleted molecule of 6.4 kb of the type shown in panel C will be linearized by Pvu2, but will not be cut by SnaB1 and so in the latter case will run as a circular molecule (depicted as a *broken line*, as the mobility of circular molecules can be very different according to the gel electrophoresis conditions). In contrast, a partially duplicated molecule containing the exact same 6.4 kb of mtDNA will yield linear fragments of 16.6 and 6 kb with Pvu2, but a single linear of 23 kb with SnaB1.

mitochondria with deletions were at a distinct disadvantage. The difficulty of maintaining deleted mtDNAs in a population of dividing cells was graphically illustrated when myoblasts from patients with deletions in muscle were cultivated and found to rapidly lose the population of deleted molecules [3]. Moreover, the patients with a specific form of deleted mtDNA were often sporadic cases, thereby indicating that obstacles exist to the transmission of both somatic and germ-line deleted mtDNAs. That said, in a cohort of 226 patients with a specific partially deleted mtDNA in muscle, the risk of an affected mother transmitting mutant mtDNA to her offspring was 1 in 24 [4]. However, it is not clear what number of these was transmitted in the form of other mtDNA rearrangements (see partial duplications below).

Heteroplasmy and the marked differences in mutant load between blood and muscle appeared to provide a straightforward explanation for the tissue-specific aspects of mtDNA disorders: affected tissues have high levels of mutant mtDNA. However, as we shall see, later studies of other defects in mtDNA indicated that the situation is more complicated than this hypothesis suggests.

To younger researchers, it is difficult to convey how much of a challenge manual DNA sequencing represented at that time, even with the advent of the recently invented technique of DNA amplification via PCR. Hence, it was a year or more before the precise deletion junctions were identified. The most frequent junction was found to constitute a residual 13 base pair direct repeat, thereby suggesting that the deletion arose via replication slippage or DNA recombination [5,6]. Other deletions also involved direct repeats; however, there was little or no homology in a substantial minority of deletions [7], and work many years later suggested that these could be the result of breakage and rejoining events, analogous to non-homologous end-joining (NHEJ) of nuclear DNA [7]. The puzzle is: why would human mtDNA operate a system akin to NHEJ when it is so gene-dense that almost every such event will result in a deleterious form of mtDNA? On the other hand, deleted mtDNAs are relatively rare in the human population (<1 in 100,000) and these rejoined disparate ends of mtDNA constitute a minority of the sporadic deletions; thus, NHEJ-like DNA repair could be seen as an occasional mishap rather than a routinely employed mechanism of DNA repair in mammalian mitochondria. Comparison of the sequence of the mitochondrial genome with the location of the deletion junctions suggests that G-quadruplexes may play a role in their formation [8,9], perhaps because they could function as replication pause sites, by analogy with nuclear DNA replication [10]. Interrupted DNA replication has also been implicated in the formation of the so-called common deletion, $mtDNA^{4977}$, based on an analysis of replicating mtDNA from cells expressing a defective form of the mtDNA helicase, Twinkle [11].

Another form of mutant mtDNA closely related to the deletions was discovered the following year, these were partial duplications [12] (Fig.9.1C). Duplications could form by the identical mechanism to deletions, either by the deletion event occurring on a mtDNA dimer, or a singleton deleted molecule recombining with a wild-type mtDNA [13]. The interchangeable nature of the two types of rearrangement is strikingly illustrated by some patients featuring both duplications and deletions of mtDNA [14]. Because the partially deleted and duplicated mtDNAs have identical junctions, they cannot be distinguished by

PCR using primers across the deletion/duplication junction. Hence, it is necessary to detect the mtDNA by Southern hybridization, which is considerably more time-consuming than PCR, and consequently is a dying art. The key is to cut the mtDNA with a restriction enzyme that has no recognition site *within* the putative deleted region—SnaB1 in the illustrated example (Fig. 9.1D). In the case of a partial duplication, this will produce a fragment considerably larger than that of wild-type mtDNA (Fig. 9.1D) [2]. Although few centers distinguish between the two types of rearrangement, it is expected that the risk of maternal transmission will be markedly higher for duplications than deletions [13]. Moreover, disease diagnosis is generally made on the basis of a muscle biopsy, which may contain exclusively deleted mtDNAs in the case of an affected mother, yet there may be partial duplications in other tissues (the germ-line being the critical one for transmission). This was likely the case in a mother and son who carried the same mtDNA deletion, as the son also had a small population of partial duplicated mtDNA in his muscle [15]. Thus the 1 in 24 risk of affected mothers with a single mtDNA deletion transmitting pathological mtDNA rearrangements to their offspring [4] probably reflects, at least in part, the incidence of partial duplications in germ cells rather than deletions per se.

Partial duplications of mtDNA are still considered, by some, to be benign, as they do not lack any genetic information [4]. Moreover, because deletions and duplications are not only theoretically interchangeable but also both present in differing proportions in some patients [14], it was suggested that the deleted molecules might be the only ones contributing to the pathology [4]. However, this perspective ignores the fact that no deleted molecules were detected in blood or muscle of a patient with partially duplicated mtDNA [2]. While the duplication was associated with mild mitochondrial dysfunction, a triplication that formed ex vivo provoked severe respiratory deficiency [16]. Hence, these findings refute the idea that duplications (and triplications) are benign and that only the deleted forms are deleterious.

As mentioned above, the existence of deletions and duplications in the same family and individual is most readily explained by DNA recombination. There is unequivocal evidence of mtDNA recombination enzymes and products in some organisms, such as yeast and plants [17,18]. However, the putative mammalian homolog of the yeast Holliday junction resolvase functions primarily as an ancillary factor of the mitochondrial RNA polymerase [19], and one study found no evidence of mtDNA recombination products across 50 generations in mice [20]. Moreover, it is possible that the mixtures of deleted, wild-type, and partially duplicated mtDNAs could reflect a single founder event followed by competition between the three distinct forms (see Section 9.5.3), rather than an active conversion process mediated by DNA recombination. On the other hand, interconnected multicopy mtDNA molecules have been detected in a number of contexts in mammals, most notably adult human heart [21]. It is unlikely that these arrangements of mtDNA exist without an accompanying DNA recombination apparatus to create and resolve them. Moreover, mammalian mtDNA recombination products may be rarely seen because the segregation of daughter molecules is carefully orchestrated, and after segregation, mtDNA molecules are generally kept apart. Certainly, there was little evidence of mixing when two different mtDNA genotypes were introduced into the same cell [22].

9.1.1.2 Mitochondrial DNA point mutants

The first pathological point mutation of mtDNA was discovered later in the same year deletions were reported [23], and with hindsight, it is something of an outlier. The point mutation, m.11778G was located in one of the genes encoding a subunit of respiratory chain complex I (ND4) and the resulting disease is largely restricted to the retinal ganglion cells of the optic nerve, which manifests as a rapid deterioration in vision (subacute bilateral visual impairment). Blindness is not a typical feature of mitochondrial diseases, although optic atrophy is not uncommon, and the loss of vision in a matter of weeks after decades without symptoms is the opposite of the slow progressive deterioration that typifies many other mitochondrial (DNA) disorders. The disease is named Leber's Hereditary Optic Neuropathy (LHON) [Online Mendelian Inheritance in Man (OMIM) # 535000], after its discoverer, although it would be more accurate and informative to call it Leber's Maternally Inherited Optic Neuropathy. In most [23], but not all cases [24], every copy of the mtDNA carries the mutation. Onset is usually in the third or fourth decade, and considerably more males than females manifest the disease. Issues of selection and distribution of the mutant mtDNA do not apply to homoplasmic mutants, and so neither the tissue specificity nor the variable penetrance is explained by differences in mutant load. In 2014, it was reported that heightened mitochondrial biogenesis reduces the penetrance of Leber's disease, although the genetic traits responsible remain to be determined [25]. Two other homoplasmic mutations in the mtDNA cause LHON, and all three affect genes encoding components of respiratory complex I. The m.14484C variant is associated with a less severe visual impairment and does not cause complex I deficiency in fibroblasts, though it may produce complex I deficiency in the affected tissue [26]. On the other hand, it is puzzling that m.3460 causes a marked complex I deficiency in fibroblasts equivalent to some severe paediatric complex I disorders, yet, symptoms are restricted to the eye. All three LHON mutants are "organ specific", following the broad rule that homoplasmic mutations usually affect one organ or tissue. Over the course of three decades, point mutations in other protein-encoding genes have been documented, although most of them are heteroplasmic, including ones in ATP synthase subunit *a* or A6 [27,28]. However, mutations in protein-encoding genes are far outnumbered by pathological point mutations in transfer RNA genes.

In 1990 two mutations in mitochondrial transfer RNA genes were linked to two specific syndromes MELAS (**M**itochondrial **E**ncephalomyopathy, **L**actic **A**cidosis, and **S**troke-like episodes OMIM #540000) and MERRF (**M**yoclonic **E**pilepsy with **R**agged-**R**ed **F**iber OMIM # 545000) [29,30]. Subsequently, mutations in all the tRNA genes have been reported and 10 pathological variants have been assigned to the $tRNA^{Leu(UUR)}$ gene. As expected, high levels of these mutant mtDNAs are associated with impaired mitochondrial protein synthesis [31,32]. This, in turn, decreases respiratory chain capacity of complexes I, III, and IV; however, at lower mutant load, complex I can be the only affected respiratory chain enzyme [33]. Mitochondrial tRNA gene mutants are maternally transmitted and, in the vast majority of cases, are heteroplasmic.

There are very few pathological mutations in the ribosomal RNA genes; however, a notable exception is a point mutation in the 12S rRNA gene m.1555G. This mutant is strictly associated with hearing impairment, but only rarely causes disease in normal circumstances [34]. It came to light chiefly because m.1555G confers heightened sensitivity to aminoglycoside antibiotics, and these were in the past widely, and sometimes, casually prescribed, in various parts of the world, leading to many cases of antibiotic-induced deafness.

To our knowledge, there are no established pathological mutations in the major noncoding region. This includes not only two hypervariable regions but also some critically important *cis*-elements. Thus, one could argue that variants in this region will be either benign or devastating. Although heteroplasmy allows highly deleterious variants to be sustained, as evidenced by the aforementioned deletions, considering the entire mitochondrial genome a clear spectrum emerges: tRNAs are diffusible and thus most readily complemented by the products of sister wild-type mtDNAs and these are the most numerous and varied class of mtDNA mutants. At the other extreme, deleterious mutations in the control region and the rRNAs will affect all the protein products, and these are particularly rare, while point mutations in individual protein encoding genes occupy the middle ground in both respects.

9.2 Criteria to designate a primary mtDNA mutation as pathological

Mitochondrial DNA is highly polymorphic and so designation of a variant as pathogenic, rather than a neutral polymorphism, is based on a number of criteria:

First, the variant must be absent from a substantial number of geographically and genetically matched controls; second, the mutation should affect a conserved site, as this is an indication of functional importance; third, the variant must segregate with the disease, and any transmission must be through the maternal line. Fourth, where the mutant coexists with wild-type molecules, there must be a correlation between the mutant load and the mitochondrial dysfunction. Although these criteria are valuable for ascertaining the likelihood that a mutation is disease-causing, not all the pathogenic mutations dutifully comply. For example, one mtDNA mutant is effectively "dominant" [35] and one well-established point mutant, m.8344G, affects a site that is not conserved in mammals. Moreover, some mutants produce only mild impairment of OXPHOS, so respiratory chain analysis may not reveal any defect; nevertheless, at least some of these milder mutants display increased glycolytic ATP production (e.g., Ref. [16]).

9.3 Clinical and biochemical correlates

Two clinical camps formed in the 1970s and 1980s, comprising those who favored classifying mitochondrial diseases according to specific syndromes and those who argued instead that it was better to recognize them as forming a spectrum of clinical features (e.g., muscle pathology with, in some cases, additional features such as neurological manifestation (e.g., ataxia, peripheral neuropathy) and/or deafness). The identification of the molecular basis of the disorders showed that both perspectives were legitimate. For example, mtDNA deletions were associated with a spectrum of disease that encompassed specific conditions: Pearson syndrome in neonates (OMIM # 557000) and Kearns-Sayre syndrome (OMIM # 530000), as well as impaired ocular muscle function in adult onset cases. MELAS syndrome was frequently, but not exclusively, associated with the m.3243G mutation; yet the exact same mutation also causes a quite different phenotypes featuring deafness and diabetes MIDD (OMIM # 520000). There is a clear benefit to patients and clinicians to be able to assign a name to a disease, even if it is not entirely specific, but the chief problem with the syndrome approach to classifying mitochondrial diseases is that it has given rise to a cornucopia of cumbersome acronyms that even specialists in the field struggle to commit to memory.

Mitochondrial respiration is often low in muscle homogenates or purified mitochondria from patients with deletions, and disorders such as MELAS, but this is not always the case and an obvious problem is that these measurements, whether enzyme assay or polarography based, assess the total population of mitochondria. Histochemical assays are therefore very helpful as they often show a mosaic of respiratory chain-deficient (COX-negative) fibers and ones where a huge expansion in mitochondrial number has occurred (ragged red fibers). Only a few percentage abnormal fibers marks a sample as quite different from healthy control subjects.

9.4 Mitochondrial genetic rules

Mutations in nuclear genes follow Mendelian genetics, where a mutant variant can be dominant or recessive. In contrast, the 1000s of mtDNAs in a typical cell create a situation more akin to population genetics, with mutant load for a given variant ranging from 0.01% to 100%, and for the most part, mtDNA is maternally transmitted. Nevertheless, some considerations apply to both nuclear and mtDNA. Highly deleterious variants that cause early-onset disorders are invariably recessive in the case of nuclear DNA, and most mtDNA mutants are recessive in that a disease only manifests if the mutant load exceeds 50% of the total. Only one pathological variant has been reported as dominant [35], whereas a number of mtDNA maintenance disorders due to nuclear defects (see below) are dominant. Variable penetrance is another feature of Mendelian genetics that is mirrored in mitochondrial diseases. In one striking case, a woman with a homoplasmic mutant mtDNA was sufficiently healthy to have children with several different fathers, yet all the offspring died in early infancy, indicating that this particular mutant mtDNA produces a catastrophic mitochondrial dysfunction on many, but not all, nuclear backgrounds [36].

Heteroplasmy often adds further complexity and provides a clear explanation for the clinical variation seen with the m.8993G pathological mutant: it can be benign up to 75% of the total mtDNA, produce NARP (OMIM # 551500) at 75%–90% and the more severe Leigh's syndrome (OMIM # 551500) at 90%–100% mutant load [27,28]. Although the situation is more complicated for m.3243G, differences in heteroplasmy levels between tissues are thought to be at least part of the explanation for the m.3243G mutant causing quite different clinical phenotypes [30,37].

9.5 Selection and counterselection of deleterious mtDNA variants

Transmission and selection of mtDNA variants occur not only in germ-line cells but also in cells and tissues, a phenomenon known as somatic mtDNA segregation. While there are differences between the two, the biggest being the restriction in mitochondrial (DNA) number during oogenesis (the mitochondrial bottleneck), which does not occur during somatic cell division, important factors that influence the competition between functional and deleterious variants apply in both contexts, as we will explain.

9.5.1 Phenotypic selection of fully functional mtDNAs

The first report of pathological mtDNA deletions [1]. Provided compelling, albeit indirect, evidence that severely deleterious mtDNA variants are lost from the population via Darwinian selection. Partially deleted mtDNAs are mostly sporadic and are rarely transmitted from mother to offspring. Moreover, the deleted mtDNAs are not maintained in blood (beyond the early years of life), which implies that they are lost via intercellular competition. Thus, there are obstacles to germ-line and somatic transmission of partially deleted mtDNAs, with the obvious explanation being that counterselection is attributable to respiratory incompetence of these mtDNAs. Selection on the basis of functional competence was shown to be more generally the case years later, when female mice carrying a broad range of mtDNA point mutants were found to transmit chiefly synonymous mutations (those that have no effect on the amino acid sequence) [38]. Conversely, Drosophila much more readily permit deleterious variants to pass through the germ-line [39], perhaps because of the pooling of cell contents that occurs in this metazoan during gametogenesis.

9.5.2 Propagation of dysfunctional mitochondria—misuse of the natural process of coupling mitochondrial mass to energy demand

Given the evident capacity for selection based on fitness, coupled with the fact that this can, at least in theory, act against deleterious mtDNA variants both at the subcellular level (intracellular selection between individual organelles) and between cells, one would expect primary mtDNA disorders to be vanishingly rare. On the contrary, they occur at a frequency of approximately 1 in 5000 live births making them one of the most common causes of genetic disease [40]. Two insights provide a partial explanation for the appearance and persistence of deleterious variants. First, mitochondrial biogenesis is coupled to the demand for energy production. Tissues such as muscle have a much higher mtDNA copy number and more OXPHOS complexes than, say, blood cells or fibroblasts; thus mitochondrial numbers increase in response to the energy demand. Unfortunately, this can, in the case of mutant mitochondria, lead to a futile expansion of dysfunctional organelles, as widely seen in the muscle of patients with mitochondrial diseases, with the resultant ragged red fibers (RRFs) being typically respiratory chain deficient and may be stuffed with mutant mtDNA [41]. That said, although RRFs have served for decades as a standard marker of mitochondrial dysfunction in muscle, we should not lose sight of the fact that only a small percentage of muscle fibers (rarely more than 20%) are affected after decades of living with the disorder, and so it is not the case that proliferation is a universal response that is triggered at the first hint of impaired respiratory capacity. Second, mtDNA faces the constant battle of any genetic element to avoid functionally deleterious variants arising via selfish mechanisms.

9.5.3 Selfish mechanisms

Many people struggle with the abstract concept of a selfish mechanism. Put simply, DNA has no awareness of what it encodes; hence, what is good for the DNA may be harmful for its products. A simple example would be a point mutation in the mtDNA that creates a new and highly active origin of replication and produces an amino acid change in subunit I of complex IV that abolishes cytochrome *c* oxidase activity. Without phenotypic selection (for cytochrome *c* oxidase activity), such a variant would be positively selected and cells would be left with mitochondria that were incapable of producing energy.

Striking examples of this principle are a frequent occurrence in yeasts that are facultative aerobes. *Saccharomyces cerevisiae* spontaneously produces mtDNA variants known as rho minus (ρ^-) genomes—they contain a fraction of the 80 kb of the wild-type mitochondrial genome, which is invariably unable to produce the mitochondrially encoded OXPHOS components and so the ρ^- yeasts are respiratory incompetent. Nevertheless, the residual fragment of mtDNA outcompetes wild-type mtDNA. The most dramatic examples are ρ^- mtDNAs as short as a few hundred base pairs that may be no more than a (pseudo-) origin and produce a hypersuppressive phenotype—in competition with wild-type mtDNA up to 98% of the progeny carry the ρ^- variant [42]. The physician faced with his/her patients with mitochondrial disease in the clinic might not immediately see the relevance of this exotic behavior of mtDNA in a single-celled organism that can happily grow without oxygen. However, a study of human mtDNA rearrangements 20 years ago indicated striking parallels between the two distant eukaryotes [16]. The earlier mentioned partial tandem duplications of human mtDNA [12] have an obvious potential replicative advantage as the additional DNA includes the control region and a segment of mtDNA that functions as a replication initiation zone [43]. The theoretical advantage of duplicating this region was demonstrated when cells carrying one such partial duplication spontaneously selected a partially triplicated mtDNA [16], and later still underwent further rearrangement to partially quadruplicated mtDNA (Vergani and Holt unpublished findings). Thus the control region and adjacent sequences can function as the "driver" of selection of this mtDNA rearrangement, with the only limit being phenotypic counterselection. In this regard, the triplicated mtDNA was associated with a more severe OXPHOS deficiency than the duplicated molecules [16], and quadruplicated mtDNA was highly unstable (unpublished data). Therefore phenotypic selection is active, albeit weakly in culture conditions similar to those used to maintain respiratory incompetent cells without mtDNA [44]. Although one or more extra origins explains the selection of duplications and related rearrangements, partially deleted mtDNAs are thought to repopulate organelles faster than full-length molecules by the simple expedient that they will complete the replication cycle quicker and thus be the first to become available for subsequent rounds of replication. It remains to be demonstrated whether this is attributable to the deleted molecules sequestering a scarce initiation factor at the end of the replication cycle, but in any case, replicative advantage has been demonstrated empirically by multiple groups using different cell types [45–47].

The 1997 analysis of rearranged human mtDNAs indicated that partially duplicated mtDNA could thrive in one cell type, and even go on to produce still more deleterious variants, but be rapidly lost in another cell type [16]. This distinct behavior of mutant mtDNA according to the cell background (i.e., the accompanying nuclear DNA) reinforced and extended an earlier study of the most common pathological mtDNA point mutant (m.3243G), which revealed selection of mutant or wild-type mtDNA according to the nuclear background [48]. These studies established the important principal that *mtDNA heteroplasmy* could be altered by changing *nuclear* DNA. The additional origin(s) hypothesis provides a wholly credible explanation for the partial duplication [16], and a direct replicative advantage (which is the essence of all selfish mechanisms) was also posited for the m.3243G pathological variant. The point mutation at m.3243 is within the binding site for the founding member of the mTERF family. Hence, mTERF binding to mtDNA can cause replication pausing at the tridecamer sequence that includes nucleotide position 3243, and this was confirmed experimentally by transgenically expressing mTERF in a human cell line [49]. Because the pathological variant m.3243G decreases mTERF binding in vitro [50], we propose that it alleviates replication pausing, thereby conferring an advantage to the mutant molecules [49] (Fig. 9.2). Finally, a mtDNA variant could confer an advantage by decreasing either mitochondrial or mtDNA turnover, as opposed to enhancing replication.

The study of mtDNA of nematode worms has provided further striking evidence that deleterious variants (rearrangements) can behave as selfish genetic elements [51].

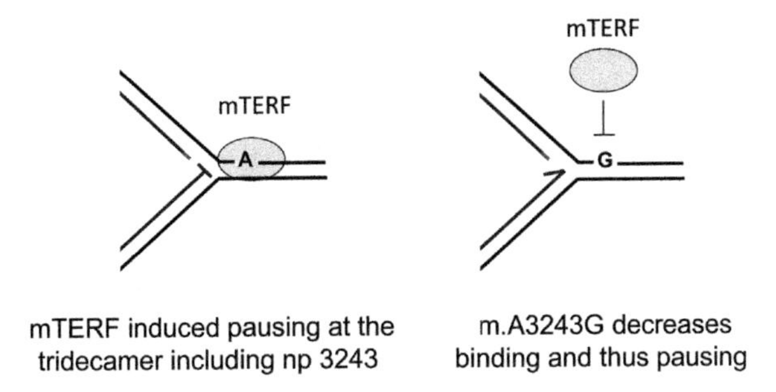

Figure 9.2: Repressed replication pausing owing to a point mutation can confer a direct replicative advantage on the mtDNA.
In the case of the pathological m.3243G mutant, the altered nucleotide resides in a tridecamer binding site for the mTERF protein and weakens binding of the protein to the DNA, which is expected to reduce replication pausing, as illustrated.

Mechanistically, this involves induction of the so-called mitochondrial unfolded protein response (UPR^{mt}), although there are differing views on whether it stimulates mitochondrial biogenesis [52], or inhibits mitochondrial turnover [51]. The latter study also invoked the idea of the deleted mtDNA "hitch-hiking" on the wild-type mtDNA, but unfortunately the literal version of this idea, namely that the variants are partial duplications of mtDNA, was not tested [51]. Indeed, unless mtDNA transmission in worms is quite different to humans, then the very fact that the rearranged mtDNA was transmitted across 100 generations points to, at least some of, it being in the form of a partial duplication, as does the lack of overt phenotypes [53]. Work in Drosophila has also highlighted the selfish behavior of deleterious mtDNA variants [54].

9.5.4 Metabolic configuration and nutrient availability

Another more obvious relevant external factor, nutrient availability, has been used to manipulate mtDNA heteroplasmy in human cells. One simple approach is to reduce glucose availability; often, this is achieved by replacing glucose with galactose, as this decreases the glycolytic flux. In cells with m.3243G, a switch from glucose to galactose produced a decrease in mutant load of six percentage points (from 99% to 93%); however, this was accompanied by extensive cell death, and so almost certainly reflected intercellular selection [55]. That is, the few cells with slightly lower mutant load were presumably the only ones capable of surviving the more demanding growth conditions—clearly the prospects for using such an approach in patients is limited. More promising was the replacement of glucose with ketone bodies in one study of cells carrying partially deleted mtDNAs. Ketone bodies can only be used to produce energy via the mitochondria, and the result was a functionally significant decrease in mutant load of 9% within 5 days that moreover appeared to occur via intracellular selection and did not involve cell death [56]. On the other hand, as earlier described, counterselection against partial deletions is very strong so these nonfunctional rearranged mtDNAs are in theory the easiest to select against. In our hands, ketone bodies did not induce a decrease in mutant load in cells carrying m.3243G, even on a cell background that occasionally spontaneously selects the wild-type mtDNAs (our unpublished data). Nor are we aware of any other study where ketone bodies were used to reduce the mutant load for a pathological point mutant. Notwithstanding this, a more recent study highlighted the profound effects of ketone bodies on mtDNA organization in controls as well as cell disease models [57], and so we expect that nutraceuticals can form a part of the management and treatment of mtDNA disorders.

Another factor sensitive to external stimuli and conditions is free radicals. Most discussions of reactive oxygen species (ROS) and mitochondria tend to focus on the former's capacity to cause damage to the latter, but the role of ROS in mtDNA segregation more likely reflects its ability to act as a signal transducer [58]. ROS stimulates mtDNA replication in yeasts via the action of Ntg1 [59], although there is no known animal homolog this line of research merits further study, especially as RNase H1, one of the critical enzymes for mtDNA replication, is potentially redox regulated [60].

9.6 Genetic drift

In the absence of selective pressure, DNA variants increase and decrease randomly in a population and, over time, there is an inevitable tendency for particular variants to become fixed, or in mtDNA terms to reach homoplasmy. Such genetic drift is conducive to mathematical modeling and several studies of human mtDNA disorders found that the spread of mutant load in multiple pedigrees was compatible with genetic drift [61]. Mathematical modeling has also been used to show that clonal expansion of a single deletion (or point mutation) could account for the accumulation of mtDNA variants seen *in vivo* [62]. For many years, these studies were taken by many to mean that selfish mechanisms were an unnecessary complication that was not required to explain the occurrence of mutant mtDNAs in patients, and it was often said that the underpinning data on biased segregation relied too heavily on studies in aneuploid cancer cells that were of dubious physiological relevance. However, later studies of worms, flies, and mice fully corroborated the human cell cybrid studies (as detailed earlier in this chapter), and the proponents of genetic drift made some assumptions that were at least open to question and did not consider some potentially relevant parameters. First, Elson et al. [62] adopted a relatively high theoretical mutation rate (five orders of magnitude above that of *Escherichia coli* DNA polymerase I, even before DNA repair systems set to work). Second, in respect of the rate of mtDNA replication, if mtDNA molecules compete for limited factors required for the initiation of replication, then shorter molecules, by completing the replication cycle earlier, can sequester such factors and thereby ensure they are selected for the next round of replication.[1] Third, a point mutant can also produce a direct replicative advantage, for instance by affecting the number or activity of origins of replication, or the strength of replication pause sites. Fourth, further modeling showed that even if one accepted the parameters set by the authors, genetic drift could not explain the accumulation of mutant mtDNAs in short-lived animals [63], to say nothing of cultured cells displaying substantial increases in mutant load within a matter of days or weeks (e.g., Refs. [64,65]). None of this is to say that random genetic drift is unimportant for the fixation of mutant mtDNA variants, but it is evidently far from the only parameter determining the level of mutant mtDNA.

9.7 Mitochondrial DNA selection—more or less?

Mitochondrial biogenesis is a two-edged sword; it clearly has the capacity to compensate for impaired mitochondrial function [25,66–68]; but, by mitigating decreased mitochondrial respiratory chain capacity, it can potentially sustain mutant mtDNA variants. Mitochondrial mass is also a function of turnover and mitochondria and their DNA can be recycled via autophagic pathways involving Parkin/PINK1 and Beclin [69]. In one study, this process of mitophagy was demonstrated to remove aberrant forms of respiratory complex I [70], that are the first "victim" of the m.3243G mutant tRNA [33][2], which suggests mitophagy can favor the maintenance of *mutant* mtDNA by clearing its defective products. On the other hand, artificially high levels of Parkin counterselected one of two mutant mtDNA variants [71], and a cell type capable of selecting wild-type over mutant mtDNA [16,48] has higher mitophagy than one that selects mutant mtDNA [72]. Thus, the effect of mitophagy on mutant load is variable and must be influenced by other factors. Selective turnover of mutant or wild-type mtDNA is another potential mechanism of biased segregation, but while the mtDNA polymerase γ (POLG) and the endonuclease MGME1 both have established roles in mtDNA degradation as well as replication [73], their contribution to the selection of mutant variants is unknown.

[1] If this sounds complicated, it is not. Consider passengers arriving at a bus station over the course of 24 hours and their bus leaves the next day. There would appear to be no advantage to arriving early, but if the early arrivals buy up all the tickets (initiation factors), the advantage they gain is obvious—the late arrivals miss the bus as there are no more tickets available!

[2] The particular sensitivity of respiratory complex I to the m.3243G mutation has been attributed to the mitochondrially encoded subunit ND6, which proportionally has the highest number of leucine residues that are decoded by tRNA-LeuUUR, and whose synthesis is most vulnerable to the mutant tRNA [33].

9.8 Stable heteroplasmy—the persistence of a fixed proportion of mutant and wild-type mtDNA

Another behavior of mtDNA that cannot be explained by genetic drift is the long-term persistence of fixed proportions of two genotypes (stable heteroplasmy) (Ref. [55] and references therein). The phenomenon suggests that the mitochondrial genotype can be "locked" in some way and all variants are excluded from selection—whether good or bad. A comprehensive study of individual cells with m.3243G found evidence that such stable periods of heteroplasmy were interspersed with temporary shifts in the mutant load [74]. These changes could reflect the formation and dissolution of physical connections between mitochondrial nucleoids (mtDNA packaged with proteins)—a concept that is described in more detail later in this chapter in relation to nuclear-coded proteins that maintain the mtDNA.

Combining biogenesis, recycling, and selfish mechanisms versus phenotypic selection and stable heteroplasmy produces a highly complex set of parameters contributing to the mutant load that is tissue and genetic background dependent, and can additionally alter in response to a range of external stimuli. Hence, many pedigrees, even for one particular type of mutant, could well give the appearance of randomness. And of course random genetic drift itself can be an important factor. Often, the solution to problems of this type is simply to ramp up the size of the test population, and recently the 1000 genome data collection was used to assess mtDNA inheritance patterns. This approach for the first time provided *in vivo* evidence of nuclear background effects on mtDNA transmission and selection in the human population [75], thereby fully substantiating the nuclear background effects documented for pathological mtDNA variants in the 1990s carried out in cultured cells. Moreover it strengthens the thesis propounded here that there is considerable overlap between the factors determining germ-line and somatic cell transmission of mtDNA, notwithstanding the marked contraction in mtDNA numbers that is unique to germ-line transmission.

9.9 Mitochondrial DNA maintenance disorders

This topic is covered only briefly here to compare and contrast with the primary mtDNA mutants; readers can find more details elsewhere in this book. Secondary mtDNA mutants are the result of defects in nuclear genes encoding factors that contribute to the maintenance of the mtDNA. At the molecular level, the mtDNA abnormalities manifest as low copy number (depletion) or multiple deletions, which are indicative of slow replication or stalling.

The core of the mtDNA maintenance system comprises the replication machinery and factors that produce and regulate the supply of DNA building blocks (deoxynucleotide triphosphates, dNTPs—A, G, C, and T). The former group includes the presumed mtDNA replicative polymerase and helicase (Twinkle). Equally essential is the dedicated mitochondrial RNA polymerase POLRMT, as it makes the several key primers for replication, which are subsequently processed by RNase H1. Pathological variants in many components of the mtDNA replication machinery have been reported, save (thus far) POLRMT.

All four precursors of DNA need to be balanced, as well as present in sufficient amount, for the replication and repair of both nuclear and mtDNA. Since mtDNA replicates in nondividing and postmitotic cells, it is particularly vulnerable to changes in dNTP concentration, which can be detrimental to mtDNA integrity, and can cause a mtDNA disorder regardless of the location of the enzyme or factor. Indeed, the first identified cause of mtDNA depletion syndrome, known as Mitochondrial GastroIntestinal NeuroEncephalomyopathy or MNGIE, was the catabolic and cytosolic enzyme thymidine phosphorylase [76]. Loss of function of thymidine phosphorylase results in high levels of thymidine [77], which in turn leads to an imbalance of dNTPs in mitochondria and mtDNA depletion or multiple mtDNA deletions. Likewise, cytosolic ribonucleotide reductase, p53r2, which supports the *de novo* synthesis of deoxynucleotides in nondividing cells, is needed for mtDNA maintenance [78] and defects in the mitochondrial salvage pathways enzymes, deoxyguanosine kinase, and thymidine kinase also cause mtDNA depletion or multiple deletions [79]. Another protein in this category is MPV17: although its precise function is still not defined, MPV17 loss-of-function results in low mitochondrial dNTP levels, and mtDNA depletion or multiple deletions [80–82]. MPV17 deficiency also played an important role in identifying a hitherto unrecognized mtDNA abnormality, misincorporation of ribonucleotides in mtDNA.

9.10 Ribonucleotide incorporation—a new mtDNA abnormality and a potential precursor or mitigator of mtDNA deletions and depletion

Forty-four years elapsed between the first demonstrations of ribonucleotides embedded in mammalian mtDNA [83,84] and the recognition that this is much more frequent in mtDNAs of solid tissues than cultured cells [85]. However, it could have been predicted from first principles. Despite a marked preference for dNTPs, DNA polymerases occasionally insert a ribonucleotide in the nascent DNA chain. Clearly, the ratio of dNTPs to NTPs will be an important factor in how many ribonucleotides are incorporated in the DNA, as well as the discriminatory properties of the DNA polymerase (e.g., UTP is inserted much less than other ribonucleotides because it differs most from the corresponding dNTP) [86]. In mitochondria of solid tissues, the major activity is the production of ATP, to provide the energy source for a host of cellular reactions, but this is also one of the four ribonucleotides required for RNA synthesis. Thus, the major ribonucleotide of mtDNA will logically be AMP from ATP—and so it proved [85], although to our knowledge, no one had predicted this in the preceding five decades. As mentioned above, one of the largest group of mtDNA disorders is caused by disturbed nucleotide homeostasis, in particular shortages of dNTPs, which will slow replication in and of itself, but also can produce a problem that has hitherto been overlooked—elevated or altered ribonucleotide incorporation in mtDNA. For example, if the level of dGTP in mitochondria is low, this will increase the GTP/dGTP ratio and thereby increase the presence of GMP in mtDNA. This occurs to such an extent in Mpv17 deficiency that GMP becomes the major embedded ribonucleotide in the mtDNA of liver [85]. That said, in the Mpv17-deficient mouse, elevated GMP levels in mtDNA are evident in tissues with normal mtDNA and dNTP levels, leading us to suggest that the local concentration of usable dNTPs may be altered. Extrapolating from the idea that the fidelity of DNA replication will be influenced by the nucleotide levels in the mitochondria of different tissues [87], we have noted that equalizing the dNTP pools could reduce the error rate of the DNA polymerase, which would fit with the slowing of replication being an adaptation to another underlying problem. The incorporation of GTP slows DNA synthesis [88] and furthermore, rGMPs in G-quadruplexes may alter the rate of replication. Thus nucleotide dyshomeostasis has diverse consequences for mtDNA metabolism and it will take some time to elucidate all the ramifications of the normal (rAMP) and abnormal (e.g., rGMP) ribonucleotides in mammalian mtDNA, including its impact on the selection of mtDNA variants.

9.11 Overlaps between nuclear defects in the mtDNA maintenance system and primary mtDNA mutants

Rearrangements and point mutants have been seen to cause mtDNA depletion in particular cell types cultured in the laboratory [16,89,90] and in patients' leukocytes carrying m.3243G [91]. Moreover, the MELAS point mutant m.3243G is associated with gastrointestinal (GI) problems reminiscent of MNGIE, and MNGIE features mtDNA depletion [76]. Hence, mtDNA depletion owing to m.3243G may occur in the small intestine and in leukocytes, among other cell types, and it should be determined whether this is attributable to a disturbance of deoxynucleotide homeostasis, or other side effect. We need also to consider that some disorders may not fit neatly into one category or another and should be classifed as dual nuclear and mtDNA disorders. For instance, polymorphisms in mtDNA maintenance factors, such as mTERF, could well affect the segregation bias in a primary mtDNA disorder to the extent of cosegregating with the disease.

9.12 A mitochondrial DNA network and its implications for heteroplasmy

Numerous factors impinge on mtDNA maintenance, including mitochondrial morphology and dynamics, protein synthesis, biogenesis and turnover. Furthermore, mtDNA is tightly attached to the inner mitochondrial membrane [92–94] and so perturbations of membrane cholesterol [95,96] and lipid content can affect the mtDNA. In yeasts, defects in mitochondrial protein synthesis cause mtDNA loss, and although this strict coupling is not evident in mammals, there is growing evidence of physical and functional ties between mtDNA and the translation machinery, which are likely to have implications for the physical and genetic segregation of mtDNA. Translation factors are among the most highly enriched proteins when mtDNA binding proteins are used to purify mtDNA [97], and the small subunit of the mitochondrial ribosome appears to be assembled on the mtDNA [98], which makes the mitochondrial nucleoid vulnerable to defects in mitochondrial ribosome assembly [99]. Perhaps this is one reason that pathological mutations in the mitochondrial ribosomal RNA genes are particularly rare, in that they can disrupt multiple aspects of mtDNA maintenance and expression. On the other hand, ATAD3 protein appears to contribute to diverse aspects of mitochondrial structure and function, including mtDNA segregation and organization and protein synthesis in mitochondria, as well as (possibly indirectly) membrane cholesterol content and architecture and cristae maintenance [93,95,100], and yet mutant forms of the protein are an increasingly recognized cause of human disease [95,101,102]. The behavior and properties of ATAD3, FARS2 (an amino acyl tRNA synthetase that in mutant form can cause mtDNA aggregation [103]), C14orf14/NOA1 [98], and LETM1 support a model in which mitochondrial nucleoids are linked through some form of scaffold to mitochondrial ribosomes [57] (Fig. 9.3). This proposed network within a network will inevitably affect mtDNA transmission and segregation in the genetic as well as the physical sense. The mitochondrial nucleoid and ribosome scaffold could be disassembled in particular conditions or at certain stages of development (e.g., in germ cells); conversely, extensively connected mitochondrial nucleoids could underpin stable heteroplasmy (see above). The nucleoid network also gives a physical form to the idea that there is a coupling between genotype and phenotype, to maintain mitochondrial integrity by selection [108]. And Kowald and Kirkwood's idea that a feedback mechanism acts via transcriptional repression of replication is interchangeable with activation of replication owing to aberrant translation. That is, the production of functional OXPHOS products leads to continued expression of the mtDNA, whereas dysfunctional proteins could cause the translation machinery to uncouple from the mtDNA, which may be necessary to permit replication to commence. In this way, deleterious variants would replicate more often than their wild-type counterparts.

The interplay between different facets of mitochondrial maintenance is nicely illustrated by the protein DRP1. The primary role of DRP1 is to mediate mitochondrial fission [109], yet its repression causes mtDNA aggregation [110] and loss [111]. This is because mitochondrial fission can be coupled to mtDNA replication and segregation, and involves contacts with the endoplasmic reticulum [112,113]. Actin is another protein that has links to mtDNA [105] and appears to participate in ER-mediated and DRP1-dependent mitochondrial division [114]. Hence, we should not be surprised that elevated expression of DRP1 can induce biased mtDNA segregation, causing the mutant load to increase in osteosarcoma cells

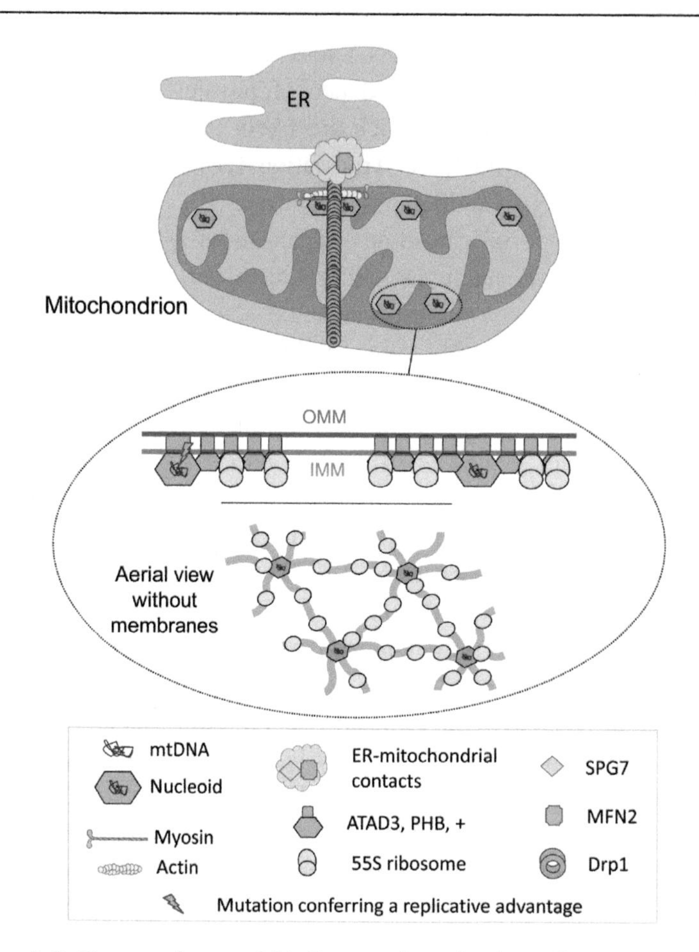

Figure 9.3: Factors that could influence the selection of mtDNA variants.

The upper part of the illustration shows the mitochondria in close proximity with the endoplasmic reticulum (ER). Note that ER mitochondrial contacts formed to mediate mtDNA segregation or to facilitate replication may be distinct from those involved in other processes such as calcium exchange [104]. The *gray* cloud surrounding the ER-mitochondrial connections is by way of indicating that little is known about these structures. MFN2 and SPG7 are shown as both have been linked to the ER and mutants cause mtDNA abnormalities (see main text). The role of actomyosin in mtDNA segregation is also uncertain, although both beta-actin and nonmuscle myosin type II-C are enriched with highly purified mtDNA, and gene silencing of either gene produces mtDNA abnormalities [105]. Proteins involved in mtDNA organization, replication, and segregation (i.e., nucleoid proteins) that could affect the selection of mutant variants include TFAM, POLG, PEO1, RNASEH1, MGME1, TOP3A, SSBP1, POLRMT, ATAD3, and Prohibitin (see main text). Mitochondrial fission via DRP1 is not only linked to the physical separation of mtDNA molecules, at high abundance, it can induce biased segregation of mtDNA [106]. The lightning flash represents changes to the primary sequence, which can create one or more additional origins or point mutants that have a replicative advantage. Trans-acting nucleic acids, in particular the D-loop and

(*Continued*)

carrying m.3243G [106]. Other nuclear-encoded proteins that affect mitochondrial morphology and mtDNA maintenance, and thus may also influence the segregation of mtDNA variants include OPA1, MFN2, and SPG7 [115–117]. Further concrete evidence of mitochondrial fusion and fission influencing mtDNA heteroplasmy levels comes from studies of flies, as less Mitofusin or more DRP1 than normal both selected against mitochondria with a mutant form of cytochrome *c* oxidase subunit 1 [118].

Research perspectives

A recent study in mice carrying two types of mtDNA indicates that the same forces described throughout this chapter influence mtDNA segregation *in vivo*: from random drift to strong selection of one or other haplotype, via mechanisms that are dependent upon mitochondrial–nuclear interactions affecting metabolism and cell fitness [119]. The major challenge is to determine to what extent selection can be manipulated by interventions. Ideally, this will prove to be possible using small molecules, as these are much easier to bring to the clinic than the genetic manipulation of mtDNA via DNA-modifying proteins. Finally, the world's attention is being drawn to the possibility of preventing the transmission of primary mtDNA mutants. Plans are well advanced to implement mitochondrial replacement in eggs of mothers carrying high levels of pathological mtDNA, in an effort to enable them to have healthy children of their own [120]. Clearly, it is critical to understand how any residual mutant mtDNA behaves in competition with the donor wild-type molecules, as biased segregation of the type described in this chapter could potentially confound attempts to spare the child a mitochondrial disease.

Acknowledgments

IJH is supported by the Basque Government (2018111043; 2018222031 and PRE_2018_1_0253), and the Carlos III Health Program (PI17/00380). AS is supported by the UK Medical Research Council with a Senior Non-Clinical Fellowship (MC_PC_13029) and Muscular dystrophy UK and the Lily foundation.

◀ R-loop are also implicated in the physical separation/distribution of mtDNAs [107], which could influence the segregation of mutant variants. Because the mitochondrial nucleoid is intimately associated with the mitochondrial translation apparatus, proteins involved in mitochondrial protein synthesis can impact mtDNA distribution, notably FARS2 [103] (and potentially other amino acyl tRNA synthetases), ATAD3 [97], C4ORF14 [98] (bacterial ortholog YqeH), and MPV17L2 [99]. Other factors to consider, but not illustrated, are the mitochondrial RNA granules; phenotypic selection based on ATP production; mitochondrial biogenesis and turnover; the environment (nutrient availability/starvation, redox state); and finally germ-line specific considerations—especially the mitochondrial (DNA) bottleneck.

References

[1] Holt IJ, Harding AE, Morgan-Hughes JA. Deletions of muscle mitochondrial DNA in patients with mitochondrial myopathies. Nature 1988;331:717–19.

[2] Dunbar DR, Moonie PA, Swingler RJ, Davidson D, Roberts R, Holt IJ. Maternally transmitted partial direct tandem duplication of mitochondrial DNA associated with diabetes mellitus. Hum Mol Genet 1993;2:1619–24.

[3] Moraes CT, Schon EA, DiMauro S, Miranda AF. Heteroplasmy of mitochondrial genomes in clonal cultures from patients with Kearns-Sayre syndrome. Biochem Biophys Res Commun 1989;160:765–71.

[4] Chinnery PF, DiMauro S, Shanske S, Schon EA, Zeviani M, Mariotti C, et al. Risk of developing a mitochondrial DNA deletion disorder. Lancet 2004;364:592–6.

[5] Schon EA, Rizzuto R, Moraes CT, Nakase H, Zeviani M, DiMauro S. A direct repeat is a hotspot for large-scale deletion of human mitochondrial DNA. Science 1989;244:346–9.

[6] Holt IJ, Harding AE, Morgan-Hughes JA. Deletions of muscle mitochondrial DNA in mitochondrial myopathies: sequence analysis and possible mechanisms. Nucleic Acids Res 1989;17:4465–9.

[7] Mita S, Rizzuto R, Moraes CT, Shanske S, Arnaudo E, Fabrizi GM, et al. Recombination via flanking direct repeats is a major cause of large-scale deletions of human mitochondrial DNA. Nucleic Acids Res 1990;18:561–7.

[8] Dong DW, Pereira F, Barrett SP, Kolesar JE, Cao K, Damas J, et al. Association of G-quadruplex forming sequences with human mtDNA deletion breakpoints. BMC Genomics 2014;15:677.

[9] Bharti SK, Sommers JA, Zhou J, Kaplan DL, Spelbrink JN, Mergny JL, et al. DNA sequences proximal to human mitochondrial DNA deletion breakpoints prevalent in human disease form G-quadruplexes, a class of DNA structures inefficiently unwound by the mitochondrial replicative Twinkle helicase. J Biol Chem 2014;289:29975–93.

[10] Lopes J, Piazza A, Bermejo R, Kriegsman B, Colosio A, Teulade-Fichou MP, et al. G-quadruplex-induced instability during leading-strand replication. EMBO J 2011;30:4033–46.

[11] Phillips AF, Millet AR, Tigano M, Dubois SM, Crimmins H, Babin L, et al. Single-molecule analysis of mtDNA replication uncovers the basis of the common deletion. Mol Cell 2017;65:527–38 e526.

[12] Poulton J, Deadman ME, Gardiner RM. Tandem direct duplications of mitochondrial DNA in mitochondrial myopathy: analysis of nucleotide sequence and tissue distribution. Nucleic Acids Res 1989;17:10223–9.

[13] Poulton J, Holt IJ. Mitochondrial DNA: does more lead to less? Nat Genet 1994;8:313–15.

[14] Poulton J, Deadman ME, Bindoff L, Morten K, Land J, Brown G. Families of mtDNA re-arrangements can be detected in patients with mtDNA deletions: duplications may be a transient intermediate form. Hum Mol Genet 1993;2:23–30.

[15] Chapman TP, Hadley G, Fratter C, Cullen SN, Bax BE, Bain MD, et al. Unexplained gastrointestinal symptoms: think mitochondrial disease. Digestive liver Dis Off J Italian Soc Gastroenterol Ital. Assoc Study Liver 2014;46:1–8.

[16] Holt IJ, Dunbar DR, Jacobs HT. Behaviour of a population of partially duplicated mitochondrial DNA molecules in cell culture: segregation, maintenance and recombination dependent upon nuclear background. Hum Mol Genet 1997;6:1251–60.

[17] Boltin-Fukuhara M, Fukuhara H. Modified recombination and transmission of mitochondrial genetic markers in rho minus mutants of *Saccharomyces cerevisiae*. Proc Natl Acad Sci USA 1976;73:4608–12.

[18] Pring DR, Levings CS. Heterogeneity of maize cytoplasmic genomes among male-sterile cytoplasms. Genetics 1978;89:121–36.

[19] Minczuk M, He J, Duch AM, Ettema TJ, Chlebowski A, Dzionek K, et al. TEFM (c17orf42) is necessary for transcription of human mtDNA. Nucleic Acids Res 2011;39:4284–99.

[20] Hagstrom E, Freyer C, Battersby BJ, Stewart JB, Larsson NG. No recombination of mtDNA after heteroplasmy for 50 generations in the mouse maternal germline. Nucleic Acids Res 2014;42:1111–16.

[21] Pohjoismaki JL, Goffart S, Tyynismaa H, Willcox S, Ide T, Kang D, et al. Human heart mitochondrial DNA is organized in complex catenated networks containing abundant four-way junctions and replication forks. J Biol Chem 2009;284:21446–57.

[22] Gilkerson RW, Schon EA, Hernandez E, Davidson MM. Mitochondrial nucleoids maintain genetic autonomy but allow for functional complementation. J Cell Biol 2008;181:1117–28.

[23] Wallace DC, Singh G, Lott MT, Hodge JA, Schurr TG, Lezza AM, et al. Mitochondrial DNA mutation associated with Leber's hereditary optic neuropathy. Science 1988;242:1427–30.

[24] Holt IJ, Miller DH, Harding AE. Genetic heterogeneity and mitochondrial DNA heteroplasmy in Leber's hereditary optic neuropathy. J Med Genet 1989;26:739–43.

[25] Giordano C, Iommarini L, Giordano L, Maresca A, Pisano A, Valentino ML, et al. Efficient mitochondrial biogenesis drives incomplete penetrance in Leber's hereditary optic neuropathy. Brain J Neurol 2014;137:335–53.

[26] Cock HR, Cooper JM, Schapira AH. The 14484 ND6 mtDNA mutation in Leber hereditary optic neuropathy does not affect fibroblast complex I activity. Am J Hum Genet 1995;57:1501–2.

[27] Tatuch Y, Christodoulou J, Feigenbaum A, Clarke JT, Wherret J, Smith C, et al. Heteroplasmic mtDNA mutation (T----G) at 8993 can cause Leigh disease when the percentage of abnormal mtDNA is high. Am J Hum Genet 1992;50:852–8.

[28] Holt IJ, Harding AE, Petty RK, Morgan-Hughes JA. A new mitochondrial disease associated with mitochondrial DNA heteroplasmy. Am J Hum Genet 1990;46:428–33.

[29] Shoffner JM, Lott MT, Lezza AM, Seibel P, Ballinger SW, Wallace DC. Myoclonic epilepsy and ragged-red fiber disease (MERRF) is associated with a mitochondrial DNA tRNA(Lys) mutation. Cell 1990;61:931–7.

[30] Goto Y, Nonaka I, Horai S. A mutation in the tRNA(Leu)(UUR) gene associated with the MELAS subgroup of mitochondrial encephalomyopathies. Nature 1990;348:651–3.

[31] Chomyn A, Meola G, Bresolin N, Lai ST, Scarlato G, Attardi G. In vitro genetic transfer of protein synthesis and respiration defects to mitochondrial DNA-less cells with myopathy-patient mitochondria. Mol Cell Biol 1991;11:2236–44.

[32] King MP, Koga Y, Davidson M, Schon EA. Defects in mitochondrial protein synthesis and respiratory chain activity segregate with the tRNA(Leu(UUR)) mutation associated with mitochondrial myopathy, encephalopathy, lactic acidosis, and strokelike episodes. Mol Cell Biol 1992;12:480–90.

[33] Dunbar DR, Moonie PA, Zeviani M, Holt IJ. Complex I deficiency is associated with 3243G:C mitochondrial DNA in osteosarcoma cell cybrids. Hum Mol Genet 1996;5:123–9.

[34] Estivill X, Govea N, Barcelo E, Badenas C, Romero E, Moral L, et al. Familial progressive sensorineural deafness is mainly due to the mtDNA A1555G mutation and is enhanced by treatment of aminoglycosides. Am J Hum Genet 1998;62:27–35.

[35] Sacconi S, Salviati L, Nishigaki Y, Walker WF, Hernandez-Rosa E, Trevisson E, et al. A functionally dominant mitochondrial DNA mutation. Hum Mol Genet 2008;17:1814–20.

[36] McFarland R, Clark KM, Morris AA, Taylor RW, Macphail S, Lightowlers RN, et al. Multiple neonatal deaths due to a homoplasmic mitochondrial DNA mutation. Nat Genet 2002;30:145–6.

[37] van den Ouweland JM, Lemkes HH, Ruitenbeek W, Sandkuijl LA, de Vijlder MF, Struyvenberg PA, et al. Mutation in mitochondrial tRNA(Leu)(UUR) gene in a large pedigree with maternally transmitted type II diabetes mellitus and deafness. Nat Genet 1992;1:368–71.

[38] Stewart JB, Freyer C, Elson JL, Wredenberg A, Cansu Z, Trifunovic A, et al. Strong purifying selection in transmission of mammalian mitochondrial DNA. PLoS Biol 2008;6:e10.

[39] Samstag CL, Hoekstra JG, Huang CH, Chaisson MJ, Youle RJ, Kennedy SR, et al. Deleterious mitochondrial DNA point mutations are overrepresented in Drosophila expressing a proofreading-defective DNA polymerase gamma. PLoS Genet 2018;14:e1007805.

[40] Chinnery PF, Elliott HR, Hudson G, Samuels DC, Relton CL. Epigenetics, epidemiology and mitochondrial DNA diseases. Int J Epidemiol 2012;41:177–87.

[41] Petruzzella V, Moraes CT, Sano MC, Bonilla E, DiMauro S, Schon EA. Extremely high levels of mutant mtDNAs co-localize with cytochrome *c* oxidase-negative ragged-red fibers in patients harboring a point mutation at nt 3243. Hum Mol Genet 1994;3:449–54.
[42] Lockshon D, Zweifel SG, Freeman-Cook LL, Lorimer HE, Brewer BJ, Fangman WL. A role for recombination junctions in the segregation of mitochondrial DNA in yeast. Cell 1995;81:947–55.
[43] Bowmaker M, Yang MY, Yasukawa T, Reyes A, Jacobs HT, Huberman JA, et al. Mammalian mitochondrial DNA replicates bidirectionally from an initiation zone. J Biol Chem 2003;278:50961–9.
[44] King MP, Attardi G. Human cells lacking mtDNA: repopulation with exogenous mitochondria by complementation. Science 1989;246:500–3.
[45] Russell OM, Fruh I, Rai PK, Marcellin D, Doll T, Reeve A, et al. Preferential amplification of a human mitochondrial DNA deletion in vitro and in vivo. Sci Rep 2018;8:1799.
[46] Spelbrink JN, Zwart R, Van Galen MJ, Van den Bogert C. Preferential amplification and phenotypic selection in a population of deleted and wild-type mitochondrial DNA in cultured cells. Curr Genet 1997;32:115–24.
[47] Diaz F, Bayona-Bafaluy MP, Rana M, Mora M, Hao H, Moraes CT. Human mitochondrial DNA with large deletions repopulates organelles faster than full-length genomes under relaxed copy number control. Nucleic Acids Res 2002;30:4626–33.
[48] Dunbar DR, Moonie PA, Jacobs HT, Holt IJ. Different cellular backgrounds confer a marked advantage to either mutant or wild-type mitochondrial genomes. Proc Natl Acad Sci USA 1995;92:6562–6.
[49] Hyvarinen AK, Pohjoismaki JL, Reyes A, Wanrooij S, Yasukawa T, Karhunen PJ, et al. The mitochondrial transcription termination factor mTERF modulates replication pausing in human mitochondrial DNA. Nucleic Acids Res 2007;35:6458–74.
[50] Hess JF, Parisi MA, Bennett JL, Clayton DA. Impairment of mitochondrial transcription termination by a point mutation associated with the MELAS subgroup of mitochondrial encephalomyopathies. Nature 1991;351:236–9.
[51] Gitschlag BL, Kirby CS, Samuels DC, Gangula RD, Mallal SA, Patel MR. Homeostatic responses regulate selfish mitochondrial genome dynamics in *C. elegans*. Cell Metab 2016;24:91–103.
[52] Lin YF, Schulz AM, Pellegrino MW, Lu Y, Shaham S, Haynes CM. Maintenance and propagation of a deleterious mitochondrial genome by the mitochondrial unfolded protein response. Nature 2016;533 (7603):416–19. Available from: https://doi.org/10.1038/nature17989.
[53] Tsang WY, Lemire BD. Stable heteroplasmy but differential inheritance of a large mitochondrial DNA deletion in nematodes. Biochem Cell Biol 2002;80:645–54.
[54] Hill JH, Chen Z, Xu H. Selective propagation of functional mitochondrial DNA during oogenesis restricts the transmission of a deleterious mitochondrial variant. Nat Genet 2014;46:389–92.
[55] Lehtinen SK, Hance N, El Meziane A, Juhola MK, Juhola KM, Karhu R, et al. Genotypic stability, segregation and selection in heteroplasmic human cell lines containing np 3243 mutant mtDNA. Genetics 2000;154:363–80.
[56] Santra S, Gilkerson RW, Davidson M, Schon EA. Ketogenic treatment reduces deleted mitochondrial DNAs in cultured human cells. Ann Neurol 2004;56:662–9.
[57] Durigon R, Mitchell AL, Jones AW, Manole A, Mennuni M, Hirst EM, et al. LETM1 couples mitochondrial DNA metabolism and nutrient preference. EMBO Mol Med 2018;10.
[58] Moreno-Loshuertos R, Acin-Perez R, Fernandez-Silva P, Movilla N, Perez-Martos A, Rodriguez de Cordoba S, et al. Differences in reactive oxygen species production explain the phenotypes associated with common mouse mitochondrial DNA variants. Nat Genet 2006;38:1261–8.
[59] Hori A, Yoshida M, Shibata T, Ling F. Reactive oxygen species regulate DNA copy number in isolated yeast mitochondria by triggering recombination-mediated replication. Nucleic Acids Res 2009;37:749–61.
[60] Holt IJ. The Jekyll and Hyde character of RNase H1 and its multiple roles in mitochondrial DNA metabolism. DNA Repair 2019;102630.

[61] Chinnery PF, Thorburn DR, Samuels DC, White SL, Dahl HM, Turnbull DM, et al. The inheritance of mitochondrial DNA heteroplasmy: random drift, selection or both? Trends Genet: TIG 2000;16:500–5.
[62] Elson JL, Samuels DC, Turnbull DM, Chinnery PF. Random intracellular drift explains the clonal expansion of mitochondrial DNA mutations with age. Am J Hum Genet 2001;68:802–6.
[63] Kowald A, Kirkwood TB. Mitochondrial mutations and aging: random drift is insufficient to explain the accumulation of mitochondrial deletion mutants in short-lived animals. Aging Cell 2013;12:728–31.
[64] Yoneda M, Chomyn A, Martinuzzi A, Hurko O, Attardi G. Marked replicative advantage of human mtDNA carrying a point mutation that causes the MELAS encephalomyopathy. Proc Natl Acad Sci USA 1992;89:11164–8.
[65] Hayashi J, Ohta S, Kikuchi A, Takemitsu M, Goto Y, Nonaka I. Introduction of disease-related mitochondrial DNA deletions into HeLa cells lacking mitochondrial DNA results in mitochondrial dysfunction. Proc Natl Acad Sci USA 1991;88:10614–18.
[66] Srivastava S, Barrett JN, Moraes CT. PGC-1alpha/beta upregulation is associated with improved oxidative phosphorylation in cells harboring nonsense mtDNA mutations. Hum Mol Genet 2007;16:993–1005.
[67] Viscomi C, Bottani E, Civiletto G, Cerutti R, Moggio M, Fagiolari G, et al. In vivo correction of COX deficiency by activation of the AMPK/PGC-1alpha axis. Cell Metab 2011;14:80–90.
[68] Dillon LM, Williams SL, Hida A, Peacock JD, Prolla TA, Lincoln J, et al. Increased mitochondrial biogenesis in muscle improves aging phenotypes in the mtDNA mutator mouse. Hum Mol Genet 2012;21:2288–97.
[69] Pickles S, Vigie P, Youle RJ. Mitophagy and quality control mechanisms in mitochondrial maintenance. Curr Biol 2018;28:R170–85.
[70] Hamalainen RH, Manninen T, Koivumaki H, Kislin M, Otonkoski T, Suomalainen A. Tissue- and cell-type-specific manifestations of heteroplasmic mtDNA 3243A > G mutation in human induced pluripotent stem cell-derived disease model. Proc Natl Acad Sci USA 2013;110:E3622–3630.
[71] Suen DF, Narendra DP, Tanaka A, Manfredi G, Youle RJ. Parkin overexpression selects against a deleterious mtDNA mutation in heteroplasmic cybrid cells. Proc Natl Acad Sci USA 2010;107:11835–40.
[72] Malena A, Pantic B, Borgia D, Sgarbi G, Solaini G, Holt IJ, et al. Mitochondrial quality control: cell-type-dependent responses to pathological mutant mitochondrial DNA. Autophagy 2016;12:2098–112.
[73] Nissanka N, Minczuk M, Moraes CT. Mechanisms of mitochondrial DNA deletion formation. Trends Genet: TIG 2019;35:235–44.
[74] Raap AK, Jahangir Tafrechi RS, van de Rijke FM, Pyle A, Wahlby C, Szuhai K, et al. Non-random mtDNA segregation patterns indicate a metastable heteroplasmic segregation unit in m.3243A > G cybrid cells. PLoS One 2012;7:e52080.
[75] Wei W, Tuna S, Keogh MJ, Smith KR, Aitman TJ, Beales PL, et al. Germline selection shapes human mitochondrial DNA diversity. Science 2019;364:eaau6520.
[76] Nishino I, Spinazzola A, Hirano M. Thymidine phosphorylase gene mutations in MNGIE, a human mitochondrial disorder. Science 1999;283:689–92.
[77] Spinazzola A, Marti R, Nishino I, Andreu AL, Naini A, Tadesse S, et al. Altered thymidine metabolism due to defects of thymidine phosphorylase. J Biol Chem 2002;277:4128–33.
[78] Bourdon A, Minai L, Serre V, Jais JP, Sarzi E, Aubert S, et al. Mutation of RRM2B, encoding p53-controlled ribonucleotide reductase (p53R2), causes severe mitochondrial DNA depletion. Nat Genet 2007;39:776–80.
[79] Almannai M, El-Hattab AW, Scaglia F. Mitochondrial DNA replication: clinical syndromes. Essays Biochem 2018;62:297–308.
[80] Spinazzola A, Viscomi C, Fernandez-Vizarra E, Carrara F, D'Adamo P, Calvo S, et al. MPV17 encodes an inner mitochondrial membrane protein and is mutated in infantile hepatic mitochondrial DNA depletion. Nat Genet 2006;38:570–5.
[81] Dalla Rosa I, Camara Y, Durigon R, Moss CF, Vidoni S, Akman G, et al. MPV17 loss causes deoxynucleotide insufficiency and slow DNA replication in mitochondria. PLoS Genet 2016;12:e1005779.

[82] Blakely EL, Butterworth A, Hadden RD, Bodi I, He L, McFarland R, et al. MPV17 mutation causes neuropathy and leukoencephalopathy with multiple mtDNA deletions in muscle. Neuromuscul Disord 2012;22:587–91.

[83] Wong-Staal F, Mendelsohn J, Goulian M. Ribonucleotides in closed circular mitochondrial DNA from HeLa cells. Biochem Biophys Res Commun 1973;53:140–8.

[84] Grossman LI, Watson R, Vinograd J. The presence of ribonucleotides in mature closed-circular mitochondrial DNA. Proc Natl Acad Sci USA 1973;70:3339–43.

[85] Moss CF, Dalla Rosa I, Hunt LE, Yasukawa T, Young R, Jones AWE, et al. Aberrant ribonucleotide incorporation and multiple deletions in mitochondrial DNA of the murine MPV17 disease model. Nucleic Acids Res 2017;45:12808–15.

[86] Kasiviswanathan R, Copeland WC. Ribonucleotide discrimination and reverse transcription by the human mitochondrial DNA polymerase. J Biol Chem 2011;286:31490–500.

[87] Song S, Pursell ZF, Copeland WC, Longley MJ, Kunkel TA, Mathews CK. DNA precursor asymmetries in mammalian tissue mitochondria and possible contribution to mutagenesis through reduced replication fidelity. Proc Natl Acad Sci USA 2005;102:4990–5.

[88] Forslund JME, Pfeiffer A, Stojkovic G, Wanrooij PH, Wanrooij S. The presence of rNTPs decreases the speed of mitochondrial DNA replication. PLoS Genet 2018;14:e1007315.

[89] Turner CJ, Granycome C, Hurst R, Pohler E, Juhola MK, Juhola MI, et al. Systematic segregation to mutant mitochondrial DNA and accompanying loss of mitochondrial DNA in human NT2 teratocarcinoma cybrids. Genetics 2005;170:1879–85.

[90] Vergani L, Rossi R, Brierley CH, Hanna M, Holt IJ. Introduction of heteroplasmic mitochondrial DNA (mtDNA) from a patient with NARP into two human rho degrees cell lines is associated either with selection and maintenance of NARP mutant mtDNA or failure to maintain mtDNA. Hum Mol Genet 1999;8:1751–5.

[91] Pyle A, Taylor RW, Durham SE, Deschauer M, Schaefer AM, Samuels DC, et al. Depletion of mitochondrial DNA in leucocytes harbouring the 3243A->G mtDNA mutation. J Med Genet 2007;44:69–74.

[92] Albring M, Griffith J, Attardi G. Association of a protein structure of probable membrane derivation with HeLa cell mitochondrial DNA near its origin of replication. Proc Natl Acad Sci USA 1977;74:1348–52.

[93] He J, Mao CC, Reyes A, Sembongi H, Di Re M, Granycome C, et al. The AAA + protein ATAD3 has displacement loop binding properties and is involved in mitochondrial nucleoid organization. J Cell Biol 2007;176:141–6.

[94] Rajala N, Gerhold JM, Martinsson P, Klymov A, Spelbrink JN. Replication factors transiently associate with mtDNA at the mitochondrial inner membrane to facilitate replication. Nucleic Acids Res 2014;42:952–67.

[95] Desai R, Frazier AE, Durigon R, Patel H, Jones AW, Dalla Rosa I, et al. ATAD3 gene cluster deletions cause cerebellar dysfunction associated with altered mitochondrial DNA and cholesterol metabolism. Brain: J Neurol 2017;140:1595–610.

[96] Gerhold JM, Cansiz-Arda S, Lohmus M, Engberg O, Reyes A, van Rennes H, et al. Human mitochondrial DNA-protein complexes attach to a cholesterol-rich membrane structure. Sci Rep 2015;5:15292.

[97] He J, Cooper HM, Reyes A, Di Re M, Sembongi H, Litwin TR, et al. Mitochondrial nucleoid interacting proteins support mitochondrial protein synthesis. Nucleic Acids Res 2012;40:6109–21.

[98] He J, Cooper HM, Reyes A, Di Re M, Kazak L, Wood SR, et al. Human C4orf14 interacts with the mitochondrial nucleoid and is involved in the biogenesis of the small mitochondrial ribosomal subunit. Nucleic Acids Res 2012;40:6097–108.

[99] Dalla Rosa I, Durigon R, Pearce SF, Rorbach J, Hirst EM, Vidoni S, et al. MPV17L2 is required for ribosome assembly in mitochondria. Nucleic Acids Res 2014;42:8500–15.

[100] Peralta S, Goffart S, Williams SL, Diaz F, Garcia S, Nissanka N, et al. ATAD3 controls mitochondrial cristae structure in mouse muscle, influencing mtDNA replication and cholesterol levels. J Cell Sci 2018;131.

[101] Cooper HM, Yang Y, Ylikallio E, Khairullin R, Woldegebriel R, Lin KL, et al. ATPase-deficient mitochondrial inner membrane protein ATAD3A disturbs mitochondrial dynamics in dominant hereditary spastic paraplegia. Hum Mol Genet 2017;26:1432–43.

[102] Harel T, Yoon WH, Garone C, Gu S, Coban-Akdemir Z, Eldomery MK, et al. Recurrent de novo and biallelic variation of ATAD3A, encoding a mitochondrial membrane protein, results in distinct neurological syndromes. Am J Hum Genet 2016;99:831–45.

[103] Almalki A, Alston CL, Parker A, Simonic I, Mehta SG, He L, et al. Mutation of the human mitochondrial phenylalanine-tRNA synthetase causes infantile-onset epilepsy and cytochrome *c* oxidase deficiency. Biochim Biophys Acta 2014;1842:56–64.

[104] Bartok A, Weaver D, Golenar T, Nichtova Z, Katona M, Bansaghi S, et al. IP3 receptor isoforms differently regulate ER-mitochondrial contacts and local calcium transfer. Nat Commun 2019;10:3726.

[105] Reyes A, He J, Mao CC, Bailey LJ, Di Re M, Sembongi H, et al. Actin and myosin contribute to mammalian mitochondrial DNA maintenance. Nucleic Acids Res 2011;39:5098–108.

[106] Malena A, Loro E, Di Re M, Holt IJ, Vergani L. Inhibition of mitochondrial fission favours mutant over wild-type mitochondrial DNA. Hum Mol Genet 2009;18:3407–16.

[107] Holt IJ. The mitochondrial R-loop. Nucleic Acids Res 2019;47:5480–9.

[108] Kowald A, Kirkwood TB. Transcription could be the key to the selection advantage of mitochondrial deletion mutants in aging. Proc Natl Acad Sci USA 2014;111:2972–7.

[109] Bleazard W, McCaffery JM, King EJ, Bale S, Mozdy A, Tieu Q, et al. The dynamin-related GTPase Dnm1 regulates mitochondrial fission in yeast. Nat Cell Biol 1999;1:298–304.

[110] Ban-Ishihara R, Ishihara T, Sasaki N, Mihara K, Ishihara N. Dynamics of nucleoid structure regulated by mitochondrial fission contributes to cristae reformation and release of cytochrome *c*. Proc Natl Acad Sci USA 2013;110:11863–8.

[111] Parone PA, Da Cruz S, Tondera D, Mattenberger Y, James DI, Maechler P, et al. Preventing mitochondrial fission impairs mitochondrial function and leads to loss of mitochondrial DNA. PLoS One 2008;3:e3257.

[112] Lewis SC, Uchiyama LF, Nunnari J. ER-mitochondria contacts couple mtDNA synthesis with mitochondrial division in human cells. Science 2016;353:aaf5549.

[113] Murley A, Lackner LL, Osman C, West M, Voeltz GK, Walter P, et al. ER-associated mitochondrial division links the distribution of mitochondria and mitochondrial DNA in yeast. Elife 2013;2:e00422.

[114] Korobova F, Ramabhadran V, Higgs HN. An actin-dependent step in mitochondrial fission mediated by the ER-associated formin INF2. Science 2013;339:464–7.

[115] De la Casa-Fages B, Fernandez-Eulate G, Gamez J, Barahona-Hernando R, Moris G, Garcia-Barcina M, et al. Parkinsonism and spastic paraplegia type 7: expanding the spectrum of mitochondrial Parkinsonism. Mov Disord Off J Mov Disord Soc 2019;34:1547–61.

[116] Amati-Bonneau P, Valentino ML, Reynier P, Gallardo ME, Bornstein B, Boissiere A, et al. OPA1 mutations induce mitochondrial DNA instability and optic atrophy 'plus' phenotypes. Brain: J Neurol 2008;131:338–51.

[117] Rouzier C, Bannwarth S, Chaussenot A, Chevrollier A, Verschueren A, Bonello-Palot N, et al. The MFN2 gene is responsible for mitochondrial DNA instability and optic atrophy 'plus' phenotype. Brain: J Neurol 2012;135:23–34.

[118] Lieber T, Jeedigunta SP, Palozzi JM, Lehmann R, Hurd TR. Mitochondrial fragmentation drives selective removal of deleterious mtDNA in the germline. Nature 2019;570:380–4.

[119] Latorre-Pellicer A, Lechuga-Vieco AV, Johnston IG, Hamalainen RH, Pellico J, Justo-Mendez R, et al. Regulation of mother-to-offspring transmission of mtDNA heteroplasmy. Cell Metab 2019;30:1120–1130.e5.

[120] Herbert M, Turnbull D. Progress in mitochondrial replacement therapies. Nat Rev Mol Cell Biol 2018;19:71–2.

CHAPTER 10

Mitochondrial DNA mutations and aging

Karolina Szczepanowska[1,2] and Aleksandra Trifunovic[1,2]

[1]Cologne Excellence Cluster on Cellular Stress Responses in Ageing-Associated Diseases (CECAD) and Center for Molecular Medicine (CMMC), University of Cologne, Cologne, Germany, [2]Institute for Mitochondrial Diseases and Ageing, Medical Faculty, University of Cologne, Cologne, Germany

10.1 Introduction

Corrupted integrity of the mitochondrial genome (mtDNA) accompanied by a gradual decline of mitochondrial function is a hallmark of aging. The majority of people carry detectable levels of mtDNA mutations in somatic tissues, and the burden of those mutations builds up considerably with age. Hence, the causative nature of the relationship between the mtDNA mutations and aging emerges, awaiting further rigorous testing. Although over several decades numerous studies have been conducted and many experimental models developed, the origin and a precise time of occurrence of mtDNA mutations in the course of aging remain controversial. Here, we review the possible sources of somatic mtDNA mutations in our cells, and the mechanism underlying their expansion during aging. We also discuss whether in light of currently available evidence the link between mtDNA mutations and aging is credible.

10.2 Old and new mitochondrial theories of aging—how changes in mtDNA contribute to aging?

Aging is a slowly progressing collapse of organism induced by gradually deteriorating cellular functions. Although multiple molecular pathways in our cells seem to contribute to aging, mitochondrial dysfunction is undoubtedly one of the brightest hallmarks of age-related decline of the organism. The implication of mitochondria in aging was initially suggested by Harman in his Free Radical Theory of Aging that postulates mitochondrial-derived oxidative damage as a major force behind the age-related deterioration [1,2]. This theory was redefined by others who suggested the mutational component of the entire process highlighting the mtDNA as the most critical target of free radicals in the course of aging [3,4]. It was further extended by the self-accelerating mechanism, known as a "vicious cycle," that stimulates the oxidative catastrophe as a consequence of

The Human Mitochondrial Genome.
DOI: https://doi.org/10.1016/B978-0-12-819656-4.00010-3

exponentially accumulating mtDNA lesions (Fig. 10.1) [5]. In light of those theories, it seems that mitochondria, and in particular age-associated changes of the mitochondrial genome, could serve as a sort of "aging clock" for our cells. Indeed, mtDNA replicates independently and more frequently than the nuclear genome (the phenomenon known as "relaxed replication"). Thus the mitochondrial genome might serve as a better indicator of

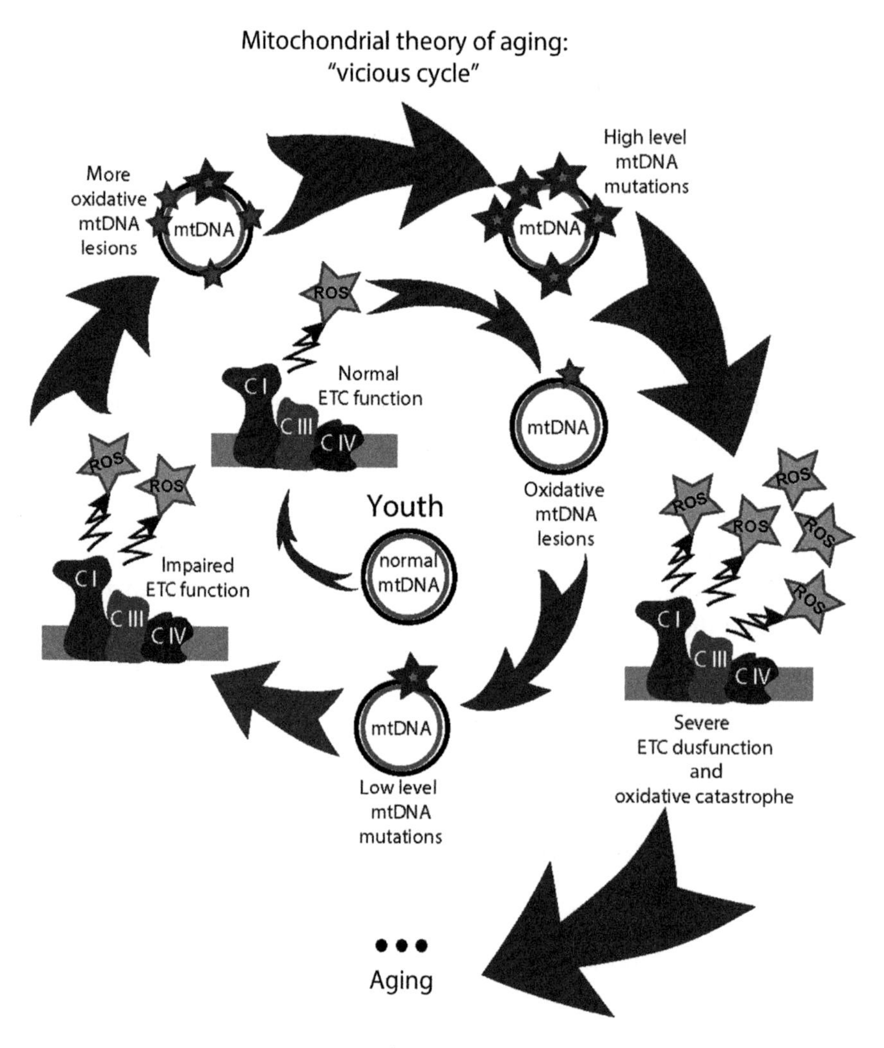

Figure 10.1
Old mitochondrial theory of aging—vicious cycle theory.
Over the course of aging, free radicals produced by the mitochondrial electron transport chain (ETC) cause minor mtDNA oxidative lesions and, consequently, some mtDNA mutations. This, in turn, causes production of imperfect ETC complexes that further stimulate ROS production and accelerate generation of mtDNA mutations leading to ultimate mitochondrial dysfunction, energy crisis, and oxidative catastrophe.

"wear and tears" events[1] in the postmitotic cells than the genetic information enclosed in the nucleus. In agreement, the frequency of somatic mtDNA mutations appears several magnitudes higher than the one detected in the nuclear genome [6]. Therefore the mitochondrial mutational hypothesis of aging remains attractive even today, although some of its concepts did not find much support in experimental data. For example, the clear causality between mitochondrial reactive oxygen species (ROS) and aging lacks evidences [7,8], though the traces of accumulating oxidative damage were repeatedly detected in aging tissues and senescent cells [9,10]. The oxidative damage in mtDNA also does not determine the lifespan of nematode mitochondrial mutants [11]. Furthermore, the physiological levels of ROS were shown to execute a rather beneficial effect on longevity, acting as signaling molecules through the phenomenon known as mitohormesis[2] [12,13]. Nevertheless, a body of evidence still supports the causative relation of somatic (spontaneous) mtDNA mutations with the aging process.

10.3 Mitochondrial genetics from the perspective of aging

Several basic principles of mitochondrial genetics rule the occurrence of mtDNA mutations during aging. Age-associated propagation of mtDNA mutations entails the *clonal expansion* and *mitotic segregation*, while translation of mutational changes into phenotype obeys the principles of *threshold effect* and *heteroplasmy/homoplasmy*. These concepts are thoroughly explained elsewhere in this book.

Two general types of mutations affect mtDNA, namely large-scale genome rearrangements and point mutations. Genome rearrangements arise as a consequence of prominent mtDNA deletions that lead to the formation of smaller circular mtDNA molecules. Point mutations include nucleotide substitutions, base deletions, or insertions. Essentially, both types of changes have been shown to accumulate with age in manifold of human and animal tissues

[1] "Wear and tear" concept underlies one of the most basic and long-standing theory of ageing. It states that we age because during the lifetime our bodies accumulate damage that cannot be efficiently repaired. Therefore our body "wears out" as a consequence of passing time. The principles of "wear and tear" concept are firmly established in the sociological perspective on ageing. Furthermore, many components of the living systems follow the rules of thermodynamics as their entropy increases during ageing supporting the "wear and tears" concept.

[2] "Mitohormesis" is a type of dose-dependent, biphasic response of biological systems, and an important antiaging paradigm. The hormetic concept suggests that low levels of stress will stimulate adaptive response of the cell (organism) that in turn will increase the stress resistance and propagate overall longevity. In the case of mitohormesis, stress is defined mostly as the reactive oxygen species, and adaptive stress responses, as a buildup of cellular antioxidant defense. However, beside upregulating antioxidant response, ROS, H_2O_2 in particular might activate additional signaling pathways due to its prominent role as secondary messenger. Mitohormesis phenomenon underlies the prolongevity mechanisms implicated in caloric restriction and physical exercise, and is also suggested to be the major cause of longevity induced by mild-to-moderate mitochondrial dysfunction.

including skeletal muscle, heart, brain, liver, and colon [14–19]. Given the distinct nature of mtDNA mutations, their propagation during the aging process as well as contribution to age-related pathologies may implement different mechanisms. Thus they will be discussed separately below.

10.4 mtDNA deletions and aging

10.4.1 The origin of mtDNA deletions

Large-scale deletions are the most frequently detected somatic mtDNA mutations, possibly due to technical difficulties in detecting individual point mutations. It is well documented that the load of mtDNA deletions increases substantially in aged individuals, particularly in tissues and cell types with high energy demand as skeletal muscle, heart, liver, brain, retina, ovaries, and sperm [19–27]. Deletions in the mitochondrial genome are most likely formed spontaneously as a consequence of aberrant mtDNA replication due to a slippage or stalling of replication forks, resulting in a formation of double-strand breaks (DSBs) [28–31]. A recent update in the mechanisms underlying formation of deletions reveals high complexity of the entire process [32]. The highly unstable linear by-products of DSBs seem not to play any crucial role in aging [33], albeit recent findings suggest that they can further propagate the formation of mtDNA deletions [32,34]. Although artificial and potent mutagens can stimulate DSBs in mtDNA, the direct involvement of oxidative damage in a generation of mtDNA deletions under physiological conditions is speculative [32,35,36].

10.4.2 How do mtDNA deletions expand during aging?

MtDNA deletions are often characterized by patchy and cell-type-specific distribution, reaching high levels in individual cells while having rather low general abundance in the entire tissue [19,23,37]. This mosaic character can be attributed to the mechanisms involved in expansions of mtDNA deletions in the cell population. Noticeably, it is still not clear how mtDNA deletions are propagated during aging, and whether this process involves any positive selection or acts randomly. From one side, a clonal expansion might be favored by the replicative advantage of smaller, deleted mtDNA molecules, as shorter DNA could replicate faster and consequently more often than the wild type genome [38,39]. This hypothesis was however challenged by the observation that time spent by mtDNA on replication is irrelevant when compared to the average half-life of mtDNA molecule [40], and further disproved by computational modeling [39]. Indeed, no correlation between the length of deletion and level of OXPHOS deficiency in muscle fibers has been observed [37]. More recently, a new selection-based mechanism for clonal expansion of mtDNA deletions has been proposed. It grounds on the close link between replication and transcription and implies that higher transcription rates will lead to the higher rates of replication initiation

[41]. If a surplus of products negatively inhibits mitochondrial transcription, then lacking some genes in mtDNA can give a replicational advantage to deleted molecules, and propagate them over the wild type nucleoids [41]. However, this mechanism still awaits further experimental evaluation.

Alternative hypotheses postulate that mtDNA deletions can expand clonally as a result of random genetic drift that does not require any positive selection [42]. In this case, a sufficient number of generations granted by "relaxed mtDNA replication" can allow the particular deletion mutation to dominate the entire population by chance [42] (Fig. 10.2). Such an explanation can be particularly correct for long-lived species, including humans, as confirmed by computational simulations [42]. However, the short-lived species as rodents, that also accumulate mtDNA deletions during aging, rather escape the principles of genetic drift when tested with computer models [43].

Above, we have described only several putative mechanisms that might be implicated in expansion and selection of mtDNA variants, particularly those contributing to increased mutational load cooccurring with aging. Further mechanisms of mutations expansion were comprehensively presented and discussed elsewhere in this book.

10.4.3 Do mtDNA deletions play a role in aging?

It is still debated whether the link between accumulation of mtDNA deletions and progression of aging-related deterioration of tissue function is genuinely causative. Clonal accumulation of mtDNA deletions seems to be involved in progressive loss of muscle fibers, a common manifestation of aging [19,23,44,45]. Similarly, clonal propagation of mtDNA deletions accompanies age-related pathologies of the substantia nigra [46,47]. A high accumulation of mtDNA deletions is also detected in neurons of patients with Parkinson's disease but absent in patients with Alzheimer's [48,49]. Instead, in the hippocampus of patients with Alzheimer's diseases, high levels of mtDNA point mutations were detected [50,51].

The correlative data obtained from patient samples are not supported by causative studies in animal models. Although different animal models prove that accumulation of mtDNA deletions is inseparably associated with a progressive decline of respiratory capacity leading to a strong energetic crisis [30,52], they do not show any signs of accelerated aging phenotype. Furthermore, a recently described mouse model of MGME1 deficiency that has high levels of mtDNA depletion, carrying multiple deleted mtDNA molecules and linear mtDNA fragments, does not show signs of progeria [53]. Despite the opposing data from the animal models, and in the light of evidence from human tissues, the contribution of mtDNA deletions to the aging process should be still considered, especially regarding the age-related deterioration of particular postmitotic tissues as muscle fibers and certain brain regions.

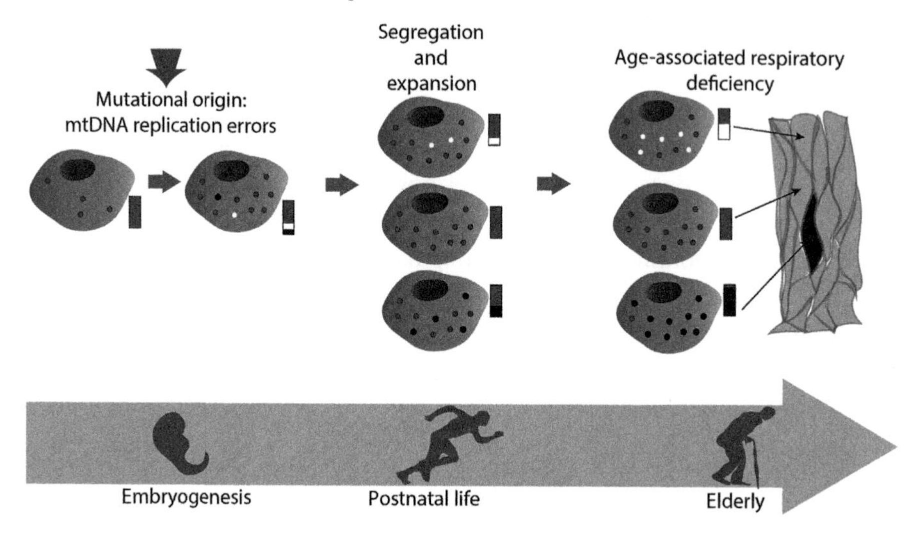

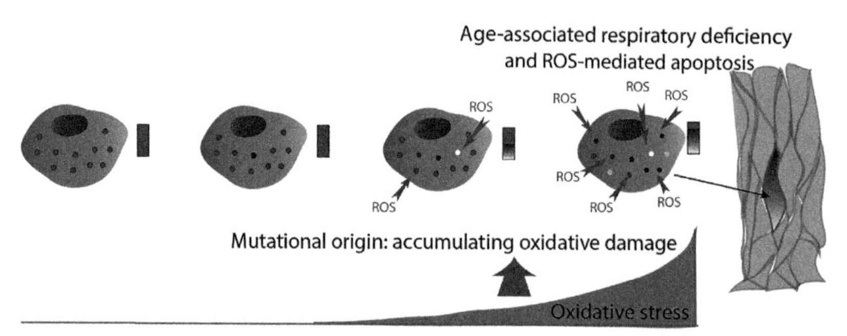

Figure 10.2

Two opposing views on the mitochondrial mutational theory of aging.

MtDNA point mutation can originate from replicative errors (top panel) or oxidative DNA damage (bottom panel). Top panel: replicative errors (*red and yellow dots*) arise early in life and are propagated via mitotic segregation and clonal expansion during the lifetime to reach the mutational threshold (*red dots*) and associated mosaic mitochondrial deficiency in aged individuals. Bottom panel: mtDNA mutations arise in the course of life as a consequence of DNA oxidative lesions. Oxidative stress increases with age aggravating the mutational burden substantially. The population of mtDNA mutations in aged cells represents a high load of mutations with low levels of individual heteroplasmy (*multicolored dots*). Dominant mutations are responsible for mitochondrial deficiency and apoptosis in aged individuals.

Bars represent the levels of heteroplasmy.

10.5 mtDNA point mutations

10.5.1 MtDNA point mutations occur during aging

Recent breakthroughs in technology that led to increased sensitivity of sequencing methods provided evidence that mtDNA point mutations are universal, and that most of us carry low levels of heteroplasmy [54–56]. Remarkably, it seems that at least 20% of individuals are carrying an mtDNA mutation that has been already implicated in different diseases [54]. As in case of mtDNA deletions, point mutations also tend to accumulate in an age-dependent manner in tissues primarily affected by age-related diseases [15,17,57–60].

10.5.2 The origin of somatic mtDNA point mutations during aging: oxidative stress versus replication errors?

The source of mtDNA point mutations in somatic cells and their contribution to aging is a matter of vivid debate for the last decades. Considering that the causal relationship between mtDNA point mutations and aging is real, the appropriate allocation of the mutational origin is essential for understanding the mechanism that governs their expansion during aging, and for selection of relevant antiaging therapies and potential prophylactic interventions.

The overall mutational burden in cells is determined by several factors including execution and accuracy of DNA replication events, ample DNA protection against mutagens, an adaptation of DNA repair apparatus to mutagenic features of the local environment, and rates of hazardous insults in DNA surroundings. The evaluation of which of those factors are safeguarded in the mitochondrial milieu can help to estimate whether mtDNA is more prone to mutagenesis through oxidation damage or rather undergoes spontaneous replicational errors. The accurate assessment of mutational source can be challenging. Nevertheless, the majority of mutagenic insults imprint the specific mutational signatures in DNA that can be used to determine the true origin of each point mutation. In general, single nucleotide substitutions can be divided into transversions and transitions. Transversions represent exchanges of pyrimidines with purines and vice versa, while transitions refer to the replacement of one pyrimidine or purine by another nucleotide of the same type.

10.5.3 Oxidative damage

Oxidative damage in DNA is produced via the action of free radical on the base or sugar moiety leading to their modification. Such a modified base may not pair appropriately during the following round of DNA replication, ultimately leading to persistent nucleotide substitution, thereby point mutation (Fig. 10.3). Although some oxidative lesions can lead to the generation of transitions, it is generally viewed that most of them lead to transversion changes [61]. The 8-oxoguanine (8-oxoG) is the most frequently reported oxidative lesion

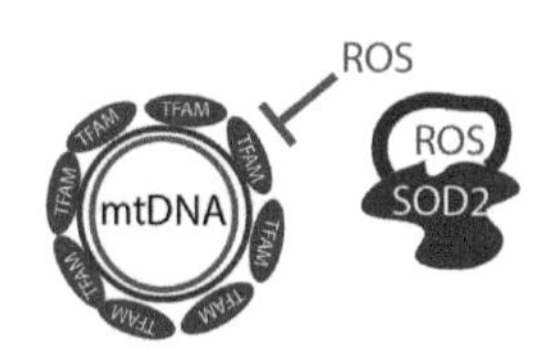

I. Prevention: ROS scavengers and mtDNA coating

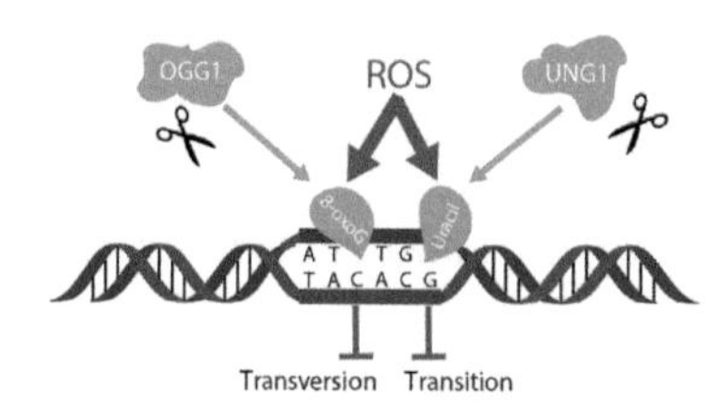

II. Repair: BER-mediated correction of oxidative lesions

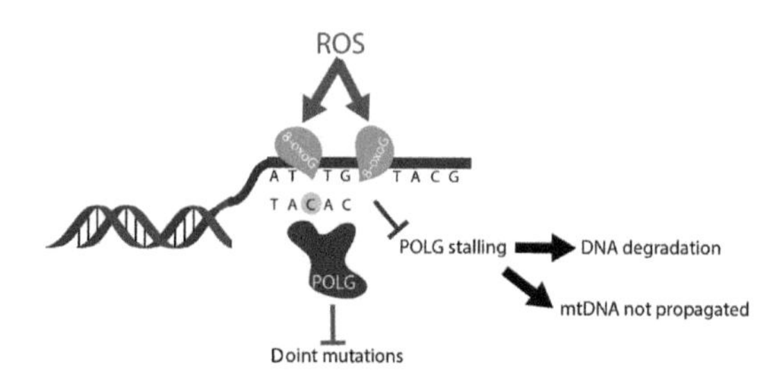

III. Overcoming: Error-ignoring translesion DNA synthesis or replication stalling

Figure 10.3

Defense against oxidative mtDNA lesions in mitochondria.

Mitochondria developed several layers of defense against oxidative mtDNA lesions and their mutational consequences. (I) The mitochondrial genome is protected from oxidative damage by robust protein coating (TFAM), and a set of potent ROS scavengers (including SOD2). (II) Although the importance of base-excision repair (BER) machinery to the removal of oxidative damage remains ambiguous, mammalian mitochondria own several BER enzymes (e.g., OGG1 and UNG1) that are capable of eliminating a broad spectrum of oxidative mtDNA adducts (e.g., 8-oxoG or cytosine deamination product uracil). (III) Once appeared, the oxidative lesion does not have to be translated into mutation but can be ignored by replication machinery (POLG) instead, or lead to replication stalling and eventual loss of the corrupted mtDNA molecule.

in DNA, including mtDNA, used commonly as a universal marker for DNA oxidative damage [62]. If not repaired, the 8-oxoG adducts lead to the preferential formation of transversions in vivo [63]. Other common oxidative DNA lesions, as 2-hydroxyadenine or formamidopyrimidines, also lead to the generation of transversions [61,64].

10.5.4 DNA polymerases

Similar to oxidative lesions, DNA polymerases also have their mutational signature, meaning that they tend to make specific mistakes due to their structural and biochemical properties. Noteworthy, the fidelity of DNA replication is a crucial determinant of the stability of genetic information in our cells, and DNA polymerases dictate how accurate this process is (Fig. 10.4). The DNA polymerase gamma (POLG) is a unique mtDNA replicase in the entire animal kingdom responsible for all possible DNA synthesis events inside the mitochondria. The overall accuracy of POLG results from three basal features: selectivity toward dNTPs, a preferential extension of the well-paired nascent chain, and the ability to excise wrongly paired bases known as proofreading activity. The latter improves the accuracy of human POLG of at least 20-fold, suggesting that ~95% of polymerase errors are repaired thanks to the self-correction mechanism [65]. A number of studies showed that the mutational signature of POLG displays the overwhelming prevalence of transitions over transversions [66–68]. This is strongly consistent with the observations made on brain samples obtained from young versus aged human individuals, where transitions were the most frequent mutations in those biopsies, including predominant C:G to T:A substitutions [17,57]. Similar patterns were observed in tissues of aged rodents [69] and the hippocampus of patients with Alzheimer's disease [50]. Furthermore, naturally occurring mtDNA polymorphisms and disease-causing somatic mutations in the global human population are also strongly biased toward transitions (www.mitomap.org). In light of those findings, only marginal detection of transversions in the mitochondrial genome suggests that the most common oxidative lesions have minimal contribution to overall mtDNA mutational load also in the course of aging. In contrast to bacterial ancestors or yeast mitochondria, mammalian mitochondria do not have canonical mismatch repair (MMR) machinery [70]. Instead, the initial reports suggested that mitochondrial MMR activity could be mediated by YB-1 factor, previously implicated in the nuclear base excision repair (BER) pathway [71]. Nevertheless, further evidence for the robust MMR in mammalian mitochondria is still missing. Therefore it is possible that mtDNA mutations originating from imperfect mtDNA replication cannot be efficiently recognized and corrected, increasing their chances of further propagation in mtDNA pool (Fig. 10.4).

10.5.5 Is mtDNA susceptible to oxidative damage?

Do transversion-promoting oxidative lesions occur in a lifetime but are effectively removed by the dedicated repair machinery? Are oxidative lesions in mtDNA translated into point mutations? Although for decades it was believed that rates of oxidative damage in mitochondrial genome exceed by 10-fold, the ones in nuclear DNA, more rigorous studies showed that previous calculations were highly overestimated and a real difference is minor [7,72]. The same studies indicate that age of tested individuals does not influence the levels

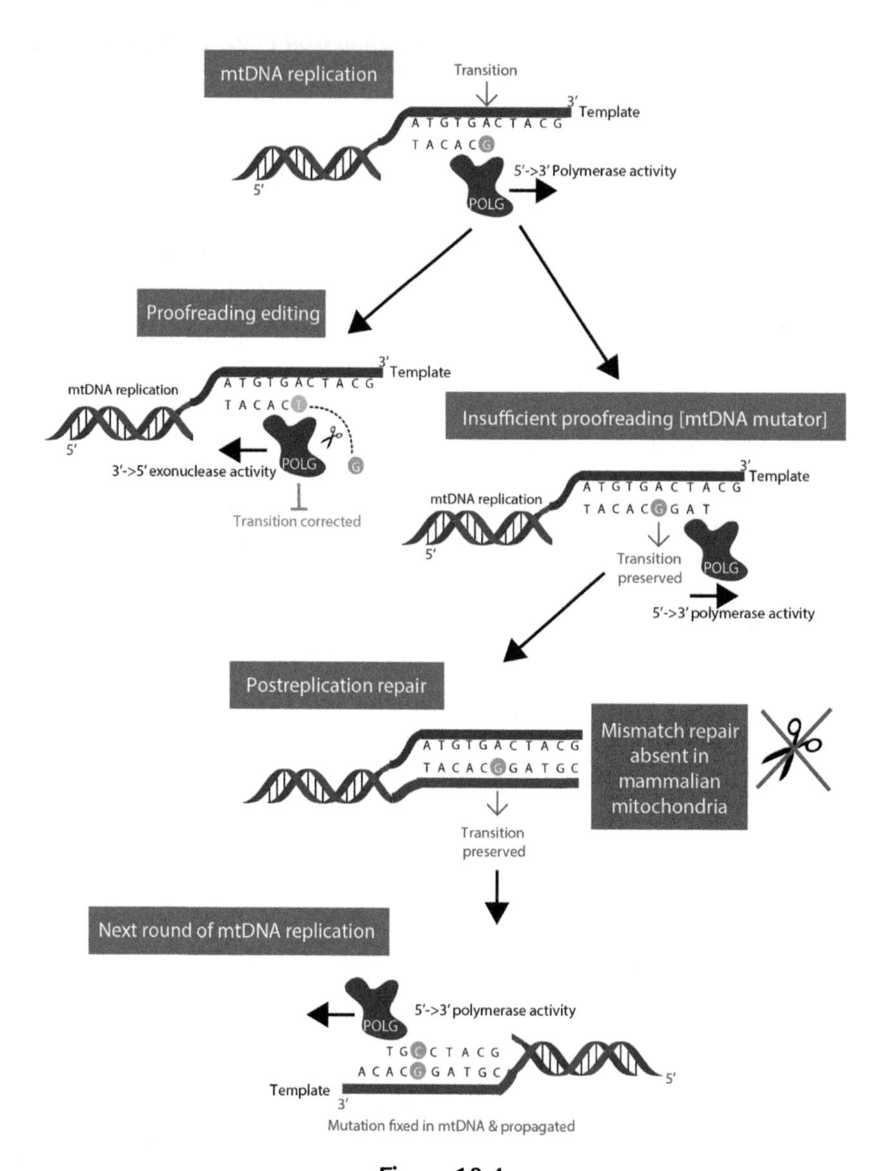

Figure 10.4

Imperfect fidelity of mtDNA replication can be a source of mtDNA mutations during aging. DNA polymerases are a major determinant of replication fidelity. Mitochondrial DNA polymerase gamma (POLG) exhibits intrinsic 3′- > 5′ exonuclease activity that can proofread the errors produced during mtDNA replication. Due to the imperfectness of polymerase, some types of nucleotide mismatches (predominantly transitions) can escape the POLG-mediated proofreading editing cooccurring with replication. The canonical MMR machinery is absent in mammalian mitochondria, questioning whether mismatches produced by POLG can be efficiently recognized and removed. If true, lack of MMR will lead to fixation of mtDNA mutation during the next round of relaxed mtDNA replication. MtDNA undergoes extensive replication early in life (oogenesis and embryogenesis). Those frequent DNA synthesis events can provide a robust source of mutation due to imperfect mtDNA synthesis fidelity, and contribute to the pool of mutations that clonally expand during aging (see Fig. 10.2).

of oxidative mtDNA lesions. A first layer of defense against oxidative damage is provided by set of potent ROS scavengers that reside in the mitochondrial matrix and efficiently neutralize free radicals not allowing them to get in touch with mtDNA (Fig. 10.3). Additionally, and in contrast to a persistent dogma, mtDNA is not naked but carefully coated with proteins (TFAM) and packed into nucleoids, providing a histone-like shielding of genetic information against the direct action of free radicals [73] (Fig. 10.3). Furthermore, mitochondria possess functional BER machinery (shared with nuclear compartment), which would allow repair of a high spectrum of oxidatively damaged bases, thus preventing eventual base substitution [74] (Fig. 10.3). However, the indispensability of mitochondrial BER machinery to mtDNA integrity has been challenged by recent studies. The genetic inactivation of OGG1 and MUTYH components that remove 8-oxyG lesions and prevent transversions, as well as inactivation of UNG1 that removes cytosine deamination and prevents transitions, did not enhance mtDNA mutagenesis in mice [75–77]. This indicates that independently of the nature of the oxidative lesion, there is a little chance that they interfere with mtDNA integrity in vivo. Finally, several in vitro studies suggest that when mitochondrial DNA polymerase gamma meets the oxidative lesion during DNA replication, it tends to stall, or it introduces the correct nucleotide opposite to the lesions [78–80] (Fig. 10.3). Overall, it seems that mtDNA does not accumulate extensive oxidative damage during aging and, if some oxidative lesions occur, mitochondria have developed several strategies to mitigate possible mtDNA mutagenesis.

10.5.6 Origin of mtDNA mutations—evidence from animal models

The free radical theory of aging found support in a study showing that targeting of catalase, an antioxidant enzyme, into mitochondria extends the lifespan of mice and is associated with decreased oxidative damage, reduced levels of mtDNA deletions, and delayed deterioration of heart [81]. Surprisingly, overexpression of many other antioxidant enzymes in mice, including mitochondrial isoforms of superoxide dismutase (SOD2) and glutathione peroxidase (GPX4), failed to recapitulate the lifespan extension despite the pronounced reduction in oxidative stress and several beneficial effects on cell physiology [82].

In contrast, a dramatic stimulation of oxidative stress in haploinsufficient MCKL1 (*Mclk1*$^{+/-}$) mutants of mitochondrial ubiquinone biosynthesis increased the lifespan of animals despite no changes in oxidative lesions in cellular DNA [83]. Similarly, although the heart-specific SOD2 deficiency manifests dilated cardiomyopathy accompanied by severe signs of oxidative stress, no increase in mutational load, either in transversions or in transitions, has been detected in the mtDNA of those animals [76]. Further suppression of mitochondrial repair of oxidative DNA lesions in the same model by inactivation of OGG1 did not stimulate the mtDNA mutagenesis in mice [76]. Similar findings were obtained for genetic inactivation of many other components of mitochondrial BER, additionally revealing no substantial relationship between lack of oxidative lesion repair and age-related phenotypes in mammals (reviewed in Ref. [84]).

In contrast to models of oxidative damage, mice carrying the proofreading-deficient version of mitochondrial DNA polymerase, so-called mtDNA mutators, accumulate high amounts of mtDNA mutations associated with an apparent progeroid phenotype [85,86]. MtDNA mutator mice display reduced lifespan, osteoporosis, sarcopenia, progressive hearing loss, anemia, reduced fertility, hair graying, and alopecia—a set of features that strikingly recapitulate signs of human aging. Consistent with the principles of clonal expansion, the majority of point mutation in mtDNA mutator mice occurs during embryonic development, while phenotypes are observed much later and only upon clonal expansion of individual mutations, over the certain threshold, leading to a clear respiratory chain deficiency [85]. Strikingly, mtDNA mutations that accumulate in the oocytes of heterozygous mothers carrying mutator alleles can be transmitted to progeny and result in mild aging phenotypes in wild type offspring [87]. Besides point mutations, mtDNA mutator mice also carry large linear deleted mtDNA fragments that likely do not contribute to aging phenotypes [32,33]. The major critics of mtDNA mutator mouse as a model for the mtDNA theory of aging indicate that the burden of mutations in mtDNA mutators is much higher than typically observed in aged animals, and that heterozygous mutator mice do not manifest accelerated aging despite relatively high levels of mutations [88]. Nevertheless, mtDNA mutator has been instrumental to many important discoveries regarding the mode of mtDNA replication, and the role of mtDNA mutations in different pathological phenotypes, as well as in stem cells, as will be discussed later. It also cannot be disputed that the complexity of phenotypes observed in mtDNA mutator mice resembles strongly normal aging phenotypes in both mice and humans.

10.5.7 When are point mutations generated, and how do they expand?

In general, mechanisms driving the expansion of point mutations are similar as for deletions. The principles of mtDNA mutation generation and expansion according to two opposing mitochondrial hypotheses of aging were presented in Fig. 10.2. Assuming that point mutations are mostly generated as random errors arising from inaccurate replication events, they can be produced through the entire lifetime, hence will accumulate significantly during aging. Therefore mutations produced early in our life might be able to reach very high levels of heteroplasmy in single cells due to random genetic drift. This will lead to the patchy, mosaic distribution of cells with high rates of individual, potentially hazardous mutations. Levels of heteroplasmy are extremely important for the phenotypic output of mtDNA mutations. The overwhelming majority of mtDNA point mutations need to be present in a high proportion of cellular mtDNA molecules (high heteroplasmy) to elicit harmful effects to the cell. Therefore it is not clear if mutations produced late in life, and thus likely present at low heteroplasmy, will have a detrimental impact on cellular wellbeing. Experimental evidence supports these notions, as it was shown that early to midlife somatic point mutations are much more critical to the age-associated decline of body function, than late-life mtDNA changes [15,42]. The mtDNA mutations seem not to contribute to aging of short-lived metazoans as fruit fly [89] or roundworms [90]. This indicates that aging of long-lived species undergoes distinct regulatory mechanism and also reflects the amount of time needed for the clonal expansion of mutations.

There are also differences in the mechanisms governing point mutation expansion in actively dividing versus postmitotic cells. In proliferating tissues, as in blood cells or embryo, changes in heteroplasmy will likely emerge mostly through the partitioning of mitochondria to the daughter cell, a process known as vegetative segregation. This partitioning of organelle pool can be random or driven by selective pressure, giving an opportunity of fast selection for or against the particular mutation type as shown for blood stem cells [91]. In postmitotic cells, like neurons or myocytes, changes in heteroplasmy occur mostly through the processes associated with relaxed mtDNA replication. They can undergo the random genetic drift or stay under selection pressure, as was shown for mutations in the mtDNA control region that succumb positive replication selection [92–94]. Therefore the time for clonal expansion of point mutation can be very distinctive and depend not only on a tissue type but also on the properties of a particular mutation.

Finally, the mosaic distribution of cells with OXPHOS defect is evident in tissues primarily affected by age-dependent accumulation of mtDNA point mutations, as muscle fibers or colonic crypts [95,96], providing an essential link between the presence of high heteroplasmy mutations and their physiological effect. Moreover, a recent population study revealed that elevated heteroplasmy of mtDNA genetic variant m3243A > G transition manifests with several aging outcomes in affected individuals including reduced strength, cognitive, metabolic and cardiovascular functioning, and increased risk of mortality [97]. Together, those findings emphasize the importance of clonal expansion in age-associated deterioration of tissues.

10.6 How do somatic mtDNA mutations lead to aging?

Mammalian mtDNA encodes a set of components of respiratory chain complexes (RC) and all rRNA and tRNA species needed for the expression of those subunits in mitochondria. Therefore primary defects of any mtDNA mutation, including the ones associated with aging, will reflect on a function of the respiratory chain and ultimately lead to respiratory deficiency. In this regard, as in many other aspects, somatic mtDNA mutations that contribute to aging mimic the mutations associated with mitochondrial diseases. In both cases, mtDNA mutations undergo clonal expansion, leading to different levels of heteroplasmy and phenotypic manifestation dictated by different thresholds specific for each mutation. The prominent difference between mtDNA mutations causing mitochondrial diseases and aging-associated mutations is that formers are usually caused by a single mutation present at high level, while later show high variety in the type and levels of different mtDNA mutations. Therefore upon reaching a specific threshold, mtDNA mutations associated with specific disease will have a much more uniform effect on cell physiology. In contrast, aging-associated mtDNA mutations will accumulate in some, but not all cells of the affected tissues, thus creating highly mosaic pattern of mitochondrial dysfunction. Additionally, results from mtDNA mutator mice argue that high levels of deleterious mutations in protein coding genes are a likely cause of premature aging phenotypes. Remarkably, these kinds of mutations are rarely observed in mitochondrial disease patients because of a strong negative selective pressure against them [98,99]. Point mutations in protein-encoded subunits can strongly decrease or completely abolish the activity of a particular respiratory complex, leading to the instability of affected subunit and specific RC complex, or even result in a loss of other complexes through destabilization of the supercomplex formation. In some rare cases, these point mutations can also stimulate an extensive ROS production through respiratory complexes and result in rapid loss of affected cells [100]. In contrast, point mutations in RNA-encoding genes often cause only mild destabilization of affected RNA species, and only in rare cases can lead to loss of translation fidelity or even the entire mitochondrial gene expression. The most dramatic consequences could be attributed to point mutations in the D-loop, which can affect mtDNA replication and lead to mtDNA depletion [17]. Remarkably, it has been suggested that some of the point mutations in the mtDNA control region can potentially drive a positive replication selection and speed up the clonal expansion of the mutated molecule [101]. Overall, mtDNA point mutations not always result in milder phenotype than mtDNA deletions, so their role in age-related phenotypes should not be underestimated.

Circular mtDNA deleted molecules usually lack several genes encoding for both subunits of respiratory complexes and different RNA molecules. Therefore the effect of large mtDNA deletions is often dramatic, as a loss of any mitochondrial rRNA or tRNA will abolish mitochondrial gene expression and lead to a decline of all mtDNA-encoded complexes (I, III, IV, and V).

10.6.1 Tissue-specific consequences of mtDNA mutations during aging

As mentioned at the beginning of this chapter, mtDNA mutations accumulate during aging in number of organs (including muscles or neurons) and might have very peculiar consequences dependently on the tissue they affect. Although respiratory defects might occur in only a small fraction of aged fibers, their implications on the entire muscle physiology can be harmful and range from common age-related muscle weakness till sarcopenia [23]. Similarly, mosaic distribution of respiratory deficient cardiomyocytes can lead to arrhythmia and cardiac failure, which are also frequent age-associated pathologies [102,103]. Furthermore, it has been proposed that mtDNA deletions and depletion found in the substantia nigra neurons of aged individuals, might play a significant role in the pathogenesis of Parkinson's disease [48,104]. It has also been proposed that a mosaic respiratory deficiency in some parts of the brain can lead to local energy deprivation and problems with signal transduction through the neuronal network, contributing in turn to age-related dementia [30]. The clear mosaic respiratory deficiency has been frequently observed in colonic crypts of aged individuals. In this case, the phenomenon is explained by the aging of stem cells responsible for the rejuvenation of colon epithelium, and in contrast to previous examples, attributed to aging of proliferative cells [15,96].

10.7 mtDNA mutations and aging of stem cells

Stem cells are essential for rejuvenation of cell pools and play a central role in the regeneration and maintenance of tissues. The ability of an organism to regenerate declines considerably with progressing age and stem cells decline contributes to aging of both proliferative and postmitotic tissues [105]. The hypothesis that mtDNA mutations can be propagated through stem cells, and that mtDNA integrity plays a role in age-associated decline of the stem cell function drew much attention in recent years. Mitochondrial respiratory function and adequate partitioning of the mitochondria during stem cell division were shown to greatly contribute to stem cell fitness, and their ability to differentiate [106,107].

Stem cells can significantly contribute to the clonal expansion of mtDNA variants in proliferative tissues as shown for human colonic crypts and stomach [108,109]. The mtDNA mutational load increases with age in induced pluripotent stem cells (iPSCs) and plays a role in the expansion of somatic mtDNA mutations in human blood and skin cells [110]. Studies on mtDNA mutator mice, particularly on their hematopoietic system, allowed uncovering of several mechanisms that control the aging of the stem cells. Dramatic failure of the hematopoiesis, manifested with progressing anemia and lymphopenia, is one of the hallmarks of the progeroid phenotype in mtDNA mutator mice [85,86]. Curiously, deterioration of hematopoiesis in those animals initiates already during embryogenesis [111]. Stem cell defects were also described in the intestine of mtDNA mutator mice, where the accumulation of mtDNA point mutations resulted in problems with nutrient absorption, aberrant morphology of intestine, dysregulation of the cell cycle in crypts, and prominent apoptosis [112]. Similar observations have been made for neural and cardiac stem cells of mtDNA mutator mice, indicating that the rejuvenating capacity of postmitotic tissues undergoes regulation driven by mtDNA mutational load [59,111]. The increased mutational burden in cardiac stem cells disrupted mitochondrial function and compromised cell differentiation leading to massive cell death [59]. Intriguingly, the redox balance inside the mitochondria might play a critical role in a regulation of stem cell aging. Treatment with N-acetyl-L-cysteine, a molecule that increases cellular thiols (e.g., GSH) thereby decreasing overall oxidation level in the cell, improved the fitness of the stem cells in mtDNA mutator mice, suggesting that the accumulation of point mutations in hematopoietic stem cells exerts a small but essential effect on their redox balance that could be therapeutically manipulated [111]. Similar findings were obtained in pluripotent stem cells, where increased mtDNA mutagenesis affected the reprogramming and stemness of the cells, and the effect could be substantially reversed by pharmacologically increasing the redox capacity [113]. Remarkably, redox balance appears to be very tightly regulated in stem cells, as the elevated levels of specific antioxidants (MitoQ) were proven to be toxic to pluripotent stem cells, in particular to the neural ones [113]. Together, these evidences suggest that mtDNA mutations play an essential role in the age-progressing decline in cell stemness and consequent deterioration of the rejuvenation capacity of the organism. MtDNA integrity appears crucial particularly for the differentiation ability of stem cells and this seems to be highly dependent on a tightly controlled redox balance.

10.8 Conclusions

Although mtDNA encodes only a handful of all genes present in humans, an intimate link exists between the integrity of the mitochondrial genome and age-associated decline of our organism. Remarkably, even after years of intensive research, it is yet to be determined whether this relationship is genuinely causative or merely correlative. Here, we outlined major hypotheses aimed at explaining the complex interplay between mtDNA mutational load and aging, and confronted them with experimental evidence. The current state of knowledge does not allow us to make a final conclusion as to whether mtDNA point mutations and deletion contribute to aging to the same extent. However, it becomes increasingly clear that most mtDNA point mutations likely originate from replication errors rather than from oxidative damage. Strong experimental data also indicate that new somatic mtDNA mutations need to arise early enough to be able to expand to levels apt for the phenotypic manifestation and consequent mitochondrial dysfunction. The expansion of mtDNA mutations can undergo several layers of regulation, and this makes the entire phenomenon even more complicated. We also do not know yet if environmental factors can modulate the dynamics of mtDNA mutation expansion or process as the whole is only intrinsically regulated. Nevertheless, mitochondrial failure seems to substantially limit our health-span expectations, hence being an attractive target for antiaging interventions.

Acknowledgments

The study was supported by grants of the European Research Council (ERC-StG-2012–310700 and ERC-2018-PoC-813169), Deutsche Forschungsgemeinschaft (DFG, German Research Fundation) - SFB 1218 - Projektnummer 269925409, and Center for Molecular Medicine Cologne, University of Cologne (54-RP). The authors declare that they have no competing financial interests related to this manuscript.

Research perspectives

Although the first two decades of 21st century beget extensive studies in the field of mitochondrial genetics, we are still far from a complete understanding of whether and how mtDNA mutations contribute to aging. The mtDNA mutator mice not only provided a fresh look at the mitochondrial theory of aging but also allowed multiple new questions to emerge, while inflaming the long-lasting debate on the role of mtDNA oxidative damage in the age-related decline of the organism. Notwithstanding this, we are still missing direct evidence for oxidative damage being a straightforward cause of mtDNA mutations and underlying defects. Indeed, the grasp of such mechanisms would highly benefit from studies on animal models that avert the age-associated mtDNA mutation accumulation by improving the replicative fidelity in mitochondria in vivo, for example, by the generation of mtDNA antimutator mice. Eventually, alternative models displaying increased early versus late-onset mtDNA mutagenesis could shed light on the complexity of the entire process. Although the occurrence of aging-eliciting mtDNA mutations early in life is credible and consistent with computational modeling, the definitive experimental evidence for a higher relevance of those mutations over the once appearing later in life is still missing. We also do not know whether the expansion of mutations in the course of aging occurs randomly, or is prone to selective pressure, or combines both mechanisms depending on the mutational or tissue-specific context. Improvement of mtDNA integrity in stem cells (e.g., via targeted modulation of redox capacity) may bring new therapeutic perspectives for age-associated disorders. Finally, it would be essential to evaluate how the intrinsic and environmental changes can modulate mtDNA mutational load in long-lived species.

References

[1] Harman D. Aging: a theory based on free radical and radiation chemistry. J Gerontol 1956;11:298–300.

[2] Harman D. The biologic clock: the mitochondria? J Am Geriatr Soc 1972;20:145–7.

[3] Linnane AW, Marzuki S, Ozawa T, Tanaka M. Mitochondrial DNA mutations as an important contributor to ageing and degenerative diseases. Lancet 1989;642–5.

[4] Miquel J. An update on the oxygen stress-mitochondrial mutation theory of aging: genetic and evolutionary implications. Exp Gerontol 1998;33:113–26.

[5] Bandy B, Davison AJ. Mitochondrial mutations may increase oxidative stress: implications for carcinogenesis and aging? Free Radic Biol Med 1990;8:523–39.

[6] Nachman MW, Crowell SL. Estimate of the mutation rate per nucleotide in humans. Genetics 2000;156:297–304.

[7] Lim KS, Jeyaseelan K, Whiteman M, Jenner A, Halliwell B. Oxidative damage in mitochondrial DNA is not extensive. Ann N Y Acad Sci 2005;1042:210–20.

[8] Sanz A, Fernandez-Ayala DJ, Stefanatos RK, Jacobs HT. Mitochondrial ROS production correlates with, but does not directly regulate lifespan in Drosophila. Aging (Albany NY) 2010;2:200–23.

[9] Halliwell B. Biochemistry of oxidative stress. Biochem Soc Trans 2007;35:1147–50.

[10] Cadenas E, Davies KJ. Mitochondrial free radical generation, oxidative stress, and aging. Free Radic Biol Med 2000;29:222–30.

[11] Ng LF, Ng LT, van Breugel M, Halliwell B, Gruber J. Mitochondrial DNA damage does not determine *C. elegans* lifespan. Front Genet 2019;10:311.

[12] Yang W, Hekimi S. A mitochondrial superoxide signal triggers increased longevity in *Caenorhabditis elegans*. PLoS Biol 2010;8:e1000556.

[13] Lapointe J, Hekimi S. When a theory of aging ages badly. Cell Mol Life Sci 2009;67:1–8.
[14] Cortopassi GA, Arnheim N. Using the polymerase chain reaction to estimate mutation frequencies and rates in human cells. Mutat Res 1992;277:239–49.
[15] Greaves LC, Nooteboom M, Elson JL, Tuppen HA, Taylor GA, Commane DM, et al. Clonal expansion of early to mid-life mitochondrial DNA point mutations drives mitochondrial dysfunction during human ageing. PLoS Genet 2014;10:e1004620.
[16] Katayama M, Tanaka M, Yamamoto H, Ohbayashi T, Nimura Y, Ozawa T. Deleted mitochondrial DNA in skeletal muscle of aged individuals. Biochem Int 1991;25:47–56.
[17] Kennedy SR, Salk JJ, Schmitt MW, Loeb LA. Ultra-sensitive sequencing reveals an age-related increase in somatic mitochondrial mutations that are inconsistent with oxidative damage. PLoS Genet 2013;9:e1003794.
[18] Piko L, Hougham AJ, Bulpitt KJ. Studies of sequence heterogeneity of mitochondrial DNA from rat and mouse tissues: evidence for an increased frequency of deletions/additions with aging. Mech Ageing Dev 1988;43:279–93.
[19] Bua E, Johnson J, Herbst A, Delong B, McKenzie D, Salamat S, et al. Mitochondrial DNA-deletion mutations accumulate intracellularly to detrimental levels in aged human skeletal muscle fibers. Am J Hum Genet 2006;79:469–80.
[20] Yen TC, Pang CY, Hsieh RH, Su CH, King KL, Wei YH. Age-dependent 6kb deletion in human liver mitochondrial DNA. Biochem Int 1992;26:457–68.
[21] Corral-Debrinski M, Shoffner JM, Lott MT, Wallace DC. Association of mitochondrial DNA damage with aging and coronary atherosclerotic heart disease. Mutat Res 1992;275:169–80.
[22] Cortopassi GA, Arnheim N. Detection of a specific mitochondrial DNA deletion in tissues of older humans. Nucleic Acids Res 1990;18:6927–33.
[23] Herbst A, Wanagat J, Cheema N, Widjaja K, McKenzie D, Aiken JM. Latent mitochondrial DNA deletion mutations drive muscle fiber loss at old age. Aging Cell 2016;15:1132–9.
[24] Kao SH, Chao HT, Wei YH. Mitochondrial deoxyribonucleic acid 4977-bp deletion is associated with diminished fertility and motility of human sperm. Biol Reprod 1995;52:729–36.
[25] Brossas JY, Barreau E, Courtois Y, Treton J. Multiple deletions in mitochondrial DNA are present in senescent mouse brain. Biochem Biophys Res Commun 1994;202:654–9.
[26] Simonetti S, Chen X, DiMauro S, Schon EA. Accumulation of deletions in human mitochondrial DNA during normal aging: analysis by quantitative PCR. Biochim Biophys Acta 1992;1180:113–22.
[27] Taylor SD, Ericson NG, Burton JN, Prolla TA, Silber JR, Shendure J, et al. Targeted enrichment and high-resolution digital profiling of mitochondrial DNA deletions in human brain. Aging Cell 2014;13:29–38.
[28] Damas J, Samuels DC, Carneiro J, Amorim A, Pereira F. Mitochondrial DNA rearrangements in health and disease—a comprehensive study. Hum Mutat 2014;35:1–14.
[29] Dong DW, Pereira F, Barrett SP, Kolesar JE, Cao K, Damas J, et al. Association of G-quadruplex forming sequences with human mtDNA deletion breakpoints. BMC Genomics 2014;15:677.
[30] Fukui H, Moraes CT. Mechanisms of formation and accumulation of mitochondrial DNA deletions in aging neurons. Hum Mol Genet 2009;18:1028–36.
[31] Krishnan KJ, Reeve AK, Samuels DC, Chinnery PF, Blackwood JK, Taylor RW, et al. What causes mitochondrial DNA deletions in human cells? Nat Genet 2008;40:275–9.
[32] Nissanka N, Minczuk M, Moraes CT. Mechanisms of mitochondrial DNA deletion formation. Trends Genet 2019;35:235–44.
[33] Kauppila TE, Kauppila JH, Larsson NG. Mammalian mitochondria and aging: an update. Cell Metab 2017;25:57–71.
[34] Nissanka N, Moraes CT. Mitochondrial DNA damage and reactive oxygen species in neurodegenerative disease. FEBS Lett 2018;592:728–42.
[35] Dumont P, Burton M, Chen QM, Gonos ES, Frippiat C, Mazarati JB, et al. Induction of replicative senescence biomarkers by sublethal oxidative stresses in normal human fibroblast. Free Radic Biol Med 2000;28:361–73.

[36] Prithivirajsingh S, Story MD, Bergh SA, Geara FB, Ang KK, Ismail SM, et al. Accumulation of the common mitochondrial DNA deletion induced by ionizing radiation. FEBS Lett 2004;571:227–32.
[37] Campbell G, Krishnan KJ, Deschauer M, Taylor RW, Turnbull DM. Dissecting the mechanisms underlying the accumulation of mitochondrial DNA deletions in human skeletal muscle. Hum Mol Genet 2014;23:4612–20.
[38] Wallace DC. Mitochondrial genetics: a paradigm for aging and degenerative diseases? Science 1992;256:628–32.
[39] Kowald A, Dawson M, Kirkwood TB. Mitochondrial mutations and ageing: can mitochondrial deletion mutants accumulate via a size based replication advantage? J Theor Biol 2014;340:111–18.
[40] Kowald A, Kirkwood TB. Transcription could be the key to the selection advantage of mitochondrial deletion mutants in aging. Proc Natl Acad Sci USA 2014;111:2972–7.
[41] Kowald A, Kirkwood TBL. Resolving the enigma of the clonal expansion of mtDNA deletions. Genes (Basel) 2018;9.
[42] Elson JL, Samuels DC, Turnbull DM, Chinnery PF. Random intracellular drift explains the clonal expansion of mitochondrial DNA mutations with age. Am J Hum Genet 2001;68:802–6.
[43] Kowald A, Kirkwood TB. Mitochondrial mutations and aging: random drift is insufficient to explain the accumulation of mitochondrial deletion mutants in short-lived animals. Aging Cell 2013;12:728–31.
[44] Wanagat J, Cao Z, Pathare P, Aiken JM. Mitochondrial DNA deletion mutations colocalize with segmental electron transport system abnormalities, muscle fiber atrophy, fiber splitting, and oxidative damage in sarcopenia. FASEB J 2001;15:322–32.
[45] Cao Z, Wanagat J, McKiernan SH, Aiken JM. Mitochondrial DNA deletion mutations are concomitant with ragged red regions of individual, aged muscle fibers: analysis by laser-capture microdissection. Nucleic Acids Res 2001;29:4502–8.
[46] Kraytsberg Y, Kudryavtseva E, McKee AC, Geula C, Kowall NW, Khrapko K. Mitochondrial DNA deletions are abundant and cause functional impairment in aged human substantia nigra neurons. Nat Genet 2006;38:518–20.
[47] Reeve AK, Krishnan KJ, Elson JL, Morris CM, Bender A, Lightowlers RN, et al. Nature of mitochondrial DNA deletions in substantia nigra neurons. Am J Hum Genet 2008;82:228–35.
[48] Bender A, Krishnan KJ, Morris CM, Taylor GA, Reeve AK, Perry RH, et al. High levels of mitochondrial DNA deletions in substantia nigra neurons in aging and Parkinson disease. Nat Genet 2006;38:515–17.
[49] Nido GS, Dolle C, Flones I, Tuppen HA, Alves G, Tysnes OB, et al. Ultradeep mapping of neuronal mitochondrial deletions in Parkinson's disease. Neurobiol Aging 2018;63:120–7.
[50] Hoekstra JG, Hipp MJ, Montine TJ, Kennedy SR. Mitochondrial DNA mutations increase in early stage Alzheimer disease and are inconsistent with oxidative damage. Ann Neurol 2016;80:301–6.
[51] Lin MT, Simon DK, Ahn CH, Kim LM, Beal MF. High aggregate burden of somatic mtDNA point mutations in aging and Alzheimer's disease brain. Hum Mol Genet 2002;11:133–45.
[52] Tyynismaa H, Mjosund KP, Wanrooij S, Lappalainen I, Ylikallio E, Jalanko A, et al. Mutant mitochondrial helicase Twinkle causes multiple mtDNA deletions and a late-onset mitochondrial disease in mice. Proc Natl Acad Sci USA 2005;102:17687–92.
[53] Matic S, Jiang M, Nicholls TJ, Uhler JP, Dirksen-Schwanenland C, Polosa PL, et al. Mice lacking the mitochondrial exonuclease MGME1 accumulate mtDNA deletions without developing progeria. Nat Commun 2018;9:1202.
[54] Ye K, Lu J, Ma F, Keinan A, Gu Z. Extensive pathogenicity of mitochondrial heteroplasmy in healthy human individuals. Proc Natl Acad Sci USA 2014;111:10654–9.
[55] Payne BA, Wilson IJ, Yu-Wai-Man P, Coxhead J, Deehan D, Horvath R, et al. Universal heteroplasmy of human mitochondrial DNA. Hum Mol Genet 2013;22:384–90.
[56] Stewart JB, Larsson NG. Keeping mtDNA in shape between generations. PLoS Genet 2014;10:e1004670.
[57] Williams SL, Mash DC, Zuchner S, Moraes CT. Somatic mtDNA mutation spectra in the aging human putamen. PLoS Genet 2013;9:e1003990.

[58] Coxhead J, Kurzawa-Akanbi M, Hussain R, Pyle A, Chinnery P, Hudson G. Somatic mtDNA variation is an important component of Parkinson's disease. Neurobiol Aging 2016;38 217 e211–216.
[59] Orogo AM, Gonzalez ER, Kubli DA, Baptista IL, Ong SB, Prolla TA, et al. Accumulation of mitochondrial DNA mutations disrupts cardiac progenitor cell function and reduces survival. J Biol Chem 2015;290:22061–75.
[60] Pinto M, Moraes CT. Mechanisms linking mtDNA damage and aging. Free Radic Biol Med 2015;85:250–8.
[61] Evans MD, Dizdaroglu M, Cooke MS. Oxidative DNA damage and disease: induction, repair and significance. Mutat Res 2004;567:1–61.
[62] Radak Z, Boldogh I. 8-Oxo-7,8-dihydroguanine: links to gene expression, aging, and defense against oxidative stress. Free Radic Biol Med 2010;49:587–96.
[63] Ohno M, Sakumi K, Fukumura R, Furuichi M, Iwasaki Y, Hokama M, et al. 8-oxoguanine causes spontaneous de novo germline mutations in mice. Sci Rep 2014;4:4689.
[64] Yasui M, Kanemaru Y, Kamoshita N, Suzuki T, Arakawa T, Honma M. Tracing the fates of site-specifically introduced DNA adducts in the human genome. DNA Repair (Amst) 2014;15:11–20.
[65] Longley MJ, Nguyen D, Kunkel TA, Copeland WC. The fidelity of human DNA polymerase gamma with and without exonucleolytic proofreading and the p55 accessory subunit. J Biol Chem 2001;276:38555–62.
[66] Zheng W, Khrapko K, Coller HA, Thilly WG, Copeland WC. Origins of human mitochondrial point mutations as DNA polymerase gamma-mediated errors. Mutat Res 2006;599:11–20.
[67] Itsara LS, Kennedy SR, Fox EJ, Yu S, Hewitt JJ, Sanchez-Contreras M, et al. Oxidative stress is not a major contributor to somatic mitochondrial DNA mutations. PLoS Genet 2014;10:e1003974.
[68] Spelbrink JN, Toivonen JM, Hakkaart GA, Kurkela JM, Cooper HM, Lehtinen SK, et al. In vivo functional analysis of the human mitochondrial DNA polymerase POLG expressed in cultured human cells. J Biol Chem 2000;275:24818–28.
[69] Stewart JB, Chinnery PF. The dynamics of mitochondrial DNA heteroplasmy: implications for human health and disease. Nat Rev Genet 2015;16:530–42.
[70] Foury F, Hu J, Vanderstraeten S. Mitochondrial DNA mutators. Cell Mol Life Sci 2004;61 (22):2799–811. Available from: https://doi.org/10.1007/s00018-004-4220-y.
[71] de Souza-Pinto NC, Maynard S, Hashiguchi K, Hu J, Muftuoglu M, Bohr VA. The recombination protein RAD52 cooperates with the excision repair protein OGG1 for the repair of oxidative lesions in mammalian cells. Mol Cell Biol 2009;29(16):4441–54. Available from: https://doi.org/10.1128/MCB.00265-09.
[72] Anson RM, Hudson E, Bohr VA. Mitochondrial endogenous oxidative damage has been overestimated. FASEB J 2000;14:355–60.
[73] Kukat C, Wurm CA, Spahr H, Falkenberg M, Larsson NG, Jakobs S. Super-resolution microscopy reveals that mammalian mitochondrial nucleoids have a uniform size and frequently contain a single copy of mtDNA. Proc Natl Acad Sci USA 2011;108:13534–9.
[74] Szczepanowska K, Trifunovic A. Origins of mtDNA mutations in ageing. Essays Biochem 2017;61:325–37.
[75] Halsne R, Esbensen Y, Wang W, Scheffler K, Suganthan R, Bjoras M, et al. Lack of the DNA glycosylases MYH and OGG1 in the cancer prone double mutant mouse does not increase mitochondrial DNA mutagenesis. DNA Repair (Amst) 2012;11:278–85.
[76] Kauppila JHK, Bonekamp NA, Mourier A, Isokallio MA, Just A, Kauppila TES, et al. Base-excision repair deficiency alone or combined with increased oxidative stress does not increase mtDNA point mutations in mice. Nucleic Acids Res 2018;46:6642–69.
[77] Nilsen H, Steinsbekk KS, Otterlei M, Slupphaug G, Aas PA, Krokan HE. Analysis of uracil-DNA glycosylases from the murine Ung gene reveals differential expression in tissues and in embryonic development and a subcellular sorting pattern that differs from the human homologues. Nucleic Acids Res 2000;28:2277–85.

[78] Graziewicz MA, Bienstock RJ, Copeland WC. The DNA polymerase gamma Y955C disease variant associated with PEO and parkinsonism mediates the incorporation and translesion synthesis opposite 7,8-dihydro-8-oxo-2′-deoxyguanosine. Hum Mol Genet 2007;16:2729–39.

[79] Pinz KG, Shibutani S, Bogenhagen DF. Action of mitochondrial DNA polymerase gamma at sites of base loss or oxidative damage. J Biol Chem 1995;270:9202–6.

[80] Stojkovic G, Makarova AV, Wanrooij PH, Forslund J, Burgers PM, Wanrooij S. Oxidative DNA damage stalls the human mitochondrial replisome. Sci Rep 2016;6:28942.

[81] Schriner SE, Linford NJ, Martin GM, Treuting P, Ogburn CE, Emond M, et al. Extension of murine life span by overexpression of catalase targeted to mitochondria. Science 2005;308:1909–11.

[82] Lei XG, Zhu JH, Cheng WH, Bao Y, Ho YS, Reddi AR, et al. Paradoxical roles of antioxidant enzymes: basic mechanisms and health implications. Physiol Rev 2016;96:307–64.

[83] Liu X, Jiang N, Hughes B, Bigras E, Shoubridge E, Hekimi S. Evolutionary conservation of the clk-1-dependent mechanism of longevity: loss of mclk1 increases cellular fitness and lifespan in mice. Genes Dev 2005;19:2424–34.

[84] Szczepanowska K, Trifunovic A. Different faces of mitochondrial DNA mutators. Biochim Biophys Acta 2015;1847:1362–72.

[85] Trifunovic A, Wredenberg A, Falkenberg M, Spelbrink JN, Rovio AT, Bruder CE, et al. Premature ageing in mice expressing defective mitochondrial DNA polymerase. Nature 2004;429:417–23.

[86] Kujoth GC, Hiona A, Pugh TD, Someya S, Panzer K, Wohlgemuth SE, et al. Mitochondrial DNA mutations, oxidative stress, and apoptosis in mammalian aging. Science 2005;309:481–4.

[87] Ross JM, Coppotelli G, Hoffer BJ, Olson L. Maternally transmitted mitochondrial DNA mutations can reduce lifespan. Sci Rep 2014;4:6569.

[88] Khrapko K, Turnbull D. Mitochondrial DNA mutations in aging. Prog Mol Biol Transl Sci 2014;127:29–62.

[89] Kauppila TES, Bratic A, Jensen MB, Baggio F, Partridge L, Jasper H, et al. Mutations of mitochondrial DNA are not major contributors to aging of fruit flies. Proc Natl Acad Sci USA 2018;115: E9620–9.

[90] Lakshmanan LN, Yee Z, Ng LF, Gunawan R, Halliwell B, Gruber J. Clonal expansion of mitochondrial DNA deletions is a private mechanism of aging in long-lived animals. Aging Cell 2018;17: e12814.

[91] Rajasimha HK, Chinnery PF, Samuels DC. Selection against pathogenic mtDNA mutations in a stem cell population leads to the loss of the 3243A-- > G mutation in blood. Am J Hum Genet 2008;82:333–43.

[92] He Y, Wu J, Dressman DC, Iacobuzio-Donahue C, Markowitz SD, Velculescu VE, et al. Heteroplasmic mitochondrial DNA mutations in normal and tumour cells. Nature 2010;464:610–14.

[93] Li M, Schroder R, Ni S, Madea B, Stoneking M. Extensive tissue-related and allele-related mtDNA heteroplasmy suggests positive selection for somatic mutations. Proc Natl Acad Sci USA 2015;112:2491–6.

[94] Samuels DC, Li C, Li B, Song Z, Torstenson E, Boyd Clay H, et al. Recurrent tissue-specific mtDNA mutations are common in humans. PLoS Genet 2013;9:e1003929.

[95] Mueller-Hocker J. Cytochrome-c-oxidase deficient cardiomyocytes in the human heart—an age-related phenomenon. Am J Pathol 1989;134:1167–73.

[96] Taylor RW, Barron MJ, Borthwick GM, Gospel A, Chinnery PF, Samuels DC, et al. Mitochondrial DNA mutations in human colonic crypt stem cells. J Clin Invest 2003;112:1351–60.

[97] Tranah GJ, Katzman SM, Lauterjung K, Yaffe K, Manini TM, Kritchevsky S, et al. Mitochondrial DNA m.3243A > G heteroplasmy affects multiple aging phenotypes and risk of mortality. Sci Rep 2018;8:11887.

[98] Edgar D, Shabalina I, Camara Y, Wredenberg A, Calvaruso MA, Nijtmans L, et al. Random point mutations with major effects on protein-coding genes are the driving force behind premature aging in mtDNA mutator mice. Cell Metab 2009;10:131–8.

[99] Stewart JB, Freyer C, Elson JL, Wredenberg A, Cansu Z, Trifunovic A, et al. Strong purifying selection in transmission of mammalian mitochondrial DNA. PLoS Biol 2008;6:e10.

[100] Fan W, Waymire KG, Narula N, Li P, Rocher C, Coskun PE, et al. A mouse model of mitochondrial disease reveals germline selection against severe mtDNA mutations. Science 2008;319:958–62.

[101] Michikawa Y, Mazzucchelli F, Bresolin N, Scarlato G, Attardi G. Aging-dependent large accumulation of point mutations in the human mtDNA control region for replication. Science 1999;286:774–9.

[102] Zhang D, Mott JL, Farrar P, Ryerse JS, Chang SW, Stevens M, et al. Mitochondrial DNA mutations activate the mitochondrial apoptotic pathway and cause dilated cardiomyopathy. Cardiovasc Res 2003;57:147–57.

[103] Baris OR, Ederer S, Neuhaus JF, von Kleist-Retzow JC, Wunderlich CM, Pal M, et al. Mosaic deficiency in mitochondrial oxidative metabolism promotes cardiac arrhythmia during aging. Cell Metab 2015;21:667–77.

[104] Dolle C, Flones I, Nido GS, Miletic H, Osuagwu N, Kristoffersen S, et al. Defective mitochondrial DNA homeostasis in the substantia nigra in Parkinson disease. Nat Commun 2016;7:13548.

[105] Goodell MA, Rando TA. Stem cells and healthy aging. Science 2015;350:1199–204.

[106] Anso E, Weinberg SE, Diebold LP, Thompson BJ, Malinge S, Schumacker PT, et al. The mitochondrial respiratory chain is essential for haematopoietic stem cell function. Nat Cell Biol 2017;19:614–25.

[107] Katajisto P, Dohla J, Chaffer CL, Pentinmikko N, Marjanovic N, Iqbal S, et al. Stem cells. Asymmetric apportioning of aged mitochondria between daughter cells is required for stemness. Science 2015;348:340–3.

[108] Gutierrez-Gonzalez L, Deheragoda M, Elia G, Leedham SJ, Shankar A, Imber C, et al. Analysis of the clonal architecture of the human small intestinal epithelium establishes a common stem cell for all lineages and reveals a mechanism for the fixation and spread of mutations. J Pathol 2009;217:489–96.

[109] McDonald SA, Greaves LC, Gutierrez-Gonzalez L, Rodriguez-Justo M, Deheragoda M, Leedham SJ, et al. Mechanisms of field cancerization in the human stomach: the expansion and spread of mutated gastric stem cells. Gastroenterology 2008;134:500–10.

[110] Kang E, Wang X, Tippner-Hedges R, Ma H, Folmes CD, Gutierrez NM, et al. Age-related accumulation of somatic mitochondrial DNA mutations in adult-derived human iPSCs. Cell Stem Cell 2016;18:625–36.

[111] Ahlqvist KJ, Hamalainen RH, Yatsuga S, Uutela M, Terzioglu M, Gotz A, et al. Somatic progenitor cell vulnerability to mitochondrial DNA mutagenesis underlies progeroid phenotypes in Polg mutator mice. Cell Metab 2012;15:100–9.

[112] Fox RG, Magness S, Kujoth GC, Prolla TA, Maeda N. Mitochondrial DNA polymerase editing mutation, PolgD257A, disturbs stem-progenitor cell cycling in the small intestine and restricts excess fat absorption. Am J Physiol Gastrointest Liver Physiol 2012;302:G914–24.

[113] Hamalainen RH, Ahlqvist KJ, Ellonen P, Lepisto M, Logan A, Otonkoski T, et al. mtDNA mutagenesis disrupts pluripotent stem cell function by altering redox signaling. Cell Rep 2015;11:1614–24.

CHAPTER 11

Methods for the identification of mitochondrial DNA variants

Claudia Calabrese*[,1,2], Aurora Gomez-Duran*[,1,2], Aurelio Reyes[2] and Marcella Attimonelli[3]

[1]Department of Clinical Neurosciences, University of Cambridge, Cambridge Biomedical Campus, Cambridge, United Kingdom, [2]MRC Mitochondrial Biology Unit, University of Cambridge, Cambridge, United Kingdom, [3]Department of Biosciences, Biotechnology and Biopharmaceutics, University "Aldo Moro", Bari, Italy

11.1 Introduction to human mtDNA variants detection

Human mitochondrial DNA (mtDNA) is present in multiple copies within mitochondria, varying its number between cells and tissues depending upon the cellular energy demand [1,2]. Due to this physiological polyploidy, more than one mtDNA species, defined by their sequences, can be present within a single mitochondrion or single cell [3]. This condition is known as *heteroplasmy*, as opposed to *homoplasmy*, in which all mtDNA copies are identical [4]. mtDNA heteroplasmy was long considered rare and unique to mitochondrial disorders [5]. Some early evidence, however, challenged this hypothesis by showing the presence of heteroplasmic variants in the control region of mtDNA in hair [6] and brain [7] from healthy individuals. Furthermore, in order to be disease-causing, the abundance of a given pathogenic mutation has to reach a certain threshold [8], that is variable for different types of mutations [9].

Detection of mtDNA variants (either populational or pathogenic) and heteroplasmy has quickly evolved since the discovery of the mtDNA in the early 1960s [10]. Traditionally, mitochondrial variants were studied through a combination of molecular biology techniques (Sections 11.2.1). However, these methods are often constrained to a single and specific variant detection [11] and, hence, unable to detect the presence of other mtDNA pathogenic mutations elsewhere in the mtDNA [12]. The discovery of a second mutation in a patient already diagnosed with mitochondrial disease [13], together with the growing interest in mitochondrial population genetics [14], instigated the vast majority of laboratories to

---

* These authors contributed equally.

The Human Mitochondrial Genome.
DOI: https://doi.org/10.1016/B978-0-12-819656-4.00011-5

perform comprehensive screening of the entire mtDNA molecule using Sanger sequencing (Section 11.2.2.3). Most of these techniques (including Sanger sequencing) are time-consuming, nonquantitative, and prone to errors [15]. As a consequence, while they provide valuable information, their application is limited to small cohorts of samples and do not ensure accurate estimates of the level of mtDNA heteroplasmy.

Advances in robotics and nanotechnologies led to the development of modern DNA microarrays [16]. Microarrays have propelled mtDNA clinical research and population genetics into large-scale genome studies, allowing a high-throughput interrogation of panels with single-nucleotide variants in multiple individuals (Section 11.2.2.4). Likewise, microarrays are designed to detect known sequence variants only, limiting their application to hypothesis-driven studies.

The application of next-generation sequencing (NGS) technologies (Section 11.2.2.5) has greatly improved the detection, characterization, and quantification of not only mitochondrial point mutations [17] but also mtDNA rearrangements, including large and small deletions and duplications [18,19]. Furthermore, the analysis of an ever-increasing number of sequences generated by NGS has confirmed the initial observation from the 1990s that mtDNA heteroplasmy is "universal," as it is present in several tissues from healthy individuals [20–22] and it is not exclusively related to mitochondrial diseases. In addition, NGS technologies have also been used in several other mtDNA-related applications, such as mtDNA carryover quantification in mtDNA replacement technologies [23,24], mitochondrial genomic variation profiling in cancer [25,26], neurodegenerative diseases characterization [27–29], large-scale population studies [30,31], induced pluripotent stem cell generation and differentiation [32,33], and single-cell lineage tracing [34].

A further revolution in genomics, with a predictable impact also on mtDNA analysis, is currently caused by the so-called third (3rd) generation sequencing techniques. These new technologies are able to sequence long tracts, encompassing entire small genomes and long DNA fragments (>10 kb) [35], making them particularly suitable for the detection and quantification of large genome rearrangements, such as large mtDNA deletions and duplications (Section 11.2.2.6). These advances in sequencing technologies have been calling for more sophisticated laboratory techniques to extract mtDNA with lower nuclear DNA (nDNA) carryover (Section 11.3.1) and for the development of ad hoc bioinformatic methods in order to ensure an accurate detection of mtDNA variants and quantification of heteroplasmy (Section 11.4). Both aspects of great importance will be discussed later in this chapter.

11.2 Techniques for detecting mitochondrial variants

The complete human mtDNA sequence was published in 1981 [36,37] and only two years later, the first mtDNA heteroplasmy was reported [38]. Shortly afterward (Section 11.3.1), the first mtDNA variants associated with a pathological phenotype were identified. Within a couple of years, the presence of deletions in muscle mtDNA from patients with mitochondrial myopathies [39] and the variants mt.11778A > G and m.8344A > G causing Leber's Hereditary Optic Neuropathy (LHON) [40] and Myoclonic Epilepsy with Ragged Red Fibers [5,41], respectively, were described. Since then, an ever-increasing number of mtDNA variants have been identified as a result, in part, of the application of fast-evolving new methodologies for mtDNA analysis. Here, we present a small compendium of the most widely used techniques to detect mitochondrial variants in wet laboratories around the world.

11.2.1 Polymerase chain reaction–based methods and mtDNA rearrangements detection

The polymerase chain reaction (PCR) is a method that allows the synthesis of specific regions of DNA and their accumulation in billions of copies and that has revolutionized science since it was developed by Kary Mullis in the 1980s [42,43]. Approaches to detect single and large mtDNA variants employing PCR include restriction fragment length polymorphism PCR [39], long-range PCR (LR-PCR) [44–46], real-time quantitative PCR (qPCR), amplification refractory mutation system qPCR (ARMS-qPCR) [47], single-molecule PCR (smPCR) [48], pyrosequencing [49,50], and digital PCR (dPCR) [51,52]. RFLP, LR-PCR, and qPCR are the most frequently used approaches due to their low cost, plasticity of implementation, and small equipment requirements. Yet, others like pyrosequencing, smPCR, and dPCR have higher sensitivity, as briefly discussed below.

11.2.1.1 Polymerase chain reaction restriction fragment length polymorphism

RFLP technology makes use of the fact that some genetic variants cause the loss or appearance of a restriction site that, when digested with the appropriate restriction endonuclease, generates allele-specific DNA fragments (Fig. 11.1A). These digested DNA products are then loaded onto agarose or acrylamide gels and separated by length using electrophoresis. Visualization of DNA requires staining with either fluorescent (e.g., ethidium bromide, SYBR Green, or Eva Green) or eye-visible dyes such as methylene blue or silver staining. Next, the electrophoretic gel image is captured and the relative proportions of mutant and wild-type mtDNA are quantified by densitometric analysis [5,39]. RFLP has been traditionally the most frequently employed method in the detection of mtDNA point mutations. However, it has limited sensitivity to 5%–10% in heteroplasmy quantification [53], it relies on the creation or loss of a new restriction enzyme site, it often has problems associated with incomplete restriction enzyme digestion, and it does not exclude the formation of heteroduplex. Duplex DNA commonly originates by annealing of single-stranded PCR products from different templates during the PCR amplification, which limits the use of RFLP for the quantification of mitochondrial heteroplasmy [5]. In order to overcome the formation

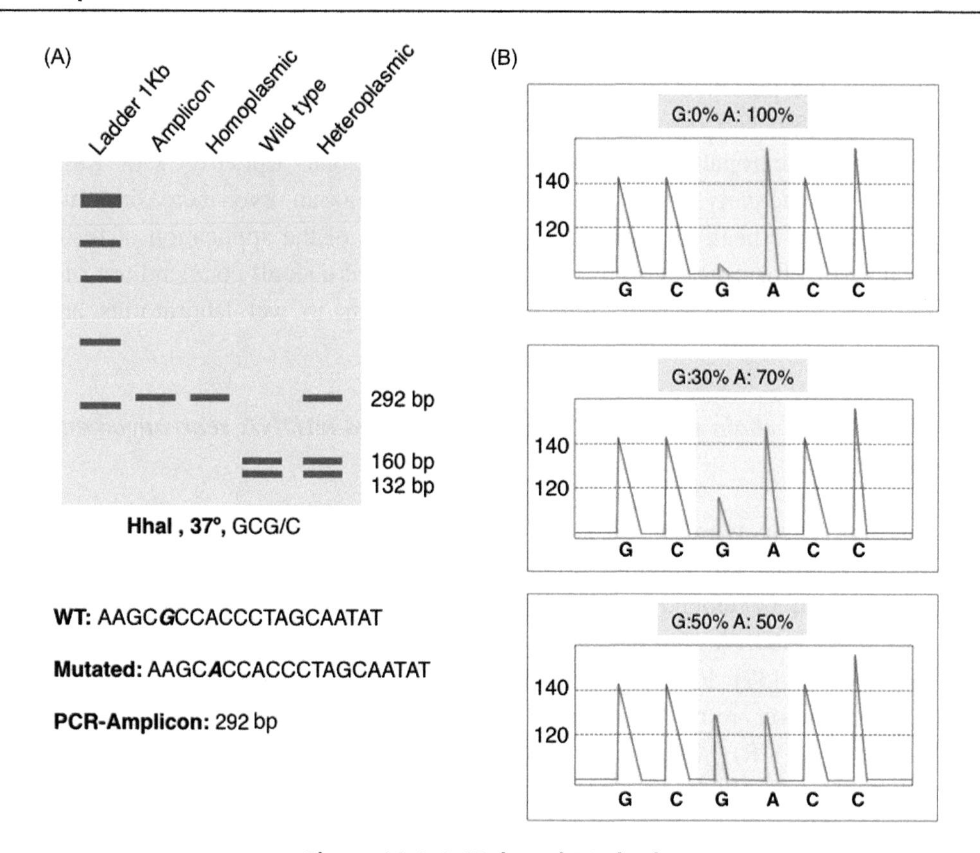

Figure 11.1: PCR-based Methods.
(A) PCR-RFLP. Example of detection of the m.9055G > A/MT-ATP6 variant. Samples were amplified with forward primer (GCCCTAGCCCACTTCTTAC) and reverse primer (AGAGGCTTACTAGAAGTGTG) following the standard conditions (95 degrees (45 seconds), 95 degrees (45 seconds), 63 degrees (45 seconds), 72 degrees (1 minute) for 35 circles, and 72 degrees (5 minutes of elongation time)) generating an amplicon of 292 bp. The alternative allele m.9055A generates the loss of a restriction site for the enzyme HhaI (37 degrees, GCG/C). If a heteroplasmic variant is present, only one part of the amplicon is digested. (B) Pyrosequencing. Example of detection of homoplasmy and heteroplasmy for the m.9055G > A variant.

of heteroduplex and to increase the sensitivity of detection, alternatives such as the Last Hot-Cycle PCR [54] and Fluorescent PCR [55] were designed. In *last Hot-Cycle PCR*, [α-^{32}P]-dCTP is added to the last PCR cycle to avoid the underestimation of heteroplasmy [56]. Next, the PCR fragments are digested and separated by length using electrophoresis, gels are dried and exposed to X-ray films in a cassette at -80°C for 24 h before being developed. Similarly to the common RFLP [5,39], the relative proportions of mutant and wild-type mtDNA are quantified by densitometric analysis. In the case of Fluorescent PCR,

3′ fluorescently labeled oligos are used in the last cycle of PCR and products are directly quantified after digestion and gel electrophoresis as above.

11.2.1.2 Southern blotting and long-range polymerase chain reaction

Restriction enzyme and *PCR*-based methods such as Southern blotting and LR-PCR are commonly employed to detect mtDNA rearrangements rather than single-nucleotide variants. In Southern blot, genomic DNA from nondividing cells such as muscle biopsies [39] is linearized with restriction enzymes that cut only once in the human mtDNA (e.g., *Pvu* II and *BamH* I). The resulting digested DNA fragments are separated by electrophoresis on a 0.6%–0.8% agarose gel and transferred onto a membrane. Hybridization is then performed to identify specific DNA products using radiolabeled, fluorescent, or chromogenic-dye probes, derived from random primer radiolabeling of purified whole mtDNA or from PCR fragments amplified mainly from the D-loop-containing region [39,56]. If rearrangements are present, the pattern of migration in the mutant samples will show differences compared to the wild type in the form of either smaller (deletions) or bigger (duplications) fragments. Southern blot can be used semiquantitatively, but when applied for accurate quantification of heteroplasmic deletions, it is important to ensure that the probe hybridizes solely to a nondeleted or duplicated region of the region of the genome, in order to allow the detection of the ratio between the two genomes (e.g., wild type and deleted). Southern blot requires large amounts of DNA (1–5 μg), and has low resolution for small deletions like those observed in neurodegeneration and aging. Recently, a high-resolution approach has been described, in which a 5% polyacrylamide instead of 0.6% agarose gel is used in order to ensure better separation of variants [57].

LR-PCR was first described in the late 1980s and exploits the fact that the hypervariable region is highly preserved in partially deleted and duplicated mtDNA molecules. Thus primers that bind to that region can be used to amplify almost the whole mtDNA sequence [44,45]. As PCR preferentially amplifies shorter over longer genomes, amplification of deleted genomes will be favored over full-length wild-type mtDNA, making their identification much easier. Detection of genomes carrying duplications by contrast is also possible by LR-PCR but less effective than deleted genomes. For visualization, PCR products are separated according to their size by electrophoresis and stained as previously described. If deletions or duplications are present, smaller or bigger fragments, respectively, than those present in the wild-type DNA will be observed in the gel. LR-PCR is frequently used to amplify smaller deleted genomes; however, it requires good-quality mtDNA to avoid PCR-mediated false-positive deletions that can be created during amplification. In addition, it is limited by the number of primers and the variability in the location of the deletions (for an extensive review, see Ref. [56]). Recently, an advanced version of single-molecule PCR (smPCR) has been developed to allow LR-PCR analysis (Section 11.2.2.5).

11.2.1.3 Pyrosequencing

Pyrosequencing enables real-time DNA sequencing based on the detection of pyrophosphate released during DNA elongation. A specific pair of PCR primers, one of them biotin labeled, is used to generate a locus-specific amplicon of approximately 200 base pairs (bp) while a sequencing primer is used to sequence the region of interest and quantify heteroplasmy levels. The procedure requires denaturation of the double-stranded PCR products and isolation of biotin-linked single strands with streptavidin-coated beads to be used as template for pyrosequencing. After annealing of the sequencing primer, nucleotides are added one at a time into the reaction mix. If the complementary nucleotide is incorporated into the nascent elongating strand, pyrophosphate is released and converted to ATP by ATP sulfurylase. This ATP along with oxygen is used by the luciferase to convert luciferin into oxyluciferin in a reaction that generates light and releases pyrophosphate Light production is proportional to the amount of incorporated nucleotide, and it is represented as a pyrogram peak that provides information that can be used to quantify heteroplasmy levels (Fig. 11.1B). The main drawbacks of this approach are sensitivity of detection is only about 5% heteroplasmy [58], it is PCR based, and it needs very specific equipment, which makes reactions quite expensive. Another limitation of the pyrosequencing is its low fidelity in homopolymeric stretches where high undercalling has been described [59]. For this reason and its high costs, this technique is not commonly used in diagnostics labs with some described exceptions [23,58,60]. In studies on segregation of heteroplasmic variants in mice, however, pyrosequencing is frequently employed [61–63].

11.2.1.4 Quantitative polymerase chain reaction

qPCR, also known as real-time polymerase chain reaction as it monitors the amplification of the DNA in real time, is a quantitative PCR-based technology to measure DNA/allele amount. Quantification is based on the use of unspecific fluorescent dyes (e.g., SYBR Green) and/or sequence-specific DNA probes (e.g., Taqman). Design and guidelines for qPCR experiments are extensively detailed elsewhere [64]. For accurate mtDNA variant quantification by qPCR, amplification must be within the linear exponential increase phase in each reaction and a parallel amplification of a standard curve containing known copy numbers of mutant and wild-type DNA is required. This artificial DNA standard is normally generated by cloning each allele [5,56] and mixing them together in different known quantities. The level of heteroplasmy is then calculated, using the cycle threshold (Ct) value, by comparing this value to the one obtained from the artificial standard curve [47]. qPCR has been successfully applied to detect mtDNA heteroplasmy. Kurelac and colleagues employed established the limit of detection of heteroplasmy with this method at about 75% [59]. Other groups have used a similar approach for the detection of nucleotide polymorphisms [65] and to quantify haplotype competition in dividing cells [66].

ARMS-qPCR is a simple method for detecting variants such as single base changes or small deletions. ARMS-qPCR employs a primer with one or two mismatched nucleotides immediately 5′ to the variant of interest. This modification increases the binding specificity to the target variant but not to the alternative allele during PCR amplification [67]. Thus two different forward primers (one for wild type and one for alternative variant) are used in combination with a common reverse primer, and quantification is performed by qPCR, as previously described. The heteroplasmic detection limit for ARMS-qPCR has been set at 0.5% heteroplasmy for one variant [68], but this may not be applicable to every variant, as the design of the primers and the amplification is highly dependent on the sequence context. Likewise, the use of two independent reactions and the low specificity of the SYBR Green dye limit its applications.

Quantification of mitochondrial rearrangements can also be performed by *qPCR*. Indeed, qPCR is the most common application for the quantification of mtDNA copy number [18,69]. In the case of deletion quantification, qPCR is normally carried out by amplification of the commonly deleted mitochondrial gene, in 85% of the cases in the major arc of the mtDNA [70] (e.g., *MT-ND4*), a nondeleted mitochondrial gene (e.g., *MT-ND1*) and/or a nuclear gene (e.g., *B2M*). Linear exponential amplification is verified in each gene reaction and quantification is based on parallel detection of a standard curve using serial diluted (1:10 to 1:10,000) genomic DNA, purified PCR fragments, or cloned DNA for each gene. The level of the deletion is then expressed as absolute number of deleted copies (inferred from the standard) or as a ratio deleted/nondeleted [69]. qPCR has many advantages over other technologies, such as Southern blot as it requires smaller amount of DNA (about 10 ng) and it can detect both single and multiple mtDNA deletions, through the combination of probes for different genes. In addition, it has higher sensitivity for low abundance mtDNA deletions [51]. Recently, an advanced version of single-molecule qPCR, dPCR (Section 11.2.2.5), has been developed to allow the analysis of mtDNA deletions.

These techniques, although providing a quantification of the amount of deleted molecules, do not provide information on the exact breakpoint. Deletion breakpoint determination is still quite time-consuming, as it normally implies the use of different sets of primers for PCR amplification, followed by Sanger sequencing [71,72]. High-throughput mtDNA sequencing of long reads (Section 10.1.2.6) currently represents a valid alternative for detecting large mtDNA rearrangements, especially when coupled with ad hoc computational approaches, as described later in this chapter (Section 10.3).

11.2.1.5 Single molecule–based detection techniques

dPCR is a more advanced qPCR that allows the absolute quantification of mtDNA copy number and deletions without the need of an arbitrary standard curve. dPCR is based on the use of *smPCR*. In *smPCR*, the DNA templates are diluted to the point that each amplification reaction receives either 0 or 1 DNA molecules of mtDNA, reducing the number of errors [48,73]. The amplified single molecule is then quantified by fluorescence detection, similar to the qPCR [51,52] (Section 11.2.2.4).

11.2.2 Broad-spectrum techniques for detection of variants in whole mtDNA genomes

The techniques described so far provide very useful information. However, many of them are limited to a small number of samples, prone to errors, constrained to a single "known" variant; they require artificial standard curves for establishing the levels of heteroplasmy and they do not provide a precise estimate of the error. In order to overcome these problems and increase the number of variants that can be detected, several technologies have been developed. Some examples include multiplex PCR/allele-specific oligonucleotide [74], multiplex competitive primer extension [75] and its modified version with multiplex PCR combined with mass spectroscopy detection [76,77], single-strand conformation polymorphisms (SSCPs) [78], denaturing gradient gel electrophoresis [79], denaturing high-performance liquid chromatography (DHPLC) [80], and Surveyor Nuclease assay [81]. Among them, SSCP and DHPLC are widely used in population and forensic studies and will be briefly described.

11.2.2.1 Single-strand conformation polymorphism

SSCP allows simultaneous detection of several genomic variants in a large number of samples depending on electrophoretic mobility differences. SSCP is frequently used to detect base substitutions, small deletions, or insertions in the DNA. Radioactive labeled PCR fragments of approximately 200–350 bp are denatured and single strands are separated by non-denaturing polyacrylamide gel electrophoresis. Detection is performed by changes in the DNA mobility [82]. A modified SSCP for mtDNA polymorphism assessment has also been described using a semiautomatic electrophoretic system followed by silver staining [78]. This method is, however, not very informative as it requires complementary confirmation by DNA sequencing for variant identification and has low sensitivity.

11.2.2.2 Denaturing high-performance liquid chromatography

DHPLC is an advanced version of the denaturing gradient gel electrophoresis [53] employing chromatography for the detection of base substitutions, small deletions, or insertions in the DNA. The key features of DHPLC are the differential affinity of the solid phase for double- and single-stranded DNA and the automated processing of the data [53]. While DHPLC is widely used in population genetics [83], it has been less frequently used in other fields. In the case of mtDNA analysis, the entire or partial mtDNA is amplified and digested with restriction enzymes. These restriction fragments are denatured and renatured to allow heteroduplex (mismatched duplex DNA) formation and then digested again into smaller fragments, ranging from 90 to 560 bp, depending on the aim of the experiment. The resulting digestion products are then resolved by liquid chromatography through a hydrophobic column using a set of optimized temperatures to separate heteroduplex and homoduplex DNAs. As heteroduplexes are thermally less stable than their corresponding homoduplexes and the column has lower affinity for single-stranded DNA, their peak will be resolved within a lower retention time compared to the homoduplexes. Each eluted DNA fraction is detected by ultraviolet light and generates a chromatogram with different peaks depending on the presence of the variants and the number of fragments. Next, alternative variants and the percentage heteroplasmy are determined by quantitative analysis of the resulting peaks [80,84]. The limit of detection of heteroplasmy for DHPLC has been set at 2% [59].

11.2.2.3 Sanger sequencing

The gold standard technique for whole mtDNA sequencing is *Sanger sequencing*, which was developed by Sanger et al. in 1967 and is based on the idea of in vitro synthesis of many copies of single-stranded DNA from a specific template (nowadays normally achieved by PCR) [85]. The DNA synthesis is carried out in vitro by a DNA polymerase added to the reaction along with all the necessary cofactors, including deoxynucleotides (dNTPs) (Fig. 11.2A). The DNA polymerase incorporates the dNTPs to the free 3′-hydroxyl (3′-OH) of a radiolabeled primer complementary to the single-stranded DNA template. A small quantity (100-fold less than dNTPs) of 2′,3′-dideoxynucleotides (ddNTPs), chemically modified dNTPs lacking the 3′-OH (Fig. 11.2B), are also added to the reaction to terminate the chain elongation by preventing the formation of the phosphodiester bonds with the incoming dNTP (Fig. 11.2Cand D). The newly synthesized DNA fragments are heat denatured and separated by size using gel electrophoresis. Gels are dried and radioactive fragments are detected by autoradiography [86] (Fig. 11.2E). DNA sequence is inferred directly from the fragments generated in each of the four reactions (one per nucleotide/ddNTP). This technique was used to sequence the whole human mtDNA sequence [36]. Shortly afterward, the laboratory of Leroy Hood modified this method by using fluorescently labeled ddNTPs, each with a different emission wavelength, so they can be mixed in a single reaction and the sequence information acquired directly by a computer [87]. Since then, Sanger sequencing has extremely evolved. Yet, the technique shows limitations in the termination, dye affinity, which can result in unequal peak heights in the chromatogram (read-outs of the sequencing) (Fig. 11.2F) that could subsequently affect the heteroplasmy calling. This aspect has been improved with the use of brighter dyes, such as Big Dye chemistry, that allow to obtain good results from low DNA quantity and better-defined electrophoretic peaks which, in turn, eases the data analysis and the heteroplasmy detection [88]. Nowadays, DNA Sanger sequencing is performed by capillary electrophoresis [89], which enables a further level of high throughput and can produce sequences beyond 1000 bases in 1–3 hours for up to 96 samples. The majority of whole mtDNA and control region sequences present in GenBank and other nucleotide databases have been generated using this technology. Those sequences have been uploaded in databases using *FASTA* files [90], which lack the information about the sequencing quality and mtDNA heteroplasmy.

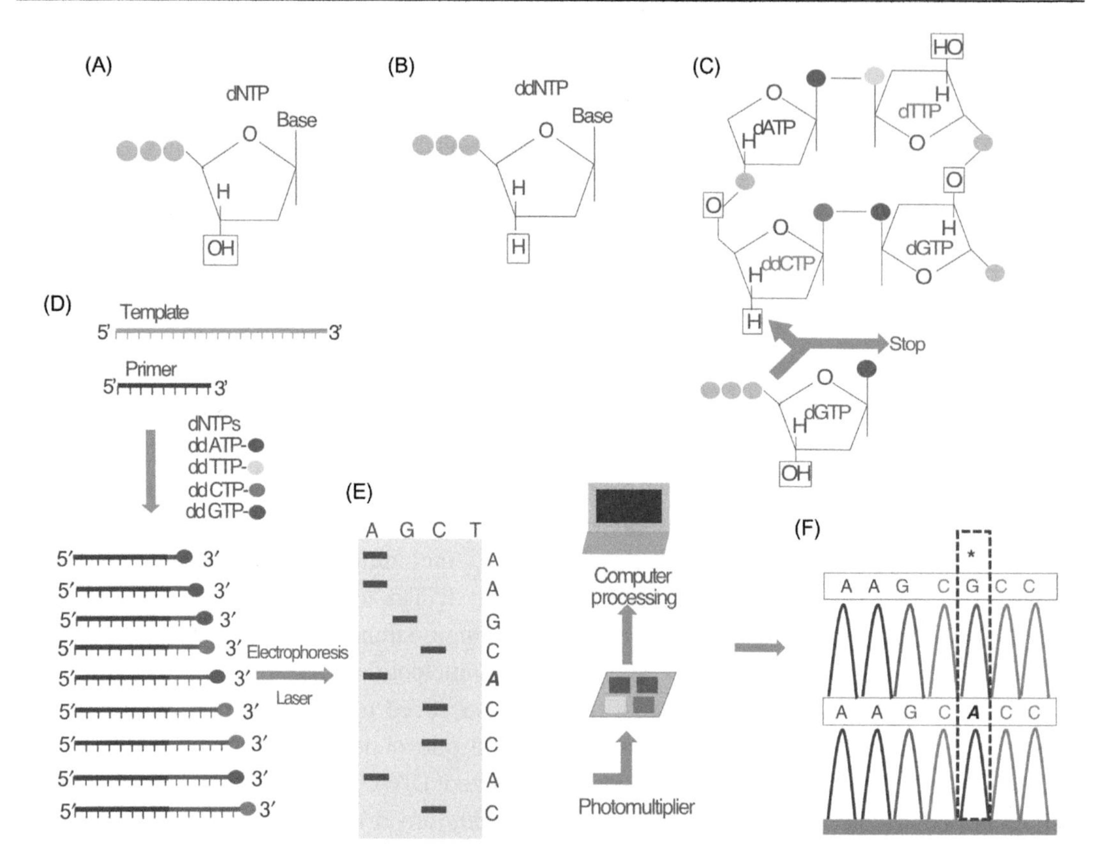

Figure 11.2: Sanger sequencing.

(A) Deoxynucleotides (dNTPs). (B) Chemically modified dNTPs lacking the 3′-OH di-deoxy-nucleotides (ddNTPs). (C) and (D) Incorporation of nucleotides during DNA synthesis. Incorporation of a ddNTP prevents the formation of the phosphodiester bonds with the incoming dNTP, due to the lack of 3′-OH. (E) Electrophoresis detection of the homoplasmic mtDNA variant m.9055G > A/MT-ND6 defining the haplogroup U. (F) Electropherogram detection of the variant m.9055G > A. Upper sequence track and electropherogram represents the rCRS mitochondrial reference genome (m.9055G), lower electropherogram corresponds to the DNA of interest carrying the variant m.9055A.

However, analysis of the raw data (electrophoretic chromatograms) sets the limit of detection of heteroplasmy using Sanger sequencing at about 10%–20% [91].

11.2.2.4 Microarrays

Microarrays allow a high-throughput interrogation of DNA by hybridizing target DNA to a set of nucleotide probes of known sequence immobilized onto either a solid support (like glass or silicon, Affymetrix) or microscopic beads (Illumina). Base calling is possible by quantification of probe intensities that reflect the abundance of the hybridized alleles using either fluorescence, chemiluminescence, or silver stain. There are few arrays for human mtDNA analysis and they are mainly fluorescence-based. In 2004, the MitoChip array (Affymetrix), that captures both strands of almost the whole human mtDNA, was developed [92]. An updated version, the Affymetrix GeneChip Mitochondrial Resequencing Array [93], allows interrogation of the whole human mitochondrial genome by hybridization with oligonucleotide probes complementary to the rCRS (revised Cambridge Reference Sequence) (Fig. 11.3A). Illumina arrays use BeadChip technology instead, and thus oligonucleotides are attached to silica beads allocated within microwells on an array support, hybridizing target DNA fragments as they pass over the array.

Commercial software can be used for mitochondrial base calling analysis and quality filtering, as well as custom bioinformatics pipelines [94,95]. Although some studies have attempted to develop algorithms to quantify heteroplasmy, using relative intensities from allele-specific probes [94,96], microarray-based techniques remain unsuitable for accurate heteroplasmy quantification, being unable to distinguish between the noise of fluorescent signal and true low-level heteroplasmy (<40%) [96] (Fig. 11.3B). Likewise, this technology cannot be used to call for small insertions, deletions, homopolymeric regions, or closely spaced polymorphisms [97].

Microarrays can detect a variable number of known common or rare homoplasmic mtDNA single variants (approximately between 140 and 300) in multiple individuals, depending on the type of application required (e.g., clinical or population-scale studies). As a consequence, several studies have implemented this technology in a wide range of applications, from mtDNA population classification [98–100] for screening of nonsyndromic hearing loss [101], schizophrenia [102], HIV biomarkers [103], or to study the prevalence of mtDNA variants in mitochondrial disorders and common diseases [104].

11.2.2.5 Second-generation sequencing

High-throughput detection techniques have substantially improved since 2005, with the advent of NGS (also known as second-generation sequencing) and the commercialization of Roche/454, Solexa, and Illumina sequencing, followed later on by other technologies, like the Ion Torrent sequencing. NGS provides the opportunity to achieve an extraordinary yield in sequencing, exploiting massive parallelization of reactions, at a relatively small cost [35,105,106]. While chemistry can be quite different, all NGS technologies rely on similar workflows, that is, they generate spatially clustered amplicons with clonal amplifications from each single-target DNA molecule. These clusters of molecules are recovered on a solid support and then subjected to sequencing-by-synthesis approach [107], as shown in Fig. 11.3C for the Illumina technology. The methods used to read the sequence of each cluster are different depending on the chemistry adopted by each provider, that is, fluorescent light from labeled dNTPs (Illumina and 454 technologies) or changes in pH due to

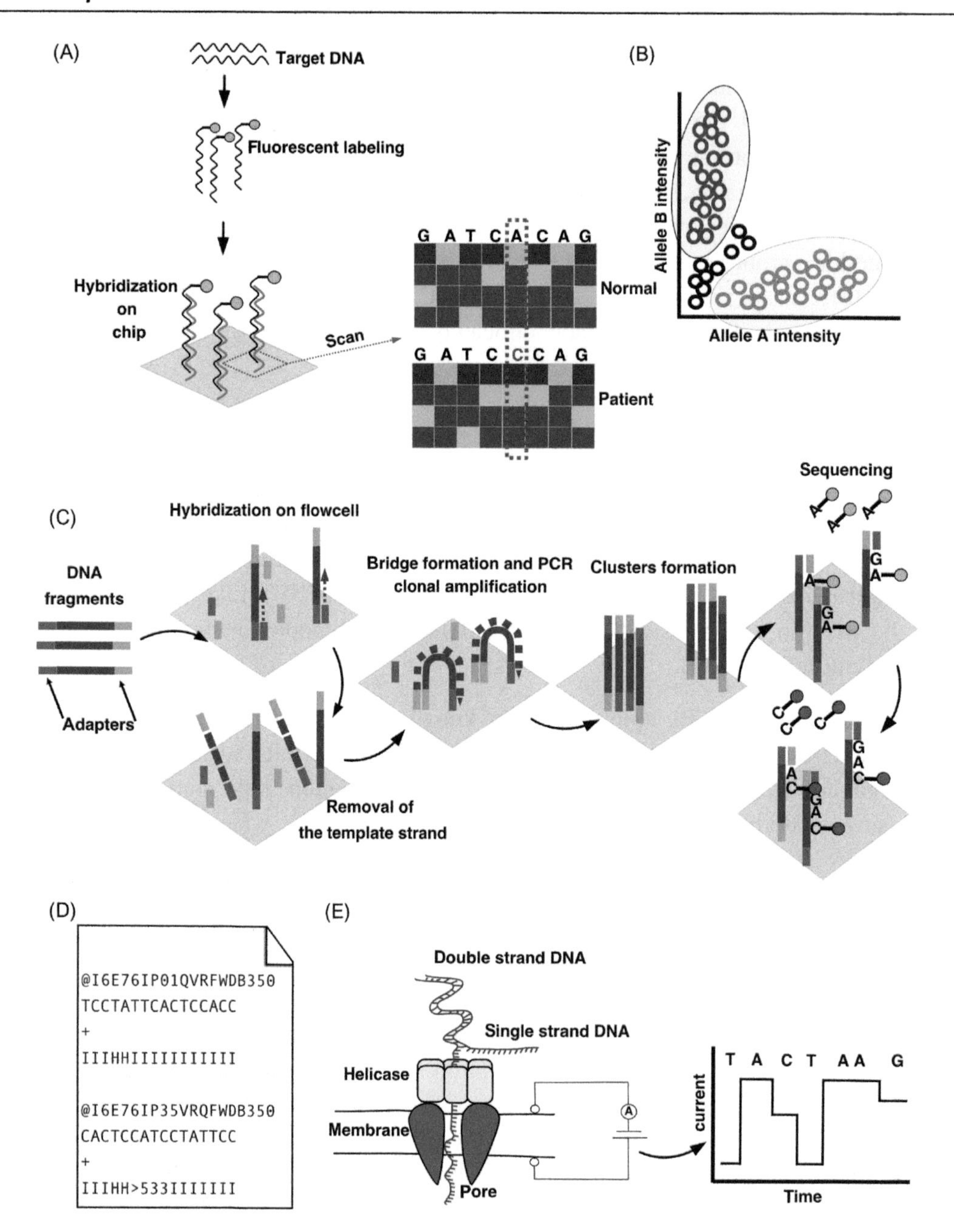

Figure 11.3: Microarray, Illumina, and Nanopore technologies

(A) and (B) Microarray detection of mtDNA variants. (A) Two magnified sections of the chip are depicted: one querying alleles from a healthy control (normal; *top rectangle*) and one from a patient (*bottom rectangle*). In the MitoChip and GeneChip arrays, each single mtDNA site is queried by four oligonucleotides tiled on the chip (corresponding to the four rows of each rectangle), which differ

(Continued)

◀ by one position, where they carry all the four possible nucleotides. The *lighter squares* indicate the fluorescent light emitted, when target mtDNA hybridizes to the complementary probe. The *darker squares* indicate no hybridization. A different mitochondrial allele is detected in the fifth position of the patient's mtDNA. (B) Example of a cluster plot of mitochondrial allele intensities. Axes show the fluorescent intensity for allele A and B (within clouds), of a mitochondrial biallelic position. Each *dot* represents a homoplasmic allele detected in every individual screened on the array. Dots outside the two clouds indicate fluorescent signal that cannot be attributed to any of the two alleles because of heteroplasmy or noisy signal. (C) Illumina sequencing. Fragments of target DNA are ligated to Illumina adapters and hybridized on a flow cell. A synthesis of complementary fragments using the hybridized DNA as template is performed and afterward the hybridized DNA is washed away (*dashed bars*). The remaining single strands are clonally amplified by double-strand bridge formation. Clusters of clonal copies of the initial template are created and used as templates for the sequencing-by-synthesis chemistry, based on fluorescently tagged nucleotides. (D) Example of a *FASTQ* file where two reads are shown. Each entry consists of four lines: the read id, the nucleotide sequence, a separator (+), and a string of per-base quality scores represented by ASCII characters. (E) Nanopore sequencing. A helicase protein attached to the DNA fragment binds to the nanopore embedded in a synthetic membrane, causing denaturation of the double-stranded DNA. The single-stranded DNA passes through the pore, causing a disruption of an ion flow, recorded as an electric current over time. The latter is interpreted by the software as a nucleotide sequence.

release of protons (Ion Torrent technologies), every time a new nucleotide is incorporated [106]. The outcome of NGS are digital short sequences (also known as "reads") with variable lengths (between 35 and 700 bp), depending on the technology [106] in *FASTQ* format. *FASTQ* is a text-based format for storing both nucleotide sequence and corresponding quality score expressed with a Phred-like formula [108] (Fig. 11.3D). The Phred score is logarithmically related to the base calling error probability P and is inversely related to the error rate, that is, higher scores correspond to more accurate base calls. This is one of the greatest achievements of this technology, as a measure of the quality of the sequencing data is provided for the first time. In addition, NGS guarantees high scalability, enabling multiplexing of large sample numbers that can be simultaneously sequenced in a single experiment. To accomplish this, individual "barcode" sequences are added to each sample so that they can be differentiated during the data analysis. The most commonly used NGS applications for DNA sequencing are Whole Exome Sequencing (WES) and Whole Genome Sequencing (WGS).

WES involves the capture of coding regions by hybridizing genomic DNA to biotinylated oligonucleotide baits that cover the whole human exome, followed by enrichment of targeted regions by pull-down with magnetic streptavidin beads and then sequencing. There are several commercial exome enrichment platforms that differ in their target choice, bait lengths, density, and type of molecules used for capture [109–111]. Due to the cross-hybridization of nuclear baits with Nuclear mitochondrial Sequences (NumtSs) or random nonspecific entrapment of NumtS, WES has proven to retrieve mtDNA as an off-target sequence [25,112–114]. Indeed, even if not specific for mtDNA, the read depth and coverage provided by off-target WES results to be enough ($\sim 100\times$) to perform mtDNA single-variant calling and heteroplasmy quantification [30,115], depending on the array platform. Studies have shown that low-density platforms ensure the highest enrichment of mtDNA over nDNA [113,114]. Moreover, WES read depth also depends on the relative number of mtDNA molecules present in a particular sample, with tissues rich in mitochondria (like muscle, heart, and liver) ensuring higher mtDNA fold enrichment [114].

While WES is targeted to specific regions of the nDNA, *WGS* delivers a comprehensive view of the entire nuclear genome. It does not require a prior knowledge of the target sequence and the starting material is whole genomic DNA subjected to random fragmentation or amplified with random primers [116]. WGS studies have demonstrated that it is possible to retrieve sufficient mtDNA to perform variant identification and heteroplasmy calling [30]. However, a more cost-effective strategy to obtain high read depth ($> 1000\times$) in mtDNA sequencing is to prepare PCR amplicon libraries targeted directly to mtDNA. Targeted libraries coupled with NGS technologies (e.g., Illumina) are indeed able to accurately estimate the level of heteroplasmy (down to 1%) [117,118], with a better error estimate, depending on the sequencing read depth [21].

NGS techniques present an intrinsic error rate, due to the chemistry adopted or to the sequence context of the target DNA, which might limit their accuracy in correctly quantifying mtDNA heteroplasmy. For example, the sequencing-by-synthesis technology used by Illumina Miseq is more prone to substitution errors for A and C bases, as both are labeled by fluorophores that have the highest intensities and are identified through the same channel [119]. In ion semiconductor technologies (used in Torrent platform) instead, the signal is directly proportional to the number of incorporated bases during sequence elongation. Therefore these methods tend to be highly error-prone in sequencing DNA stretches of repeated nucleotides (homopolymers), resulting in spurious insertions and deletions [59,107,120]. Besides sequencing errors, a significant fraction of miscalls is also expected to arise during the PCR amplification steps, due to failed proofreading activity of DNA polymerases. These can be introduced either before sequencing during library preparation or cluster generation as both of them envisage PCR-based steps of clonal amplification or during the sequencing itself, as demonstrated by their presence in only one of the DNA strands [21]. Currently, the estimated error rate of NGS techniques is between 1 per 100 and 1 per 1000 base pairs sequenced, varying depending on the sequencing technology [121]. To control the sequencing and PCR errors in second-generation sequencing, it is possible to use ad hoc bioinformatics strategies, as discussed in Section 11.4.

11.2.2.6 Third-generation sequencing

More recently, third-generation technologies (like *PacBio* and *Oxford Nanopore Technologies* (*ONT*)) represent a valid alternative to short reads NGS, enabling the detection of single molecules with real-time sequencing [106]. These technologies offer the extraordinary opportunity to provide full-length mtDNA sequencing in a single reaction. Long-read techniques are more likely to increase the specificity for mitochondrial variant by capturing the whole mtDNA sequence instead of NumtS. However, NumtS covering the entire mtDNA have also been mapped (as detailed in Chapter 6: Human nuclear mitochondrial sequences (NumtS)) and hence the risk of contamination still exists with long-read sequencing technologies.

Besides, these methods enable the sequencing of PCR-free libraries, thus promising an increased accuracy in mtDNA variant detection, especially of low-level heteroplasmy, as they are devoid of PCR errors. Nonetheless, PCR-free protocols require a considerable amount of input DNA (1–5 μg) to be able to achieve a sensible read depth. It is worth highlighting that third-generation techniques can also ensure an accurate phasing of variants [116] without the need of genome assembly (necessary with short NGS reads); thus they could be used to study possible linkages between heteroplasmic variants.

The *PacBio* core technology is based on single-molecule real-time sequencing (SMRT), where DNA target molecules are ligated to a hairpin adapter and loaded onto nanoscale observation chambers [122,123]. Each chamber captures a single DNA molecule where it is sequenced by a DNA polymerase elongating a primer annealed to the adapter and incorporating fluorescently labeled nucleotides [122,123]. The *SMRT* technology has demonstrated very high sensitivity (91%) in detecting low-frequency mtDNA variants (down to 0.1% heteroplasmy) in fresh frozen tumor tissues [124]. *ONT* sequencing is based instead on electrical current changes caused by nucleic acids passing through a protein nanopore [125] (Fig. 11.3E). This method yet suffers from high error rate, especially in homopolymeric regions, where it exhibits an increased number of false short insertions and deletions [126]. Therefore at the moment, ONT is less suitable for mitochondrial single-variant detection and heteroplasmy quantification, but rather more useful to identify long mtDNA rearrangements [127].

11.3 Challenges in mitochondrial variant studies

In order to achieve an accurate detection of mtDNA variants and heteroplasmy quantification, it is fundamental to first enrich the mitochondrial genome over nDNA [128]. Residual nDNA contamination, due to inefficient mtDNA extraction and/or to coamplification of NumtS, represents an important challenge to mitochondrial variant studies and a problem that has to be tackled. Here, we briefly discuss different strategies to isolate mtDNA and to reduce the risk of NumtS contamination.

11.3.1 Mitochondrial DNA isolation

Several methods can be used for mtDNA isolation and the choice of the most appropriate one quite often depends on the amount and type of starting material. Crude mitochondrial fractions can be extracted from tissues or cell homogenates through a series of differential centrifugations, where contaminants are washed away [129,130]. These protocols allow to obtain, with considerably high yield, intact and highly functional mitochondria from different mammalian tissues and cells, that can be further used for biogenetical studies [129]. However, due to the harsh nature of high-speed centrifugation, disruptions of the nuclear and mitochondrial membranes can lead to less robust mtDNA enrichment and contamination with nDNA [131]. In improved protocols, crude mitochondria are further purified in sucrose gradients and/or treated with DNase I prior to mitochondrial lysis and mtDNA extraction resulting in higher mtDNA enrichment [132]. These methods are quite laborious, costly, and usually present side effects, like the need of large quantities of starting material [128].

Some methods instead have been designed to purify mtDNA from genomic DNA. Density gradient enrichment using cesium chloride (CsCl) was the first method used to separate mtDNA from its nuclear counterpart [133,134]. CsCl establishes a gradient by centrifugation where the DNA molecules are separated depending on their buoyant densities [133]. Alternatively, removal of nDNA can be achieved by enzymatic methods in which enzymes like Exonuclease V preferentially cut linear nDNA, leaving circular mtDNA intact [135]. This method was successfully used for the isolation of mtDNA from fresh frozen tumor tissues [124]. Another growing trend in mtDNA enrichment is direct mtDNA genome capture by means of DNA extraction kits originally designed for plasmid supercoiled DNA (The Plasmid Miniprep kits slightly modified and rebranded as mtDNA isolation kits) [136]. To achieve a more high-throughput enrichment, it is also possible to use capture methods based on the hybridization of DNA or RNA probes with complementary mitochondrial fragments, which are then captured in solution or on a solid surface [128]. The use of capture methods is very effective in ancient DNA and forensic studies, where the DNA can be degraded and present as short fragments (<100 bp) [137,138]. However, probe capture strategies can show low specificity [137], which could be increased by isolation with PCR amplification.

PCR-based methods use variable numbers of overlapping primer pairs, complementary to the mtDNA genome sequence, to trigger amplification reactions and generate mitochondrial double-stranded amplicons (typically 100–2000 bp long) [114,139]. Yet, nucleotide composition of the target sequence, PCR reaction conditions, and presence of polymorphisms in primer binding sites can lead to an uneven coverage across the mtDNA molecule. An alternative strategy to overcome this problem is LR-PCR (Section 11.2.1.2). Originally, LR-PCR method was implemented to amplify the whole mtDNA with one pair of primers [140], while more recent methods also envisage the use of two long and slightly overlapping amplicons [141]. Overall, LR-PCR methods provide a more evenly distributed coverage, which is particularly useful when combined with high-throughput sequencing, especially with long-read sequencing techniques [142] (Section 11.2.2.6).

Finally, a recent technology developed to overcome errors introduced by PCR-based methods is the rolling circle amplification (also known as MDA, for Multiple Displacement Amplification), which uses the Phi29 polymerase, a highly processive enzyme, to trigger a single priming event to generate a single DNA molecule containing several head-to-tail copies of the mtDNA (circular template). This strategy has proven to highly enrich the mtDNA over the nDNA and increase the accuracy of the mtDNA amplification compared to PCR-based methods [143].

11.3.2 NumtS contamination

Methods to isolate mtDNA, like capture arrays and PCR amplification, can suffer from copurification of NumtS, which can confound heteroplasmy quantification leading to false positives. Given the presence of more than 700 NumtS in the nDNA [144–148], mapping of sequencing data is highly prone to incorrect assignment of nuclear reads to mtDNA. Therefore a critical step in mtDNA analysis is to accurately design probes and primers sets, taking into account the nucleotide similarity of these sequences to NumtS by browsing the UCSC NumtS tracks [147] or by checking the nucleotide similarity of the set to the reference genome, by applying Primer-Blast [149]. Also, primer pairs targeted to mtDNA should be first tested on cell lines lacking mtDNA (rho0) [150] to ensure they do not cross-hybridize with NumtS. Santibanez-Koref and colleagues observed an increased risk of false heteroplasmy in protocols adopting short PCR amplicons and with pooled cells, rather than LR-PCR amplicons and tissue homogenates [142]. The low performance of short amplicons can be due to differences in starting material, inaccuracy of the PCR leading to NumtS coamplification, or to an increased ambiguity in mapping short reads to the reference genome. To overcome this, a further optimization of the bioinformatics analyses is necessary, as discussed in Section 11.4.

11.4 Bioinformatics strategies to detect mitochondrial variants and heteroplasmy

11.4.1 Reads mapping and genome assembly

The first step in mtDNA variant analysis is reads alignment (also known as reads mapping) and genome assembly. Aligning short reads either just on mtDNA or nDNA sequences is a strategy prone to generate misaligned reads and false positives in mitochondrial variant calling [96,142]. Instead, it is recommended to map reads (especially if generated via short amplicons) on both mtDNA and nDNA reference genomes simultaneously, in order to detect and remove putative NumtS [113,142] (Fig. 11.4A). Such strategies are implemented in several bioinformatics pipelines for mtDNA sequencing analysis such as MToolBox [151].

Different strategies should be considered to map large mtDNA deletions and structural rearrangements. The MitoDel pipeline [152] uses the BLAT algorithm [153] to identify "split" reads, that is, reads split into segments and aligned to separate positions on the reference mtDNA (Fig. 11.4B). However, unlike BWA [154] and other aligners specifically designed for NGS data, the use of BLAT comes at a great computational cost, which is, however, compensated by the small size of mtDNA [152]. An alternative pipeline to detect large mtDNA deletions is eKLIPse [155], where BLAT analysis is used to identify breakpoint positions starting from "soft-clipped" reads, that is, reads with trimmed extremities that do not align to the reference (Fig. 11.4B). Large soft clipping can indeed be the result of NGS designed algorithms attempting to align reads encompassing the deletion, at the cost of trimming one of their extremities.

All the above-mentioned strategies are considered reference-guided approaches, where the assignment of reads to the mtDNA is achieved by mapping to a reference sequence. Alternatively, de novo genome assembly can be implemented to reconstruct the mitochondrial sequence without a reference support (Fig. 11.4A). The Novoplasty pipeline [156,157], for example, starts the assembly from multiple seeds, which are iteratively extended bidirectionally until the circular genome is reconstructed. While de novo strategies are useful to assemble genomes that are very different from the reference sequence, where too many mismatches could interfere with a reference-guided mapping, the process itself can be halted by the presence of sequencing errors in nucleotides repeats and homopolymeric regions [156,157].

The NGS data input of both reference-guided and de novo strategies are usually files with raw reads, called *FASTQ* (Fig. 11.3D), including sequences and per-base quality scores, as mentioned above (Section 11.2.2.5). Output files typically include aligned reads stored in BAM (Binary Alignment Map) or SAM (Sequence Alignment Map) files [158]. An overview of the principal tools for mitochondrial genome mapping and assembly is provided in Table 11.1.

11.4.2 Mitochondrial variant calling

Variant calling is usually the step that follows reads mapping and it is aimed at identifying mtDNA variants and quantify their heteroplasmy (Fig. 11.4C). There are several bioinformatic tools to perform variant calling, as summarized in Table 11.1. These pipelines implement different quality procedures for mitochondrial reads alignment and calling of a valid

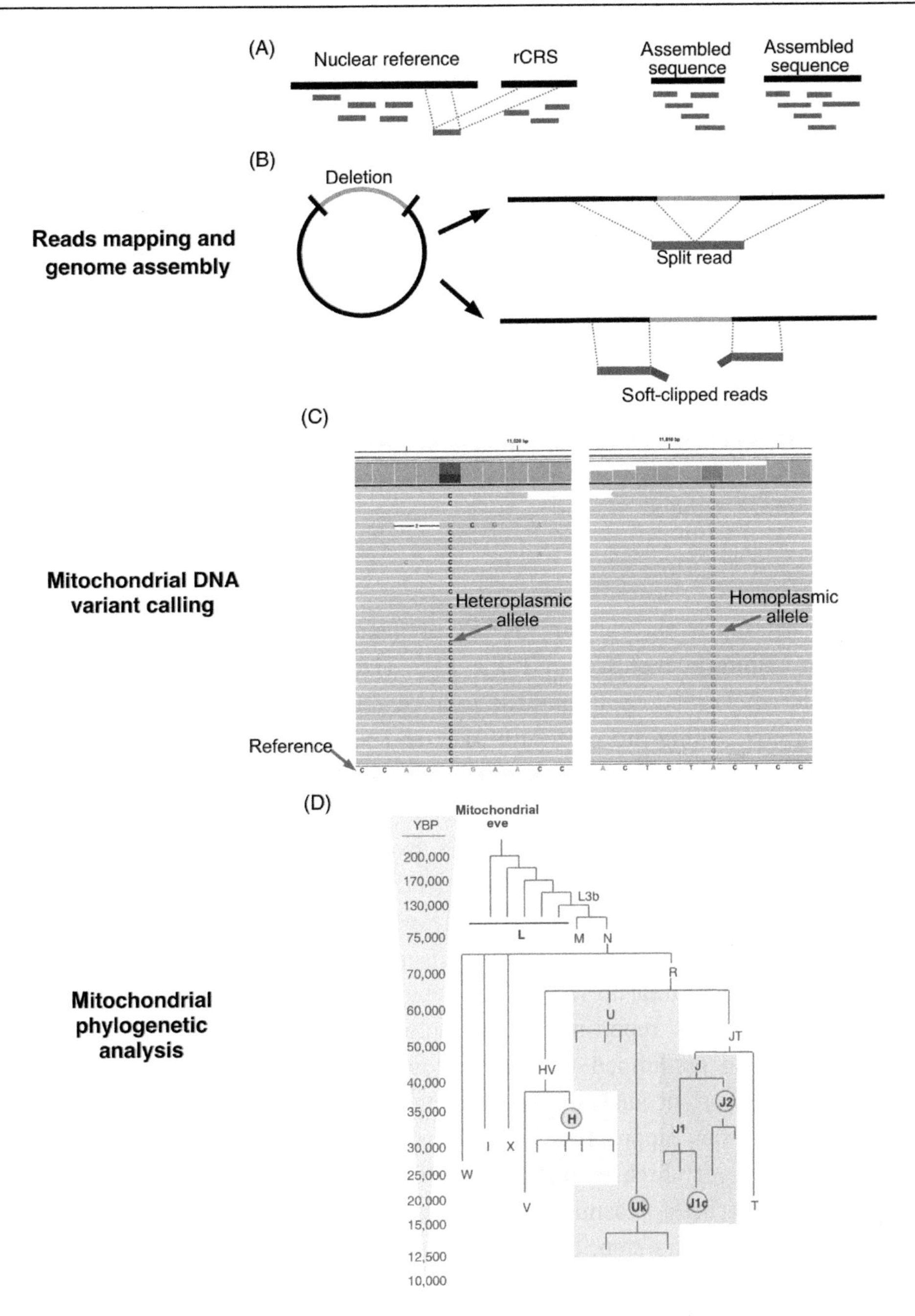

Figure 11.4: Bioinformatic workflow of analysis of mtDNA

(A) Reference-guided assembly, where reads are aligned on both the nuclear and the mitochondrial (rCRS) reference sequence (left figure). This allows the identification of reads with multiple mapping locations that likely belong to NumtS (rectangle connected with dashed lines to the

(*Continued*)

reference sequences) rather than true mitochondrial DNA (*gray rectangles*). Overlapping reads (*gray rectangles*) can be instead used for de novo assembly, without the need of references, to reconstruct the correct original DNA sequence (right figure). (B) Examples of the alignment of split reads and soft-clipped reads on the mitochondrial reference sequence when the read encompasses a large mitochondrial deletion (lighter portion of the circular and linearized genome). Overhanging extremities of soft-clipped reads could not be mapped on the reference sequence. (C) Track visualization of sequencing reads encompassing a heteroplasmic allele (m.11017T>C/*MT-ND4*; left plot) and a homoplasmic allele (m.11812A>G/*MT-ND4*; right plot) present in a human mtDNA sample, using the Integrative Genome Viewer software [166]. The reference sequence used to align the reads (rCRS) is shown below each track. (D) Representation of the human mitochondrial phylogeny including the arising age (YBP) of the different mtDNA haplogroups. The major European haplogroups H, U, and J are highlighted in rectangles. As an example, some of their subhaplogroups are highlighted with *green circles*. Source: *Adapted from Chinnery PF, Gomez-Duran A. Oldies but goldies mtDNA population variants and neurodegenerative diseases. Front Neurosci 2018;12:682. https://doi.org/10.3389/fnins.2018.00682 [167].*

set of mtDNA heteroplasmic variants. A list of general recommendations to perform an accurate variant calling is enclosed in Table 11.2. Heteroplasmy is usually calculated as a ratio between the allele read depth and the total read depth per position (Fig. 11.4C). Cut-offs used for variant calling are quite arbitrary and vary between pipelines. For example, the mtDNA-Server tool [159] removes mtDNA variant calls at sites showing read depth <10 per strand. Then variants are further retained only if the heteroplasmic ratio of the variant is $\geq 1\%$ and each variant is supported by at least three bases per strand. The MToolBox pipeline [151] adopts some default read depth values and per-base quality score above which the variant is retained (5 and 25, respectively). MToolBox also excludes small insertions and deletions occurring close to read ends (<5 bp or more), where sequencing quality can be lower. A similar to MToolBox pipeline, MitoSeek [160], allows to specify a heteroplasmy cut-off, thus reporting in the output only the filtered variants identified. Although there are no predefined rules to follow on the use of a specific heteroplasmy threshold in mtDNA variant analysis, sequencing metrics, like the sequencing read depth, can help choose the most appropriate heteroplasmy cut-off. Usually, a greater sensitivity can be achieved with very high read depth ($>1000\times$) and therefore it would be possible to detect sensibly low heteroplasmic values (e.g., 1%). However, rare undetected NumtS, not excluded in the mapping step, and/or sequencing errors that failed the quality filtering might still be present as false low heteroplasmy, even after stringent quality checks [142,161]. Therefore higher heteroplasmy cut-offs (e.g., $>1\%$ or $>5\%$) would be preferred to reduce false positives, despite inevitably leading to the exclusion of true low-frequency variants [96]. Studies that skip this further quality control step have, in fact, shown to be more susceptible to flawed data [162].

Table 11.1: List of the principal bioinformatic tools for the in silico analysis of mtDNA sequencing data.

			Read mapping and genome assembly		**Mitochondrial variant Calling**			**Phylogenetic analysis**
Tool (on-line availability)	Input	Output	Reference-guided assembly	De novo assembly	Single mtDNA variant[a]	MtDNA rearrangements	Heteroplasmy quantification	Haplogroup predictions
LoFreq [168] https://csb5.github.io/lofreq/	BAM	• VCF with identified variants and variant allele fractions			✓		✓	
MitoSeek [160] https://github.com/riverlee/MitoSeek	BAM	• TEXT file with identified variants and heteroplasmy loads • MPILEUP with identified variants • PNG plots to display results			✓	✓	✓	✓
MToolBox [151] https://github.com/mitoNGS/MToolBox	• FASTA • FASTQ • BAM • SAM	• VCF with identified variants and heteroplasmic loads • CSV tables with identified variants and functional annotations • FASTA with assembled *consensus* contigs • TEXT file with haplogroup predictions	✓		✓		✓	✓
Phy-Mer [169] https://github.com/MEEIBioinformaticsCenter/phy-mer	• FASTA • FASTQ • BAM	• CSV file haplogroup defining variants • FASTA with assembled *consensus* sequence						✓
mtDNA-Server [159] https://mtdna-server.uibk.ac.at/index.html	• FASTQ • SAM • BAM	• Interactive HTML report summarizing all findings: identified variants, heteroplasmic loads, haplogroups predicted	✓		✓		✓	✓

(*Continued*)

Table 11.1: (Continued)

			Read mapping and genome assembly		Mitochondrial variant Calling			Phylogenetic analysis
Novoplasty [156,157] https://github.com/ndierckx/NOVOPlasty	• FASTA • FASTQ	• FASTA or TEXT files with assembled *consensus* contigs • VCF with identified variants and heteroplasmic ratios • FASTA or TEXT files with putative NumtS • TEXT file for circos plots		✓	✓	✓	✓	
MitoDel [152] http://mendel.gene.cwru.edu/laframboiselab/.	FASTQ	• AXT. file including coverage, breakpoint nucleotide annotations, and quality filtering score	✓		✓	✓		
eKLIPse [155] https://github.com/dooguypapua/eKLIPse	• BAM • SAM	• CIRCOS plot • CSV tables with breakpoint nucleotide annotation, deletion sizes, heteroplasmic loads	✓			✓	✓	
Haplogrep2 [170,171] http://haplogrep.uibk.ac.at/about.html	VCF	• TEXT files with haplogroup QC reports • Graphical phylogenetic tree of all input samples						✓

[a]Point mutations, small insertions, and small deletions.

Table 11.2: Steps usually adopted in mitochondrial variant calling analysis of high-throughput sequencing data.

Steps
Removal of duplicate reads (reads mapping to the same position with the same orientation on the reference)
Removal of mtDNA calls with poor Phred-like quality score (e.g., <20–30)
Removal of mtDNA variants showing strand bias, while keeping only variants supported by both strands
Removal of variants close to reads ends
Optimization of read alignments in complex regions and/or eventually exclusion of variants occurring in error-prone motifs (like homopolymeric stretches or repeats)
Removal of heteroplasmic variants occurring in multiple samples (when cohort of multiple individuals are available)
Removal of heteroplasmic variants below a certain heteroplasmy threshold (e.g., <1% or <5%) depending on sequencing metrics (e.g., read depth)

Reporting of variant calling in NGS studies can be done using different outputs. These are text-based formats, usually PILEUP files [158], that contain information of each variant in each read, or VCF (Variant Call Format) files, that report all the positions where at least one minor allele variant has been observed. VCFs also contain other metainformation (e.g., number of genotypes, read depth, and average quality score observed per position). Bioinformatic tools for identifying mtDNA variants and heteroplasmy provide different output file formats. Table 11.1 offers an overview of outputs generated by the different variant calling pipelines.

Research perspectives

While traditional approaches for genomic variants identification set the stage for the discovery of mtDNA variation in human evolution and disease, deep sequencing has steered a "genomic revolution" in mitochondrial research, where modern detection techniques have sufficient sensitivity to detect low heteroplasmic mtDNA changes. This has revealed the existence of a "universal heteroplasmy," present in healthy and pathological conditions, thus changing the prism of mtDNA pathogenic variant criteria and emphasizing the importance of sequencing the entire mitochondrial genome in diagnostic applications. Likewise, long-reads sequencing technologies promise great progress in the detection and quantification of large mtDNA genome rearrangements, as well as in heteroplasmic allele phasing. Nonetheless, the challenges raised by mtDNA variant detection and heteroplasmy quantification here discussed call for standardized workflow of analysis and established criteria for heteroplasmy calling. The increasing amount of sequencing data also requires continuous development of specialized databases and resources to collect and share mitochondrial genomics data, in order to improve the understanding of the mitochondrial genomic variability in pathology and population.

11.4.3 Mitochondrial phylogenetic analysis

Another important step in mitochondrial variant studies is to perform phylogenetic analysis of each analyzed genome, using homoplasmic variants [163], with the aim to predict the individual haplogroups (Fig. 11.4D). Usually, haplogroup analysis can be performed using the Phylotree human phylogeny [164] (Chapter 5: Haplogroups and the history of human evolution through mtDNA). This phylogenetic analysis can help identifying contamination during sample preparation [163,165], as in cases in which more than one haplotype is detected within the same individual or mismatched haplogroups are detected between mother–child pairs. In addition, there are several in silico automated workflows for haplogroup analysis from high-throughput sequencing or Sanger sequencing data as shown in Table 11.1.

References

[1] Lightowlers RN, Chinnery PF, Turnbull DM, Howell N. Mammalian mitochondrial genetics: heredity, heteroplasmy and disease. Trends Genet: TIG 1997;13(11):450–5. Available from: https://doi.org/10.1016/s0168-9525(97)01266-3.

[2] Chinnery PF, Hudson G. Mitochondrial genetics. Br Med Bull 2013;106:135–59. Available from: https://doi.org/10.1093/bmb/ldt017.

[3] DiMauro S, Schon EA, Carelli V, Hirano M. The clinical maze of mitochondrial neurology. Nat Rev Neurol 2013;9(8):429–44. Available from: https://doi.org/10.1038/nrneurol.2013.126.

[4] Stewart JB, Chinnery PF. The dynamics of mitochondrial DNA heteroplasmy: implications for human health and disease. Nat Rev Genet 2015;16(9):530–42. Available from: https://doi.org/10.1038/nrg3966.

[5] Shoffner JM, Lott MT, Lezza AM, Seibel P, Ballinger SW, Wallace DC. Myoclonic epilepsy and ragged-red fiber disease (MERRF) is associated with a mitochondrial DNA tRNA(Lys) mutation. Cell 1990;61(6):931–7. Available from: https://doi.org/10.1016/0092-8674(90)90059-n.

[6] Comas D, Pääbo S, Bertranpetit J. Heteroplasmy in the control region of human mitochondrial DNA. Genome Res 1995;5(1):89–90. Available from: https://doi.org/10.1101/gr.5.1.89.

[7] Jazin EE, Cavelier L, Eriksson I, Oreland L, Gyllensten U. Human brain contains high levels of heteroplasmy in the noncoding regions of mitochondrial DNA. Proc Natl Acad Sci U S A 1996;93(22):12382–7.

[8] Rossignol R, Faustin B, Rocher C, Malgat M, Mazat J-P, Letellier T. Mitochondrial threshold effects. Biochemical J 2003;370(Pt 3):751–62. Available from: https://doi.org/10.1042/BJ20021594.

[9] Wilson IJ, Carling PJ, Alston CL, et al. Mitochondrial DNA sequence characteristics modulate the size of the genetic bottleneck. Hum Mol Genet 2016;25(5):1031–41. Available from: https://doi.org/10.1093/hmg/ddv626.

[10] Nass MMK, Nass S. Intramitochondrial fibers with DNA characteristics: I. Fixation and electron staining reactions. J Cell Biol 1963;19(3):593–611. Available from: https://doi.org/10.1083/jcb.19.3.593.

[11] Finsterer J, Harbo HF, Baets J, et al. EFNS guidelines on the molecular diagnosis of mitochondrial disorders. Eur J Neurol 2009;16(12):1255–64.

[12] Montoya J, López-Gallardo E, Díez-Sánchez C, López-Pérez MJ, Ruiz-Pesini E. 20 years of human mtDNA pathologic point mutations: carefully reading the pathogenicity criteria. Biochim Biophys Acta 2009;1787(5):476–83. Available from: https://doi.org/10.1016/j.bbabio.2008.09.003.

[13] McFarland R, Elson JL, Taylor RW, Howell N, Turnbull DM. Assigning pathogenicity to mitochondrial tRNA mutations: when "definitely maybe" is not good enough. Trends Genet: TIG 2004;20(12):591–6. Available from: https://doi.org/10.1016/j.tig.2004.09.014.

[14] Kogelnik AM, Lott MT, Brown MD, Navathe SB, Wallace DC. MITOMAP: a human mitochondrial genome database. Nucleic Acids Res 1996;24(1):177–9. Available from: https://doi.org/10.1093/nar/24.1.177.

[15] Forster P. To err is human. Ann Hum Genet 2003;67(Pt 1):2–4. Available from: https://doi.org/10.1046/j.1469-1809.2003.00002.x.

[16] Bumgarner R. Overview of DNA microarrays: types, applications, and their future. Curr Protoc Mol Biol 2013;. Available from: https://doi.org/10.1002/0471142727.mb2201s101 Chapter 22:Unit 22.1.

[17] Cui H, Li F, Chen D, et al. Comprehensive next-generation sequence analyses of the entire mitochondrial genome reveal new insights into the molecular diagnosis of mitochondrial DNA disorders. Genet Med J Am Coll Med Genet 2013;15(5):388–94. Available from: https://doi.org/10.1038/gim.2012.144.

[18] Rygiel KA, Tuppen HA, Grady JP, et al. Complex mitochondrial DNA rearrangements in individual cells from patients with sporadic inclusion body myositis. Nucleic Acids Res 2016;44(11):5313–29. Available from: https://doi.org/10.1093/nar/gkw382.

[19] Bosworth CM, Grandhi S, Gould MP, LaFramboise T. Detection and quantification of mitochondrial DNA deletions from next-generation sequence data. BMC Bioinforma 2017;18(Suppl 12):407. Available from: https://doi.org/10.1186/s12859-017-1821-7.

[20] Payne BAI, Wilson IJ, Yu-Wai-Man P, et al. Universal heteroplasmy of human mitochondrial DNA. Hum Mol Genet 2013;22(2):384–90. Available from: https://doi.org/10.1093/hmg/dds435.

[21] Li M, Schönberg A, Schaefer M, Schroeder R, Nasidze I, Stoneking M. Detecting heteroplasmy from high-throughput sequencing of complete human mitochondrial DNA genomes. Am J Hum Genet 2010;87 (2):237–49. Available from: https://doi.org/10.1016/j.ajhg.2010.07.014.

[22] He Y, Wu J, Dressman DC, et al. Heteroplasmic mitochondrial DNA mutations in normal and tumour cells. Nature 2010;464(7288):610–14. Available from: https://doi.org/10.1038/nature08802.

[23] Hyslop LA, Blakeley P, Craven L, et al. Towards clinical application of pronuclear transfer to prevent mitochondrial DNA disease. Nature 2016;534(7607):383–6. Available from: https://doi.org/10.1038/nature18303.

[24] Hudson G, Takeda Y, Herbert M. Reversion after replacement of mitochondrial DNA. Nature 2019;574 (7778):E8–11. Available from: https://doi.org/10.1038/s41586-019-1623-3.

[25] Larman TC, DePalma SR, Hadjipanayis AG, et al. Spectrum of somatic mitochondrial mutations in five cancers. Proc Natl Acad Sci U S A 2012;109(35):14087–91. Available from: https://doi.org/10.1073/pnas.1211502109.

[26] Ju YS, Alexandrov LB, Gerstung M, et al. Origins and functional consequences of somatic mitochondrial DNA mutations in human cancer. eLife 2014;3. Available from: https://doi.org/10.7554/eLife.02935.

[27] Coxhead J, Kurzawa-Akanbi M, Hussain R, Pyle A, Chinnery P, Hudson G. Somatic mtDNA variation is an important component of Parkinson's disease. Neurobiol Aging 2016;38(217):e1–217.e6. Available from: https://doi.org/10.1016/j.neurobiolaging.2015.10.036.

[28] Dölle C, Fløhnes I, Nido GS, et al. Defective mitochondrial DNA homeostasis in the substantia nigra in Parkinson disease. Nat Commun 2016;7:13548. Available from: https://doi.org/10.1038/ncomms13548.

[29] Wei W, Keogh MJ, Wilson I, et al. Mitochondrial DNA point mutations and relative copy number in 1363 disease and control human brains. Acta Neuropathol Commun 2017;5(1):13. Available from: https://doi.org/10.1186/s40478-016-0404-6.

[30] Diroma MA, Calabrese C, Simone D, et al. Extraction and annotation of human mitochondrial genomes from 1000 Genomes Whole Exome Sequencing data. BMC Genomics 2014;15(Suppl 3):S2. Available from: https://doi.org/10.1186/1471-2164-15-S3-S2.

[31] Wei W, Tuna S, Keogh MJ, et al. Germline selection shapes human mitochondrial DNA diversity. Science (New York, NY) 2019;364(6442). Available from: https://doi.org/10.1126/science.aau6520.

[32] Deuse T, Hu X, Agbor-Enoh S, et al. De novo mutations in mitochondrial DNA of iPSCs produce immunogenic neoepitopes in mice and humans. Nat Biotechnol 2019;37(10):1137–44. Available from: https://doi.org/10.1038/s41587-019-0227-7.

[33] Perales-Clemente E, Cook AN, Evans JM, et al. Natural underlying mtDNA heteroplasmy as a potential source of intra-person hiPSC variability. EMBO J 2016;35(18):1979–90. Available from: https://doi.org/10.15252/embj.201694892.

[34] Ludwig LS, Lareau CA, Ulirsch JC, et al. Lineage tracing in humans enabled by mitochondrial mutations and single-cell genomics. Cell 2019;176(6):1325–1339.e22. Available from: https://doi.org/10.1016/j.cell.2019.01.022.

[35] van Dijk EL, Jaszczyszyn Y, Naquin D, Thermes C. The third revolution in sequencing technology. Trends Genet: TIG 2018;34(9):666–81. Available from: https://doi.org/10.1016/j.tig.2018.05.008.

[36] Anderson S, Bankier AT, Barrell BG, et al. Sequence and organization of the human mitochondrial genome. Nature 1981;290(5806):457–65. Available from: https://doi.org/10.1038/290457a0.

[37] Andrews RM, Kubacka I, Chinnery PF, Lightowlers RN, Turnbull DM, Howell N. Reanalysis and revision of the Cambridge reference sequence for human mitochondrial DNA. Nat Genet 1999;23(2):147. Available from: https://doi.org/10.1038/13779.

[38] Greenberg BD, Newbold JE, Sugino A. Intraspecific nucleotide sequence variability surrounding the origin of replication in human mitochondrial DNA. Gene 1983;21(1-2):33–49. Available from: https://doi.org/10.1016/0378-1119(83)90145-2.

[39] Holt IJ, Harding AE, Morgan-Hughes JA. Deletions of muscle mitochondrial DNA in patients with mitochondrial myopathies. Nature 1988;331(6158):717–19. Available from: https://doi.org/10.1038/331717a0.

[40] Wallace DC, Singh G, Lott MT, et al. Mitochondrial DNA mutation associated with Leber's hereditary optic neuropathy. Science (New York, NY) 1988;242(4884):1427–30. Available from: https://doi.org/10.1126/science.3201231.

[41] Wallace DC, Zheng XX, Lott MT, et al. Familial mitochondrial encephalomyopathy (MERRF): genetic, pathophysiological, and biochemical characterization of a mitochondrial DNA disease. Cell 1988;55(4):601–10. Available from: https://doi.org/10.1016/0092-8674(88)90218-8.

[42] Saiki RK, Scharf S, Faloona F, et al. Enzymatic amplification of beta-globin genomic sequences and restriction site analysis for diagnosis of sickle cell anemia. Science (New York, NY) 1985;230(4732):1350–4. Available from: https://doi.org/10.1126/science.2999980.

[43] Saiki RK, Gelfand DH, Stoffel S, et al. Primer-directed enzymatic amplification of DNA with a thermostable DNA polymerase. Science (New York, NY) 1988;239(4839):487–91. Available from: https://doi.org/10.1126/science.2448875.

[44] Tengan CH, Moraes CT. Detection and analysis of mitochondrial DNA deletions by whole genome PCR. Biochemical Mol Med 1996;58(1):130–4. Available from: https://doi.org/10.1006/bmme.1996.0040.

[45] Fromenty B, Manfredi G, Sadlock J, Zhang L, King MP, Schon EA. Efficient and specific amplification of identified partial duplications of human mitochondrial DNA by long PCR. Biochim Biophys Acta 1996;1308(3):222–30. Available from: https://doi.org/10.1016/0167-4781(96)00110-8.

[46] Zeviani M, Bresolin N, Gellera C, et al. Nucleus-driven multiple large-scale deletions of the human mitochondrial genome: a new autosomal dominant disease. Am J Hum Genet 1990;47(6):904–14.

[47] Bai R-K, Wong L-JC. Detection and quantification of heteroplasmic mutant mitochondrial DNA by real-time amplification refractory mutation system quantitative PCR analysis: a single-step approach. Clin Chem 2004;50(6):996–1001. Available from: https://doi.org/10.1373/clinchem.2004.031153.

[48] Van Haute L, Spits C, Geens M, Seneca S, Sermon K. Human embryonic stem cells commonly display large mitochondrial DNA deletions. Nat Biotechnol 2013;31(1):20–3. Available from: https://doi.org/10.1038/nbt.2473.

[49] Ronaghi M, Uhlén M, Nyrén P. A sequencing method based on real-time pyrophosphate. Science (New York, NY) 1998;281(5375):363–5. Available from: https://doi.org/10.1126/science.281.5375.363.

[50] Andréasson H, Asp A, Alderborn A, Gyllensten U, Allen M. Mitochondrial sequence analysis for forensic identification using pyrosequencing technology. Biotechniques 2002;32(1):124–6. Available from: https://doi.org/10.2144/02321rr01 128, 130-133.

[51] Belmonte FR, Martin JL, Frescura K, et al. Digital PCR methods improve detection sensitivity and measurement precision of low abundance mtDNA deletions. Sci Rep 2016;6:25186. Available from: https://doi.org/10.1038/srep25186.
[52] Trifunov S, Pyle A, Valentino ML, et al. Clonal expansion of mtDNA deletions: different disease models assessed by digital droplet PCR in single muscle cells. Sci Rep 2018;8(1):11682. Available from: https://doi.org/10.1038/s41598-018-30143-z.
[53] Wong L-JC, Boles RG. Mitochondrial DNA analysis in clinical laboratory diagnostics. Clin Chim Acta Int J Clin Chem 2005;354(1-2):1–20. Available from: https://doi.org/10.1016/j.cccn.2004.11.003.
[54] Blok RB, Gook DA, Thorburn DR, Dahl HH. Skewed segregation of the mtDNA nt 8993 (T-- > G) mutation in human oocytes. Am J Hum Genet 1997;60(6):1495–501. Available from: https://doi.org/10.1086/515453.
[55] Gigarel N, Ray PF, Burlet P, et al. Single cell quantification of the 8993T > G NARP mitochondrial DNA mutation by fluorescent PCR. Mol Genet Metab 2005;84(3):289–92. Available from: https://doi.org/10.1016/j.ymgme.2004.10.008.
[56] Moraes CT, Atencio DP, Oca-Cossio J, Diaz F. Techniques and pitfalls in the detection of pathogenic mitochondrial DNA mutations. J Mol Diagn 2003;5(4):197–208.
[57] Nicholls TJ, Zsurka G, Peeva V, et al. Linear mtDNA fragments and unusual mtDNA rearrangements associated with pathological deficiency of MGME1 exonuclease. Hum Mol Genet 2014;23(23):6147–62. Available from: https://doi.org/10.1093/hmg/ddu336.
[58] White HE, Durston VJ, Seller A, Fratter C, Harvey JF, Cross NCP. Accurate detection and quantitation of heteroplasmic mitochondrial point mutations by pyrosequencing. Genet Test 2005;9(3):190–9. Available from: https://doi.org/10.1089/gte.2005.9.190.
[59] Kurelac I, Lang M, Zuntini R, et al. Searching for a needle in the haystack: comparing six methods to evaluate heteroplasmy in difficult sequence context. Biotechnol Adv 2012;30(1):363–71. Available from: https://doi.org/10.1016/j.biotechadv.2011.06.001.
[60] Ng YS, Hardy SA, Shrier V, et al. Clinical features of the pathogenic m.5540G > A mitochondrial transfer RNA tryptophan gene mutation. Neuromuscul Disord 2016;26(10):702–5. Available from: https://doi.org/10.1016/j.nmd.2016.08.009.
[61] Kauppila JHK, Baines HL, Bratic A, et al. A phenotype-driven approach to generate mouse models with pathogenic mtDNA mutations causing mitochondrial disease. Cell Rep 2016;16(11):2980–90. Available from: https://doi.org/10.1016/j.celrep.2016.08.037.
[62] Gammage PA, Viscomi C, Simard M-L, et al. Genome editing in mitochondria corrects a pathogenic mtDNA mutation in vivo. Nat Med 2018;24(11):1691–5. Available from: https://doi.org/10.1038/s41591-018-0165-9.
[63] Pan J, Wang L, Lu C, et al. Matching mitochondrial DNA haplotypes for circumventing tissue-specific segregation bias. iScience 2019;13:371–9. Available from: https://doi.org/10.1016/j.isci.2019.03.002.
[64] Bustin SA, Benes V, Garson JA, et al. The MIQE guidelines: minimum information for publication of quantitative real-time PCR experiments. Clin Chem 2009;55(4):611–22. Available from: https://doi.org/10.1373/clinchem.2008.112797.
[65] Niederstätter H, Coble MD, Grubwieser P, Parsons TJ, Parson W. Characterization of mtDNA SNP typing and mixture ratio assessment with simultaneous real-time PCR quantification of both allelic states. Int J Leg Med 2006;120(1):18–23. Available from: https://doi.org/10.1007/s00414-005-0024-3.
[66] Gómez-Durán A, Pacheu-Grau D, López-Gallardo E, et al. Unmasking the causes of multifactorial disorders: OXPHOS differences between mitochondrial haplogroups. Hum Mol Genet 2010;19(17):3343–53. Available from: https://doi.org/10.1093/hmg/ddq246.
[67] Newton CR, Graham A, Heptinstall LE, et al. Analysis of any point mutation in DNA. The amplification refractory mutation system (ARMS). Nucleic Acids Res 1989;17(7):2503–16. Available from: https://doi.org/10.1093/nar/17.7.2503.
[68] Biffi S, Bortot B, Carrozzi M, Severini GM. Quantification of heteroplasmic mitochondrial DNA mutations for DNA samples in the low picogram range by nested real-time ARMS-qPCR. Diagn Mol Pathol

Am J Surg Pathol Part B 2011;20(2):117–22. Available from: https://doi.org/10.1097/PDM.0b013e3181efe2c6.

[69] He L, Chinnery PF, Durham SE, et al. Detection and quantification of mitochondrial DNA deletions in individual cells by real-time PCR. Nucleic Acids Res 2002;30(14):e68. Available from: https://doi.org/10.1093/nar/gnf067.

[70] Pitceathly RDS, Rahman S, Hanna MG. Single deletions in mitochondrial DNA—molecular mechanisms and disease phenotypes in clinical practice. Neuromuscul Disord 2012;22(7):577–86. Available from: https://doi.org/10.1016/j.nmd.2012.03.009.

[71] Moslemi AR, Melberg A, Holme E, Oldfors A. Autosomal dominant progressive external ophthalmoplegia: distribution of multiple mitochondrial DNA deletions. Neurology 1999;53(1):79–84. Available from: https://doi.org/10.1212/wnl.53.1.79.

[72] Yakes FM, Van Houten B. Mitochondrial DNA damage is more extensive and persists longer than nuclear DNA damage in human cells following oxidative stress. Proc Natl Acad Sci U S A 1997;94(2):514–19. Available from: https://doi.org/10.1073/pnas.94.2.514.

[73] Osborne A, Reis AH, Bach L, Wangh LJ. Single-molecule LATE-PCR analysis of human mitochondrial genomic sequence variations. PLoS One 2009;4(5):e5636. Available from: https://doi.org/10.1371/journal.pone.0005636.

[74] Wong LJ, Senadheera D. Direct detection of multiple point mutations in mitochondrial DNA. Clin Chem 1997;43(10):1857–61.

[75] Fauser S, Wissinger B. Simultaneous detection of multiple point mutations using fluorescence-coupled competitive primer extension. Biotechniques 1997;22(5):964–8. Available from: https://doi.org/10.2144/97225rr05.

[76] Elliott HR, Samuels DC, Eden JA, Relton CL, Chinnery PF. Pathogenic mitochondrial DNA mutations are common in the general population. Am J Hum Genet 2008;83(2):254–60. Available from: https://doi.org/10.1016/j.ajhg.2008.07.004.

[77] Cerezo M, Bandelt H-J, Martín-Guerrero I, et al. High mitochondrial DNA stability in B-cell chronic lymphocytic leukemia. PLoS One 2009;4(11). Available from: https://doi.org/10.1371/journal.pone.0007902.

[78] Barros F, Lareu MV, Salas A, Carracedo A. Rapid and enhanced detection of mitochondrial DNA variation using single-strand conformation analysis of superposed restriction enzyme fragments from polymerase chain reaction-amplified products. Electrophoresis 1997;18(1):52–4. Available from: https://doi.org/10.1002/elps.1150180110.

[79] Sternberg D, Danan C, Lombès A, et al. Exhaustive scanning approach to screen all the mitochondrial tRNA genes for mutations and its application to the investigation of 35 independent patients with mitochondrial disorders. Hum Mol Genet 1998;7(1):33–42. Available from: https://doi.org/10.1093/hmg/7.1.33.

[80] van Den Bosch BJ, de Coo RF, Scholte HR, et al. Mutation analysis of the entire mitochondrial genome using denaturing high performance liquid chromatography. Nucleic Acids Res 2000;28(20):E89. Available from: https://doi.org/10.1093/nar/28.20.e89.

[81] Bannwarth S, Procaccio V, Paquis-Flucklinger V. Surveyor Nuclease: a new strategy for a rapid identification of heteroplasmic mitochondrial DNA mutations in patients with respiratory chain defects. Hum Mutat 2005;25(6):575–82. Available from: https://doi.org/10.1002/humu.20177.

[82] Suomalainen A, Ciafaloni E, Koga Y, Peltonen L, DiMauro S, Schon EA. Use of single strand conformation polymorphism analysis to detect point mutations in human mitochondrial DNA. J Neurological Sci 1992;111(2):222–6. Available from: https://doi.org/10.1016/0022-510x(92)90074-u.

[83] Underhill PA, Jin L, Lin AA, et al. Detection of numerous Y chromosome biallelic polymorphisms by denaturing high-performance liquid chromatography. Genome Res 1997;7(10):996–1005. Available from: https://doi.org/10.1101/gr.7.10.996.

[84] Lim KS, Naviaux RK, Wong S, Haas RH. Pitfalls in the denaturing high-performance liquid chromatography analysis of mitochondrial DNA mutation. J Mol Diagn 2008;10(1):102–8. Available from: https://doi.org/10.2353/jmoldx.2008.070081.

[85] Sanger F, Nicklen S, Coulson AR. DNA sequencing with chain-terminating inhibitors. Proc Natl Acad Sci U S A 1977;74(12):5463−7. Available from: https://doi.org/10.1073/pnas.74.12.5463.
[86] Maxam AM, Gilbert W. A new method for sequencing DNA. Proc Natl Acad Sci U S A 1977;74 (2):560−4. Available from: https://doi.org/10.1073/pnas.74.2.560.
[87] Smith LM, Sanders JZ, Kaiser RJ, et al. Fluorescence detection in automated DNA sequence analysis. Nature 1986;321(6071):674−9. Available from: https://doi.org/10.1038/321674a0.
[88] Rosenblum BB, Lee LG, Spurgeon SL, et al. New dye-labeled terminators for improved DNA sequencing patterns. Nucleic Acids Res 1997;25(22):4500−4.
[89] Karger BL, Guttman A. DNA sequencing by CE. Electrophoresis 2009;30(Suppl 1):S196−202. Available from: https://doi.org/10.1002/elps.200900218.
[90] Pearson WR, Lipman DJ. Improved tools for biological sequence comparison. Proc Natl Acad Sci U S A 1988;85(8):2444−8. Available from: https://doi.org/10.1073/pnas.85.8.2444.
[91] Just RS, Irwin JA, Parson W. Mitochondrial DNA heteroplasmy in the emerging field of massively parallel sequencing. Forensic Sci Int Genet 2015;18:131−9. Available from: https://doi.org/10.1016/j.fsigen.2015.05.003.
[92] Maitra A, Cohen Y, Gillespie SED, et al. The Human MitoChip: a high-throughput sequencing microarray for mitochondrial mutation detection. Genome Res 2004;14(5):812−19. Available from: https://doi.org/10.1101/gr.2228504.
[93] GeneChip® Human Mitochondrial Resequencing Array 2.0:2.
[94] Xie HM, Perin JC, Schurr TG, et al. Mitochondrial genome sequence analysis: a custom bioinformatics pipeline substantially improves Affymetrix MitoChip v2.0 call rate and accuracy. BMC Bioinforma 2011;12(402). Available from: https://doi.org/10.1186/1471-2105-12-402.
[95] Zhao S, Jing W, Samuels DC, Sheng Q, Shyr Y, Guo Y. Strategies for processing and quality control of Illumina genotyping arrays. Brief Bioinforma 2017;19(5):765−75. Available from: https://doi.org/10.1093/bib/bbx012.
[96] Zhang P, Samuels DC, Lehmann B, et al. Mitochondria sequence mapping strategies and practicability of mitochondria variant detection from exome and RNA sequencing data. Brief Bioinforma 2016;17 (2):224−32. Available from: https://doi.org/10.1093/bib/bbv057.
[97] Vallone PM, Just RS, Coble MD, Butler JM, Parsons TJ. A multiplex allele-specific primer extension assay for forensically informative SNPs distributed throughout the mitochondrial genome. Int J Leg Med 2004;118(3):147−57. Available from: https://doi.org/10.1007/s00414-004-0428-5.
[98] Sigurdsson S, Hedman M, Sistonen P, Sajantila A, Syvänen A-C. A microarray system for genotyping 150 single nucleotide polymorphisms in the coding region of human mitochondrial DNA. Genomics 2006;87(4):534−42. Available from: https://doi.org/10.1016/j.ygeno.2005.11.022.
[99] Hartmann A, Thieme M, Nanduri LK, et al. Validation of microarray-based resequencing of 93 worldwide mitochondrial genomes. Hum Mutat 2009;30(1):115−22. Available from: https://doi.org/10.1002/humu.20816.
[100] Bybjerg-Grauholm J, Hagen CM, Gonçalves VF, et al. Complex spatio-temporal distribution and genomic ancestry of mitochondrial DNA haplogroups in 24,216 Danes. PLoS One 2018;13(12):e0208829. Available from: https://doi.org/10.1371/journal.pone.0208829.
[101] Lévêque M, Marlin S, Jonard L, et al. Whole mitochondrial genome screening in maternally inherited non-syndromic hearing impairment using a microarray resequencing mitochondrial DNA chip. Eur J Hum Genet 2007;15(11):1145−55. Available from: https://doi.org/10.1038/sj.ejhg.5201891.
[102] Rollins B, Martin MV, Sequeira PA, et al. Mitochondrial variants in schizophrenia, bipolar disorder, and major depressive disorder. PLoS One 2009;4(3):e4913. Available from: https://doi.org/10.1371/journal.pone.0004913.
[103] Samuels DC, Kallianpur AR, Ellis RJ, et al. European mitochondrial DNA haplogroups are associated with cerebrospinal fluid biomarkers of inflammation in HIV infection. Pathog Immun 2016;1 (2):330−51. Available from: https://doi.org/10.20411/pai.v1i2.156.

[104] Mitchell AL, Elson JL, Howell N, Taylor RW, Turnbull DM. Sequence variation in mitochondrial complex I genes: mutation or polymorphism? J Med Genet 2006;43(2):175–9. Available from: https://doi.org/10.1136/jmg.2005.032474.

[105] Metzker ML. Sequencing technologies—the next generation. Nat Rev Genet 2010;11(1):31–46. Available from: https://doi.org/10.1038/nrg2626.

[106] Goodwin S, McPherson JD, McCombie WR. Coming of age: ten years of next-generation sequencing technologies. Nat Rev Genet 2016;17(6):333–51. Available from: https://doi.org/10.1038/nrg.2016.49.

[107] Shendure J, Ji H. Next-generation DNA sequencing. Nat Biotechnol 2008;26(10):1135–45. Available from: https://doi.org/10.1038/nbt1486.

[108] Ewing B, Hillier L, Wendl MC, Green P. Base-calling of automated sequencer traces using phred. I. Accuracy assessment. Genome Res 1998;8(3):175–85. Available from: https://doi.org/10.1101/gr.8.3.175.

[109] Chilamakuri CSR, Lorenz S, Madoui M-A, et al. Performance comparison of four exome capture systems for deep sequencing. BMC Genomics 2014;15:449. Available from: https://doi.org/10.1186/1471-2164-15-449.

[110] Clark MJ, Chen R, Lam HYK, et al. Performance comparison of exome DNA sequencing technologies. Nat Biotechnol 2011;29(10):908–14. Available from: https://doi.org/10.1038/nbt.1975.

[111] Sulonen A-M, Ellonen P, Almusa H, et al. Comparison of solution-based exome capture methods for next generation sequencing. Genome Biol 2011;12(9):R94. Available from: https://doi.org/10.1186/gb-2011-12-9-r94.

[112] Samuels DC, Han L, Li J, et al. Finding the lost treasures in exome sequencing data. Trends Genet 2013;29(10):593–9. Available from: https://doi.org/10.1016/j.tig.2013.07.006.

[113] Picardi E, Pesole G. Mitochondrial genomes gleaned from human whole-exome sequencing. Nat Methods 2012;9(6):523–4. Available from: https://doi.org/10.1038/nmeth.2029.

[114] Griffin HR, Pyle A, Blakely EL, et al. Accurate mitochondrial DNA sequencing using off-target reads provides a single test to identify pathogenic point mutations. Genet Med J Am Coll Med Genet 2014;16 (12):962–71. Available from: https://doi.org/10.1038/gim.2014.66.

[115] Wei W, Keogh MJ, Aryaman J, et al. Frequency and signature of somatic variants in 1461 human brain exomes. Genet Med J Am Coll Med Genet 2019;21(4):904–12. Available from: https://doi.org/10.1038/s41436-018-0274-3.

[116] Bentley DR, Balasubramanian S, Swerdlow HP, et al. Accurate whole human genome sequencing using reversible terminator chemistry. Nature 2008;456(7218):53–9. Available from: https://doi.org/10.1038/nature07517.

[117] Floros VI, Pyle A, Dietmann S, et al. Segregation of mitochondrial DNA heteroplasmy through a developmental genetic bottleneck in human embryos. Nat Cell Biol 2018;20(2):144–51. Available from: https://doi.org/10.1038/s41556-017-0017-8.

[118] Liu C, Fetterman JL, Liu P, et al. Deep sequencing of the mitochondrial genome reveals common heteroplasmic sites in NADH dehydrogenase genes. Hum Genet 2018;137(3):203–13. Available from: https://doi.org/10.1007/s00439-018-1873-4.

[119] Schirmer M, Ijaz UZ, D'Amore R, Hall N, Sloan WT, Quince C. Insight into biases and sequencing errors for amplicon sequencing with the Illumina MiSeq platform. Nucleic Acids Res 2015;43(6):e37. Available from: https://doi.org/10.1093/nar/gku1341.

[120] Feng W, Zhao S, Xue D, et al. Improving alignment accuracy on homopolymer regions for semiconductor-based sequencing technologies. BMC Genomics 2016;17(Suppl 7). Available from: https://doi.org/10.1186/s12864-016-2894-9.

[121] Salk JJ, Schmitt MW, Loeb LA. Enhancing the accuracy of next-generation sequencing for detecting rare and subclonal mutations. Nat Rev Genet 2018;19(5):269–85. Available from: https://doi.org/10.1038/nrg.2017.117.

[122] Ardui S, Ameur A, Vermeesch JR, Hestand MS. Single molecule real-time (SMRT) sequencing comes of age: applications and utilities for medical diagnostics. Nucleic Acids Res 2018;46(5):2159–68. Available from: https://doi.org/10.1093/nar/gky066.

[123] Eid J, Fehr A, Gray J, et al. Real-time DNA sequencing from single polymerase molecules. Science (New York, NY) 2009;323(5910):133–8. Available from: https://doi.org/10.1126/science.1162986.

[124] Weerts MJA, Timmermans EC, Vossen RHAM, et al. Sensitive detection of mitochondrial DNA variants for analysis of mitochondrial DNA-enriched extracts from frozen tumor tissue. Sci Rep 2018;8(1):1–12. Available from: https://doi.org/10.1038/s41598-018-20623-7.

[125] Branton D, Deamer DW, Marziali A, et al. The potential and challenges of nanopore sequencing. Nat Biotechnol 2008;26(10):1146–53. Available from: https://doi.org/10.1038/nbt.1495.

[126] Bowden R, Davies RW, Heger A, et al. Sequencing of human genomes with nanopore technology. Nat Commun 2019;10(1):1869. Available from: https://doi.org/10.1038/s41467-019-09637-5.

[127] Wood E, Parker MD, Dunning MJ, et al. Clinical long-read sequencing of the human mitochondrial genome for mitochondrial disease diagnostics. bioRxiv 2019;597187. Available from: https://doi.org/10.1101/597187 April.

[128] Duan M, Tu J, Lu Z. Recent advances in detecting mitochondrial DNA heteroplasmic variations. Mol Basel Switz 2018;23(2). Available from: https://doi.org/10.3390/molecules23020323.

[129] Fernández-Vizarra E, López-Pérez MJ, Enriquez JA. Isolation of biogenetically competent mitochondria from mammalian tissues and cultured cells. Methods San Diego Calif 2002;26(4):292–7. Available from: https://doi.org/10.1016/S1046-2023(02)00034-8.

[130] Lang BF, Burger G. Purification of mitochondrial and plastid DNA. Nat Protoc 2007;2(3):652–60. Available from: https://doi.org/10.1038/nprot.2007.58.

[131] Gould MP, Bosworth CM, McMahon S, Grandhi S, Grimerg BT, LaFramboise T. PCR-free enrichment of mitochondrial DNA from human blood and cell lines for high quality next-generation DNA sequencing. PLoS One 2015;10(10). Available from: https://doi.org/10.1371/journal.pone.0139253.

[132] Reyes A, He J, Mao CC, et al. Actin and myosin contribute to mammalian mitochondrial DNA maintenance. Nucleic Acids Res 2011;39(12):5098–108. Available from: https://doi.org/10.1093/nar/gkr052.

[133] Hudson B, Vinograd J. Sedimentation velocity properties of complex mitochondrial DNA. Nature 1969;221(5178):332–7. Available from: https://doi.org/10.1038/221332a0.

[134] Tobler H, Gut C. Mitochondrial DNA from 4-cell stages of *Ascaris lumbricoides*. J Cell Sci 1974;16(3):593–601.

[135] Jayaprakash AD, Benson EK, Gone S, et al. Stable heteroplasmy at the single-cell level is facilitated by intercellular exchange of mtDNA. Nucleic Acids Res 2015;43(4):2177–87. Available from: https://doi.org/10.1093/nar/gkv052.

[136] Quispe-Tintaya W, White RR, Popov VN, Vijg J, Maslov AY. Fast mitochondrial DNA isolation from mammalian cells for next-generation sequencing. Biotechniques 2013;55(3):133–6. Available from: https://doi.org/10.2144/000114077.

[137] Shih SY, Bose N, Gonçalves ABR, Erlich HA, Calloway CD. Applications of probe capture enrichment next generation sequencing for whole mitochondrial genome and 426 nuclear SNPs for forensically challenging samples. Genes 2018;9(1). Available from: https://doi.org/10.3390/genes9010049.

[138] Hofreiter M, Serre D, Poinar HN, Kuch M, Pääbo S. Ancient DNA. Nat Rev Genet 2001;2(5):353–9. Available from: https://doi.org/10.1038/35072071.

[139] Fendt L, Zimmermann B, Daniaux M, Parson W. Sequencing strategy for the whole mitochondrial genome resulting in high quality sequences. BMC Genomics 2009;10(1):139. Available from: https://doi.org/10.1186/1471-2164-10-139.

[140] Zhang W, Cui H, Wong LJC. Comprehensive one-step molecular analyses of mitochondrial genome by massively parallel sequencing. Clin Chem 2012;58(9):1322–31. Available from: https://doi.org/10.1373/clinchem.2011.181438.

[141] Kang E, Wu J, Gutierrez NM, et al. Mitochondrial replacement in human oocytes carrying pathogenic mitochondrial DNA mutations. Nature 2016;540(7632):270–5. Available from: https://doi.org/10.1038/nature20592.

[142] Santibanez-Koref M, Griffin H, Turnbull DM, Chinnery PF, Herbert M, Hudson G. Assessing mitochondrial heteroplasmy using next generation sequencing: a note of caution. Mitochondrion 2019;46:302–6. Available from: https://doi.org/10.1016/j.mito.2018.08.003.
[143] Marquis J, Lefebvre G, Kourmpetis YAI, et al. MitoRS, a method for high throughput, sensitive, and accurate detection of mitochondrial DNA heteroplasmy. BMC Genomics 2017;18(1):326. Available from: https://doi.org/10.1186/s12864-017-3695-5.
[144] Mishmar D, Ruiz-Pesini E, Brandon M, Wallace DC. Mitochondrial DNA-like sequences in the nucleus (NUMTs): insights into our African origins and the mechanism of foreign DNA integration. Hum Mutat 2004;23(2):125–33. Available from: https://doi.org/10.1002/humu.10304.
[145] Lascaro D, Castellana S, Gasparre G, Romeo G, Saccone C, Attimonelli M. The RHNumtS compilation: features and bioinformatics approaches to locate and quantify Human NumtS. BMC Genomics 2008;9:267. Available from: https://doi.org/10.1186/1471-2164-9-267.
[146] Hazkani-Covo E, Zeller RM, Martin W. Molecular poltergeists: mitochondrial DNA copies (NumtS) in sequenced nuclear genomes. PLoS Genet 2010;6(2):e1000834. Available from: https://doi.org/10.1371/journal.pgen.1000834.
[147] Simone D, Calabrese FM, Lang M, Gasparre G, Attimonelli M. The reference human nuclear mitochondrial sequences compilation validated and implemented on the UCSC genome browser. BMC Genomics 2011;12:517. Available from: https://doi.org/10.1186/1471-2164-12-517.
[148] Li M, Schroeder R, Ko A, Stoneking M. Fidelity of capture-enrichment for mtDNA genome sequencing: influence of NUMTs. Nucleic Acids Res 2012;40(18):e137. Available from: https://doi.org/10.1093/nar/gks499.
[149] Ye J, Coulouris G, Zaretskaya I, Cutcutache I, Rozen S, Madden TL. Primer-BLAST: a tool to design target-specific primers for polymerase chain reaction. BMC Bioinforma 2012;13:134. Available from: https://doi.org/10.1186/1471-2105-13-134.
[150] King MP, Attardi G. Human cells lacking mtDNA: repopulation with exogenous mitochondria by complementation. Science (New York, NY) 1989;246(4929):500–3. Available from: https://doi.org/10.1126/science.2814477.
[151] Calabrese C, Simone D, Diroma MA, et al. MToolBox: a highly automated pipeline for heteroplasmy annotation and prioritization analysis of human mitochondrial variants in high-throughput sequencing. Bioinforma Oxf Engl 2014;30(21):3115–17. Available from: https://doi.org/10.1093/bioinformatics/btu483.
[152] Bosworth CM, Grandhi S, Gould MP, LaFramboise T. Detection and quantification of mitochondrial DNA deletions from next-generation sequence data. BMC Bioinforma 2017;18(Suppl 12):407. Available from: https://doi.org/10.1186/s12859-017-1821-7.
[153] Kent WJ. BLAT—the BLAST-like alignment tool. Genome Res 2002;12(4):656–64. Available from: https://doi.org/10.1101/gr.229202.
[154] Li H, Durbin R. Fast and accurate long-read alignment with Burrows-Wheeler transform. Bioinforma Oxf Engl 2010;26(5):589–95. Available from: https://doi.org/10.1093/bioinformatics/btp698.
[155] Goudenège D, Bris C, Hoffmann V, et al. eKLIPse: a sensitive tool for the detection and quantification of mitochondrial DNA deletions from next-generation sequencing data. Genet Med J Am Coll Med Genet 2019;21(6):1407–16. Available from: https://doi.org/10.1038/s41436-018-0350-8.
[156] Dierckxsens N, Mardulyn P, Smits G. Unraveling heteroplasmy patterns with NOVOPlasty. NAR Genomics Bioinforma 2020;2(1). Available from: https://doi.org/10.1093/nargab/lqz011.
[157] Dierckxsens N, Mardulyn P, Smits G. NOVOPlasty: de novo assembly of organelle genomes from whole genome data. Nucleic Acids Res 2017;45(4):e18. Available from: https://doi.org/10.1093/nar/gkw955.
[158] Li H, Handsaker B, Wysoker A, et al. The sequence alignment/map format and SAMtools. Bioinforma (Oxford, Engl) 2009;25(16):2078–9. Available from: https://doi.org/10.1093/bioinformatics/btp352.
[159] Weissensteiner H, Forer L, Fuchsberger C, et al. mtDNA-Server: next-generation sequencing data analysis of human mitochondrial DNA in the cloud. Nucleic Acids Res 2016;44:W64–9. Available from: https://doi.org/10.1093/nar/gkw247 Web Server issue.

[160] Guo Y, Li J, Li C-I, Shyr Y, Samuels DC. MitoSeek: extracting mitochondria information and performing high-throughput mitochondria sequencing analysis. Bioinforma (Oxford, Engl) 2013;29(9):1210–11. Available from: https://doi.org/10.1093/bioinformatics/btt118.

[161] Pyle A, Hudson G, Wilson IJ, et al. Extreme-depth re-sequencing of mitochondrial DNA finds no evidence of paternal transmission in humans. PLoS Genet 2015;11(5):e1005040. Available from: https://doi.org/10.1371/journal.pgen.1005040.

[162] Just RS, Irwin JA, Parson W. Questioning the prevalence and reliability of human mitochondrial DNA heteroplasmy from massively parallel sequencing data. Proc Natl Acad Sci U S A 2014;111(43): E4546–4547. Available from: https://doi.org/10.1073/pnas.1413478111.

[163] Bandelt HJ, Lahermo P, Richards M, Macaulay V. Detecting errors in mtDNA data by phylogenetic analysis. Int J Leg Med 2001;115(2):64–9.

[164] van Oven M, Kayser M. Updated comprehensive phylogenetic tree of global human mitochondrial DNA variation. Hum Mutat 2009;30(2):E386–394. Available from: https://doi.org/10.1002/humu.20921.

[165] Salas A, Carracedo A, Macaulay V, Richards M, Bandelt H-J. A practical guide to mitochondrial DNA error prevention in clinical, forensic, and population genetics. Biochemical Biophysical Res Commun 2005;335(3):891–9. Available from: https://doi.org/10.1016/j.bbrc.2005.07.161.

[166] Robinson JT, Thorvaldsdóttir H, Winckler W, et al. Integrative genomics viewer. Nat Biotechnol 2011;29(1):24–6. Available from: https://doi.org/10.1038/nbt.1754.

[167] Chinnery PF, Gomez-Duran A. Oldies but goldies mtDNA population variants and neurodegenerative diseases. Front Neurosci 2018;12:682. Available from: https://doi.org/10.3389/fnins.2018.00682.

[168] Wilm A, Aw PPK, Bertrand D, et al. LoFreq: a sequence-quality aware, ultra-sensitive variant caller for uncovering cell-population heterogeneity from high-throughput sequencing datasets. Nucleic Acids Res 2012;40(22):11189–201. Available from: https://doi.org/10.1093/nar/gks918.

[169] Navarro-Gomez D, Leipzig J, Shen L, et al. Phy-Mer: a novel alignment-free and reference-independent mitochondrial haplogroup classifier. Bioinforma Oxf Engl 2015;31(8):1310–12. Available from: https://doi.org/10.1093/bioinformatics/btu825.

[170] Weissensteiner H, Pacher D, Kloss-Brandstätter A, et al. HaploGrep 2: mitochondrial haplogroup classification in the era of high-throughput sequencing. Nucleic Acids Res 2016;44(W1):W58–63. Available from: https://doi.org/10.1093/nar/gkw233.

[171] Kloss-Brandstätter A, Pacher D, Schönherr S, et al. HaploGrep: a fast and reliable algorithm for automatic classification of mitochondrial DNA haplogroups. Hum Mutat 2011;32(1):25–32. Available from: https://doi.org/10.1002/humu.21382.

CHAPTER 12

Bioinformatics resources, databases, and tools for human mtDNA

Marcella Attimonelli[1], Roberto Preste[1], Ornella Vitale[1], Marie T. Lott[2], Vincent Procaccio[3], Zhang Shiping[2] and Douglas C. Wallace[2,4]

[1]Department of Biosciences, Biotechnology and Biopharmaceutics, University "Aldo Moro", Bari, Italy, [2]Center for Mitochondrial and Epigenomic Medicine, Children's Hospital of Philadelphia, Philadelphia, PA, United States, [3]Biochemistry and Genetics Department, MitoVasc Institute, UMR CNRS 6015 – INSERM U1083, CHU Angers, Angers, France, [4]Perelman School of Medicine, University of Pennsylvania, Philadelphia, PA, United States

12.1 Introduction to human mtDNA variability (Wallace DC and Attimonelli M)

A unique feature of human mitochondrial DNA (mtDNA) is its high mutation rate relative to comparably functional nuclear DNA (nDNA) coded genes [1–5]. There are three clinically relevant classes of mtDNA mutations: ancient polymorphisms, recent deleterious mutations, and somatic mutations [3]. Since the mtDNA is exclusively maternally inherited [6,7], mutations can only accumulate along radiating maternal lineages of mtDNAs that create a global phylogenetic tree. This tree originated in Africa in the order of 150,000–200,000 years before the present (YBP), with two mtDNA lineages leaving Africa about 65,000 YBP to colonize the rest of the world [8]. While the majority of ancient mitochondrial variants are neutral, about 1/3 of the population polymorphisms are functionally relevant, with each new branch of the mtDNA tree being founded by one of several functional variants that were regionally selected and gave rise to regional groups of related haplotypes, known as haplogroups [9–11] (Chapter 5: Haplogroups and the history of human evolution through mtDNA). These ancient "adaptive" mtDNA variants can become maladaptive in alternative environments and can be predisposed to a wide range of common metabolic and degenerative diseases, cancer, and aging [10]. In addition to these global mtDNA variants, new mutations are constantly arising in the mtDNA, a subset of which significantly impair mitochondrial function and result in maternally inherited diseases. Because of the high ploidy of the mtDNA, new mutations are initially heteroplasmic. Through mitotic or meiotic replication, the percentage of mutant mtDNAs can drift, and if a

The Human Mitochondrial Genome.
DOI: https://doi.org/10.1016/B978-0-12-819656-4.00012-7

deleterious mutation reaches a high enough heteroplasmic level, it can impair function and result in disease [1,3,12–14]. The frequency of clinically ascertained mtDNA diseases is of the order of 1 in 5000 [15,16]. Finally, mtDNA mutations arise during development and in somatic tissues. These are generally heteroplasmic and have been implicated in aging (Chapter 10: Mitochondrial DNA mutations and ageing) and the delayed onset and progressive course of adult diseases [3].

Because of the range of mtDNA variants with potential clinical and evolutionary involvement, difficulties can be encountered in deciding whether or not a particular mtDNA variant is relevant for either clinical or evolutionary arguments [9,10,12,17]. Thus in order to prioritize mitochondrial variations, there are four generally accepted canonical criteria that support the role of human mitochondrial variants:

1. the knowledge about the neutral nature of a polymorphism,
2. the effects caused by the variant in a conserved or functionally important site,
3. the heteroplasmic traits of a certain variant,
4. the degree of heteroplasmy linked to the severity of symptoms [12].

In this scenario, the advent of sequencing technology has permitted a detailed analysis of mtDNA variation in populations, patients, and tissues. The first mtDNA sequence was produced by the Sanger laboratory in Cambridge, England in 1981 [18]. This sequence was subsequently revisited and corrected to result in the revised Cambridge reference sequence (rCRS) [19]. The rCRS has been extensively used as a reference sequence for analyzing and describing new mtDNA sequences. However, the rCRS was from a Western European, and thus arose after humans migrated out of Africa. To permit a more logical phylogenetic perspective on mtDNA variation, a hypothetical founder mtDNA sequence was devised: the "Reconstructed Sapiens Reference Sequence" (RSRS) [20]. Because of the historical precedence of the rCRS, it continues to be commonly used as the reference sequence in clinical studies [21]. However, RSRS has gained acceptance as a root sequence for human mtDNA evolutionary studies.

Since the publication of the first mtDNA sequence almost 30 years ago, an enormous amount of information on mtDNA sequence variation has been accumulated in connection with human origins [22,23] (Chapter 5: Haplogroups and the history of human evolution through mtDNA), forensic sample analysis [24,25], and clinical mtDNA classification [10,26].

Several bioinformatics systems have been developed to manage and analyze this enormous volume of information with the aim of annotating and classifying genomes and variants.

In this chapter, our goal is to assist the physician, scientist, historian, and individual to penetrate the complexities of mtDNA sequence variation. We provide a comprehensive overview of concepts underlying the criteria to assess mitochondrial variant pathogenicity, as well as bioinformatics databases and tools.

12.2 Human mtDNA genomes and variants

12.2.1 Primary databases: GenBank/ENA/DDBJ (Attimonelli M)

As reported in the introduction of the present chapter, the advent of sequencing technologies has allowed the production of complete or partial human mtDNA sequences that, according to international agreements, may and must be submitted by the authors to the primary nucleotide databases GenBank [27], EMBL/ENA [28], and DDBJ [29] altogether associated as the "International Nucleotide Sequence Database Collaboration" (INSDC) (http://www.insdc.org). The resulting submission is organized in the database as a record or otherwise named "Entry" whose format allows annotation of primary information regarding the sequence, description of the sample, authorship of the sequence, location of the genes and of any other functional feature mapped onto the entire submitted sequence. Data are under the responsibility of the authors who receive the Accession Number as guarantee of authorship and also of false or erroneous results. Errors may be always corrected through further submissions. However, considering that the aim of sequencing human mtDNA is to recognize site-specific alleles with a role as a marker of a population or disease, the information regarding the variability of each genome is not stored in the human mtDNA entries available in primary databases; only the pairwise alignment of the sequence, extracted in FASTA format [30], toward the reference genome (rCRS or RSRS) allows variants to be annotated. To this aim, specialized databases and resources reporting sequence data, variants, and their features have been designed and implemented. Here, below we describe the most updated and informative human mitochondrial genome databases, that is, MitoMap (Section 12.2.2) and HmtDB together with its derived HmtVar and HmtPhenome databases, namely the Human MitoCompendium (Section 12.3). The reference resource for genome's haplotyping, Phylotree [31], is described in Chapter 5: Haplogroups and the history of human evolution through mtDNA. Other resources are briefly reported in Section 12.5.

12.2.2 MITOMAP (Lott MT, Procaccio V, and Zhang S)

MITOMAP (https://mitomap.org) contains key background information to reinforce mitochondrial-specific essentials before a user embarks upon variant analysis. In 1996 the MITOMAP database became the first curated online compilation of human mtDNA variants, containing 582 general variants and 55 mutations with possible or confirmed disease associations from published reports [32]. Almost 25 years later, MITOMAP now contains over 14,000 observed variants, including 730 reported disease mutations. In 2013, MITOMAP was expanded to include a repository of 18,000 GenBank mtDNA sequences. As of September 2019, this number has grown to 49,135 full length and 72,235 control region human mtDNA sequences in the database. These sequences represent almost every mitochondrial haplotype known to date (https://mitomap.org/MITOMAP/GBFreqInfo).

Careful curation of specialized databases is essential to ensure value and accuracy. In MITOMAP, current peer-reviewed publications and clinical reports are collected on a weekly basis and the reported variants are manually indexed into the database. Any published variant reported with possible disease association is noted. The GenBank sequences are updated on a regular basis throughout the year. Sequences are sorted into broad categories [general, ancient DNA (aDNA), cancer/tumor/abnormal tissues, and cell lines] as to their source based on information contained in their sequence header or indexed publication. All new sequences are haplotyped and their variants extracted before being added to the database. MITOMAP contains MITOMASTER, a comprehensive sequence analysis tool for human mtDNA. All of MITOMAP's 120,000 + sequences and variants are preloaded in the MITOMASTER tool to provide a large comparison set for user variant and sequence analysis. Amino acid translation and gene loci tables are provided as well as an annotated rCRS, a summary of the RSRS variants, and easy links to pull down mitochondrial sequences from GenBank. Classic illustrations include the morbid map of the mtDNA, a world map of ancestral haplogroup migrations, and a tree of basic phylogenetic relationships.

The heart of MITOMAP is its annotated listing of over 14,000 mtDNA variants accumulated over the years of curation, both in the general population and in patients. All variants display the gene locus, nucleotide change, translation product, GenBank frequency and sequences, and references (Fig. 12.1). A variant's GenBank frequency is linked to a full table of sequences, complete with Accession ID links, PubMed links, haplotype, and haplogroup-specific frequency. Sequences from aDNA, cancers, or cell lines are flagged

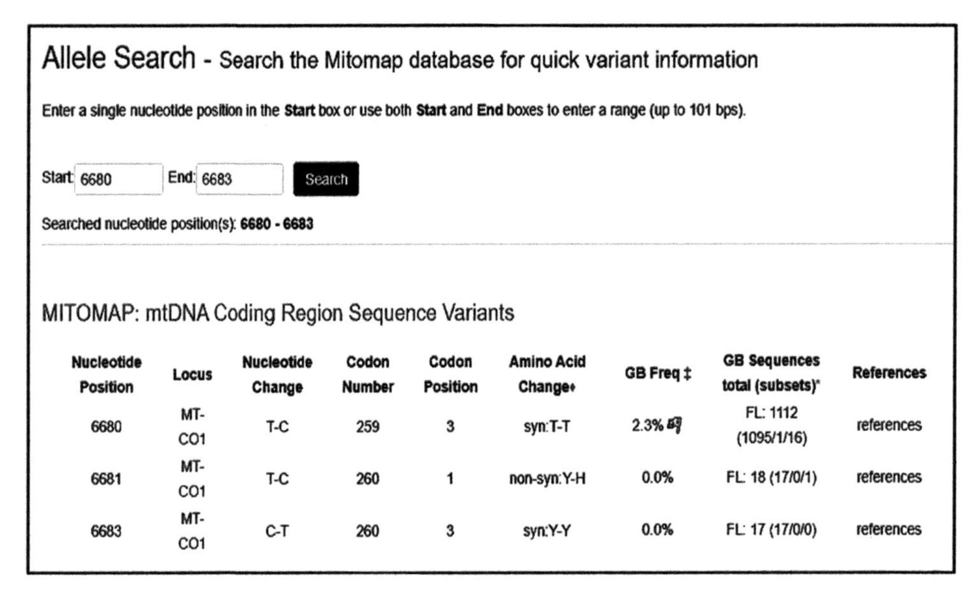
Allele Search - Search the Mitomap database for quick variant information

Enter a single nucleotide position in the **Start** box or use both **Start** and **End** boxes to enter a range (up to 101 bps).

Start: 6680 End: 6683 Search

Searched nucleotide position(s): **6680 - 6683**

MITOMAP: mtDNA Coding Region Sequence Variants

Nucleotide Position	Locus	Nucleotide Change	Codon Number	Codon Position	Amino Acid Change•	GB Freq ‡	GB Sequences total (subsets)*	References
6680	MT-CO1	T-C	259	3	syn:T-T	2.3%	FL: 1112 (1095/1/16)	references
6681	MT-CO1	T-C	260	1	non-syn:Y-H	0.0%	FL: 18 (17/0/1)	references
6683	MT-CO1	C-T	260	3	syn:Y-Y	0.0%	FL: 17 (17/0/0)	references

Figure 12.1
Example of Allele Search results in MITOMAP.

(Fig. 12.2). MITOMAP's MITOMASTER engine is used to analyze these SNPS and sequences (cf. Sections 12.2.2.1–12.2.2.5). An API (Application Programming Interface) is provided for high volume users to access MITOMAP data directly from their own applications, and a full database dump is available to the public.

12.2.2.1 Variant status in MITOMAP

Variants reported in the literature with possible disease associations carry a simple "Reported" status until the time when further evaluation by MITOMAP warrants a change to confirmed pathogenic status (Fig. 12.3). For MITOMAP to assign a confirmed (Cfrm) status to a variant, there must be sufficient reports, which address the criteria outlined in the literature [33–36]. These criteria include the following: (1) independent reports of two or more unrelated families with evidence of similar disease; (2) evolutionary conservation of the nucleotide (for RNA variants) or amino acid (for coding variants); (3) presence of heteroplasmy; (4) correlation of variant with phenotype/segregation of the mutation with the disease related to the mutant load within a family; (5) biochemical defects in complexes I,

GenBank Record for Coding Variant G->A at rCRS position 3380

click hyperlink to view GenBank Record, PubMed Reference, and Mitomaster running results

GenBank ID	Seq Type*	PubMed Reference	Predicted haplogroup (HG) branch	# in HG branch with variant	Total # HG branch seqs	Frequency in HG branch (%)	MITOMASTER Results
AY195771.1	FL.Main	12509511	A5b	1	25	4.00	view
KC990651.1	FL.Main		M3c	1	46	2.17	view
MF588827.1	FL.aDNA	29033326	D2a	1	63	1.59	view

* **Footnotes:**

Full Length (**FL**) sequences are now classified into three subsets: **aDNA** for ancient DNA, **CancerCL** for cancer, tumor, other abnormal tissues, or cell line studies, and **Main** for the rest.

The current sequence counts are from two sets of human mitochondrial sequences collected from GenBank on **Sep 01, 2019**. These sets consist of:

- **49,135** Full Length (**FL**) sequences (>15.4 kb)
  - FL.Main: **47,487**
  - FL.aDNA: **954**
  - FL.CancerCL: **693** (**602** cancer + **91** cell lines)
- **72,235** Control Region (**CR**) only sequences (0.4-1.6 kb)

Figure 12.2

Example of a linked sequence detail page in MITOMAP.

Mitomap Frequency Annotation Example, with and without a High Frequency Haplogroup Flag

Position	Locus	Disease	Allele	Nucleotide Change	Amino Acid Change	Homo-plasmy	Hetero-plasmy	Status	GB Freq	GB Seqs	References
14568	MT-ND6	LHON	C14568T	C-T	G-S	+	-	Cfrm	0.0%	6	9
14577	MT-ND6	MIDM	T14577C	T-C	I-V	-	+	Reported	1.8%	411	1

Figure 12.3

An example of variants with "Confirmed" and "Reported" pathogenic status in MITOMAP, as well as GB frequency annotation with and without a High Frequency Haplogroup flag.

III, or IV of the respiratory chain in affected or multiple tissues; (6) functional studies showing differential defects segregating with the mutation (cybrid or single-fiber studies); (7) histochemical evidence of a mitochondrial disorder; and (8) for fatal or severe clinical phenotypes, the absence or extremely rare occurrence of the variant in large mtDNA sequence databases. As of the publication date, only 88 out of 730 reported disease variants in MITOMAP carry a confirmed pathogenic status. A current listing of MITOMAP's confirmed pathogenic mutations is available at https://mitomap.org/MITOMAP/ConfirmedMutations.

12.2.2.2 Haplogroup assignment in MITOMAP

MITOMAP has put an emphasis on enhanced annotations of frequency and haplogroup distribution to encourage the use of haplotyping in patient mtDNA analysis. An awareness of variant frequency will avoid the occasional mistaken attribution of pathogenicity to near ubiquitous variants [37–40]. Most of these extremely common variants are ancestral and occur throughout the phylogenetic tree. However, some may be found at up to 99% in one or two of the three major lineages (L, M, N) but be virtually absent in another. MITOMAP has compiled the most common variants and their distribution at https://mitomap.org/MITOMAP/TopVariants. It is also important to be aware that while variants found at high frequency in certain haplogroups are usually considered benign, occasionally such variants can modulate patient phenotypes when the background haplogroup is different [41–43].

To assist in evaluating variant pathogenicity, MITOMAP has recently added a prominent annotation to each variant in its database found at high frequency in a top-level haplogroup or branch. A variant is flagged with a "High Frequency Alert" when it has an overall frequency of 0.5% or higher among MITOMAP's 49,000 + GenBank sequences and is seen at 50% or higher in haplogroup branches (flag seen in Figs. 12.1 and 12.3, with haplogroup details in Fig. 12.4). All haplogroup assignments are calculated in MITOMASTER using HaploGrep2 with Phylotree 17.

High Frequency Haplogroups

m.10086A>G: 418 sequences (0.9% overall)

Lineage	Top Level HG	Top Level HG Branch (ltr-num)	HG Branch (ltr-num-ltr)
L 396 (6.6%)	L3	L3	L3b 396 (100.0%)
N 19 (0.1%)	W	W1	W1b 17 (81.0%)

Figure 12.4
An example of a High Frequency Haplogroup detail page in MITOMAP.

In addition, the frequency of markers for all top-level haplogroups (A-Z, L0-L6, HV) is assessed on a regular basis using MITOMAP's full length sequence set. Marker variants at $>80\%$ are reported, as well as those found at $>50\%$. These marker frequencies are available for easy reference through MITOMAP's home page or may be accessed directly at https://mitomap.org/MITOMAP/HaplogroupMarkers. A "Marker Finder" tool is also provided for user-generated frequency queries of specific haplogroups. Other related tools are under development (https://mitomap.org/MITOMAP/ToolLaunchpad).

12.2.2.3 Allele search function in MITOMAP

One of the most used tools for quick and easy information about a variant is Allele Search (http://mitomap.org/MITOMAP/SearchAllele) (Fig. 12.1). After entering the position or a position range up to 100 bp, all variant alleles in the MITOMAP database at that position are displayed. Basic data such as nucleotide change, amino acid change if coding, homo or/ and heteroplasmy status and curated references are included. The count of full-length GenBank sequences carrying that variant will also be displayed; the sequence count is further linked to a full table of accession IDs, sequence haplotypes, and the frequency within each haplogroup. The High Frequency Alert flag, described earlier, has recently been incorporated into Allele Search. When evaluating a patient variant for possible pathogenicity, its frequency, in both the patient's haplogroup and others around the globe, plays an important role in pathogenicity assessment.

12.2.2.4 Analyzing mtDNA variability using MITOMASTER

At the core of MITOMAP is the sequence analysis engine, MITOMASTER (https://mitomap.org/foswiki/bin/view/MITOMASTER/WebHome). This powerful tool analyzes single nucleotide variants relative to the rCRS, user-supplied sequences, and GenBank identifiers. MITOMASTER performs haplotyping using the Phylotree-based Haplogrep2 engine [44,45]. MITOMASTER haplotypes the different sequences that carry the variant and calculates the number, frequency, and distribution within the total full-length sequences in its database. Variant conservation among a user-selected species set is also calculated.

12.2.2.5 MitoTIP

Integrated into MITOMASTER is MitoTIP (Mitochondrial tRNA Informatics Predictor), an in silico analysis tool predicting the pathogenicity of novel tRNA variants [46]. The MitoTIP score for a tRNA variant is a starting point for pathogenicity assessment. It is intended as a preliminary guide for novel variants, creating a score from the variant's conservation, structural disruption, and its location within the cloverleaf structure of the tRNA. Quartile ranks are assigned as "likely benign," "possibly benign," "possibly pathogenic," and "likely pathogenic." Once a variant's pathogenicity is confirmed by MITOMAP, its MitoTIP label changes to "pathogenic." More information about MitoTIP scoring is available at https://mitomap.org/MITOMAP/MitoTipInfo

12.3 The Human MitoCompendium: HmtDB, HmtVar, and HmtPhenome (Attimonelli M, Preste R, and Vitale O)

12.3.1 HmtDB

HmtDB [47,48] is a web-based human mitochondrial resource collecting all the human mtDNA sequences derived from data available through the primary nucleotide databases, further annotated and structured according to criteria allowing the user to browse and extract information regarding human mitochondrial phylogeny and disease. Differently from MITOMAP, from its inception, the main aim of HmtDB was to collect available Human mitochondrial genomes and hence to annotate their variants. Vice versa MITOMAP started with the collection of reported mtDNA mutations.

The first release of HmtDB was published in 2005 with the aim of reporting the list of variants detected after the comparison with the reference mitochondrial genome, for any available human mitochondrial genome (at that time 1255). As of July 2019, HmtDB annotates 50,871 mitochondrial genomes, including 1427 mitochondrial genomes reconstructed as off-target sequences from exome data [49] generated by the 1000 Genomes Project [50].

The sequences are organized in two datasets representing individuals with healthy and disease phenotype. Samples reported as "healthy" mainly derive from population studies or control in clinical studies, while "pathologic" genomes come from individuals affected by mitochondrial diseases or other stated clinical conditions. It must be underlined here that the "healthy" feature merely refers to what is reported in the study that produced that specific sequence, meaning that subjects referred to as controls in clinical studies are reported as healthy on the basis of not having *that* specific pathology, but not of not having *any* disease. Similarly, subjects included in population studies are not necessarily free of disease, if this has not been taken into account in the study, and are therefore reported as "healthy" in the absence of more specific details. Both sets are grouped in continent-specific subsets (AF: Africa, AM: America, AS: Asia, EU: Europe, OC: Oceania, XX: Undefined Continent).

An automatic system allowing the database to be updated is available and is applied routinely, thus ensuring a timely updated database.

Each genome is annotated with information regarding the sample reported both in primary databases and in the associated publications, as well as with data obtained by the application of specific tools implemented in the resource aimed at estimating variability and phylogenicity (Table 12.1).

Variant assignment in HmtDB is based on the comparison with RSRS. The entire sets of healthy and pathologic genomes, as well as the continent-specific subsets, all including RSRS are multialigned by applying an automatic process. Nucleotide site-specific

Table 12.1: Information associated with each HmtDB genome.

Sample information
- Population data (continent, country, ethnicity > subgroups)
- Age, sex, health state (healthy or pathologic both distinct according to continent)
- Haplogroup assigned by the authors
- Authors and institution submitters
- PubMed reference

Variants and variability
- Position and variation with respect to RSRS
- Nucleotide site-specific variability
  - Variant's allele frequencies obtained through SiteVar
  - Intrahuman and intermammalia amino-acidic variability
- Haplogroup prediction

site	nucleotide	variability	A	C	G	T	gap	others
3243	A	0.000438	0.936111	0.000000	0.000095	0.000000	0.063699	0.000095
3308	T	0.019271	0.000000	0.008536	0.000143	0.927408	0.063699	0.000214

Figure 12.5
Example of variability and allele frequencies of the 3243 and 3308 reference sites after the application of the SiteVar algorithm to the multialignment of the complete genomes stored in the last HmtDB update (July 2019).

variability is estimated starting from the multialignment of each dataset calculated by applying the algorithm SiteVar [51]: the variability score ranges between 0 and 1, the higher the score the more variable is the site. Fig. 12.5 reports an example of output of the SiteVar application where, in addition to the variability value, the frequency of each nucleotide in the specific position is reported. Moreover, intrahuman and intermammalia amino-acidic variability data estimated on the 13 coding for proteins genes of the available human genomes by applying the MitVarProt algorithm [52] are also reported. Furthermore, the higher the score (range 0–1), the less functionally constrained the site should be.

Although genomes in primary databases may report the author's assigned haplogroup, with the aim of reporting in HmtDB the up-to-date haplogroup, the haplotyping of each genome is performed through an ad hoc implemented pipeline based on the most updated Phylotree built. In the case of not completely assembled genomes, the tool suggests the equally probable haplogroups.

HmtDB can be accessed at https://www.hmtdb.uniba.it and its information can be accessed using its Query page (https://www.hmtdb.uniba.it/query) (Table 12.2).

Table 12.2: List of the HmtDB query criteria.

Query criteria	Description
HmtDB genome identifier	Select a genome whose HmtDB genome identifier is known
Reference DB identifier	Select a genome whose INSDC accession number is known
Geographical origin	Select genomes whose associated subject belongs to a specific continent/country
Haplogroup	Select genomes matching a specific macrohaplogroup as it has been assigned in the associated paper
Complete genomes/coding regions only	Select either complete genomes or genomes not inclusive of the D-loop region
SNP position	Position of the rCRS reference sequence in which the selected genomes present a mutation
Variation type	Search for genomes with specific variations (transitions, transversions, insertions, deletions)
Age and sex	Select genomes whose related subject had a specific age at sampling time and/or a given sex
DNA source	Select genomes sequenced from samples extracted from a specific tissue
Individual type	Select genomes from healthy or pathologic datasets or from a phenotype related to a specific disease
References	Select genomes related to a specific paper or to papers published from a specific author, in a specific journal of with a given PMID

The result of the query function application is a list of HmtDB genome identifiers together with the Primary database Accession number (hyperlinked vs GenBank) and the Complete reference, if the genome has been published (hyperlink to PubMed). By clicking on the HmtDB identifier the Genome card is displayed (https://www.hmtdb.uniba.it/genomeCard/20726).

The multialignment of the selected genomes may be downloaded.

The number of items of each data set is available through the *Statistics* window https://www.hmtdb.uniba.it/stats, where the world map is shown, offering a straightforward view of the number of genomes and variants annotated in the database for each continent, followed by detailed information about the genome type.

12.3.1.1 HmtDB API

HmtDB also offers an API that can be used to access data in a programmatic manner and to integrate them into analysis pipelines. Users can access the same information available through the Query web page, but instead of using a graphical interface the request is made through an http address and results are returned in a dictionary-like format, which is both human- and machine-readable. HmtDB's API was also employed by RD-Connect, a platform that connects bioinformatics databases, registries, and biobanks involved in rare disease research, to augment their data using information regarding mitochondrial disease derived from HmtDB.

12.3.2 HmtVar (Preste R, Vitale O, and Attimonelli M)

HmtVar is an online database of human mitochondrial variants [53]. These variants are primarily collected from variations found in human mitochondrial genomes, which have been completely sequenced (as opposed to those lacking the D-loop or other sections of the genome) available within HmtDB [47]. This set of observed variants is further enriched with potential variations (not detected in genomes from HmtDB), considering every possible single nucleotide variant for each site of the rCRS reference sequence regarding CDS and tRNA loci.

Each variant in HmtVar is annotated with various information, such as its variability and allele frequency as estimated in HmtDB (Fig. 12.5), the participation of the variant in specific haplogroups as reported in the last Phylotree build, pathogenicity assessment estimated according to the criteria described in Section 12.3.2.1, and additional data gathered from third-party resources such as ClinVar, MITOMAP, OMIM, and dbSNPs.

HmtVar can be accessed at https://www.hmtvar.uniba.it and its information can be searched using its Query page (https://www.hmtvar.uniba.it/query) (Table 12.3), which allows users to query the database using many search parameters, from broader criteria to narrower ones, thus producing a list of variants satisfying the user request.

When a specific variant is selected from the query results, its details are shown in a Variant Card, which gathers all the information available about that variant. These data are arranged in different tabs based on the type of information provided:

- Main Info: basic information such as location, codon position, and amino acid change due to the variation (if applicable), associated haplogroups (if any), and pathogenicity.
- Variability: nucleotide and amino acid variability and allele frequency in both healthy and diseased individuals as well as continent specific.
- Pathogenicity Predictions estimated according to the criteria described in Section 12.5 as well as derived from different predictors: MutPred [54], Panther [55], PhD-SNP [56], SNPs & GO [57], Polyphen-2 [58].
- External Resources: information from third-party resources focused on diseases and population studies: that is, dbSNP (https://www.ncbi.nlm.nih.gov/snp/) [59], OMIM, the Online Mendelian Inheritance in Man database (https://www.omim.org) [60], MITOMAP (cf. Section 12.2), and ClinVar (https://www.ncbi.nlm.nih.gov/clinvar/) [61]. ClinVar (https://www.ncbi.nlm.nih.gov/clinvar/) [61] is a database reporting relationships between human variations and phenotypes, together with the available experimental evidence supporting them. dbSNP, OMIM, and ClinVar host data regarding polymorphic (dbSNP) and clinical variants (OMIM and ClinVar) from both nuclear and mitochondrial DNA.

All the information in HmtVar related to a variant can be downloaded through the *Download Data tab.*

Table 12.3: Description of the query criteria implemented in HmtVar.

Query criteria	Description
Locus type	Query the entire DB or restrict to CDS, rRNA, tRNA, or regulatory sequences
Variation	Search for a specific mutation in different formats
Position	Search for variants occurring in one or more specific positions
Locus	Search for variants occurring in a specific mitochondrial locus
Codon	Search for all variants or for variants in a specific codon position
Amino acid change	Search for synonymous, stop-gain, or nonsynonymous variants
Disease score	Search variants with a given disease score
tRNA model	Search for variants belonging to a specific tRNA model
Pathogenicity	Search for variants predicted with a certain pathogenicity prediction
Nt variability	Search for variants with a defined healthy or patient nucleotide variability
Aa variability	Search for variants with a defined healthy or patient amino acid variability

Applications wanting to exploit data from HmtVar can take advantage of its API, which allows access to information programmatically and in a standardized format for further analysis.

HmtVar aims at building a unique aggregated database specialized in human mitochondrial variants, to fulfill researchers and clinicians' needs when looking for pathogenicity information related to human mtDNA dysfunctions.

12.3.2.1 HmtVar variants pathogenicity assessment (Attimonelli M and Vitale O)

The knowledge about prioritization has grown over time and in the literature, there are different tools and methods for the interpretation of novel and rare variants related to mitochondrial loci [62]. Variants with an importance for clinical care are divided into two main groups: those affecting mitochondrial protein synthesis, and those mapping in protein-coding genes [12] where their estimation of pathogenicity is principally based on the application of prediction algorithms and scores of morbidity [9,62] (Sections 12.2.2.1 and 12.2.2.5). In the literature, the general approach applied for pathogenicity assessment of protein-coding genes is based on the stand-alone usage of pathogenicity predictors or on the combination of these algorithms with nucleotide site-specific variability [63] or allele frequencies [53]. The main computational predictors, such as MutPred [64], Polyphen-2 [58], SNP&GO, PhD-SNP [57], and PANTHER [65], assess pathogenicity by considering several features reflecting protein structural effects and dynamics, biochemical properties, and evolutionary conservation of sites assessed by analyzing the multialignments [66]. As a result, these algorithms release as output a quantitative score and/or a qualitative mark. Considering that these predictors are based on mitochondrial datasets that are no longer updated and moreover that do not take into account the real variability of mitochondrial variants, the stand-alone use of each predictor is prone to errors, so the best practice proposed is to use a consensus of predictors [66]. With respect to the evaluation of the effects of variants on mitochondrial protein synthesis, it will be necessary to distinguish between tRNA and rRNA mitochondrial variants. The mitochondrial tRNA point mutations are known to be more deleterious than the protein-coding ones because of their involvement in transcriptional and translational pathways. Considering this, the pathogenicity assessment of mitochondrial variants for these loci can be made by considering both canonical predictors of pathogenicity integrated into specific tools, for instance MitoTIP [46] or Pon-tRNA (Niroula et al., 2016), and pathogenicity scoring systems [34,67]. Even though tRNA variant tools take into account features such as conservation, structure, and biochemical effect of variants, the pathogenicity scoring systems are based on additional criteria. These criteria include functional studies such as molecular genetics assays, biochemical and histochemical testing, transmitochondrial hybrids, and single-fiber cell studies [34,49]. In line with this concept, in Ref. [68], the pathogenicity of nonsynonymous variants is evaluated with a robust statistical approach that combines the information derived from a consensus of pathogenicity predictors, known as the disease score, with the allele frequency values up to date in relation with the HmtDB and HmtVar updating [47,53]. This assessment consists of an improvement of the method described in Santorsola et al. [63], where the site-specific nucleotide variability was combined with the disease score values instead of the allele frequencies that are more specific in evaluating the effect of a given variant [53]. Therefore the disease score values consist of the weighted mean of probabilities that an amino acid substitution may affect gene or protein function and take into account the prediction of the six above-mentioned algorithms. The evaluation of the disease score, made on a training dataset of 15,385 complete genomes from healthy individuals, led to the derivation of a disease score threshold (DS_T), fixed at 0.43, that allows the potential deleterious mitochondrial variants to be distinguished from neutral polymorphisms. In addition, a statistical approach, based on empirical cumulative distribution of allele frequency values belonging to variants with a disease score that exceeds the DS_T, leads to the determination of the allele frequency threshold, fixed at 0.003264 (AF_T). As reported in Table 12.4, starting from these two thresholds, the tiers of pathogenicity have been assessed for protein-coding variants, in order to assign a clinical significance and classify them in a specific tier. A similar statistical approach has been applied to derive both thresholds and tiers regarding tRNA variants, where in this case, the disease score values are estimated by taking into account the criteria of pathogenicity reported in Ref. [34].

12.3.3 HmtPhenome (Preste R and Attimonelli M)

HmtPhenome [69] is a web resource that integrates a knowledge network involving genes and related variants located both on the nuclear and on the mitochondrial genome, together with phenotypes and diseases having any level of involvement in the mitochondrial function. This can be extremely useful to fully understand the pathological mechanisms of mitochondrial syndromes and diseases; users can start their query from a variant position, gene,

Table 12.4: Tiers of pathogenicity of nonsynonymous and tRNA mtDNA variants.

Tier	Disease score range	Allele frequency range
General rules		
Polymorphic	DS < DS_T	AF > AF_T
Likely polymorphic	DS < DS_T	AF ≤ AF_T
Likely pathogenic	DS ≥ DS_T	AF > AF_T
Pathogenic	DS ≥ DS_T	AF ≤ AF_T
Nonsynonymous variants		
Polymorphic	DS < 0.43	AF > 0.003264
Likely polymorphic	DS < 0.43	AF ≤ 0.003264
Likely pathogenic	DS ≥ 0.43	AF > 0.003264
Pathogenic	DS ≥ 0.43	AF ≤ 0.003264
tRNA variants		
Polymorphic	DS < 0.35	AF > 0.005020
Likely polymorphic	DS < 0.35	AF ≤ 0.005020
Likely pathogenic	DS ≥ 0.35	AF > 0.005020
Pathogenic	DS ≥ 0.35	AF ≤ 0.005020

phenotype or disease, and retrieve all the related information through an integrated network of these biological entities.

Due to the high number of resources involved and the great amount of data considered in each query, all the necessary information is collected from external resources at the time of the query, and only a small amount of data is actually stored in the ad hoc designed local database. Third-party resources exploited to gather these data include the Human Phenotype Ontology (HPO, https://hpo.jax.org/app/), the Experimental Factor Ontology (EFO, https://www.ebi.ac.uk/efo/) [70], Ensembl (https://www.ensembl.org/), BioMart (https://www.ensembl.org/biomart/martview) [71], OMIM (https://www.omim.org) [60], and Orphanet (https://www.orpha.net/consor/cgi-bin/index.php) [72]. While in other canonical resources, such as HmtVar, these data are periodically dumped and stored in the database with each update, in HmtPhenome this information is retrieved at query time using external APIs, ensuring that the provided information is always up to date. Not all external resources, however, provide a suitable API; in such cases, an automated protocol was set up to obtain the required data and store them in a local database.

HmtPhenome can be accessed at https://www.hmtphenome.uniba.it. An example of the output is reported in Fig. 12.6. In addition to this Network view, resulting data can be further explored through hyperlinks to third-party services reported in the Table view; these include dbSNP (https://www.ncbi.nlm.nih.gov/snp) [59] and HmtVar (Roberto [53]) for variants, Ensembl [73] for genes, UMLS (https://www.nlm.nih.gov/research/umls/index.html) [74] for diseases, and HPO [75] for phenotypes. Query results can also be exported as tabular files, so that they can be further investigated and analyzed using downstream software.

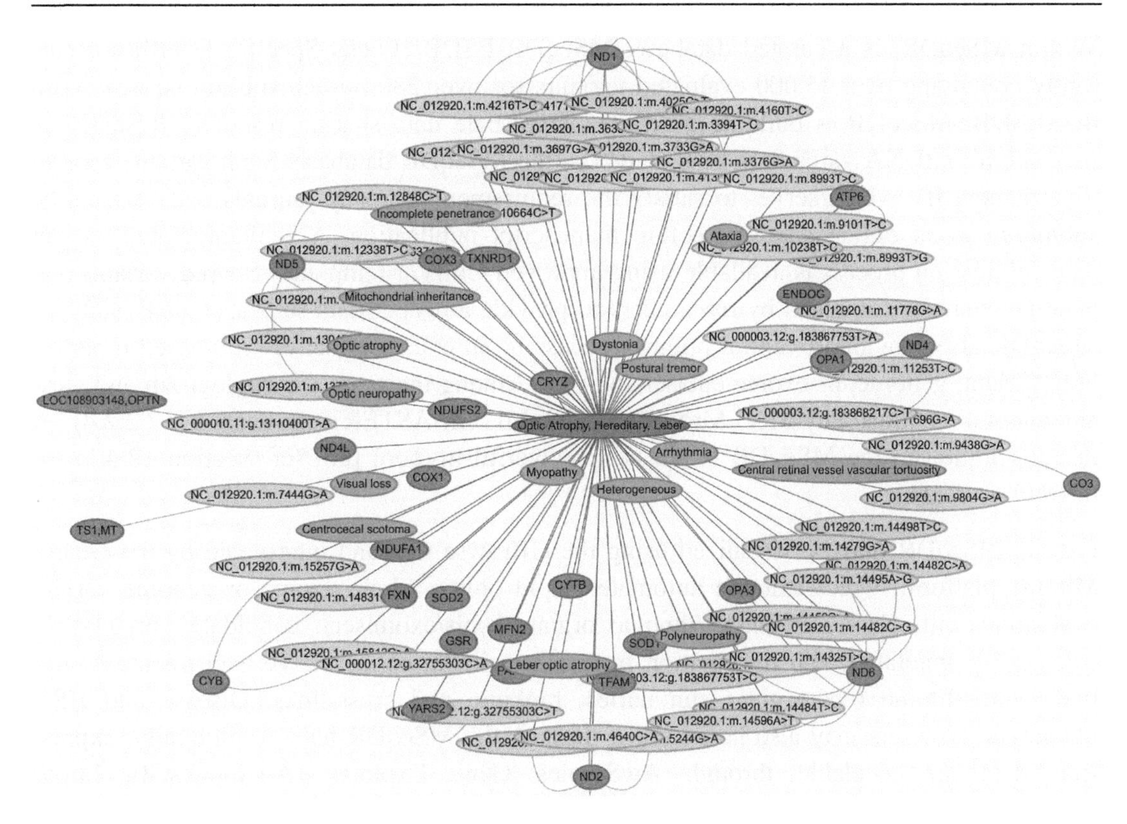

Figure 12.6
Network returned after a query by disease name "Leber Optic Atrophy" OMIM: 535000.

12.4 MSeqDR—Mitochondrial Disease Sequence Data Resource (MSeqDR) Consortium (Lott M)

MSeqDR (https://MSeqDR.org) is a collaborative data hub focused on mitochondrial disease. Its extensive data resources and bioinformatics tools support mtDNA variant annotation and mining, as well as phenotype-based exome data analysis tuned for mitochondrial diseases [76,77]. The MSeqDR portal aggregates input from expert-curated databases such as MITOMAP (https://www.mitomap.org), HmtDB (http://www.hmtdb.uniba.it), ClinVar (https://www.ncbi.nlm.nih.gov/clinvar/), ClinGen (https://clinicalgenome.org), LeighMap (https://www.vmh.life/#leighmap), Genesis [78], and PhenoTips (https://phenotips.org). MSeqDR's reference dataset includes approximately 90,000 mitochondrial genomes from MITOMAP, GeneDx (https://www.genedx.com), peer-reviewed publications, and sequencing projects such as 1000 Genomes [79]. As part of its collaborative efforts, MSeqDR also includes "pseudo" case registries of Leigh syndrome and Leigh-like syndrome (LS/LLS), and LS.

Within MSeqDR, is a curated database, MSeqDR-LSDB (Locus Specific DataBase), currently containing over 15,000 evaluated variants for over 280 mitochondrial diseases. With this LSDB, MSeqDR is building a secure, large-scale data sharing resource. Researchers may submit mtDNA variants to be included in the MSeqDR database. Such variants receive a permanent ID, which serves to satisfy the requirement of many journals that variants be submitted to an online database before manuscript publication. A semiautomated variant data submission process is available using approved ClinVar templates. Shared variants and genomes may be combined by researchers to provide a deeper data set for clinical analysis. MSeqDR also provides a central point of access to an extensive set of web-based tools for user variant, genome, or exome dataset analysis. Among these are the haplogroup and variant annotation tools Phy-Mer, MtoolBox, and MITOMASTER (cf. Section 12.2.2.4). A recent addition to the MSeqDR is the new, powerful mvTool [80] (cf. Section 12.4.1) for mtDNA variants.

Data in MSeqDR may be visualized using the GBrowse tool and interpreted by the Quick-Mitome platform. Quick-Mitome automates rapid online whole exome or genome variant evaluation with Exomiser (https://hpo.jax.org/app/tools/exomiser); candidate variants are extensively annotated with information from all available collaborative data tracks. A unified QuickMitome report gives summaries, ranking, and cross-links. Disease and HPO Phenotype browsers may also be accessed through the MSeqDR hub. Additionally, expression data are available through AwSomics Gene Explorer (AwSomicsGE, http://52.90.192.24:3838). A new platform, MSeq-OpenCGA (http://mseq.org), is currently under development by MSeqDR for large-scale interactive variant warehousing, analysis, and prioritization.

12.4.1 MvTool

Shen et al. [80]—https://mseqdr.org/mvtool.php. The MSeqDR mvTool accepts mtDNA variants in a wide range of possible formats and then, through the application of a specific converter, transforms them into the desired output format, such as HGVS, Ensembl and Mutalyzer nomenclatures. mvTool can also be used to annotate one or more variants with the support of the MSeqDR infrastructure and data from external resources like MitoMap, HmtDB, GeneDx (https://www.genedx.com) and the 1000 Genomes Project; annotations include pathogenicity data, population frequencies, and haplogroup assignment. mvTool can be used through either a web interface or an API. The web interface allows users to submit a list of variants and obtain annotations as an HTML table, which can also be downloaded as an Excel-formatted file; using the API, it is possible to submit variants in classical and HGVS format as well as from a VCF file, in order to retrieve results in JSON format or as an annotated VCF file.

12.5 Other specialized human mitochondrial databases (Attimonelli M and Preste R)

Various databases storing data about human mtDNA have been published worldwide. Some are no longer updated while others report specific and nonredundant information complementing the databases described above. A short description with hyperlinks to access them and references is reported in Table 12.5.

12.6 Tools for variant annotations (Attimonelli M and Preste R)

As already reported above, the main goal of sequencing human mtDNA is to have an overview of its variability both in health and disease. To this aim, once mtDNA has been sequenced, the next steps are to detect the variants through the application of specific tools and thereafter to annotate them. To achieve this goal, it is important to understand how a variant may affect the functionality of a gene and therefore the related metabolic processes and how much a variant contributes to mtDNA haplotyping.

The objective of sequencing the human mtDNA is to detect sample-specific variants with the aim of characterizing their role in health as well as in disease. Hence, the process regarding variant annotations is the main core of the human mtDNA studies. To annotate mtDNA implies the addition of information to the variant regarding the functional role of the locus where the variant is mapped, its frequency in various populations, its inter- and intraspecies conservation, as well as its pathogenicity and disease-associated scores.

Table 12.5: List of minor but focused human mtDNA databases.

DB name	DB description	DB hyperlink	References
AmtDB—the ancient mtDNA database	*Published mtDNA from ancient DNAs (aDNA)	https://amtdb.org/	[81]
MitoBreak—the mtDNA rearrangements database	*List of breakpoints from: • circular deleted mtDNAs (deletions) • circular partially duplicated mtDNAs (duplications) • linear mtDNAs	http://mitobreak.portugene.com/cgi-bin/Mitobreak_home.cgi	[82]
MitoAge	*Extensive collection of animalia mtDNA integrated with longevity records. Basic tools for comparative analysis of mtDNA, focused on animal longevity	http://www.mitoage.info/	[83]
Mamit-tRNA	Compilation of mammalian mitochondrial tRNA genes reporting structural features helpful in the frame of human diseases linked to point mutations	http://mamit-trna.u-strasbg.fr/	[84]

Notes: The asterisk (*) marks up-to-date status.

Annotation may be performed by applying one of the various available tools. The most efficient widely used and updated tools are reported below.

12.6.1 HmtNote

HmtNote software annotates human mitochondrial variants by exploiting data available on the HmtVar database [53]. Starting from an input VCF file with human mitochondrial variants, which can be produced from any variant calling software, each variant is searched for on HmtVar to retrieve all the available data. These annotations are distinguished into four groups, based on the type of information they provide:

- basic data about a variant (locus, amino acid change, HmtVar pathogenicity assessment, and disease score);
- cross-reference information derived from ClinVar ID [61], dbSNP ID [59], OMIM ID [60], associated diseases according to MITOMAP [85];
- variability and allele frequency data (nucleotide and amino acid variability and allele frequency in both healthy and diseased individuals);
- pathogenicity predictions from external resources (cf. Section 12.3.2).

Users can decide to use all these annotations or choose one or more specific annotation groups if preferred; HmtNote also allows the annotation database to be downloaded locally to perform offline annotation. The annotated VCF file can also be converted into a CSV format for better visual inspection of the data.

12.6.2 HaploGrep

HaploGrep [45] is a web application (http://haplogrep.uibk.ac.at/index.html) based on Phylotree, that haplotypes entire or partial mtDNAs. For every input sample, the top 10 results and the phylogenetic position of the corresponding haplogroup are displayed, thus providing a detailed explanation for how and why a haplogroup was ranked best. HaploGrep generates an interactive data visualization of the results and provides recommendations as to which polymorphisms should be analyzed additionally to obtain a more accurate result. HaploGrep is free and can be used without any login. It is worth mentioning that the last Phylotree update is at September 2016. However, considering the large number of genomes already haplotyped and considered within Phylotree, the accuracy of the results obtained through Haplogrep as well as through other tools based on Phylotree is sufficient to evaluate the mtDNA sequence quality. Haplogrep may also be locally applied by downloading it through GitHub.

12.6.3 MitImpact3D

MitImpact (http://mitimpact.css-mendel.it/description) is a collection of precomputed pathogenicity predictions for all possible nucleotide changes that cause nonsynonymous substitution in human mitochondrial protein-coding genes. The new version 3 greatly updates the tool published in Castellana et al. [86] (PMID: 25516408) with new predictors and information regarding the protein. Details regarding the he list of predictors and databases are available at http://mitimpact.css-mendel.it/description#annotations.

12.6.4 PON-mt-tRNA [87]

PON-mt-tRNA (http://structure.bmc.lu.se/PON-mt-tRNA/) is a posterior probability-based method for classification of mitochondrial tRNA variations. It integrates machine learning-based probability of pathogenicity and evidence-based likelihood of pathogenicity to predict the posterior probability of pathogenicity. In the absence of evidence, it classifies the variations based on the machine learning-based probability of pathogenicity. It has been trained and tested on variants classified as definitely pathogenic and definitely neutral according to Yarham et al. [34].

12.7 Nuclear encoded mitochondrial genes databases (Vitale O and Attimonelli M)

Mitochondria are involved in an increasing number of cellular conditions; hence, mitochondrial proteins are interest for clinicians and researchers because of their implications [88]. Proteins in mitochondria are produced by both nuclear and mitochondrial genomes where the low number of mitochondrial protein-coding genes is supplied by a large number of nuclear proteins that are imported into this organelle [89]. Therefore this interconnection between nucleus and mitochondria implies an intense cross-talk between the two eukaryotic genomes, thus generating the mitochondrial interactome. The conformation and interactions of this network are not fully understood even if knowledge about these crucial aspects could elucidate the role and involvement of mitochondria in health and disease. Data about the mitochondrial proteome and interactome are stored in several databases focused on the mitochondrial-related network of proteins. The main databases regarding the human mitochondrial proteome are Mitocarta [90], MitoMiner [88], and MitoProteome [91], while the interactome-related databases are principally MitoInteractome [89], PSICQUIC [92], and Mentha [93].

MitoCarta [90] is a comprehensive inventory of genes encoding for mitochondrial localized proteins, which in version MitoCarta2.0 consists of a human mitochondrial network of 1158 genes to date. These genes were derived from evidence mined from different sources such as literature, ascorbate peroxidase-based mass spectrometry, green fluorescent-tagging microscopy, and integrated by applying a Bayesian method. MitoCarta [90] is a good resource to investigate mitochondrial proteins despite the fact that it presents several limitations, such as its static nature due to the absence of a regular updating, the mitochondrial centricity that means the absence of proteins interacting with mitochondria under specific conditions and finally the knowledge about the protein only within double membrane.

MitoMiner [88] is a mitochondrial database for the storage and analysis of proteomics data integrating data and biological information. This database hosts the Integrated Mitochondrial Protein Index, also called IMPI, which consists of a collection of genes encoding for proteins with mitochondrial localization. The IMPI list integrates data from multiple kinds of resources and evidence including: Mitocarta [90], novel data about human mitochondrial proteome, homology relationships, metabolic models and pathway from KEGG [94], antibody data derived from the Human Protein Atlas [95], data from Gene Ontology [96], clinical information from OMIM [97], and experimental evidence based on green fluorescence protein tagging and mass spectrometry. All these data are combined by using a machine learning classifier, InterMine, that integrates, stores, and classifies evidence to produce the IMPI dataset consisting of 1626 human genes with a known or predicted mitochondrial localization.

MitoProteome [91] is a comprehensive mitochondrial database and annotation system of 847 human mitochondrial protein sequences derived from literature curation, mass spectrometry studies and extensive cross-linked data derived from several external resources such as Entrez, KEGG [94], OMIM [97], MINT [98], DIP [99], PFAM [100], InterPro [102], PRINTS [101], SWISS-PROT [103], PDB [104], PMD [105], and taxonomic data. In addition to proteome databases that list mitochondrial-localization genes and proteins, MitoInteractome [89], PSICQUIC [92], and Mentha [93] are resources focused on protein--protein interactions (PPIs). MitoInteractome [89], the main mitochondrial protein interactome database, contains 6549 protein sequences extracted from SWISS-PROT [102], MitoP2 [105], MitoProteome [91], and HPRD [107] and annotated through Gene Ontology [96] database. In addition, MitoInteractome [89] integrates information about protein–protein interactions (PPI), physical–chemical properties, polymorphism in genes encoding from proteins from dbSNP [59] where the effect of these SNPs is estimated through the Polyphen [58] algorithm, mitochondrial disease in which alterations of mitochondrial genes from OMIM are involved [97], metabolic pathway from KEGG [94], and mitochondrial enzymes from BRENDA [108] database. The integration of all these data is based on a computational approach based on the use of two algorithms: PSIMAP (protein structural interactome MAP) and PEIMAP (protein experimental interactome MAP). PSIMAP is a structure-based interaction prediction method based on the information from the PDB database to evaluate the Euclidean distance between protein domains. PEIMAP is an experimental-based predictor that combines interaction data available from MINT [98] and

HPDR [107], under the concept of "homologous interaction." PSICQUIC [92] is an EMBL web service able to access multiple protein interaction resources at the same time and that contains approximately 16 million of interactions derived from different providers among which 2,750,019 are binary interactions found for Homo Sapiens interactome. The advantage of this database is the standardized access to molecular interaction datasets based on IMEx policy and a controlled vocabulary [109]. On the same line, Mentha [93] is a protein–protein interaction resource that collects data from different curated primary protein–protein interaction databases linked to the IMEx Consortium. Mentha contains approximately 19,300 human protein and 340,916 related networks of interactions that are regularly updated by applying the PSICQUIC standard protocol and based on a controlled dictionary.

Research perspectives

In the future, improvements in NGS software, used to detect human mtDNA variants, will be required in order to avoid sequencing artifacts and to detect true variants also when the heteroplasmic fraction is very low. Furthermore, in the context of variant annotation, the usage of a standard nomenclature and reference sequence must be considered to avoid misunderstanding in variant interpretation. Finally, considering the importance of information about functional studies to assess pathogenicity, it will be important to increase the validation in certified laboratories by applying several techniques regarding the acquisition of empirical evidences of pathogenicity of human mitochondrial variants.

References

[1] Tuppen HAL, Blakely EL, Turnbull DM, Taylor RW. Mitochondrial DNA mutations and human disease. Biochim Biophys Acta Bioenerg 2010;1797:113–28. Available from: https://doi.org/10.1016/j.bbabio.2009.09.005.

[2] Wallace DC, et al. Sequence analysis of cDNAs for the human and bovine ATP synthase b-subunit: mitochondrial DNA genes sustain seventeen times more mutations. Curr Genet 1987;12(2):81–90.

[3] Brown WM, George M, Wilson AC. Rapid evolution of animal mitochondrial DNA. Proc Natl Acad Sci USA 1979;76(4):1967–71.

[4] Brown WM, Prager EM, Wan A, Wilson AC. Mitochondrial DNA sequences in primates: tempo and mode of evolution. J Mol Evol 1982;18(4):225–39.

[5] Neckelmann N, Li K, Wade RP, Shuster R, Wallace DC. cDNA sequence of a human skeletal muscle ADP/ATP translocator: lack of a leader peptide, divergence from a fibroblast translocator cDNA, and coevolution with mitochondrial DNA genes. Proc Natl Acad Sci USA 1987;84(21):7580–4.

[6] Giles RE, Blanc H, Cann HM, Wallace DC. Maternal inheritance of human mitochondrial DNA. Proc Natl Acad Sci USA 1980;77(11):6715–19.

[7] Sato M, Sato K. Maternal inheritance of mitochondrial DNA by diverse mechanisms to eliminate paternal mitochondrial DNA. Biochim Biophys Acta Mol Cell Res 2013;1833:1979–84. Available from: https://doi.org/10.1016/j.bbamcr.2013.03.010.

[8] Mishmar D, et al. Natural selection shaped regional mtDNA variation in humans. Proc Natl Acad Sci USA 2003;100(1):171–6.
[9] Wang J, Schmitt ES, Landsverk ML, Zhang VW, Li F-Y, Graham BH, et al. An integrated approach for classifying mitochondrial DNA variants: one clinical diagnostic laboratory's experience. Genet Med 2012;14:620–6. Available from: https://doi.org/10.1038/gim.2012.4.
[10] Wallace DC. Mitochondrial DNA variation in human radiation and disease. Cell 2015;163:33–8. Available from: https://doi.org/10.1016/j.cell.2015.08.067.
[11] Ruiz-Pesini E, Wallace DC. Evidence for adaptive selection acting on the tRNA and rRNA genes of the human mitochondrial DNA. Hum Mutat 2006;27(11):1072–81.
[12] DiMauro S, Schon EA. Mitochondrial DNA mutations in human disease. Am J Med Genet 2001;106:18–26. Available from: https://doi.org/10.1002/ajmg.1392.
[13] Chinnery PF, Hudson G. Mitochondrial genetics. Br Med Bull 2013;106:135–59. Available from: https://doi.org/10.1093/bmb/ldt017.
[14] Wallace DC, Ruiz-Pesini E, Mishmar D. mtDNA variation, climatic adaptation, degenerative diseases, and longevity. Cold Spring Harb Symp Quant Biol 2003;68:471–8. Available from: https://doi.org/10.1101/sqb.2003.68.471.
[15] Gorman GS, et al. Prevalence of nuclear and mitochondrial DNA mutations related to adult mitochondrial disease. Ann Neurol 2015;77(5):753–9.
[16] Lightowlers RN, Taylor RW, Turnbull DM. Mutations causing mitochondrial disease: what is new and what challenges remain? Science 2015;349(6255):1494–9.
[17] Wallace DC. Mitochondrial genetic medicine. Nat Genet 2018;50(12):1642–9.
[18] Anderson S, Bankier AT, Barrell BG, de Bruijn MH, Coulson AR, Drouin J, et al. Sequence and organization of the human mitochondrial genome Nature 1981;290(5806):457–65Apr 9; PubMed PMID. Available from: 7219534.
[19] Andrews RM, Kubacka I, Chinnery PF, Lightowlers RN, Turnbull DM, Howell N. Reanalysis and revision of the Cambridge reference sequence for human mitochondrial DNA Nat Genet 1999;23 (2):147PubMed PMID. Available from: 10508508.
[20] Behar DM, van Oven M, Rosset S, Metspalu M, Loogväli EL, Silva NM, et al. "Copernican" reassessment of the human mitochondrial DNA tree from its root Am J Hum Genet 2012;90(4):675–84. Available from: https://doi.org/10.1016/j.ajhg.2012.03.002Erratum in: Am J Hum Genet. 2012 May 4;90(5):936. PubMed PMID. Available from: 22482806 PubMed Central PMCID: PMC3322232.
[21] Bandelt HJ, Kloss-Brandstätter A, Richards MB, Yao YG, Logan I. The case for the continuing use of the revised Cambridge reference sequence (rCRS) and the standardization of notation in human mitochondrial DNA studies J Hum Genet 2014;59(2):66–77. Available from: https://doi.org/10.1038/jhg.2013.120Epub 2013 Dec 5. Review. PubMed PMID. Available from: 24304692.
[22] Wallace DC. The mitochondrial genome in human adaptive radiation and disease: on the road to therapeutics and performance enhancement. Gene 2005;354:169–80.
[23] Pakendorf B, Stoneking M. Mitochondrial DNA and human evolution. Annu Rev Genomics Hum Genet 2005;6:165–83. Available from: https://doi.org/10.1146/annurev.genom.6.080604.162249.
[24] Parson W, Brandstätter A, Alonso A, Brandt N, Brinkmann B, Carracedo A, et al. The EDNAP mitochondrial DNA population database (EMPOP) collaborative exercises: organisation, results and perspectives. Forensic Sci Int 2004;139:215–26. Available from: https://doi.org/10.1016/j.forsciint.2003.11.008.
[25] Salas A, Carracedo A, Macaulay V, Richards M, Bandelt H-J. A practical guide to mitochondrial DNA error prevention in clinical, forensic, and population genetics. Biochem Biophys Res Commun 2005;335:891–9. Available from: https://doi.org/10.1016/j.bbrc.2005.07.161.
[26] Chinnery PF, Gomez-Duran A. Oldies but goldies mtDNA population variants and neurodegenerative diseases. Front Neurosci 2018;12:682. Available from: https://doi.org/10.3389/fnins.2018.00682.
[27] Sayers EW, Cavanaugh M, Clark K, Ostell J, Pruitt KD, Karsch-Mizrachi I. GenBank. Nucleic Acids Res 2019;47(D1):D94–9. Available from: https://doi.org/10.1093/nar/gky989.

[28] Harrison PW, Alako B, Amid C, Cerdeño-Tárraga A, Cleland I, Holt S, et al. The European nucleotide archive in 2018. Nucleic Acids Res 2019;47(D1):D84–8. Available from: https://doi.org/10.1093/nar/gky1078.

[29] Kodama Y, Mashima J, Kosuge T, Ogasawara O. DDBJ update: the Genomic Expression Archive (GEA) for functional genomics data. Nucleic Acids Res 2019;47(D1):D69–73. Available from: https://doi.org/10.1093/nar/gky1002.

[30] Pearson WR, Lipman DJ. Improved tools for biological sequence comparison. Proc Natl Acad Sci U S A 1988;85(8):2444–8.

[31] van Oven M, Kayser M. Updated comprehensive phylogenetic tree of global human mitochondrial DNA variation. Hum Mutat 2009;30(2):E386–94. Available from: https://doi.org/10.1002/humu.20921. Available from: http://www.phylotree.org.

[32] Kogelnik AM, Lott MT, Brown MD, Navathe SB, Wallace DC. MITOMAP: a human mitochondrial genome database. Nucleic Acids Res 1996;24(1):177–9.

[33] Wong LJ. Pathogenic mitochondrial DNA mutations in protein-coding genes. Muscle Nerve 2007;36 (3):279–93.

[34] Yarham JW, Al-Dosary M, Blakely EL, Alston CL, Taylor RW, Elson JL, et al. A comparative analysis approach to determining the pathogenicity of mitochondrial tRNA mutations. Hum Mutat 2011;32:1319–25. Available from: https://doi.org/10.1002/humu.21575.

[35] González-Vioque E, Bornstein B, Gallardo ME, Fernández-Moreno MÁ, Garesse R. The pathogenicity scoring system for mitochondrial tRNA mutations revisited. Mol Genet Genomic Med 2014;2:107–14. Available from: https://doi.org/10.1002/mgg3.47.

[36] Mitchell AL, Elson JL, Howell N, Taylor RW, Turnbull DM. Sequence variation in mitochondrial complex I genes: mutation or polymorphism? J Med Genet 2006;43(2):175–9.

[37] Aikhionbare FO, Khan M, Carey D, Okoli J, Go R. Is cumulative frequency of mitochondrial DNA variants a biomarker for colorectal tumor progression? Mol Cancer 2004;3(30):7.

[38] Houshmand M, et al. Is 8860 variation a rare polymorphism or associated as a secondary effect in HCM disease? Arch Med Sci 2011;7(2):242–6.

[39] Koh H, et al. Mitochondrial mutations in cholestatic liver disease with biliary atresia. Sci Rep 2018;8(1):905.

[40] Roshan M, et al. Analysis of mitochondrial DNA variations in Indian patients with congenital cataract. Mol Vis 2012;18:181–93.

[41] Chalkia D, et al. Mitochondrial DNA associations with East Asian metabolic syndrome. Biochim Biophys Acta 2018;1859(9):878–92.

[42] Ji F, et al. Mitochondrial DNA variant associated with Leber hereditary optic neuropathy and high-altitude Tibetans. Proc Natl Acad Sci USA 2012;109(19):7391–6.

[43] Kang L, et al. MtDNA analysis reveals enriched pathogenic mutations in Tibetan highlanders. Sci Rep 2016;6:31083.

[44] van Oven M. PhyloTree Build 17: growing the human mitochondrial DNA tree. Forensic Sci Int: Genet Suppl Ser 2015;5:e392–4.

[45] Weissensteiner H, Pacher D, Kloss-Brandstätter A, Forer L, Specht G, Bandelt H-J, et al. HaploGrep 2: mitochondrial haplogroup classification in the era of high-throughput sequencing. Nucleic Acids Res 2016;. Available from: https://doi.org/10.1093/nar/gkw233 2016 Apr 15.

[46] Sonney S, Leipzig J, Lott MT, Zhang S, Procaccio V, Wallace DC, et al. Predicting the pathogenicity of novel variants in mitochondrial tRNA with MitoTIP. PLoS Comput Biol 2017;13. Available from: https://doi.org/10.1371/journal.pcbi.1005867.

[47] Clima R, Preste R, Calabrese C, Diroma MA, Santorsola M, Scioscia G, et al. HmtDB 2016: data update, a better performing query system and human mitochondrial DNA haplogroup predictor. Nucleic Acids Res 2017;45:D698–706. Available from: https://doi.org/10.1093/nar/gkw1066.

[48] Attimonelli M, Accetturo M, Santamaria M, Lascaro D, Scioscia G, Pappadà G, et al. HmtDB, a human mitochondrial genomic resource based on variability studies supporting population genetics and biomedical research. BMC Bioinform 2005;6:S4. Available from: https://doi.org/10.1186/1471-2105-6-S4-S4.

[49] Diroma MA, Lubisco P, Attimonelli M. A comprehensive collection of annotations to interpret sequence variation in human mitochondrial transfer RNAs. BMC Bioinform 2016;17:338. Available from: https://doi.org/10.1186/s12859-016-1193-4.
[50] Consortium, T. 1000 G.P. An integrated map of genetic variation from 1,092 human genomes. Nature 2012;491:56–65.
[51] Pesole G, Saccone C. A novel method for estimating substitution rate variation among sites in a large dataset of homologous DNA sequences. Genetics 2001;157(2):859–65.
[52] Horner DS, Pesole G. The estimation of relative site variability among aligned homologous protein sequences. Bioinformatics 2003;19:600–6. Available from: https://doi.org/10.1093/bioinformatics/btg063.
[53] Preste R, Clima R, Attimonelli M, 2019. Human mitochondrial variant annotation with HmtNote. bioRxiv 600619. 10.1101/600619
[54] Pejaver V, Urresti J, Lugo-Martinez J, Pagel KA, Lin GN, Nam H-J, et al., 2017. MutPred2: inferring the molecular and phenotypic impact of amino acid variants. bioRxiv 134981. 10.1101/134981
[55] Mi H, Muruganujan A, Thomas PD. PANTHER in 2013: modeling the evolution of gene function, and other gene attributes, in the context of phylogenetic trees. Nucleic Acids Res 2013;41:D377–86. Available from: https://doi.org/10.1093/nar/gks1118.
[56] Capriotti E, Calabrese R, Casadio R. Predicting the insurgence of human genetic diseases associated to single point protein mutations with support vector machines and evolutionary information. Bioinformatics 2006;22:2729–34. Available from: https://doi.org/10.1093/bioinformatics/btl423.
[57] Capriotti E, Calabrese R, Fariselli P, Martelli PL, Altman RB, Casadio R. WS-SNPs&GO: a web server for predicting the deleterious effect of human protein variants using functional annotation. BMC Genom 2013;14:S6. Available from: https://doi.org/10.1186/1471-2164-14-S3-S6.
[58] Adzhubei I, Jordan DM, Sunyaev SR. 2013. Predicting functional effect of human missense mutations using PolyPhen-2. Curr Protoc Hum Genet 0 7, Unit7.20. 10.1002/0471142905.hg0720s76
[59] Sherry ST, Ward M-H, Kholodov M, Baker J, Phan L, Smigielski EM, et al. dbSNP: the NCBI database of genetic variation. Nucleic Acids Res 2001;29:308–11.
[60] Amberger JS, Bocchini CA, Schiettecatte F, Scott AF, Hamosh A. OMIM.org: Online Mendelian Inheritance in Man (OMIM®), an online catalog of human genes and genetic disorders. Nucleic Acids Res 2015;43:D789–98. Available from: https://doi.org/10.1093/nar/gku1205.
[61] Landrum MJ, Lee JM, Benson M, Brown G, Chao C, Chitipiralla S, et al. ClinVar: public archive of interpretations of clinically relevant variants. Nucleic Acids Res 2016;44:D862–8. Available from: https://doi.org/10.1093/nar/gkv1222.
[62] Bris C, Goudenege D, Desquiret-Dumas V, Charif M, Colin E, Bonneau D, et al. Bioinformatics tools and databases to assess the pathogenicity of mitochondrial DNA variants in the field of next generation sequencing. Front Genet 2018;9. Available from: https://doi.org/10.3389/fgene.2018.00632.
[63] Santorsola M, Calabrese C, Girolimetti G, Diroma MA, Gasparre G, Attimonelli M. A multi-parametric workflow for the prioritization of mitochondrial DNA variants of clinical interest. Hum Genet 2016;135:121–36. Available from: https://doi.org/10.1007/s00439-015-1615-9.
[64] Li B, Krishnan VG, Mort ME, Xin F, Kamati KK, Cooper DN, et al. Automated inference of molecular mechanisms of disease from amino acid substitutions. Bioinformatics 2009;25:2744–50. Available from: https://doi.org/10.1093/bioinformatics/btp528.
[65] Mi H, Huang X, Muruganujan A, Tang H, Mills C, Kang D, et al. PANTHER version 11: expanded annotation data from Gene Ontology and Reactome pathways, and data analysis tool enhancements. Nucleic Acids Res 2017;45:D183–9. Available from: https://doi.org/10.1093/nar/gkw1138.
[66] Ohanian M, Otway R, Fatkin D. Heuristic methods for finding pathogenic variants in gene coding sequences. J Am Heart Assoc 2012;1:e002642. Available from: https://doi.org/10.1161/JAHA.112.002642.
[67] McFarland R, Elson JL, Taylor RW, Howell N, Turnbull DM. Assigning pathogenicity to mitochondrial tRNA mutations: when "definitely maybe" is not good enough. Trends Genet 2004;20:591–6. Available from: https://doi.org/10.1016/j.tig.2004.09.014.

[68] Preste R, Vitale O, Clima R, Gasparre G, Attimonelli M. HmtVar: a new resource for human mitochondrial variations and pathogenicity data. Nucleic Acids Res 2019;47:D1202–10. Available from: https://doi.org/10.1093/nar/gky1024.

[69] Preste R, Attimonelli M, 2019. Integration of genomic variation and phenotypic data using HmtPhenome. bioRxiv 660282. 10.1101/660282

[70] Malone J, Holloway E, Adamusiak T, Kapushesky M, Zheng J, Kolesnikov N, et al. Modeling sample variables with an Experimental Factor Ontology. Bioinformatics 2010;26:1112–18. Available from: https://doi.org/10.1093/bioinformatics/btq099.

[71] Kinsella RJ, Kähäri A, Haider S, Zamora J, Proctor G, Spudich G, et al. Ensembl BioMarts: a hub for data retrieval across taxonomic space. Database 2011;2011. Available from: https://doi.org/10.1093/database/bar030 bar030.

[72] Weinreich SS, Mangon R, Sikkens JJ, Teeuw MEEN, Cornel MC. Orphanet: a European database for rare diseases. Ned Tijdschr Geneeskd 2008;152:518–19.

[73] Zerbino DR, Achuthan P, Akanni W, Amode MR, Barrell D, Bhai J, et al. Ensembl 2018. Nucleic Acids Res 2018;46:D754–61. Available from: https://doi.org/10.1093/nar/gkx1098.

[74] Bodenreider O. The Unified Medical Language System (UMLS): integrating biomedical terminology. Nucleic Acids Res 2004;32:D267–70. Available from: https://doi.org/10.1093/nar/gkh061.

[75] Köhler S, Carmody L, Vasilevsky N, Jacobsen JOB, Danis D, Gourdine J-P, et al. Expansion of the Human Phenotype Ontology (HPO) knowledge base and resources. Nucleic Acids Res 2019;47: D1018–27. Available from: https://doi.org/10.1093/nar/gky1105.

[76] Shen L, et al. MSeqDR: a centralized knowledge repository and bioinformatics web resource to facilitate genomic investigations in mitochondrial disease. Hum Mutat (Online) 2016;37(6):540–8.

[77] Falk MJ, et al. Mitochondrial Disease Sequence Data Resource (MSeqDR): a global grass-roots consortium to facilitate deposition, curation, annotation, and integrated analysis of genomic data for the mitochondrial disease clinical and research communities. Mol Genet Metab 2015;114 (3):388–96.

[78] Gonzalez M, et al. Innovative genomic collaboration using the GENESIS (GEM.app) platform. Hum Mutat 2015;36(10):950–6.

[79] 1000 Genomes Project C, et al. A global reference for human genetic variation. Nature 2015;526 (7571):68–74.

[80] Shen L, et al. MSeqDR mvTool: a mitochondrial DNA web and API resource for comprehensive variant annotation, universal nomenclature collation, and reference genome conversion. Hum Mutat 2018;39 (6):806–10.

[81] Ehler E, Novotný J, Juras A, Chylenski M, Moravcík O, Paces J. AmtDB: a database of ancient human mitochondrial genomes. Nucleic Acids Res 2019;47(D1):D29–32. Available from: https://doi.org/10.1093/nar/gky843.

[82] Damas J, Carneiro J, Amorim A, Pereira F. MitoBreak: the mitochondrial DNA breakpoints database. Nucleic Acids Res 2014;42(D1):D1261–8. Available from: https://doi.org/10.1093/nar/gkt982.

[83] Toren D, Barzilay T, Tacutu R, Lehmann G, Muradian KK, Fraifeld VE. MitoAge: a database for comparative analysis of mitochondrial DNA, with a special focus on animal longevity. Nucleic Acids Res 2016;44(D1):D1262–5.

[84] Pütz J, Dupuis B, Sissler M, Florentz C. Mamit-tRNA, a database of mammalian mitochondrial tRNA primary and secondary structures. RNA 2007;13(8):1184–90.

[85] Lott MT, Leipzig JN, Derbeneva O, Xie HM, Chalkia D, Sarmady M, et al., 2013. mtDNA variation and analysis using MITOMAP and MITOMASTER. Curr. Protoc. Bioinforma. Ed. Board Andreas Baxevanis Al 1, 1.23.1-1.23.26. 10.1002/0471250953.bi0123s44

[86] Castellana S, Rónai J, Mazza T. MitImpact: an exhaustive collection of pre-computed pathogenicity predictions of human mitochondrial non-synonymous variants. Hum Mutat 2015;36(2):E2413–22. Available from: https://doi.org/10.1002/humu.22720.

[87] Niroula A, Vihinen M. PON-mt-tRNA: a multifactorial probability-based method for classification of mitochondrial tRNA variations. Nucleic Acids Res 2016;44:2020–7. Available from: https://doi.org/10.1093/nar/gkw046.

[88] Smith AC, Robinson AJ. MitoMiner v3.1, an update on the mitochondrial proteomics database. Nucleic Acids Res 2016;44:D1258–61. Available from: https://doi.org/10.1093/nar/gkv1001.

[89] Reja R, Venkatakrishnan A, Lee J, Kim B-C, Ryu J-W, Gong S, et al. MitoInteractome: mitochondrial protein interactome database, and its application in "aging network" analysis. BMC Genom 2009;10: S20. Available from: https://doi.org/10.1186/1471-2164-10-S3-S20.

[90] Calvo SE, Clauser KR, Mootha VK. MitoCarta2.0: an updated inventory of mammalian mitochondrial proteins. Nucleic Acids Res 2016;44(D1):D1251–7. Available from: https://doi.org/10.1093/nar/gkv1003 Epub 2015 Oct 7.

[91] Cotter D. MitoProteome: mitochondrial protein sequence database and annotation system. Nucleic Acids Res 2004;32:463D–7D. Available from: https://doi.org/10.1093/nar/gkh048.

[92] Aranda B, Blankenburg H, Kerrien S, Brinkman FSL, Ceol A, Chautard E, et al. PSICQUIC and PSISCORE: accessing and scoring molecular interactions. Nat Methods 2011;8:528–9. Available from: https://doi.org/10.1038/nmeth.1637.

[93] Calderone A, Castagnoli L, Cesareni G. mentha: a resource for browsing integrated protein-interaction networks. Nat Methods 2013;10:690–1. Available from: https://doi.org/10.1038/nmeth.2561.

[94] Kanehisa M, Goto S. KEGG: kyoto encyclopedia of genes and genomes. Nucleic Acids Res 2000;28:27–30.

[95] Pontén F, Jirström K, Uhlen M. The Human Protein Atlas—a tool for pathology. J Pathol 2008;216:387–93. Available from: https://doi.org/10.1002/path.2440.

[96] Ashburner M, Ball CA, Blake JA, Botstein D, Butler H, Cherry JM, et al. Gene ontology: tool for the unification of biology. Nat Genet 2000;25:25–9. Available from: https://doi.org/10.1038/75556.

[97] Hamosh A, Scott AF, Amberger JS, Bocchini CA, McKusick VA. Online Mendelian Inheritance in Man (OMIM), a knowledgebase of human genes and genetic disorders. Nucleic Acids Res 2005;33:D514–17. Available from: https://doi.org/10.1093/nar/gki033.

[98] Chatr-aryamontri A, Ceol A, Palazzi LM, Nardelli G, Schneider MV, Castagnoli L, et al. MINT: the Molecular INTeraction database. Nucleic Acids Res 2007;35:D572–4. Available from: https://doi.org/10.1093/nar/gkl950.

[99] Xenarios I, Rice DW, Salwinski L, Baron MK, Marcotte EM, Eisenberg D. DIP: the database of interacting proteins. Nucleic Acids Res, 28. 2000. p. 289–91.

[100] Finn RD, Bateman A, Clements J, Coggill P, Eberhardt RY, Eddy SR, et al. Pfam: the protein families database. Nucleic Acids Res 2014;42:D222–30. Available from: https://doi.org/10.1093/nar/gkt1223.

[101] Attwood TK, Coletta A, Muirhead G, et al. The PRINTS database: a fine-grained protein sequence annotation and analysis resource-its status in 2012. Database (Oxford) 2012;2012. Available from: https://doi.org/10.1093/database/bas019. Published 2012 Apr 15.

[102] Hunter S, Apweiler R, Attwood TK, Bairoch A, Bateman A, Binns D, et al. InterPro: the integrative protein signature database. Nucleic Acids Res 2009;37:D211–15. Available from: https://doi.org/10.1093/nar/gkn785.

[103] Bairoch A, Apweiler R. The SWISS-PROT protein sequence database and its supplement TrEMBL in 2000. Nucleic Acids Res 2000;28:45–8.

[104] Parasuraman S. Protein data bank. J Pharmacol Pharmacother 2012;3:351–2. Available from: https://doi.org/10.4103/0976-500X.103704.

[105] Xu Z, Huang L, Zhang H, Li Y, Guo S, Wang N, et al. PMD: a resource for archiving and analyzing protein microarray data. Sci Rep 2016;6. Available from: https://doi.org/10.1038/srep19956.

[106] Prokisch H, Ahting U. MitoP2, an integrated database for mitochondrial proteins. Methods Mol Biol 2007;372:573–86. Available from: https://doi.org/10.1007/978-1-59745-365-3_39.

[107] Prasad AS. Zinc: role in immunity, oxidative stress and chronic inflammation. Curr Opin Clin Nutr Metab Care 2009;12:646–52. Available from: https://doi.org/10.1097/MCO.0b013e3283312956.

[108] Schomburg I, Chang A, Schomburg D. BRENDA, enzyme data and metabolic information. Nucleic Acids Res 2002;30:47–9.

[109] Orchard S, et al. Protein interaction data curation: the International Molecular Exchange (IMEx) consortium. Nat Methods 2012;9:345–50.

Further reading

Bannwarth S, Procaccio V, Lebre AS, Jardel C, Chaussenot A, Hoarau C, et al. Prevalence of rare mitochondrial DNA mutations in mitochondrial disorders. J Med Genet 2013;50:704–14. Available from: https://doi.org/10.1136/jmedgenet-2013-101604.

Barros F, Lareu MV, Salas A, Carracedo A. Rapid and enhanced detection of mitochondrial DNA variation using single-strand conformation analysis of superposed restriction enzyme fragments from polymerase chain reaction-amplified products. Electrophoresis 1997;18:52–4. Available from: https://doi.org/10.1002/elps.1150180110.

Brownlee GG, Sanger F, Barrell BG. Nucleotide sequence of 5S-ribosomal RNA from Escherichia coli Nature 1967;215(5102):735–6PubMed PMID:. Available from: 4862513.

Diroma MA, Calabrese C, Simone D, Santorsola M, Calabrese FM, Gasparre G, et al. Extraction and annotation of human mitochondrial genomes from 1000 Genomes Whole Exome Sequencing data. BMC Genom 2014;15(Suppl. 3):S2.

Keshava Prasad TS, Goel R, Kandasamy K, Keerthikumar S, Kumar S, Mathivanan S, et al. Human Protein Reference Database—2009 update. Nucleic Acids Res 2009;37:D767–72. Available from: https://doi.org/10.1093/nar/gkn892.

Maxam AM, Gilbert W. A new method for sequencing DNA Proc Natl Acad Sci USA 1977;74(2):560–4. Available from: https://doi.org/10.1073/pnas.74.2.560PubMed PMID. Available from: 265521 PubMed Central.

McCormick EM, et al. Standards and guidelines for mitochondrial DNA variant interpretation. In: United Mitochondrial Disease Foundation (UMDF) Symposium—Mitochondrial Medicine 2019 (Washington, D. C.), 2019a.

McCormick EM, et al. Specifications of the ACMG/AMP standards and guidelines for mitochondrial DNA variant interpretation. Preprint server TBD, 2019b, in preparation.

Murakami K, Sugita M. Evaluation of database annotation to determine human mitochondrial proteins. IJBBB 2018;8:210–17. Available from: https://doi.org/10.17706/ijbbb.2018.8.4.210-217.

Parson W, Dür A. EMPOP—a forensic mtDNA database. Forensic Sci Int Genet 2007;1:88–92. Available from: https://doi.org/10.1016/j.fsigen.2007.01.018.

Richards S, et al. Standards and guidelines for the interpretation of sequence variants: a joint consensus recommendation of the American College of Medical Genetics and Genomics and the Association for Molecular Pathology. Genet Med 2015;17(5):405–24.

Rossignol R, Faustin B, Rocher C, Malgat M, Mazat J-P, Letellier T. Mitochondrial threshold effects. Biochem J 2003;370:751–62. Available from: https://doi.org/10.1042/BJ20021594.

Ruiz-Pesini E, Mishmar D, Brandon M, Procaccio V, Wallace DC. Effects of purifying and adaptive selection on regional variation in human mtDNA. Science 2004;303(5655):223–6.

Smith PM, Elson JL, Greaves LC, Wortmann SB, Rodenburg RJT, Lightowlers RN, et al. The role of the mitochondrial ribosome in human disease: searching for mutations in 12S mitochondrial rRNA with high disruptive potential. Hum Mol Genet 2014;23:949–67. Available from: https://doi.org/10.1093/hmg/ddt490.

Torroni A, Huoponen K, Francalacci P, Petrozzi M, Morelli L, Scozzari R, et al. Classification of European mtDNAs from an analysis of three European populations. Genetics 1996;144:1835–50.

Torroni A, Wallace DC. Mitochondrial DNA variation in human populations and implications for detection of mitochondrial DNA mutations of pathological significance. J Bioenerg Biomembr 1994;26:261–71. Available from: https://doi.org/10.1007/bf00763098.

Wallace DC, Chalkia D. Mitochondrial DNA genetics and the heteroplasmy conundrum in evolution and disease. Cold Spring Harb Perspect Biol 2013;5:a021220. Available from: https://doi.org/10.1101/cshperspect.a021220.

Wallace DC. Bioenergetics in human evolution and disease: implications for the origins of biological complexity and the missing genetic variation of common diseases. Philos Trans R Soc Lond B Biol Sci 2013;368:20120267. Available from: https://doi.org/10.1098/rstb.2012.0267.

CHAPTER 13

Methods and models for functional studies on mtDNA mutations

Luisa Iommarini[1], Anna Ghelli[1] and Francisca Diaz[2]

[1]Department of Pharmacy and Biotechnology, University of Bologna, Bologna, Italy, [2]Department of Neurology, University of Miami, Miller School of Medicine, Miami, FL, United States

13.1 Introduction

The age of mitochondrial medicine began in the late 1980s, when mitochondrial DNA (mtDNA) deletions and point mutations were identified as the genetic cause of neuromuscular disorders, which are now referred as mitochondrial diseases [1,2]. From these early studies, the involvement of mtDNA alterations (point mutations, insertions, deletions, or copy number variations) has expanded not only to rare mitochondrial disorders but also to other human pathologies, including diabetes, cardiovascular and neurodegenerative diseases, cancer, and aging [3]. The discovery of human diseases with a mitochondrial etiology enormously prompted the research in the field. This led to the creation of several *in vitro* and *in vivo* models and to the optimization of different methods to assess the functional implications of mtDNA alterations. Early studies mainly relied on patient-derived bioptic tissues and primary cell cultures. Histological analyses combined with classic enzymology methods and molecular biology techniques allowed to demonstrate that certain mtDNA alterations were responsible for dysfunctional oxidative phosphorylation (OXPHOS) and caused the pathological phenotype. These experiments were mainly performed on skeletal muscle specimens or on mitochondria isolated from affected tissues. Patients' derived cell cultures were largely used to obtain the amount of mitochondria necessary for biochemical analyses, especially when skeletal muscle biopsies were not available or for those mitochondrial disorders in which skeletal muscle was not affected. As research proceeded, such first-generation approaches evolved into more sophisticated and precise techniques, and models were developed to mirror human pathology. Having appropriate models to study mitochondrial defects is critical to replace invasive diagnostic procedures and to reduce the amount of required biological material. Many efforts were made to clarify all the aspects that were and still remain unclear in the pathogenesis of mitochondrial disorders, to generate faithful experimental models and to identify possible therapeutic options. Transcytoplasmic hybrid

The Human Mitochondrial Genome.
DOI: https://doi.org/10.1016/B978-0-12-819656-4.00013-9

cells (cybrids) and induced pluripotent stem cells (iPSCs) from patients' fibroblasts advanced our understanding on the pathogenetic role of mtDNA mutations and the tissue specificity in mitochondrial diseases [4,5]. *In vivo* models have also been generated, but they are relatively rare since the direct manipulation of mtDNA is affected by technical hurdles [6,7]. In this chapter, we summarize the current knowledge regarding *in vitro* and *in vivo* models and techniques used to investigate the impact of mtDNA alterations to OXPHOS complexes function and structure.

13.2 Models for the study of mtDNA mutations: in vitro *models*

Low amount of material derived from muscle biopsies or peripheral blood cells from patients limits the study of the effect of mtDNA alterations on cell metabolism. Therefore it was necessary to create cell models that were highly proliferative and provided a renewable source of biological samples. A solution to this problem was to create *in vitro* models using cell lines derived from patients, cybrids, or yeast.

13.2.1 Human primary cell lines

Since the demonstration of the first pathogenic mtDNA mutation associated with human disease, cultured lymphoblasts, myoblasts, and skin fibroblasts have been utilized for the diagnosis of OXPHOS defects. Routine assays include the examination of the redox state of the cell, the enzymatic activity of the respiratory chain complexes, and ATP synthesis using either permeabilized cells or isolated mitochondria [8–11]. However, primary cells have a finite life span and decreased proliferation ability when cultured *in vitro*. After a limited number of divisions, they enter into a permanent quiescence [12]. To avoid this problem, immortalized lymphoblastoid cell lines can be established by the transformation of peripheral blood B lymphocytes with the Epstein–Barr virus [13]. Similarly, patients' myoblasts or fibroblasts can be immortalized through the introduction of viral oncogenes/oncoproteines [14–16].

13.2.2 Cybrids

Cells harboring the same pathogenic mtDNA variants derived from the same pedigree may exhibit different mitochondrial function resulting from their different nuclear DNA (nDNA). In order to eliminate the interference of nuclear background, in 1989, King and Attardi created the "cybrid model" in which a patient-derived enucleated mitochondrial donor cell line was fused with immortalized cells lacking mtDNA (Rho0 cells) [4] (Fig. 13.1A). In the original publication, Rho0 cells were obtained by exposure to ethidium bromide, a DNA intercalating compound that inhibits mtDNA replication without affecting the replication of nDNA [17]. Rho0 cells were generated from human osteosarcoma cell line 143B.TK^{-} on a

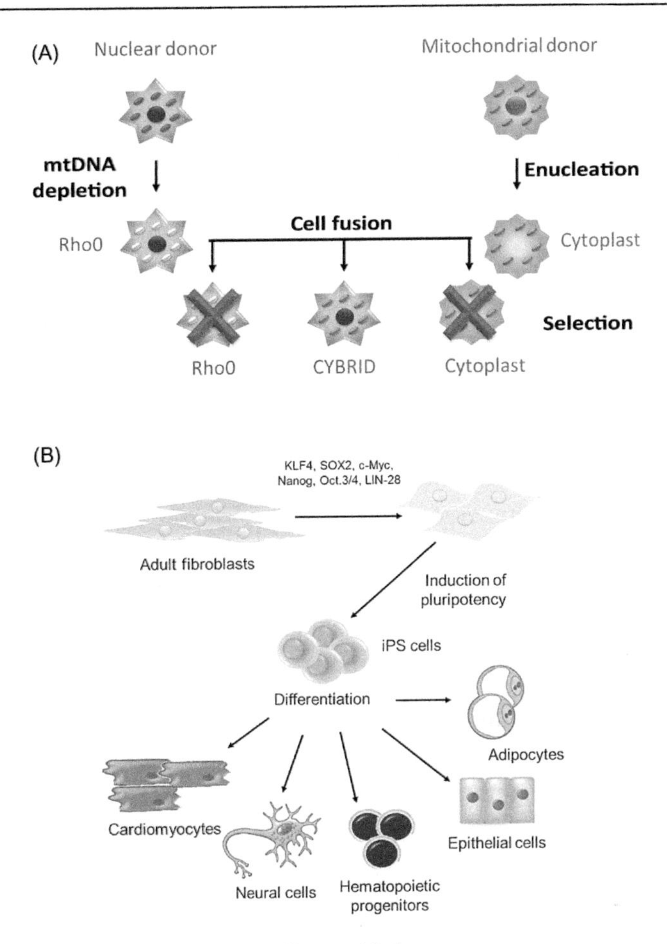

Figure 13.1

(A) Experimental design of cybrid generation. Nuclear donor cell line (eptagonal cell) is subjected to mtDNA depletion by treatment with ethidium bromide producing Rho0 cells. Mitochondrial donor cell line (decagonal cell) is enucleated as described in the main text and then subjected to cell fusion with Rho0 cells. After an appropriate selection, parental cell lines are eliminated and only cybrids (eptagonal cell with dark mitochondria) are maintained and expanded for cell culture. (B) Experimental design of induced pluripotent stem cells (iPSCs) generation and differentiation. Cells from differentiated tissues (i.e., fibroblasts) undergo introduction of specific genes encoding transcription factors that convert adult cells into iPSCs. After induction with appropriate stimuli, iPSCs can be differentiated in several types of cells using specific combinations of cytokines and growth factors.

growth medium supplemented with 50 ng/mL ethidium bromide and 50 μg/mL uridine. Under these conditions, cells were depleted of mtDNA and became auxotrophic for uridine and pyruvate after 10–15 days. The biochemical basis for this dependence on uridine resides in a crucial step of *de novo* pyrimidine synthesis pathway catalyzed by the enzyme dihydroorotate dehydrogenase. This enzyme oxidizes orotate to dihydroorotate reducing coenzyme Q, which in turn is oxidized by the mitochondrial respiratory chain. Depletion of mtDNA and subsequent respiratory chain failure disrupts such pyrimidine synthesis pathway [18]. Furthermore, cells lacking mtDNA rely on glycolysis for ATP production and need supplementation of pyruvate to push forward the formation of lactate generating NAD^+ for glycolysis [4].

The generation of Rho0 cells requires two steps: the blocking of mtDNA replication and its dilution due to cell replication. Based on this rationale, other techniques have been developed to deplete mtDNA using other mtDNA replication inhibitors [19–21] or by inactivating mtDNA polymerase γ [22]. Using the cybrid approach, mtDNA mutations can be studied under different nuclear DNA backgrounds [23], but it should be noted that concentration and exposure time of such inhibitors may vary even for cells of the same origin and that this method requires optimization [24]. To confirm the absence of mtDNA in Rho0 cells, molecular techniques (i.e., PCR or quantitative real-time PCR) or cell proliferation assays in selective medium must be applied [23]. In addition, Rho0 cells require a recessive nuclear marker such as a thymidine kinase mutation (TK^-) to select for true cybrids after repopulation of Rho0 cells with exogenous mitochondria [4,17].

Cytoplasts from mitochondrial donor cells can be obtained by treating cells with cytochalasin B that disrupts the cytoskeleton and allows mechanical enucleation by centrifugation [17]. An alternative method is the chemical enucleation with actinomycin D, which irreversibly inactivates nuclear transcription and replication in mtDNA donor cells [25]. Chomyn and colleagues demonstrated the possibility of using human blood platelets as mitochondrial donors for the repopulation of Rho0 cells. Using platelets simplifies the mitochondria-transfer procedure because platelets do not contain nuclei and the enucleation procedure is not required [26].

Among the various methods developed to fuse Rho0 cells with cytoplasts, the most frequently applied is based on polyethylene glycol 1500 (PEG 1500) [4]. Direct microinjection of mitochondria and cell electrofusion are also used [17,24]. After fusion, cells undergo a selection in a medium lacking uridine and pyruvate to eliminate Rho0 cells associated with a treatment to exclude mtDNA donor cells. The treatment with 5-bromo-2′-deoxyuridine (BrdU) is one of the most used to select cybrids deriving from TK^- cells. Thymidine kinase (TK) activity is present in mtDNA donors and the enzyme is able to phosphorylate BrdU causing its intercalation into nDNA that, in turn, inhibits nDNA replication and prevents proliferation of mitochondrial donor cells without affecting TK-defective cybrids proliferation [4,17]. Another method for selecting cybrids is the use of Rho0 cells lacking guanine-hypoxanthine phosphoribosyl transferase activity and thus resistant to treatment with 6-thioguanine or 8-azaguanine [27]. Further, if drug resistance is absent in nuclear donors, the resistance to another drug (i.e., neomycin) can be artificially introduced before cell fusion by the use of commercially available transfection vectors containing the gene for resistance [28].

The primary goal of cybrids generation was to understand the pathogenic mechanisms underlying disease-associated mtDNA mutations [23]. However, they also contributed to shed light on the threshold effect of pathogenic mutations, to study the relationship between different mitochondrial haplogroups and their susceptibility to diseases, drug response, aging, and climate adaptation [29–32]. Moreover, cybrids have also been exploited to understand the mitochondrial contribution to neurodegenerative diseases such as Parkinson's and Alzheimer's disease [23] or to investigate the role of mtDNA mutations in tumor progression, chemoresistance, and metastasis formation (described in Chapter 17: MtDNA mutations in cancer).

13.2.3 Patient-specific induced pluripotent stem cells

Despite the great contribution that the above described cell models have given to the understanding of the molecular bases of mitochondrial disorders, it remains debated if they reflect the real function and characteristics of affected tissues. Indeed, mtDNA mutations often show a high tissue specificity and these cell models may have significant differences in morphology, structure, and function compared to cells from affected tissues. New techniques developed for reprogramming differentiated cells into induced pluripotent stem cells (iPSCs) attracted great attention as new models for mitochondrial diseases for their ability to differentiate into any cell type maintaining the original nDNA and mtDNA [11,33]. A scheme of the procedure is shown in Fig. 13.1B while the standard protocol for the reprogramming of patients' cells is described elsewhere [34].

The first study regarding the use of iPSCs from a patient carrying a mtDNA mutation reported the generation and characterization of cells with heteroplasmic m.3243A > G/*MT-TL1* mutation from two diabetic patients [35]. The ability to establish iPSCs with different degrees of heteroplasmy suggests that they are useful for investigating the correlation between heteroplasmy and mitochondrial dysfunction [35]. Subsequently, Cherry and colleagues reported the generation and characterization of iPSCs carrying a heteroplasmic 2.5 kb mtDNA deletion from a patient affected by Pearson syndrome (PS) [5]. After differentiation in hematopoietic progenitor cells, iPSCs maintained mitochondrial defects recapitulating the pathological features of PS. This study demonstrated that the process of reprogramming and differentiation modifies heteroplasmy in cells and highlights the power of iPSCs as model to investigate the molecular bases of heteroplasmic shift. Other studies reported the generation and characterization of iPSCs carrying heteroplasmic mutations causing MELAS (mitochondrial encephalomyopathy, lactic acidosis, and stroke-like episodes), such as m.13513G > A/*MT-ND5* [36], m.3242A > G/*MT-TL1*, and m.5541C > T/*MT-TW* [37–41]. In general, these studies suggest that iPSCs are an excellent tool to study tissue specificity of mitochondrial diseases also in relation with the mutation load. Furthermore, it was observed that during reprogramming iPSCs undergo a reduction of mtDNA copy number similar to what occurs during early embryogenesis, known as the "bottleneck effect." Hence, iPSCs may also represent a useful tool to study this phenomenon, which is crucial to understand the mechanisms of mutant mtDNA inheritance and why different tissues present with different heteroplasmy [37]. iPSCs have been also generated from LHON (Leber's hereditary optic neuropathy) patients' cells and differentiated into retinal ganglion cells, those affected in human pathology, allowing to confirm the biochemical defects already observed in other LHON cell models [42,43]. The establishment of patient-derived isogenic iPSC-carrying mtDNA mutations appears to be a promising approach to clarify the molecular mechanisms underlying mitochondrial diseases in specifically affected tissues and a tool for personalized drug discovery. It has also been suggested that iPSC could be used in the future for autologous cell transplantation therapies although further studies on safety of these approaches are required.

13.2.4 Yeast

The baker's yeast *Saccharomyces cerevisiae* constitutes a convenient model for large genetic and drug screening studies. This model counts with several advantages including the high evolutionary conservation of OXPHOS complexes and mitochondrial function to the mammalian counterpart; the easy and relatively cheap handling of the microorganism; the extensive annotation of its mtDNA; and the almost unique possibility to directly manipulate the mitochondrial genome [44]. Yeast is also suitable for modeling mtDNA mutations by direct manipulation of its genome through a biolistic approach, which consists in shooting cells with DNA-coated particles followed by a selection via specific genetic markers [45,46]. Despite the low yield of mitochondrial transformants, this approach allowed to generate strains carrying homoplasmic mtDNA mutations, which are accumulated by mtDNA molecules segregation during budding [47]. Moreover, yeast possesses the ability to grow on fermentable carbon sources, such as glucose when OXPHOS is dysfunctional; thus pathogenic mutations that lead to severe dysfunction can be maintained.

Multiple tRNA pathogenic mutations causing different mitochondrial diseases have been modeled in yeast in homoplasmy [48–50]. The defective phenotypes of these mutants include reduced growth on non-fermentable carbon sources and impaired respiration, which depend on the strain used and the mutation type. Mutations corresponding to pathogenic MELAS mutations in yeast share the same defect in amino acylation as that observed in human cells [49,51]. Yeast has been also exploited to define the impact of mtDNA mutations affecting the Q_i and Q_o sites of cytochrome *bc1* in terms of complex III (CIII) assembly, structural stability, catalytic function, and sensitivity to inhibitors [52–55]. These models were also used to identify compensatory mutations suppressing the pathogenic phenotype [56] or the functional impact of mtDNA variants considered polymorphisms [57]. Similarly, yeast has been used to investigate mutations in mtDNA-encoded genes for respiratory complex IV (CIV) allowing to discriminate pathogenic variants from silent polymorphisms [58–60]. Pathogenic mutations occurring in *atp6* gene causing neuropathy, ataxia, and retinitis pigmentosa (NARP), Leigh syndrome (LS), or bilateral striatal lesions were modeled in yeast [61–66]. These studies revealed the heterogeneity of these mutations, as some of them clearly affect the structural stability of the mitochondrial F_1F_o ATPase (CV) [62,64], while others are responsible for a functional defect of the complex [61,63,66]. These mutated strains have been exploited to clarify the molecular mechanisms behind the enzymatic dysfunctions [67,68] or to rapidly screen drug libraries to identify possible therapeutic agents for mitochondrial disorders [69–71].

Overall, these studies highlighted the value of yeast as a model for the study of the functional effects of mtDNA mutations. A major drawback of baker's yeast as an ideal model for mammalian mitochondrial diseases is the lack of CI, which is indeed a mutational hotspot for several mitochondrial disorders, in particular LHON. In this context, the development of new models using yeast species containing CI such as *Yarrowia lipolytica* is desirable.

13.3 Animal models

Studies *in vitro* led to significant understanding of the biochemical defects produced by mtDNA alterations, but investigation on animal models is required to get more insights into their phenotypic consequences. Advances in gene editing techniques greatly improved the production of animal models for mitochondrial diseases but among them, only few carry mtDNA alterations. There are reports describing naturally occurring mutations in animals, that is, Shetland sheepdogs and Australian cattle dogs. These two breeds were described to have inherited spongiform leukoencephalomyelopathy caused by the m.G14474A/*MT-CYB* mutation producing the p.V98M change in cytochrome *b* [72]. In this section, we briefly describe animal models generated to study mtDNA deletions and mutations (summarized in Table 13.1). A comparison of mtDNA organization of the species used to model human pathologies is shown in Fig. 13.2.

13.3.1 *Caenorhabditis elegans*

The garden worm *Caenorhabditis elegans* is a nematode widely used for neurobiology, aging, and development studies. The benefits of this model organism include the presence of differentiated organs and tissues, a short life cycle, easiness to genetic manipulation, and high reproductive rate. Hermaphrodite worms use self-fertilization permitting an easy

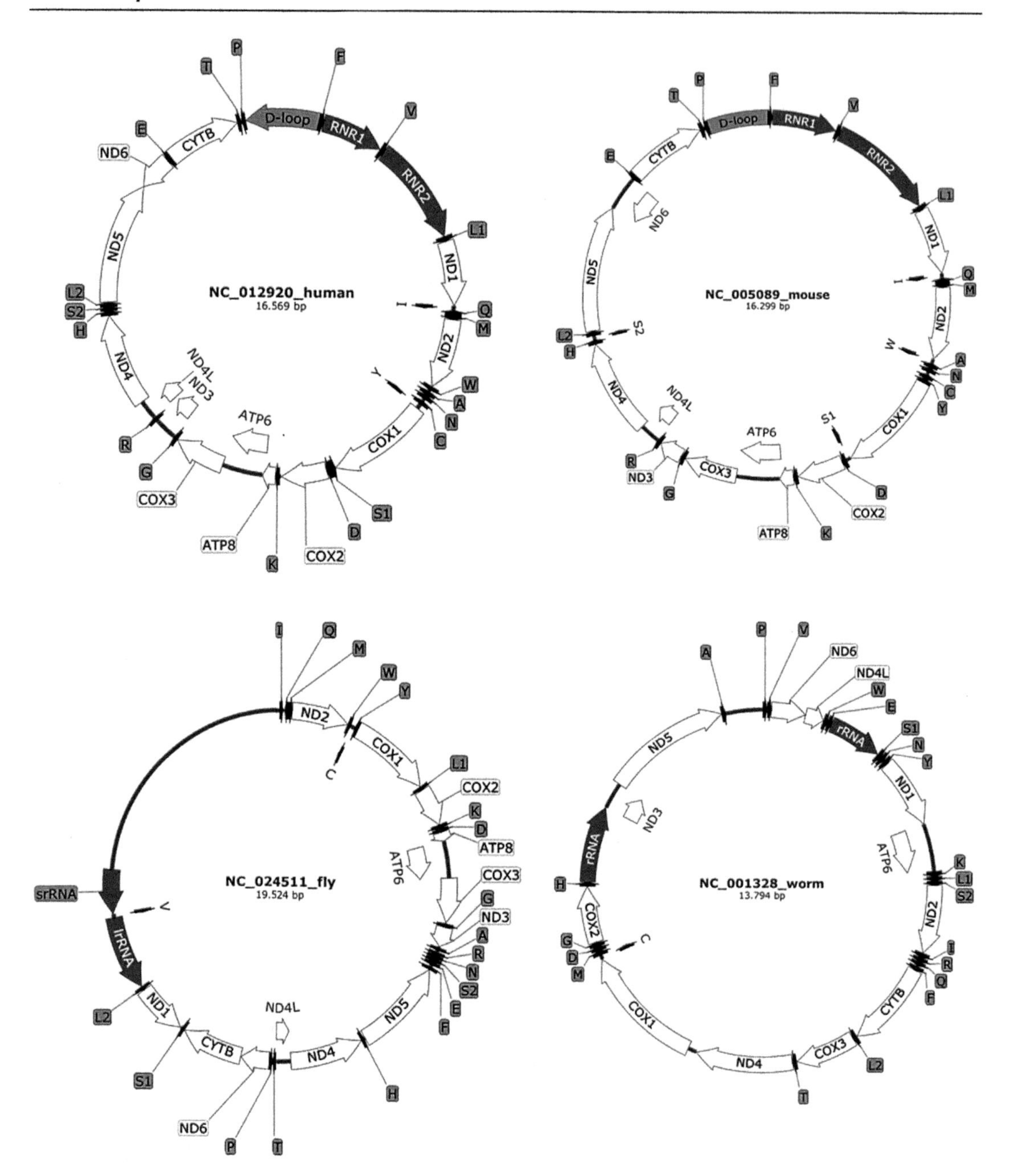

Figure 13.2
Mitochondrial genome organization in the human, mouse, fly and worm, animals used as models for mtDNA mutation-related disorders. Reference sequences have been used to construct the maps and are indicated by the GenBank accession number. Coding sequences are depicted as *white* arrows, tRNAs as *dark gray* arrows, rRNAs as *light gray* arrows, and regulatory regions in *gray*. Maps are generated using SnapGene v.4.3.11.

breeding of homozygous organism and aiding isolation and maintenance of mutant strains. Similar to yeast, this nematode represents an excellent tool for large drug or genetic screenings. It also allows the study of the mechanism of action at the whole organism level in identical individuals and under controlled conditions. *C. elegans* has been widely utilized to model human pathologies, including mitochondrial diseases [73]. Mutant strains in nuclear encoded genes for mitochondrial proteins have been produced and investigated leading to a significant improvement in the knowledge of the molecular bases of mitochondrial disorders [73,74], but to date no mtDNA-mutated model is available. Interestingly, mitochondrial dysfunction has been linked to increased life span in nematodes despite the bioenergetic failure testified by the lower ATP content and oxygen consumption [75,76]. According to the free radical theory of aging, long-living worms with mitochondrial dysfunction should produce and thus be exposed to reduced levels of reactive oxygen species (ROS). However, strains with mitochondrial dysfunction and slow physiological rates failed to show reduced oxidative stress, indicating that the prolonged life span is mostly related to reduced metabolic activity rather than to ROS production [75,77]. The age-dependent accumulation of mtDNA deletions was also taken into account. Interestingly, mtDNA deletions and mitochondrial function decline were reported in aging nematodes [78], but a recent study demonstrated that such alterations are rare and not related with aging, indicating that they are unlikely to drive the age-dependent functional decline and dampened the worth of *C. elegance* as a model for the study of such phenotype [79].

13.3.2 Drosophila melanogaster

Recently, *Drosophila melanogaster* gained much attention as a possible animal model for mtDNA-related disorders, thanks to the possibility to manipulate mtDNA in a tissue-specific manner by expressing mitochondrial-targeted restriction endonucleases recognizing unique cleavage sites. This method has been developed in human cells by Moraes' lab and then applied to flies [80,81]. Despite the extremely high toxicity of this approach, some escapers carrying mutations in the restriction site survived leading to the generation of mutant strains. The limitations of the approach reside in the presence of a few number of unique restriction sites in the mitochondrial genome and in the fact that the generated nucleotide variants may not cause a phenotype [82]. Besides the general benefits of easy handling, short life cycle, and high reproduction rate, *D. melanogaster* presents several advantages as animal model for mitochondrial disorders. In particular, mtDNA-encoded genes are highly conserved between the two species as well as mtDNA replication, transcription and translation, mitochondrial biogenesis, and dynamics [83]. Flies have all the organs and tissues affected by mitochondrial diseases and nuclear mutants mirror their symptoms, such as brain degeneration, cardiomyopathy, and deafness [84]. To date, four mtDNA-mutated *D. melanogaster* strains have been described (Table 13.1) and several models for nDNA-encoded genes are available including proteins involved in the replication and maintenance of the mtDNA [110]. A strain of *Drosophila suboscura* carrying a spontaneous heteroplasmic mtDNA large deletion of about 5 kb was described in the early 1990s and has been used to study the segregation of mtDNA genomes. Because this fly strain lacks any pathological phenotype, it has not been considered as a model for mitochondrial disorders [111].

13.3.2.1 ATP6 mutant

The first Drosophila mtDNA mutant model was discovered as a maternally inherited enhancer of $sesB^{1}$, the ortholog of ANT1, and harbors a spontaneous point mutation inducing the amino acid change p.G116E in the ATP6 subunit of CV [88]. The $mt{:}ATP6^{1}$ mutation arose and was stable at very high heteroplasmy in $sesB^{1}$ mutant background, while shifted to wild type and is purified in $sesB^{+}$ background. Mutants mirror the key features of LS as they displayed short life span; progressive muscular degeneration, and locomotor impairment; conditional paralysis; seizures; and neural dysfunction [88]. $mt{:}ATP6^{1}$ mutant flies showed mitochondrial respiration rate comparable to wild type flies but their CV activity was almost abolished. The mutation was predicted to affect ATP synthase dimerization and indeed mitochondrial *cristae* were disorganized. This fly model showed a compensatory metabolic shift toward glycolysis and ketogenesis in young individuals and was exploited to demonstrate that a high fat/ketogenic diet reduces the occurrence of seizures [112,113].

13.3.2.2 CoI mutants

By expressing the mitochondrial-targeted *Xho*I (mito-*Xho*I) enzyme, four strains carrying homoplasmic variants in *mt:CoI* gene were isolated, namely $mt{:}CoI^{A302T}$, $mt{:}CoI^{R301L}$, $mt{:}CoI^{R301S}$, and $mt{:}CoI^{T300I}$ [81,86]. While the first two mutant flies were healthy, the $mt{:}CoI^{R301S}$ strain showed growth retardation, reduced life span, age-related neurodegeneration, and myopathy caused by CIV deficiency. Conversely, the $mt{:}CoI^{T300I}$ mutant was a temperature sensitive strain that fails to eclose at 29°C displaying a severe CIV defect [114]. This model has been used to investigate the propagation of mtDNA variants during oogenesis and to dissect the molecular mechanism leading to neurodegeneration when the variant is expressed specifically in the eye [84].

13.3.2.3 ND2 mutants

By applying the same approach with a mito-*Bgl*II restriction endonuclease, two mutants in CI subunit ND2 were obtained [81,85]. The $mt{:}ND2^{ins1}$ strain harbored an in-frame insertion of three nucleotides at position 189 and was phenotypically comparable to controls while the $mt{:}ND2^{del1}$ mutant carried a nine-nucleotide deletion that removed three amino acids at positions 186–188. The latter mutant strain presented with reduced CI activity and inefficient coupling of electron transfer to proton pumping, resulting in decreased mitochondrial membrane potential ($\Delta\psi_{m}$) and diminished energy production [85]. Mutant flies also resembled several symptoms of LS including stress-induced seizures, progressive neurodegeneration, and reduced life span.

13.3.2.4 CoII mutant

A Drosophila model carrying the homoplasmic *mt:CoII*^{G177S} hypomorph mutation showed a decreased CIV activity in the sperm. This bioenergetic defect was correlated with male infertility caused by an impairment of sperm development and motility without affecting the general physiology of the flies [87]. The pathological phenotype was fully rescued in certain nuclear genetic backgrounds indicating that nuclear genomes can modulate the manifestation of certain mtDNA mutations.

13.3.3 Mice

Over the years, several mice models of mitochondrial disorders were generated greatly improving the knowledge of their pathogenesis and providing handful tools to test therapies [115–117]. Most of the models are based on defects of nuclear encoded genes because of the already mentioned technical difficulties in the manipulation of mammalian mtDNA. In this section, we summarize the mice models developed carrying mtDNA alterations (see Table 13.1).

13.3.3.1 mtDNA deletions

The first mouse with a large-scale mtDNA deletion was the "mito-mouse" created by electrofusion of cytoplasts from enucleated cybrids containing a 4696 bp deletion (nt 7759–12454) with a pronuclear stage mouse embryo and then transplanted into a pseudopregnant female [89]. The chimeric progeny carrying the heteroplasmic deletion was selected for breeding and transmitted the deletion to the F2 generation. The accumulation of mtDNA deletions reached about 85%–94% in some of the tissues causing mitochondrial dysfunction in muscle and kidney and mice died prematurely of renal failure [89].

Another approach taken by the Moraes' group to create animal models of mtDNA deletions was to produce mtDNA double-strand breaks. The first model was a transgenic mouse expressing in skeletal muscle the mitochondrial-targeted *Pst*I (mito-*Pst*I) restriction endonuclease that cleaves the mouse mtDNA in two sites [90]. Transgenic founders showed growth retardation and by 6–7 months of age, they developed a severe muscle myopathy. Analysis of muscle sections showed ragged-red fibers (RRFs), a mosaic pattern of COX negative fibers with increased SDH staining, swollen mitochondria, and disrupted *cristae*. At the molecular level, double-strand breaks created by mito-*Pst*I induced a significant mtDNA depletion but some of the remaining molecules formed large-scale deletions resulting from the recombination of one of the *Pst*I sites and the 3′ end of the D-loop control region [90]. Because accumulation of mtDNA deletions during aging has been associated with the development of neurodegenerative diseases, Moraes' group created an inducible Tet-Off system to express mito-*Pst*I in adult neurons [91]. Under different induction paradigms, transgenic mice preferentially accumulated mtDNA large deletions in neurons suggesting the presence of a replicative advantage of the small mtDNA. This model demonstrated the presence of a mitochondrial double-strand breaks repair system [91] and allowed to prove that certain neuronal populations were more susceptible to OXPHOS dysfunction caused by mtDNA deletions [92,93], that nervous system inflammation and axonal injury were caused by mtDNA deletions in oligodendrocytes [94] and that systemic damage of mtDNA produced a progeroid phenotype leading to muscle sarcopenia [95].

13.3.3.2 mt-Co1

In 2006, the first transmitochondrial model harboring the m.6589T > C/*mt-Co1* missense mutation was generated by fusing mtDNA-depleted female embryonic stem (ES) cells with an enucleated mouse cell line carrying the homoplasmic mutation [96]. After selection for successful fusion, ES cells were introduced into eight-cell embryos and chimeric founders were bred to obtain transmission of the mutant mtDNA to next generations. The F6 generation had 100% mutant mtDNA in all tissues analyzed and displayed decreased CIV activity and increased lactate in blood that resulted in growth retardation.

13.3.3.3 mt-Nd6

Using the same technique, Wallace's group created another transmitochondrial mouse model to test the fate of severe mtDNA mutations [97]. The mitochondrial donor was a mouse cell line harboring in homoplasmy the previously described m.6589T > C/*mt-Co1* that caused a decrease in CIV activity and the new m.13885InsC/*mt-Nd6* frame-shift mutation that exposes a premature stop codon leading to a severe CI deficiency. These cells were enucleated and fused to the LMEB4 Rho0 cells to generate cybrids that were subsequently enucleated and fused to Rho0 ES cells. Multiple clones were obtained and analyzed. Clone EC77 that was homoplasmic for *mt-Co1* and had 96% heteroplasmy for *mt-Nd6* mutations was injected into blastocyst to obtain chimeras. Only three chimeras were obtained and bred to determine transmission of mutant mtDNA over successive maternal generations. In F1, only one female was homoplasmic for the *mt-Co1* mutation and 47% heteroplasmic for the *mt-Nd6* mutation in the mtDNA extracted from tail. The mouse did not show any overt phenotype but at 11 months of age, molecular analysis revealed that brain had accumulated high levels of *mt-Nd6* mutation. Brain, heart, skeletal muscle, and liver showed a decrease in CI activity of 10%–33% and a decrease in CIV activity of 19%–56%. This female was mated and transmission of mutations was analyzed in the progeny in her six pregnancies. Interestingly, the mutation decreased in each litter until it disappeared at the fourth pregnancy. Females from her progeny were also mated and again the *mt-Nd6* mutation was lost by the second pregnancy indicating that the severe CI mutation is purified during oogenesis. Conversely, the *mt-Co1* mutation was preserved over generations even though it caused myopathy and cardiomyopathy [97].

Wallace's group also created a mouse harboring the m.13997G > A/*mt-Nd6* corresponding to the human m.14600G > A/*MT-ND6* found homoplasmic in a child with severe encephalomyopathy and heteroplasmic in her maternal aunt affected by LHON [99]. Despite the systemic reduction of CI activity, in mice, the mutation induced a pathologic phenotype similar to LHON restricted to the optic nerve with loss of small retinal fibers, swollen axons of retinal ganglion cells (RGCs), and increased oxidative stress. To date, this remains a unique model to study the pathogenesis of mitochondrial optic neuropathies and to test antioxidant therapies. The same mutation modeled in another mouse strain was created to study tumor development and metastasis formation [98]. Young Nd6 mutant mice displayed only a mild reduction of CI + III activity and a moderate increase of lactate in blood but were healthy. With time, animals did not show any neurological or ophthalmological alterations but developed age-associated disorders such as increased glycemia and B-cell lymphoma.

13.3.3.4 mt-tK ($tRNA^{Lys}$)

By subcloning multiple times the P29 mouse lung carcinoma cell line, Shimizu and colleagues identified a clone carrying the heteroplasmic m.7731G > A/*mt-tK* mutation [100], corresponding to the m.8328G > A/*MT-TK* reported in patients with mitochondrial disorders. Transmitochondrial mice were created as previously described and F5 generations were produced carrying different levels of heteroplasmy but none of the offspring had more than 85% heteroplasmy of the mutation because at higher heteroplasmy, it was purified during oogenesis. Mice with high mutation load were smaller and had muscle weakness. At the molecular level, they presented with decreased mitochondrial respiration. Mice were analyzed only at 4 months of age and how the phenotype progressed with age was not investigated [100].

13.3.3.5 mt-tA ($tRNA^{Ala}$)

Using a relatively easy and clever approach, Stewart and Larsson's group produced a heteroplasmic mice model carrying the m.5024C > T/*mt-tA* mutation [101]. The model was created by taking advantage of the heterozygous mutator mouse (described later) to produce and select somatic mtDNA mutations after a series of strategic breeding. They crossed a male $PolgA^{+/mut}$ with a wild type female. Then first-generation heterozygous females $PolgA^{+/mut}$ were crossed with wild type males. From this second generation, females will have wild type nuclear genome but are expected to inherit mtDNA mutations from the maternal lineage. Twelve lineages were established and then founders were screened for OXPHOS deficiency using the COX/SDH staining in colonic crypts. If the somatic mtDNA mutation clonally expanded above its mutational threshold, it should produce a COX deficiency in affected cells. Through this approach, authors identified one founder with high levels of heteroplasmy containing the m.5024C > T/*mt-tA* mutation, corresponding to the human m.5650G > A/*MT-TA* found in one patient with pure myopathy [118] and in one with encephalomyopathy [119]. The human mutation generates a wobble base pair in the acceptor stem of the $tRNA^{Ala}$, whereas in mouse it causes a mismatch that affects stability of the tRNA molecule causing mitochondrial protein synthesis impairment [101]. Female mutant mice did not have any overt phenotype, while male showed reduced weight and fat content when compared to controls. CIV deficiency was observed in epithelial cells of colon crypts at 5 months and by 10 months of age the deficiency was observed also in the colonic smooth muscle and in some cardiomyocytes. Although this mouse model does not have high heteroplasmy levels, it is very useful to understand the segregation of mtDNA mutations and to test therapeutic interventions.

13.3.3.6 PolgA

The mammalian mtDNA replication is driven by polymerase γ, a heterodimer composed of a catalytic subunit (*PolgA*), and a DNA binding accessory subunit (*Polg2*). Polγ is a proofreading polymerase with $3'-5'$ exonuclease activity in its catalytic subunit. Mutations in *POLG* are one of the most common causes of mitochondrial disorders [120]. To induce accumulation of mtDNA mutations, transgenic mouse models were created by expressing in specific tissues a *PolgA* mutant transgene with the D181A mutation that abolished the exonuclease activity without altering the DNA-polymerase activity [102–104]. The first transgenic model was designed to express the D181A mutation under the cardiac-specific α-myosin heavy chain [102]. A rapid accumulation of mtDNA mutations and deletions was observed in the heart of the transgenic mice without affecting mtDNA copy number. Mitochondrial respiration was not impaired, but the accumulation of mtDNA mutations led to the development of progressive cardiomyopathy starting at 4–5 weeks. A neuron-specific model expressing the D181A mutation under the CaMKIIα promoter displayed accumulation of mtDNA deletions and mutations in brain and developed behavioral abnormalities resembling mood disorders found in some patients with chronic progressive external ophthalmoplegia (PEO) [104]. In the third transgenic model, the *PolgA* D181A transgene was under the rat insulin promoter expressing the mutant polymerase in pancreatic β-cells. This model also accumulated mtDNA deletions and mutations that led to cell death causing insulin insufficiency and development of a diabetic phenotype [103].

Other mouse models of $3'-5'$ exonuclease proofreading deficient PolgA were created independently by Larsson's and Prolla's groups by knocking in the gene [106,107]. To create the KI mouse, an allele carrying the D257A mutation in the second exonuclease domain was introduced in ES cells to obtain homologous recombination. After intercrossing the resulting mice, they generated the homozygous $PolgA^{mut/mut}$ KI mouse named "mutator" [107]. The mutator mouse developed normally until about 6 months of age when started to shown signs of premature aging [107]. A similar progeroid phenotype was observed in the

D257A mouse model in which hearing loss and sarcopenia were also described to occur at 9 months of age [106]. Both mice died prematurely between 12 and 15 months of age. At the molecular level, the accumulation of mtDNA deletions and somatic point mutations was observed in brain, liver, and heart. Overall, these mice are considered excellent models to study the role of mtDNA alterations in aging.

13.3.3.7 Twnk

In mammals, the mitochondrial helicase Twinkle unwinds mtDNA giving access to Polγ and single-stranded mtDNA binding proteins during replication and is required for mtDNA maintenance and copy number control [121]. Mutations in *TWNK* lead to autosomal dominant PEO as a result of mtDNA instability caused by defective replication [122]. Two transgenic mice overexpressing different mutant forms of Twinkle ($Twnk^{A360T}$ and the $Twnk^{dup353-365}$) driven by the ubiquitous β-actin promoter have been created and are commonly called "deletor mice" [108]. The Twinkle mutations produced mtDNA depletion and accumulation of multiple deletions in mouse older than 1 year but no point mutations were observed. Unlike the mutator, the deletor mice did not display a progeroid phenotype although a late onset and progressive mitochondrial dysfunction was observed in muscle and brain [108]. A transgenic mouse overexpressing the mutant $Twnk^{dup353-365}$ in dopaminergic neurons was also developed to investigate the effects of accumulation of mtDNA deletions in Parkinson's disease [123]. Transgenic mice did not show any overt phenotype but displayed increased levels of age-related mtDNA deletions and signs of neurodegeneration and myopathy.

13.3.3.8 Mgme1

MGME1 is a nuclease that processes single-stranded mtDNA 5′ ends produced during replication. Mutations in this protein cause severe mitochondrial multisystem disorders in humans. Clinical presentations include PEO, dilated cardiomyopathy, microcephalus, mental retardation, muscle wasting, and gastrointestinal symptoms [124,125]. These patients displayed mtDNA depletion, multiple deletions, and accumulation of high levels of a 11 kb mtDNA linear fragment [125]. Systemic KO and heart and skeletal muscle-specific conditional KO of *Mgme1* mouse models were created [109]. At 18 months of age, these mice had no obvious phenotype, looked healthy, and were fertile. Similar to the deletor mouse, loss of Mgme1 produced mtDNA depletion and deletions due to a replication stall without accumulation of point mutations. The most abundant deletion was again a linear fragment of 11 kb that did not contain the minor arc region of the mtDNA (region between the origin of replication of the heavy strand O_H and the origin of replication of the light strand O_L). Interestingly, despite all the alterations described above, OXPHOS complexes structure and function seemed to be not affected.

Table 13.1: Animal models of mtDNA alterations

Gene	Mitochondrial function	Genetic manipulation	Phenotype		References
Drosophila melanogaster					
ND2	Structural subunit of CI	Expression of the mitochondrial-targeted *Bgl*II (mito-*Bgl*II)	*mt: ND2*^{ins1}	Healthy	[85]
			mt: ND2^{del1}	Reduced CI activity, $\Delta\psi_m$, and energy production. Stress-induced seizures, neurodegeneration and reduced life span	
CoI	Structural and catalytic subunit of CIV	Expression of the mitochondrial-targeted *Xho*I (mito-*Xho*I)	*mt: CoI*^{A302T}	Healthy	[81,86]
			mt: CoI^{R301L}	Healthy	
			mt: CoI^{R301S}	Growth retardation, reduced life span, age-related neurodegeneration and myopathy, reduced CIV activity	
			mt: CoI^{T300I}	Fails to eclose at 29°C, severe CIV defect	
CoII	Structural and catalytic subunit of CIV	Spontaneous mutant	Reduced CIV activity in the sperm, male infertility		[87]
ATP6	Structural subunit of CV	Spontaneous mutant found in *sesB*^{1} background	Short life span, progressive muscular degeneration and locomotor impairment, seizures. Abolished CV activity, disorganized mitochondrial *cristae*		[88]
Mus musculus					
Common deletion	Encompasses tRNAs and 7 structural genes of CI, CIV, and CV	Electrofusion of pronucleus-stage embryos with several enucleated cytoplasts of the ΔmtDNA cybrids	Mosaic pattern of CIV negative fibers in heart and muscle; lactic acidosis in blood; kidney failure		[89]
mtDNA depletion and large deletions	Multiple subunits of OXPHOS complexes	Expression of the mitochondrial-targeted *Pst*I (mito-*Pst*I)	RRFs, mosaic pattern of CIV, swollen mitochondria and disrupted *cristae*		[90–95]
mt-Co1	Structural and catalytic subunit of CIV	Fusion of cytoplasts carrying the 6589T > C mutation with Rho0 ES cells	Growth retardation and CIV deficiency in brain, heart, liver, and skeletal muscles. No motor or neurological phenotype		[96]
mt-Co1 *mt-Nd6*	Structural subunits of CIV and CI	Fusion of cytoplasts carrying the 6589T > C and the 13885InsC mutations with Rho0 ES cells	Mitochondrial myopathy and cardiomyopathy; decreased CIV activity in brain, liver, heart, and skeletal muscle; RRFs. No motor or neurological phenotype		[97]

(*Continued*)

Table 13.1: (Continued)

Gene	Mitochondrial function	Genetic manipulation	Phenotype	References
Nd6	Structural subunit of CI	Fusion of cells containing the 13997G > A mutation with Rho0 ES cells.	CI + III defect in multiple tissues. Age-associated disorders at longer observation	[98]
		Enucleated cell lines carrying the 13997G > A mutation fused with Rho0 ES cells	Decreased retinal response and swollen axons in the optic nerve; CI deficiency in liver and brain; high levels of ROS in the brain	[99]
mt-tK	$tRNA^{Lys}$	Somatic 7731G > A mutation introduced in female Rho0 ES cells	Heteroplasmic mice (85% mutant) displayed short body length, muscle weakness, and RRFs. Respiratory chain defect and elevated ROS production in skeletal muscle.	[100]
mt-tA	$tRNA^{Ala}$	Spontaneous mutant found in mutator mouse and subsequently selected	Female healthy, males reduced weight and fat content. Age-dependent CIV deficiency in colon crypts epithelia, smooth muscle, and cardiomyocytes	[101]
Polg	mtDNA replication (DNA polymerase)	Expression of $Polg^{D181A}$ in heart	Cardiomegaly, mtDNA mutations and deletions	[102]
		Expression of $Polg^{D181A}$ in pancreatic β-cells	Early onset diabetes, mtDNA mutations and deletions	[103]
		Expression of $Polg^{D181A}$ in neurons	Neuronal dysfunction after 5 m of age, mtDNA mutations and deletions	[104,105]
		Knock-in $Polg^{D257A}$	Reduced life span, premature aging, mtDNA mutations and deletions	[106,107]
Twnk	mtDNA replication (helicase)	Expression of $Twinkle^{dup353-365}$	Late-onset myopathy, mtDNA deletions and depletion	[108]
		Expression of $Twinkle^{A360T}$	Late-onset myopathy, mtDNA deletions and depletion	
Mgme1	mtDNA replication (nuclease)	Systemic and heart and skeletal muscle-specific conditional KO	mtDNA deletions and depletion, no overt phenotype	[109]

13.4 *Methods for assessment of functional defects induced by mtDNA alterations*

The mitochondrial genome encodes for subunits belonging to the complexes of the OXPHOS system that provides the function of generating ATP from reducing equivalents (NADH and $FADH_2$) produced in catabolic reactions. The mammalian OXPHOS system comprises five multiprotein complexes (complexes I to V) and two mobile electron carriers (ubiquinone and cytochrome *c*) embedded in the lipid bilayer of the inner mitochondrial membrane. Complexes I to IV allow the electron transfer from NADH and $FADH_2$ to molecular oxygen while generating a proton gradient across the inner mitochondrial membrane that is dissipated to synthesize ATP by CV. Hence, alterations in mtDNA may affect the function of the OXPHOS system, and several biochemical methodologies have been set up to establish its functional and structural status. A variety of reports detailing principles and protocols of such assays can be found in the scientific literature. In this section, we will describe the most popular, highlighting their benefits and limitations, but we will not describe detailed protocols for which we cite specific references.

13.4.1 OXPHOS complexes activity

13.4.1.1 Spectrophotometric methods

Oxidative phosphorylation complexes activity can be measured in homogenates, mitochondria-enriched fractions, or purified mitochondria isolated from tissues and cells [126]. One problem in the study of the biochemical effects of mitochondrial diseases is that in practice each laboratory uses its own protocols to prepare samples and to perform enzymatic assays. Variations in buffers composition, methods of sample permeabilization, redox acceptors, and concentration of reagents make almost impossible to compare results [127]. Some technical details are reported in Table 13.2 and in Box 13.1.

13.4.1.1.1 Sample preparation

An exhaustive description of methods to isolate mitochondria from animal tissues and cultured cell lines was published by Pallotti and Lenaz [137]. The general steps for isolating mitochondria include the methods of rupture of intercellular connection, cell walls and/or plasma membranes, the choice of isolation medium, the differential centrifugation, and the storage of samples. The most used methods for sample homogenization are mechanical methods such as the use of tissue blender (Waring, UltraTurrax), Dounce/Potter-Elvehjem homogenizer, or chemical methods (i.e., lysozyme or digitonin) [137]. Tissue or cell disruption must be carried out in an appropriate isolation solution, which usually contains 0.25 M of sugars, such as sucrose, mannitol, or sorbitol in different proportions, 100–150 mM KCl to maintain the cell ionic force and a buffer system to maintain the physiological pH (usually 5–20 mM Tris–HCl or Tris-acetate). In addition, the isolation solution also contains EDTA or EGTA to bind Ca^{+2} ions (1 mM), 0.1%–1% bovine serum albumin (BSA), which quenches proteolytic activity of cellular proteases and removes free fatty acids preserving mitochondrial membrane integrity, and a cocktail of protease inhibitors. After tissue and cell disruption, mitochondria are separated and collected by differential centrifugation that consists of a two-step centrifugation. The low-speed centrifugation (600–1000 *g*) is necessary to remove intact cells, cellular debris, and nuclei. Then the supernatant is centrifuged at high speed (8000–10,000 *g*) to obtain the mitochondrial fraction, called "crude mitochondria." This pellet is considered suitable for enzymatic assays but can also undergo a further step of purification using discontinuous gradients (sucrose, Ficoll, etc.). Then, mitochondrial preparations can be resuspended in a minimal volume of isolation buffer to maintain a high concentration of protein and stored frozen at −80°C.

13.4.1.1.2 NADH:ubiquinone oxidoreductase activity (CI activity)

The measurement of CI spectrophotometric activity is considered difficult and not always reliable because the site of oxidation of NADH is on the side of mitochondrial matrix; thus it is required to completely permeabilize the inner mitochondrial membrane to allow the access of the substrate. Further, the endogenous acceptor of electrons coenzyme Q is insoluble in the aqueous medium and a more soluble ubiquinone analog is necessary, that is, decylbenzoquinone (DB). Depending on the mitochondrial preparation and source, CI assay may suffer from non-specific activity due to the presence of other NADH dehydrogenases [128]. Despite the differences among the methods described in literature, the real critical factors that must be considered for a reliable assay are the electron acceptor substrate, the method of determination of rotenone sensitivity, and the mitochondrial permeabilization procedure. Few studies have dissected these issues and compared the differences among distinct methods [128–131]. The most popular procedures for mitochondrial permeabilization are based on one or two cycles of freeze/thawing that can be followed by a hypotonic shock. In some cases, 0.1%–0.5% digitonin may be used for permeabilizing membranes or freshly prepared mitochondria are directly added in the assay hypotonic buffer [126,128–136]. Most of the assays are based on the measurement of NADH oxidation at 340 nm; however, some methods are based on the measurement of the reduction of quinone at 272 nm [126] or of 2,6-dichloroindophenol (DCIP) at 600 nm [130,136]. Reduction of DCIP by electrons originated from reduced quinone is particularly useful, since DCIP extinction coefficient is three time greater than NADH one. Hence, the signal is amplified allowing the measurements of very low CI activity [130,136]. The best way to calculate the rotenone sensitivity is to perform a parallel assay with rotenone in the sample before the addiction of substrates. In fact, rotenone inhibits better the enzyme when substrates are not present probably because it competes with quinone. However, several methods indicate the use of rotenone by adding it in the same assay particularly when diagnostic samples are limited [128].

13.4.1.1.3 Succinate:ubiquinone oxidoreductase activity (CII activity)

Complex II (CII) is the only enzyme of the respiratory chain entirely encoded by nuclear genes; thus along with citrate synthase (CS), the measure of its activity is often used as marker of the amount of mitochondria. The reaction is started by adding 50–100 μM DB after 10 minutes of preincubation in assay medium to activate CII. Reduction of DB can be directly followed measuring the signal of produced quinol at 280 nm [135] or following the reduction of 50–80 μM DCIP at 600 nm [126,133]. The non-specific reaction can be measured and subtracted to the total activity adding 5 mM malonate, a specific inhibitor of CII.

13.4.1.1.4 Ubiquinol:cytochrome c oxidoreductase activity (CIII activity)

CIII activity is measured following the increase in absorbance at 550 nm caused by cytochrome *c* reduction. The antimycin-sensitive specific activity is calculated after addition of 1–2 μM antimycin A (aA). As described for CI, aA sensitivity should be determined in a parallel assay with the inhibitor at the beginning of the reaction. DBH_2 can be obtained by reducing DB with borohydride crystals [126] or sodium dithionite [133,138].

13.4.1.1.5 Cytochrome *c* oxidoreductase activity (CIV activity)

The reaction is measured following the disappearance of reduced cytochrome *c* at 550 nm [126,132–135]. The assay is generally performed in a 10–50 mM KH_2PO_4 buffer at pH 7.4 in a cuvette maintained at the controlled temperature of 30°C or 37°C. In some protocols, the buffer also contains 0.1% BSA [133] and the specificity of the reaction is measured subtracting the KCN insensitive activity after addition of 1–2 mM KCN. The protocol for reduction of cytochrome *c* is accurately described elsewhere [126].

13.4.1.1.6 NADH:cytochrome *c* oxidoreductase (CI + III activity)

The measurement of CI + III activity suffers of high levels of rotenone insensitivity (between 30% and 80% of the total activity) in cell homogenates, while it is considered reliable on mitochondrial fractions. This measurement provides information about the integrated activity of the two complexes in which electrons are transported by endogenous coenzyme Q and not by exogenous quinones. The rate of enzymatic activity is measured following the reduction of cytochrome *c* at 550 nm. The specificity of the reaction is assessed subtracting the residual activity measured in a parallel assay in the presence of 1 μM rotenone and 1 μM aA incubated with the sample before the addition of substrates.

13.4.1.1.7 Succinate:cytochrome *c* oxidoreductase (CII + III activity)

CII + III activity is measured by following the reduction of cytochrome *c* after addition of sodium succinate. The classical buffer is described in Table 13.2. Some adjustments include 0.1% BSA, 1 mM ATP (activator of CII), and 1 mM EDTA [135]. The reaction is started with 30–50 μM cytochrome *c* and the specificity of reaction is calculated subtracting the residual activity measured in a parallel assay in presence of 1 μM aA and 5 mM malonate.

13.4.1.1.8 Hydrolytic activity of ATP synthase (CV activity)

The assay for CV is measured in the reverse direction (ATP hydrolysis) by linking the hydrolysis of ATP to the oxidation of NADH using the coupled reaction of pyruvate kinase (PK) and lactate dehydrogenase (LDH) as described in Ref. [135]. The rate of NADH oxidation is monitored by measuring the absorbance at 340 nm. The specificity of the reaction is assessed adding 1–10 μM oligomycin A in a parallel assay. Alternatively, ATP hydrolysis activity can be monitored using an end point method in which the reaction is carried out in a buffer containing 125 mM sucrose, 65 mM KCl, 2.5 mM $MgCl_2$, 50 mM Hepes pH 7.2, and 2.5 mM ATP at 30°C for 10 minutes. Then, the reaction is stopped by adding an equal volume of 40% TCA in 0.5 mM KH_2PO_4, 10% molybdate dissolved in 10 N H_2SO_4. The inorganic phosphate (Pi) derived from ATP hydrolysis reacts with molybdate developing a colored solution with a maximum of absorbance at 600 nM. The amount of Pi produced is quantified using a standard curve of Pi in a range of 0–500 nmol [136].

13.4.1.1.9 Citrate synthase activity

Like CII, CS activity is used to normalize OXPHOS complexes activities on the mitochondrial content of the sample. The assay is performed following the reduction of 5,5′-dithiobis(2-nitrobenzoic acid) (DTNB) coupled to the reduction of coenzyme-A (CoA) produced by CS during the condensation of acetyl-CoA and oxaloacetate to citrate. The assay buffer used for CS activity varies between 125 and 200 mM Tris–Cl [126,136] and 50 mM KH_2PO_4 [135]. While the type of buffer system seems to be not crucial for the assay, it should be noted that different methods use a pH ranging from 7.4 to 8. The later pH value is closer to the one for the mitochondrial matrix and it is optimal for the enzymatic activity. Some methods also include in the assay buffer 0.1% Triton-X100 to ensure a complete solubilization of mitochondrial membranes and the release of CS [135,136]. The reaction mixture also contains 50–300 μM acetyl-CoA, 0.1 mM DTNB, and starts with 0.1–0.5 mM oxaloacetate. The reduction rate of DTNB is measured at 30°C–37°C with a wavelength set at 412 nm [126,135,136].

13.4.1.2 Immunocapture-based assays

Recently, commercially available kits that measure the enzymatic activity of respiratory complexes based on their immunocapture onto a coated microplate have been developed. They can be used with samples from cell culture, isolated mitochondria, or tissues from various sources including human, mouse, rat, and bovine. These kits provide a

Table 13.2: Experimental setting of spectrophotometric assays of OXPHOS complexes activities

Assay	Buffer	Substrates	Conditions	References
CI	20–25 mM KH_2PO_4 or 20–25 mM Tris–Cl pH 7.2–8 Supplemented with: 2–5 mM $MgCl_2$ 0.2%–0.5% BSA 1–2 mM KCN 1–3 μM Antimycin A	50–200 μM NADH 10–100 μM DB or CoQ1	Temperature: 30°C–37°C Constant stirring λ = 340 nm; 380 nm as reference	[126,128–136]
CII	10–50 mM KH_2PO_4 pH 7.2 1 mM EDTA 2 μM Antimycin A 1–3 μM Rotenone 1–2 mM KCN	5–20 mM Na-succinate 50–100 μM DB	Temperature: 30°C–37°C Constant stirring λ = 280 nm for DB or 600 nm for DCPIP	[126,133,135,136]
CIII	10–50 mM KH_2PO_4 or 250 mM sucrose 50 mM Tris–Cl pH 7.4 Supplemented with: 1 mM EDTA 0.1% BSA 1–2 mM KCN 1 μM rotenone 5 mM malonate	15–50 μM cyt *c* 50 μM DBH_2	Temperature: 30°C–37°C Constant stirring λ = 550 nm	[126,132–136]
CIV	10–50 KH_2PO_4 pH 7.4 Optional 0.1% BSA		Temperature: 30°C–37°C Constant stirring λ = 550 nm	[133]
CI + III	50 mM KH_2PO_4 pH 7.4 2 mM KCN	80 μM cyt *c* 100 μM NADH	Temperature: 30°C–37°C Constant stirring λ = 550 nm	[126]
CII + III	40–50 mM KH_2PO_4 pH 7.4 0.5–1 mM KCN 10–20 mM succinate incubated for 10 min	30–50 μM cyt *c*	Temperature: 30°C–37°C Constant stirring λ = 550 nm	[126]
CV	40–50 mM Tris–Cl pH 8 0.2–0.25 mM NADH 2–2.5 mM PEP 5 mM $MgCl_2$ 20UI or 50 μg/mL LDH 20UI or 50 μg/mL of PK	0.5–2.5 mM ATP	λ = 340 nm	[133,135]

BOX 13.1 Data analysis of spectrophotometric assays for OXPHOS complexes activity.

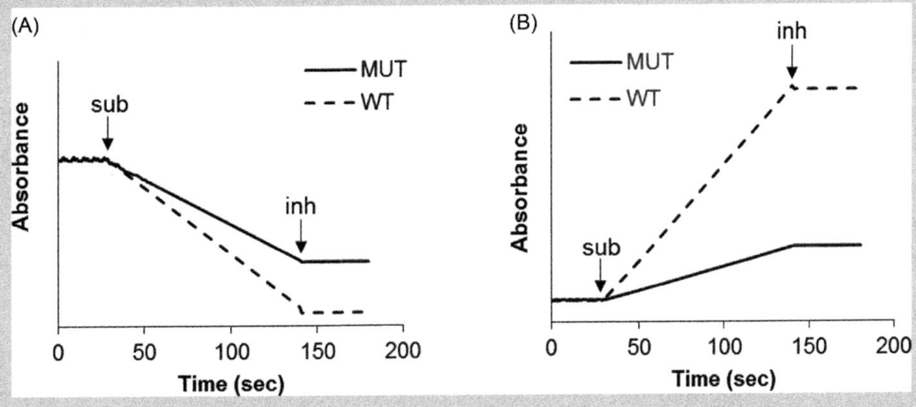

(A) The graph represents a hypothetical enzymatic assay of CI, CII, CIV, or CV in which the chromophore that is reduced (i.e., DCIP) or oxidized (i.e., NADH or reduced cytochrome *c*) decreases its absorbance at a fixed wavelength during time. Before the addition of substrate that starts the reaction, the absorbance should be stable. After the addition of substrate, the absorbance decreases until the reaction is stopped with a specific inhibitor (i.e., rotenone, malonate, KCN, or oligomycin); for assay details, see Table 13.2. The rate of reaction is calculated as variation of concentration of substrate in a minute applying the Lambert–Beer law. As described in main text, the best choice to subtract the non-enzymatic activity is to perform a parallel assay in presence of the specific inhibitor at the beginning of the reaction. However, when the amount of the sample is scarce, the only possibility is to add the inhibitor during the reaction and to subtract the residual inhibited activity. The *dotted line* indicates the rate of enzymatic activity in a control sample (WT), while the *full line* refers to the enzymatic activity of a sample deriving from a hypothetical mtDNA mutant (MUT) affecting the enzyme considered for the assay.

(B) The graph represents a hypothetical enzymatic assay of CIII or CS in which the chromophore reduced (i.e., cytochrome *c* or DTNB) increases its absorbance at a fixed wavelength during time. CIII activity is measured following the reduction of cytochrome *c* after the addition of reduced DB (DBH_2). The non-specific activity of CIII assay is measured in the presence of antimycin A in a parallel experiment or after the addition of inhibitor during the assay as showed in the graph. The CS activity assay is usually performed considering as non-specific CS activity the slope of the absorbance variation for a minute before the addition of oxaloacetate. The rates of specific enzymatic activity are calculated subtracting the activity before the addition of oxaloacetate to the rate in presence of the substrate. As in graph A, the *dotted line* corresponds to the rate of enzymatic activity of a control sample (WT), while the *full line* represents the activity of a hypothetical mutated sample (MUT).

convenient, cost-effective way and are less time-consuming when having a large number of samples to analyze, when screening for compounds that could affect OXPHOS function, or when these measurements are performed sporadically. The downside is that they are limited to measure only CI, CII, or CIV enzymatic activity. The overall procedure is very simple: proteins are extracted from samples with a detergent solution, incubated with the precoated microplate for 2–3 hours, and washed to eliminate the unbound material. Only the respiratory complex of interest remains immobilized on the microplate. Then a solution containing substrates and dye is added and changes in absorbance over time are recorded by using a plate reader. It is important to include always positive and negative controls in the assay and to ensure preservation of enzymatic activities by properly collecting and storing samples.

For CI assay, the substrates provided allow to measure its diaphorase-type activity by following NADH oxidation and the reduction of the provided dye is measured as an increase in absorbance at 450 nm. The assay does not depend on ubiquinone and therefore it is not inhibited by rotenone. In CII assay, the reaction is based on the production of ubiquinol and the reduction of DCIP, which is measured by a decrease in the absorbance at 600 nm. 2-thenoyltrifluoroacetone (TTFA) is a CII inhibitor and can be included in the assay to determine specificity of signal. The CIV assay is based on the oxidation of reduced cytochrome *c* measured by decrease in absorbance at 550 nm. As for classic spectrophotometric assays, the results of respiratory complexes activities should be normalized on protein content and/or mitochondrial content or use the activity of CS as a reference. There is a modality of the immunocapture kits called dipstick assay available for CI and CIV where antibodies against the particular complex are placed in a strip of nitrocellulose membrane and a wick pad draws the sample toward the antibody. Then the stick is submerged into a solution containing substrates and dye, color is developed and precipitated on the stick strip. Subsequently, sticks can be analyzed with an imaging system to determine signal intensity of the precipitated dye.

13.4.2 Oxygen consumption

13.4.2.1 Classic Clark-type electrode methods

Several techniques are available to measure oxygen consumption but the classical Clark-type electrode is still widely used and considered reliable. The Clark-electrode consists of a temperature-controlled incubation chamber with a magnetic stirrer and an oxygen electrode inserted and separated from the sample by an oxygen-permeable Teflon membrane [139]. The chamber contains a well-defined volume of media and is closed by a stopper to maintain the sample isolated from air. The addition of substrates/inhibitors is done through a small port inside the stopper using a syringe [139]. The electrode continuously measures the O_2 concentration in the chamber and generates an electric signal that is registered. Several methods have been published for oxygen consumption measurements in different biological samples such as isolated mitochondria, intact or permeabilized cells, bacteria, and even whole organisms [139–143].

Oxygen consumption in intact cells is usually performed in culture medium without serum and supplemented with glucose (1–30 mM), glutamine (0.5–2 mM), pyruvate (0.5–2 mM), or palmitate (100–600 nM) depending on the metabolic pathway associated with mitochondrial respiration. In these conditions, the respiration is regulated by intracellular substrate's production from the different catabolic pathways and by cellular energy demands. After cell permeabilization or mitochondria isolation, the addition of substrates and stimulators/inhibitors allows to modulate respiration and to obtain several parameters that may measure defects in the respiratory chain, coupling between substrate oxidation and phosphorylation (respiratory control ratio, RCR) or efficiency of the mitochondrial OXPHOS (ADP/O index) [139–142]. The most popular method for cell permeabilization consists in the use of detergents (such as digitonin and saponin) in properly optimized conditions [139,140,142]. The composition of the assay buffer may vary but it is crucial to maintain the osmolarity of the media in the range of 250–300 mOsm to keep the mitochondrial membranes intact.

Using permeabilized cells or isolated mitochondria, it is possible to measure respiration by dissecting the different contribution of respiratory complexes to oxygen consumption. The respiration buffer contains 1 mM ADP to maintain fully active ATP synthase. Respiration driven by CI + III + IV can be measured adding 5 mM malate and 5 mM glutamate or 10 mM pyruvate, being these substrates oxidized by dehydrogenases that produce NADH. Conversely, after inhibiting CI with 1 μM rotenone, the addition of succinate or glycerol-3-phosphate (G3P) leads to reduction of quinone bypassing CI and allowing to measure respiration depending on CII + III + IV or G3P dehydrogenase + CIII + IV. After inhibition of CIII with 1 μM aA, the addition of N,N,N′, N′-tetramethyl-*p*-phenylenediamine (TMPD) and ascorbate directly reduces cytochrome *c* providing electrons to CIV. In this way, it is possible to reveal whether and where a defect in the respiratory chain occurs [140].

In permeabilized cells or isolated mitochondria, it is possible to measure the total RCR and ADP/O index. RCR is defined as the ratio between state 3 and state 4 respiration. State 3 is the rate of maximal respiration stimulated by ADP in presence of substrates, such as malate and glutamate or pyruvate, succinate, and ascorbate/TMPD for CI-, CII-, and CIV-driven respiration, respectively. State 4 is the rate of oxygen consumption measured in the same sample after the complete transformation of ADP into ATP in which respiration slows down and the respiratory chain is used only to maintain the mitochondrial membrane potential dissipated by carriers and proton leak. Usually, RCR values reflect the quality of mitochondrial preparation. RCR values around 6–8 for CI substrates, 2–4 for CII substrates, and 1.5–2 for ascorbate/TMPD indicate a good mitochondrial preparation. The ADP/O index indicates the efficiency of mitochondrial phosphorylation system. The index is calculated as the number of nmol of ADP added to the system divided by the number of atoms of oxygen consumed during state 3 respiration [139–142].

13.4.2.2 High-resolution respirometry

The Oroboros O2K instrument is a high-resolution respirometer designed to study mitochondrial physiology using either intact or permeabilized cells, isolated mitochondria, tissue homogenates, or even permeabilized tissues [144]. The Oroboros O2K is very sensitive and can detect oxygen flux as low as 0.5 pmol oxygen/sec/mL in a temperature-controlled fashion. Updated version of the instrument couples the high-resolution respirometer with a fluorometer providing the capability to measure simultaneously oxygen consumption, and either hydrogen peroxide production, mitochondrial membrane potential, Ca^{2+} flux, or ATP production using the respective fluorescent probes Amplex Red, tetramethylrhodamine methyl ester (TMRM), Calcium Green, or Magnesium Green dyes, respectively [145–148]. Additional modules containing ion-selective electrodes can be integrated allowing to measure changes in membrane potential and pH. The module containing an amperometer allows to measure nitric oxide, hydrogen sulfide, and hydrogen peroxide production. The wide range of capabilities and its sensitivity makes the Oroboros O2K the instrument of choice to study mitochondrial bioenergetics and function. However, it comes with the limitation of having only two sample chambers. This restricts the streamline provided by high-throughput approaches like those based on multiwell plate formats, which are suitable for screening large number of samples but have low sensitivity and limited flexibility in terms of experimental protocols (number of injections and measurement type).

13.4.2.3 Microrespirometry on multiwell plate

One of the major limits of electrode-based approaches to measure oxygen consumption is the elevated quantity of sample necessary to perform the experiments and the necessity of a prolonged stirring of the sample which may interfere with cell and mitochondrial fitness. However, the advance in fluorescence- and phosphorescence-based oxygen sensors prompted the development of methods to measure oxygen consumption in adherent cell monolayers in a convenient multiwell plate format [149,150]. These methods present the advantage to reduce the required amount of biological material, to permit multiple measurements in different experimental conditions, and to maintain cells in a status closer to their physiological state. Moreover, they can be coupled with other measurements such as pH quantification, metabolic flux evaluation [151], or proteomics approaches [152] to define the metabolic setting of the cell. However, they are less sensitive than high-resolution approaches and designed for high-throughput experiments such as evaluation of multiple mutants or drug screenings. Interestingly, these approaches have been developed to measure oxygen consumption in intact cells or isolated mitochondria [153], but they have been applied also to 3D cell models, microorganisms, whole tissues, and even small organisms such as worms or zebrafish embryos

[154–158]. These sensors are commercially available from Agilent in soluble form (MitoXpress) or in the more popular format fixed on plastic support (Seahorse XF). In particular, Seahorse XF system has reached a tremendous popularity thanks to the optimized protocols and reagents that allow a rapid and easy assessment of mitochondrial and cell metabolism by bioenergetic newbies. The Seahorse XF Analyzer measures respiration (oxygen consumption rate; OCR) and extracellular pH (extracellular acidification rate; ECAR) to provide in real-time information on mitochondrial function and metabolic status of viable cells [153]. Technical details, working principles, and OCR and ECAR mathematical corrections are described on the manufacturer website and by Gerencser and colleagues [149]. Briefly, fluorescent biosensors for pH and O_2 are immobilized onto a disposable cartridge and coupled to an optic fiber waveguide. OCR and ECAR measurements are performed in real time into a transient microchamber formed by the multiwell plate and the cartridge in a medium lacking the carbonate/bicarbonate buffer. During the experiment, up to four injections can be performed allowing the addition of increasing concentrations of the same compound or different treatments. Several ready-to-use kits have been designed to cover various aspects of energy metabolism. The MitoStress Test kit exploits OXPHOS complexes inhibitors (oligomycin, rotenone, and antimycin) and the uncoupling agent carbonyl cyanide-p-trifluoromethoxyphenylhydrazone (FCCP) to determine basal respiration, ATP production-coupled respiration, maximal and reserve capacities, and non-mitochondrial respiration. This is the classic assay used to identify differences in mitochondrial function among cells and to test the effects of drugs or genetic manipulation on respiration. More details are provided in Box 13.2. Two kits called Glycolysis Stress Test and Glycolytic Rate Assay have been designed to study glycolysis. The first one measures the cell glycolytic capacity and reserve and investigates the ability of cells to boost their glycolytic activity in response to bioenergetic requirements. The second determines the glycolytic rate under basal conditions and when the respiratory chain is inhibited, a parameter that has been showed directly correlated with extracellular lactate accumulation, providing valuable insights on cell responsiveness to metabolic stress. Other optimized protocols and reagents permit the determination of fatty acid oxidation and ATP synthesis or the preferential fuel source of the respiratory chain [159–161]. Noteworthy, these kits represent only some of the possible tests that can be performed. Experimental design can be modified to test any compound that does not interfere with the oxygen and pH sensors. However, correct data interpretation, proper OCR/ECAR normalization, and strong statistics are always necessary to obtain informative and reliable results [153,162]. When using cells for these experiments, one important parameter is to set up seeding, as cells must be confluent but not overgrown in the well. Moreover, the optimal concentration of permeabilizing agents (digitonin or recombinant perfringolysin O, provided by Agilent as XF Plasma Membrane Permeabilizer) must be experimentally established to investigate substrate utilization to permit the mitochondrial substrate uptake without detaching cells from the well [163,164]. Detailed protocols are described for permeabilized cells [163] or mitochondria isolated from different mouse tissues [165].

13.4.3 Determination of mitochondrial membrane potential

According to Mitchell's chemiosmotic theory, the mitochondrial electron transport chain is coupled to the proton gradient generated by CI, CIII, and CIV and dissipated by CV to catalyze ATP production. The electrochemical potential is composed by a ΔpH given by the pH difference across the inner mitochondrial membrane and by the electrical component of mitochondrial membrane potential ($\Delta\psi_m$). At 37°C, the relation between these two components is defined by the following equation:

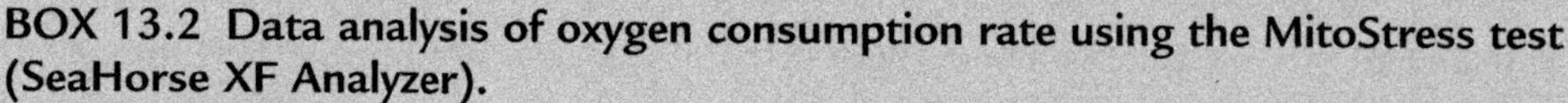

BOX 13.2 Data analysis of oxygen consumption rate using the MitoStress test (SeaHorse XF Analyzer).

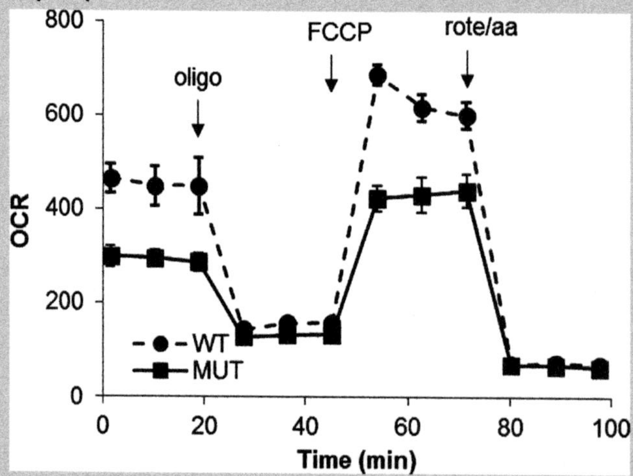

Oxygen consumption rate (OCR) can be determined in intact cells, isolated mitochondrial, and other specimens in basal conditions or upon stimulation with specific compounds. In this graph is represented a hypothetical experiment performed with SeaHorse XF Analyzer by using either wild type (*dotted line, round*) (WT) or mtDNA-mutated (*full line, square*) (MUT) intact cells. Ideally, a mutation in mtDNA can induce an OCR reduction already in basal conditions, as shown in this example. However, in this type of experiment, several parameters can be calculated and multiple information regarding mitochondrial activity can be extrapolated. After addition of oligomycin (oligo), the difference between the residual and the basal respiration is an indication of ATP production, while the residual respiration subtracted for the non-specific OCR (after rotenone and antimycin A injection) is a measurement of proton leak. Moreover, by adding the uncoupler FCCP, the maximal respiration and the spare respiratory capacity (maximal subtracted for basal) can be calculated. The residual respiration after the inhibition of CI with rotenone (Rote) and CIII with antimycin A allows to estimate the non-mitochondrial OCR for data normalization. Data must always be corrected for cell density that can be determined by using several methods, including cell counting, protein, or DNA amount or fluorescence detection of cells.

$$\Delta p(\text{mV}) = \Delta\psi - 61\Delta\text{pH}$$

where Δp represents the proton motive force [166]. In physiological conditions, $\Delta\psi_m$ value is 150 mV, while ΔpH is -0.5 units and thus Δp value is about 180 mV (mitochondrial matrix negatively charged) [166]. Several lipophilic cell-permeant cationic fluorescent dyes have been developed to determine $\Delta\psi_m$ in different protocols. Due to their positive charge, they accumulate in the polarized mitochondria according to the Nernst equation [166]. These molecules include the already mentioned TMRM, tetramethylrhodamine ethyl (TMRE) ester, rhodamine 123 (Rh123), 3,30-dihexyloxacarbocyanine iodide (DiOC6(3)), and 5,50,6,60-tetrachloro-1,10,3,30-tetraethylbenzimidazolylcarbocyanine iodide (JC-1). Insightful hints and considerations on the semiquantitative nature of these assays, on experimental set up, and on data analysis and interpretation can be found elsewhere [167,168]. JC-1 is a cationic ratiometric dye that is accumulated within the mitochondria in a $\Delta\psi_m$-dependent manner. At low concentration, the dye exists as monomer and emits a green fluorescence (λ_{em} = 525 nm) while when accumulated produces J-aggregates and the fluorescence shifts to red (λ_{em} = 590 nm) [169]. This probe is suitable for confocal microscopy and flow cytometry and permits the analysis of both intact cells and isolated mitochondria [170]. The carbocyanine DiOC6(3) is a green-fluorescent lipophilic dye selectively accumulated within the mitochondria when used at low concentrations ($<$100 nM) but stains also the endoplasmic reticulum when used at higher concentrations. Rh123, TMRM, and TMRE are also widely used for determination of $\Delta\psi_m$ and can be used in a "quenching" mode (high concentrations 1–20 μM) or in non-quenching mode (nM concentration range) [167]. Among these, TRMR has been found to display the lowest non-specific localization, to be relatively non-toxic, and to not affect mitochondrial respiration [171]. This dye has been used in flow cytometry and fluorescence microscopy including high-throughput screenings [172] and can be used simultaneously with other fluorescent probes [173–175]. An example of interpretation of a TRMR-based assay in nonquencing mode at fluorescent microscope is reported in Box 13.3. New potential-sensitive fluorescent probes and methodologies have been reported to ameliorate the specificity and the sensitivity of these assays and to expand the determination of $\Delta\psi_m$ also in *in vivo* models [176–180].

13.4.4 ATP production

Several methods have been developed to quantify ATP produced by intact or permeabilized cells or isolated mitochondria. Some methods are based on fluorometry [26] or alternatively on [^{32}P]P_i incorporation into ADP [181] and have been applied to isolated mitochondria. Another rapid and reliable way for measuring ATP not only in isolated mitochondria but also in permeabilized cells is the use of the firefly luciferin-luciferase. Using this system, the rate of ATP production can be measured by stopping parallel reactions at different end points and subsequently measuring ATP content in the sample [182] or by measuring

BOX 13.3 Data analysis of TRMR-based assay for mitochondrial membrane potential.

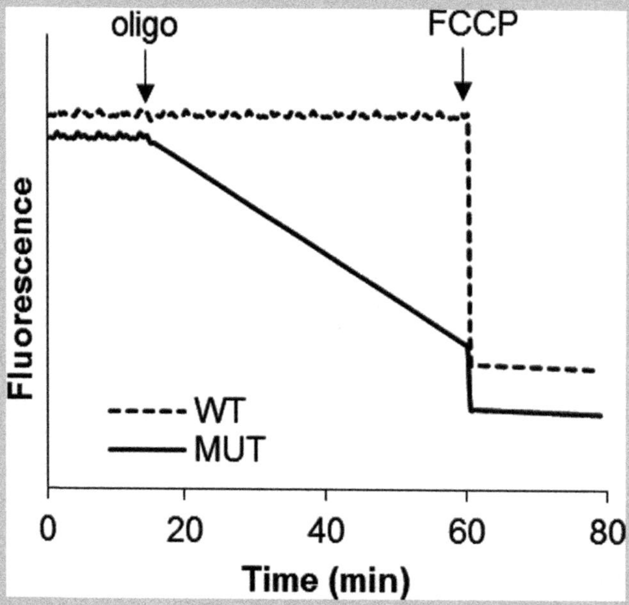

The graph represents a hypothetical result of an experiment to determine mitochondrial membrane potential in intact cells using a fluorescent microscope (i.e., TRMR in non-quenching mode). Fluorescence intensity is plotted during time. In this example, the *dotted line* represents wild type cells (WT), while the *full line* is a putative mtDNA mutant (MUT) in respiratory complexes genes that induces an impairment in the generation of mitochondrial membrane potential. In this type of experiment, cells are seeded onto a glass-bottomed dish or a glass in growth medium. After 24 hours, the medium is eliminated, cells are incubated in a buffer solution containing 20 nM TRMR and incubated for 30 minutes at 37°C to permit the dye to reach the equilibrium between mitochondria and cytosol depending on mitochondrial membrane polarization. Then cells are observed at the microscope, a field containing few cells (depends on the cell type, usually 2–6) is chosen, and one picture corresponding to time 0 is acquired. Mitochondrial areas are selected together with a background area not containing cells to normalize data. Fluorescence intensity of the areas is then measured during time upon the addition of inhibitors or other compounds. The baseline fluoresce can be the same between two cell lines or it can differ; this is not a direct indication of differences in mitochondrial membrane potential and a calibration of the dye must be performed (see the description below). To assess the impact of a mtDNA mutation on the generation of mitochondrial membrane potential, a simple experiment can be performed by recording the baseline fluorescence of the two cell lines (wild type and mutant) for at least 5 minutes, then by incubating cells with 2 μM oligomycin (oligo) to inhibit CV activity and following the fluorescence intensity during time (at least 30 minutes). If the mitochondrial membrane potential is generated by the proton translocation mediated by respiratory CI, CIII, and CIV, the fluorescence intensity

(Continued)

BOX 13.3 Data analysis of TRMR-based assay for mitochondrial membrane potential. (Continued)

will remain constant or will increase during time since CV is not able to dissipate the gradient to produce ATP. Conversely, a reduction of the TRMR intensity after oligomycin injection indicates that mitochondrial membrane potential is sustained by the reverse activity of CV that hydrolyze ATP to ADP in order to maintain the electrochemical gradient. The injection of uncoupler FCCP (usually 1 μM) at the end of the experiment is necessary to calibrate the system. In fact, it helps to normalize data especially when the initial intensity is different between the analyzed cell lines. This type of experiment is semiquantitative, as it can be used to compare multiple experimental conditions but do not allow the measurement of the electric component of the electrochemical gradient. An independent anionic indicator of plasma membrane potential can be used in parallel (see main text).

mitochondrial ATP synthesis continuously [183]. One of the most used methods applied to permeabilized fibroblasts or cybrids carrying mtDNA mutation has been published by Manfredi et al. [183]. In this method, the rate of ATP synthesis is continuously measured in digitonin-permeabilized cells and the rates of ATP synthesis driven by CI (malate and pyruvate) or CII (succinate) substrates are determined. The same method has been used and implemented to also measure ATP synthesis driven by G3P and DBH_2 to assess the contribution to ATP synthesis by CIII and CIV independently form CI or CII [184,185]. An example of interpretation of this type of experiment is reported in Box 13.4.

13.4.5 Reactive oxygen species measurement

Mitochondria are one of the main sources of reactive oxygen and nitrogen species in the cell. Various mitochondrial enzymes can produce free radicals including CI, CII, CIII, α-ketoglutarate dehydrogenase, monoamine oxidases, aconitase, and $p66^{SHC}$ and NADP oxidases [186,187]. Free radicals are short lived and have an important role in cell signaling; however, in excess, they produce oxidative stress and damage. One way to assess mitochondrial dysfunction is by determining the production of free radicals [188]. Hydrogen peroxide, superoxide, and peroxynitrite are the most abundant free radical species and can be detected using small-molecule probes. The most widely used small-molecule fluorescent probes are Amplex Red, dihydroethidium, MitoSox, dichlorodihydrofluorescein diacetate (DCFH-DA), and dihydrorhodamine [189,190]. The use of these fluorescent probes requires a lot of standardization, careful experimental design, and appropriate controls to ensure specificity of the fluorescent signal. There are multiple parameters to be considered such as loading, light sensitivity of the probe, oxidation or photoreduction, and level of endogenous peroxidases that could affect the probe's fluorescence independent of ROS production. Therefore other approaches are required to confirm the results obtained [189,191]. ROS measurements are

BOX 13.4 Data analysis of ATP synthesis assay.

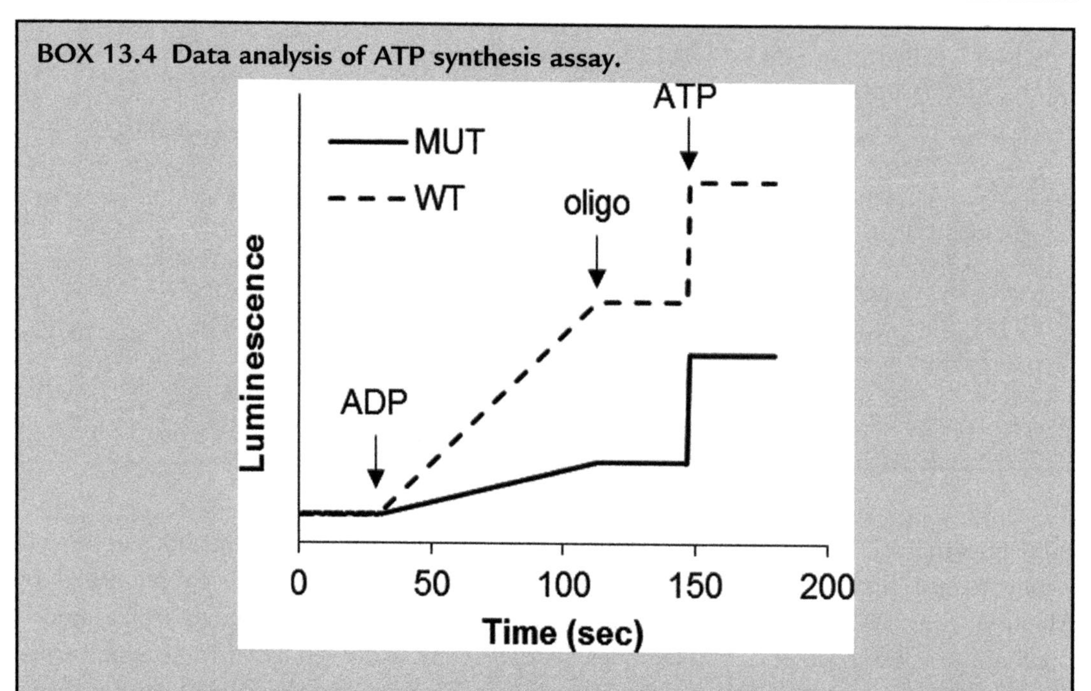

Kinetic measurement of ATP synthesis in digitonin-permeabilized cells. Luminescence curve is recorded during time in a single tube luminometer using a luciferin/luciferase assay (as described in text) in the presence of different substrates for the respiratory chain: malate/pyruvate for CI, succinate for CII, glycerol-3-phosphate for G3P dehydrogenase or DBH_2 for CIII. The reaction starts with the addition of ADP and stops in presence of oligomycin (oligo). The calibration of luminescence signal with the corresponding concentration of ATP is performed adding in the same cuvette a known amount of ATP as standard. The *dotted line* refers to a typical experiment of ATP synthesis in control cells (WT), while the *full line* corresponds to ATP synthesis in cells carrying mtDNA mutations (MUT) affecting oxidative phosphorylation.

usually performed in culture cells incubated with one of the above mentioned dyes and then changes in fluorescence over time determined either by using a plate reader, a fluorescence spectrometer, HPLC, flow cytometry, or fluorescent microscopy. More recently, fluorescent protein-based ROS indicators have been developed. These genetically encoded biosensors are extremely useful for *in vivo* applications and can be expressed in a cell type-specific fashion depending on the promoter and targeted to a specific cell compartment. Some examples of these biosensors are the redox sensitive green fluorescent protein (RoGFP) and the chimeric protein named HyPer for H_2O_2 determination [189]. Another approach to assess mitochondrial matrix H_2O_2 levels *in vivo* has been developed using the probe MitoB conjugated to triphenylphosphonium that drives its accumulation within mitochondria. Reacting with H_2O_2, MitoB forms a molecule called MitoP; thus by measuring the MitoP/MitoB ratio using liquid chromatography/mass spectrometry, it is possible to evaluate the production of

H_2O_2 inside the mitochondria, in living cells, in tissues, and in organs [192]. Alternatively, surrogate markers of ROS production and oxidative stress have also been used by many scientists. These alternatives include the detection of protein oxidation using Oxyblot, detection of protein adducts of lipid peroxidation measuring 4-hydroxynonenal and malondialdehyde levels as well as levels of nitrotyrosine. Additionally, alterations in the steady-state level or activity of cellular antioxidants such as glutathione, SOD1, SOD2, glutathione peroxidase, and glutathione reductase can serve as indicators of oxidative stress.

13.4.6 Blue Native Polyacrylamide Gel Electrophoresis

One of the methods that have gained popularity in the last decade to study samples of patients with mitochondrial disorders and to complement other analysis such as mitochondrial respiration and enzymatic activity is Blue Native PolyAcrylamide Gel Electrophoresis (BN-PAGE). BN-PAGE is used to determine the steady-state level of respiratory complexes, their assembly, enzymatic activity, and their interactions to form supercomplexes. The technique was initially described by Shägger and von Jagow to separate respiratory complexes on the basis of their native molecular weight by gel electrophoresis [193]. Proteins are extracted with mild detergents and exposed to Coomassie Blue dye, which binds proteins conferring a negative charge and in that way, proteins can migrate through an electric field without changes in their native conformation. Depending on the detergent used, one can detect individual respiratory complexes or their association into supercomplexes. Lauryl maltoside is commonly used to detect individual complexes, while digitonin is used when it is desired to preserve supercomplex structures. Lauryl maltoside is a non-ionic glycosidic detergent efficient for hydrophobic protein solubilization that allows to solubilize and disrupt the interaction between the respiratory complexes, but mild enough to maintain their multimeric composition. Digitonin is a non-ionic steroidal glycoside detergent that does not cause denaturation of proteins during solubilization and is known to interact with cholesterol in membranes. Digitonin is a much milder detergent than lauryl maltoside and does not disrupt the interaction between respiratory complexes, leaving supercomplexes intact. The respiratory complexes can be detected by either staining gels with Coomassie or other protein stain, by western blot if proteins are transferred onto a nitrocellulose or PVDF membranes or by In-Gel Activity (IGA) stain. Because the complexes retain their native conformation after electrophoresis, gels can be incubated with appropriate substrates and dye to detect the enzymatic activity by the formation of a color precipitate. IGA assays have been developed to measure CI, CII, CIV, and CV, but these stains are semiquantitative [194]. Individual components of each complex can be analyzed running a second dimension in an SDS-PAGE. For the first dimension, BN-PAGE is run, then each gel lane is cut, incubated with SDS and reducing agents to denature native proteins, placed on top of a SDS-polyacrylamide gel and stacking gel poured around it. In the second dimension, proteins are separated under denaturing conditions and individual subunit of complexes analyzed by western blot. One of the advantages of BN-PAGE is that it allows to screen for defects on various

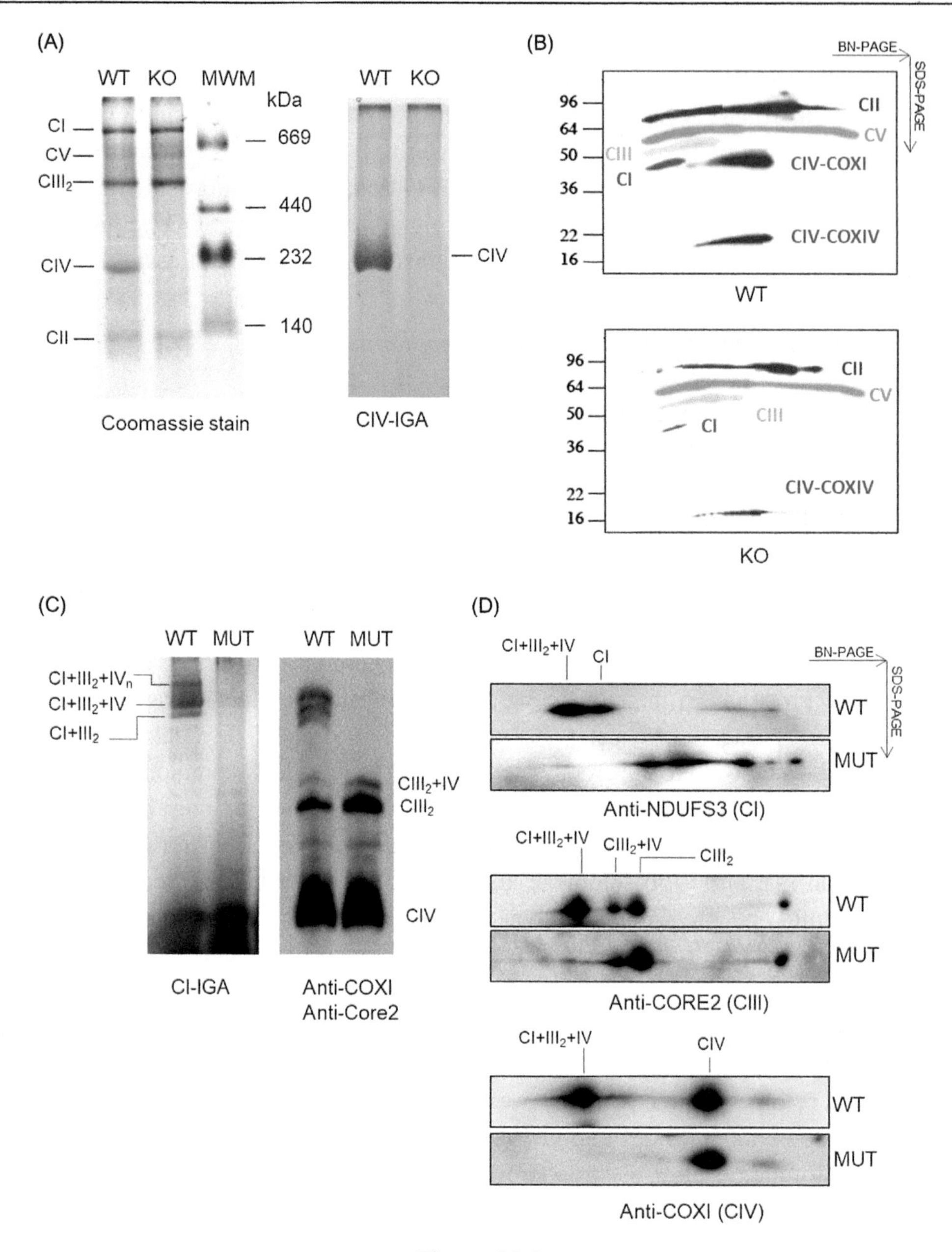

Figure 13.3

Examples of results of BN-PAGE and 2D-PAGE for isolated OXPHOS complexes and supercomplexes. (A) Analysis of CIV deficiency using BN-PAGE in skeletal muscle from wild type (WT) and *Cox10*^{KO} (KO) mouse models [198]. Cox10 is a CIV assembly factor required for maturation and stability of Cox1. Muscle mitochondria were isolated from WT and KO mice and then treated with lauryl maltoside to extract respiratory complexes. Samples were separated by BN-PAGE in a 3%–13% acrylamide native gel and proteins were stained with Coomassie blue or for CIV activity (CIV-IGA) as reported in Ref. [194]. Note the absence of fully assembled CIV in the KO compared to WT. (*Continued*) respiratory complexes at the same time and can provide information if an enzymatic activity defect detected by other methods is due to decreased steady-state levels of the respiratory complex. Detailed protocols for BN-PAGE can be found in the following references [194–197] and an example of results is shown in Fig. 13.3.

Research perspectives

The role of mtDNA mutations in cell biology is still not completely understood. Several issues remain unsolved such as the regulation of heteroplasmic shift and the cellular metabolic reprogramming induced by mtDNA mutations. Understanding these mechanisms would provide novel knowledge of fundamental biological process and would enhance our ability to discover new therapies for diseases in which mtDNA mutations are involved. In this scenario, iPSCs technology may strongly contribute to shed light on the involvement and the impact of mtDNA mutations during the different phases of cell differentiation. Furthermore, the increasing ability to obtain organoids will pave the way to understand the specific function/disfunction of mitochondria in organs specially affected by mtDNA mutations. Lastly, novel approaches in gene editing are still necessary to manipulate mtDNA with the aim to generate more faithful animal models and to test gene therapy strategies for mitochondrial disorders.

◀ (B) Mitochondria isolated from the same WT and KO mouse skin fibroblasts extracted with lauryl maltoside and then separated by BN-PAGE as in (A) followed by a second dimension. The second dimension was performed by cutting gel strips of BN-PAGE and separating proteins in a denaturing SDS-PAGE. Proteins were transferred onto a PVDF membrane and immunodetected with antibodies against various subunits of OXPHOS complexes. Signals were developed by chemiluminescence. For CI, we used anti-NDUFA9, for CII anti-SDHA, for CIII anti-Core2, for CV anti-ATPaseβ, and for CIV anti-COXI and anti-COXIV antibodies, respectively. Antibodies were added sequentially to the same blots. Signals of blots were color-coded and merged to create the figure. Note the absence of COXI signal and decreased levels of COXIV (in *blue*) in the KO sample. (C) Analysis of supercomplexes using BN-PAGE in human cybrids carrying the homoplasmic frameshift mutation m.3571insC/*MT-ND1* (MUT) and corresponding control (WT). The mutation generates a premature stop codon in ND1 causing a lack of the protein [199]. Mitochondria were isolated from WT and MUT cells and then treated with digitonin to extract respiratory complexes and supercomplexes ($CI + III_2 + IV$; $CI + III_2$; $CIII_2 + IV$). Samples were separated by BN-PAGE in a 3%–13% acrylamide native gel and proteins were stained for CI activity (CI-IGA) or transferred onto a PVDF membrane and immunodetected by using antibodies against COXI (CIV) and Core2 (CIII), which are able to highlight all the supercomplexes and isolated $CIII_2$ and CIV. Note the absence of supercomplexes containing CI in the mutant cell line (MUT). (D) Mitochondria isolated from the same WT and MUT cybrids as in (C). Respiratory supercomplexes were separated BN-PAGE followed by a denaturing SDS-PAGE. Proteins were transferred onto a PVDF membrane and immunodetected with antibodies against various subunits of OXPHOS complexes. Signals were developed by chemiluminescence. For CI, we used anti-NDUFS3, for CIII anti-Core2, and for CIV anti-COXI antibodies, respectively. Antibodies were added sequentially to the same blots. Note the absence of supercomplexes containing CI and the presence of low molecular weight staining for NDUFS3 (subcomplexes) in the MUT sample.

References

[1] Wallace DC, Singh G, Lott MT, Hodge JA, Schurr TG, Lezza AM, et al. Mitochondrial DNA mutation associated with Leber's hereditary optic neuropathy. Science 1988;242:1427–30.

[2] Holt IJ, Harding AE, Morgan-Hughes JA. Deletions of muscle mitochondrial DNA in patients with mitochondrial myopathies. Nature 1988;331:717–19.

[3] Picard M, Wallace DC, Burelle Y. The rise of mitochondria in medicine. Mitochondrion 2016;30:105–16.

[4] King MP, Attardi G. Human cells lacking mtDNA: repopulation with exogenous mitochondria by complementation. Science 1989;246:500–3.

[5] Cherry ABC, Gagne KE, McLoughlin EM, Baccei A, Gorman B, Hartung O, et al. Induced pluripotent stem cells with a mitochondrial DNA deletion. Stem Cell 2013;31:1287–97.

[6] Wallace DC, Fan W. The pathophysiology of mitochondrial disease as modeled in the mouse. Genes Dev 2009;23:1714–36.

[7] Farrar GJ, Chadderton N, Kenna PF, Millington-Ward S. Mitochondrial disorders: aetiologies, models systems, and candidate therapies. Trends Genet 2013;29:488–97.

[8] King MP, Attardi G. Isolation of human cell lines lacking mitochondrial DNA. Methods Enzymol 1996;264:304–13.

[9] Robinson BH. Use of fibroblast and lymphoblast cultures for detection of respiratory chain defects. Methods Enzymol 1996;264:454–64.

[10] Dumoulin R, Mandon G, Collombet JM, Blond JL, Carrier H, Godinot C, et al. Human cultured myoblasts: a model for the diagnosis of mitochondrial diseases. J Inherit Metab Dis 1993;16:545–7.

[11] Hu S-Y, Zhuang Q-Q, Qiu Y, Zhu X-F, Yan Q-F. Cell models and drug discovery for mitochondrial diseases. J Zhejiang Univ Sci B 2019;20:449–56.

[12] Hayflick L, Moorhead PS. The serial cultivation of human diploid cell strains. Exp Cell Res 1961;25:585–621.

[13] Amoli MM, Carthy D, Platt H, Ollier WER. EBV Immortalization of human B lymphocytes separated from small volumes of cryo-preserved whole blood. Int J Epidemiol 2008;37(Suppl 1):i41–5.

[14] Robin JD, Wright WE, Zou Y, Cossette SC, Lawlor MW, Gussoni E. Isolation and immortalization of patient-derived cell lines from muscle biopsy for disease modeling. J Vis Exp 2015;52307.

[15] Ozer HL, Banga SS, Dasgupta T, Houghton J, Hubbard K, Jha KK, et al. SV40-mediated immortalization of human fibroblasts. Exp Gerontol 1996;31:303–10.

[16] Wang Y, Chen S, Yan Z, Pei M. A prospect of cell immortalization combined with matrix microenvironmental optimization strategy for tissue engineering and regeneration. Cell Biosci 2019;9:7.

[17] King MP, Attadi G. Mitochondria-mediated transformation of human rho(0) cells. Methods Enzymol 1996;264:313–34.

[18] Grégoire M, Morais R, Quilliam MA, Gravel D. On auxotrophy for pyrimidines of respiration-deficient chick embryo cells. Eur J Biochem 1984;142:49–55.

[19] Nelson I, Hanna MG, Wood NW, Harding AE. Depletion of mitochondrial DNA by ddC in untransformed human cell lines. Somat Cell Mol Genet 1997;23:287–90.

[20] Ashley N, Harris D, Poulton J. Detection of mitochondrial DNA depletion in living human cells using PicoGreen staining. Exp Cell Res 2005;303:432–46.

[21] Inoue K, Takai D, Hosaka H, Ito S, Shitara H, Isobe K, et al. Isolation and characterization of mitochondrial DNA-less lines from various mammalian cell lines by application of an anticancer drug, ditercalinium. Biochem Biophys Res Commun 1997;239:257–60.

[22] Jazayeri M, Andreyev A, Will Y, Ward M, Anderson CM, Clevenger W. Inducible expression of a dominant negative DNA polymerase-gamma depletes mitochondrial DNA and produces a rho0 phenotype. J Biol Chem 2003;278:9823–30.

[23] Wilkins HM, Carl SM, Swerdlow RH. Cytoplasmic hybrid (cybrid) cell lines as a practical model for mitochondriopathies. Redox Biol 2014;2:619–31.

[24] Sazonova MA, Sinyov VV, Ryzhkova AI, Galitsyna EV, Melnichenko AA, Postnov AY, et al. Cybrid models of pathological cell processes in different diseases. Oxid Med Cell Longev 2018;2018:4647214.

[25] Bayona-Bafaluy MP, Manfredi G, Moraes CT. A chemical enucleation method for the transfer of mitochondrial DNA to rho(o) cells. Nucleic Acids Res 2003;31:e98.

[26] Chomyn A. Platelet-mediated transformation of human mitochondrial DNA-less cells. Methods Enzymol 1996;264:334–9.

[27] Hayashi J, Ohta S, Kikuchi A, Takemitsu M, Goto Y, Nonaka I. Introduction of disease-related mitochondrial DNA deletions into HeLa cells lacking mitochondrial DNA results in mitochondrial dysfunction. Proc Natl Acad Sci U S A 1991;88:10614–18.

[28] Ishikawa K, Hayashi J-I. Generation of mtDNA-exchanged cybrids for determination of the effects of mtDNA mutations on tumor phenotypes. Methods Enzymol 2009;457:335–46.

[29] Rossignol R, Faustin B, Rocher C, Malgat M, Mazat J-P, Letellier T. Mitochondrial threshold effects. Biochem J 2003;370:751–62.

[30] Kenney MC, Chwa M, Atilano SR, Falatoonzadeh P, Ramirez C, Malik D, et al. Molecular and bioenergetic differences between cells with African versus European inherited mitochondrial DNA haplogroups: implications for population susceptibility to diseases. Biochim Biophys Acta 1842;2014:208–19.

[31] Wallace DC. Bioenergetics in human evolution and disease: implications for the origins of biological complexity and the missing genetic variation of common diseases. Philos Trans R Soc Lond B Biol Sci 2013;368:20120267.

[32] Wallace DC. Mitochondrial DNA variation in human radiation and disease. Cell 2015;163:33–8.

[33] Prigione A. Induced pluripotent stem cells (iPSCs) for modeling mitochondrial DNA disorders. Methods Mol Biol 2015;1265:349–56.

[34] Hämäläinen RH, Suomalainen A. Generation and characterization of induced pluripotent stem cells from patients with mtDNA mutations. Methods Mol Biol 2016;1353:65–75.

[35] Fujikura J, Nakao K, Sone M, Noguchi M, Mori E, Naito M, et al. Induced pluripotent stem cells generated from diabetic patients with mitochondrial DNA A3243G mutation. Diabetologia 2012;55:1689–98.

[36] Folmes CDL, Nelson TJ, Martinez-Fernandez A, Arrell DK, Lindor JZ, Dzeja PP, et al. Somatic oxidative bioenergetics transitions into pluripotency-dependent glycolysis to facilitate nuclear reprogramming. Cell Metab 2011;14:264–71.

[37] Hämäläinen RH, Manninen T, Koivumäki H, Kislin M, Otonkoski T, Suomalainen A. Tissue- and cell-type-specific manifestations of heteroplasmic mtDNA 3243A > G mutation in human induced pluripotent stem cell-derived disease model. Proc Natl Acad Sci U S A 2013;110:E3622–30.

[38] Kodaira M, Hatakeyama H, Yuasa S, Seki T, Egashira T, Tohyama S, et al. Impaired respiratory function in MELAS-induced pluripotent stem cells with high heteroplasmy levels. FEBS Open Bio 2015;5:219–25.

[39] Hatakeyama H, Katayama A, Komaki H, Nishino I, Goto Y-I. Molecular pathomechanisms and cell-type-specific disease phenotypes of MELAS caused by mutant mitochondrial tRNA(Trp). Acta Neuropathol Commun 2015;3:52.

[40] Ma H, Folmes CDL, Wu J, Morey R, Mora-Castilla S, Ocampo A, et al. Metabolic rescue in pluripotent cells from patients with mtDNA disease. Nature 2015;524:234–8.

[41] Lin D-S, Huang Y-W, Ho C-S, Hung P-L, Hsu M-H, Wang T-J, et al. Oxidative insults and mitochondrial DNA mutation promote enhanced autophagy and mitophagy compromising cell viability in pluripotent cell model of mitochondrial disease. Cells 2019;8.

[42] Wu Y-R, Wang A-G, Chen Y-T, Yarmishyn AA, Buddhakosai W, Yang T-C, et al. Bioactivity and gene expression profiles of hiPSC-generated retinal ganglion cells in MT-ND4 mutated Leber's hereditary optic neuropathy. Exp Cell Res 2018;363:299–309.

[43] Lu H-E, Yang Y-P, Chen Y-T, Wu Y-R, Wang C-L, Tsai F-T, et al. Generation of patient-specific induced pluripotent stem cells from Leber's hereditary optic neuropathy. Stem Cell Res 2018;28:56–60.

[44] Lasserre J-P, Dautant A, Aiyar RS, Kucharczyk R, Glatigny A, Tribouillard-Tanvier D, et al. Yeast as a system for modeling mitochondrial disease mechanisms and discovering therapies. Dis Model Mech 2015;8:509–26.

[45] Butow RA, Henke RM, Moran JV, Belcher SM, Perlman PS. Transformation of *Saccharomyces cerevisiae* mitochondria using the biolistic gun. Methods Enzymol 1996;264:265–78.
[46] Bonnefoy N, Remacle C, Fox TD. Genetic transformation of *Saccharomyces cerevisiae* and *Chlamydomonas reinhardtii* mitochondria. Methods Cell Biol 2007;80:525–48.
[47] Okamoto K, Perlman PS, Butow RA. The sorting of mitochondrial DNA and mitochondrial proteins in zygotes: preferential transmission of mitochondrial DNA to the medial bud. J Cell Biol 1998;142:613–23.
[48] Feuermann M, Francisci S, Rinaldi T, De Luca C, Rohou H, Frontali L, et al. The yeast counterparts of human 'MELAS' mutations cause mitochondrial dysfunction that can be rescued by overexpression of the mitochondrial translation factor EF-Tu. EMBO Rep 2003;4:53–8.
[49] Montanari A, Besagni C, De Luca C, Morea V, Oliva R, Tramontano A, et al. Yeast as a model of human mitochondrial tRNA base substitutions: investigation of the molecular basis of respiratory defects. RNA 2008;14:275–83.
[50] De Luca C, Zhou Y, Montanari A, Morea V, Oliva R, Besagni C, et al. Can yeast be used to study mitochondrial diseases? Biolistic tRNA mutants for the analysis of mechanisms and suppressors. Mitochondrion 2009;9:408–17.
[51] Börner GV, Zeviani M, Tiranti V, Carrara F, Hoffmann S, Gerbitz KD, et al. Decreased aminoacylation of mutant tRNAs in MELAS but not in MERRF patients. Hum Mol Genet 2000;9:467–75.
[52] Blakely EL, Mitchell AL, Fisher N, Meunier B, Nijtmans LG, Schaefer AM, et al. A mitochondrial cytochrome b mutation causing severe respiratory chain enzyme deficiency in humans and yeast. FEBS J 2005;272:3583–92.
[53] Fisher N, Castleden CK, Bourges I, Brasseur G, Dujardin G, Meunier B. Human disease-related mutations in cytochrome b studied in yeast. J Biol Chem 2004;279:12951–8.
[54] Fisher N, Bourges I, Hill P, Brasseur G, Meunier B. Disruption of the interaction between the Rieske iron-sulfur protein and cytochrome b in the yeast bc1 complex owing to a human disease-associated mutation within cytochrome b. Eur J Biochem 2004;271:1292–8.
[55] Kessl JJ, Ha KH, Merritt AK, Lange BB, Hill P, Meunier B, et al. Cytochrome b mutations that modify the ubiquinol-binding pocket of the cytochrome bc1 complex and confer anti-malarial drug resistance in *Saccharomyces cerevisiae*. J Biol Chem 2005;280:17142–8.
[56] Meunier B, Fisher N, Ransac S, Mazat J-P, Brasseur G. Respiratory complex III dysfunction in humans and the use of yeast as a model organism to study mitochondrial myopathy and associated diseases. Biochim Biophys Acta 2013;1827:1346–61.
[57] Song Z, Laleve A, Vallières C, McGeehan JE, Lloyd RE, Meunier B. Human mitochondrial cytochrome *b* variants studied in yeast: not all are silent polymorphisms. Hum Mutat 2016;37:933–41.
[58] Bratton M, Mills D, Castleden CK, Hosler J, Meunier B. Disease-related mutations in cytochrome c oxidase studied in yeast and bacterial models. Eur J Biochem 2003;270:1222–30.
[59] Meunier B. Site-directed mutations in the mitochondrially encoded subunits I and III of yeast cytochrome oxidase. Biochem J 2001;354:407–12.
[60] Meunier B, Taanman J-W. Mutations of cytochrome c oxidase subunits 1 and 3 in *Saccharomyces cerevisiae*: assembly defect and compensation. Biochim Biophys Acta 2002;1554:101–7.
[61] Rak M, Tetaud E, Duvezin-Caubet S, Ezkurdia N, Bietenhader M, Rytka J, et al. A yeast model of the neurogenic ataxia retinitis pigmentosa (NARP) T8993G mutation in the mitochondrial ATP synthase-6 gene. J Biol Chem 2007;282:34039–47.
[62] Kucharczyk R, Rak M, di Rago J-P. Biochemical consequences in yeast of the human mitochondrial DNA 8993T>C mutation in the ATPase6 gene found in NARP/MILS patients. Biochim Biophys Acta 1793;2009:817–24.
[63] Kucharczyk R, Ezkurdia N, Couplan E, Procaccio V, Ackerman SH, Blondel M, et al. Consequences of the pathogenic T9176C mutation of human mitochondrial DNA on yeast mitochondrial ATP synthase. Biochim Biophys Acta Bioenerg 2010;1797:1105–12.
[64] Kucharczyk R, Salin B, di Rago J-P. Introducing the human Leigh syndrome mutation T9176G into *Saccharomyces cerevisiae* mitochondrial DNA leads to severe defects in the incorporation of

Atp6p into the ATP synthase and in the mitochondrial morphology. Hum Mol Genet 2009;18:2889–98.
[65] Kabala AM, Lasserre J-P, Ackerman SH, di Rago J-P, Kucharczyk R. Defining the impact on yeast ATP synthase of two pathogenic human mitochondrial DNA mutations, T9185C and T9191C. Biochimie 2014;100:200–6.
[66] Kucharczyk R, Giraud M-F, Brèthes D, Wysocka-Kapcinska M, Ezkurdia N, Salin B, et al. Defining the pathogenesis of human mtDNA mutations using a yeast model: the case of T8851C. Int J Biochem Cell Biol 2013;45:130–40.
[67] Kucharczyk R, Dautant A, Gombeau K, Godard F, Tribouillard-Tanvier D, di Rago J-P. The pathogenic MT-ATP6 m8851T > C mutation prevents proton movements within the n-side hydrophilic cleft of the membrane domain of ATP synthase. Biochim Biophys Acta Bioenerg 1860;2019:562–72.
[68] Skoczeń N, Dautant A, Binko K, Godard F, Bouhier M, Su X, et al. Molecular basis of diseases caused by the mtDNA mutation m8969G > A in the subunit a of ATP synthase. Biochim Biophys Acta Bioenerg 2018;1859:602–11.
[69] Couplan E, Aiyar RS, Kucharczyk R, Kabala A, Ezkurdia N, Gagneur J, et al. A yeast-based assay identifies drugs active against human mitochondrial disorders. Proc Natl Acad Sci U S A 2011;108:11989–94.
[70] Aiyar RS, Bohnert M, Duvezin-Caubet S, Voisset C, Gagneur J, Fritsch ES, et al. Mitochondrial protein sorting as a therapeutic target for ATP synthase disorders. Nat Commun 2014;5:5585.
[71] Garrido-Maraver J, Cordero MD, Moñino ID, Pereira-Arenas S, Lechuga-Vieco AV, Cotán D, et al. Screening of effective pharmacological treatments for MELAS syndrome using yeasts, fibroblasts and cybrid models of the disease. Br J Pharmacol 2012;167:1311–28.
[72] Li FY, Cuddon PA, Song J, Wood SL, Patterson JS, Shelton GD, et al. Canine spongiform leukoencephalomyelopathy is associated with a missense mutation in cytochrome b. Neurobiol Dis 2006;21:35–42.
[73] Maglioni S, Ventura N. *C. elegans* as a model organism for human mitochondrial associated disorders. Mitochondrion 2016;30:117–25.
[74] Rea SL, Graham BH, Nakamaru-Ogiso E, Kar A, Falk MJ. Bacteria, yeast, worms, and flies: exploiting simple model organisms to investigate human mitochondrial diseases. Dev Disabil Res Rev 2010;16:200–18.
[75] Lee SS, Lee RYN, Fraser AG, Kamath RS, Ahringer J, Ruvkun G. A systematic RNAi screen identifies a critical role for mitochondria in *C. elegans* longevity. Nat Genet 2003;33:40–8.
[76] Dillin A, Hsu A-L, Arantes-Oliveira N, Lehrer-Graiwer J, Hsin H, Fraser AG, et al. Rates of behavior and aging specified by mitochondrial function during development. Science 2002;298:2398–401.
[77] Van Raamsdonk JM, Meng Y, Camp D, Yang W, Jia X, Bénard C, et al. Decreased energy metabolism extends life span in *Caenorhabditis elegans* without reducing oxidative damage. Genetics 2010;185:559–71.
[78] Melov S, Lithgow GJ, Fischer DR, Tedesco PM, Johnson TE. Increased frequency of deletions in the mitochondrial genome with age of *Caenorhabditis elegans*. Nucleic Acids Res 1995;23:1419–25.
[79] Lakshmanan LN, Yee Z, Ng LF, Gunawan R, Halliwell B, Gruber J. Clonal expansion of mitochondrial DNA deletions is a private mechanism of aging in long-lived animals. Aging Cell 2018;17:e12814.
[80] Bayona-Bafaluy MP, Blits B, Battersby BJ, Shoubridge EA, Moraes CT. Rapid directional shift of mitochondrial DNA heteroplasmy in animal tissues by a mitochondrially targeted restriction endonuclease. Proc Natl Acad Sci U S A 2005;102:14392–7.
[81] Xu H, DeLuca SZ, O'Farrell PH. Manipulating the metazoan mitochondrial genome with targeted restriction enzymes. Science 2008;321:575–7.
[82] Sen A, Cox RT. Fly models of human diseases: Drosophila as a model for understanding human mitochondrial mutations and disease. Curr Top Dev Biol 2017;121:1–27.
[83] Garesse R, Kaguni LS. A Drosophila model of mitochondrial DNA replication: proteins, genes and regulation. IUBMB Life 2005;57:555–61.
[84] Chen Z, Zhang F, Xu H. Human mitochondrial DNA diseases and Drosophila models. J Genet Genomics 2019;46:201–12.
[85] Burman JL, Itsara LS, Kayser E-B, Suthammarak W, Wang AM, Kaeberlein M, et al. A Drosophila model of mitochondrial disease caused by a complex I mutation that uncouples proton pumping from electron transfer. Dis Model Mech 2014;7:1165–74.

[86] Hill JH, Chen Z, Xu H. Selective propagation of functional mitochondrial DNA during oogenesis restricts the transmission of a deleterious mitochondrial variant. Nat Genet 2014;46:389–92.
[87] Patel MR, Miriyala GK, Littleton AJ, Yang H, Trinh K, Young JM, et al. A mitochondrial DNA hypomorph of cytochrome oxidase specifically impairs male fertility in *Drosophila melanogaster*. eLife 2016;5:e16923.
[88] Celotto AM. Mitochondrial encephalomyopathy in Drosophila. J Neurosci 2006;26:810–20.
[89] Inoue K, Nakada K, Ogura A, Isobe K, Goto Y, Nonaka I, et al. Generation of mice with mitochondrial dysfunction by introducing mouse mtDNA carrying a deletion into zygotes. Nat Genet 2000;26:176–81.
[90] Srivastava S, Moraes CT. Double-strand breaks of mouse muscle mtDNA promote large deletions similar to multiple mtDNA deletions in humans. Hum Mol Genet 2005;14:893–902.
[91] Fukui H, Moraes CT. Mechanisms of formation and accumulation of mitochondrial DNA deletions in aging neurons. Hum Mol Genet 2009;18:1028–36.
[92] Pickrell AM, Pinto M, Hida A, Moraes CT. Striatal dysfunctions associated with mitochondrial DNA damage in dopaminergic neurons in a mouse model of Parkinson's disease. J Neurosci 2011;31:17649–58.
[93] Pickrell AM, Fukui H, Wang X, Pinto M, Moraes CT. The striatum is highly susceptible to mitochondrial oxidative phosphorylation dysfunctions. J Neurosci 2011;31:9895–904.
[94] Madsen PM, Pinto M, Patel S, McCarthy S, Gao H, Taherian M, et al. Mitochondrial DNA double-strand breaks in oligodendrocytes cause demyelination, axonal injury, and CNS inflammation. J Neurosci 2017;37:10185–99.
[95] Wang X, Pickrell AM, Rossi SG, Pinto M, Dillon LM, Hida A, et al. Transient systemic mtDNA damage leads to muscle wasting by reducing the satellite cell pool. Hum Mol Genet 2013;22:3976–86.
[96] Kasahara A, Ishikawa K, Yamaoka M, Ito M, Watanabe N, Akimoto M, et al. Generation of trans-mitochondrial mice carrying homoplasmic mtDNAs with a missense mutation in a structural gene using ES cells. Hum Mol Genet 2006;15:871–81.
[97] Fan W, Waymire KG, Narula N, Li P, Rocher C, Coskun PE, et al. A mouse model of mitochondrial disease reveals germline selection against severe mtDNA mutations. Science 2008;319:958–62.
[98] Hashizume O, Shimizu A, Yokota M, Sugiyama A, Nakada K, Miyoshi H, et al. Specific mitochondrial DNA mutation in mice regulates diabetes and lymphoma development. Proc Natl Acad Sci U S A 2012;109:10528–33.
[99] Lin CS, Sharpley MS, Fan W, Waymire KG, Sadun AA, Carelli V, et al. Mouse mtDNA mutant model of Leber hereditary optic neuropathy. Proc Natl Acad Sci U S A 2012;109:20065–70.
[100] Shimizu A, Mito T, Hayashi C, Ogasawara E, Koba R, Negishi I, et al. Transmitochondrial mice as models for primary prevention of diseases caused by mutation in the tRNA(Lys) gene. Proc Natl Acad Sci U S A 2014;111:3104–9.
[101] Kauppila JHK, Baines HL, Bratic A, Simard ML, Freyer C, Mourier A, et al. A phenotype-driven approach to generate mouse models with pathogenic mtDNA mutations causing mitochondrial disease. Cell Rep 2016;16:2980–90.
[102] Zhang D, Mott JL, Chang S-W, Denniger G, Feng Z, Zassenhaus HP. Construction of transgenic mice with tissue-specific acceleration of mitochondrial DNA mutagenesis. Genomics 2000;69:151–61.
[103] Bensch KG, Mott JL, Chang SW, Hansen PA, Moxley MA, Chambers KT, et al. Selective mtDNA mutation accumulation results in beta-cell apoptosis and diabetes development. Am J Physiol Endocrinol Metab 2009;296:E672–80.
[104] Kasahara T, Kubota M, Miyauchi T, Noda Y, Mouri A, Nabeshima T, et al. Mice with neuron-specific accumulation of mitochondrial DNA mutations show mood disorder-like phenotypes. Mol Psychiatry 2006;11:577–93.
[105] Kong YXG, Van Bergen N, Trounce IA, Bui BV, Chrysostomou V, Waugh H, et al. Increase in mitochondrial DNA mutations impairs retinal function and renders the retina vulnerable to injury: mitochondrial DNA mutations lead to neuronal vulnerability. Aging Cell 2011;10:572–83.
[106] Kujoth GC. Mitochondrial DNA mutations, oxidative stress, and apoptosis in mammalian aging. Science 2005;309:481–4.

[107] Trifunovic A, Wredenberg A, Falkenberg M, Spelbrink JN, Rovio AT, Bruder CE, et al. Premature ageing in mice expressing defective mitochondrial DNA polymerase. Nature 2004;429:417–23.
[108] Tyynismaa H, Mjosund KP, Wanrooij S, Lappalainen I, Ylikallio E, Jalanko A, et al. Mutant mitochondrial helicase Twinkle causes multiple mtDNA deletions and a late-onset mitochondrial disease in mice. Proc Natl Acad Sci U S A 2005;102:17687–92.
[109] Matic S, Jiang M, Nicholls TJ, Uhler JP, Dirksen-Schwanenland C, Polosa PL, et al. Mice lacking the mitochondrial exonuclease MGME1 accumulate mtDNA deletions without developing progeria. Nat Commun 2018;9:1202.
[110] Foriel S, Willems P, Smeitink J, Schenck A, Beyrath J. Mitochondrial diseases: *Drosophila melanogaster* as a model to evaluate potential therapeutics. Int J Biochem Cell Biol 2015;63:60–5.
[111] Alziari S, Petit N, Lefai E, Beziat F, Lecher P, Touraille S, et al. A heteroplasmic strain of *D. subobscura* an animal model of mitochondrial genome rearrangement. In: Lestienne P, editor. Mitochondrial diseases: models and methods. Berlin, Heidelberg: Springer Berlin Heidelberg; 1999. p. 197–208.
[112] Celotto AM, Chiu WK, Van Voorhies W, Palladino MJ. Modes of metabolic compensation during mitochondrial disease using the drosophila model of ATP6 dysfunction. PLoS One 2011;6:e25823.
[113] Fogle KJ, Hertzler JI, Shon JH, Palladino MJ. The ATP-sensitive K channel is seizure protective and required for effective dietary therapy in a model of mitochondrial encephalomyopathy. J Neurogenet 2016;30:247–58.
[114] Chen Z, Qi Y, French S, Zhang G, Covian Garcia R, Balaban R, et al. Genetic mosaic analysis of a deleterious mitochondrial DNA mutation in Drosophila reveals novel aspects of mitochondrial regulation and function. Mol Biol Cell 2015;26:674–84.
[115] Torraco A, Peralta S, Iommarini L, Diaz F. Mitochondrial diseases part I: Mouse models of OXPHOS deficiencies caused by defects in respiratory complex subunits or assembly factors. Mitochondrion 2015;21:76–91.
[116] Iommarini L, Peralta S, Torraco A, Diaz F. Mitochondrial diseases part II: Mouse models of OXPHOS deficiencies caused by defects in regulatory factors and other components required for mitochondrial function. Mitochondrion 2015;22:96–118.
[117] Peralta S, Torraco A, Iommarini L, Diaz F. Mitochondrial diseases part III: Therapeutic interventions in mouse models of OXPHOS deficiencies. Mitochondrion 2015;23:71–80.
[118] McFarland R, Swalwell H, Blakely EL, He L, Groen EJ, Turnbull DM, et al. The m5650G > A mitochondrial tRNAAla mutation is pathogenic and causes a phenotype of pure myopathy. Neuromuscul Disord 2008;18:63–7.
[119] Finnilä S, Tuisku S, Herva R, Majamaa K. A novel mitochondrial DNA mutation and a mutation in the Notch3 gene in a patient with myopathy and CADASIL. J Mol Med 2001;79:641–7.
[120] Rahman S, Copeland WC. POLG-related disorders and their neurological manifestations. Nat Rev Neurol 2019;15:40–52.
[121] Tyynismaa H, Sembongi H, Bokori-Brown M, Granycome C, Ashley N, Poulton J, et al. Twinkle helicase is essential for mtDNA maintenance and regulates mtDNA copy number. Hum Mol Genet 2004;13:3219–27.
[122] Spelbrink JN, Li FY, Tiranti V, Nikali K, Yuan QP, Tariq M, et al. Human mitochondrial DNA deletions associated with mutations in the gene encoding Twinkle, a phage T7 gene 4-like protein localized in mitochondria. Nat Genet 2001;28:223–31.
[123] Song L, Shan Y, Lloyd KCK, Cortopassi GA. Mutant Twinkle increases dopaminergic neurodegeneration, mtDNA deletions and modulates Parkin expression. Hum Mol Genet 2012;21:5147–58.
[124] Kornblum C, Nicholls TJ, Haack TB, Scholer S, Peeva V, Danhauser K, et al. Loss-of-function mutations in MGME1 impair mtDNA replication and cause multisystemic mitochondrial disease. Nat Genet 2013;45:214–19.
[125] Nicholls TJ, Zsurka G, Peeva V, Scholer S, Szczesny RJ, Cysewski D, et al. Linear mtDNA fragments and unusual mtDNA rearrangements associated with pathological deficiency of MGME1 exonuclease. Hum Mol Genet 2014;23:6147–62.

[126] Trounce IA, Kim YL, Jun AS, Wallace DC. Assessment of mitochondrial oxidative phosphorylation in patient muscle biopsies, lymphoblasts, and transmitochondrial cell lines. Methods Enzymol 1996;264:484–509.

[127] Connolly NMC, Theurey P, Adam-Vizi V, Bazan NG, Bernardi P, Bolaños JP, et al. Guidelines on experimental methods to assess mitochondrial dysfunction in cellular models of neurodegenerative diseases. Cell Death Differ 2018;25:542–72.

[128] Oliveira KK, Kiyomoto BH, Rodrigues ADS, Tengan CH. Complex I spectrophotometric assay in cultured cells: detailed analysis of key factors. Anal Biochem 2013;435:57–9.

[129] Chretien D, Bénit P, Chol M, Lebon S, Rötig A, Munnich A, et al. Assay of mitochondrial respiratory chain complex I in human lymphocytes and cultured skin fibroblasts. Biochem Biophys Res Commun 2003;301:222–4.

[130] Janssen AJM, Trijbels FJM, Sengers RCA, Smeitink JAM, van den Heuvel LP, Wintjes LTM, et al. Spectrophotometric assay for complex I of the respiratory chain in tissue samples and cultured fibroblasts. Clin Chem 2007;53:729–34.

[131] de Wit LEA, Sluiter W. Chapter 9 Reliable assay for measuring complex I activity in human blood lymphocytes and skin fibroblasts. Methods Enzymol 2009;456:169–81.

[132] Barrientos A, Fontanesi F, Díaz F. Evaluation of the mitochondrial respiratory chain and oxidative phosphorylation system using polarography and spectrophotometric enzyme assays. Curr Protoc Hum Genet 2009; Chapter 19, Unit19.3.

[133] Bénit P, Goncalves S, Philippe Dassa E, Brière J-J, Martin G, Rustin P. Three spectrophotometric assays for the measurement of the five respiratory chain complexes in minuscule biological samples. Clin Chim Acta 2006;374:81–6.

[134] Kirby DM, Thorburn DR, Turnbull DM, Taylor RW. Biochemical assays of respiratory chain complex activity. Methods Cell Biol 2007;80:93–119.

[135] Frazier AE, Thorburn DR. Biochemical analyses of the electron transport chain complexes by spectrophotometry. Methods Mol Biol 2012;837:49–62.

[136] Teodoro JS, Palmeira CM, Rolo AP. Determination of oxidative phosphorylation complexes activities. Methods Mol Biol 2015;1241:71–84.

[137] Pallotti F, Lenaz G. Isolation and subfractionation of mitochondria from animal cells and tissue culture lines. Methods Cell Biol 2007;80:3–44.

[138] Rieske JS. [44] Preparation and properties of reduced coenzyme Q-cytochrome c reductase (complex III of the respiratory chain). Methods Enzymol 1967;10:239–45.

[139] Silva AM, Oliveira PJ. Evaluation of respiration with Clark-type electrode in isolated mitochondria and permeabilized animal cells. Methods Mol Biol 1782;2018:7–29.

[140] Hofhaus G, Shakeley RM, Attardi G. Use of polarography to detect respiration defects in cell cultures. Methods Enzymol 1996;264:476–83.

[141] Li Z, Graham BH. Measurement of mitochondrial oxygen consumption using a Clark electrode. Methods Mol Biol 2012;837:63–72.

[142] Simonnet H, Vigneron A, Pouysségur J. Conventional techniques to monitor mitochondrial oxygen consumption. Methods Enzymol 2014;542:151–61.

[143] Palikaras K, Tavernarakis N. Measuring oxygen consumption rate in *Caenorhabditis elegans*. Bio Protoc 2016;6.

[144] Doerrier C, Garcia-Souza LF, Krumschnabel G, Wohlfarter Y, Meszaros AT, Gnaiger E. High-resolution FluoRespirometry and OXPHOS protocols for human cells, permeabilized fibers from small biopsies of muscle, and isolated mitochondria. Methods Mol Biol 1782;2018:31–70.

[145] Makrecka-Kuka M, Krumschnabel G, Gnaiger E. High-resolution respirometry for simultaneous measurement of oxygen and hydrogen peroxide fluxes in permeabilized cells, tissue homogenate and isolated mitochondria. Biomolecules 2015;5:1319–38.

[146] Krumschnabel G, Fontana-Ayoub M, Sumbalova Z, Heidler J, Gauper K, Fasching M, et al. Simultaneous high-resolution measurement of mitochondrial respiration and hydrogen peroxide production. Methods Mol Biol 2015;1264:245–61.

[147] Elustondo PA, Negoda A, Kane CL, Kane DA, Pavlov EV. Spermine selectively inhibits high-conductance, but not low-conductance calcium-induced permeability transition pore. Biochim Biophys Acta 1847;2015:231–40.

[148] Chinopoulos C, Kiss G, Kawamata H, Starkov AA. Measurement of ADP-ATP exchange in relation to mitochondrial transmembrane potential and oxygen consumption. Methods Enzymol 2014;542:333–48.

[149] Gerencser AA, Neilson A, Choi SW, Edman U, Yadava N, Oh RJ, et al. Quantitative microplate-based respirometry with correction for oxygen diffusion. Anal Chem 2009;81:6868–78.

[150] Dmitriev RI, Papkovsky DB. Optical probes and techniques for O_2 measurement in live cells and tissue. Cell Mol Life Sci 2012;69:2025–39.

[151] Nonnenmacher Y, Palorini R, Hiller K. Determining compartment-specific metabolic fluxes. Methods Mol Biol 2019;1862:137–49.

[152] Walheim E, Wiśniewski JR, Jastroch M. Respiromics—an integrative analysis linking mitochondrial bioenergetics to molecular signatures. Mol Metab 2018;9:4–14.

[153] Divakaruni AS, Paradyse A, Ferrick DA, Murphy AN, Jastroch M. Analysis and interpretation of microplate-based oxygen consumption and pH data. Methods Enzymol 2014;547:309–54.

[154] Leek R, Grimes DR, Harris AL, McIntyre A. Methods: using three-dimensional culture (spheroids) as an *in vitro* model of tumour hypoxia. Adv Exp Med Biol 2016;899:167–96.

[155] Dwyer DJ, Belenky PA, Yang JH, MacDonald IC, Martell JD, Takahashi N, et al. Antibiotics induce redox-related physiological alterations as part of their lethality. Proc Natl Acad Sci U S A 2014;111:E2100–9.

[156] Eichenlaub T, Villadsen R, Freitas FCP, Andrejeva D, Aldana BI, Nguyen HT, et al. Warburg effect metabolism drives neoplasia in a Drosophila genetic model of epithelial cancer. Curr Biol 2018;28:3220–8 e6.

[157] Koopman M, Michels H, Dancy BM, Kamble R, Mouchiroud L, Auwerx J, et al. A screening-based platform for the assessment of cellular respiration in *Caenorhabditis elegans*. Nat Protoc 2016;11:1798–816.

[158] Bond ST, McEwen KA, Yoganantharajah P, Gibert Y. Live metabolic profile analysis of zebrafish embryos using a seahorse XF 24 extracellular flux analyzer. Methods Mol Biol 2018;1797:393–401.

[159] Rogers GW, Nadanaciva S, Swiss R, Divakaruni AS, Will Y. Assessment of fatty acid beta oxidation in cells and isolated mitochondria. Curr Protoc Toxicol 2014;60:1–19 25.3.

[160] Divakaruni AS, Hsieh WY, Minarrieta L, Duong TN, Kim KKO, Desousa BR, et al. Etomoxir inhibits macrophage polarization by disrupting CoA homeostasis. Cell Metab 2018;28:490–503 e7.

[161] Mookerjee SA, Gerencser AA, Nicholls DG, Brand MD. Quantifying intracellular rates of glycolytic and oxidative ATP production and consumption using extracellular flux measurements. J Biol Chem 2017;292:7189–207.

[162] Yépez VA, Kremer LS, Iuso A, Gusic M, Kopajtich R, Koňaříková E, et al. OCR-Stats: robust estimation and statistical testing of mitochondrial respiration activities using seahorse XF analyzer. PLoS One 2018;13:e0199938.

[163] Divakaruni AS, Rogers GW, Murphy AN. Measuring mitochondrial function in permeabilized cells using the seahorse XF analyzer or a Clark-type oxygen electrode. Curr Protoc Toxicol 2014;60:1–16 25.2.

[164] Clerc P, Polster BM. Investigation of mitochondrial dysfunction by sequential microplate-based respiration measurements from intact and permeabilized neurons. PLoS One 2012;7:e34465.

[165] Iuso A, Repp B, Biagosch C, Terrile C, Prokisch H. Assessing mitochondrial bioenergetics in isolated mitochondria from various mouse tissues using seahorse XF96 analyzer. Methods Mol Biol 2017;1567:217–30.

[166] Nicholls DG, Ward MW. Mitochondrial membrane potential and neuronal glutamate excitotoxicity: mortality and millivolts. Trends Neurosci 2000;23:166–74.

[167] Nicholls DG. Fluorescence measurement of mitochondrial membrane potential changes in cultured cells. Methods Mol Biol 1782;2018:121–35.

[168] Perry SW, Norman JP, Barbieri J, Brown EB, Gelbard HA. Mitochondrial membrane potential probes and the proton gradient: a practical usage guide. BioTechniques 2011;50:98–115.

[169] Smiley ST, Reers M, Mottola-Hartshorn C, Lin M, Chen A, Smith TW, et al. Intracellular heterogeneity in mitochondrial membrane potentials revealed by a J-aggregate-forming lipophilic cation JC-1. Proc Natl Acad Sci U S A 1991;88:3671–5.

[170] De Biasi S, Gibellini L, Cossarizza A. Uncompensated polychromatic analysis of mitochondrial membrane potential using JC-1 and multilaser excitation. Curr Protoc Cytom 2015;72:1–11 7.32.

[171] Scaduto RC, Grotyohann LW. Measurement of mitochondrial membrane potential using fluorescent rhodamine derivatives. Biophys J 1999;76:469–77.

[172] Iannetti EF, Smeitink JAM, Beyrath J, Willems PHGM, Koopman WJH. Multiplexed high-content analysis of mitochondrial morphofunction using live-cell microscopy. Nat Protoc 2016;11:1693–710.

[173] Zhang X, Lemasters JJ. Translocation of iron from lysosomes to mitochondria during ischemia predisposes to injury after reperfusion in rat hepatocytes. Free Radic Biol Med 2013;63:243–53.

[174] McKenzie M, Duchen MR. Impaired cellular bioenergetics causes mitochondrial calcium handling defects in MT-ND5 mutant cybrids. PLoS One 2016;11:e0154371.

[175] Nicholls DG. Simultaneous monitoring of ionophore- and inhibitor-mediated plasma and mitochondrial membrane potential changes in cultured neurons. J Biol Chem 2006;281:14864–74.

[176] Wang C, Wang G, Li X, Wang K, Fan J, Jiang K, et al. Highly sensitive fluorescence molecular switch for the ratio monitoring of trace change of mitochondrial membrane potential. Anal Chem 2017;89:11514–19.

[177] Chen Y, Qi J, Huang J, Zhou X, Niu L, Yan Z, et al. A nontoxic, photostable and high signal-to-noise ratio mitochondrial probe with mitochondrial membrane potential and viscosity detectivity. Spectrochim Acta A Mol Biomol Spectrosc 2018;189:634–41.

[178] Li J, Kwon N, Jeong Y, Lee S, Kim G, Yoon J. Aggregation-induced fluorescence probe for monitoring membrane potential changes in mitochondria. ACS Appl Mater Interfaces 2018;10:12150–4.

[179] Logan A, Pell VR, Shaffer KJ, Evans C, Stanley NJ, Robb EL, et al. Assessing the mitochondrial membrane potential in cells and *in vivo* using targeted click chemistry and mass spectrometry. Cell Metab 2016;23:379–85.

[180] Springett R. Novel methods for measuring the mitochondrial membrane potential. Methods Mol Biol 2015;1264:195–202.

[181] Vázquez-Memije ME, Shanske S, Santorelli FM, Kranz-Eble P, DeVivo DC, DiMauro S. Comparative biochemical studies of ATPases in cells from patients with the T8993G or T8993C mitochondrial DNA mutations. J Inherit Metab Dis 1998;21:829–36.

[182] Ouhabi R, Boue-Grabot M, Mazat JP. Mitochondrial ATP synthesis in permeabilized cells: assessment of the ATP/O values in situ. Anal Biochem 1998;263:169–75.

[183] Manfredi G, Yang L, Gajewski CD, Mattiazzi M. Measurements of ATP in mammalian cells. Methods 2002;26:317–26.

[184] Zanna C, Ghelli A, Porcelli AM, Karbowski M, Youle RJ, Schimpf S, et al. OPA1 mutations associated with dominant optic atrophy impair oxidative phosphorylation and mitochondrial fusion. Brain 2008;131:352–67.

[185] Ghelli A, Tropeano CV, Calvaruso MA, Marchesini A, Iommarini L, Porcelli AM, et al. The cytochrome b p278Y > C mutation causative of a multisystem disorder enhances superoxide production and alters supramolecular interactions of respiratory chain complexes. Hum Mol Genet 2013;22:2141–51.

[186] Giorgio M, Migliaccio E, Orsini F, Paolucci D, Moroni M, Contursi C, et al. Electron transfer between cytochrome c and p66Shc generates reactive oxygen species that trigger mitochondrial apoptosis. Cell 2005;122:221–33.

[187] Bedard K, Krause KH. The NOX family of ROS-generating NADPH oxidases: physiology and pathophysiology. Physiol Rev 2007;87:245–313.

[188] Cadenas E, Davies KJ. Mitochondrial free radical generation, oxidative stress, and aging. Free Radic Biol Med 2000;29:222–30.

[189] Wang X, Fang H, Huang Z, Shang W, Hou T, Cheng A, et al. Imaging ROS signaling in cells and animals. J Mol Med (Berl) 2013;91:917–27.
[190] Dikalov SI, Harrison DG. Methods for detection of mitochondrial and cellular reactive oxygen species. Antioxid Redox Signal 2014;20:372–82.
[191] Dikalov S, Griendling KK, Harrison DG. Measurement of reactive oxygen species in cardiovascular studies. Hypertension 2007;49:717–27.
[192] Cochemé HM, Quin C, McQuaker SJ, Cabreiro F, Logan A, Prime TA, et al. Measurement of H2O2 within living Drosophila during aging using a ratiometric mass spectrometry probe targeted to the mitochondrial matrix. Cell Metab 2011;13:340–50.
[193] Schagger H, von Jagow G. Blue native electrophoresis for isolation of membrane protein complexes in enzymatically active form. Anal Biochem 1991;199:223–31.
[194] Diaz F, Barrientos A, Fontanesi F. Evaluation of the mitochondrial respiratory chain and oxidative phosphorylation system using blue native gel electrophoresis. Curr Protoc Hum Genet 2009; Chapter 19, Unit19 4.
[195] Wittig I, Braun HP, Schagger H. Blue native PAGE. Nat Protoc 2006;1:418–28.
[196] Nijtmans LG, Henderson NS, Holt IJ. Blue Native electrophoresis to study mitochondrial and other protein complexes. Methods 2002;26:327–34.
[197] Calvaruso MA, Smeitink J, Nijtmans L. Electrophoresis techniques to investigate defects in oxidative phosphorylation. Methods 2008;46:281–7.
[198] Diaz F, Thomas CK, Garcia S, Hernandez D, Moraes CT. Mice lacking COX10 in skeletal muscle recapitulate the phenotype of progressive mitochondrial myopathies associated with cytochrome c oxidase deficiency. Hum Mol Genet 2005;14:2737–48.
[199] Iommarini L, Ghelli A, Tropeano CV, Kurelac I, Leone G, Vidoni S, et al. Unravelling the effects of the mutation m3571insC/MT-ND1 on respiratory complexes structural organization. Int J Mol Sci 2018;19.

PART 4

mtDNA-determined diseases and therapies

CHAPTER 14

Mitochondrial DNA-related diseases associated with single large-scale deletions and point mutations

Robert D.S. Pitceathly[1] and Shamima Rahman[2]

[1]*Department of Neuromuscular Diseases, UCL Queen Square Institute of Neurology and The National Hospital for Neurology and Neurosurgery, London, United Kingdom,* [2]*UCL Great Ormond Street Institute of Child Health, London, United Kingdom*

14.1 Clinical syndromes of mitochondrial DNA-related diseases associated with single large-scale deletions and point mutations

The existence of mitochondrial DNA (mtDNA) in vertebrates was first identified in 1963 [1]. Subsequently, the demonstration of maternal inheritance of human mtDNA [2] and the observation of large pedigrees with apparent maternal inheritance of Leber hereditary optic neuropathy (LHON) and myoclonic epilepsy with ragged red fibers (MERRF) led to the suggestion that mtDNA mutations may be responsible for human disease. This suspicion was confirmed in 1988 with the detection of mtDNA point mutations responsible for MERRF and LHON [3,4]. In the same year, single large-scale mtDNA deletions (SLSMDs) were discovered in the muscle of individuals with mitochondrial myopathy [5]. Since these initial discoveries, more than 250 different potentially pathogenic variants of mtDNA have been reported (https://www.mitomap.org/), in association with a huge array of diverse phenotypes. In this chapter, we will discuss mtDNA mutations that cause "canonical" syndromes in which there are characteristic and often stereotyped constellations of clinical features. However, it should be noted that many affected individuals have symptoms overlapping two or more mitochondrial syndromes, some patients have complex phenotypes that do not closely match any of the known syndromes, and others may have isolated involvement of a single organ. For each of the clinical syndromes associated with mtDNA point mutations discussed in this chapter (Table 14.1), we report the genes involved, most prevalent mutation(s) and whether the mutations are typically heteroplasmic or homoplasmic, followed by a brief clinical description. Under "Mutations" we list those variants that are listed as confirmed pathogenic in MitoMap (https://www.mitomap.org/); the reader may

The Human Mitochondrial Genome.
DOI: https://doi.org/10.1016/B978-0-12-819656-4.00014-0

Table 14.1: Clinical syndromes of mitochondrial DNA-related diseases associated with single large-scale deletions and point mutations.

Clinical syndrome/disease	Gene(s)	Mutation(s)
Exercise intolerance	*MT-CYB*	m.14849T>C
Kearns-Sayre syndrome	Multiple genes deleted	SLSMDs
Leber-dystonia	*MT-ND6*	m.14459G>A
Leber hereditary optic neuropathy	*MT-ND1, MT-ND4, MT-ND6*	m.11778G>A, m.3460G>A and m.14484T>C (~95% of cases), m.3376G>A, m.3635G>A, m.3697G>A, m.3700G>A, m.3733G>A, m.4171C>A, m.10197G>A, m.10663T>C, m.13051G>A, m.13094T>C, m.14459G>A, m.14482C>A, m.14482C>G, m.14495A>G, m.14502T>C, m.14568C>T
Maternally inherited diabetes and deafness	*MT-TL1*	m.3243A>G
Maternally inherited Leigh syndrome	*MT-ATP6, MT-ND1, MT-ND3, MT-ND4, MT-ND5, MT-ND6, MT-TL1, MT-TK, MT-TV, MT-TW*	m.8993T>G, m.8993T>C, m.9176T>C, m.9176T>G, m.9185T>C, m.3697G>A, m.10158T>C, m.10191T>C, m.10197G>A, m.11777C>A, m.12706T>C, m.13513G>A, m.14459G>A, m.14487T>C
Mitochondrial encephalomyopathy with lactic acidosis and stroke-like episodes	*MT-TL1, MT-TF, MT-TQ, MT-ND1, MT-ND5*	m.3243A>G (~80% of cases), m.3256C>T, m.3271T>C, m.3291T>C, m.583G>A, m.4332G>A, m.3697G>A, m.13513G>A, m.13514A>G
Mitochondrial myopathy and cardiomyopathy	*MT-TL1, MT-TI*	m.3260A>G, m.3303C>T, m.4300A>G
Myoclonic epilepsy with ragged red fibers	*MT-TK, MT-TH*	m.8344A>G (~80% of cases), m.8356T>C, m.8363G>A, m.12147G>A
Neurogenic muscle weakness, ataxia, retinitis pigmentosa	*MT-ATP6*	m.8993T>G, m.8993T>C, m.9185T>C
Nonsyndromic sensorineural hearing loss	*MT-RNR1, MT-TS1*	m.1555A>G, m.1494C>T, m.7445A>G, m.7511T>C
Pearson marrow pancreas syndrome	Multiple genes deleted	SLSMDs
Progressive external ophthalmoplegia/progressive external ophthalmoplegia plus	Multiple genes deleted, *MT-TL1, MT-TI, MT-TN, MT-TL2*	SLSMDs, m.3243A>G, m.3243A>T, m.4298G>A, m.4308G>A, m.5690A>G, m.5703G>A, m.12276G>A, m.12294G>A, m.12315G>A, m.12316G>A
Reversible infantile mitochondrial myopathy	*MT-TE*	m.14674T>C, m.14674T>G

SLSMDs, Single large-scale mitochondrial DNA deletions.

wish to refer to MitoMap for other variants that have been reported in association with particular phenotypes but not yet been confirmed as pathogenic. Fig. 14.1 illustrates the positions of the confirmed mutations, including the high-frequency zone for SLSMDs, discussed in this chapter, together with the associated clinical syndromes, on a cartoon representation of the human mtDNA molecule.

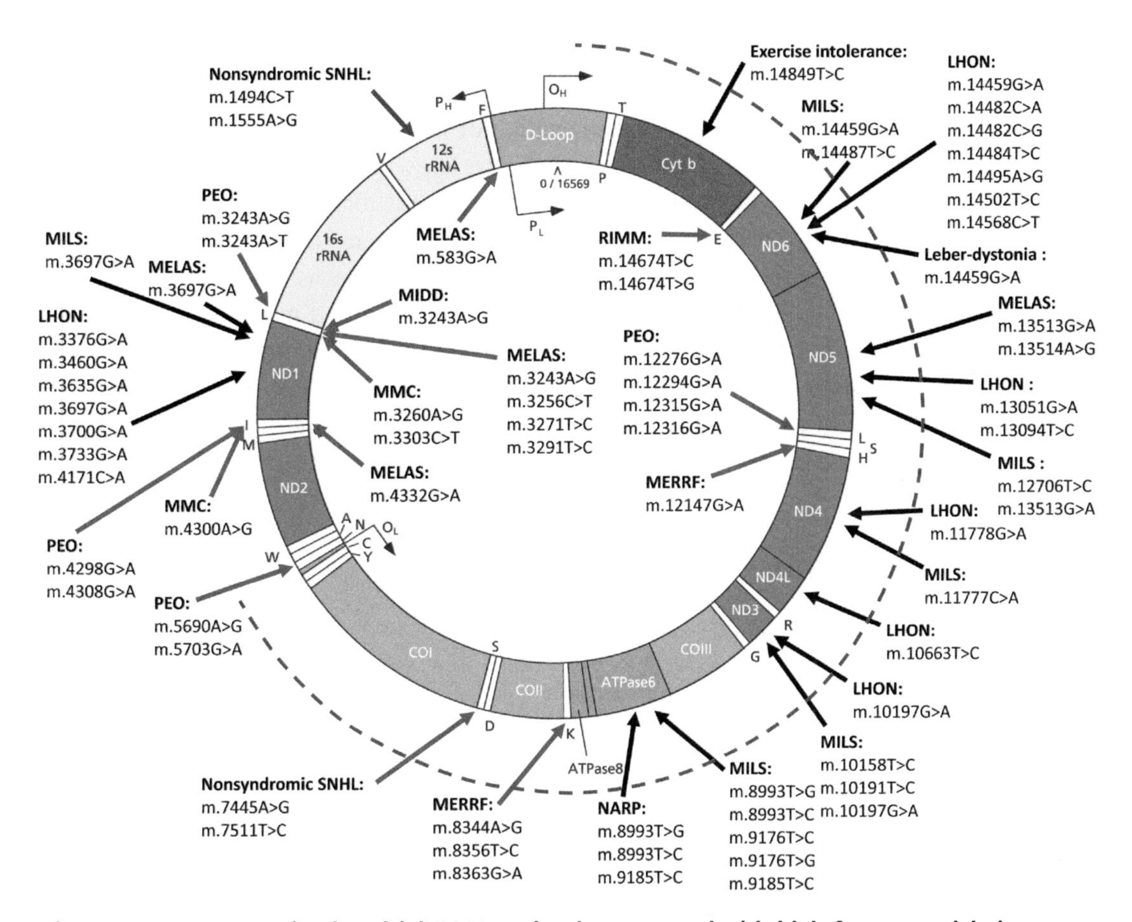

Figure 14.1: Human mitochondrial DNA molecule annotated with high frequency deletion zone and confirmed pathogenic point mutations, as listed in MitoMap (https://www.mitomap.org/), and associated clinical syndromes/diseases (*arrows*).

The letters around the outside perimeter or on the inside circle indicate related amino acids of the tRNA genes. *Black arrows* = protein-coding gene mutations; *Red arrows* = tRNA gene mutations; *Purple arrow* = rRNA gene mutations; *Dashed line* = high frequency deletion zone. *LHON*, Leber hereditary optic neuropathy; *MELAS*, mitochondrial encephalomyopathy with lactic acidosis and stroke-like episodes; *MERRF*, myoclonic epilepsy with ragged red fibers; *MIDD*, maternally inherited diabetes and deafness; *MILS*, maternally inherited Leigh syndrome; *MMC*, maternally inherited myopathy with cardiomyopathy; *NARP*, Neurogenic muscle weakness, ataxia, retinitis pigmentosa; *PEO*, progressive external ophthalmoplegia; *RIMM*, reversible infantile mitochondrial myopathy; *SNHL*, sensorineural hearing loss.

14.1.1 Exercise intolerance

Gene(s): *MT-CYB*

Mutation(s): m.14849T > C

Heteroplasmic

Prevalence: Rare

Exercise intolerance (EI) is a common manifestation of mitochondrial myopathy; more than 20% of patients with mitochondrial disease experience EI, with higher frequencies observed when cytochrome *c* oxidase (COX) negative and ragged red fibers (RRFs) are present, and with specific genotypes [6]. Pathogenic variants in *MT-CYB* were first reported in 1996 as a cause of EI [7], and subsequently in sporadic cases when the EI was associated with proximal, muscle weakness, and rhabdomyolysis [8]. The only confirmed mutation in *MT-CYB* is m.14849T > C, which was detected an 85% heteroplasmy in muscle mtDNA of an 18-year-old man with EI, fatigue, myalgia, resting lactic acidosis [9]. COX positive RRFs and reduced complex III enzyme activity were detected in muscle tissue. However, additional private, de novo mutations are reported within individual families [10].

14.1.2 Kearns-Sayre syndrome

Gene(s): Multiple genes deleted

Mutation(s): Single large-scale mtDNA deletions

Heteroplasmic

Prevalence: Rare

Kearns-Sayre syndrome (KSS) is defined by the triad of progressive external ophthalmoplegia (PEO) with onset prior to 20 years and one or more of the following clinical features: pigmentary retinopathy, cerebellar ataxia, heart block, and/or elevated cerebrospinal fluid (CSF) protein (> 100 mg/dL) [11]. Additional signs include sensorineural hearing loss (SNHL), cognitive impairment, seizures, renal tubular acidosis, short stature, and endocrine disturbance (diabetes mellitus, hypoparathyroidism, and growth hormone deficiency). There may also be an overlap between KSS and mitochondrial encephalomyopathy with lactic acidosis and stroke-like episodes (MELAS) or MERRF. The clinical course is progressive and patients frequently develop a progressive skeletal myopathy and require a pacemaker because of cardiac conduction defects. Those with central nervous system (CNS) and/or cardiac complications often die in the third or fourth decade. KSS is usually sporadic and caused by SLSMDs that arise in the mother's oocyte.

14.1.3 Leber-dystonia

Gene(s): *MT-ND6*

Mutation(s): m.14459G>A

Heteroplasmic

Prevalence: Rare

Some of the point mutations initially reported to cause LHON, particularly affecting the ND6 subunit of respiratory chain complex I, were subsequently identified to cause a LHON-dystonia overlap syndrome. Maternally inherited LHON with dystonia was first reported by Doug Wallace's group in 1994, associated with a heteroplasmic m.14459A>G variant in *MT-ND6* [12]. Clinical features are variable but the dystonia usually starts in early childhood, and may be particularly disabling and difficult to treat. Management options include baclofen, trihexyphenidyl, and botulinum toxin injections. Deep brain stimulation has not been reported for this disorder. Other clinical features include ptosis and ophthalmoplegia. Neuroimaging reveals bilateral symmetrical T2-weighted hyperintense lesions in the putamina and caudate nuclei. Nowadays this phenotype would be considered part of the Leigh syndrome spectrum, as described in more detail in Section 14.1.6.

14.1.4 Leber hereditary optic neuropathy

Gene(s): *MT-ND1*, *MT-ND4*, *MT-ND6*

Mutation(s): m.11778G>A, m.3460G>A, and m.14484T>C account for 95% of all cases. Rarer confirmed mutations include m.3376G>A, m.3635G>A, m.3697G>A, m.3700G>A, m.3733G>A, m.4171C>A, m.10197G>A, m.10663T>C, m.13051G>A, m.13094T>C. m.14459G>A, m.14482C>A, m.14482C>G, m.14495A>G, m.14502T>C, m.14568C>T.

Homoplasmic

Prevalence: 3.7 per 100,000 [13]

LHON was originally defined by Theodore Leber in 1871 [14]. Clinically, LHON is characterized by a subacute, painless loss of central vision in one eye, with both eyes affected in the majority at 6 months. Visual loss is bilateral at onset in ~25%. The peak age of onset is in the second and third decades, and 95% of those who lose vision do so by the sixth decade. Three mutations account for 95% of LHON: m.3460G>A in *MT-ND1* [15,16], m.11778G>A in *MT-ND4* [4], and m.14484T>C in *MT-ND6* [17], which are usually present at homoplasmic levels. Although visual loss is often severe and permanent, the underlying genotype does influence recovery; the m.3460G>A mutation is associated with the worst, and the m.14484T>C mutation with the best long-term visual outcome [18].

LHON exhibits a marked gender bias and reduced penetrance; 50% of males and 90% of females harboring pathogenic mutations do not develop blindness. Tobacco smoking and alcohol have been proposed as potential precipitants in patients harboring pathogenic LHON mutations. Patients with LHON may also develop an MS-like illness with disseminated CNS demyelination, typical radiological white matter lesions and unmatched CSF oligoclonal bands (so-called "Harding syndrome") [19].

14.1.5 Maternally inherited diabetes and deafness

Gene(s): *MT-TL1*

Mutation(s): m.3243A > G

Heteroplasmic

Prevalence: 3.5 per 100,000 for m.3243A > G [13] (30% with m.3243A > G mutation have maternally inherited diabetes and deafness [20])

Maternally inherited diabetes and deafness (MIDD) is the most frequent phenotypic manifestation of the m.3243A > G mutation that was initially reported to cause MELAS syndrome (see above) [20]. MIDD is defined by SNHL and diabetes in adulthood, but may also be accompanied by other neurological and nonneurological system features including PEO, ptosis, myopathy, cerebellar ataxia, vestibular impairment, migraine, cardiomyopathy, retinopathy, and chronic kidney disease.

14.1.6 Maternally inherited Leigh syndrome

Gene(s): MT-ATP6, MT-ND1, MT-ND3, MT-ND4, MT-ND5, MT-ND6, MT-TL1, MT-TK, MT-TV, MT-TW

Mutation(s): m.8993T > G, m.8993T > C, m.9176T > C, m.9176T > G, m.9185T > C, m.3697G > A, m.10158T > C, m.10191T > C, m.10197G > A, m.11777C > A, m.12706T > C, m.13513G > A, m.14459G > A, m.14487T > C

Heteroplasmic (but typically at very high mutant load, >95%)

Prevalence: Approximately 1:300,000 (~10% of Leigh syndrome) [18]

Maternally inherited Leigh syndrome (MILS) is most frequently caused by two mutations at the same nucleotide position in the *MT-ATP6* gene encoding a subunit of the ATP synthase (complex V): m.8993T > G and m.8993T > C. These mutations are estimated to account for approximately 10% of Leigh syndrome [21]. Leigh syndrome is a subacute neurodegenerative disorder, originally defined neuropathologically by symmetrical focal lesions in the brainstem, midbrain, and basal ganglia characterized by spongiosis with vacuolation of the neuropil, relative preservation of neurons, demyelination, gliosis, and capillary proliferation [22]. The Leigh syndrome spectrum is now thought of as a clinical continuum encompassing developmental delay and/or regression with neurological signs related to basal ganglia and/or brainstem dysfunction, abnormal neuroimaging (symmetrical focal T2 hyperintense lesions in the brainstem, midbrain, thalami, and basal ganglia) and elevated lactates in blood and/or CSF, or other biochemical evidence of mitochondrial dysfunction. There is a huge genotypic spectrum. Pathogenic variants in at least 89 genes have been linked to this clinical entity, mostly nuclear-encoded, but including 14 mitochondrial genes (10 confirmed—see list above) [23]. Modes of inheritance of Leigh syndrome include maternal, recessive, X-linked and sporadic. MILS may be caused by a range of *MT-ATP6* mutations, and also by mutations in mitochondrial genes encoding tRNAs and complex I subunits [24]. The mutations at m.8993 are particularly interesting because of skewed segregation in oocytes; mothers are often unaffected but may have eggs with either very high or very low mutant loads.

Clinical presentation of Leigh syndrome is typically in infancy or early childhood with developmental regression and lactic acidosis following a metabolic decompensation, which in some cases may be triggered by intercurrent infection or other metabolic stressor such as fasting, anesthesia, or surgery. Adult onset of Leigh syndrome is also recognized, although this occurs much less frequently. Neurological features include hypotonia, ataxia, dystonia, movement disorder, and seizures. Variable multisystem disease complications may occur, including cardiomyopathy and renal tubulopathy.

14.1.7 Mitochondrial encephalomyopathy with lactic acidosis and stroke-like episodes

Gene(s): *MT-TL1, MT-TF, MT-TQ, MT-ND1, MT-ND5*

Mutation(s): m.3243A > G (~80% of cases), m.3256C > T, m.3271T > C, m.3291T > C, m.583G > A, m.4332G > A, m.3697G > A, m.13513G > A, m.13514A > G

Heteroplasmic

Prevalence: 3.5 per 100,000 for m.3243A > G [13] (10% with m.3243A > G mutation have mitochondrial encephalomyopathy with lactic acidosis and stroke-like episodes [20])

MELAS is one of the canonical mitochondrial syndromes and was first recognized as a clinical entity as early as 1984 [25]. That 80% of cases are caused by the same mtDNA point mutation, m.3243A > G in the *MT-TL1* gene, is a tribute to the fine art of clinical phenotyping. However, it should be acknowledged that most individuals with this mtDNA mutation do not have MELAS. In fact, only 10% of clinically manifesting patients with the m.3243A > G mutation have MELAS syndrome [20]. The m.3243A > G mutation is one of the most prevalent pathogenic variants in the general population, with an estimated frequency of ~1 in 400 [26]. The majority of people carrying this mutation have MIDD (see above), a variety of other clinical manifestations (e.g., isolated nephropathy, gastrointestinal dysmotility, sudden unexpected death caused by cardiac arrhythmia) or are asymptomatic. Another conundrum is the overlap between MELAS and other canonical mitochondrial syndromes, especially MERRF (see below) [27,28].

The clinical presentation of MELAS is typically with a migraine headache associated with focal motor seizures evolving to generalized seizures and a stroke-like episode that may include hemianopsia, hemiplegia, or aphasia, but resolves within weeks to months. The stroke-like episodes typically start toward the end of the first decade of life, but adult onset is relatively frequent and the first stroke-like episode may be at any time up to 40 years. Clinical evolution of MELAS is with recurrent stroke-like episodes, although there may be periods of stability that last for several years. Associated clinical features include SNHL, optic atrophy, anorexia, short stature, ataxia, seizures, and progressive cognitive decline. Neuroimaging reveals parieto-occipital stroke-like lesions that are not confined to vascular territories, and which later disappear on subsequent imaging. Muscle biopsy features include RRFs and isolated complex I deficiency or a combined defect of complexes I and IV.

14.1.8 Mitochondrial myopathy and cardiomyopathy

Gene(s): *MT-TL1*, *MT-TI*

Mutation(s): m.3260A > G, m.3303C > T, m.4300A > G

Heteroplasmic (occasionally homoplasmic)

Prevalence: Extremely rare

Three mutations in the *MT-TL1* and *MT-TI* genes (see above), encoding the mitochondrial tRNAs for leucine and isoleucine respectively, have been associated particularly with a syndrome of maternally inherited myopathy with cardiomyopathy (MMC) presenting in adult life [29–31]. Several mutations in *MT-TI* and genes encoding other mitochondrial tRNAs have been reported to cause infantile onset cardiomyopathy, but their pathogenicity has never been confirmed (https://www.mitomap.org/). It is important to note that infantile mitochondrial cardiomyopathy is more typically associated with biallelic mutations in a range of nuclear genes, including genes encoding assembly factors of complexes I (e.g., NDUFAF1) [32] and IV (e.g., SCO2) [33], and mitochondrial translation factors. A peculiar syndrome of histiocytoid cardiomyopathy has been associated with mutations in the X-linked nuclear gene *NDUFB11* encoding a subunit of complex I [34].

14.1.9 Myoclonic epilepsy with ragged red fibers

Gene(s): *MT-TK, MT-TH*

Mutation(s): m.8344A > G (~80% of cases), m.8356T > C, m.8363G > A, m.12147G > A

Heteroplasmic

Prevalence: 0.7/100,000 [13]

MERRF is another canonical mitochondrial syndrome that was clinically defined decades before its genetic basis was understood [3,35]. Onset is typically in the second decade with variable clinical features including ataxia, myoclonus, and SNHL. An unusual feature present in some individuals with MERRF is multiple symmetrical lipomatosis, the pathogenic basis of which remains unclear. Other clinical features in individuals with MERRF may include optic atrophy, pigmentary retinopathy, ophthalmoparesis, exercise intolerance, lactic acidosis, cardiomyopathy, and psychiatric manifestations. Some individuals with mutations more typically associated with MERRF may develop clinical features overlapping those of MELAS syndrome, including stroke-like episodes [36,37], and vice versa—those with the m.3243A > G mutation associated with MELAS may present with features of MERRF and never develop stroke-like episodes [28]. As with the m.3243A > G mutation, many people with the m.8344A > G mutation have atypical clinical features, or may be oligo- or asymptomatic [38].

14.1.10 Neurogenic muscle weakness, ataxia, retinitis pigmentosa

Gene(s): *MT-ATP6*

Mutation(s): m.8993T > G, m.8993T > C, m.9185T > C

Heteroplasmic

Prevalence: Rare

Neurogenic muscle weakness, ataxia, and retinitis pigmentosa (NARP) is caused by three mutations in *MT-ATP6* (m.8993T > C/G, m.9185T > C), all of which also cause MILS (see above). Generally speaking, for the most common m.8993T > G mutation, a mutant load of 70%–90% causes NARP, while those individuals with >90% mutant mtDNA develop MILS [39]. The m.8993T > C is a milder variant that only tends to be associated with disease when there is >90% abnormal mtDNA. NARP is characterized by proximal neurogenic muscle weakness, sensory neuropathy, cerebellar ataxia, short stature, SNHL, PEO, cardiac conduction defects, pigmentary retinopathy, seizures, learning difficulties, and dementia [24]. Onset of symptoms, in particular the cerebellar ataxia and learning difficulties, is in childhood. The disease may follow a relapsing-remitting course, with decompensation triggered by metabolic stressors, for example, viral illnesses.

14.1.11 Nonsyndromic sensorineural hearing loss

Gene(s): *MT-RNR1, MT-TS1*

Mutation(s): m.1555A > G, m.1494C > T, m.7445A > G, m.7511T > C

Homoplasmic (occasionally heteroplasmic)

Prevalence: 1 in 500 for m.1555A > G [40,41]

Maternally inherited SNHL is commonly associated with mutations in several mitochondrial genes: *MT-RNR1*, *MT-TS1*, *MT-TL1*, and *MT-TK*. Mitochondrial hearing loss may be syndromic when part of a canonical syndrome, such as MELAS or MERRF, or a rarer symptom cluster, such as hearing loss-ataxia-myoclonus (HAM) associated with mutations in *MT-TS1* [42]. However, it may also be nonsyndromic, that is, isolated hearing loss without other clinical features. The latter is especially true for the m.1555A > G variant in *MT-RNR1* associated with exquisite sensitivity to aminoglycoside-induced hearing loss. This mutation occurs particularly frequently, with an estimated prevalence of at least 1 in 500 in White European populations [40,41]. It has been suggested that the phenotypic severity associated with the m.1555A > G mutation may be mediated by nuclear gene variants, including heterozygous variants in *TRMU* encoding a tRNA modifying factor and *SSBP1* encoding the mitochondrial single-stranded DNA binding protein [43,44].

14.1.12 Pearson marrow pancreas syndrome

Gene(s): Multiple genes deleted

Mutation(s): Single large-scale mtDNA deletions

Heteroplasmic

Prevalence: Rare

The Pearson marrow pancreas syndrome was first reported in 1979 as a constellation of refractory sideroblastic anaemia with vacuolization of marrow precursors and exocrine pancreatic dysfunction [45]. However, the genetic basis was not discovered for another 10 years, when SLSMDs were found in affected individuals [46]. Babies with Pearson syndrome present in infancy (mean 2.5 months, range birth to 16 months) with a transfusion-dependent anaemia, usually associated with lactic acidosis and faltering growth, with variable additional clinical features that may include neutropaenia, thrombocytopaenia, faltering growth, developmental delay, renal tubulopathy, and exocrine pancreatic insufficiency [47]. Bone marrow examination reveals the presence of ringed sideroblasts and vacuolated myeloid precursors. There is a very high mortality in Pearson syndrome, with the most severely affected individuals succumbing to liver failure or metabolic decompensation in the first few years of life (approaching 60% die by 4 years, and only 22% are alive at 18 years) [47,48]. Surviving individuals usually have resolution of the transfusion requirement by around the age of 2 years, but invariably go on to develop the multisystem features of KSS [49]. The reasons for this transition between phenotypes is believed to be because of the clearance of the SLSMDs from rapidly dividing blood cells, whereas SLSMDs progressively accumulate in nondividing tissues such as muscle and brain.

14.1.13 Progressive external ophthalmoplegia/progressive external ophthalmoplegia plus

Gene(s): Multiple genes deleted, *MT-TL1, MT-TI, MT-TN, MT-TL2*

Mutation(s): Single large-scale mtDNA deletions, m.3243A > G/T, m.4298G > A, m.4308G > A, m.5690A > G, m.5703G > A, m.12276G > A, m.12294G > A, m.12315G > A, m.12316G > A

Heteroplasmic

Prevalence: Rare (especially in isolation) [50]

Isolated PEO is a genetically heterogeneous mitochondrial syndrome characterized by slowly progressive bilateral ptosis and limitation of eye movements secondary to extraocular muscle weakness [11]. SLSMDs are the most frequent cause of PEO, accounting for two thirds [50] of cases. Mutations in nuclear-encoded mitochondrial genes involved with mtDNA maintenance are also an important cause of autosomal dominant and recessive forms of PEO, and are frequently associated with impaired mtDNA replication resulting in multiple (polyclonal) mtDNA deletions. Point mutations in mtDNA are a relatively uncommon cause of PEO accounting for approximately 10% of cases, the majority of which are due to the m.3243A > G mutation in *MT-TL1* [50]. Conversely, 12% of patients with m.3243A > G have PEO as part of their clinical presentation [20]. Unlike single mtDNA deletions, isolated PEO is rarely seen with mtDNA point mutations without other neurological and non-neurological system involvement [20] or as part of a canonical mitochondrial syndrome, such as MELAS. PEO is frequently associated with additional muscle-related symptoms, including proximal myopathy, exercise intolerance, and fatigue. The most prevalent extramuscular manifestations include SNHL, diabetes mellitus, cerebellar ataxia, parkinsonism, and peripheral neuropathy [50]; when these exist the syndrome is termed PEO plus.

14.1.14 Reversible infantile mitochondrial myopathy

Gene(s): *MT-TE*

Mutation(s): m.14674T>C, m.14674T>G

Homoplasmic

Prevalence: Rare

An infantile mitochondrial myopathy that improved with age was first reported in the early 1980s and initially termed "benign reversible COX deficiency" [51,52]. Later this was recognized to be a reversible myopathy associated with multiple respiratory chain enzyme deficiencies, not isolated COX deficiency, and subsequently was linked to two homoplasmic mtDNA point mutations in the gene encoding the tRNA for glutamate [53,54]. Affected infants typically present at a few weeks of age, after a symptom-free interval, with progressive muscle weakness and lactic acidosis leading to feeding difficulties and, in some cases, respiratory failure requiring prolonged invasive ventilation for as long as 18 months. Thereafter there is a progressive recovery of muscle strength, although mild weakness persists into adult life. Even within families, not all individuals harboring the homoplasmic *MT-TE* mutation have muscle weakness, and so it is likely that other genetic variants modify the phenotype.

14.2 Molecular genetics of mitochondrial DNA single large-scale deletions and point mutations

The (almost) universal consensus is that mtDNA is maternally inherited. There is one perplexing case of apparent paternal inheritance of a mtDNA mutation [55] and a more recent report of biparental inheritance of mtDNA in humans [56]. However, the latter phenomenon has not been observed in other studies [57] and may be explained by the presence of so-called mega NUMTs [58].

MtDNA point mutations are maternally inherited but may also arise spontaneously in the affected individual. Conversely, SLSMDs are usually sporadic events that are not inherited, although one retrospective review suggested possible recurrence risk within families of 1 in 24 births, with wide confidence intervals (95% CI 1 in 117 to 1 in 9) [59]. Homoplasmic mtDNA mutations are typically maternally inherited through multiple generations, but frequently have variable penetrance. For example, the risk of vision loss across all LHON mutations is 46% in men [60], which is approximately five times greater than females. For many years it was assumed that there must be an X chromosomal gene encoding a modifying factor. However gene mapping strategies to identify such a modifier gene proved fruitless, and it appears that the gender differential may be explained by a protective effect of oestrogen [61].

Heteroplasmic mtDNA mutations may also be maternally inherited and may be associated with huge variability in heteroplasmy levels (mutant loads), even between first-degree maternal relatives. In some cases, the mtDNA point mutation appears to have arisen de novo. This has most frequently been reported with the m.8993T > G/C mutations associated with MILS, but has also been noted with some of the mutations in the *MT-CO1–3* and *MT-CYB* genes associated with exercise intolerance [8,10]. Most mtDNA point mutations are functionally recessive. Thus, in the case of heteroplasmic mtDNA mutations, a clinical phenotype is observed when the mutant load exceeds a critical threshold above which the biochemical function of the mitochondrion is impaired. However some mtDNA mutations appear to exert dominant negative effects, for example, a heteroplasmic m.5545C > T mutation in *MT-TW* appeared to have a biochemical threshold of only 4%–8% in cybrids [62].

One major contributor to the varied clinical presentation of mtDNA-related diseases is the unpredictable, and seemingly random, segregation of mutant mtDNA between different tissue types; mutant mtDNA segregates both in the female oocytes and in somatic tissues. Rapid segregation of mtDNA heteroplasmy may also occur via a genetic bottleneck, when a temporary reduction in the population of mtDNA molecules (~200 mtDNA molecules per oogonium) in the female primordial germ cells precedes a rapid replication during oocyte maturation, thus potentially greatly increasing the rate of genetic drift between generations [63]. The genetic drift of mtDNA heteroplasmy is also an important factor to consider during adult life when investigating the presence of mutant mtDNA. In rapidly dividing tissues (e.g., blood leukocytes) a reduction in mutant mtDNA is observed over time, particularly with certain pathogenic mutations [64]. The mechanisms underpinning this phenomenon are unclear, but selection pressure toward wild-type mtDNA and/or the general fitness of the cell to divide in the absence of a respiratory chain defect have been suggested. It is therefore important that the diagnosis of mtDNA diseases is based on genetic analysis of more than one tissue type (and if necessary a postmitotic tissue such as skeletal muscle) to avoid false-negative results.

14.3 Diagnostic approach to mitochondrial DNA-related diseases associated with single large-scale deletions and point mutations

Diagnosis of mitochondrial disease rests ultimately on clinical suspicion, followed by targeted investigations as listed below. Traditionally a muscle biopsy has been an essential component of the diagnostic work-up, but the increased sensitivity of next generation sequencing methods allows detection of low levels of heteroplasmy, so it may be reasonable to perform genetic investigations in blood first and a muscle biopsy for confirmation, if needed, as a second tier investigation. However, it should be emphasized that the absence of mtDNA mutations in blood does not exclude mtDNA-related mitochondrial disease, and analysis of mtDNA extracted from urinary epithelial cells, bone marrow aspirate (SLSMDs in Pearson marrow pancreas syndrome) and/or muscle tissue may be required to exclude the diagnosis.

14.3.1 Laboratory tests

Mitochondrial metabolic testing includes serum creatine kinase (CK), lactate and pyruvate, plasma acylcarnitines, plasma and urine amino acids, urine organic acids, and CSF lactate and pyruvate if there is CNS involvement. CSF protein and 5-methyltetrahydrofolate (5MTHF) levels should be measured in KSS to support the diagnosis and confirm the presence of cerebral folate deficiency prior to folinic acid supplementation. Serum CK is usually normal or mildly elevated in patients with mitochondrial myopathies. Furthermore, normal values for lactate and lactate/pyruvate ratio do not exclude mitochondrial disease, particularly in adults; mtDNA-related diseases are often associated with normal or only mildly elevated resting blood lactate levels but significant elevation occurs with exercise. Blood levels of free carnitine may be decreased in mtDNA-related disorders, with a relative increase in some acylcarnitine species.

14.3.2 Neuroimaging

The use of neuroimaging, in particular brain magnetic resonance imaging (MRI), has greatly facilitated the detection of CNS involvement in mitochondrial disorders. Brain atrophy is common in children with mitochondrial disease. The diagnosis of MELAS can be aided by the clinical association of stroke-like episodes with radiological lesions that do not conform to the anatomical vascular territories of blood vessels, typically in the parietal-occipital region. Leigh syndrome characteristically shows bilateral hyperintense signals on T2-weighted and fluid-attenuated inversion recovery (FLAIR) MRIs in the basal ganglia, thalamus, and/or brainstem. The most common MRI findings in KSS include cerebral and cerebellar atrophy with bilateral, symmetrical hyperintense lesions in the subcortical white matter, thalamus, basal ganglia, and brainstem. Basal ganglia calcification is a relatively frequent feature of mitochondrial disease, particularly in KSS, but may not be apparent on the MRI scan; computed tomography (CT) may be needed for confirmation of this finding. Extraocular muscle T2 signal is elevated in PEO and correlates negatively with ocular movements, thus providing a potential quantitative measure of disease severity [65]. ^{1}H MRS can detect lactate accumulation in the CSF and in specific areas of the brain, while PET, which measures metabolic flux, has identified several metabolic abnormalities in mitochondrial disease patients using radio-isotopically labeled metabolites relevant to the study of bioenergetics.

14.3.3 Skeletal muscle histochemistry, electron microscopy, and respiratory chain biochemistry

If mtDNA mutations are undetectable in blood, muscle biopsy may be a helpful diagnostic test to investigate further the possibility of mtDNA-related mitochondrial disease.

Histochemical techniques include Gomori trichrome, the red staining of which highlights subsarcolemmal accumulation of mitochondria, so-called RRFs, while succinate dehydrogenase (SDH) enzyme staining highlights a similar phenomenon, so-called ragged blue fibers. Combined staining with COX and SDH (COX/SDH) is a useful tool to highlight COX-negative fibers, which are highly suggestive of an underlying mtDNA mutation, particularly when a mosaic pattern is present indicative of heteroplasmy. Fig. 14.2 illustrates histochemical staining of skeletal muscle sections and demonstrates findings consistent with mitochondrial myopathy.

In our hands, electron microscopy (EM) of muscle tissue does not add significantly to diagnostic yield when histochemical, biochemical, and genetic studies are combined. However, EM may reveal subsarcolemmal and intermyofibrillar proliferation of mitochondria and the presence of abnormal mitochondria in muscle fibers. Enlarged, elongated, irregular, and dumbbell-shaped mitochondria with hypoplastic and dystrophic cristae and paracrystalline inclusions suggest mitochondrial dysfunction. However, these findings are nonspecific and can be present in other neuromuscular disorders.

Muscle homogenate (fresh or frozen) can be used to undertake enzymatic assays of specific respiratory chain complexes. This is particularly useful in children with mtDNA-related disease in whom muscle histochemistry may appear normal. Multiple respiratory chain enzyme deficiencies are seen with mtDNA tRNA mutations (e.g., m.3243A > G), while isolated defects suggest dysfunction of protein-coding genes, although these rules are not universally adhered to.

14.3.4 Genetic testing

Genetic investigation of suspected mtDNA disease may be performed in DNA extracted from blood, urinary sediment, skeletal muscle, or another affected tissue. Long-range polymerase chain reaction (PCR) and Southern blot genetic techniques are used to detect SLSMDs of the mitochondrial genome. Genetic testing for mtDNA point mutations may include screening for the common mutations at nucleotide positions 3243, 8344, and 8993, or next generation sequencing of the whole mitochondrial genome. Owing to the high copy number of mtDNA, mtDNA sequences are very well captured by both exome and genome next generation sequencing, so in some cases, a mtDNA mutation may be detected by one of these methods when a mtDNA disorder had not been suspected clinically. Confirming pathogenesis of a novel mtDNA variant can be challenging. Criteria for pathogenicity include segregation of

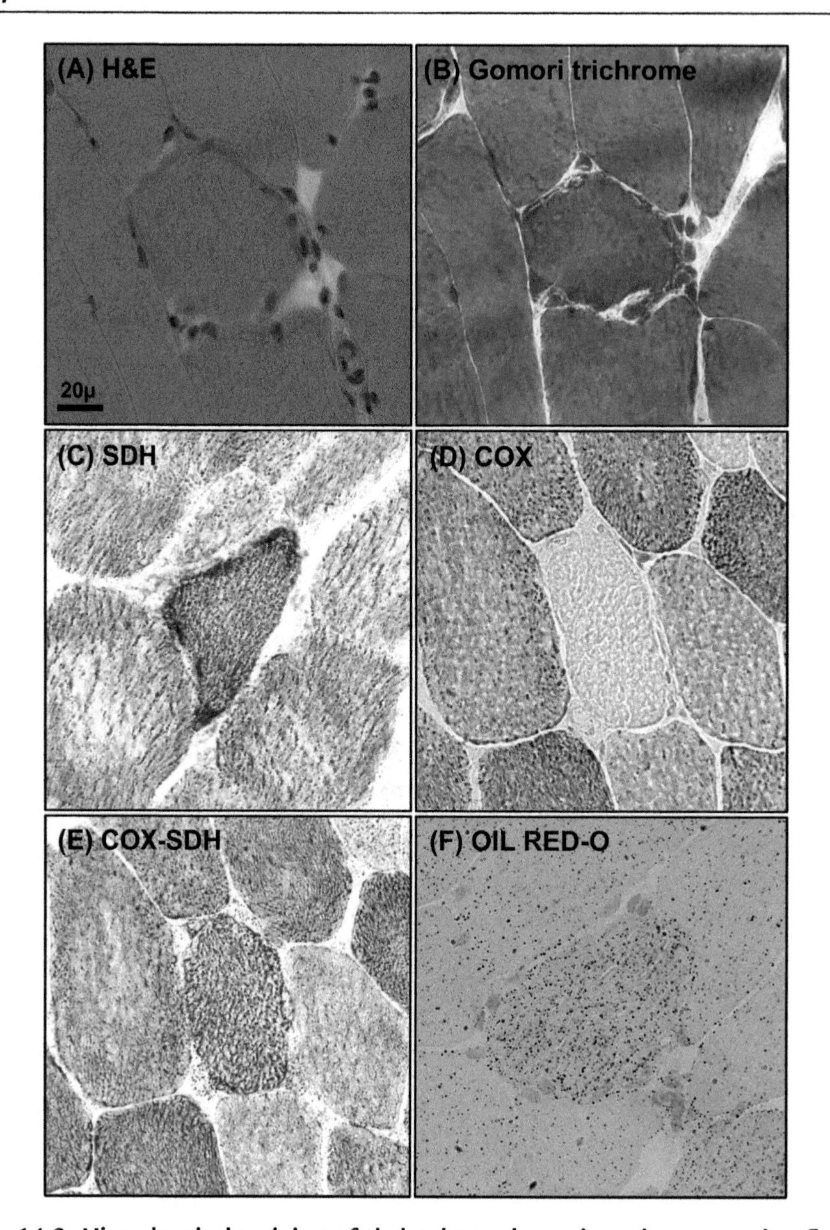

Figure 14.2: Histochemical staining of skeletal muscle sections demonstrating findings consistent with mitochondrial myopathy.

(A) Prominent internal mitochondria including subsarcolemmal mitochondrial aggregates. These fibers are referred to as "ragged red" fibers and are usually most clearly visible with Gomori trichrome histochemistry (B). On oxidative stains, particularly succinate dehydrogenase (SDH) staining, these fibers show "ragged blue" changes (C). Deficiency of cytochrome c oxidase (COX) is an important diagnostic feature (D) that is further highlighted using combined COX-SDH staining (E). The fibers may also show excess internal lipid droplets as seen on Oil red-O stain (F).

phenotype with mutant load (so testing of maternal relatives may be helpful), absence in healthy controls, evolutionary conservation and evidence of a functional effect of the variant on OXPHOS function. Functional studies may include single-fiber PCR to determine whether the mutation is at higher levels in COX-negative fibers or RRFs or demonstration that a biochemical defect is transmitted with the mitochondrial genome in transmitochondrial cybrids. The latter is considered the "gold standard" for functional confirmation of the pathogenicity of a novel mtDNA mutation but is a laborious and time-consuming test that is only offered in specialist research laboratories.

14.4 Management of mitochondrial DNA-related diseases associated with single large-scale deletions and point mutations

14.4.1 Supportive therapies

Patients and families with mitochondrial disease require supportive care in a multidisciplinary clinical team setting. This is often coordinated by a clinical expert in mitochondrial diseases (adult or pediatric) with close links to other medical disciplines, including neurologists, cardiologists, respiratory physicians, endocrinologists, ophthalmologists, gastroenterologists, audiovestibular physicians, rehabilitation physicians, physiotherapists, occupational therapists, speech and language therapists, and psychologists. Although no specific therapy exists to halt or reverse the progression of the underlying OXPHOS disorder, monitoring and treatment of complications arising from the disease is crucial to maintain independence, improve quality of life and reduce morbidity.

14.4.2 Vitamins and cofactors

Numerous vitamins and cofactors have been proposed for use in mitochondrial disease, aiming to support mitochondrial function. These vitamins/cofactors include coenzyme Q_{10}, riboflavin and nicotinamide, but evidence for efficacy of these agents is currently lacking [66], so they will not be discussed in detail here. CSF 5MTHF deficiency is observed relatively frequently in KSS, variably associated with white matter disease and seizures, and clinical responses to folinic acid supplementation have been reported in some patients [67]. Folic acid is contraindicated in this disorder as it competes with 5MTHF for transport across the blood-CSF barrier and thus exacerbates the cerebral folate deficiency.

14.4.3 Emerging therapies

There remain no curative therapies for mtDNA point mutation diseases. Developing novel therapies is the subject of intense research focus across the globe. Emerging therapies may be divided broadly into pharmacological and genetic approaches [68,69]. The former include antioxidants (such as coenzyme Q_{10} and vitamin E derivatives), agents to boost mitochondrial biogenesis, small molecules that can stabilize the mitochondrial lipid membranes, and rapamycin to target mitophagy by mTOR inhibition. Genetic approaches specifically targeting the mtDNA include selective destruction of mutant mtDNA using transcription activator-like effector nucleases (TALENs) or zinc finger nucleases. These methods have shown promising early results in cell and animal models [70,71], and some of these approaches are discussed in this book. Another option, for mutations in mtDNA protein-coding genes, is to recode the gene using the nuclear genetic code so that it can be delivered using a viral vector and be expressed from the nucleus. The encoded protein is then synthesized on cytosolic ribosomes and, since the vector includes a mitochondrial targeting signal, imported into the mitochondrion. This approach is currently in clinical trial for LHON [72].

14.4.4 Reproductive options

MtDNA point mutations can only be transmitted by females carrying the mutations in their oocytes; these include affected women and asymptomatic carriers. These individuals should all receive expert genetic counseling from geneticists experienced in mitochondrial genetics. Reproductive options available to these women include (1) prenatal diagnosis for certain heteroplasmic mutations, where oocyte mutation load is known to be markedly skewed toward homoplasmy for wild-type or mutant mtDNA (e.g., the m.8993T > G mutation associated with MILS and NARP) [73,74]; (2) preimplantation genetic diagnosis (only suitable for heteroplasmic mtDNA mutations where there is the possibility of some eggs with low mutation load) [75]; (3) mtDNA replacement [76]; (4) egg donation; (5) adoption; and (6) voluntary childlessness.

Research perspectives in mitochondrial DNA-related diseases associated with single large-scale deletions and point mutations

Despite significant advances in genomic medicine, current understanding of the causal relationship between the many different mtDNA point mutations and their clinical manifestations remains incomplete, while the number of reported mtDNA point mutations linked to human diseases continues to expand. However, national and international efforts to develop cohorts comprising patients with genetically confirmed mtDNA-related diseases is helping to address these fundamental questions, while facilitating accurate genotype-phenotype correlations, better understanding of natural history and the development of evidence-based management guidelines. Furthermore, genetic-based approaches that target mtDNA deletions and point mutations are in the preclinical phase of development, raising the real possibility of licensed therapies that treat the underlying molecular defects, rather than their downstream consequences.

Acknowledgments

We are very grateful to Dr. Ashirwad Merve and Dr. Rahul Phadke for providing representative images of muscle section staining for the chapter.

References

[1] Nass MM, Nass S. Intramitochondrial fibers with DNA characteristics. I. Fixation and electron staining reactions. J Cell Biol 1963;19:593–611.

[2] Giles RE, Blanc H, Cann HM, Wallace DC. Maternal inheritance of human mitochondrial DNA. Proc Natl Acad Sci U S A 1980;77:6715–19.

[3] Wallace DC, et al. Familial mitochondrial encephalomyopathy (MERRF): genetic, pathophysiological, and biochemical characterization of a mitochondrial DNA disease. Cell 1988;55:601–10.

[4] Wallace DC, et al. Mitochondrial DNA mutation associated with Leber's hereditary optic neuropathy. Science 1988;242:1427–30.

[5] Holt IJ, Harding AE, Morgan-Hughes JA. Deletions of muscle mitochondrial DNA in patients with mitochondrial myopathies. Nature 1988;331:717–19.

[6] Mancuso M, et al. Fatigue and exercise intolerance in mitochondrial diseases. Literature revision and experience of the Italian Network of mitochondrial diseases. Neuromuscul Disord 2012;22(Suppl. 3): S226–9.

[7] Dumoulin R, et al. A novel gly290asp mitochondrial cytochrome b mutation linked to a complex III deficiency in progressive exercise intolerance. Mol Cell Probes 1996;10:389–91.

[8] Andreu AL, et al. Exercise intolerance due to mutations in the cytochrome b gene of mitochondrial DNA. N Engl J Med 1999;341:1037–44.

[9] Massie R, Wong L-JC, Milone M. Exercise intolerance due to cytochrome b mutation. Muscle Nerve 2010;42:136–40.

[10] Rahman S, et al. A missense mutation of cytochrome oxidase subunit II causes defective assembly and myopathy. Am J Hum Genet 1999;65:1030–9.

[11] Pitceathly RDS, Rahman S, Hanna MG. Single deletions in mitochondrial DNA—molecular mechanisms and disease phenotypes in clinical practice. Neuromuscul Disord 2012;22:577–86.

[12] Jun AS, Brown MD, Wallace DC. A mitochondrial DNA mutation at nucleotide pair 14459 of the NADH dehydrogenase subunit 6 gene associated with maternally inherited Leber hereditary optic neuropathy and dystonia. Proc Natl Acad Sci U S A 1994;91:6206–10.

[13] Gorman GS, et al. Prevalence of nuclear and mitochondrial DNA mutations related to adult mitochondrial disease. Ann Neurol 2015;77:753–9.

[14] Leber T. Ueber hereditäre und congenital-angelegte Sehnervenleiden. Albrecht Von Graefes Arch Für Ophthalmol 1871;17:249–91.

[15] Huoponen K, Vilkki J, Aula P, Nikoskelainen EK, Savontaus ML. A new mtDNA mutation associated with Leber hereditary optic neuroretinopathy. Am J Hum Genet 1991;48:1147–53.

[16] Howell N, et al. Leber hereditary optic neuropathy: identification of the same mitochondrial ND1 mutation in six pedigrees. Am J Hum Genet 1991;49:939–50.

[17] Johns DR, Neufeld MJ, Park RD. An ND-6 mitochondrial DNA mutation associated with Leber hereditary optic neuropathy. Biochem Biophys Res Commun 1992;187:1551–7.

[18] Spruijt L, et al. Influence of mutation type on clinical expression of Leber hereditary optic neuropathy. Am J Ophthalmol 2006;141:676–82.

[19] Harding AE, et al. Occurrence of a multiple sclerosis-like illness in women who have a Leber's hereditary optic neuropathy mitochondrial DNA mutation. Brain J Neurol 1992;115(Pt 4):979–89.

[20] Nesbitt V, et al. The UK MRC Mitochondrial Disease Patient Cohort Study: clinical phenotypes associated with the m.3243A > G mutation—implications for diagnosis and management. J Neurol Neurosurg Psychiatry 2013;84:936–8.

[21] Rahman S, et al. Leigh syndrome: clinical features and biochemical and DNA abnormalities. Ann Neurol 1996;39:343–51.

[22] Leigh D. Subacute necrotizing encephalomyelopathy in an infant. J Neurol Neurosurg Psychiatry 1951;14:216–21.

[23] Rahman J, Noronha A, Thiele I, Rahman S. Leigh map: a novel computational diagnostic resource for mitochondrial disease. Ann Neurol 2017;81:9–16.

[24] Thorburn DR, Rahman J, Rahman S. Mitochondrial DNA-associated Leigh syndrome and NARP. In: Adam MP, et al., editors. GeneReviews®. Seattle: University of Washington; 1993.

[25] Pavlakis SG, Phillips PC, DiMauro S, De Vivo DC, Rowland LP. Mitochondrial myopathy, encephalopathy, lactic acidosis, and stroke-like episodes: a distinctive clinical syndrome. Ann Neurol 1984;16:481–8.

[26] Manwaring N, et al. Population prevalence of the MELAS A3243G mutation. Mitochondrion 2007;7:230–3.

[27] Liolitsa D, Rahman S, Benton S, Carr LJ, Hanna MG. Is the mitochondrial complex I ND5 gene a hot-spot for MELAS causing mutations? Ann Neurol 2003;53:128–32.

[28] Verma A, Moraes CT, Shebert RT, Bradley WG. A MERRF/PEO overlap syndrome associated with the mitochondrial DNA 3243 mutation. Neurology 1996;46:1334–6.

[29] Zeviani M, et al. Maternally inherited myopathy and cardiomyopathy: association with mutation in mitochondrial DNA tRNA(Leu)(UUR). Lancet Lond Engl 1991;338:143–7.

[30] Sweeney MG, Brockington M, Weston MJ, Morgan-Hughes JA, Harding AE. Mitochondrial DNA transfer RNA mutation Leu(UUR)A-- > G 3260: a second family with myopathy and cardiomyopathy. Q J Med 1993;86:435–8.

[31] Casali C, et al. A novel mtDNA point mutation in maternally inherited cardiomyopathy. Biochem Biophys Res Commun 1995;213:588–93.

[32] Fassone E, et al. Mutations in the mitochondrial complex I assembly factor NDUFAF1 cause fatal infantile hypertrophic cardiomyopathy. J Med Genet 2011;48:691–7.

[33] Papadopoulou LC, et al. Fatal infantile cardioencephalomyopathy with COX deficiency and mutations in SCO2, a COX assembly gene. Nat Genet 1999;23:333–7.

[34] Shehata BM, et al. Exome sequencing of patients with histiocytoid cardiomyopathy reveals a de novo NDUFB11 mutation that plays a role in the pathogenesis of histiocytoid cardiomyopathy. Am J Med Genet A 2015;167A:2114–21.

[35] Fukuhara N, Tokiguchi S, Shirakawa K, Tsubaki T. Myoclonus epilepsy associated with ragged-red fibres (mitochondrial abnormalities): disease entity or a syndrome? Light-and electron-microscopic studies of two cases and review of literature. J Neurol Sci 1980;47:117–33.

[36] Zeviani M, et al. A MERRF/MELAS overlap syndrome associated with a new point mutation in the mitochondrial DNA tRNA(Lys) gene. Eur J Hum Genet 1993;1:80–7.

[37] Miyahara H, et al. Autopsied case with MERRF/MELAS overlap syndrome accompanied by stroke-like episodes localized to the precentral gyrus. Neuropathol J Jpn Soc Neuropathol 2019;39:212–17.

[38] Mancuso M, et al. Phenotypic heterogeneity of the 8344A > G mtDNA 'MERRF' mutation. Neurology 2013;80:2049–54.

[39] Claeys KG, et al. Novel genetic and neuropathological insights in neurogenic muscle weakness, ataxia, and retinitis pigmentosa (NARP). Muscle Nerve 2016;54:328–33.

[40] Bitner-Glindzicz M, et al. Prevalence of mitochondrial 1555A-- > G mutation in European children. N Engl J Med 2009;360:640–2.

[41] Vandebona H, et al. Prevalence of mitochondrial 1555A-- > G mutation in adults of European descent. N Engl J Med 2009;360:642–4.

[42] Pulkes T, et al. New phenotypic diversity associated with the mitochondrial tRNA(SerUCN) gene mutation. Neuromuscul Disord 2005;15:364–71.

[43] Guan M-X, et al. Mutation in TRMU related to transfer RNA modification modulates the phenotypic expression of the deafness-associated mitochondrial 12S ribosomal RNA mutations. Am J Hum Genet 2006;79:291–302.

[44] Kullar PJ, et al. Heterozygous SSBP1 start loss mutation co-segregates with hearing loss and the m.1555A > G mtDNA variant in a large multigenerational family. Brain J Neurol 2018;141:55–62.

[45] Pearson HA, et al. A new syndrome of refractory sideroblastic anemia with vacuolization of marrow precursors and exocrine pancreatic dysfunction. J Pediatr 1979;95:976–84.
[46] Rotig A, et al. Mitochondrial DNA deletion in Pearson's marrow/pancreas syndrome. Lancet Lond Engl 1989;1:902–3.
[47] Broomfield A, et al. Paediatric single mitochondrial DNA deletion disorders: an overlapping spectrum of disease. J Inherit Metab Dis 2015;38:445–57.
[48] Rötig A, Bourgeron T, Chretien D, Rustin P, Munnich A. Spectrum of mitochondrial DNA rearrangements in the Pearson marrow-pancreas syndrome. Hum Mol Genet 1995;4:1327–30.
[49] McShane MA, et al. Pearson syndrome and mitochondrial encephalomyopathy in a patient with a deletion of mtDNA. Am J Hum Genet 1991;48:39–42.
[50] Horga A, et al. Peripheral neuropathy predicts nuclear gene defect in patients with mitochondrial ophthalmoplegia. Brain J Neurol 2014;137:3200–12.
[51] DiMauro S, et al. Benign infantile mitochondrial myopathy due to reversible cytochrome c oxidase deficiency. Trans Am Neurol Assoc 1981;106:205–7.
[52] DiMauro S, et al. Benign infantile mitochondrial myopathy due to reversible cytochrome c oxidase deficiency. Ann Neurol 1983;14:226–34.
[53] Horvath R, et al. Molecular basis of infantile reversible cytochrome c oxidase deficiency myopathy. Brain J Neurol 2009;132:3165–74.
[54] Chen T-H, Tu Y-F, Goto Y-I, Jong Y-J. Benign reversible course in infants manifesting clinicopathological features of fatal mitochondrial myopathy due to m.14674 T > C mt-tRNAGlu mutation. QJM Mon J Assoc Physicians 2013;106:953–4.
[55] Schwartz M, Vissing J. Paternal inheritance of mitochondrial DNA. N Engl J Med 2002;347:576–80.
[56] Luo S, et al. Biparental inheritance of mitochondrial DNA in humans. Proc Natl Acad Sci U S A 2018;115:13039–44.
[57] Rius R, et al. Biparental inheritance of mitochondrial DNA in humans is not a common phenomenon. Genet Med J Am Coll Med Genet 2019;21:2823–6. Available from: https://doi.org/10.1038/s41436-019-0568-0.
[58] Balciuniene J, Balciunas D. A nuclear mtDNA concatemer (Mega-NUMT) could mimic paternal inheritance of mitochondrial genome. Front Genet 2019;10:518.
[59] Chinnery PF, et al. Risk of developing a mitochondrial DNA deletion disorder. Lancet Lond Engl 2004;364:592–6.
[60] Fraser JA, Biousse V, Newman NJ. The neuro-ophthalmology of mitochondrial disease. Surv Ophthalmol 2010;55:299–334.
[61] Giordano C, et al. Oestrogens ameliorate mitochondrial dysfunction in Leber's hereditary optic neuropathy. Brain J Neurol 2011;134:220–34.
[62] Sacconi S, et al. A functionally dominant mitochondrial DNA mutation. Hum Mol Genet 2008;17:1814–20.
[63] Jenuth JP, Peterson AC, Fu K, Shoubridge EA. Random genetic drift in the female germline explains the rapid segregation of mammalian mitochondrial DNA. Nat Genet 1996;14:146–51.
[64] Rahman S, Poulton J, Marchington D, Suomalainen A. Decrease of 3243 A-- > G mtDNA mutation from blood in MELAS syndrome: a longitudinal study. Am J Hum Genet 2001;68:238–40.
[65] Pitceathly RDS, et al. Extra-ocular muscle MRI in genetically-defined mitochondrial disease. Eur Radiol 2016;26:130–7.
[66] Pfeffer G, Majamaa K, Turnbull DM, Thorburn D, Chinnery PF. Treatment for mitochondrial disorders. Cochrane Database Syst Rev 2012;. Available from: https://doi.org/10.1002/14651858.CD004426.pub3 CD004426.
[67] Quijada-Fraile P, et al. Follow-up of folinic acid supplementation for patients with cerebral folate deficiency and Kearns-Sayre syndrome. Orphanet J Rare Dis 2014;9:217.
[68] Rahman J, Rahman S. Mitochondrial medicine in the omics era. Lancet Lond Engl 2018;391:2560–74.

[69] Hirano M, Emmanuele V, Quinzii CM. Emerging therapies for mitochondrial diseases. Essays Biochem 2018;62:467–81.
[70] Bacman SR, et al. MitoTALEN reduces mutant mtDNA load and restores tRNAAla levels in a mouse model of heteroplasmic mtDNA mutation. Nat Med 2018;24:1696–700.
[71] Gammage PA, et al. Genome editing in mitochondria corrects a pathogenic mtDNA mutation in vivo. Nat Med 2018;24:1691–5.
[72] Guy J, et al. Gene therapy for Leber hereditary optic neuropathy: low- and medium-dose visual results. Ophthalmology 2017;124:1621–34.
[73] Harding AE, Holt IJ, Sweeney MG, Brockington M, Davis MB. Prenatal diagnosis of mitochondrial DNA8993 T----G disease. Am J Hum Genet 1992;50:629–33.
[74] Blok RB, Gook DA, Thorburn DR, Dahl HH. Skewed segregation of the mtDNA nt 8993 (T-- > G) mutation in human oocytes. Am J Hum Genet 1997;60:1495–501.
[75] Sallevelt SCEH, et al. Preimplantation genetic diagnosis for mitochondrial DNA mutations: analysis of one blastomere suffices. J Med Genet 2017;54:693–7.
[76] Pickett SJ, et al. Mitochondrial donation—which women could benefit? N Engl J Med 2019;380:1971–2.

CHAPTER 15

Nuclear genetic disorders of mitochondrial DNA gene expression

Ruth I.C. Glasgow[1,2], Albert Z. Lim[1,2], Thomas J. Nicholls[1,3], Robert McFarland[1,2], Robert W. Taylor[1,2] and Monika Oláhová[1,3]

[1]Wellcome Centre for Mitochondrial Research, Newcastle University, Newcastle upon Tyne, United Kingdom, [2]Newcastle University Translational and Clinical Research Institute, Newcastle University, Newcastle upon Tyne, United Kingdom, [3]Newcastle University Biosciences Institute, Newcastle University, Newcastle upon Tyne, United Kingdom

15.1 Introduction

While the mitochondrial genome encodes 13 oxidative phosphorylation (OXPHOS) proteins and all of the RNA molecules required for mitochondrial translation, the majority of mitochondrial proteins are encoded by the nuclear genome. The current estimate of mitochondrial proteins of nuclear genetic origin lies at ~1158 (MitoCarta 2.0) and the number of pathogenic variants identified within nuclear genes involved in mitochondrial DNA (mtDNA) gene expression continues to grow [1]. The clinical features associated with defects of mtDNA gene expression, and the consequent OXPHOS deficiency are diverse. The spectrum of clinical disease includes multisystem and isolated organ dysfunction, resulting in a range of "syndromic" presentations. However, many patients do not fit these classic syndromic archetypes, presenting at various stages of life, with markedly different disease severity and outcome. Tissues with high energy demands are considered the most sensitive to defects in ATP synthesis and consequently heart, brain, and skeletal muscle are the most commonly affected tissues. However, patients carrying identical pathogenic variants in the same gene can present with quite different pathology, while those with similar clinical features may have distinct genetic etiologies. This extensive clinical and genetic heterogeneity makes the identification, characterization, and diagnosis of mitochondrial disease challenging. Further complicating matters, the clinical features of mitochondrial disease often overlap with other neurological or systemic diseases.

Increased implementation of high-throughput next-generation sequencing (NGS) to diagnostic pathways is leading to the widespread application of whole-exome sequencing (WES)

The Human Mitochondrial Genome.
DOI: https://doi.org/10.1016/B978-0-12-819656-4.00015-2

and whole-genome sequencing (WGS) in rare disease, thus revolutionizing the genetic diagnosis of mitochondrial diseases. Where candidate or targeted gene panel approaches to genetic diagnosis rely on neat phenotype-genotype correlations, WES/WGS are unbiased sequencing tools, that when used in conjunction with bioinformatic filtering pipelines, have proven to be extremely effective in the diagnosis of such a complex phenotypic spectrum, with some centers achieving successful diagnosis in a high percentage of cases [2,3]. Upon identification of novel, or previously unreported, variants, functional characterization within a research setting is vital in the assignment of pathogenicity and confirmation of genetic diagnosis [4].

Over one-third of reported mitochondrial disease-causing nuclear defects are in genes encoding proteins with a role in mtDNA gene expression, encompassing mtDNA maintenance and replication through to mitochondrial transcription and translation [4].

What are the consequences when these mechanisms of mtDNA expression (Fig. 15.1; see also Chapter 1: MtDNA replication, maintenance, and nucleoid organization and Chapter 2: Human mitochondrial transcription and translation in this book) are perturbed? Outlined below are some of the pertinent examples of the functional effects of reported mutations and the clinical phenotypes associated with each disease gene. Fig. 15.2 provides a list of the mtDNA expression disease genes identified to date.

15.2 Mechanisms of mtDNA replication

In the matrix of all mitochondria, there are an estimated 1–15 mtDNA molecules; however, this number is dynamic and responsive to cell type-specific mechanisms of copy number control [5,6]. At 16,569 bp in length, the circular double-stranded mitochondrial genome encodes 37 genes. Of these genes, 13 encode polypeptides, all of which are core subunits of OXPHOS complexes I, III, IV, and V. The remaining genes encode 22 transfer RNAs (tRNAs) and 2 ribosomal RNAs (rRNAs), 16S and 12S, that are integral parts of the large and small mitoribosomal subunits, respectively. The two strands of mtDNA are termed the heavy strand (H strand) and light strand (L strand).

A faithful system of mtDNA replication is essential for the maintenance of the mitochondrial genome and the expression of its genes. Two main origins of replication have been described: the origin of H-strand replication (O_H), which lies within the noncoding region (NCR) of the mitochondrial genome, and the origin of L-strand replication (O_L), which lies in a cluster of tRNAs approximately two-thirds of the genome downstream from O_H. The RNA primers necessary for mtDNA replication are synthesized by the mitochondrial RNA polymerase (POLRMT) [7,8]. DNA synthesis is driven by the mitochondria-specific DNA polymerase γ (Pol γ), a heterotrimer composed of one catalytic subunit encoded by *POLGA* and a homodimeric accessory subunit, encoded by *POLGB*, which confers processivity to

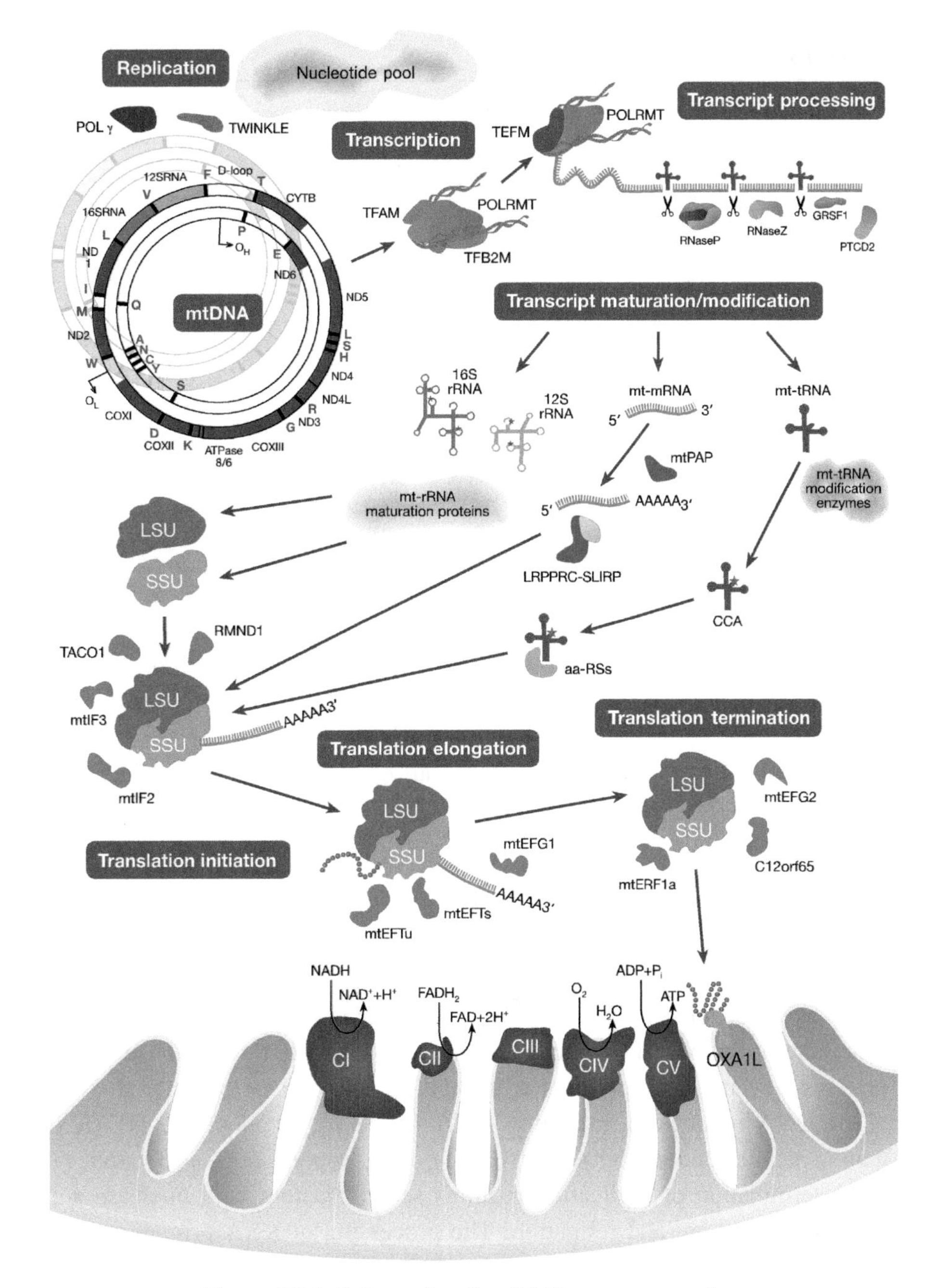

Figure 15.1: Schematic of mtDNA gene expression.

Shown are some of the key nuclear-encoded mitochondrial gene products involved in each major step. Replication and transcription of the mitochondrial genome depend on a pool of available nucleotides. Following transcription and processing, mt-rRNAs, mt-mRNAs, and mt-tRNAs undergo a number of posttranscriptional modification steps facilitated by numerous maturation and modification enzymes. The mitoribosomal large subunit (LSU) and small subunit (SSU) assemble to synthesize each of the 13 mitochondrial-encoded OXPHOS polypeptides, followed by their insertion into the inner mitochondrial membrane as constituents of complexes I, III, IV, and V (CI, CIII, CIV and CV).

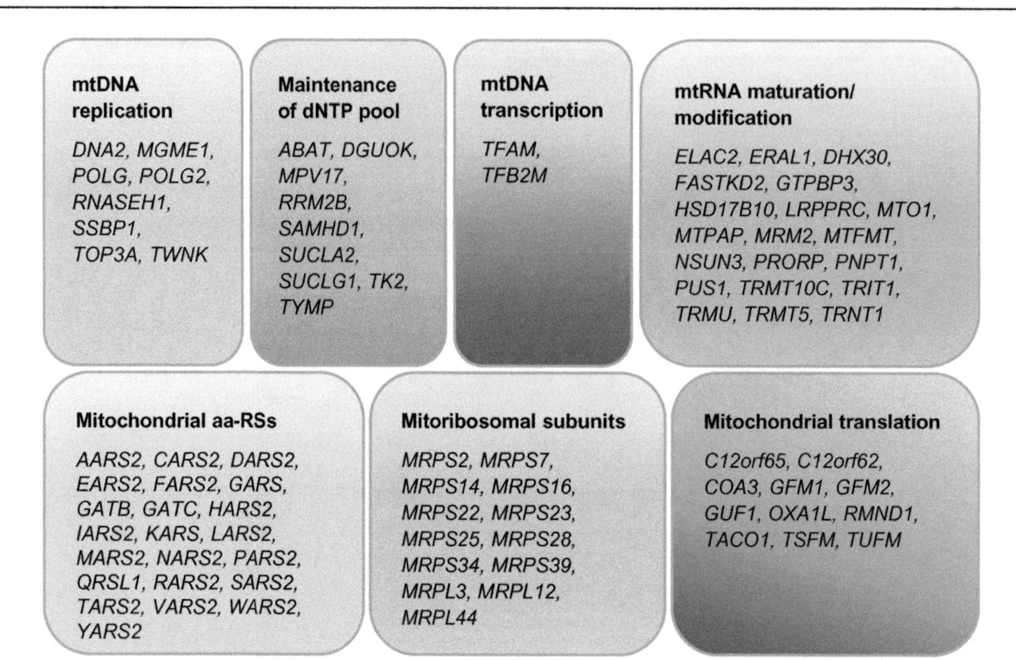

Figure 15.2
List of nuclear-encoded mitochondrial disease genes linked to defects of mitochondrial DNA gene expression to date. Genes are categorised according to their function. Accurate as of April 2020.

Pol γ by increasing its affinity to DNA [9]. In order for Pol γ to proceed the two mtDNA strands must be unwound at the replication fork, a role carried out by the hexameric mitochondrial helicase Twinkle (TWNK) through disruption of the hydrogen bonds between strands [10]. The mitochondrial single-stranded DNA-binding protein (mtSSB) additionally enhances the activity of both Pol γ and Twinkle [11]. In addition to these mitochondria-specific proteins, several other proteins with dual nuclear and mitochondrial functions are involved in mtDNA replication, including DNA ligase III, ribonuclease H1 (RNase H1), flap structure-specific endonuclease 1 (FEN1), and topoisomerase 3 alpha (TOP3α) [12–14]. The termination of replication requires several factors with nuclease activity for primer removal, processing, and religation. The processing of RNA primers can take place in a number of ways, usually involving the partial degradation of annealed primers by RNase H1 and subsequent displacement of remaining nucleotides by Pol γ to generate a "flap" of varying length. This flap can then be cleaved and cleared by FEN1, DNA2, or MGME1 [15]. To complete each replication event, separation of the two newly synthesized mtDNA molecules must take place. It has been demonstrated that the protein topoisomerase 3α (TOP3α) carries out decatenation of interlinked mtDNA, releasing each mtDNA molecule to form their final compact nucleoid structure [14]. Mitochondria maintain continuous cycles of DNA replication, running independently of the broader cell cycle in both actively proliferating and postmitotic cells [16,17]. The replication of the mtDNA L- and H-strands is highly asynchronous, with contention between whether the lagging-strand template is coated by mtSSB (the "strand displacement model") or hybridized to RNA (the "bootlace model"). Fully double-stranded replication intermediates resembling the products of strand-coupled replication have also been observed in mitochondria [18–20]. For a description of the finer details of the maintenance and replication of mtDNA, see Chapter 1: MtDNA replication, maintenance, and nucleoid organization.

15.3 Defects of mtDNA replication

15.3.1 Mutations in POLG

Pathogenic variants in *POLG*, the gene encoding the catalytic subunit of DNA polymerase γ (Pol γ), are the most common single gene causes of inherited mitochondrial disorders [21]. In one Australian cohort, *POLG* mutations accounted for 10% of adult cases [22]. At present, over 300 different pathogenic *POLG* variants have been deposited into the Human DNA Polymerase Gamma Mutation Database (https://tools.niehs.nih.gov/polg/). These mutations span the entirety of the amino acid sequence of *POLG*, some with autosomal recessive and others with autosomal dominant inheritance patterns. Functionally, these *POLG* defects can cause decreased activity of Pol γ and stalling at the replication fork. The result of this mtDNA synthesis defect can be mtDNA depletion, mtDNA multiple deletion, or a combination of both [23]. Mutations in the *POLG* gene can manifest in a group of disorders that comprise a continuum of overlapping phenotypes [24,25].

Alpers-Huttenlocher syndrome (AHS), characterized by progressive neurodegeneration, refractory seizures, movement disorder, neuropathy, and hepatic failure, is the most commonly reported phenotype of *POLG*-related disease in children and has a high mortality rate [24,26]. Disease-onset in carriers of *POLG* mutations is most commonly between 2 and 4 years [27,28] with a second peak onset between 17 and 24 years [29,30]. Progression is rapid and death usually occurs within 4 years of onset [29,31]. Homozygous variants located in the linker domain of the *POLG* gene correlate with a later age of onset and longer survival compared to compound heterozygous variants [32,33].

Another phenotype of neurodegenerative disease secondary to *POLG* mutations is the childhood myocerebrohepatopathy spectrum, characterized by hypotonia, myopathy, developmental regression, liver failure, pancreatitis, cyclical vomiting, and renal tubular acidosis [34]. These features typically become apparent from a few months of age and are distinct from AHS in that seizures are a less prominent feature [32,34].

Myoclonic epilepsy myopathy sensory ataxia (MEMSA), another *POLG*-related mitochondrial disorder, refers to a spectrum of disorders with epilepsy, myopathy, and ataxia without ophthalmoplegia and now also encapsulates what was previously known as spinocerebellar ataxia with epilepsy (SCAE) [24]. Ataxia neuropathy spectrum (ANS) is a group of *POLG*-related disorders in which patients have difficulties with coordination along with nerve dysfunction [24]. This spectrum also includes sensory ataxia, neuropathy, dysarthria, and ophthalmoplegia (SANDO), and mitochondrial recessive ataxia syndrome (MIRAS) [35].

The autosomal dominant PEO (AdPEO) *POLG*-related disorder typically presents in adult life with progressive weakness of extraocular muscles resulting in bilateral ptosis and/or double vision. Most of the affected patients also have generalized myopathy. Sensorineural hearing loss, axonal neuropathy, ataxia, parkinsonism, cataracts, premature ovarian failure, cardiomyopathy, and gastrointestinal dysmotility have been reported [36,37]. The autosomal recessive PEO, in contrast, rarely has systemic involvement [38].

15.3.2 Mutations in TWNK

As with *POLG* defects, biallelic mutations in *TWNK*, the gene encoding the mtDNA helicase Twinkle, result in the stalling of replication and can cause qualitative defects in the form of mtDNA deletions or quantitative mtDNA depletion defects [39]. Recent investigation into the molecular basis for *TWNK*-related disease suggests that mutant Twinkle is unable to oligomerize through the normal conformational changes made by *wild-type* Twinkle when interacting with DNA [40].

The most common phenotype in *TWNK*-related mitochondrial DNA maintenance defects is autosomal-dominant PEO. Almost all affected individuals with monoallelic pathogenic variants in *TWNK* develop ptosis and ophthalmoplegia. Apart from these near, universal ophthalmic findings, other frequently reported, though generally mild, features include fatigue, proximal myopathy, myalgia, and exercise intolerance [41–43]. The onset of disease could vary from childhood to late adulthood with the mean in the fourth decade [42].

Biallelic pathogenic variants in *TWNK* are not common. Affected individuals present in early life with progressive and life-limiting conditions. The infantile-onset hepatocerebral type mitochondrial DNA depletion secondary to *TWNK* mutations typically presents with hepatopathy, lactic acidosis, psychomotor delay, and hypotonia [44–46]. A genetically isolated population in Finland have reported *TWNK*-related spinocerebellar ataxia with disease-onset around 1 year [44,47]. These infants present with extrapyramidal signs, low tone, and depressed reflexes associated with white matter changes in the cerebellum. Another rare association with *TWNK* mutations is Perrault syndrome, characterized by hearing loss and ovarian dysfunction [48,49].

15.3.3 Mutations in DNA2

Both autosomal-dominant and autosomal-recessive mutations have been identified in the mitochondrial DNA maintenance helicase *DNA2* gene, resulting in two distinct disease presentations. Autosomal dominant *DNA2* disease results in a clear mtDNA maintenance defect of multiple mtDNA deletions and a myopathic phenotype, which usually presents during adolescence with ptosis, ophthalmoplegia, proximal muscle weakness, muscle atrophy, and fatigability. The heterozygous missense variants within the nuclease domain of DNA2 are believed to disturb RNA primer and flap intermediate removal and impede mtDNA synthesis [50]. Recessive *DNA2* variants are associated with Seckel syndrome, which is characterized by intrauterine growth retardation, dwarfism, microcephaly with mental retardation, and a characteristic "bird-headed" facial appearance [51,52].

15.3.4 Mutations in MGME1

Autosomal recessive mutations (homozygous nonsense and homozygous missense variants) in the *MGME1* gene have been reported in three families, resulting in both mtDNA depletion and multiple deletions. The mtDNA rearrangements identified in all affected patients are significantly larger than deletions characteristic of *POLG*-related mtDNA maintenance defects and include numerous duplications [53]. Abnormally increased levels of 7S DNA, the linear component of the D-loop region, were also identified in fibroblasts and muscle tissues of patients, indicating an important role of MGME1 in 7S DNA turnover [54].

A homozygous frameshift deletion in *MGME1* leading to a premature termination of the gene product has been identified by WES in a child presenting with early onset cerebellar ataxia, speech delay, microcephaly, and cerebellar atrophy [39]. The *MGME1* frameshift deletion variant is characterized by a more severe phenotype and expands the clinical spectrum of the previously reported mtDNA depletion syndrome 11 patients. These patients usually present during childhood or early adulthood with ptosis, ophthalmoplegia, easy fatigability, exercise intolerance, progressive proximal muscular weakness, muscle wasting, and weight loss [54]. The involvement of chest wall muscles can result in breathlessness, which can progress to respiratory failure resulting in the need for noninvasive ventilation. Other common reported features are kyphosis, intellectual disability, and ataxia [54].

15.4 Maintenance of dNTP pool

The replication and repair of both the nuclear and mitochondrial genomes require a steadily available pool of deoxyribonucleoside triphosphates (dNTPs). The two major pathways that regulate dNTP supply to mitochondria are the salvage and the *de novo* synthesis pathways. The latter operates in the cytosol where the key regulatory enzymes involved in dNTP synthesis are ribonucleotide reductase (RNR) and thymidylate synthase (TS). RNR is responsible for the reduction of ribonucleotides to deoxyribonucleotides [55,56]. Specific mitochondrial carriers, such as MPV17, then facilitate the transport of cytosolic dNTPs into the matrix, as mitochondria do not possess *de novo* nucleotide synthesis pathways, nor can such molecules permeate the inner mitochondrial membrane (IMM) due to their charge [57].

The unique mitochondrial dNTP salvage pathway constantly converts former deoxynucleosides, already within the mitochondrial matrix as a result of DNA turnover, into dNTPs [58]. Mitochondrial thymidine kinase 2 (TK2) is a key driver of the pyrimidine nucleotide salvage pathway as it carries out the initial phosphorylation of pyrimidine precursors [59]. Similarly, the primary phosphorylation step in purine nucleotide salvage is carried out by mitochondrial deoxyguanosine kinase (DGK) [60]. Two additional phosphorylation steps follow, carried out by nucleotide monophosphate kinase (NMPK) and then nucleotide diphosphate kinase (NDPK), resulting in the conversion of each of the deoxyribonucleoside monophosphates (dNMPs) into dNTPs [61]. For NMPK to execute the final phosphorylation of each nucleotide, it must form a complex with 4-aminobutyrate transaminase (ABAT) and succinyl CoA ligase (SUCL) [62,63].

15.5 Defects of the dNTP salvage pathway and nucleotide metabolism

15.5.1 Mutations in TK2

Loss-of-function mutations in the *TK2* gene are now a well-described cause of mtDNA maintenance defects. Due to the role of mitochondrial thymidine kinase 2 in the first step of the pyrimidine nucleotide salvage pathway, these mutations drastically affect the nucleotide pool within mitochondria and result in severe mtDNA depletion [64]. *TK2*-related mitochondrial disease presents in infancy, childhood (juvenile), or adult life with prominent myopathic features [65–67]. Infantile-onset *TK2*-related disease with mtDNA depletion is the most severe form often associated with additional central nervous system (CNS) involvement, and if untreated, it progresses rapidly to early death from respiratory failure. Carriers of this genetic mutation who experience juvenile-onset of symptoms have better outcomes, but still develop progressive moderate-to-severe generalized weakness with mean survival around age 13 years [66]. Adult-onset myopathy of *TK2*-related mitochondrial disease usually takes a slower progression with mean survival beyond the second decade [66–68]. These adult patients typically have ophthalmoparesis and facial weakness. Their muscle biopsies often demonstrate multiple mtDNA deletions rather than depletion [69]. There is growing evidence that nucleotide or nucleoside treatment can alter the natural history of *TK2*-related mitochondrial disease and early commencement of therapy seems to be advantageous [70–72].

15.5.2 *Mutations in* RRM2B

Pathogenic mutations in the *RRM2B* gene, encoding the p53-inducible small subunit of the cytosolic dNTP salvage enzyme RNR, have been implicated in both mtDNA depletion and deletion phenotypes. Many of the reported *RRM2B* mutations are located within a highly conserved alpha helical region and are thought to interrupt intramolecular interactions [55]. *RRM2B*-related mitochondrial disease manifests with three typical phenotypes—mitochondrial DNA depletion syndrome (encephalomyopathic form with renal tubulopathy), mitochondrial neurogastrointestinal encephalopathy, and chronic PEO with multiple mtDNA deletions [73–79]. Autosomal dominant variants usually present with PEO, mild myopathy, bulbar dysfunction, and fatigue associated with the finding of multiple mtDNA deletions in a skeletal muscle biopsy [73,80]. Homozygous or compound heterozygous carriers tend to have a more severe multisystemic disorder and present much earlier than the single heterozygous carriers (mean age 7 years vs 46 years) [73,81]. Infants with recessive *RRM2B* mutations typically present with profound myopathy, hypotonia, gastrointestinal dysmotility, proximal renal tubulopathy, and severe CNS involvement with epileptic encephalopathy and diffuse cerebral atrophy. A later onset in adolescence or early adult life presents with PEO, ptosis, and proximal myopathy, followed by sensorineural hearing loss, bulbar dysfunction, sensory ataxia, and gastrointestinal dysmotility. To date, nearly 100 individuals with a confirmed genetic diagnosis of *RRM2B* have been reported [81].

15.5.3 *Mutations in* MPV17

IMM protein MPV17 has been demonstrated to form a nonselective channel believed to play a role in dNTP homeostasis through the nucleotide salvage pathway [82,83]. Autosomal recessive mutations in the *MPV17* gene are a reported cause of infantile-onset multisystemic mtDNA depletion disease with liver and neurological involvement [84,85]. *MPV17* mutations have also been identified in juvenile- and adult-onset mtDNA depletion cases causing both multisystemic disease and isolated axonal sensorimotor neuropathy [86–88]. Rescue of mtDNA depletion in quiescent patient fibroblasts through deoxynucleoside supplementation points to dNTP insufficiency as the driver of pathogenicity in these mtDNA depletion syndromes [83].

15.6 *Mechanism of mitochondrial transcription*

The mitochondrial transcription machinery is entirely nuclear encoded and it is well demonstrated that mitochondrial transcription can be reconstituted *in vitro* in the presence of just three proteins: mitochondrial RNA polymerase (POLRMT), mitochondrial transcription factor A (TFAM), and mitochondrial transcription factor B2 (TFB2M) [89,90]. POLRMT, a single-subunit RNA polymerase that is related to the bacteriophage T7 RNA polymerase (T7 RNAP), is the driver of mitochondrial transcription. The transcription activity of POLRMT is exclusive to the mitochondrial genome [91]. POLRMT requires the two additional cofactors TFAM and TFB2M for promoter recognition and melting during transcription initiation [92].

Mammalian TFAM is a multifunctional protein with essential roles in mitochondrial transcription initiation, packaging of mtDNA into compact nucleoids, and regulating mtDNA copy number. TFAM has the ability to bind to high-affinity binding sites ~10–15 bp upstream of the promoter sites and introduce a 180 degrees bend in the bound mtDNA, thus allowing for the recruitment of POLRMT to promoter DNA to form the preinitiation complex [92]. Finally, the recruitment of TFB2M to the preinitiation complex introduces structural changes in POLRMT that drive promoter melting and allow the entry of the initiating nucleotide into the catalytic site of POLRMT, thus completing initiation and allowing elongation to commence.

During the transition from initiation to elongation, TFB2M dissociates and a new combination of proteins, the elongation complex, assembles. The mitochondrial transcription elongation factor TEFM dimerizes and binds to the POLRMT site previously occupied by TFB2M, forming a sliding clamp structure [93]. TEFM drives elongation by increasing POLRMT–mtDNA interactions, greatly enhancing POLRMT processivity along the mtDNA strand and preventing premature termination or stalling [89,94].

The exact mechanism by which mitochondrial transcription is terminated remains somewhat unresolved. One protein known to play a key role in the termination of transcription is MTERF1 [95]. MTERF1 binds along the major groove of mtDNA at the 3′ end of the 16S rRNA gene and induces a bend, partial unwinding of the double helix and the eversion of three nucleotides to stabilize this interaction [96]. MTERF1 exhibits polar transcription termination activity consistent with a role in terminating transcription originating only from LSP [95]. Equivalent proteins involved in the termination of transcription originating from HSP are yet to be identified. See Chapter 2: Human mitochondrial transcription and translation for a more in-depth description of mitochondrial transcription.

15.7 Defects of mitochondrial transcription

Mutations in the nuclear-encoded mtDNA maintenance machinery represent a frequent cause of inherited mitochondrial disease. For instance, more than 300 disease-causing mutations have been identified in the mtDNA polymerase *POLG* alone [25]. In contrast, pathogenic variants in the mitochondrial transcription machinery remain relatively unknown and to date only single pathogenic variants in the mitochondrial transcription factors *TFAM* and *TFB2M* have been identified.

15.7.1 Mutations in TFAM

Pathogenic mutations in *TFAM*, the gene encoding the mitochondrial transcription factor A, required for mitochondrial transcription initiation and mtDNA packaging of nucleoids, have been identified in only one family with two affected siblings each carrying a homozygous missense c.533C > T, p.Pro178Leu variant (RefSeq: NM_003201.2). These two siblings presented with intrauterine growth restriction and hypoglycemia with abnormal liver function tests. They progressed rapidly to liver failure and death in infancy. Samples from liver biopsies demonstrated evidence of cirrhosis, cholestasis, and steatosis. Depletion of mtDNA was present in cultured fibroblasts and affected tissues, along with a combined OXPHOS defect of complexes I, II, III, and IV. The number of nucleoids was decreased in fibroblasts, and abnormal nucleoid aggregates could be observed [97].

15.7.2 Mutations in TFB2M

Only one homozygous missense variant in the *TFB2M* gene c.790C > T, p.His264Tyr (RefSeq: NM_022366.3) has been reported in two Korean siblings with an unusual presentation of mitochondrial disease pathology. The variant does not appear to affect the expression of TFB2M, but instead may cause disease through a gain of function effect. In patient fibroblasts, a significant increase was observed in the level of *MT-ND4L*, *MT-ND5*, *MT-CYB*, *MT-CO1*, and *MT-ATP6* transcripts, along with a dramatic increase in relative mitochondrial membrane potential, ATP, and reactive oxygen species (ROS). This overall enhancement of mitochondrial function is suggested to raise oxidative stress through ROS during brain development leading to a clinical presentation of features consistent with the diagnosis of an autistic spectrum disorder. These two Korean siblings had significant delay in language and social development along with stereotyped repetitive speech, as well as intellectual disability [98].

15.8 Transcript processing

The long polycistronic products of mitochondrial transcription undergo essential post-transcriptional processing and modification steps prior to mitochondrial translation. The majority of protein-coding mtDNA genes, along with both mt-rRNA genes, are separated by individual mt-tRNA genes. The mt-tRNAs act as guides for endonucleolytic excision and thus the release of each flanking mRNA and rRNA transcript is a structured system termed the "tRNA punctuation model" [99]. Cleavage of each tRNA is carried out by ribonuclease P (RNaseP) and ElaC ribonuclease Z 2 (ELAC2) at the 5′ and 3′ ends, respectively. Though this model can account for the release of the majority of individual transcripts, there are some protein-coding genes not flanked by tRNAs on either side. It has been suggested that the remaining three precursors that are not flanked by tRNAs require one of two proteins for their processing, GRSF1 or PTCD2 [100–102].

15.9 Defects of maturation of pre mt-RNA

15.9.1 Mutations in RNase P complex (MRPP1, 2, 3)

The mitochondrial RNase P complex is composed of mitochondrial RNase P proteins I–III (MRPP1, 2, and 3), all three of which have been implicated in mitochondrial disease. Along with their role as RNase P constituents, MRPP1 and MRPP2 also form a complex responsible for the methylation of position 9 of mt-tRNAs [103].

Missense variants affecting conserved residues in the *TRMT10C* gene, encoding MRPP1, have been identified in two unrelated subjects who presented at birth with lactic acidosis, hypotonia, deafness, and feeding difficulties [104]. The compound heterozygous c.542G > T, p.Arg181Leu; c.814A > G, p.Thr272Ala; and homozygous c.542G > T, p.Arg181Leu (RefSeq: NM_017819.3) changes caused a decrease in steady-state levels of the MRPP1 protein with consequent impairment of mt-tRNA processing. This resulted in a defect in mitochondrial translation and a combined OXPHOS deficiency affecting complexes I and IV observed in patient fibroblasts [104].

Missense variants in the *HSD17B10* gene, encoding the MRPP2 subunit of RNaseP, have been identified in over 20 families to date. MRPP2 acts as a dehydrogenase in isoleucine metabolism independently of its role in tRNA modification and 5′ transcript processing. Reported variants have been demonstrated to disrupt both RNase P activity and the methylation of mt-tRNAs carried out by the MRPP1/2 complex as well as the dehydrogenase activity of MRPP2 to varying extents [105,106]. The described clinical course of *HSD17B10*-related mitochondrial disease is characterized by a period of normal development in the first year of life followed by a relentlessly progressive neurodegenerative disease course. Developmental regression with loss of speech and motor skills in conjunction with ataxia and movement disorders is typical. The deterioration may be further complicated by development of dilated cardiomyopathy, leading to death at around the age of 2 years. Metabolic acidosis, hyperlactatemia, hypoglycemia, and hyperammonemia are a common biochemical profile shared by these patients. Neuroimaging usually shows frontotemporal or generalized cortical atrophy and lesions in the basal ganglia [105].

The third component of RNase P, MRPP3, is encoded by the *PRORP* gene. A homozygous missense c.1454C > T, p.Ala485Val variant in *PRORP* (RefSeq: NM_014672.3) has been identified in a single family case of Perrault syndrome. The variant affected the stability of MRPP3 protein, reducing the amount of 5′ mt-tRNA processing carried out by the RNase P complex by ~35%–45% leading to an accumulation of unprocessed transcripts and a combined OXPHOS defect [107]

15.9.2 Mutations in RNase Z (ELAC2)

The exonuclease responsible for 3′ mt-tRNA processing is encoded by the *ELAC2* gene, in which numerous disease-causing variants have been reported [108–112]. *In vitro* modeling of ELAC2 with patient mutations demonstrated impairment of RNase Z activity. Almost all of the reported cases had neonatal and early childhood-onset cardiomyopathy, both hypertrophic and dilated types. Orthotropic heart transplant had been successful in some cases [112]. In most cases, an accumulation of unprocessed mt-tRNAs is observed in patient tissues along with defects of mitochondrial translation and combined OXPHOS deficiencies [112].

15.10 mt-mRNA maturation and turnover

Following the nucleolytic processing of each polycistron, all but one of the mt-mRNA transcripts undergoes adenylation on their 3′ ends. The majority of mt-mRNAs are polyadenylated with poly(A) tails with the exception of the *MT-ND6* transcript [113]. Poly(A) tail synthesis is carried out by the poly(A) polymerase (mtPAP), which localizes to mitochondrial RNA granules [114]. An essential function of poly(A) tails is to provide mt-mRNA transcripts with a complete "stop" codon. Seven of the thirteen mitochondrial open reading frames have incomplete stop codons ("U" or "UA") that, without poly- or oligo-adenylation into "UAA," would not cause termination of mitochondrial translation [113].

An important regulator of mtDNA gene expression is the "mitochondrial degradosome" dedicated to the turnover of RNAs within the matrix. The RNA degradosome apparatus consists of a mitochondria-specific helicase hSuv3p that can unwind double-stranded DNA (dsDNA), dsRNA and DNA–RNA hybrids [115], and polynucleotide phosphorylase (PNPase), encoded by the *PNPT1* gene, that has poly(A) polymerase (PAP) and 3′–5′ exoribonuclease activities. It has been demonstrated that hSuv3p and PNPase form a stable complex that is believed to play a role in the removal of RNA transcripts that are antisense, aberrant, or have undergone damage as well as normally processed mature transcripts in the mitochondrial matrix [116,117]. PNPase shows additional localization to the intermembrane space (IMS) where it is thought to facilitate nuclear-encoded RNA import into the matrix [118]. A third protein with a proposed role in mitochondrial RNA degradation is RNA exonuclease 2 (REXO2). The substrates of REXO2, degraded in the 3′ to 5′ direction, are dinucleotides generated during the degradation of RNA transcripts by hSuv3p-PNPase or by other nucleases [119,120].

The leucine-rich PPR-containing protein (LRPPRC) and stem-loop interacting RNA-binding protein (SLIRP) are both mitochondrial RNA binding proteins involved in the regulation of mt-mRNA stability. It has been demonstrated that the LRPPRC–SLIRP complex conveys stability to mt-mRNAs by blocking PNPase access and aiding in the polyadenylation of the 3′ terminus of bound mRNAs [121].

15.11 Defects of mt-mRNA maturation and turnover

At present *PNPT1*, the gene encoding PNPase, is the only member of the mitochondrial degradosome to be implicated in disease. A growing cohort of patients with biallelic variants in *PNPT1* present with a broad spectrum of clinical phenotypes thought to result from accumulation of double-stranded RNA molecules [122].

15.11.1 Mutations in MTPAP

A rare founder mutation c.1432A > G, p.Asn478Asp (RefSeq: NM_018109.3) resulting in a homozygous missense change at a highly conserved residue of the *MTPAP* gene has been identified in a large Old Order Amish family as the pathogenic variant responsible for a progressive neurodegeneration phenotype in multiple affected children. Encoded by *MTPAP*, the mtPAP protein is responsible for polyadenylation of mt-mRNA transcripts. RNA extracted from patient blood revealed severe truncation of mt-mRNA transcript poly(A) tails [123]. Patient fibroblasts demonstrate a profound OXPHOS deficiency affecting complexes I and IV along with a pattern of transcript-specific effects on mt-mRNA stability [124].

15.11.2 Mutations in LRPPRC

The first pathogenic variant identified in the *LRPRRC* gene, encoding a gene product involved in mt-mRNA maintenance and stability, was a founder mutation discovered in a unique type of Leigh syndrome, characterized by tissue-specific cytochrome *c* oxidase (COX) deficiency within the French-Canadian population. In a cohort of Leigh Syndrome patients, the French-Canadian type (LSFC) was found to be present in a homozygous state in 55/56 patients carrying the c.1061C > T, p.Ala354Val variant [125,126]. More recently, WES has identified novel disease-causing *LRPPRC* variants in 10 further patients from different ethnic backgrounds presenting with multiple OXPHOS defects in fibroblasts and skeletal muscle tissue [127]. Although in initial studies the biochemical defect in LSFC patients had been confined to Complex IV, further studies have shown that the *LRPPRC* variants cause multiple biochemical OXPHOS defects that manifest in a tissue-specific manner, and the levels of LRPPRC appear to determine the nature of the OXPHOS defect [127].

15.12 mt-tRNA maturation

Following the release of mt-tRNAs from primary polycistronic transcripts, they are subjected to a broad range of different posttranscriptional maturation and modification steps, necessary for the formation of stable and fully functioning tRNA structures [128,129]. A number of different nuclear-encoded tRNA modifying enzymes are responsible for the chemical modifications of different sites of each mitochondrial tRNA, some of which also modify nuclear tRNAs [130]. The final maturation steps involve the addition of CCA nucleotides to the 3′ terminus of mt-tRNAs followed by aminoacylation by the correct aminoacyl-tRNA synthetase (aa-RS). 19 aa-RSs are required for mitochondrial translation [131]. The aa-RSs that produce aminoacyl-tRNA conjugates in the cytosol and in mitochondria are generally encoded by two distinct genes. However, two aa-RSs, glycyl-ARS (GARS) and lysyl-ARS (LARS), act within both the cytosolic and mitochondrial systems, which is achieved via the inclusion or exclusion of a mitochondrial targeting sequence as a result of alternative initiation sites or splicing [132,133].

15.13 Defects of mt-tRNA maturation and modification

15.13.1 Mutations in TRNT1

The enzyme tRNA nucleotidyl transferase, CCA-adding 1 (TRNT1), catalyzes the addition of a 3′ CCA trinucleotide to tRNA molecules. The 3′-CCA is an essential modification as it facilitates correct aminoacylation of all tRNAs, both cytoplasmic and mitochondrial [134]. The clinical presentation of patients with mutations in the *TRNT1* gene is varied; however, some features, such as autoinflammation and immunodeficiency, are common. *TRNT1* disease may present in childhood with a severe sideroblastic anemia, B-cell immunodeficiency, fever, and developmental delay (SIFD), often in the presence of additional multiorgan involvement such as retinitis pigmentosa and sensorineural deafness [135,136]. *TRNT1* mutations are also reported in patients with a milder disease phenotype characterized by adult-onset retinitis pigmentosa as the sole phenotype or in the presence of mild immunological abnormalities or anemia [137]. Defective 3′-CCA addition to mt-tRNA^{Cys} and mt-tRNA^{His} was observed in the fibroblasts of two patients with compound heterozygous frameshift and missense *TRNT1* mutations [138]. Patient fibroblasts carrying missense *TRNT1* mutations demonstrate decreased steady-state levels of both nuclear and mtDNA-encoded OXPHOS proteins, along with an observable defect in oxygen consumption [139].

15.13.2 Mutations in PUS1

The *PUS1* gene, encoding pseudouridylate synthase 1 (PUS1), has been implicated in the rare disorder characterized by mitochondrial myopathy, lactic acidosis, and sideroblastic anemia (MLASA). PUS1 is a tRNA modification enzyme that carries out the conversion of uridine into pseudouridine in tRNAs of both nuclear and mitochondrial origins [140]. To date, nine families with pathogenic *PUS1* variants have been identified with missense, splice-site, and truncating mutations reported [140–146]. All but one *PUS1* patient has presented with MLASA, the only exception being an adult patient with chronic sideroblastic anemia, diarrhea, microcephaly, and failure to thrive [143]. However, some degree of clinical variability can be observed between different cases of MLASA, with features such as microcephaly, cognitive impairment, and growth hormone deficiency manifesting variably even within families [140,144]. Defects of mitochondrial translation and combined OXPHOS deficiencies have been demonstrated in the fibroblasts of some patients carrying *PUS1* mutations. Interestingly, these OXPHOS deficiencies affected the entirely nuclear-encoded complex II, indicating the presence of a cytosolic translation defect. It is possible that the extent to which cytosolic translation is affected, alongside a mitochondrial translation defect, could be partly responsible for some of the clinical heterogeneity observed in different cases of *PUS1* disease [140].

15.13.3 Mutations in MTO1 *and* GTPBP3

Encoded by the *MTO1* gene, the mitochondrial translation optimization 1 (MTO1) protein forms one subunit of a heterodimeric enzyme responsible for the 5-taurinomethylation of five mitochondrial tRNAs at their wobble uridine base. Patients with *MTO1* mutations are widely reported, with the most common clinical features of disease being lactic acidosis, hypertrophic cardiomyopathy, and global developmental delay. A more severe, sometimes fatal, phenotype is observed in patients who harbor one truncating variant rather than two missense mutations. Defects in the activities of multiple OXPHOS respiratory enzymes (complexes I, III, and IV) are characteristic of *MTO1* disease [147–149]. Also involved in the generation of the same 5-taurinomethyluridine modification is the mitochondrial GTP-binding protein 3 (GTPBP3). Mutations in the *GTPBP3* gene have been reported in patients with a Leigh-like syndrome characterized by hypertrophic cardiomyopathy, lactic acidosis, and encephalopathy [150].

15.13.4 Mutations in TRMU

The enzyme tRNA 5-methylaminomethyl-2-thiouridylate methyltransferase (TRMU), encoded by the gene of the same name, catalyzes the thiouridylation at the wobble base of mt-tRNA^{Lys}, mt-tRNA^{Glu}, and mt-tRNA^{Gln} [151]. This process is dependent on sulfur from L-cysteine, an amino acid essential during early infancy when levels of cystathionase, the enzyme required for cysteine production, are low [152]. Autosomal recessive mutations in the *TRMU* gene have been reported in at least 30 patients to date, predominantly presenting with infantile liver failure. The disease phenotype of patients with TRMU deficiency has been shown to be transient, reversing in the majority of patients who survive the period of acute hepatopathy. L-cysteine supplementation from birth has recently been demonstrated to alleviate the liver dysfunction in a single patient with compound heterozygous missense mutations in the *TRMU* gene [153].

15.13.5 Mutations in TRIT1

Encoded by the *TRIT1* gene, tRNA iso-pentenyltransferase (TRIT1) catalyzes the i^6A modification at position A37 of both cytosolic and mitochondrial tRNAs, believed to play an important role in correct codon–anticodon interactions [154]. Autosomal recessive variants in *TRIT1* have been reported in five unrelated families in which affected individuals presented with very similar clinical features including microcephaly, developmental delay, epilepsy, and brain abnormalities. Neuroimaging showed generalized mild atrophy with frontal lobes predominantly affected, prominent extra-axial fluid, partial absence of the corpus callosum, broad septum pellucidum, and small anterior horns of lateral ventricles [155]. Patient fibroblasts exhibited decreased i^6A modification, translation defects, and combined OXPHOS deficiencies at the steady-state level [155–157].

15.13.6 Mutations in mitochondrial aminoacyl-tRNA synthetases

Pathologies caused by mutations in all 19 mitochondrial aminoacyl-tRNA synthetases (mt-tRNA aa-RS) have now been described, with a variety of different clinical presentations. The first mitochondrial aminoacyl-tRNA synthetase to be reported as a disease gene was *DARS2*, encoding aspartyl tRNA synthetase. Patients from 30 families with leukoencephalopathy with brain stem and spinal cord involvement and lactate elevation (LBSL) underwent linkage mapping and candidate gene sequencing to identify a range of pathogenic variants in the *DARS2* gene, although patient fibroblasts did not exhibit any clear mitochondrial dysfunction [158]. Diseases resulting from mutations in mt-tRNA aa-RS genes are often hallmarked by features of CNS involvement such as leukodystrophy (*AARS2*, *DARS2*, *EARS2*, *MARS2*), encephalopathy (*RARS2*, *NARS2*, *CARS2*, *IARS2*, *FARS2*, *PARS2*, *TARS2*, *VARS2*), deafness or hearing loss (*NARS2*, *PARS2*, *MARS2*), or Perrault syndrome (*HARS2*, *LARS2*) [159,160]. For review please see [131,161]. However, non-CNS and isolated pathologies also occur in mt-tRNA aa-RS disease. Cardiomyopathy has been reported in patients with pathogenic variants in *GARS*, *KARS*, *YARS2*, and *AARS2* [162–165], while two distinct syndromes, MLASA (mitochondrial Myopathy, Lactic acidosis, and Sideroblastic Anemia) and HUPRA (Hyperuricemia, Pulmonary hypertension, Renal failure in infancy, and Alkalosis), can be caused by mutations in the *YARS2* and *SARS2* genes, respectively [166,167]. Along with the clinical heterogeneity between different mt-tRNA aa-RS disease genes, vast variability in disease presentation also exists within disorders of single aa-RS genes, such as *AARS2*. Through WES, pathogenic mutations in the *AARS2* gene were first reported in an infant with fatal hypertrophic cardiomyopathy [168] and then in two further families with cardiomyopathic clinical presentation in infancy [165], consistent with the original report of *AARS2* disease. However, a second distinct disease phenotype is seen in patients with *AARS2* mutations, characterized by leukoencephalopathy with, in female patients, premature ovarian failure [169]. The mutations reported across all 19 mt-tRNA aa-RS enzymes do not appear to follow any general trends in regard to the location of variants within core domains or at evolutionarily conserved residues. This, together with the variability in clinical presentation, both within and between the mt-tRNA aa-RS group, suggests a number of different pathomechanisms underlying this broad group of diseases [131].

15.14 mt-rRNA maturation

Like mt-tRNAs, both the 12S and 16S mt-rRNAs undergo a number of nucleotide modifications believed to convey stability and promote mitoribosome biogenesis. The 12S mt-rRNA is subject to methylation of a cytidine at position 841 and dimethylation of neighboring adenines at positions 936 and 937 by the methyltransferase NSUN4 and dimethyltransferase TFB1M proteins, respectively [170,171]. The recently identified METTL15 protein introduces an N-4 methylcytidine at position 839 of the 12S mt-rRNA believed to stabilize folding [172].

The 16S mt-rRNA undergoes one pseudouridinylation, carried out by the pseudouridine synthase RPUSD4 at position 1397 [173]. There are also several methylation sites on the 16S mt-rRNA. The protein TRMT61B, a known tRNA modifying enzyme, has been demonstrated to introduce a methyladenosine at position 947 of the 16S mt-rRNA. Structurally, this modification sits at the interface of the mitochondrial large (mt-LSU) and small subunit (mt-SSU) and may be required to maintain mitoribosomal structure and function [174]. The three proteins MRM1, MRM2, and MRM3 are a group of 2′-O-ribose methyltransferases, each believed to modify specific bases (Gm1145, Um1369, and Gm1370) within the peptidyl transferase center (PTC) of the 16S mt-rRNA [175–177].

15.15 Defects of mt-rRNA maturation, modification, and stability

15.15.1 Mutations in MRM2

To date, the only mt-rRNA modifying enzyme linked to human disease is the 2′-O-ribose methyltransferase MRM1, which is responsible for the modification of the 16S mt-rRNA at position U1369. A homozygous missense mutation c.567G>A, p. Gly189Arg; (RefSeq: NM_013393) affecting a highly conserved residue was identified in one patient with childhood-onset encephalopathy and stroke-like episodes. Cultured patient fibroblasts did not demonstrate any loss of the U1369 methylation nor a decrease in mitochondrial translation or OXPHOS deficiency. This absence of mitochondrial dysfunction in patient cells was attributed to the strong tissue specificity of the patient's disease presentation and pathogenicity of the identified variant was supported by a yeast model of the same mutation [178].

15.15.2 Mutations in FASTKD2

The FASTKD2 protein is one member of the FAS-activated serine/threonine kinase family of RNA-binding proteins and has a role in 16S rRNA and *MT-ND6* mRNA stability along with mitoribosome assembly [179]. A homozygous missense mutation c.1246C > T, p. Arg416*; (RefSeq: NM_014929) in *FASTKD* was first reported as the cause of early onset encephalopathy and decreased cytochrome oxidase *c* activity in two affected siblings [180]. More recently, compound heterozygous mutations [c.613C > T, p.Arg205* and c.764T > C, p.Leu255Pro; (RefSeq: NM_014929.3)] have been reported in a much milder presentation of late-onset MELAS-like disease [181].

15.16 Mechanism of mitochondrial translation

Reflecting their alphaproteobacterial ancestral origins, the mechanism of translation employed by mitochondria is more akin to prokaryotic protein synthesis systems than to the cytosolic translation of eukaryotes [182]. Unlike transcription, mitochondrial translation is yet to be successfully reconstituted *in vitro* and as such is not as well characterized as eubacterial or eukaryotic cytosolic translation. The core driver of mitochondrial translation is the mitoribosome. Comprised of a small (28S) subunit (SSU) and large (39S) subunit (LSU), the human mitoribosome contains 80 nuclear-encoded proteins, two mt-rRNAs (12S and 16S), and a mt-tRNA^{Val} that assemble to form the 55S monosome [183,184]. The process of mitochondrial translation can be divided into three major stages: initiation, elongation, and termination.

Initiation of mitochondrial translation requires the recruitment of a mitochondrial mRNA transcript to the SSU of the mitoribosome. When not active in translation, the SSU is bound by mitochondrial initiation factor 3 (mtIF3), which blocks its association with the LSU preventing 55S formation [185]. The binding of an mRNA transcript at the mRNA entry channel of the SSU may be facilitated by the PPR-containing mS39 protein [186]. Upon the entry of an initiating codon of an mt-mRNA into the SSU entry site, formylated tRNA^{Met} (tRNA^{fMet}) is recruited by mitochondrial initiation factor 2 in its GTP-bound state (mtIF2: GTP) to form an initiation complex with the SSU [187]. Binding of tRNA^{fMet} through a codon–anticodon interaction triggers the recruitment of the LSU to form the monosome, and the subsequent hydrolysis of mtIF2:GTP to mtIF2:GDP results in the release of both mtIF2 and mtIF3 from the SSU. The monosome can then enter the elongation phase of mitochondrial translation.

The elongation phase begins with the binding of GTP-bound mitochondrial elongation factor Tu (mtEFTu:GTP) to an aa-tRNA which directs the aa-tRNA into the A-site of the mitoribosome. The formation of a correctly matched codon:anticodon between the bound mt-mRNA transcript and the aa-tRNA in the A-site triggers hydrolysis of the mtEFTu:GTP to mtEFTu:GDP that is then released from the mitoribosome [188]. This GTP hydrolysis catalyzes the formation of a peptide bond at the PTC between the aa-tRNA within the A-site and the amino acid sitting at the adjacent peptidyl tRNA site (P-site). The tRNA occupying the P-site becomes deacylated as a result and is then displaced by the translocation of the bi-peptidyl tRNA from the A-site to the P-site driven by the hydrolysis of mitochondrial elongation factor G1 (mtEFG1). Regeneration of mtEFTu:GDP is carried out by mitochondrial elongation factor Ts (mtEFTS), which allows this process to cycle and the peptide chain to grow [189].

Upon entry of a "stop" codon into the A-site of the mitoribosome, mitochondrial translation release factor A (mtRF1a) is recruited [190]. mtRF1a catalyzes the hydrolysis of the ester bond between the peptidyl-tRNA occupying the A-site and the terminal amino acid of the complete nascent polypeptide. This GTP-dependent cleavage results in the release of the full length polypeptide through the exit tunnel of the LSU [191]. Following this release, the mitoribosome must undergo a recycling process to return to independent large and small subunits available for the initiation of a new translation event, also releasing the mRNA template and final terminating tRNA. Two proteins, mitochondrial ribosome release factor (mtRRF) and mitochondrial elongation factor G2 (mtEFG2), promote the dissociation of the mitoribosomal subunits in a GTP-dependent manner [192]. A ribosome-dependent peptidyl-tRNA hydrolase, ICT1, thought to be a structural component of the LSU, and C12orf65 are believed to facilitate the hydrolysis and release of prematurely terminated peptidyl chains from any stalled mitoribosomes [193]. The mechanism and machinery underlying mitochondrial translation are explored in depth in Chapter 2: Human mitochondrial transcription and translation.

15.17 Mutations in mitoribosomal proteins

The first mitoribosomal protein to be associated with disease was *MRPS16*, in a patient with neonatal lactic acidosis, agenesis of the corpus callosum, and dysmorphism. Patient fibroblasts exhibited a severe mitochondrial translation defect and a combined OXPHOS deficiency, the latter also observed in patient muscle and liver homogenates [194]. In the past 15 years, disease causing variants have been reported in genes for a further nine proteins of the mitoribosomal SSU: *MRPS2*, *MRPS7*, *MRPS14*, *MRPS22*, *MRPS23*, *MRPS25*, *MRPS28* *MRPS34*, and *MRPS39* (also known as *PTCD3*) [195–203]; and three of the LSU: *MRPL3*, *MRPL12*, *MRPL44* [204–206], taking the total number of reported MRP disease genes to 13. Although there is significant variation in clinical presentation, some common features are seen across the MRP cases. Disease-onset consistently occurs very early in life (neonatal or infantile) and almost all patients exhibit mild to severe/fatal lactic acidosis [195,197], which may explain the small number of patients carrying pathogenic variants from the 80 mitoribosomal proteins that have been identified to date.

15.18 Defects of translation initiation

15.18.1 Mutations in MTFMT

Initiation of mitochondrial translation is dependent on fMet-tRNA^{Met} for the recruitment of mtIF2. The protein methionyl-tRNA formyltransferase (MTFMT), encoded by the *MTFMT* gene, carries out formylation of a tightly regulated proportion of the general Met-tRNA^{Met} pool to provide Met-tRNAs for both mitochondrial translation initiation and elongation [207]. Autosomal recessive variants in the *MTFMT* gene were first identified as pathogenic in two cases of Leigh syndrome [208], but have since been implicated in a range of Leigh-like encephalomyopathic presentations, along with one case of relapsing-remitting attacks of neurological dysfunction more reminiscent of a de-myelinating disease [209,210]. *MTFMT*-related Leigh syndrome has an age of onset around one year, presenting with developmental delay affecting gross motor function. Cranial MRI findings in patients with *MTFMT* mutations have included symmetrical basal ganglia changes, periventricular and subcortical white matter abnormalities, and brain stem lesions. Almost all investigated patient fibroblasts demonstrate a severe decrease or complete loss of steady-state MTFMT protein, resulting in a translation defect and either an isolated complex I deficiency or combined complex I and IV deficiencies [209,211].

15.18.2 Mutations in RMND1

Located on the IMM, RMND1 is responsible for the anchoring and stabilization of the mitoribosome in close proximity to sites of mt-mRNA maturation. Combined and isolated OXPHOS defects have both been observed in *RMND1* disease patients, along with decreased steady-state levels of mitoribosomal proteins and defects of mitochondrial translation [212,213]. The clinical disease spectrum of *RMND1* variants ranges from fatal encephalomyopathy with lactic acidosis to developmental delay, sensorineural deafness, hypotonia, and renal disease [212,214,215]. The disease-onset varies from severe infantile encephalomyopathy culminating in death to mild, childhood-onset nephropathy with longer survival [212,214].

15.19 Defects of translation elongation

15.19.1 Mutations in GFM1

The first nuclear disease gene to be identified associated with defective mitochondrial translation [216], *GFM1* encodes the elongation factor mtEFG1. Recessive variants in *GFM1* have been reported in a total of 17 patients with early onset mitochondrial disease. Many of the early cases of *GFM1* disease were rapidly progressive and fatal before the age of two and a half years [216–221]. The brain imaging typically showed hypoplasia of corpus callosum, symmetrical cystic lesions in the white matter, and involvement of basal ganglia. However, more recent cases of *GFM1* disease demonstrate long-term survival further into childhood, with a less severe clinical disease presentation [222,223]. The loss-of-function mutations described in *GFM1* result in generalized defects of mitochondrial translation, resulting in combined OXPHOS deficiencies. The severity of OXPHOS deficiency in each case appears to correlate with the residual amount of expressed mtEFG1. Patients with a higher residual steady-state level of mtEFG1 tend to exhibit a less severe OXPHOS defect [222].

15.19.2 Mutations in TSFM

Recessive mutations in the *TSFM* gene, encoding the elongation factor EF-Ts, have been described in a number of cases of early onset mitochondrial disease resulting in death in early infancy, or a more slowly progressing childhood-onset cardiomyopathy with ataxia and a neurological phenotype [224–227]. The majority of reported *TSFM* variants are located within the C-subdomain of the core domain, a region that mediates the proper interaction of EF-Ts with EF-Tu [228]. Compound heterozygous variants in *TSFM* were recently identified in a unique case of adult-onset mitochondrial cardiomyopathy. A decrease in EF-Ts expression was observed in explanted cardiac tissue, which appears to result in a decrease in EF-Tu stability along with a combined OXPHOS defect [229].

15.20 Defects of translation termination and mitoribosome recycling

15.20.1 Mutations in C12orf65

The protein encoded by the *C12orfF65* gene is believed to be a peptide release factor required for the release of peptides from stalled mitoribosomes during the elongation phase of mitochondrial translation. Mutations in the *C12orf65* gene disturb mitochondrial translation and result in combined OXPHOS deficiencies. Variants have been reported in a number of patients presenting in early childhood with mitochondrial disease that for the most part align with a clinical diagnosis of Behr syndrome (optic atrophy, spastic paraparesis, and peripheral neuropathy) [230]. However, clinical variability in the presentation of *C12orf65* defects has been demonstrated. An adult presenting with Leigh Syndrome in the fifth decade of life was found to harbor the same frameshift mutation as a previously reported pediatric patient (c.210delA, p.Gly72Alafs*13) despite exhibiting a much milder phenotype [231,232].

15.20.2 *Mutations in* GFM2

Recessive variants in the *GFM2* gene, encoding the mitoribosome recycling factor mtEFG2, have been reported in four families to date with various clinical presentations including Leigh syndrome, microcephaly with simplified brain gyral pattern and insulin-dependent diabetes, and cases of global developmental delay [233–235]. Both missense and frameshift mutations affecting the stability of mtEFG2 have been demonstrated to result in combined OXPHOS defects in patient fibroblasts and skeletal muscle [235].

15.21 *Defects of translational activation and coupling*

15.21.1 *Mutations in* TACO1

The *TACO1* gene encodes the translational activator of COXI protein believed to specifically promote translation of *MT-CO1* either through the promotion of start codon recognition or through stabilization of the COXI polypeptide during its synthesis. To date, just one pathogenic variant in the *TACO1* gene has been reported [c.472insC, p.His158Pfs*8; (RefSeq: NM_016360.4)], identified in five patients within one large consanguineous Turkish family presenting with late-onset Leigh syndrome. The frameshift mutation causes the introduction of a premature stop codon and results in a complete loss of TACO1 protein. Consequently, patient fibroblasts exhibit a translation defect affecting only COXI, leading to instability of the COXII subunit along with a severe decrease in the level of fully assembled complex IV [236,237].

15.21.2 *Mutations in* COA3 *and* C12orf62

Two further proteins, COX assembly factor 3 (COA3) and C12orf62, also play a role in the regulation of the specific synthesis of COXI protein and its coupling with complex IV assembly. Pathogenic variants in both genes cause severe complex IV assembly defects, but appear to be detrimental to the translation of COXI only [238,239].

15.22 *IMM insertion of mtDNA-encoded OXPHOS proteins*

Following their synthesis by the mitoribosome, the 13 mtDNA-encoded OXPHOS proteins are inserted into the IMM. Correct insertion of nascent polypeptides into the IMM is believed to be aided by a family of insertases. A major candidate for a human insertase that carries out this role, OXA1L, is a homologue of the yeast Oxa1p, known to be important for the co-translational membrane insertion of a number of mitochondrial-encoded subunits.

15.22.1 Mutations in OXA1L

The first pathogenic variant in the *OXA1L* gene, encoding a membrane insertase of the same name, was recently described in a patient presenting with encephalopathy, hypotonia, and developmental delay, resulting in death at 5 years of age. Patient fibroblasts exhibited instability of nascent peptides synthesized by the mitoribosome and a deficiency of complexes I, IV, and V. OXA1L has been shown to directly interact with at least 9 of the 13 mtDNA-encoded polypeptides and other nuclear-encoded accessory subunits of OXPHOS complexes to aid insertion of newly synthesized proteins into the IMM. These data confirmed the long-believed role of OXA1L in the insertion of newly synthesized mtDNA-encoded OXPHOS proteins into the IMM [240].

Research perspectives

The power of NGS technologies for the identification of novel mitochondrial disease-causing genes will likely continue to grow for some years. The integration of NGS techniques and improvements in analytical approaches, database sharing of variants, alongside "omics" studies including transcriptomic and proteomic analyses will continue to reveal an array of mitochondrial disease genes, a large proportion of which have vital roles in the expression of mtDNA. The subsequent functional characterization of patient cells and tissues is essential to the identification of novel pathogenic variants in these genes and to help to elucidate the pathomechanisms driving these disorders. In some instances, these studies have resulted in detailed molecular characterization and understanding of disease pathology of genes involved in mitochondrial gene expression pathways (e.g. *TRMT10C*, *ELAC2*, or *TACO1*). In other instances, such as in *TK2*-deficient patients, it has led to the development of potential nucleoside substrate enhancement treatments to elevate the clinical symptoms of the disease [71,241]. Our overview of the mitochondrial nuclear genetic disorders highlights the vast clinical and genetic heterogeneity that characterizes defects of mtDNA maintenance and replication, transcription and translation, emphasizing the importance of unbiased genotype-driven approaches in the diagnosis of Mendelian disease.

Acknowledgments

Work in our laboratories is supported by the Wellcome Centre for Mitochondrial Research (203105/Z/16/Z), the Medical Research Council (MRC) International Centre for Genomic Medicine in Neuromuscular Disease, Newcastle University Centre for Ageing and Vitality (supported by the Biotechnology and Biological Sciences Research Council and Medical Research Council (G016354/1)), the UK NIHR Biomedical Research Centre in Age and Age Related Diseases award to the Newcastle upon Tyne Hospitals NHS Foundation, the MRC/ESPRC

Newcastle Molecular Pathology Node, the UK National Health Service Highly Specialised Service for Rare Mitochondrial Disorders, and the Lily Foundation. RICG is supported by a PhD studentship from the Lily Foundation. TJN is the recipient of a Sir Henry Dale Fellowship jointly funded by the Wellcome Trust and the Royal Society (213464/Z/18/Z) and of a Rosetrees and Stoneygate Trust Research Fellowship (M811).

References

[1] Calvo SE, Clauser KR, Mootha VK. MitoCarta2.0: an updated inventory of mammalian mitochondrial proteins. Nucleic Acids Res 2016;44(D1):D1251–7.
[2] Pronicka E, et al. New perspective in diagnostics of mitochondrial disorders: two years' experience with whole-exome sequencing at a national paediatric centre. J Transl Med 2016;14(1).
[3] Taylor RW, et al. Use of whole-exome sequencing to determine the genetic basis of multiple mitochondrial respiratory chain complex deficiencies. JAMA 2014;312(1):68.
[4] Thompson K, et al. Recent advances in understanding the molecular genetic basis of mitochondrial disease. J Inherit Metab Dis 2019;.
[5] Clay Montier LL, Deng JJ, Bai Y. Number matters: control of mammalian mitochondrial DNA copy number. J Genet Genomics 2009;36(3):125–31.
[6] Satoh M, Kuroiwa T. Organization of multiple nucleoids and DNA molecules in mitochondria of a human cell. Exp Cell Res 1991;196(1) 137-40.
[7] Fusté JM, et al. Mitochondrial RNA polymerase is needed for activation of the origin of light-strand DNA replication. Mol Cell 2010;37(1):67–78.
[8] Kuhl I, et al. POLRMT regulates the switch between replication primer formation and gene expression of mammalian mtDNA. Sci Adv 2016;2(8):e1600963.
[9] Lee Y-S, Kennedy WD, Yin YW. Structural insight into processive human mitochondrial DNA synthesis and disease-related polymerase mutations. Cell 2009;139(2):312–24.
[10] Milenkovic D, et al. TWINKLE is an essential mitochondrial helicase required for synthesis of nascent D-loop strands and complete mtDNA replication. Hum Mol Genet 2013;22(10):1983–93.
[11] Oliveira MT, Kaguni LS. Functional roles of the N- and C-terminal regions of the human mitochondrial single-stranded DNA-binding protein. PLoS One 2010;5(10):e15379.
[12] Cerritelli SM, et al. Failure to produce mitochondrial DNA results in embryonic lethality in Rnaseh1 null mice. Mol Cell 2003;11(3):807–15.
[13] Ruhanen H, Ushakov K, Yasukawa T. Involvement of DNA ligase III and ribonuclease H1 in mitochondrial DNA replication in cultured human cells. Biochim Biophys Acta Mol Basis Dis 2011;1813(12):2000–7.
[14] Nicholls TJ, et al. Topoisomerase 3alpha is required for decatenation and segregation of human mtDNA. Mol Cell 2018;69(1):9–23 e6.
[15] Uhler JP, Falkenberg M. Primer removal during mammalian mitochondrial DNA replication. DNA Repair 2015;34:28–38.
[16] Korr H, et al. Mitochondrial DNA synthesis studied autoradiographically in various cell types in vivo. Braz J Med Biol Res 1998;31(2):289–98.
[17] Magnusson J, et al. Replication of mitochondrial DNA occurs throughout the mitochondria of cultured human cells. Exp Cell Res 2003;289(1):133–42.
[18] Robberson DL, Kasamatsu H, Vinograd J. Replication of mitochondrial DNA. circular replicative intermediates in mouse L cells. Proc Natl Acad Sci U S A 1972;69(3):737–41.
[19] Clayton DA. Replication of animal mitochondrial DNA. Cell 1982;28(4):693–705.
[20] Falkenberg M. Mitochondrial DNA replication in mammalian cells: overview of the pathway. Essays Biochem 2018;62(3):287–96.
[21] Hikmat O, et al. The clinical spectrum and natural history of early-onset diseases due to DNA polymerase gamma mutations. Genet Med 2017;19(11):1217–25.

[22] Woodbridge P, et al. POLG mutations in Australian patients with mitochondrial disease. Intern Med J 2013;43(2):150–6.

[23] El-Hattab AW, Craigen WJ, Scaglia F. Mitochondrial DNA maintenance defects. Biochim Biophys Acta Mol Basis Dis 2017;1863(6):1539–55.

[24] Cohen BH, Chinnery PF, Copeland WC. POLG-related disorders. In: Adam MP, et al., editors. *GeneReviews((R))*. Seattle: WA; 1993.

[25] Rahman S, Copeland WC. POLG-related disorders and their neurological manifestations. Nat Rev Neurol 2019;15(1):40–52.

[26] Darin N, et al. The incidence of mitochondrial encephalomyopathies in childhood: clinical features and morphological, biochemical, and DNA abnormalities. Ann Neurol 2001;49(3):377–83.

[27] Harding BN. Progressive neuronal degeneration of childhood with liver disease (Alpers-Huttenlocher syndrome): a personal review. J Child Neurol 1990;5(4):273–87.

[28] Horvath R, et al. Phenotypic spectrum associated mutant mitochondrial polymerase gamma gene. Brain 2006;129(Pt 7):1674–84.

[29] Wiltshire E, et al. Juvenile Alpers disease. Arch Neurol 2008;65(1):121–4.

[30] Uusimaa J, et al. Homozygous W748S mutation in the POLG1 gene in patients with juvenile-onset Alpers syndrome and status epilepticus. Epilepsia 2008;49(6) 1038-45.

[31] Harding BN, et al. Progressive neuronal degeneration of childhood with liver disease (Alpers' disease) presenting in young adults. J Neurol Neurosurg Psychiatry 1995;58(3):320–5.

[32] Anagnostou ME, et al. Epilepsy due to mutations in the mitochondrial polymerase gamma (POLG) gene: a clinical and molecular genetic review. Epilepsia 2016;57(10):1531–45.

[33] Farnum GA, Nurminen A, Kaguni LS. Mapping 136 pathogenic mutations into functional modules in human DNA polymerase gamma establishes predictive genotype-phenotype correlations for the complete spectrum of POLG syndromes. Biochim Biophys Acta 2014;1837(7):1113–21.

[34] Wong LJ, et al. Molecular and clinical genetics of mitochondrial diseases due to POLG mutations. Hum Mutat 2008;29(9):E150–72.

[35] Fadic R, et al. Sensory ataxic neuropathy as the presenting feature of a novel mitochondrial disease. Neurology 1997;49(1):239–45.

[36] Luoma P, et al. Parkinsonism, premature menopause, and mitochondrial DNA polymerase gamma mutations: clinical and molecular genetic study. Lancet 2004;364(9437):875–82.

[37] Pagnamenta AT, et al. Dominant inheritance of premature ovarian failure associated with mutant mitochondrial DNA polymerase gamma. Hum Reprod 2006;21(10):2467–73.

[38] Lamantea E, et al. Mutations of mitochondrial DNA polymerase gamma A are a frequent cause of autosomal dominant or recessive progressive external ophthalmoplegia. Ann Neurol 2002;52(2):211–19.

[39] Hebbar M, et al. Homozygous c.359del variant in MGME1 is associated with early onset cerebellar ataxia. Eur J Med Genet 2017;60(10):533–5.

[40] Peter B, et al. Structural basis for adPEO-causing mutations in the mitochondrial TWINKLE helicase. Hum Mol Genet 2019;28(7):1090–9.

[41] Paradas C, et al. Longitudinal clinical follow-up of a large family with the R357P Twinkle mutation. JAMA Neurol 2013;70(11):1425–8.

[42] Van Hove JLK, et al. Finding twinkle in the eyes of a 71-year-old lady: a case report and review of the genotypic and phenotypic spectrum of TWINKLE-related dominant disease. Am J Med Genet Part A 2009;149A(5):861–7.

[43] Kiferle L, et al. Twinkle mutation in an Italian family with external progressive ophthalmoplegia and parkinsonism: a case report and an update on the state of art. Neurosci Lett 2013;556:1–4.

[44] Nikali K, et al. Infantile onset spinocerebellar ataxia is caused by recessive mutations in mitochondrial proteins Twinkle and Twinky. Hum Mol Genet 2005;14(20):2981–90.

[45] Hakonen AH, et al. Infantile-onset spinocerebellar ataxia and mitochondrial recessive ataxia syndrome are associated with neuronal complex I defect and mtDNA depletion. Hum Mol Genet 2008;17 (23):3822–35.

[46] Sarzi E, et al. Twinkle helicase (PEO1) gene mutation causes mitochondrial DNA depletion. Ann Neurol 2007;62(6):579–87.
[47] Hakonen AH, et al. Recessive Twinkle mutations in early onset encephalopathy with mtDNA depletion. Brain 2007;130(11):3032–40.
[48] Demain LAM, et al. Expanding the genotypic spectrum of Perrault syndrome. Clin Genet 2017;91(2):302–12.
[49] Ołdak M, et al. Novel neuro-audiological findings and further evidence for TWNK involvement in Perrault syndrome. J Transl Med 2017;15(1):25.
[50] Ronchi D, et al. Mutations in DNA2 link progressive myopathy to mitochondrial DNA instability. Am J Hum Genet 2013;92(2):293–300.
[51] Shanske A, et al. Central nervous system anomalies in Seckel syndrome: report of a new family and review of the literature. Am J Med Genet 1997;70(2):155–8.
[52] Shaheen R, et al. Genomic analysis of primordial dwarfism reveals novel disease genes. Genome Res 2014;24(2):291–9.
[53] Nicholls TJ, et al. Linear mtDNA fragments and unusual mtDNA rearrangements associated with pathological deficiency of MGME1 exonuclease. Hum Mol Genet 2014;23(23):6147–62.
[54] Kornblum C, et al. Loss-of-function mutations in MGME1 impair mtDNA replication and cause multisystemic mitochondrial disease. Nat Genet 2013;45(2):214–19.
[55] Penque BA, et al. A homozygous variant in RRM2B is associated with severe metabolic acidosis and early neonatal death. Eur J Med Genet 2018;.
[56] Pontarin G, et al. Ribonucleotide reduction is a cytosolic process in mammalian cells independently of DNA damage. Proc Natl Acad Sci U S A 2008;105(46):17801–6.
[57] Dahout-Gonzalez C. Molecular, functional, and pathological aspects of the mitochondrial ADP/ATP carrier. Physiology (Bethesda) 2006;21(4):242–9.
[58] Aaron, Minczuk M. Mitochondrial transcription and translation: overview. Essays Biochem 2018;62 (3):309–20.
[59] Johansson M, Karlsson A. Cloning of the cDNA and chromosome localization of the gene for human thymidine kinase 2. J Biol Chem 1997;272(13):8454–8.
[60] Johansson M, Karlsson A. Cloning and expression of human deoxyguanosine kinase cDNA. Proc Natl Acad Sci U S A 1996;93(14):7258–62.
[61] Wang L. Mitochondrial purine and pyrimidine metabolism and beyond. Nucleosides Nucleotides Nucleic Acids 2016;35(10-12):578–94.
[62] Besse A, et al. The GABA transaminase, ABAT, is essential for mitochondrial nucleoside metabolism. Cell Metab 2015;21(3):417–27.
[63] Kowluru A, Tannous M, Chen H-Q. Localization and characterization of the mitochondrial isoform of the nucleoside diphosphate kinase in the pancreatic β cell: evidence for its complexation with mitochondrial succinyl-CoA synthetase. Arch Biochem Biophys 2002;398(2):160–9.
[64] Martín-Hernández E, et al. Myopathic mtDNA depletion syndrome due to mutation in TK2 gene. Pediatr Dev Pathol 2017;20(5):416–20.
[65] Adam M, et al. TK2-related mitochondrial DNA maintenance defect, myopathic form. In: GeneReviews®.
[66] Garone C, et al. Retrospective natural history of thymidine kinase 2 deficiency. J Med Genet 2018;55 (8):515–21.
[67] Wang J, et al. Clinical and molecular spectrum of thymidine kinase 2-related mtDNA maintenance defect. Mol Genet Metab 2018;124(2):124–30.
[68] Wang J, El-Hattab AW, Wong LJC. TK2-related mitochondrial DNA maintenance defect, myopathic form. In: Adam MP, et al., editors. GeneReviews((R)). Seattle: University of Washington; 1993. University of Washington, Seattle. GeneReviews is a registered trademark of the University of Washington, Seattle. All rights reserved: Seattle (WA).
[69] Poulton J, et al. Collated mutations in mitochondrial DNA (mtDNA) depletion syndrome (excluding the mitochondrial gamma polymerase, POLG1). Biochim Biophys Acta 2009;1792(12):1109–12.

[70] Garone C, et al. Deoxypyrimidine monophosphate bypass therapy for thymidine kinase 2 deficiency. EMBO Mol Med 2014;6(8):1016–27.
[71] Lopez-Gomez C, et al. Deoxycytidine and deoxythymidine treatment for thymidine kinase 2 deficiency. Ann Neurol 2017;81(5):641–52.
[72] Dominguez-Gonzalez C, et al. Deoxynucleoside therapy for thymidine kinase 2-deficient myopathy. Ann Neurol 2019;86(2):293–303.
[73] Pitceathly RD, et al. Adults with RRM2B-related mitochondrial disease have distinct clinical and molecular characteristics. Brain 2012;135(Pt 11):3392–403.
[74] Takata A, et al. Exome sequencing identifies a novel missense variant in RRM2B associated with autosomal recessive progressive external ophthalmoplegia. Genome Biol 2011;12(9):R92.
[75] Spinazzola A, et al. Clinical and molecular features of mitochondrial DNA depletion syndromes. J Inherit Metab Dis 2009;32(2):143–58.
[76] Shaibani A, et al. Mitochondrial neurogastrointestinal encephalopathy due to mutations in RRM2B. Arch Neurol 2009;66(8) 1028-32.
[77] Kollberg G, et al. A novel homozygous RRM2B missense mutation in association with severe mtDNA depletion. Neuromuscul Disord 2009;19(2):147–50.
[78] Acham-Roschitz B, et al. A novel mutation of the RRM2B gene in an infant with early fatal encephalomyopathy, central hypomyelination, and tubulopathy. Mol Genet Metab 2009;98 (3):300–4.
[79] Bornstein B, et al. Mitochondrial DNA depletion syndrome due to mutations in the RRM2B gene. Neuromuscul Disord 2008;18(6):453–9.
[80] Fratter C, et al. RRM2B mutations are frequent in familial PEO with multiple mtDNA deletions. Neurology 2011;76(23):2032–4.
[81] Gorman GS, Taylor RW. RRM2B-related mitochondrial disease. In: Adam MP, et al., editors. *GeneReviews((R))*. Seattle: University of Washington; 1993. University of Washington, Seattle. GeneReviews is a registered trademark of the University of Washington, Seattle. All rights reserved.: Seattle (WA).
[82] Antonenkov VD, et al. The human mitochondrial DNA depletion syndrome gene MPV17 encodes a non-selective channel that modulates membrane potential. J Biol Chem 2015;290(22):13840–61.
[83] Dalla Rosa I, et al. MPV17 loss causes deoxynucleotide insufficiency and slow DNA replication in mitochondria. PLOS Genet 2016;12(1):e1005779.
[84] Karadimas CL, et al. Navajo neurohepatopathy is caused by a mutation in the MPV17. Gene. 2006;79 (3):544–8.
[85] Spinazzola A, et al. MPV17 encodes an inner mitochondrial membrane protein and is mutated in infantile hepatic mitochondrial DNA depletion. Nat Genet 2006;38(5):570–5.
[86] Blakely EL, et al. MPV17 mutation causes neuropathy and leukoencephalopathy with multiple mtDNA deletions in muscle. Neuromuscul Disord 2012;22(7):587–91.
[87] Garone C, et al. MPV17 mutations causing adult-onset multisystemic disorder with multiple mitochondrial DNA deletions. Arch Neurol 2012;69(12):1648.
[88] Baumann M, et al. MPV17 mutations in juvenile- and adult-onset axonal sensorimotor polyneuropathy. Clin Genet 2018;.
[89] Posse V, et al. TEFM is a potent stimulator of mitochondrial transcription elongation in vitro. Nucleic Acids Res 2015;43(5):2615–24.
[90] Falkenberg M, et al. Mitochondrial transcription factors B1 and B2 activate transcription of human mtDNA. Nat Genet 2002;31(3):289–94.
[91] Kuhl I, et al. POLRMT does not transcribe nuclear genes. Nature 2014;514(7521):E7–11.
[92] Hillen HS, Temiakov D, Cramer P. Structural basis of mitochondrial transcription. Nat Struct Mol Biol 2018;25(9):754–65.
[93] Hillen HS, et al. Mechanism of transcription anti-termination in human mitochondria. Cell 2017;171 (5):1082–93 e13.

[94] Posse V, et al. The amino terminal extension of mammalian mitochondrial RNA polymerase ensures promoter specific transcription initiation. Nucleic Acids Res 2014;42(6):3638–47.

[95] Terzioglu M, et al. MTERF1 binds mtDNA to prevent transcriptional interference at the light-strand promoter but is dispensable for rRNA gene transcription regulation. Cell Metab 2013;17(4):618–26.

[96] Asin-Cayuela J, et al. The human mitochondrial transcription termination factor (mTERF) is fully active in vitro in the non-phosphorylated form. J Biol Chem 2005;280(27):25499–505.

[97] Stiles AR, et al. Mutations in TFAM, encoding mitochondrial transcription factor A, cause neonatal liver failure associated with mtDNA depletion. Mol Genet Metab 2016;119(1-2):91–9.

[98] Park CB, et al. Identification of a rare homozygous c.790C > T variation in the TFB2M gene in Korean patients with autism spectrum disorder. Biochem Biophys Res Commun 2018;507(1-4):148–54.

[99] Ojala D, Montoya J, Attardi G. tRNA punctuation model of RNA processing in human mitochondria. Nature 1981;290(5806):470–4.

[100] Jourdain AA, et al. GRSF1 regulates RNA processing in mitochondrial RNA granules. Cell Metab 2013;17(3):399–410.

[101] Antonicka H, et al. The mitochondrial RNA-binding protein GRSF1 localizes to RNA granules and is required for posttranscriptional mitochondrial gene expression. Cell Metab 2013;17(3):386–98.

[102] Xu F, et al. Disruption of a mitochondrial RNA-binding protein gene results in decreased cytochrome b expression and a marked reduction in ubiquinol-cytochrome c reductase activity in mouse heart mitochondria. Biochem J 2008;416(1):15–26.

[103] Vilardo E, et al. A subcomplex of human mitochondrial RNase P is a bifunctional methyltransferase—extensive moonlighting in mitochondrial tRNA biogenesis. Nucleic Acids Res 2012. 40(22):11583–93.

[104] Metodi, et al., Recessive mutations in TRMT10C cause defects in mitochondrial RNA processing and multiple respiratory chain deficiencies. Am J Hum Genet 2016;98(5):993–1000.

[105] Zschocke J. HSD10 disease: clinical consequences of mutations in the HSD17B10 gene. J Inherit Metab Dis 2012;35(1):81–9.

[106] Oerum S, et al. Novel patient missense mutations in the HSD17B10 gene affect dehydrogenase and mitochondrial tRNA modification functions of the encoded protein. Biochim Biophys Acta Mol Basis Dis 2017;1863(12):3294–302.

[107] Hochberg I, et al. A homozygous variant in mitochondrial RNase P subunit PRORP is associated with Perrault syndrome characterized by hearing loss and primary ovarian insufficiency. bioRxiv 2017;168252.

[108] Tobias et al. ELAC2 mutations cause a mitochondrial RNA processing defect associated with hypertrophic cardiomyopathy. Am J Hum Genet 2013;93(2):211–23.

[109] Shinwari ZMA, et al. The phenotype and outcome of infantile cardiomyopathy caused by a homozygous ELAC2 mutation. Cardiology 2017;137(3):188–92.

[110] Akawi NA, et al. A homozygous splicing mutation in ELAC2 suggests phenotypic variability including intellectual disability with minimal cardiac involvement. Orphanet J Rare Dis 2016;11(1):139.

[111] Kim YA, et al. The First Korean case of combined oxidative phosphorylation deficiency-17 diagnosed by clinical and molecular investigation. Korean J Pediatr 2017;60(12):408–12.

[112] Saoura M, et al. Mutations in ELAC2 associated with hypertrophic cardiomyopathy impair mitochondrial tRNA 3′-end processing. Hum Mutat 2019;40(10):1731–48.

[113] Temperley RJ, et al. Human mitochondrial mRNAs—like members of all families, similar but different. Biochim Biophys Acta 2010;1797(6–7):1081–5.

[114] Bai Y, et al. Structural basis for dimerization and activity of human PAPD1, a noncanonical poly(A) polymerase. Mol Cell 2011;41(3):311–20.

[115] Minczuk M, et al. Localisation of the human hSuv3p helicase in the mitochondrial matrix and its preferential unwinding of dsDNA. Nucleic Acids Res 2002;30(23):5074–86.

[116] Borowski LS, et al. Human mitochondrial RNA decay mediated by PNPase-hSuv3 complex takes place in distinct foci. Nucleic Acids Res 2013;41(2):1223–40.

[117] Szczesny RJ, et al. Human mitochondrial RNA turnover caught in flagranti: involvement of hSuv3p helicase in RNA surveillance. Nucleic Acids Res 2010;38(1):279–98.
[118] Shepherd DL, et al. Exploring the mitochondrial microRNA import pathway through Polynucleotide Phosphorylase (PNPase). J Mol Cell Cardiol 2017;110:15–25.
[119] Bruni F, et al. REXO2 is an oligoribonuclease active in human mitochondria. PLoS One 2013;8(5): e64670.
[120] Nicholls TJ, et al. Dinucleotide degradation by REXO2 maintains promoter specificity in mammalian mitochondria. Mol Cell 2019;76(5):784–796.e6.
[121] Chujo T, et al. LRPPRC/SLIRP suppresses PNPase-mediated mRNA decay and promotes polyadenylation in human mitochondria. Nucleic Acids Res 2012;40(16):8033–47.
[122] Rius R, et al. Clinical spectrum and functional consequences associated with bi-allelic pathogenic PNPT1 variants. J Clin Med 2019;8(11).
[123] Crosby AH, et al. Defective mitochondrial mRNA maturation is associated with spastic ataxia. Am J Hum Genet 2010;87(5):655–60.
[124] Wilson WC, et al. A human mitochondrial poly(A) polymerase mutation reveals the complexities of post-transcriptional mitochondrial gene expression. Hum Mol Genet 2014;23(23):6345–55.
[125] Mootha VK, et al. Identification of a gene causing human cytochrome c oxidase deficiency by integrative genomics. Proc Natl Acad Sci U S A 2003;100(2):605–10.
[126] Debray FG, et al. LRPPRC mutations cause a phenotypically distinct form of Leigh syndrome with cytochrome c oxidase deficiency. J Med Genet 2011;48(3):183–9.
[127] Oláhová M, et al. LRPPRC mutations cause early-onset multisystem mitochondrial disease outside of the French-Canadian population. Brain 2015;138(12):3503–19.
[128] Salinas-Giege T, Giege R, Giege P. tRNA biology in mitochondria. Int J Mol Sci 2015;16(3):4518–59.
[129] Suzuki T, Suzuki T. A complete landscape of post-transcriptional modifications in mammalian mitochondrial tRNAs. Nucleic Acids Res 2014;42(11):7346–57.
[130] Suzuki T, Nagao A, Suzuki T. Human mitochondrial tRNAs: biogenesis, function, structural aspects, and diseases. Annu Rev Genet 2011;45(1):299–329.
[131] Sissler M, Gonzalez-Serrano LE, Westhof E. Recent advances in mitochondrial aminoacyl-tRNA synthetases and disease. Trends Mol Med 2017;23(8):693–708.
[132] Tolkunova E, et al. The human lysyl-tRNA synthetase gene encodes both the cytoplasmic and mitochondrial enzymes by means of an unusual alternative splicing of the primary transcript. J Biol Chem 2000;275(45):35063–9.
[133] Echevarria L, et al. Glutamyl-tRNAGln amidotransferase is essential for mammalian mitochondrial translation in vivo. Biochem J 2014;460(1):91–101.
[134] Hou Y-M, CCA addition to tRNA: implications for tRNA quality control. IUBMB Life 2010;62 (4):251–60.
[135] Chakraborty PK, et al. Mutations in TRNT1 cause congenital sideroblastic anemia with immunodeficiency, fevers, and developmental delay (SIFD). Blood 2014;124(18):2867–71.
[136] Kumaki E, et al. Atypical SIFD with novel TRNT1 mutations: a case study on the pathogenesis of B-cell deficiency. Int J Hematol 2019;109(4):382–9.
[137] Deluca AP, et al. Hypomorphic mutations in TRNT1 cause retinitis pigmentosa with erythrocytic microcytosis. Hum Mol Genet 2016;25(1):44–56.
[138] Wedatilake Y, et al. TRNT1 deficiency: clinical, biochemical and molecular genetic features. Orphanet J Rare Dis 2016;11(1):90.
[139] Liwak-Muir U, et al. Impaired activity of CCA-adding enzyme TRNT1 impacts OXPHOS complexes and cellular respiration in SIFD patient-derived fibroblasts. Orphanet J Rare Dis 2016;11 (1):79.
[140] Fernandez-Vizarra E, et al. Nonsense mutation in pseudouridylate synthase 1 (PUS1) in two brothers affected by myopathy, lactic acidosis and sideroblastic anaemia (MLASA). J Med Genet 2006;44 (3):173–80.

[141] Bykhovskaya Y, et al. Missense mutation in pseudouridine synthase 1 (PUS1) causes mitochondrial myopathy and sideroblastic anemia (MLASA). Am J Hum Genet 2004;74(6):1303–8.
[142] Zeharia A, et al. Mitochondrial myopathy, sideroblastic anemia, and lactic acidosis: an autosomal recessive syndrome in Persian Jews caused by a mutation in the PUS1 gene. J Child Neurol 2005;20 (5):449–52.
[143] Metodiev MD, et al. Unusual clinical expression and long survival of a pseudouridylate synthase (PUS1) mutation into adulthood. Eur J Hum Genet 2015;23(6):880–2.
[144] Cao M, et al. Clinical and molecular study in a long-surviving patient with MLASA syndrome due to novel PUS1 mutations. Neurogenetics 2016;17(1):65–70.
[145] Kasapkara CS, et al. A myopathy, lactic acidosis, sideroblastic anemia (MLASA) case due to a novel PUS1 mutation. Turk J Haematol 2017;34(4):376–7.
[146] Tesarova M, et al. Sideroblastic anemia associated with multisystem mitochondrial disorders. Pediatr Blood Cancer 2019;66(4):e27591.
[147] O'Byrne JJ, et al. The genotypic and phenotypic spectrum of MTO1 deficiency. Mol Genet Metab 2018;123(1):28–42.
[148] Ghezzi D, et al. Mutations of the mitochondrial-tRNA modifier MTO1 cause hypertrophic cardiomyopathy and lactic acidosis. Am J Hum Genet 2012;90(6):1079–87.
[149] Baruffini E, et al. MTO1 mutations are associated with hypertrophic cardiomyopathy and lactic acidosis and cause respiratory chain deficiency in humans and yeast. Hum Mutat 2013;34(11):1501–9.
[150] Kopajtich R, et al. Mutations in GTPBP3 cause a mitochondrial translation defect associated with hypertrophic cardiomyopathy, lactic acidosis, and encephalopathy. Am J Hum Genet 2014;95(6) 708-20.
[151] Zeharia A, et al. Acute infantile liver failure due to mutations in the TRMU gene. Am J Hum Genet 2009;85(3):401–7.
[152] Sturman JA, Gaull G, Raiha NC. Absence of cystathionase in human fetal liver: is cystine essential? Science 1970;169(3940):74–6.
[153] Soler-Alfonso C, et al. L-Cysteine supplementation prevents liver transplantation in a patient with TRMU deficiency. Mol Genet Metab Rep 2019;19:100453.
[154] Schweizer U, Bohleber S, Fradejas-Villar N. The modified base isopentenyladenosine and its derivatives in tRNA. RNA Biol 2017;14(9):1197–208.
[155] Kernohan KD, et al. Matchmaking facilitates the diagnosis of an autosomal-recessive mitochondrial disease caused by biallelic mutation of the tRNA isopentenyltransferase (TRIT1) gene. Hum Mutat 2017;38 (5):511–16.
[156] Yarham JW, et al. Defective i6A37 modification of mitochondrial and cytosolic tRNAs results from pathogenic mutations in TRIT1 and its substrate tRNA. PLoS Genet 2014;10(6):e1004424.
[157] Takenouchi T, et al. Noninvasive diagnosis of TRIT1-related mitochondrial disorder by measuring i^6A37 and ms^2i^6A37 modifications in tRNAs from blood and urine samples. Am J Med Genet Part A 2019;179 (8):1609–14.
[158] Scheper GC, et al. Mitochondrial aspartyl-tRNA synthetase deficiency causes leukoencephalopathy with brain stem and spinal cord involvement and lactate elevation. Nat Genet 2007;39(4):534–9.
[159] Webb BD, et al. Novel, compound heterozygous, single-nucleotide variants in MARS2 associated with developmental delay, poor growth, and sensorineural hearing loss. Hum Mutat 2015;36 (6):587–92.
[160] Mizuguchi T, et al. PARS2 and NARS2 mutations in infantile-onset neurodegenerative disorder. J Hum Genet 2017;62(5):525–9.
[161] Fine AS, Nemeth CL, Kaufman ML, Fatemi A. Mitochondrial aminoacyl-tRNA synthetase disorders: an emerging group of developmental disorders of myelination. J. Neurodev. Disord 2019;11:29.
[162] McMillan HJ, et al. Compound heterozygous mutations in glycyl-tRNA synthetase are a proposed cause of systemic mitochondrial disease. BMC Med Genet 2014;15(1):36.
[163] Verrigni D, et al. Novel mutations in KARS cause hypertrophic cardiomyopathy and combined mitochondrial respiratory chain defect. Clin Genet 2017;91(6):918–23.

[164] Riley LG, et al. Phenotypic variability and identification of novel YARS2 mutations in YARS2 mitochondrial myopathy, lactic acidosis and sideroblastic anaemia. Orphanet J Rare Dis 2013;8(1):193.
[165] Sommerville EW, et al. Instability of the mitochondrial alanyl-tRNA synthetase underlies fatal infantile-onset cardiomyopathy. Hum Mol Genet 2018;28(2):258–68.
[166] Nakajima J, et al. A novel homozygous YARS2 mutation causes severe myopathy, lactic acidosis, and sideroblastic anemia 2. J Hum Genet 2014;59(4):229–32.
[167] Belostotsky R, et al. Mutations in the mitochondrial seryl-tRNA synthetase cause hyperuricemia, pulmonary hypertension, renal failure in infancy and alkalosis, HUPRA syndrome. Am J Hum Genet 2011;88 (2):193–200.
[168] Gotz A, et al. Exome sequencing identifies mitochondrial alanyl-tRNA synthetase mutations in infantile mitochondrial cardiomyopathy. Am J Hum Genet 2011;88(5):635–42.
[169] Dallabona C, et al. Novel (ovario) leukodystrophy related to AARS2 mutations. Neurology 2014;82 (23):2063–71.
[170] Metodiev MD, et al. NSUN4 is a dual function mitochondrial protein required for both methylation of 12S rRNA and coordination of mitoribosomal assembly. PLoS Genet 2014;10(2):e1004110.
[171] Metodiev MD, et al. Methylation of 12S rRNA is necessary for in vivo stability of the small subunit of the mammalian mitochondrial ribosome. Cell Metab 2009;9(4):386–97.
[172] Haute LV, et al. METTL15 introduces N4-methylcytidine into human mitochondrial 12S rRNA and is required for mitoribosome biogenesis. Nucleic Acids Res 2019;.
[173] Zaganelli S, et al. The pseudouridine synthase RPUSD4 is an essential component of mitochondrial RNA granules. J Biol Chem 2017;292(11):4519–32.
[174] Bar-Yaacov D, et al. Mitochondrial 16S rRNA is methylated by tRNA methyltransferase TRMT61B in all vertebrates. PLoS Biol 2016;14(9):e1002557.
[175] Rorbach J, et al. MRM2 and MRM3 are involved in biogenesis of the large subunit of the mitochondrial ribosome. Mol Biol Cell 2014;25(17) 2542–55.
[176] Lee KW, Bogenhagen DF. Assignment of 2′-O-methyltransferases to modification sites on the mammalian mitochondrial large subunit 16S ribosomal RNA (rRNA). J Biol Chem 2014;289(36):24936–42.
[177] Lee KW, et al. Mitochondrial ribosomal RNA (rRNA) methyltransferase family members are positioned to modify nascent rRNA in foci near the mitochondrial DNA nucleoid. J Biol Chem 2013;288 (43):31386–99.
[178] Garone C, et al. Defective mitochondrial rRNA methyltransferase MRM2 causes MELAS-like clinical syndrome. Hum Mol Genet 2017;26(21):4257–66.
[179] Popow J, et al. FASTKD2 is an RNA-binding protein required for mitochondrial RNA processing and translation. RNA 2015;21(11):1873–84.
[180] Ghezzi D, et al. FASTKD2 nonsense mutation in an infantile mitochondrial encephalomyopathy associated with cytochrome c oxidase deficiency. Am J Hum Genet 2008;83(3):415–23.
[181] Yoo DH, et al. Identification of FASTKD2 compound heterozygous mutations as the underlying cause of autosomal recessive MELAS-like syndrome. Mitochondrion 2017;35:54–8.
[182] Smits P, Smeitink J, van den Heuvel L. Mitochondrial translation and beyond: processes implicated in combined oxidative phosphorylation deficiencies. J Biomed Biotechnol 2010;2010:737385.
[183] Chrzanowska-Lightowlers Z, Rorbach J, Minczuk M. Human mitochondrial ribosomes can switch structural tRNAs—but when and why? RNA Biol 2017;14(12):1668–71.
[184] Amunts A, et al. Ribosome. The structure of the human mitochondrial ribosome. Science 2015;348 (6230):95–8.
[185] Christian BE, Spremulli LL. Evidence for an active role of IF3mt in the initiation of translation in mammalian mitochondria. Biochemistry 2009;48(15):3269–78.
[186] Greber BJ, et al. Ribosome. The complete structure of the 55S mammalian mitochondrial ribosome. Science 2015;348(6232):303–8.
[187] Christian BE, Spremulli LL. Preferential selection of the 5′-terminal start codon on leaderless mRNAs by mammalian mitochondrial ribosomes. J Biol Chem 2010;285(36):28379–86.

[188] Cai YC, et al. Interaction of mitochondrial elongation factor Tu with aminoacyl-tRNA and elongation factor Ts. J Biol Chem 2000;275(27):20308–14.

[189] Mai N, Chrzanowska-Lightowlers ZMA, Lightowlers RN. The process of mammalian mitochondrial protein synthesis. Cell Tissue Res 2017;367(1):5–20.

[190] Christian BE, Spremulli LL. Mechanism of protein biosynthesis in mammalian mitochondria. Biochim Biophys Acta 2012;1819(9-10):1035–54.

[191] Lightowlers RN, Chrzanowska-Lightowlers ZM. Terminating human mitochondrial protein synthesis: a shift in our thinking. RNA Biol 2010;7(3):282–6.

[192] Tsuboi M, et al. EF-G2mt is an exclusive recycling factor in mammalian mitochondrial protein synthesis. Mol Cell 2009;35(4):502–10.

[193] Richter R, et al. A functional peptidyl-tRNA hydrolase, ICT1, has been recruited into Hum mitochondrial ribosome. EMBO J 2010;29(6):1116–25.

[194] Miller C, et al. Defective mitochondrial translation caused by a ribosomal protein (MRPS16) mutation. Ann Neurol 2004;56(5):734–8.

[195] Gardeitchik T, et al. Bi-allelic mutations in the mitochondrial ribosomal protein MRPS2 cause sensorineural hearing loss, hypoglycemia, and multiple OXPHOS complex deficiencies. Am J Hum Genet 2018;102(4):685–95.

[196] Menezes MJ, et al. Mutation in mitochondrial ribosomal protein S7 (MRPS7) causes congenital sensorineural deafness, progressive hepatic and renal failure and lactic acidemia. Hum Mol Genet 2015;24 (8):2297–307.

[197] Jackson CB, et al. A variant in MRPS14 (uS14m) causes perinatal hypertrophic cardiomyopathy with neonatal lactic acidosis, growth retardation, dysmorphic features and neurological involvement. Hum Mol Genet 2019;28(4):639–49.

[198] Saada A, et al. Antenatal mitochondrial disease caused by mitochondrial ribosomal protein (MRPS22) mutation. J Med Genet 2007;44(12):784–6.

[199] Kohda M, et al. A comprehensive genomic analysis reveals the genetic landscape of mitochondrial respiratory chain complex deficiencies. PLoS Genet 2016;12(1):e1005679.

[200] Lake NJ, et al. Biallelic mutations in MRPS34 lead to instability of the small mitoribosomal subunit and Leigh syndrome. Am J Hum Genet 2017;101(2):239–54.

[201] Borna NN, et al. Mitochondrial ribosomal protein PTCD3 mutations cause oxidative phosphorylation defects with Leigh syndrome. Neurogenetics 2019;20(1):9–25.

[202] Bugiardini E, et al. MRPS25 mutations impair mitochondrial translation and cause encephalomyopathy. Hum Mol Genet 2019;28(16):2711–19.

[203] Pulman J, et al. Mutations in the MRPS28 gene encoding the small mitoribosomal subunit protein bS1m in a patient with intrauterine growth retardation, craniofacial dysmorphism and multisystemic involvement. Hum Mol Genet 2019;28(9):1445–62.

[204] Galmiche L, et al. Exome sequencing identifies MRPL3 mutation in mitochondrial cardiomyopathy. Hum Mutat 2011;32(11):1225–31.

[205] Serre V, et al. Mutations in mitochondrial ribosomal protein MRPL12 leads to growth retardation, neurological deterioration and mitochondrial translation deficiency. Biochim Biophys Acta 2013;1832 (8):1304–12.

[206] Carroll CJ, et al. Whole-exome sequencing identifies a mutation in the mitochondrial ribosome protein MRPL44 to underlie mitochondrial infantile cardiomyopathy. J Med Genet 2013;50(3):151–9.

[207] Takeuchi N, et al. Mammalian mitochondrial methionyl-tRNA transformylase from bovine liver. Purification, characterization, and gene structure. J Biol Chem 1998;273(24):15085–90.

[208] Tucker EJ, et al. Mutations in MTFMT underlie a human disorder of formylation causing impaired mitochondrial translation. Cell Metab 2011;14(3):428–34.

[209] Haack TB, et al. Phenotypic spectrum of eleven patients and five novel MTFMT mutations identified by exome sequencing and candidate gene screening. Mol Genet Metab 2014;111(3):342–52.

[210] Pena JA, et al. Methionyl-tRNA formyltransferase (MTFMT) deficiency mimicking acquired demyelinating disease. J Child Neurol 2016;31(2):215–9.
[211] Hayhurst H, et al. Leigh syndrome caused by mutations in MTFMT is associated with a better prognosis. Ann Clin Transl Neurol 2019;6(3):515–24.
[212] Janer A, et al. RMND1 deficiency associated with neonatal lactic acidosis, infantile onset renal failure, deafness, and multiorgan involvement. Eur J Hum Genet 2015;23(10):1301–7.
[213] Ng YS, et al. The clinical, biochemical and genetic features associated with RMND1-related mitochondrial disease. J Med Genet 2016;53(11):768–75.
[214] Garcia-Diaz B, et al. Infantile encephaloneuromyopathy and defective mitochondrial translation are due to a homozygous RMND1 mutation. Am J Hum Genet 2012;91(4):729–36.
[215] Casey JP, et al. Periventricular calcification, abnormal pterins and dry thickened skin: expanding the clinical spectrum of RMND1? JIMD Rep 2016;26:13–19.
[216] Coenen MJ, et al. Mutant mitochondrial elongation factor G1 and combined oxidative phosphorylation deficiency. N Engl J Med 2004;351(20):2080–6.
[217] Smits P, et al. Mutation in subdomain G′ of mitochondrial elongation factor G1 is associated with combined OXPHOS deficiency in fibroblasts but not in muscle. Eur J Hum Genet 2011;19(3):275–9.
[218] Balasubramaniam S, et al. Infantile progressive hepatoencephalomyopathy with combined OXPHOS deficiency due to mutations in the mitochondrial translation elongation factor gene GFM1. JIMD Rep 2012;5:113–22.
[219] Galmiche L, et al. Toward genotype phenotype correlations in GFM1 mutations. Mitochondrion 2012;12(2):242–7.
[220] Antonicka H, et al. The molecular basis for tissue specificity of the oxidative phosphorylation deficiencies in patients with mutations in the mitochondrial translation factor EFG1. Hum Mol Genet 2006;15(11):1835–46.
[221] Valente L, et al. Infantile encephalopathy and defective mitochondrial DNA translation in patients with mutations of mitochondrial elongation factors EFG1 and EFTu. Am J Hum Genet 2007;80(1):44–58.
[222] Brito S, et al. Long-term survival in a child with severe encephalopathy, multiple respiratory chain deficiency and GFM1 mutations. Front Genet 2015;6.
[223] Simon MT, et al. Activation of a cryptic splice site in the mitochondrial elongation factor GFM1 causes combined OXPHOS deficiency. Mitochondrion 2017;34:84–90.
[224] Smeitink JAM, et al. Distinct clinical phenotypes associated with a mutation in the mitochondrial translation elongation factor EFTs. Am J Hum Genet 2006;79(5):869–77.
[225] Vedrenne V, et al. Mutation in the mitochondrial translation elongation factor EFTs results in severe infantile liver failure. J Hepatol 2012;56(1):294–7.
[226] Calvo SE, et al. Molecular diagnosis of infantile mitochondrial disease with targeted next-generation sequencing. Sci Transl Med 2012;4(118):118ra10.
[227] Emperador S, et al. Molecular-genetic characterization and rescue of a TSFM mutation causing childhood-onset ataxia and nonobstructive cardiomyopathy. Eur J Hum Genet 2017;25(1):153–6.
[228] Scala M, et al. Novel homozygous TSFM pathogenic variant associated with encephalocardiomyopathy with sensorineural hearing loss and peculiar neuroradiologic findings. Neurogenetics 2019;20(3):165–72.
[229] Perli E, et al. Novel compound mutations in the mitochondrial translation elongation factor (TSFM) gene cause severe cardiomyopathy with myocardial fibro-adipose replacement. Sci Rep 2019;9(1).
[230] Pyle A, et al. Behr's syndrome is typically associated with disturbed mitochondrial translation and mutations in the C12orf65 gene. J Neuromuscul Dis 2014;1(1):55–63.
[231] Wesolowska M, et al. Adult onset Leigh syndrome in the intensive care setting: a novel presentation of a C12orf65 related mitochondrial disease. J Neuromuscul Dis 2015;2(4):409–19.
[232] Antonicka H, et al. Mutations in C12orf65 in patients with encephalomyopathy and a mitochondrial translation defect. Am J Hum Genet 2010;87(1):115–22.

[233] Dixon-Salazar TJ, et al. Exome sequencing can improve diagnosis and alter patient management. Sci Transl Med 2012;4(138):138ra78.
[234] Fukumura S, et al. Compound heterozygous GFM2 mutations with Leigh syndrome complicated by arthrogryposis multiplex congenita. J Hum Genet 2015;60(9):509–13.
[235] Glasgow RIC, et al. Novel GFM2 variants associated with early-onset neurological presentations of mitochondrial disease and impaired expression of OXPHOS subunits. Neurogenetics 2017;18(4):227–35.
[236] Weraarpachai W, et al. Mutation in TACO1, encoding a translational activator of COX I, results in cytochrome c oxidase deficiency and late-onset Leigh syndrome. Nat Genet 2009;41(7):833–7.
[237] Seeger J, et al. Clinical and neuropathological findings in patients with TACO1 mutations. Neuromuscul Disord 2010;20(11):720–4.
[238] Ostergaard E, et al. Mutations in COA3 cause isolated complex IV deficiency associated with neuropathy, exercise intolerance, obesity, and short stature. J Med Genet 2015;52(3):203–7.
[239] Weraarpachai W, et al. Mutations in C12orf62, a factor that couples COX I synthesis with cytochrome c oxidase assembly, cause fatal neonatal lactic acidosis. Am J Hum Genet 2012;90(1):142–51.
[240] Thompson K, et al. OXA1L mutations cause mitochondrial encephalopathy and a combined oxidative phosphorylation defect. EMBO Mol Med 2018;e9060.
[241] Hirano M, Emmanuele V, Quinzii CM. Emerging therapies for mitochondrial diseases. Essays Biochem 2018;62(3):467–81.

CHAPTER 16

mtDNA maintenance: disease and therapy

Corinne Quadalti[1,2] and Caterina Garone[1,2]

[1]Department of Medical and Surgical Sciences, Medical Genetics Unit, University of Bologna, Bologna, Italy, [2]Center for Applied Biomedical Research, University of Bologna, Bologna, Italy

16.1 Introduction

Mitochondrial DNA (mtDNA) is essential in pluripotency, cellular differentiation, and development other than its main role in energy producing apparatus of the cell. The relative amount of mtDNA per cell varies in a tissue-specific manner, as an adequate number of mtDNA copies must be maintained to support aerobic respiration and meet cellular energetic demands. The mtDNA copy number also regulates the rate of segregation of mtDNA [1,2].

Primordial germ cell population for both sexes possesses low mtDNA copies. During oogenesis, mtDNA copy number increases throughout from approximately 200 copies in the primordial germ cells to 200,000–300,000 in mature oocytes. During early development, pluripotent stem cells possess low mtDNA copy number. They establish the mtDNA set point to promote cell proliferation in order to let every specific lineage to acquire the appropriate numbers of mtDNA copy for meeting their specific energy demand, as they become specialized cells [2]. This process is mediated by changes to DNA methylation at exon 2 of the catalytic subunit of the mitochondrial-specific polymerase (polymerase-γ, POLGA) [1]. In somatic cells, mtDNA copy number ranges from 200 to 10,000 copies/cell [2].

The maintenance of mtDNA copy number requires a set of proteins that is distinct from the nuclear replication machinery that operates the mtDNA replication, a supply of nucleotides, a number of regulatory factors, and proteins for the packaging of mtDNA into the nucleoids [3].

Hallmarks of defect in mtDNA maintenance are qualitative (point mutations, multiple deletions) and/or quantitative (copy number reduction) molecular genetic defect of mtDNA, biochemical dysfunction with multiple OXPHOS defects, and morphological alterations of the mitochondrial network. Genetically, an increasing number of autosomal dominant or

The Human Mitochondrial Genome.
DOI: https://doi.org/10.1016/B978-0-12-819656-4.00016-4

recessive pathogenic variants have been identified with large unbiased genetic sequencing by applying next generation sequences technologies. Clinically, they present as a continuum spectrum from infantile multisystemic and rapidly progressive disorders to childhood or adult tissue-specific diseases (Table 16.1).

16.2 Defects in mtDNA replisome

The mitochondrial replisome is composed by the polymerase-γ, responsible of the synthesis of the new mtDNA molecules and proof-reading base repair, and additional proteins playing role in initiation, processing or termination process of the mtDNA replication. They include Twinkle encoded by TWNK (previously designated as C10orf2), mitochondrial topoisomerase I, mitochondrial RNA polymerase (mtRNAP), RNase H1 (encoded by RNASEH1), mitochondrial genome maintenance exonuclease 1 (encoded by MGME1), single-stranded DNA-binding protein (mtSSB), DNA ligase III, DNA helicase/nuclease 2 (encoded by DNA2), and RNA and DNA flap endonuclease 1 (FEN1) [3].

Here, we present an overview of the clinical syndromes due to defects in mtDNA replisome. Defect in POLG, DNA2, MGME1, TWNK were extensively introduced, Nuclear genetic disorders of mitochondrial DNA gene expression, while only key features will be highlighted in the following section.

POLG-related disorders represent the most frequent cause of mitochondrial disorders due to pathogenic variants in nuclear encoded genes [30]. More than 1000 patients were reported in the last decade carrying pathogenic variants in the catalytic subunit encoded by *POLG* on chromosome 15q25 and additional few cases with pathogenic variants in the accessory subunits encoded by *POLG2* on chromosome 17q. Clinically, POLG-related disease recognizes five main clinical syndromes with predominant brain and liver tissue specificity [31].

POLG2 defect can present as autosomal recessive inherited early-onset fatal liver failure (#618528) or autosomal dominant adult-onset PEO (#610131) with cardiac conduction defect, myopathy and increased creatine kinase [32,33].

TWNK encodes the motor protein TWINKLE that acts as a helicase at the replication fork. Tissue specificity of Twinkle-related diseases (#271245, #609286) is characterized by prevalent extra-pyramidal and cerebellar features with parkinsonism, dyskinesia, ataxia presenting as predominantly debilitating symptoms in the infantile onset or as PEO plus syndrome in the adult onset form. Renal and liver dysfunctions are also reported in the most severe cases of the spectrum. Sporadic cases presented as Perrault syndrome (#616138) (sensorineural hearing loss and ovarian dysgenesis) with milder neurological involvement.

Table 16.1: Summarizes clinical forms of known disease causing genes in mitochondrial mtDNA maintenance.

Replisome					
Gene	**Function**	**Disease**			
POLG	Mitochondrial DNA polymerase γ	Onset	Infantile	Childhood	Adult
		Inheritance	AR	AR	AR/AD
		Clinical phenotype	Alpers–Huttenlocher hepatopathic poliodystrophy	MCHS: childhood myocerebrohepatopathy spectrum MNGIE-like	PEO SANDO: sensory ataxic neuropathy, dysarthria, ophthalmoplegia occasionally cerebellar signs, myoclonus, seizures ANS: neuropathy, ataxia without myopathy, occasionally progressive external ophthalmoplegia MEMSA: epilepsy, myopathy, and ataxia, without ophthalmoplegia
		mtDNA	Depletion	Depletion	Multiple deletions
POLG2	Polymerase γ accessory subunit	Onset	Infantile	–	Adult
		Inheritance	AR	–	AD
		Clinical phenotype	Fatal liver failure	–	PEO
		mtDNA	Depletion	–	Multiple deletions
TWNK	Mitochondrial DNA helicase	Onset	Infantile	–	Adult
		Inheritance	AR	–	AD
		Clinical phenotype	IOSCA: spinocerebellar ataxia, with athetosis, areflexia, muscle hypotonia, severe epilepsy	–	PEO
		mtDNA	Depletion	–	Multiple deletions

(*Continued*)

Table 16.1: (Continued)

Replisome					
Gene	Function	Disease			
TFAM	mtDNA synthesis priming	Onset Inheritance Clinical phenotype mtDNA	Infantile AR Hepatocerebral syndrome Depletion	– – – –	– – – –
RNASEH1	RNase H1	Onset Inheritance Clinical phenotype mtDNA	– – – –	– – – –	Adult AR Myopathy/PEO Del
MGME1	Mitochondrial DNA flappase	Onset Inheritance Clinical phenotype mtDNA	– – – –	– – – –	Adult AR Myopathy/PEO Depletion/multiple deletions
DNA2	DNA helicase/nuclease 2	Onset Inheritance Clinical phenotype mtDNA	Infantile AR Seckel syndrome NA	Childhood AD PEO Multiple deletions	Adult AD Myopathy/PEO Deletions
SSBP1	Mitochondrial single-stranded DNA-binding protein	Onset Inheritance Clinical phenotype mtDNA	– – – –	Childhood AD/AR Optic atrophy, retinal macular dystrophy, sensorineural deafness, mitochondrial myopathy, kidney failure Depletion	Adult AD/AR Optic atrophy, retinal macular dystrophy, sensorineural deafness, mitochondrial myopathy, kidney failure Depletion

Mitochondrial nucleotide pool balance					
Gene	**Function**	**Disease**			
TK2	Mitochondrial thymine kinase 2	Onset Inheritance Clinical phenotype mtDNA	Infantile AR Rapidly progressive myopathy Depletion	Childhood AR Myopathy Depletion/multiple deletions	Adult AR Slowly progressive myopathy Multiple deletions
DGUOK	Mitochondrial deoxyguanosine kinase	Onset Inheritance Clinical phenotype mtDNA	Infantile AR Hepatocerebral syndrome Depletion	– – – –	Adult AD Myopathy/PEO Multiple deletions
SUCLG1	Mitochondrial succinyl-CoA ligase [GDP-forming] subunit alpha	Onset Inheritance Clinical phenotype mtDNA	Infantile AR Hepatocerebral syndrome Depletion	– – – –	– – – –
SUCLA2	Mitochondrial succinyl-CoA ligase [GDP-forming] subunit beta	Onset Inheritance Clinical phenotype mtDNA	Infantile AR Encephalomyopathy with/without methylmalonic aciduria Depletion	Childhood AR Encephalomyopathy with/without methylmalonic aciduria Depletion	– – – –
ABAT	Encoding for mitochondrial γ-aminobutyric acid transaminase (GABAT enzyme)	Onset Inheritance Clinical phenotype mtDNA	Infantile AR Encephalomyopathy with elevated GABA Depletion	– – – –	– – – –
TYMP	Thymidine phosphorylase	Onset Inheritance Clinical phenotype mtDNA	– – –	Childhood AR Mitochondrial neurogastrointestinal encephalopathy (MNGIE) Depletion/multiple deletions	Adult AR Mitochondrial neurogastrointestinal encephalopathy (MNGIE) Depletion/multiple deletions

(*Continued*)

Table 16.1: (Continued)

Mitochondrial nucleotide pool balance					
Gene	Function	Disease			
RRM2B	Ribonucleotide-diphosphate reductase subunit M2 B	Onset Inheritance Clinical phenotype mtDNA	Infantile AR Encephalomyopathy with kidney failure MNGIE-like syndrome Depletion		Adult AD PEO Del
SLC25A4 (ANT1)	Mitochondrial muscle-specific isoform of the adenine nucleotide translocator	Onset Inheritance Clinical phenotype mtDNA	 AD Myopathy/ cardiomyopathy Depletion	 AR Myopathy/ cardiomyopathy Multiple deletions	 AD PEO Multiple deletions
AGK	Mitochondrial acylglycerol kinase	Onset Inheritance Clinical phenotype mtDNA	 AR Sengers syndrome Depletion	– – – –	– – – –
MPV17	Encoding a small protein of the inner mitochondrial membrane of unknown function	Onset Inheritance Clinical phenotype mtDNA	Infantile AR Hepatocerebral syndrome Depletion	 – – –	Adult AR Myopathy/PEO/ parkinsonism Multiple deletions
DTYMK	Nuclear-encoded deoxythymidylate kinase	Onset Inheritance Clinical phenotype mtDNA	Infantile AR Encephalopathy –	– – – –	– – – –

Mitochondrial dynamics					
Gene	Function	Disease			
OPA1	Mitochondrial dynamin-like 120 kDa protein	Onset Inheritance Clinical phenotype mtDNA	Infantile AR Encephalopathy with/ without cardiomyopathy Behr syndrome	Childhood AD Optic atrophy Optic atrophy plus Behr syndrome Del	Adult Optic atrophy

MFN2	Mitofusin 2	Onset Inheritance Clinical phenotype mtDNA	– – – –	Childhood AD Charcot-Marie-Tooth disease, axonal, type 2A2A or type 2A2B Hereditary motor and sensory neuropathy VIA Multiple deletions	– – – –
FBXL4	F-box and leucine-rich repeat protein 4	Onset Inheritance Clinical phenotype mtDNA	Infantile AR Encephalomyopathy with multisystemic features Depletion	– – – –	– – – –
DNM1L		Onset Inheritance Clinical phenotype mtDNA	Infantile AR/AD Encephalopathy Depletion	Childhood AD Optic atrophy –	Adult AD Optic atrophy –
MSTO1	Cytosolic mitochondrial fusion protein misato homolog 1	Onset Inheritance Clinical phenotype mtDNA	– – – –	Childhood AR Muscular dystrophy, corticospinal tract dysfunction early-onset nonprogressive cerebellar atrophy Depletion	– – – –
Nucleoids					
Gene	**Function**	**Disease**			
TFAM		Onset Inheritance Clinical phenotype mtDNA	Infantile AR Hepatocerebral syndrome –	– – – –	– – – –
ATAD3A		Onset Inheritance Clinical phenotype mtDNA	Infantile AR/AD Encephalopathy with cardiomyopathy Severe congenital cerebellar atrophy –	Childhood AR/AD Spastic paraplegia Cerebellar ataxia –	– – – –

MGME1 encodes a mitochondrial RecB-type exonuclease. Pathogenic variants (#615084) were identified in only six patients from three unrelated families with ptosis, PEO, muscle weakness and atrophy, severe emaciation, and respiratory insufficiency due to muscle weakness as common features [34].

DNA2 (#615156) and ***RNASEH1*** (#616479) respectively encoding nuclear/mitochondrial helicase/nuclease and endonuclease showed eye muscle tissue specificity when defective, presenting with predominant PEO and additional myopathic features [35,36]. When in homozygosity, pathogenic variants in DNA2 can cause Seckel syndrome with intrauterine growth retardation, dwarfism, microcephaly with mental retardation, and a characteristic 'bird-headed' facial appearance [37].

SSBP1 (#N/A) encodes single-stranded binding protein 1 is required to stabilize single-stranded mtDNA (ssmtDNA) and stimulate DNA synthesis by POLG. Defect in SSBP1 primarily affect optic nerve and/or retina causing optic atrophy and blindness as initial and predominant clinical phenotype. Isolated optic atrophy was described in 25 patients in 5 unrelated families while a more complex phenotype with optic atrophy and severe renal dysfunction leading to organ transplantation in 7 patients in 4 unrelated families. Homozygous recessive mutation was found in only one patient and associated with a more severe clinical phenotype with early-onset blindness and deafness complicated during the clinical course by hypertrophic cardiomyopathy, ataxia, growth failure [38–40].

16.3 Defects in mitochondrial nucleotides pool balance

MtDNA replication requires a balanced amount of deoxynucleotides triphosphates representing the building blocks for the newly synthesized mtDNA. As previously introduced, and further detailed in this section, a complex biochemical pathway regulates the synthesis and exchange of the dNTPs between the cytosolic and mitochondrial compartments via de novo synthesis or salvage route in a tissue-specific and ontogenic manner. Cytosolic and mitochondrial enzymes have overlapping substrates specificity. In addition, nucleosides and nucleotides are constantly exchanged between the two compartments via a number of dedicated transporters [41]. In physiological conditions, the nucleotides pool amount can regulate the enzyme activity by enhancing or inhibiting the product conversion. Pathogenic variants in gene encoding nucleosides/nucleotides anabolic or catabolic enzymes or carriers perturb dNTP pool balance and are responsible of mtDNA depletion, multiple deletions or point mutations. Specifically, in the salvage pathway, mitochondrial purine metabolism is perturbed by pathogenic variants in the deoxyguanosine kinase (*DGUOK*) gene, encoding deoxyguanosine (dGuo) kinase, a mitochondrial enzyme that convert dGuo and deoxyadenosine (dAdo) into their monophosphates, while mitochondrial pyrimidine metabolism is perturbed by pathogenic variants in the Thymidine Kinase 2

(***TK2***) gene encoding thymidine kinase, responsible of the conversion of deoxythymidine (dThd) and deoxycytidine (dCtd) into their monophosphates. In both cases, lack of enzyme activities causes reduction of the correspondent nucleotides triphosphates that are incorporated in the newly synthetized mtDNA. Clinically, Tk2 deficiency (#609560) present as a phenotypic spectrum including three major clinical presentation: infantile-onset (<1 years) myopathy rapidly progressing to early death (median survival of 20 months); childhood-onset (>1 and <12 years) myopathy with longer survival (>9 years); late-onset of slowly progressing myopathy and median survival of 50 years [42,43]. Similarly, dGK deficiency present as infantile onset of liver failure and encephalopathy (#251880), childhood-onset of myopathy and increases transaminases or late-onset of adult PEO, myopathy and parkinsonism (#617070) [44]. Thymidine phosphorylase (TP) encoded by *TYMP gene* is instead a cytosolic enzyme responsible for the conversion of dThd and dU into thymine and uracil. TP deficiency causes a toxic accumulation of dU and dThd that modifies mitochondrial nucleotide pool balances with excess of dTTP and secondary depletion of dCTP [45]. Clinically, pathogenic variants in *TYMP* are associated to mitochondrial neurogastrointestinal encephalomyopathy (MNGIE) characterized by onset at average age of 18.5 years of PEO, demyelinating peripheral neuropathy, leukoencephalopathy, gastrointestinal dysmotility. TP activity is reduced up to 25% in healthy heterozygous carriers while in patients is less than 15% causing toxic accumulation of dThd and dU in plasma and tissues and an abnormal excretion in the urine [14]. The excess of dThd and dU causes mitochondrial nucleotide pool unbalance and affect quantitatively and qualitatively the mtDNA during replication, causing mtDNA depletion, multiple deletions, and point mutations [46]. Mutations in ribonucleotidereductase regulatory TP53 inducible subunit (R1-p53R2), encoded by RRM2B gene, are responsible for altered metabolism of the cytosolic ribonucleotides that cannot be converted into deoxynucleosides in the de novo synthesis. Therefore lack of R1-p53R2 activity causes unbalance of both purine and pyrimidine metabolism. Clinically, RRM2B defect causes infantile onset of encephalomyopathy with renal tubulopathy (#612075), MNGIE-like syndrome (#613077), or childhood/adult onset of PEO.

Recently, compound heterozygous variants predicted to be damaging were identified in ***DTYMK***(#N/A) gene, encoding deoxythymidylate kinase, an enzyme responsible of the phosphorylation of deoxy-TMP to deoxy-TDP [47]. Patients were two siblings of a quartet family, presenting with hypotonia, microcephaly, and severe intellectual disability. However, functional studies were not performed in order to confirm the causal effect of the variants [48].

Pathogenic variants in ***SUCLG1***, ***SUCLA2***, ***ABAT***, ***ANT1***, ***AGK***, and ***MPV17*** may perturb the nucleotides pool balance and they are responsible of clinical syndromes associated with mtDNA depletions and/or multiple deletions. Details are reported in Table 16.1.

16.4 Defects in mitochondrial dynamics

Mitochondria are highly dynamic organelles undergoing coordinated cycles of fission and fusion referred to as mitochondrial dynamics in order to maintain their shape, distribution, and size. Mitochondrial fission is characterized by the division of one mitochondrion into two daughter mitochondria, whereas mitochondrial fusion is the union of two mitochondria resulting in one mitochondrion. The core machinery of the mitochondrial dynamics is composed by large GTPase proteins belonging to the dynamin family. Mitochondrial constriction and scission are carried out by the dynamin-related/like protein 1 (drp1) and dynamin 2 (dnm2). Mitochondrial fusion by mitofusin 1 and 2 and optic atrophy 1, which mediate, respectively, outer mitochondrial membrane and inner mitochondrial membrane fusion [49].

Retinal ganglion cells appear particularly affected by dysfunction in mitochondrial dynamics as demonstrated, respectively, by autosomal dominant defects of optic atrophy 1 (***OPA1***) (#165500; #125250) and dynamin-like 1 protein (***DNM1L***) (#610708) presenting with optic atrophy and peripheral nervous system involvement.

OPA1 (#165500) pathogenic variants are responsible for the most frequent autosomal dominant mitochondrial related optic atrophy due to degeneration of retinal ganglion cells presenting with bilateral visual loss, central scotoma, temporal optic disc atrophy, dyschromatopsia (ADOA) with 20% of patients presenting additional symptoms (ADOA plus, #125250) such as sensorineural deafness, ataxia, myopathy, peripheral neuropathy, and PEO [50]. A recent updated OPA1 database reported a total of 831 patients: 697 with isolated dominant optic atrophy (DOA), 47 with DOA "plus," and 83 with asymptomatic or unclassified DOA [51].

However, when pathogenic variants occur in compound heterozygosity or homozygosity, the clinical severity and complexity may vary from ataxia and optic atrophy with severe infantile cardiomyopathy [52] or childhood multiorgan failure or peripheral neuropathy [50,53] (#616896) to Behr syndrome (#210000) characterized by early-onset optic neuropathy with spinocerebellar degeneration, pyramidal signs, peripheral neuropathy, gastrointestinal dysmotility, and retarded development [54,55]. Other rare associations of OPA1 mutations have been reported with spastic paraplegia [50], multiple sclerosis-like syndrome [56], and syndromic parkinsonism and dementia [57,58].

Similarly, dominant negative mutation or autosomal recessive mutation in DRP1 can affect neurodevelopment with early-onset encephalopathy or childhood epileptic encephalopathy [49,59,60].

A tissue-specific disorder predominantly affecting peripheral nervous system has been described when *MFN2* heterozygous variant occur with a clinical phenotype of

Charcot-Marie-Tooth disease, axonal, type 2A2A (#609260) or type 2A2B (#617087), or hereditary motor and sensory neuropathy VIA(#601152) [61].

A more complex and multisystemic phenotype has been instead associated with autosomal recessive defect in Fbxl4 and MSTO1 genes.

FBXL4 encodes a f-box leucine-rich protein deemed to play a role in mitochondrial dynamics based on the observation of mtDNA depletion and mitochondrial network fragmentation in defective human cells. However, the exact function is still unknown.

Autosomal recessive pathogenic variants of FBXL4 (#615471) were described in 94 patients with a clinical syndrome characterized by failure to thrive, remarkable neurodevelopmental delays, encephalopathy, cerebral atrophy, generalized hypotonia, and persistent lactic acidosis. Other features are feeding difficulties, growth failure, microcephaly, hyperammonemia, seizures, hypertrophic cardiomyopathy, elevated liver transaminases, recurrent infections, variable distinctive facial features, white matter abnormalities, and cerebral atrophy found in neuroimaging [62,63].

MSTO1 (#617675) encodes MISATO1, a mitochondrial distribution and morphology regulator, associated to the outer membrane and playing an important role in mitochondrial fusion. When defective causes childhood-onset muscular dystrophy with proximal muscle weakness, corticospinal tract dysfunction including increased tone, spastic catch, clonus and/or increased deep tendon reflexes, and early-onset nonprogressive cerebellar atrophy presenting with dysmetria, ataxia, and dysarthria. Additional symptoms were speech delay, learning disability, and pigmentary retinopathy in some [64–68].

16.5 Defect in nucleoid proteins

Mitochondrial transcription factor-A (TFAM) (#617156) plays a role in the compaction and organization of mtDNA in the nucleoid and regulates mtDNA copy number indirectly by producing the necessary RNA primer for mtDNA replication [3]. Defect in TFAM has been described in only two siblings who presented IUGR, elevated transaminases, conjugated hyperbilirubinemia, and hypoglycemia with progression to liver failure and death in early infancy [69].

ATAD3A (ATP-ase family AAA domain containing 3A) has been deemed to be part of mtDNA nucleoids and to play role in several cellular functions including mtDNA maintenance and translation but the mechanistic details are unknown. ***ATAD3*** gene family in human includes three paralogs positioned in tandem in chromosome 1p36.33 and appears to have recently evolved by duplication on a single ancestral gene. Four distinct neurometabolic syndromes have been described in ATAD3 defect: recurrent de novo variant in *ATAD3A* (p.Arg528Trp) associated with severe developmental delay, hypotonia, optic atrophy, peripheral neuropathy, and cardiomyopathy; dominant inherited *ATAD3A* variant causing spastic paraplegia; bi-allelic variant in *ATAD3A* causing cerebellar atrophy, ataxia, congenital cataract, and seizure; large rearrangement of the three paralogs causing congenital encephalopathy with cerebellar atrophy, seizure, respiratory failure [70–72].

16.6 Experimental therapies

Advances in mitochondrial genetics have led to the design of therapies targeting disease mechanisms in mitochondrial disorders. Therapies with "general action" act on the function of key metabolic sensors, generating a plethora of downstream effects finally resulting in an increase of mitochondrial biogenesis (generation of new mitochondria) or mitophagy (selective elimination of damaged mitochondria) or a shift of energy metabolism. These therapies are potentially applicable to all mitochondrial disorders. On the other hand, "disease-tailored" therapy acts replacing the wild-type gene or protein or modifying their biochemical pathways. In both cases, the approach can be either pharmacological or gene therapy based. As "general action" treatments for mtDNA maintenance disorders, *in vivo* and *in vitro* studies have exploited both the efficacy of mtTOR inhibition, a nutrient sensing enzyme that regulate cell and organismal growth [73], and the busting of mitochondrial biogenesis through peroxisome proliferator-activated receptor alpha (PPAR)-α axis activation [74]. Instead, "disease-tailored" treatments include nucleosides supplementation, clearance of toxic metabolites, enzyme replacement therapy, and gene therapy. Preclinical evidences of efficacy and safety were strong enough for some of them to be translated in clinical trials (Figs. 16.1 and 16.2; Table 16.2).

16.7 General pharmacological approaches

16.7.1 Targeting mitochondrial biogenesis

Mitochondrial biogenesis can be activated via peroxisome proliferator-activated receptor gamma coactivator 1-alpha (PGC1α), a key regulator of cellular metabolism, that regulates the activation of a variety of transcription factors essential for mitochondrial function and energy production (e.g., estrogen-related receptor alpha—ERRα, peroxisome proliferator-activated receptor alpha—PPARα, and nuclear respiratory factors, NRF1 and NRF2) [75–77].

In particular, nicotinamide adenine dinucleotide (NAD) promotes mitochondrial biogenesis through deacetylation of PGC1α mediated by Sirtuin 1 (Sirt1) [78,79]. NAD has minimal toxicity, but has poor *in vivo* bioavailability being rapidly hydrolyzed in the small intestine [80,81]. Therefore nicotinamide riboside (NR), a NAD precursor, has been considered for his better pharmacokinetics and bioavailability [82].

NR is able to increase NAD+ in skeletal muscle and brown adipose tissue [83] and to cross the blood–brain barrier [84], increasing NAD levels in brain. Efficacy and safety of

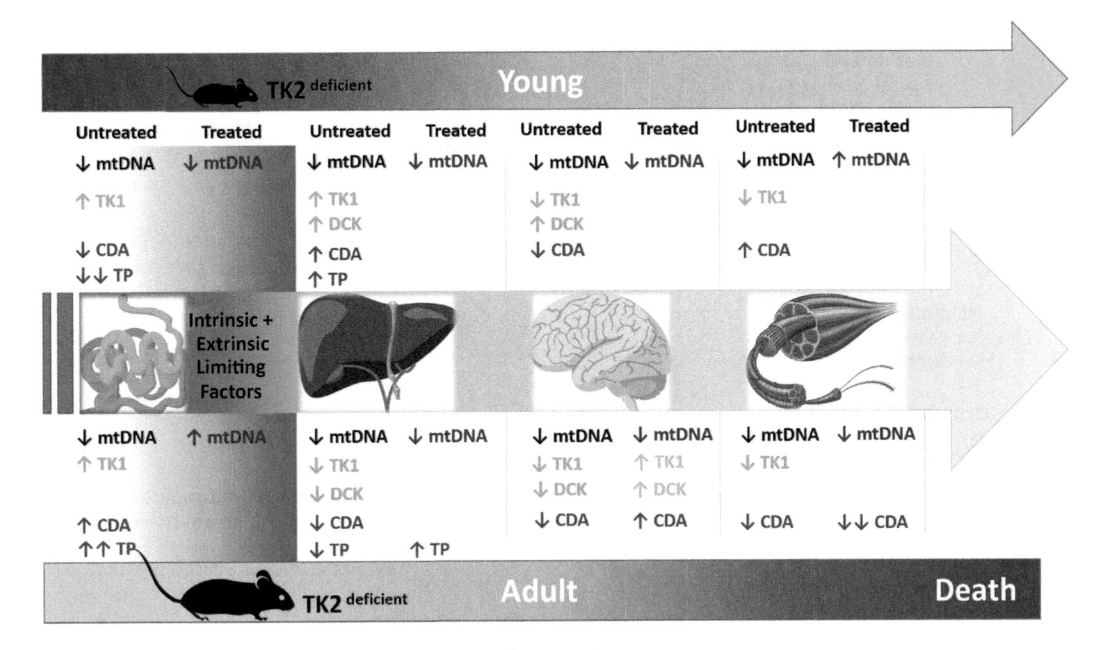

Figure 16.1
Age relation and tissue specificity of nucleoside metabolism in TK2 deficient mouse models (KO and H126N knockin): the figure is representative of intrinsic (catabolic and anabolic enzymes activities and their ontogenic and tissue specificity) and extrinsic factors (nucleosides availability and administration route) that influence the efficacy of nucleosides supplementation at different stages of life. A color-code font is used to represent anabolic enzymes (green: Tk1 and DCK) versus catabolicenzymes (purple: CDA and TP) and mtDNA level analyzed at 12 days (young) or 29 days (adult) in untreated (black: first column) or treated (blue: second column) mice.

NR have been already proved in mouse model for mitochondrial diseases of OXPHOS deficiencies and recent studies expanded NR potential application to mtDNA maintenance disorders, where it could potentially rescue muscular and neurological manifestations [4,74,83,85].

In order to identify potential new drugs for mtDNA maintenance disorders, DGUOK-deficient hepatocyte-like cells were generated using induced pluripotent stem cells (iPSCs) and tested with "SPECTRUM," a library of repurposing drugs that could improve mitochondrial ATP production and mitochondrial function. NAD was found to improve mitochondrial function in DGUOK-deficient hepatocyte-like cells by activating the PGC1α. To confirm efficacy and safety of NAD and NR, they were tested, respectively, in DGUOK-deficient iPSC-derived hepatocyte-like cells and in DGUOK KO rats. Results show increased ATP level and OXPHOS activity. In addition, in the DGUOK-deficient iPSC-derived hepatocyte-like cells, NAD was able to improve cells viability in 2-deoxy-D-glucose-stressed cell lines, to increase expression of all mitochondrial-encoded electron

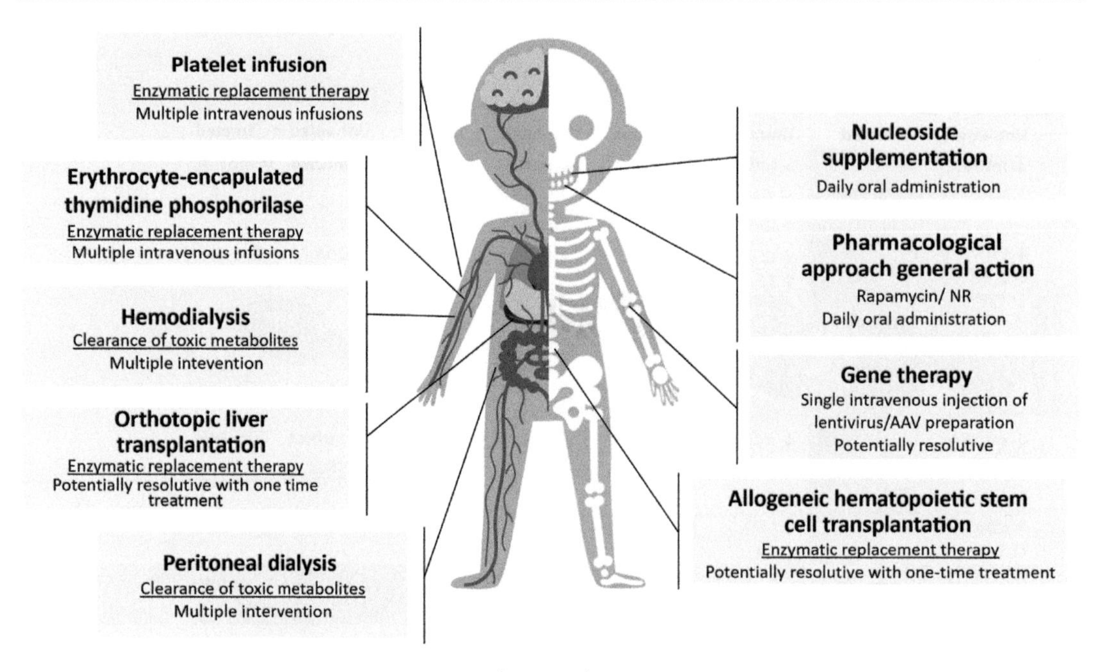

Figure 16.2
Schematic representation of the different therapeutic approaches for mtDNA maintenance diseases available to date.

transport chain genes examined, to reinstate normal membrane potential and mitochondrial morphology with increased mitochondria matrix density and normal cristae structure. Long-term exposure to NAD treatment in differentiated hepatocytes confirmed efficacy and safety. The improvement was independent by the mtDNA copy number, which in fact was still reduced, suggesting the promotion of different metabolic pathways. Similar to dGK deficient models, NAD treatment in *RRM2B*$^{-/-}$ iPSCs, generated through CRISPR/Cas9 technology, confirms the positive effect on ATP production [4].

Based on these preclinical data, NR treatment is a valid option also for mtDNA replication disorders. Currently, a clinical trial is ongoing in patients with mitochondrial myopathies (NCT03432871), opening the possibility to further translate the treatment to other mitochondrial disorders.

16.7.2 Targeting mTOR pathway

Rapamycin targets mammalian target of rapamycin (mTOR) pathway by inhibiting his activity and consequently modifying several metabolic pathways including nucleotides and lipids synthesis, protein translation, autophagy, glucose metabolism, and lysosomal biogenesis.

Table 16.2: The tables summarizes pre-clinical (*in vivo* and *in vitro*) and clinical studies on experimental therapies for mtDNA maintenance disease. Details on mechanism of action and administration route are also reported. In particular, past and/or ongoing clinical trials are indicated. Besides, the most significant limits of each approach are cited.

Therapy	Therapeutic approach	Disease	Clinical trial phase	Treatment	Limits	References
NR	Activator of mitochondrial biogenesis	dGK deficiency RRM2B defect	–	Oral	–	[4]
Rapamycin	Metabolic shift	TK2 deficiency	–	Oral	No effect on morbidity Severe cachexia	[5]
Nucleoside supplementation	Bypass molecular therapy Substrate enhancement	TK2 deficiency RRM2B dGK MNGIE Polg	Phase 1/2 NCT03639701	Oral	Unknown long-term efficacy	[6–11]
Peritoneal dialysis	Clearance of plasma nucleosides	MNGIE	–	Multiple interventions	Transient effect	[12]
Hemodialysis	Clearance of plasma nucleoside levels	MNGIE	–	Multiple interventions	Transient effect; multiple treatment	[13]
Platelet infusion	Enzyme replacement	MNGIE	–	Multiple interventions	Transient effect	[14,15]
Allogeneic hematopoietic stem cell transplantation (HSCT)	Enzyme replacement	MNGIE	Phase 1 NCT02427178	One time treatment	High mortality rate	[16]
Allogenic hematopoietic stem cell transplantation +/− Platelet infusion +/− Peritoneal dialysis	Multiple	MNGIE		Multiple approaches		[17,18]
Erythrocyte-encapsulated thymidine phosphorylase (EE-TP)	Enzyme replacement	MNGIE	Phase 2 NCT03866954	Multiple IV infusions	Possibility of immune response Limited efficacy	[19]
Orthotopic liver transplantation	Enzyme replacement Organ replacement	MNGIE dGK deficiency Alpers disease	–	One time treatment	Unknown long-term efficacy No effect on gastrointestinal symptoms	[20–24]
Gene therapy	Enzyme replacement (AAV) Gene replacement (lentivirus)	MNGIE	–	One time treatment	Risk of mutagenesis Cost for treatment production	[25–29]

The rational for using rapamycin in mitochondrial disorders is based on the ability of coupling autophagy and lysosomal biogenesis and consequently inducing a selective autophagy of dysfunctional mitochondria [86,87]. Rapamycin treatments have been tested in several mouse models producing evidence of efficacy on prolonging lifespan of the mice and ameliorating clinical phenotype [5,83,86,88–90]. However, the rapamycin signature on mitochondrial and other metabolic pathways is not univocal. First evidence of *in vivo* efficacy was obtained by daily administration of rapamycin in *Ndfus4*$^{-/-}$ knockout (KO) mouse model, whose clinical and biochemical phenotype reproduce Leigh syndrome due to Complex I deficiency. The treatment with rapamycin was able to prolong the lifespan, improve growth curve, prevent appearance of brain lesions, and delay neurological decline. Similarly, postonset oral administration delayed the development of the encephalopathy [88]. However, no effect on autophagy or mitochondrial function was observed, thus a shift toward amino acids catabolism and reduction of glycolytic catabolism was demonstrated to be responsible of the phenotypical efficacy [89]. Low dose rapamycin has been tested in *TK2* H126N knock-in mouse, a disease model for Tk2 deficiency, reproducing human infantile encephalomyopathy due to mtDNA depletion syndrome [5]. The treatment was administered daily in the pre- and postnatal periods showing increased lifespan with no effect on the morbidity or on the reduced mtDNA copy number and OXPHOS activities. Tissue-specific and systemic changes were identified in the metabolic profile by metabolomic and transcriptomic analyses with specific regard to amino acids and lipid metabolism. In the Deletor mouse, carrying a dominant mutation in the mitochondrial helicase Twinkle, rapamycin downregulated several components of the mtISR, improving mitochondrial myopathies and preventing the pathogenic progression, without inducing mitochondrial biogenesis [83,91].

Based on primary indication as immunosuppressant and encouraging preliminary data in *in vivo* and *in vitro* models, rapamycin has been first tested for human use in mitochondrial disorders in four patients with MELAS/MIDD (Mitochondrial Encephalomyopathy Lactic acidosis and stroke-like episodes/ Maternal Inherited Diabetes and Deafness) syndrome who were undergoing kidney transplant. After screening the response to rapamycin of patients' fibroblasts cell lines, patients were switched from calcineurin inhibitors to mTOR inhibitors during post-transplant immunosuppression, reporting improvement of general asthenia and anorexia with weight gain. Together with the enhancement of general conditions, metabolic markers showed reduction of glucose serum 3-nitro-tyrosine, a marker of oxidative stress, and 1- and 3-methyl-histidine, a marker of skeletal muscle breakdown, together with an increase of free fatty acid levels, tricarboxylic acid intermediates and circulating glucose level [92]. Those results confirmed previous findings in the *Ndufs4*$^{-/-}$ model.

One additional infantile patient with early-onset MELAS and one pediatric patient with Leigh syndrome were treated under compassionate use with rapamycin reporting different results. In fact, while rapamycin was able to dramatically improve the clinical status in the

Leigh patient, the treatment was not able to modify the clinical course of the patients with MELAS disease, who died at 79 months [93]. The different outcome of the use of rapamycin can have different explanations: adult versus pediatric versus infantile patients might present different responses, with greater efficacy in adult patients; the timing of the clinical course can influence response to therapy, being more effective in patients treated at earlier stages; the genetic diagnosis and therefore the underlined mitochondrial disease pathway could also have a role. The latter is sustained by the different metabolic signature observed in the preclinical studies.

Currently, a Phase 2a, open-label study with ABI-009 (Nab-sirolimus) in patients with genetically confirmed Leigh or Leigh-like syndrome is ongoing with the aim to evaluate its safety, tolerability, and clinical activity (NCT03747328). Additional preclinical studies in *in vivo* models are necessary to potentially extend the treatment in other mitochondrial disorders including mtDNA maintenance disorders.

16.8 Disease-tailored therapies

16.8.1 Nucleos(t)ide supplementation therapies

Nucleoside or nucleotide supplementation has been demonstrated as safe and effective treatment to bypass enzyme deficiency or enhance the enzymatic activity in *in vitro* and *in vivo* models.

Preliminary studies showed that supplementation with dGuo alone in *DGUOK*-deficient human fibroblasts and addition of dCtd and tetrahydrouridine in both human TP-deficient cell and *Tymp/Upp1* double knockout mouse model of MNGIE disease were able to prevent mtDNA depletion [8]. Similarly, administration of pyrimidine and purine nucleosides rescued the level of mtDNA copy number in human fibroblasts deficient for RRM2B. Instead, their monophosphates form failed to correct depletion in RRM2B deficient human myoblast [9,10]. Nucleosides supplementation was tested in *Tk2* knock-in H126N and *Tk2*-KO mouse models for *Tk2* deficiency [94,95]. Although genetically different, the two models share the same biochemical, molecular genetics, and clinical phenotype being Tk2 dramatically reduced in the affected tissues. Molecular bypass therapy was first tested in the Tk2 H126N knock-in mouse model that reproduces the human infantile encephalomyopathy: mutant mice are born alive and they have a normal development up to postnatal day 8–10 when they present grow failure, coarse tremor, ataxia with rapid progression to death at day 12–14. Biochemically, they present lack of enzyme activities in all of the tissues, with brain and muscle being the most affected. Oral administration of dThd and dCtd monophosphates was able to ameliorate the phenotype with prolonged lifespan and delayed disease onset and to increase mtDNA level and mitochondrial respiratory chain enzyme activities. Efficacy was dose-related with a mean survival of 44 days with the highest dose of 400 mg/kg/day and no side effects were reported [6]. Because of rapid catabolism of monophosphates in the intestine and in the cytosol of the target tissues, nucleosides (dCtd + dThd) were the active products as demonstrated by oral administration of dCtd + dThd to Tk2 deficient mice [7]. Molecular bypass therapy and substrate enhancement therapies confirmed to be safe and effective in the *Tk2* KO model [11]. However, they do not represent a definitive cure. Indeed, in both mouse models, mtDNA depletion was present in all analyzed tissues at 29 days except for the small intestine. Bioavailability of compounds is limited by intrinsic and extrinsic factors explaining the reduced efficacy of nucleosides supplementation therapy [11,43]. Extrinsic factors are represented by treatment route and nucleosides availability. Plasma levels of nucleosides are higher when treatment is administered by intravenous (iv) injection followed by intraperitoneal injection and oral administration. In addition, tissue concentration is higher in visceral tissues than in brain, explaining the different levels of mtDNA copy number at earlier stage. Nevertheless, intraperitoneal (ip) administration of nucleosides did neither delay disease-onset nor improve survival time, indicating that the brain is responsible for disease progression.

Moreover, studies demonstrated that the bioavailability is higher for dThd than for dCtd, independently from the administration route, prompting the idea that the use of dThd alone could be effective to modify the clinical and molecular genetics defect. Administration of dThd was able to increase survival time and positively modify level of mtDNA copy number in brain, but had no effect on liver and muscle mtDNA level, confirming that dCtd is critical for liver and also muscle. Therefore both dThd and dCtd are necessary to achieve the greatest nucleosides supplementation response.

Administration of the same doses of dThd and dCtd resulted in much lower plasma levels of these compounds in 29-day-old mice than in 12–13-day-old littermate due to intrinsic factors represented by the tissue-specific level of anabolic (thymidine kinase 1 (Tk1) and deoxycytidine kinase (dCK)) versus catabolic enzymes (TP and cytidine deaminase (CDA)) activities. Tk1 and dCK activities resulted both downregulated during the development in all the tissues preventing these tissues from responding to dCtd + dThd therapy regardless of the dose or bioavailability of each nucleoside. Tk1 levels are very low in brain since postnatal day 4 and become undetectable at day 29 explaining the appearance of mtDNA depletion in brain since postnatal day 13 in the Tk2 deficient mice. Study of Tk1 activity in human muscle from 3 adult and 1 infant patients demonstrated that downregulation of Tk1 is milder in the human compared to mouse muscle and therefore the former is more likely to respond to dCtd + dThd therapy than the latter.

Contrarily to anabolic enzymes, TP and CDA activities resulted increased in small intestine and reduced in liver while constant in brain during the development of tk2 deficient mice. In addition, CDA activity was higher than TP. In fact, although similar concentrations of dCtd and dThd were reached in the plasma after parental treatment, levels of dThd in brain were 10-fold higher than those of dCtd.

The detailed intrinsic and extrinsic factors may be target of modification in order to increase the effect of nucleosides supplementation therapy and therefore they will be subject of further studies.

Treatment with nucleotide and nucleoside monophosphates has been already translated for human compassionate use under Food and Drug Administration (FDA) emergency Investigational New Drug (IND) and local ethic committee approval in 28 patients with TK2 deficiency. Sixteen of them were retrospectively analyzed by Dominguez-Gonzalez et al. [96]. Patients were treated with up to 300–400 mg/kg of dCtd + dThd for a period of 6–36 months. Five patients presented early-onset severe myopathy as defined by Garone et al. [42], while other 11 had slower progression. Treatment was able to modify the natural history of the disease and improve motor and respirator functions and the general conditions of the patients. The range of survival age was 0.17–0.45 years for early-onset Tk2 patients untreated versus 2.1–6.3 years of treated patients. Motor function improvement was greater for patients with more severe symptoms at the baseline as demonstrated by six minute walking test (6MWT) (mean increase of 171.9 m after 1 year of treatment). Two patients with early-onset severe myopathy and one childhood onset patient who had lost the ability to walk prior to treatment gained independent ambulation on treatment. Respiratory function remained stable or improved partially after 12 months of treatment. One patient weaned off mechanical ventilation within 18 months of treatment initiation. Growth curve improved in all patients. Patients discontinued enteric feeding. Growth differentiation factor 15 (GDF15), a mitochondrial biomarker, improved in all of the patients while creatine kinase (CK) level was normalized in patients with 5–10-fold higher level. Treatment was safe, with the only adverse effect observed in 8 patients being diarrhea, which was dose-dependent and transient in most cases and did not prompt suspension of treatment in any patients. One patient experienced mild to transient abdominal pain. Prior to treatment, one patient had elevated transaminases that were attributed to Tk2 deficiency and normalized after 1 year of treatment of dCtd + dThd. In addition to the 16 patients reported, 12 patients initiated dCtd + dThd. Two adults after 3–4 months of treatment developed increased transaminases and gamma-glutamyl transferase with normal bilirubin and alkaline phosphatase levels. In both cases, 3 months after discontinuing therapy, transaminases returned to normal. Two clinical trials have been approved for the analysis of the safety and efficacy of nucleosides supplementation therapy in Tk2 deficiency: (1) RETROspective study (NCT03701568) analyzing retrospective data of patients treated under compassionate use and (2) an Open-Label study of continuation treatment (NCT03845712) analyzing prospective data with good manufacturing practice (GMP) grade nucleosides products.

Mitochondrial nucleotide pool size influences mtDNA replication rate in physiological and pathological conditions. Recently, it has been demonstrated that Polg-deficient cell lines (patients' fibroblasts) supplemented with four nucleosides plus an inhibitor of dAdo catabolism, erythro-9-(2-hydroxy-3-nonyl) adenine, are able to recover from mtDNA depletion induced by ethidium bromide [97]. Therefore nucleosides supplementation may represent a therapeutic option in other mtDNA replication disorders due to defective proteins not primarily involved in mtDNA nucleotide pool balance.

16.8.2 Clearance of toxic metabolites

Lack of enzyme activities can cause toxic accumulations of metabolites that are responsible of the pathogenesis of the disease. In MNGIE, the excess of dThd and dU causes mitochondrial nucleotide pool unbalance and affect quantitatively and qualitatively the mtDNA during replication, causing mtDNA depletion, multiple deletions, and point mutations. Therefore treatment approaches in MNGIE disease were designed to target either the enzyme deficiency or the biochemical effects of toxic circulating nucleosides. Hemodialysis and peritoneal dialysis have been attempted to clear dThd and dU from blood in four and three patients, respectively, with confirmed MNGIE [12,13,15,17]. Both approaches resulted in transient and partial effect on the level of nucleosides indicating that only continuous elimination of nucleosides will lead to the permanent reduction of their levels. Dialysis is able to reduce up to 50% both plasma and urine levels of dThd and dU, but the biochemical effect lasts only hours in both cases. Long-term efficacy of hemodialysis was evaluated in a 29-year-old patient receiving the treatment for one year (3 times/week in the first 6 months, 4 times/week in the following six). A longitudinal biochemical and clinical follow-up demonstrated progression of the disease with no improvement in neurological or gastrointestinal symptoms [13]. Clinical benefit was instead proven with long-term administration of peritoneal dialysis in two patients [12].

16.8.3 Enzyme replacement

Enzyme replacement therapy represents an option for enzyme deficiency in the de novo or salvage pathway of nucleotide pool metabolism. Theoretically, homozygous mutations are linked to severe enzyme deficiency and severe phenotype while heterozygous mutations are able to encode between 35% and 50% of active enzyme with associated healthy carrier status or very mild phenotype in the majority of disorders. Therefore a partial restoration of enzyme activity should potentially correct the biochemical and metabolic defect in the pathway and consequently ameliorate clinical phenotype. The enzyme replacement therapy has been exploited in MNGIE disease with different approaches using platelet or hematopoietic stem cells or liver as a source of enzymes.

16.8.4 Platelet infusion

Platelet infusion produced the proof of principle for enzyme replacement therapy in two patients, of 23 and 16 years of age, respectively, with confirmed diagnosis of MNGIE. Multiple platelet infusions at increasing concentrations demonstrated that the level of TP activity and the related decline in the plasma and urine levels of dU and dThd were directly related to the number of platelet infused. The metabolic effect was transient with a preinfusion level of dThd and dU in biological fluids in less than one week. Therefore the treatment approach is able to restore the catabolism of dThd and dU, but repeated scheduled platelet infusions are required to sustain normal level of TP and observe a potential clinical efficacy. In addition, it is possible to anticipate a positive effect on mtDNA replication, but it is unlikely to reverse the presence of somatic mtDNA point mutations in postmitotic cells. Therefore the treatment should be initiated at earliest disease stage to prevent mitochondrial damage [14].

16.8.5 Hematopoietic stem cell transplantation

In order to restore enzymatic activity, hematopoietic stem cells can be used as source of enzyme for transplantation. In 2006 Hirano et al. performed for the first time reduced intensity allogeneic stem cell transplantations (alloSCTs) in two patients presenting MNGIE. The first patient was a 21-year-old man, with late stage of MNGIE disease. He was wheelchair-bound and required parenteral nutrition for gastrointestinal dysmotility and cachexia. Other than peripheral neuropathy, ptosis, and ophthalmoparesis, he presented hepatomegaly. For his medical conditions and the absence of a family donor, he received risk-adapted reduced-intensity alloSCT with human leukocyte antigen (HLA) 4/6 matched as a donor source. The patient had primary nonengraftment of donor cells with spontaneous autologous recovery and died 86 days after transplant from disease progression complicated by sepsis and respiratory failure. The second patient was a 30-year-old woman with prominent borborygmi, ptosis, mild ophthalmoparesis, impaired hearing, proximal limb weakness, stocking-glove sensory loss, and areflexia. She was treated with the same alloSCT protocol with her HLA 6/6 matched brother as donor but less aggressive preconditioning and posttransplant immunosuppressive protocol. She achieved mixed donor chimerism (26% at 5 month of follow-up) with partial restore of buffy coat TP activity (181 nmol/h/mg protein) and reduced level of circulating nucleosides (dThd undetectable, dU 0.2 μM). Clinically, improvement in abdominal pain, swallowing ability, and peripheral neuropathy with decrease numbness in her hands and feet and elicitable tendon reflex were observed at 6.5 months after the transplant [98].

Guideline for HSCT in MNGIE patients was established in a consensus conference in 2011. In order to maximize the benefit of the treatment and reduce the morbidity and mortality, it was recommended: (1) All patients should undergo pre- and post-transplant evaluation; (2) transplantation should be considered early in the disease for optimal recovery and minimum risk of complications; (3) HLA identical sibling should be the first choice as donors. If not available, 10/10 HLA matched allele unrelated donor; (4) bone marrow is the recommended source because of less recurrence of GVHD; (5) autologous backup of hematopoietic stem cells should be considered before the start of conditioning for rescue in case of rejection; (6) fludarabine and BU are the recommended drug for conditioning and CYA and methotrexate as immunosuppressive regimen [99].

Twenty-four patients were treated with a success rate of 37.5% (9/24 patients) and improvement in body mass index, gastrointestinal manifestations, and peripheral neuropathy in seven patients (29%) who were engrafted and living more than 2 years after transplantation. Survival was correlated to comorbidity risk factors such as the severity of liver and gastrointestinal symptoms prior to transplant, and to the careful selection of donor. Five patients who underwent unrelated cord blood transplant died confirming that HLA identical or 10/10 HLA antigen match should be the first choice as donor. Deaths were attributed to transplant-related causes in seven patients, transplant-related toxicity after second HSCT in two patients, and disease progression such as gastrointestinal complications (intestinal perforation, liver and pancreatic failure) or infection in six patients (pneumonia, septicemia) or multiorgan failure [16].

Combination of hematopoietic stem cell transplantation (HSCT) and other treatment approaches such as platelet infusion [18] or peritoneal dialysis [17] have been considered in anecdotal case reports as options to reduce the risk related to conditioning, to improve the clinical conditions of the patient prior to transplant and to sustain the TP activity thereafter.

A clinical trial is currently recruiting for evaluating the safety of allogeneic HSCT with stem cells from an individual, who is HLA 10/10 matched (NCT02427178).

16.8.6 Erythrocytes encapsulated thymidine phosphorylase

Erythrocytes encapsulated thymidine phosphorylase (EE-TP) therapy consists of ex vivo encapsulation of a highly purified recombinant *Escherichia coli* TP enzyme into patient's autologous erythrocytes. The rationale of the therapy is to use erythrocytes as a source of TP to replace the enzyme deficiency and clear patient plasma from toxic levels of circulating dThd and dU by catabolizing them in the cytosol of the erythrocytes. Thymidine and deoxyuridine are indeed able to pass across the erythrocytes membrane via ENT1, that is, the equilibrative nucleosides transporter. The concentrated level of recombinant TP catabolizes them to uracil and thymine, which erythrocytes release into circulation where they are further catabolized. The preparation of encapsulated erythrocytes requires a predetermined volume of blood removed from the patient. Under hypoosmotic conditions, the erythrocytes swell and form pores that permit the recombinant TP to enter the cytosol. When the iso-osmotic conditions are re-established, the erythrocytes' membrane pores close and TP remains encapsulated within the cells. The advantage of the therapy is potentially low immunization effect and prolonged circulating half-life of the enzyme [100–102].

In 2013 the first patient was treated with EE-TP under compassionate use and results from this case report represented the proof of principle of safety and efficacy of EE-TP for human use. She was a 28-year-old, presenting sensorimotor polyneuropathy, external ophthalmoplegia, minimum intestinal dysmotility, and difficulties in walking (1 km walking distance). Treatment was infused at escalating doses starting from 17 UI/kg up to 46 UI/kg for 27 months and it showed remarkable biochemical effect bringing the level of plasma dThd and dU from values of 20.5 and 30.6 mmol/L, respectively, to 8.1 and 12.6 mmol/L, and urine dThd and dU from values of 421 and 324 mmol/24 hours, respectively, to 192 and 282 mmol/24 hours for dThd and 0.0–184 mmol/24 hours for dU. Clinically, the patient presented a progressive improvement in gait and balance, sensory ataxia, and fine finger function, and body weight. The patient-reported outcomes after 23 months included being able to walk 10 km and climb stairs without assistance, tie shoelaces, and feel the sensation of sand on her feet when walking on beach. Numbness in hands and feet also resolved [103]. Treatment was safe with mild side effect such as erythema of the face and the neck and coughing occurring 5 minutes after infusion. Premedication with antihistaminic and glucocorticoid and the use of highly purified enzyme prevented further appearance of those symptoms. She was treated for additional 49 months together with other two patients in a dose escalating treatment study [104]. Major clinical improvement in muscle weakness and neuropathy in the first reported patient and minor changes in a second patients were reported confirming the potential efficacy of EE-TP.

After the initial experiment performed by using EE-TP produced by the investigator team, using sterile, single-use materials, and reagents throughout, under a special license held by St. George's Healthcare Trust Pharmacy according to the Rules and Guidance for Pharmaceutical Manufacturers 2007 (MHRA), within Class A isolators contained within a Pharmacy Clean room, the EE-TP therapy is now produced under Good Manual Practice using an automated red cell loader device [104,105]. A Phase 2, multicenter, multiple dose, open-label trial without a control (NCT03866954) has been approved to investigate the application of EE-TP as an enzyme replacement therapy for MNGIE. After the first screening phase of 30 days, clinical and biochemical parameters of the patients will be observed for 90 days for determining the severity of the patient's most disabling symptom, in order to use those parameters as a control to monitor efficacy of the therapy. Patients will start 4 cycles of therapy every 21 days with escalating doses until the metabolic correction is achieved. Then, they will enter in an open-label study where the treatment dose will be administrated every 2–4 weeks at flexible dose for 24 months followed by a 90 days follow-up dose. Primary endpoints are to determine the safety, tolerability, pharmacodynamics, and efficacy of multiple doses of EE-TP. The secondary endpoints are to assess EE-TP immunogenicity after multiple dose administrations and changes in clinical assessments, and the pharmacodynamics of EE-TP on clinical assessments [19].

16.8.7 Liver transplant: tissue-specific disorder and source of enzyme replacement

Defects in mtDNA maintenance can affect liver function with mild elevation of transaminases or severe hepatic insufficiency. In a cohort of 39 patients with acute liver failure and onset before 2 years of age, the incidence of a confirmed genetic diagnosis of mtDNA depletion syndrome was 17% [106]. A predominant liver dysfunction has been described in hepatocerebral syndromes due to MPV17, TWNK, TFAM defects, in infantile or childhood dGK deficiency and in childhood myocerebrohepatopathy spectrum (MCHS) and Alpers–Huttenlocher syndrome (AHS) due to mutations in *POLG* [44]. Cirrhosis has been also documented as a rare complication in a patient with MNGIE likely due to the accumulation of toxic intermediates in the liver. Therefore liver transplant has been considered as a therapeutic option in mitochondrial diseases, as a prevalent tissue specificity occurs in this organ, thereby leading to severe hepatocerebral disorders. Liver transplant is well tolerated in mitochondrial disorders with successful rate of 90% and it might be resolutive in case of single organ failure. However, it cannot prevent the further appearance of neurological symptoms and it should be considered with caution in disorder due to *POLG* gene defects [20–22]. Other than representing a treatment for a specific liver injury, liver can represent an important source of enzyme for replacement therapy as demonstrated for TP. TP values in liver are 0.5 ± 0.07 (range 0.5–0.75) ng/μg total protein with a widespread distribution in the hepatocytes (nuclei, cytoplasm) and in the sinusoidal lining cells. Those values are 2–3 times higher with respect to normal buffy coat and intestinal mucosa and 6 times than bone marrow [107].

Therefore considering the potential hepatic complication of MNGIE disease and the high success of liver transplant in neurometabolic disorders, liver transplant has been considered as a potential source for enzyme replacement therapy to prevent the nucleoside-induced injuries and treat the neuromuscular symptoms. The therapeutic option has been administrated in two patients with an immediate biochemical effect of normalization of nucleosides level in blood. Patients were young adults of 21 and 25 years of age, observed with follow-up period of 18 months and 90 days, respectively. Clinically, an improvement in neurological features and brain lactate at MRs was reported while there was no effect on the gastrointestinal symptoms that represent the more debilitating in MNGIE disease. The latter might be explained by the timing of the orthotopic liver transplant and the presence of previous damage [23,24].

Further studies in a large cohort of patients are required to confirm this therapeutic option.

16.8.8 Gene therapy

Adeno-associated or lentivirus-mediated gene therapies represent the most promising and definitive cure for rare genetic disorders.

Lentiviruses have been considered in MNGIE for stem cell transplantation with lentiviral transduction. Despite the great clinical and biochemical results, some concerns still remain regarding the safety for potential mutagenesis. Therefore further studies are required at pre-clinical level.

Preclinical studies have been performed with Adeno-associated viruses (AAV)-mediated or lentiviral gene therapies in different mouse models for rare genetic disorders including mtDNA maintenance diseases. Other than demonstrating their efficacy, pre-clinical studies had to address risk of mutagenesis, persistence in the cell, minimum effective dose, and their translation to human, pleiotropism or tissue specificity of the gene therapy. AAV gene therapy is currently considered the best option for the low risk of insertional mutagenesis, long-term persistence in cells and the promising results obtained in the first clinical trial on spinal muscular atrophy using AAV9, showing remarkable amelioration of the treated patients compared to the natural history of the disease [74,108,109]. Liver-targeted AAV (AAV2,8 vector) have been experimented in a Mpv17 knockout mouse model of mtDNA depletion and hepatocerebral syndrome [27]. By re-expressing the wild-type gene, the vector was able to rescue the mtDNA depletion and prevent liver steatosis induced by ketogenic diet in this model. A similar approach was used in $Tp^{-/-}/Upp^{-/-}$ mouse model for MNGIE disease [25,26,28,29].

Despite the highly promising results of the preclinical study, no clinical trials are currently scheduled due to the extremely high production costs, the regulatory requirements, and the rare frequency of those disorders.

Research perspectives

Treatment options are now available for mtDNA maintenance disorders at preclinical level or for human use under compassionate use or controlled clinical trials. Although they are able to modify the natural history of the disease, they are not a definitive cure. Further studies are necessary to understand their mechanism of action, to evaluate combined treatment approach (e.g., nucleosides clearance and enzyme replacement therapies or coupled enzyme replacement therapies in MNGIE), and to identify more effective and safe therapies.

Acknowledgments

Our work was supported by Rita Levi Montalcini Programma Rientro Cervelli, Italian Minister for Education and Research.

References

[1] Kelly RD, Mahmud A, McKenzie M, Trounce IA, St John JC. Mitochondrial DNA copy number is regulated in a tissue specific manner by DNA methylation of the nuclear-encoded DNA polymerase gamma A. Nucleic Acids Res 2012;40(20):10124–38. Available from: https://doi.org/10.1093/nar/gks770.

[2] Lee WT, St John J. The control of mitochondrial DNA replication during development and tumorigenesis. Ann N Y Acad Sci 2015;1350:95–106. Available from: https://doi.org/10.1111/nyas.12873.

[3] Gustafsson CM, Falkenberg M, Larsson NG. Maintenance and expression of mammalian mitochondrial DNA. Annu Rev Biochem 2016;85:133–60. Available from: https://doi.org/10.1146/annurev-biochem-060815-014402.

[4] Jing R, Corbett JL, Cai J, Beeson GC, Beeson CC, Chan SS, et al. A screen using iPSC-derived hepatocytes reveals NAD(+) as a potential treatment for mtDNA depletion syndrome. Cell Rep 2018;25 (6):1469–1484 e1465. Available from: https://doi.org/10.1016/j.celrep.2018.10.036.

[5] Siegmund SE, Yang H, Sharma R, Javors M, Skinner O, Mootha V, et al. Low-dose rapamycin extends lifespan in a mouse model of mtDNA depletion syndrome. Hum Mol Genet 2017;26(23):4588–605. Available from: https://doi.org/10.1093/hmg/ddx341.

[6] Garone C, Garcia-Diaz B, Emmanuele V, Lopez LC, Tadesse S, Akman HO, et al. Deoxypyrimidine monophosphate bypass therapy for thymidine kinase 2 deficiency. EMBO Mol Med 2014;6(8):1016–27. Available from: https://doi.org/10.15252/emmm.201404092.

[7] Lopez-Gomez C, Levy RJ, Sanchez-Quintero MJ, Juanola-Falgarona M, Barca E, Garcia-Diaz B, et al. Deoxycytidine and deoxythymidine treatment for thymidine kinase 2 deficiency. Ann Neurol 2017;81 (5):641–52. Available from: https://doi.org/10.1002/ana.24922.

[8] Camara Y, Gonzalez-Vioque E, Scarpelli M, Torres-Torronteras J, Caballero A, Hirano M, et al. Administration of deoxyribonucleosides or inhibition of their catabolism as a pharmacological approach for mitochondrial DNA depletion syndrome. Hum Mol Genet 2014;23(9):2459–67. Available from: https://doi.org/10.1093/hmg/ddt641.

[9] Bulst S, Holinski-Feder E, Payne B, Abicht A, Krause S, Lochmuller H, et al. In vitro supplementation with deoxynucleoside monophosphates rescues mitochondrial DNA depletion. Mol Genet Metab 2012;107 (1-2):95–103. Available from: https://doi.org/10.1016/j.ymgme.2012.04.022.

[10] Pontarin G, Ferraro P, Bee L, Reichard P, Bianchi V. Mammalian ribonucleotide reductase subunit p53R2 is required for mitochondrial DNA replication and DNA repair in quiescent cells. Proc Natl Acad Sci U S A 2012;109(33):13302–7. Available from: https://doi.org/10.1073/pnas.1211289109.

[11] Blazquez-Bermejo C, Molina-Granada D, Vila-Julia F, Jimenez-Heis D, Zhou X, Torres-Torronteras J, et al. Age-related metabolic changes limit efficacy of deoxynucleoside-based therapy in thymidine kinase 2-deficient mice. EBioMedicine 2019;46:342–55. Available from: https://doi.org/10.1016/j.ebiom.2019.07.042.

[12] Yavuz H, Ozel A, Christensen M, Christensen E, Schwartz M, Elmaci M, et al. Treatment of mitochondrial neurogastrointestinal encephalomyopathy with dialysis. Arch Neurol 2007;64(3):435–8. Available from: https://doi.org/10.1001/archneur.64.3.435.

[13] Roeben B, Marquetand J, Bender B, Billing H, Haack TB, Sanchez-Albisua I, et al. Hemodialysis in MNGIE transiently reduces serum and urine levels of thymidine and deoxyuridine, but not CSF levels and neurological function. Orphanet J Rare Dis 2017;12(1):135. Available from: https://doi.org/10.1186/s13023-017-0687-0.

[14] Lara MC, Weiss B, Illa I, Madoz P, Massuet L, Andreu AL, et al. Infusion of platelets transiently reduces nucleoside overload in MNGIE. Neurology 2006;67(8):1461–3. Available from: https://doi.org/10.1212/01.wnl.0000239824.95411.52.

[15] Spinazzola A, Marti R, Nishino I, Andreu AL, Naini A, Tadesse S, et al. Altered thymidine metabolism due to defects of thymidine phosphorylase. J Biol Chem 2002;277(6):4128–33. Available from: https://doi.org/10.1074/jbc.M111028200.

[16] Halter JP, Michael W, Schupbach M, Mandel H, Casali C, Orchard K, et al. Allogeneic haematopoietic stem cell transplantation for mitochondrial neurogastrointestinal encephalomyopathy. Brain 2015;138(Pt 10):2847–58. Available from: https://doi.org/10.1093/brain/awv226.

[17] Ariaudo C, Daidola G, Ferrero B, Guarena C, Burdese M, Segoloni GP, et al. Mitochondrial neurogastrointestinal encephalomyopathy treated with peritoneal dialysis and bone marrow transplantation. J Nephrol 2015;28(1):125–7. Available from: https://doi.org/10.1007/s40620-014-0069-9.

[18] Hussein E. Non-myeloablative bone marrow transplant and platelet infusion can transiently improve the clinical outcome of mitochondrial neurogastrointestinal encephalopathy: a case report. Transfus Apher Sci 2013;49(2):208–11. Available from: https://doi.org/10.1016/j.transci.2013.01.014.

[19] Bax BE, Levene M, Bain MD, Fairbanks LD, Filosto M, Kalkan Ucar S, et al. Erythrocyte encapsulated thymidine phosphorylase for the treatment of patients with mitochondrial neurogastrointestinal encephalomyopathy: study protocol for a multi-centre, multiple dose, open label trial. J Clin Med 2019;8(8). Available from: https://doi.org/10.3390/jcm8081096.

[20] Grabhorn E, Tsiakas K, Herden U, Fischer L, Freisinger P, Marquardt T, et al. Long-term outcomes after liver transplantation for deoxyguanosine kinase deficiency: a single-center experience and a review of the literature. Liver Transpl 2014;20(4):464–72. Available from: https://doi.org/10.1002/lt.23830.

[21] Hynynen J, Komulainen T, Tukiainen E, Nordin A, Arola J, Kalviainen R, et al. Acute liver failure after valproate exposure in patients with POLG1 mutations and the prognosis after liver transplantation. Liver Transpl 2014;20(11):1402–12. Available from: https://doi.org/10.1002/lt.23965.

[22] Parikh S, Karaa A, Goldstein A, Ng YS, Gorman G, Feigenbaum A, et al. Solid organ transplantation in primary mitochondrial disease: proceed with caution. Mol Genet Metab 2016;118(3):178–84. Available from: https://doi.org/10.1016/j.ymgme.2016.04.009.

[23] De Giorgio R, Pironi L, Rinaldi R, Boschetti E, Caporali L, Capristo M, et al. Liver transplantation for mitochondrial neurogastrointestinal encephalomyopathy. Ann Neurol 2016;80(3):448–55. Available from: https://doi.org/10.1002/ana.24724.

[24] D'Angelo R, Rinaldi R, Pironi L, Dotti MT, Pinna AD, Boschetti E, et al. Liver transplant reverses biochemical imbalance in mitochondrial neurogastrointestinal encephalomyopathy. Mitochondrion 2017;34:101–2. Available from: https://doi.org/10.1016/j.mito.2017.02.006.

[25] Torres-Torronteras J, Gomez A, Eixarch H, Palenzuela L, Pizzorno G, Hirano M, et al. Hematopoietic gene therapy restores thymidine phosphorylase activity in a cell culture and a murine model of MNGIE. Gene Ther 2011;18(8):795–806. Available from: https://doi.org/10.1038/gt.2011.24.

[26] Torres-Torronteras J, Viscomi C, Cabrera-Perez R, Camara Y, Di Meo I, Barquinero J, et al. Gene therapy using a liver-targeted AAV vector restores nucleoside and nucleotide homeostasis in a murine model of MNGIE. Mol Ther 2014;22(5):901–7. Available from: https://doi.org/10.1038/mt.2014.6.

[27] Bottani E, Giordano C, Civiletto G, Di Meo I, Auricchio A, Ciusani E, et al. AAV-mediated liver-specific MPV17 expression restores mtDNA levels and prevents diet-induced liver failure. Mol Ther 2014;22(1):10–17. Available from: https://doi.org/10.1038/mt.2013.230.

[28] Torres-Torronteras J, Cabrera-Perez R, Barba I, Costa C, de Luna N, Andreu AL, et al. Long-term restoration of thymidine phosphorylase function and nucleoside homeostasis using hematopoietic gene therapy in a murine model of mitochondrial neurogastrointestinal encephalomyopathy. Hum Gene Ther 2016;27(9):656–67. Available from: https://doi.org/10.1089/hum.2015.160.

[29] Yadak R, Cabrera-Perez R, Torres-Torronteras J, Bugiani M, Haeck JC, Huston MW, et al. Preclinical efficacy and safety evaluation of hematopoietic stem cell gene therapy in a mouse model of MNGIE. Mol Ther Methods Clin Dev 2018;8:152–65. Available from: https://doi.org/10.1016/j.omtm.2018.01.001.

[30] Gorman GS, Chinnery PF, DiMauro S, Hirano M, Koga Y, McFarland R, et al. Mitochondrial diseases. Nat Rev Dis Prim 2016;2:16080. Available from: https://doi.org/10.1038/nrdp.2016.80.

[31] Stumpf JD, Saneto RP, Copeland WC. Clinical and molecular features of POLG-related mitochondrial disease. Cold Spring Harb Perspect Biol 2013;5(4):a011395. Available from: https://doi.org/10.1101/cshperspect.a011395.

[32] Varma H, Faust PL, Iglesias AD, Lagana SM, Wou K, Hirano M, et al. Whole exome sequencing identifies a homozygous POLG2 missense variant in an infant with fulminant hepatic failure and mitochondrial DNA depletion. Eur J Med Genet 2016;59(10):540–5. Available from: https://doi.org/10.1016/j.ejmg.2016.08.012.

[33] Viscomi C, Zeviani M. MtDNA-maintenance defects: syndromes and genes. J Inherit Metab Dis 2017;40(4):587–99. Available from: https://doi.org/10.1007/s10545-017-0027-5.

[34] Kornblum C, Nicholls TJ, Haack TB, Scholer S, Peeva V, Danhauser K, et al. Loss-of-function mutations in MGME1 impair mtDNA replication and cause multisystemic mitochondrial disease. Nat Genet 2013;45(2):214–19. Available from: https://doi.org/10.1038/ng.2501.

[35] Reyes A, Melchionda L, Nasca A, Carrara F, Lamantea E, Zanolini A, et al. RNASEH1 mutations impair mtDNA replication and cause adult-onset mitochondrial encephalomyopathy. Am J Hum Genet 2015;97(1):186–93. Available from: https://doi.org/10.1016/j.ajhg.2015.05.013.

[36] Ronchi D, Di Fonzo A, Lin W, Bordoni A, Liu C, Fassone E, et al. Mutations in DNA2 link progressive myopathy to mitochondrial DNA instability. Am J Hum Genet 2013;92(2):293–300. Available from: https://doi.org/10.1016/j.ajhg.2012.12.014.

[37] Shaheen R, Faqeih E, Ansari S, Abdel-Salam G, Al-Hassnan ZN, Al-Shidi T, et al. Genomic analysis of primordial dwarfism reveals novel disease genes. Genome Res 2014;24(2):291–9. Available from: https://doi.org/10.1101/gr.160572.113.

[38] Del Dotto V, Ullah F, Di Meo I, Magini P, Gusic M, Maresca A, et al. SSBP1 mutations cause mtDNA depletion underlying a complex optic atrophy disorder. J Clin Invest 2019. Available from: https://doi.org/10.1172/JCI128514.

[39] Miralles Fuste J, Shi Y, Wanrooij S, Zhu X, Jemt E, Persson O, et al. In vivo occupancy of mitochondrial single-stranded DNA binding protein supports the strand displacement mode of DNA replication. PLoS Genet 2014;10(12):e1004832. Available from: https://doi.org/10.1371/journal.pgen.1004832.

[40] Piro-Megy C, Sarzi E, Tarres-Sole A, Pequignot M, Hensen F, Quiles M, et al. Dominant mutations in mtDNA maintenance gene SSBP1 cause optic atrophy and foveopathy. J Clin Invest 2019. Available from: https://doi.org/10.1172/JCI128513.

[41] Saada A. Fishing in the (deoxyribonucleotide) pool. Biochem J 2009;422(3):e3–6. Available from: https://doi.org/10.1042/BJ20091194.

[42] Garone C, Taylor RW, Nascimento A, Poulton J, Fratter C, Dominguez-Gonzalez C, et al. Retrospective natural history of thymidine kinase 2 deficiency. J Med Genet 2018;55(8):515–21. Available from: https://doi.org/10.1136/jmedgenet-2017-105012.

[43] Lopez-Gomez C, Hewan H, Sierra C, Akman HO, Sanchez-Quintero MJ, Juanola-Falgarona M, et al. Bioavailability and cytosolic kinases modulate response to deoxynucleoside therapy in TK2 deficiency. EBioMedicine 2019;46:356–67. Available from: https://doi.org/10.1016/j.ebiom.2019.07.037.

[44] Almannai M, El-Hattab AW, Scaglia F. Mitochondrial DNA replication: clinical syndromes. Essays Biochem 2018;62(3):297–308. Available from: https://doi.org/10.1042/EBC20170101.

[45] Gonzalez-Vioque E, Torres-Torronteras J, Andreu AL, Marti R. Limited dCTP availability accounts for mitochondrial DNA depletion in mitochondrial neurogastrointestinal encephalomyopathy (MNGIE). PLoS Genet 2011;7(3):e1002035. Available from: https://doi.org/10.1371/journal.pgen.1002035.

[46] Garone C, Tadesse S, Hirano M. Clinical and genetic spectrum of mitochondrial neurogastrointestinal encephalomyopathy. Brain 2011;134(Pt 11):3326–32. Available from: https://doi.org/10.1093/brain/awr245.

[47] Huang SH, Tang A, Drisco B, Zhang SQ, Seeger R, Li C, et al. Human dTMP kinase: gene expression and enzymatic activity coinciding with cell cycle progression and cell growth. DNA Cell Biol 1994;13(5):461–71. Available from: https://doi.org/10.1089/dna.1994.13.461.

[48] Lam CW, Yeung WL, Ling TK, Wong KC, Law CY. Deoxythymidylate kinase, DTYMK, is a novel gene for mitochondrial DNA depletion syndrome. Clin Chim Acta 2019;496:93–9. Available from: https://doi.org/10.1016/j.cca.2019.06.028.

[49] Tilokani L, Nagashima S, Paupe V, Prudent J. Mitochondrial dynamics: overview of molecular mechanisms. Essays Biochem 2018;62(3):341–60. Available from: https://doi.org/10.1042/EBC20170104.

[50] Yu-Wai-Man P, Griffiths PG, Gorman GS, Lourenco CM, Wright AF, Auer-Grumbach M, et al. Multi-system neurological disease is common in patients with OPA1 mutations. Brain 2010;133(Pt 3):771–86. Available from: https://doi.org/10.1093/brain/awq007.

[51] Le Roux B, Lenaers G, Zanlonghi X, Amati-Bonneau P, Chabrun F, Foulonneau T, et al. OPA1: 516 unique variants and 831 patients registered in an updated centralized Variome database. Orphanet J Rare Dis 2019;14(1):214. Available from: https://doi.org/10.1186/s13023-019-1187-1.

[52] Spiegel R, Saada A, Flannery PJ, Burte F, Soiferman D, Khayat M, et al. Fatal infantile mitochondrial encephalomyopathy, hypertrophic cardiomyopathy and optic atrophy associated with a homozygous OPA1 mutation. J Med Genet 2016;53(2):127–31. Available from: https://doi.org/10.1136/jmedgenet-2015-103361.

[53] Schaaf CP, Blazo M, Lewis RA, Tonini RE, Takei H, Wang J, et al. Early-onset severe neuromuscular phenotype associated with compound heterozygosity for OPA1 mutations. Mol Genet Metab 2011;103(4):383–7. Available from: https://doi.org/10.1016/j.ymgme.2011.04.018.

[54] Bonneau D, Colin E, Oca F, Ferre M, Chevrollier A, Gueguen N, et al. Early-onset Behr syndrome due to compound heterozygous mutations in OPA1. Brain 2014;137(Pt 10):e301. Available from: https://doi.org/10.1093/brain/awu184.

[55] Carelli V, Sabatelli M, Carrozzo R, Rizza T, Schimpf S, Wissinger B, et al. 'Behr syndrome' with OPA1 compound heterozygote mutations. Brain 2015;138(Pt 1):e321. Available from: https://doi.org/10.1093/brain/awu234.

[56] Verny C, Loiseau D, Scherer C, Lejeune P, Chevrollier A, Gueguen N, et al. Multiple sclerosis-like disorder in OPA1-related autosomal dominant optic atrophy. Neurology 2008;70(13 Pt 2):1152–3. Available from: https://doi.org/10.1212/01.wnl.0000289194.89359.a1.

[57] Carelli V, Musumeci O, Caporali L, Zanna C, La Morgia C, Del Dotto V, et al. Syndromic parkinsonism and dementia associated with OPA1 missense mutations. Ann Neurol 2015;78(1):21–38. Available from: https://doi.org/10.1002/ana.24410.

[58] Lynch DS, Loh SHY, Harley J, Noyce AJ, Martins LM, Wood NW, et al. Nonsyndromic Parkinson disease in a family with autosomal dominant optic atrophy due to OPA1 mutations. Neurol Genet 2017;3(5):e188. Available from: https://doi.org/10.1212/NXG.0000000000000188.

[59] Vanstone JR, Smith AM, McBride S, Naas T, Holcik M, Antoun G, et al. DNM1L-related mitochondrial fission defect presenting as refractory epilepsy. Eur J Hum Genet 2016;24(7):1084–8. Available from: https://doi.org/10.1038/ejhg.2015.243.

[60] Yoon G, Malam Z, Paton T, Marshall CR, Hyatt E, Ivakine Z, et al. Lethal disorder of mitochondrial fission caused by mutations in DNM1L. J Pediatr 2016;171:313–16. Available from: https://doi.org/10.1016/j.jpeds.2015.12.060 e311-312.

[61] Stuppia G, Rizzo F, Riboldi G, Del Bo R, Nizzardo M, Simone C, et al. MFN2-related neuropathies: clinical features, molecular pathogenesis and therapeutic perspectives. J Neurol Sci 2015;356(1-2):7–18. Available from: https://doi.org/10.1016/j.jns.2015.05.033.

[62] El-Hattab AW, Dai H, Almannai M, Wang J, Faqeih EA, Al Asmari A, et al. Molecular and clinical spectra of FBXL4 deficiency. Hum Mutat 2017;38(12):1649–59. Available from: https://doi.org/10.1002/humu.23341.

[63] Ballout RA, Al Alam C, Bonnen PE, Huemer M, El-Hattab AW, Shbarou R. FBXL4-related mitochondrial DNA depletion syndrome 13 (MTDPS13): a case report with a comprehensive mutation review. Front Genet 2019;10:39. Available from: https://doi.org/10.3389/fgene.2019.00039.

[64] Donkervoort S, Sabouny R, Yun P, Gauquelin L, Chao KR, Hu Y, et al. MSTO1 mutations cause mtDNA depletion, manifesting as muscular dystrophy with cerebellar involvement. Acta Neuropathol 2019;138(6):1013–31. Available from: https://doi.org/10.1007/s00401-019-02059-z.

[65] Ardicli D, Sarkozy A, Zaharieva I, Deshpande C, Bodi I, Siddiqui A, et al. A novel case of MSTO1 gene related congenital muscular dystrophy with progressive neurological involvement. Neuromuscul Disord 2019;29(6):448–55. Available from: https://doi.org/10.1016/j.nmd.2019.03.011.
[66] Gal A, Balicza P, Weaver D, Naghdi S, Joseph SK, Varnai P, et al. MSTO1 is a cytoplasmic pro-mitochondrial fusion protein, whose mutation induces myopathy and ataxia in humans. EMBO Mol Med 2017;9(7):967–84. Available from: https://doi.org/10.15252/emmm.201607058.
[67] Li K, Jin R, Wu X. Whole-exome sequencing identifies rare compound heterozygous mutations in the MSTO1 gene associated with cerebellar ataxia and myopathy. Eur J Med Genet 2019;103623. Available from: https://doi.org/10.1016/j.ejmg.2019.01.013.
[68] Nasca A, Scotton C, Zaharieva I, Neri M, Selvatici R, Magnusson OT, et al. Recessive mutations in MSTO1 cause mitochondrial dynamics impairment, leading to myopathy and ataxia. Hum Mutat 2017;38 (8):970–7. Available from: https://doi.org/10.1002/humu.23262.
[69] Stiles AR, Simon MT, Stover A, Eftekharian S, Khanlou N, Wang HL, et al. Mutations in TFAM, encoding mitochondrial transcription factor A, cause neonatal liver failure associated with mtDNA depletion. Mol Genet Metab 2016;119(1-2):91–9. Available from: https://doi.org/10.1016/j.ymgme.2016.07.001.
[70] Cooper HM, Yang Y, Ylikallio E, Khairullin R, Woldegebriel R, Lin KL, et al. ATPase-deficient mitochondrial inner membrane protein ATAD3A disturbs mitochondrial dynamics in dominant hereditary spastic paraplegia. Hum Mol Genet 2017;26(8):1432–43. Available from: https://doi.org/10.1093/hmg/ddx042.
[71] Desai R, Frazier AE, Durigon R, Patel H, Jones AW, Dalla Rosa I, et al. ATAD3 gene cluster deletions cause cerebellar dysfunction associated with altered mitochondrial DNA and cholesterol metabolism. Brain 2017;140(6):1595–610. Available from: https://doi.org/10.1093/brain/awx094.
[72] Harel T, Yoon WH, Garone C, Gu S, Coban-Akdemir Z, Eldomery MK, et al. Recurrent de novo and biallelic variation of ATAD3A, encoding a mitochondrial membrane protein, results in distinct neurological syndromes. Am J Hum Genet 2016;99(4):831–45. Available from: https://doi.org/10.1016/j.ajhg.2016.08.007.
[73] Manning BD. Game of TOR—the target of rapamycin rules four kingdoms. N Engl J Med 2017;377 (13):1297–9. Available from: https://doi.org/10.1056/NEJMcibr1709384.
[74] Garone C, Viscomi C. Towards a therapy for mitochondrial disease: an update. Biochem Soc Trans 2018;46(5):1247–61. Available from: https://doi.org/10.1042/BST20180134.
[75] Vega RB, Huss JM, Kelly DP. The coactivator PGC-1 cooperates with peroxisome proliferator-activated receptor alpha in transcriptional control of nuclear genes encoding mitochondrial fatty acid oxidation enzymes. Mol Cell Biol 2000;20(5):1868–76. Available from: https://doi.org/10.1128/mcb.20.5.1868-1876.2000.
[76] Wu Z, Puigserver P, Andersson U, Zhang C, Adelmant G, Mootha V, et al. Mechanisms controlling mitochondrial biogenesis and respiration through the thermogenic coactivator PGC-1. Cell 1999;98 (1):115–24. Available from: https://doi.org/10.1016/S0092-8674(00)80611-X.
[77] Schreiber SN, Knutti D, Brogli K, Uhlmann T, Kralli A. The transcriptional coactivator PGC-1 regulates the expression and activity of the orphan nuclear receptor estrogen-related receptor alpha (ERRalpha). J Biol Chem 2003;278(11):9013–18. Available from: https://doi.org/10.1074/jbc.M212923200.
[78] Mouchiroud L, Houtkooper RH, Moullan N, Katsyuba E, Ryu D, Canto C, et al. The NAD(+)/sirtuin pathway modulates longevity through activation of mitochondrial UPR and FOXO Signaling. Cell 2013;154 (2):430–41. Available from: https://doi.org/10.1016/j.cell.2013.06.016.
[79] Nemoto S, Fergusson MM, Finkel T. SIRT1 functionally interacts with the metabolic regulator and transcriptional coactivator PGC-1{alpha}. J Biol Chem 2005;280(16):16456–60. Available from: https://doi.org/10.1074/jbc.M501485200.
[80] Conze DB, Crespo-Barreto J, Kruger CL. Safety assessment of nicotinamide riboside, a form of vitamin B3. Hum Exp Toxicol 2016;35(11):1149–60. Available from: https://doi.org/10.1177/0960327115626254.

[81] Billington RA, Travelli C, Ercolano E, Galli U, Roman CB, Grolla AA, et al. Characterization of NAD uptake in mammalian cells. J Biol Chem 2008;283(10):6367–74. Available from: https://doi.org/10.1074/jbc.M706204200.
[82] Trammell SA, Schmidt MS, Weidemann BJ, Redpath P, Jaksch F, Dellinger RW, et al. Nicotinamide riboside is uniquely and orally bioavailable in mice and humans. Nat Commun 2016;7:12948. Available from: https://doi.org/10.1038/ncomms12948.
[83] Khan NA, Nikkanen J, Yatsuga S, Jackson C, Wang L, Pradhan S, et al. mTORC1 regulates mitochondrial integrated stress response and mitochondrial myopathy progression. Cell Metab 2017;26(2):419–428 e415. Available from: https://doi.org/10.1016/j.cmet.2017.07.007.
[84] Spector R, Johanson CE. Vitamin transport and homeostasis in mammalian brain: focus on vitamins B and E. J Neurochem 2007;103(2):425–38. Available from: https://doi.org/10.1111/j.1471-4159.2007.04773.x.
[85] Cerutti R, Pirinen E, Lamperti C, Marchet S, Sauve AA, Li W, et al. NAD(+)-dependent activation of Sirt1 corrects the phenotype in a mouse model of mitochondrial disease. Cell Metab 2014;19(6):1042–9. Available from: https://doi.org/10.1016/j.cmet.2014.04.001.
[86] Civiletto G, Dogan SA, Cerutti R, Fagiolari G, Moggio M, Lamperti C, et al. Rapamycin rescues mitochondrial myopathy via coordinated activation of autophagy and lysosomal biogenesis. EMBO Mol Med 2018;10(11). Available from: https://doi.org/10.15252/emmm.201708799.
[87] Dai Y, Zheng K, Clark J, Swerdlow RH, Pulst SM, Sutton JP, et al. Rapamycin drives selection against a pathogenic heteroplasmic mitochondrial DNA mutation. Hum Mol Genet 2014;23(3):637–47. Available from: https://doi.org/10.1093/hmg/ddt450.
[88] Felici R, Buonvicino D, Muzzi M, Cavone L, Guasti D, Lapucci A, et al. Post onset, oral rapamycin treatment delays development of mitochondrial encephalopathy only at supramaximal doses. Neuropharmacology 2017;117:74–84. Available from: https://doi.org/10.1016/j.neuropharm.2017.01.039.
[89] Johnson SC, Yanos ME, Kayser EB, Quintana A, Sangesland M, Castanza A, et al. mTOR inhibition alleviates mitochondrial disease in a mouse model of Leigh syndrome. Science 2013;342(6165):1524–8. Available from: https://doi.org/10.1126/science.1244360.
[90] Peng M, Ostrovsky J, Kwon YJ, Polyak E, Licata J, Tsukikawa M, et al. Inhibiting cytosolic translation and autophagy improves health in mitochondrial disease. Hum Mol Genet 2015;24(17):4829–47. Available from: https://doi.org/10.1093/hmg/ddv207.
[91] Suomalainen A, Battersby BJ. Mitochondrial diseases: the contribution of organelle stress responses to pathology. Nat Rev Mol Cell Biol 2018;19(2):77–92. Available from: https://doi.org/10.1038/nrm.2017.66.
[92] Johnson SC, Martinez F, Bitto A, Gonzalez B, Tazaerslan C, Cohen C, et al. mTOR inhibitors may benefit kidney transplant recipients with mitochondrial diseases. Kidney Int 2019;95(2):455–66. Available from: https://doi.org/10.1016/j.kint.2018.08.038.
[93] Sage-Schwaede A, Engelstad K, Salazar R, Curcio A, Khandji A, Garvin Jr. JH, et al. Exploring mTOR inhibition as treatment for mitochondrial disease. Ann Clin Transl Neurol 2019;6(9):1877–81. Available from: https://doi.org/10.1002/acn3.50846.
[94] Akman HO, Dorado B, Lopez LC, Garcia-Cazorla A, Vila MR, Tanabe LM, et al. Thymidine kinase 2 (H126N) knockin mice show the essential role of balanced deoxynucleotide pools for mitochondrial DNA maintenance. Hum Mol Genet 2008;17(16):2433–40. Available from: https://doi.org/10.1093/hmg/ddn143.
[95] Bartesaghi S, Betts-Henderson J, Cain K, Dinsdale D, Zhou X, Karlsson A, et al. Loss of thymidine kinase 2 alters neuronal bioenergetics and leads to neurodegeneration. Hum Mol Genet 2010;19(9):1669–77. Available from: https://doi.org/10.1093/hmg/ddq043.
[96] Dominguez-Gonzalez C, Madruga-Garrido M, Mavillard F, Garone C, Aguirre-Rodriguez FJ, Donati MA, et al. Deoxynucleoside therapy for thymidine kinase 2-deficient myopathy. Ann Neurol 2019;86(2):293–303. Available from: https://doi.org/10.1002/ana.25506.
[97] Blazquez-Bermejo C, Carreno-Gago L, Molina-Granada D, Aguirre J, Ramon J, Torres-Torronteras J, et al. Increased dNTP pools rescue mtDNA depletion in human POLG-deficient fibroblasts. FASEB J 2019;33(6):7168–79. Available from: https://doi.org/10.1096/fj.201801591R.

[98] Hirano M, Marti R, Casali C, Tadesse S, Uldrick T, Fine B, et al. Allogeneic stem cell transplantation corrects biochemical derangements in MNGIE. Neurology 2006;67(8):1458–60. Available from: https://doi.org/10.1212/01.wnl.0000240853.97716.24.
[99] Halter J, Schupbach W, Casali C, Elhasid R, Fay K, Hammans S, et al. Allogeneic hematopoietic SCT as treatment option for patients with mitochondrial neurogastrointestinal encephalomyopathy (MNGIE): a consensus conference proposal for a standardized approach. Bone Marrow Transpl 2011;46(3):330–7. Available from: https://doi.org/10.1038/bmt.2010.100.
[100] Godfrin Y, Horand F, Franco R, Dufour E, Kosenko E, Bax BE, et al. International seminar on the red blood cells as vehicles for drugs. Expert Opin Biol Ther 2012;12(1):127–33. Available from: https://doi.org/10.1517/14712598.2012.631909.
[101] Levene M, Coleman DG, Kilpatrick HC, Fairbanks LD, Gangadharan B, Gasson C, et al. Preclinical toxicity evaluation of erythrocyte-encapsulated thymidine phosphorylase in BALB/c mice and beagle dogs: an enzyme-replacement therapy for mitochondrial neurogastrointestinal encephalomyopathy. Toxicol Sci 2013;131(1):311–24. Available from: https://doi.org/10.1093/toxsci/kfs278.
[102] Moran NF, Bain MD, Muqit MM, Bax BE. Carrier erythrocyte entrapped thymidine phosphorylase therapy for MNGIE. Neurology 2008;71(9):686–8. Available from: https://doi.org/10.1212/01.wnl.0000324602.97205.ab.
[103] Bax BE, Bain MD, Scarpelli M, Filosto M, Tonin P, Moran N. Clinical and biochemical improvements in a patient with MNGIE following enzyme replacement. Neurology 2013;81(14):1269–71. Available from: https://doi.org/10.1212/WNL.0b013e3182a6cb4b.
[104] Levene M, Bain MD, Moran NF, Nirmalananthan N, Poulton J, Scarpelli M, et al. Safety and efficacy of erythrocyte encapsulated thymidine phosphorylase in mitochondrial neurogastrointestinal encephalomyopathy. J Clin Med 2019;8(4). Available from: https://doi.org/10.3390/jcm8040457.
[105] Magnani M, Rossi L, D'Ascenzo M, Panzani I, Bigi L, Zanella A. Erythrocyte engineering for drug delivery and targeting. Biotechnol Appl Biochem 1998;28(1):1–6 Retrieved from: https://www.ncbi.nlm.nih.gov/pubmed/9693082.
[106] McKiernan P, Ball S, Santra S, Foster K, Fratter C, Poulton J, et al. Incidence of primary mitochondrial disease in children younger than 2 years presenting with acute liver failure. J Pediatr Gastroenterol Nutr 2016;63(6):592–7. Available from: https://doi.org/10.1097/MPG.0000000000001345.
[107] Boschetti E, D'Alessandro R, Bianco F, Carelli V, Cenacchi G, Pinna AD, et al. Liver as a source for thymidine phosphorylase replacement in mitochondrial neurogastrointestinal encephalomyopathy. PLoS One 2014;9(5):e96692. Available from: https://doi.org/10.1371/journal.pone.0096692.
[108] Hirano M, Emmanuele V, Quinzii CM. Emerging therapies for mitochondrial diseases. Essays Biochem 2018;62(3):467–81. Available from: https://doi.org/10.1042/EBC20170114.
[109] Mendell JR, Al-Zaidy S, Shell R, Arnold WD, Rodino-Klapac LR, Prior TW, et al. Single-dose gene-replacement therapy for spinal muscular atrophy. N Engl J Med 2017;377(18):1713–22. Available from: https://doi.org/10.1056/NEJMoa1706198.

CHAPTER 17

mtDNA mutations in cancer

Giulia Girolimetti[1,2], Monica De Luise[1,2], Anna Maria Porcelli[3,4], Giuseppe Gasparre[1,2] and Ivana Kurelac[1,2]

[1]Unit of Medical Genetics, Department of Medical and Surgical Sciences (DIMEC), University of Bologna, Bologna, Italy, [2]Center for Applied Biomedical Research (CRBA), University of Bologna, Bologna, Italy, [3]Department of Pharmacy and Biotechnology (FABIT), University of Bologna, Bologna, Italy, [4]Interdepartmental Center for Industrial Research Life Sciences and Technologies for Health, University of Bologna, Ozzano dell'Emilia, Italy

17.1 The landscape of mtDNA mutations in cancer

Mitochondrial DNA (mtDNA) mutations have traditionally been investigated within mitochondrial diseases [1] and their involvement in cancer began to be recognized only in the late 1990s with the renaissance of cancer metabolism. The first report on somatic mtDNA mutations was published in 1998, describing their occurrence in colorectal cancer cell lines [2]. Nowadays, thousands of cancer mitochondrial genomes have been sequenced, reporting frequent mtDNA mutations in virtually all cancer types [3–6].

In general, mtDNA mutations in cancer are somatic, but low heteroplasmy germline mtDNA variants have been occasionally described to accumulate in cancer tissue [7], and it is not possible to exclude that germline events may be more common than predicted, since the advent of next generation sequencing (NGS) is revealing high frequency of previously unknown low heteroplasmy variants in normal and cancer tissues [8]. MtDNA is generally up to 17 times more susceptible to mutation accumulation than nuclear DNA, due to the more modest DNA repair systems and the vicinity to the main reactive oxygen species (ROS) producing sites [9–11]. Moreover, in the context of cancer, additional mechanisms of mtDNA mutagenesis are triggered, caused by chemotherapeutics or radiation, and by microenvironmental conditions such as hypoxia [12]. Importantly, recent evidence from large cohorts surprisingly identified replication errors as the most common source of mtDNA mutations in cancer [6].

In particular, mtDNA mutations are found in almost 60% of cancers, a percentage that may vary depending on the cancer type and its stage [5,6,13]. For example, colorectal cancers present with higher frequency of mtDNA mutations (50%–60%) compared to

The Human Mitochondrial Genome.
DOI: https://doi.org/10.1016/B978-0-12-819656-4.00017-6

hematological cancers [5,6]. Most of the variants are found in the regulatory D-loop region [4,13], whereas in the rest of the mtDNA no preferential hotspot regions were reported where cancer-associated mutations would accumulate [13,14]. Inherently, genes encoding subunits of respiratory complex I (CI) are more often affected, but likely due to the fact they are the most represented in the mitochondrial genome.

Cancer-associated mtDNA variants occurring in the coding regions are often nonsynonymous (60%–80% depending on the reports) [5,6,13], suggesting they may play a modifying role in the metabolism of a cancer cell. On the other hand, highly pathogenic mutations are purified from most cancers, implying severe mitochondrial damage would be disadvantageous for cancer progression [6,15]. It is important to note that such conclusions are mainly based on data deriving from large mtDNA sequencing databases, as there are not many experimental studies corroborating these ex vivo data.

Indeed, determining the functional effect of mtDNA mutations in the context of cancer progression is not a simple task, firstly due to the peculiar properties of mtDNA genetics, namely polyploidy and the threshold effect, and secondly due to the tumor heterogeneity and ever changing selective pressures. The currently available literature, which we address in detail later in the chapter, points to two major notions regarding the role of mtDNA mutations in cancer: (1) there is no proof that mtDNA mutations are cancer-driving hits, but are rather considered to be modifiers of tumor progression and (2) there is no unique effect that can be attributed to mtDNA mutations as a whole category, as they may be either neutral, beneficial, or disadvantageous for cancer cell survival. For example, depending on the type and mutant load of mtDNA mutations, their modifying effects on cancer progression may be even opposite [16,17]. While a severely damaging mtDNA mutation may have a neutral effect on tumor progression when present in low mutant load, high level heteroplasmy or homoplasmy of the same mutation may block mitochondrial respiration and cancer cell proliferation [16,18] (Fig. 17.1).

In cancer genetics, loss of function mutations leading to cancerogenesis typically define tumor suppressors, whereas gain of function mutations define oncogenes. However, mtDNA genes may harbor loss of function mutations that are protumorigenic or neutral when heteroplasmic, or exhibit anti-tumorigenic effects when homoplasmic [16,18]. Thus mtDNA genes in cancer may not simply be defined as tumor suppressors or oncogenes since, depending on the type of mutation they harbor, they may act in a dualistic way [18]. As a consequence, a concept has recently risen based on this multifaceted role of mtDNA mutations in cancer, classifying mitochondrial genes as "oncojanus" genes, a definition which is now used also for depicting the dual role of several other nonmitochondria-related cancer-associated genes [19] (Fig. 17.2).

The final functional outcome triggered by a mitochondrial genetic lesion will depend not only on its type and mutant load, but also on the selective pressures governing the

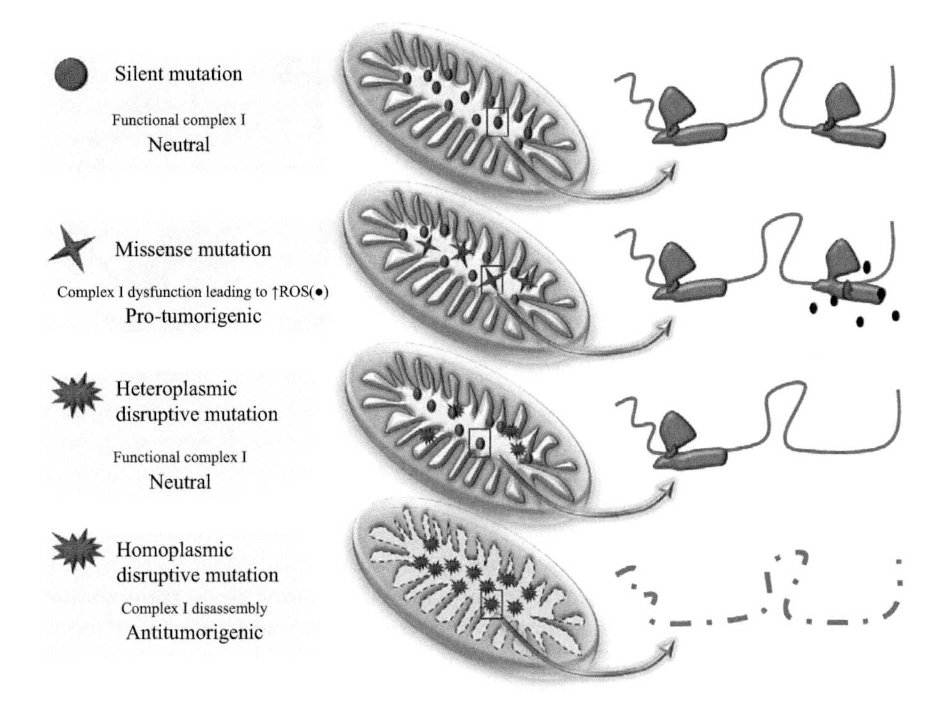

Figure 17.1

Effects of respiratory CI mtDNA mutations on tumor progression. Depending on the mutation type and mutant load level, mtDNA mutations may be neutral, beneficial, or disadvantageous for tumor growth. *Blue*: wild-type mtDNA; *Red*: mutant mtDNA. The *dashed line* indicates mitochondrial cristae remodeling, a phenotype associated with CI disassembly.

particular stage of tumor progression during which such lesion is established (see Section 17.2). For example, mtDNA mutations that slow down the efficiency of the electron transport chain (ETC) may be favorable for cancer cells in hypoxic and anoxic environment, when a decrease in oxygen consumption is required. The same lesions may be less beneficial during early metastatic stages of invasion and intravasation, when high amounts of energy are crucial to sustain cancer cell movement. This context-dependent role of mtDNA mutations has not always been clear, mainly due to a misconception influenced by Otto Warburg's observations from the beginning of the 20th century. Warburg's experiments showed that, compared to a normal tissue, the rates of glucose uptake and lactate production dramatically increase in cancer cells, even in the presence of oxygen [20,21]. This was a seminal discovery that still holds true today and is exploited in clinics to monitor neoplastic spread by positron emission tomography (PET), which may diagnose metastases by identifying areas with atypically high glucose uptake [22]. Consequently, Warburg hypothesized that the reason why cancer cells are more

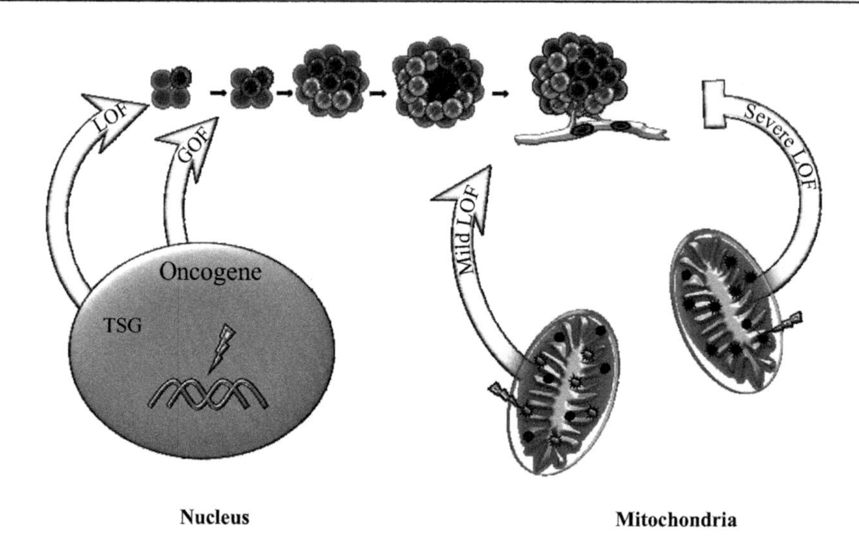

Figure 17.2
The definition of an oncojanus gene. Cell transformation and tumor progression are driven by loss of function (LOF) mutations affecting tumor suppressors (TSG) and/or gain of function (GOF) mutations affecting oncogenes. MtDNA genes are generally considered modifiers of an already existing malignant mass, but may not be classified in either of the two canonical definitions since, for example, mild LOF mutations (*gray stars*) may be advantageous while severe homoplasmic LOF mutations (*black stars*) disadvantageous for the tumor. Depending on the type and load of mutations, mtDNA genes may exhibit opposing effects on cancer progression, and be classified as "oncojanus."

glycolytic is most likely due to a functional defect in oxidative phosphorylation (OXPHOS) [23]. Therefore, at the time when mtDNA mutations were first being described in cancer, it was taken for granted their role must be protumorigenic, as they presented a perfect explanation of the Warburg's hypothesis. This even leads to the misconception that mtDNA mutations are a transforming hit involved in initiation of cell transformation, an idea which lacks concrete evidence [24]. Indeed, up to date there has been no experimental proof demonstrating transforming properties of mtDNA lesions, implying they are not sufficient to initiate cancer, but are rather mostly passenger events that in certain circumstances may act as relevant modifiers of tumor progression [25]. Therefore, Warburg's hypothesis has nowadays been revisited, demonstrating high glycolysis is mainly due to oncogene signaling, rather than a consequence of mitochondrial damage (Fig. 17.3) [26,27]. Moreover, it is generally accepted that, apart from increased glycolysis, cancer cells require functional ETC [28–30], implying that severely pathogenic mtDNA mutations would be disadvantageous for cancer cell survival, in line with the ex vivo data reporting their purification from aggressive cancers [6,13].

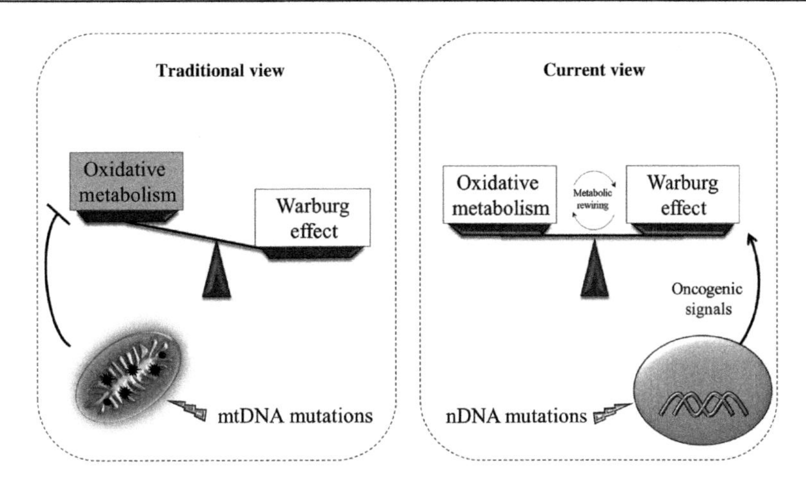

Figure 17.3

Schematic representation comparing the initial Warburg's hypothesis to the currently accepted mechanisms explaining Warburg effect. In the traditional view, the higher glycolytic flux in cancer cells (Warburg effect) was attributed to the dysfunctional oxidative metabolism, which is why mtDNA mutations found in cancer were retained to be protumorigenic. The current view recognizes that enhanced glycolysis is mainly due to oncogenic signals caused by nuclear (nDNA) mutations in oncogenes and tumor suppressors, rather than being a consequence of mtDNA damage, since both glycolytic and oxidative metabolism are maintained in cancer cells. *Gray and white squares* indicate inactive and active processes, respectively.

17.2 Functional effects of mtDNA mutations in solid cancers

Tumor progression is an evolutionary process, subjected to selective pressures and driven by somatic mutations that provide plasticity and advantage to cancer cells as they encounter rapidly changing environmental conditions [31] and acquire the hallmarks of cancer [32], resulting in turn in waves of genes expression that sustain different stages of the disease [33]. In this context, the high mtDNA mutational rates may offer a significant selective potential in driving tumor adaptation. Even though mitochondrial genes are now recognized as modifiers of tumor progression, rather than drivers of tumorigenesis, the occurrence of mtDNA mutations with a particular phenotypic effect at a specific phase of progression may play a pivotal role in the plasticity of cancer cells [25,26] (Table 17.1). Here, we will discuss the examples of mtDNA mutations that were reported to affect certain stages of cancer progression, particularly focusing on adaptation of cancer cells to high metabolic requirements, hypoxic stress, and metastatic challenge (Fig. 17.4).

17.2.1 mtDNA mutations and metabolic adaptation

In order to sustain fast proliferation, cancer cells reprogram their metabolism toward reactions that provide macromolecule synthesis, satisfying high energy and building block

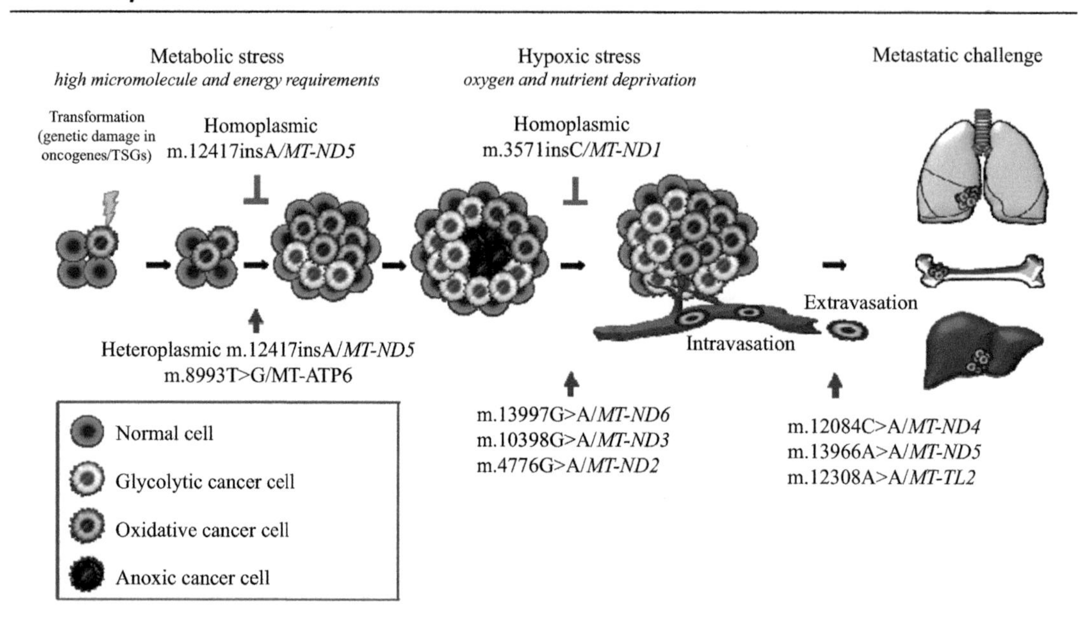

Figure 17.4
The influence of mtDNA mutations on cancer cell adaptations during solid cancer development. Following initial transformation, the fast proliferating cancer cells face metabolic stress as they require high macromolecule and energy production. Hypoxic stress sets in when the fast cell proliferation is not accompanied by adequate proangiogenic signals, resulting in scarce vasculature and development of hypoxic and anoxic areas, with consequent unavailability of nutrients causing further metabolic stress. In turn, processes of hypoxic adaptation are activated, ensuring survival and eventually leading to vasculogenesis in the solid mass. Metastatic progression includes cancer cell intravasation and extravasation, and nesting in distant tissues. In the context of these environmental selective pressures, depending from the tumor stage, mtDNA mutations may have diverse effects, either promoting (*blue*) or inhibiting (*red*) tumor progression. In this figure, mutations for which an effect has been functionally proven are reported.

requirements. For instance, glycolysis is upregulated to feed anaplerotic reactions of the Krebs cycle that lead to protein, nucleotide and lipid synthesis needed to build daughter cells (Fig. 17.5) [34,35]. Additionally, cancer cells are often more addicted to glutamine, as the latter may serve as nitrogen and carbon donor for nucleotide synthesis and anaplerotic reactions of the tricarboxylic acid (TCA) cycle [36]. Moreover, changes in lipid metabolism are common in cancer cells as they are required for new membrane formation and energy production via β-oxidation [37,38].

Apart from being driven by the type of available nutrients and the amount of oxygen present in an adverse microenvironment, metabolic reprogramming depends on the capacity of cancer cells to perform certain metabolic reactions. In this context, considering that the status of OXPHOS may dictate glycolytic and TCA cycle rates, as well as lipid metabolism, it is not surprising that mtDNA mutations have been shown to influence the

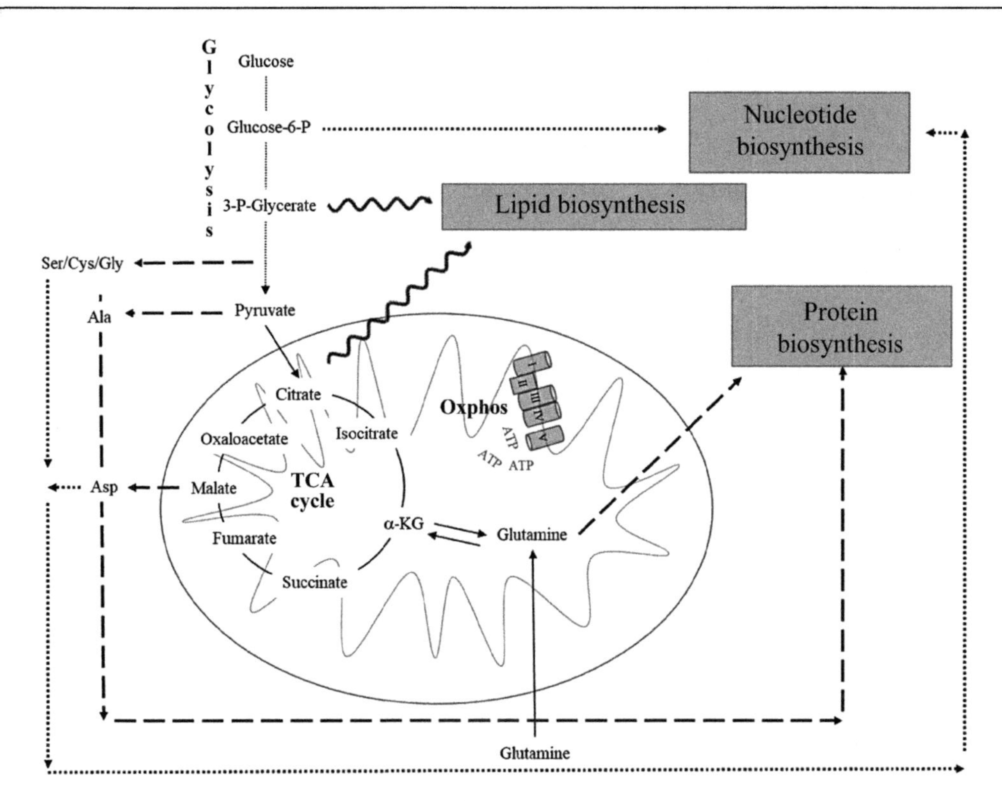

Figure 17.5

Anabolic reactions leading to macromolecule and energy (ATP) biosynthesis in proliferating cancer cells. The main metabolic axis in cells involves glycolysis, TCA cycle and OXPHOS, which may use glucose, as carbon source, and glutamine, as carbon and nitrogen source, to provide the metabolic intermediates for nucleotide (*dotted line*) and lipid biosynthesis (*waved line*); and for anaplerotic reactions of TCA cycle that support protein synthesis (*dashed line*).

metabolic adaptation of cancer cells, and in turn tumorigenic potential. For example, functional studies in trans-mitochondrial cell hybrids (cybrids) showed that 143B osteosarcoma cells carrying the heteroplasmic mutation m.12417insA in *MT-ND5* presented with an increased glycolytic metabolism, supporting tumor progression [16]. In contrast, a frameshift mutation in *MT-CYB* was shown to switch fatty acid production source from glucose to glutamine, disabling cancer cells to synthesize lipids and promoting growth under glutamine deprivation [39]. Moreover, mtDNA mutations causing severe OXPHOS damage, such as the homoplasmic m.3571insC, are generally associated with reduced tumorigenic potential when investigated in nutrient deprived conditions in vitro and especially in vivo [17,18,40,41] (Fig. 17.4). Importantly, this anti-tumorigenic effect is not only due to the energetic deficit or lipid insufficiency caused by OXPHOS dysfunction since, apart from providing energetic competence, OXPHOS ensures redox homeostasis and appropriate extracellular oxygen sensing, allowing an adequate hypoxic response [28–30].

17.2.2 MtDNA mutations and hypoxic stress

In advanced tumor stages, because of the enhanced proliferative rate of cancer cells, the central part of the mass becomes distant from the vasculature, resulting in hypoxic (<1% oxygen) and anoxic (0% oxygen) areas (Fig. 17.4). Hypoxia is one of the main selective pressures that solid tumors have to face and overcome to progress to malignancy, since it implies shortage of nutrients and oxygen, the latter required for energy production via OXPHOS.

The master regulator of adaptation to hypoxia is the Hypoxia Induced Factor 1 (HIF-1) [42], whose activity depends on constitutively expressed cytoplasmic HIF-1α, that under normoxic conditions is constantly destabilized via mechanism in which prolyl hydroxylases (PHDs) hydroxylate the protein, labeling it for Hippel-Lindau (VHL)-mediated proteasomal degradation. When oxygen tension is low, HIF-1α cannot be hydroxylated and translocates to the nucleus, where it interacts with HIF-1β as part of the HIF-1 heterodimer, in order to promote the transcription of genes dedicated to hypoxic adaptation [42,43]. Among these, HIF-1 transcribes *LDHA* to convert pyruvate in lactate and regenerate NAD^+, in order to sustain the glycolytic pathway; *GLUT1* to enable the transport of glucose inside the cell; *PKM2* which is a negative regulator of oxidative metabolism, and *VEGFA*, to promote angiogenesis [44,45].

By affecting OXPHOS functions, mtDNA mutations were shown to exert modifying effects on cancer cell metabolism and influence hypoxic adaptation (Fig. 17.4). Traditionally, it has been intuitively considered that mtDNA mutations, by slowing down respiratory chain reactions and decreasing oxygen consumption, should support cancer cells to endure extracellular hypoxia, since energy requirements may be supplied by elevated glycolysis. Indeed, the homoplasmic m.8993T>G in *MT-ATP6* (Table 17.1) was shown to confer a selective advantage by decreasing OXPHOS activity and oxygen consumption rate [46]. Moreover, the m.4776G>A mutation in *MT-ND2* was linked with HIF-1 activation through an increase in ROS production [47]. Indeed, ROS are known to trigger *HIF1A* transcription via Akt activation [48] (Fig. 17.4).

Even though cells may adapt to survive in low oxygen tension, hypoxia also implies scarcity of nutrients. Thus vasculature needs to be established to sustain proliferation, and lack of functional OXPHOS may get in the way of this process. It has been shown that hypoxic adaptation requires structural integrity of the respiratory chain [17]. In particular, severe mtDNA mutations, as demonstrated for the homoplasmic m.3571insC in *MT-ND1*, prevent HIF-1 signaling, since CI disassembly induces the accumulation of

NADH and α-ketoglutarate (α-KG). The latter increases the affinity of PHDs to bind oxygen, leading to HIF-1α degradation even in hypoxic conditions. Thus, severe damage in CI prevents hypoxic adaptation, paradoxically preventing also activation of the Warburg effect [30,49], eventually reducing tumor progression [17,18,49]. In support to this notion, the homoplasmic stop-gain m.6129G > A mutation in *MT-CO1* affecting complex IV of the respiratory chain, was associated with HIF-1 dysfunction despite a *VHL* mutation [50].

17.2.3 mtDNA mutations and metastatic progression

Metastatic progression is a complex process and only the fittest and most plastic cancer cells, which are able to quickly adapt to different environmental pressures, can succeed in this ordeal. Metastatic stages include cancer cell invasion and intravasation, survival in blood circulation, extravasation and nesting in distant tissues. Obviously, the environmental conditions surrounding a potentially metastatic cancer cell are quite different during each of these stages. For example, crossing the basement membrane and intravasation require high energy availability to sustain cell movement; survival in blood circulation requires adaptation to highly oxygenated environment, completely different from the cramped hypoxic atmosphere present in primary solid masses.

Depending on their type and mutant load, mtDNA mutations were found to either promote or block the metastatic process. Mutations leading to mild CI dysfunctions are generally considered as prometastatic, as they may induce an increase in mitochondrial ROS production that sustains extravasation of cancer cells [51,52] (Fig. 17.4). Indeed, increased ROS levels were shown to support metastatic progression, as long as they do not reach cytotoxic concentrations [53–56]. For example, the m.13997G > A in *MT-ND6* mutation described by Ishikawa et al., was shown to induce metastatic properties by boosting ROS production and promoting HIF-1 activation [52]. On the other hand, severe mtDNA mutations, leading to disassembly of the respiratory complexes and the subsequent lack of the main ROS production sites, have been associated with reduced metastatic potential [17,18].

MtDNA mutations have also been involved in promotion of metastases via ROS-independent mechanisms. For example, the m.12308A > G in *MT-TL2* was associated with hyperphosphorylated Akt and E-cadherin inhibition, crucial steps for epithelial to mesenchymal transition (EMT) of cancer cells, which is required for metastasization, particularly in the early phases related to detaching from primary tumor [57,58]. Similarly, cybrids carrying the m.10398G > A mutation in *MT-ND3* showed increased resistance to apoptosis through the Akt pathway activation, which is mandatory for the metastatic process [59,60]. On the other hand, since survival of circulating tumor cells and establishment of metastatic niche require oxidative metabolism [61], with EMT phenotype also being reported as dependent on mitochondrial function [62], it is not surprising that highly pathogenic mtDNA mutations that compromise OXPHOS, have been associated with reduced metastatic capacity [18].

Metastatic niches are the sites distant from the primary tumor, which are usually modified to favor cancer cell infiltration and nesting for metastatic colonization. This process, called "preconditioning of the niche," has been associated with metabolic changes, such as decrease of glucose uptake in the normal tissue, to obtain the increase of nutrient concentration in the niche [63]. In this context, mtDNA mutations that predispose cancer cells to a Warburg profile, may be crucial to activate mechanisms required for preconditioning of the metastatic site. Interestingly, a study on breast carcinoma cell lines revealed the presence of mtDNA mutations that confer an aggressive behavior by enhancing a highly glycolytic metabolism. In particular, the comparison between a metastatic and non-metastatic breast cancer cell line revealed the presence of two pathogenic missense mtDNA mutations in the metastatic one, namely the m.12084C>T in *MT-ND4* and the m.13966A>G in *MT-ND5*, which contribute to reduced CI activity and increased lactate production, both hallmarks of a Warburg phenotype [64].

On the other hand, the dissemination of cancer cell subpopulations facing colonization of distant tissues requires high energy consumption, making the ability of cancer cells to interconvert their energetic status an essential feature. Thus hybrid phenotypes, in which cells use both glycolysis and OXPHOS, contribute to tumor metabolic plasticity and enhanced metastatic potential [65], meaning that also in this context severe mitochondrial damage would be disadvantageous. Consistent with this conclusion, mtDNA-depleted cells, due to the lack of appropriate OXPHOS activity, were reported to have reduced capacity to disseminate metastases [66].

17.3 The fate of severely pathogenic mtDNA mutations in progressing solid tumors

As discussed above, mtDNA mutations are found in approximately 60% of solid cancers. Current meta-analyses imply that random drift and relaxed purifying selection are possible mechanisms for the accumulation of synonymous or mild missense mtDNA mutations in some types of cancers [6,67–69]. Occasionally, evidence of positive mtDNA selection was reported, as in the case of somatic missense mutations found in hepatocellular carcinoma, colorectal, and breast cancer [70,71]. Indeed, some mtDNA mutations may be selected in tumor tissue due to their beneficial effects in the context of adaptation to environmental pressures. However, most of the current literature agrees that pathogenic mtDNA mutations are purified from solid cancers. In particular, disruptive mtDNA mutations, severely affecting the ETC function, are usually negatively selected [6,13,15]. It must be noted that such mutations might persist at low, subfunctional heteroplasmic levels since in this state they do not exhibit a phenotypic effect [6,13] (Fig. 17.1). As a consequence, if purifying selection is lost due to relaxed negative selection during certain phases of constant microenvironmental changes, severe mutations may in rare cases become homoplasmic [70,72]. However, eventually the effect of such homoplasmy risks to become disadvantageous for malignant progression. Indeed, kidney chromophobe and thyroid tumors, which have been reported to accumulate a higher amount of non-synonymous mtDNA variants, are usually less malignant compared to other tumors [73,74], similarly to the example of oncocytomas (see Section 17.3.3). Here we describe the biological processes behind severe mtDNA mutation purification, tackle the mechanisms adopted by cancer cells to compensate if such damage persists, and discuss lessons learned from rare examples when highly pathogenic mtDNA mutations are accumulated in cancer cells.

17.3.1 Molecular mechanisms behind selection and purification of mtDNA mutations in cancer

In order to prevent accumulation of severe mtDNA mutations in tumors, cancer cells are selected, which adopted mechanisms upholding low heteroplasmy levels or completely purifying mitochondrial damage. One of these mechanisms regards the maintenance of high mtDNA copy number via increase of mitochondrial biogenesis orchestrated by PGC-1α. It is a general, evolutionary conserved mechanism to keep low levels of mutated mtDNA molecules and prevent their damaging phenotypic effects, and is implemented also by cancer cells [75]. However, cancer progression may involve endogenous and exogenous perturbations in processes regulating mtDNA copy number, leading to mtDNA mutation abundance. For example, mtDNA mutations within the D-loop region may predispose to decrease in mtDNA copy number, which in turn may favor the increase in the ratio between mutated and wild-type mtDNA molecules. The susceptibility of the D-loop to accumulate mutations has been associated with reduced mtDNA content in hepatocellular carcinoma [76,77]. Moreover, the abundance of the m.3234A > G homoplasmic missense mutation in *MT-TL1* has been suggested to be a consequence of the mitochondrial genome depletion in cybrid models [78]. In rare occasions, high mutant loads of mtDNA mutations have been associated with elevated mtDNA copy number, but more often than not, this was reported in benign tumors and hypothesized to be a compensatory mechanism to counteract energetic failure [79,80]. Therefore, copy number changes may be a cause, but also a consequence of somatic mtDNA mutations accumulation, in close correlation with the metabolic constraints encountered during tumorigenesis and tumor progression.

Mitochondrial biogenesis is counteracted by the autophagic targeted-mitochondrial degradation-mitophagy, a quality control mechanism that degrades defective mitochondria to preserve the energetic competence and maintain cellular homeostasis [81]. Even though the general role of autophagy is still debated in the context of cancer, there is evidence that functional mitophagy machinery seems to be required for maintenance of tumorigenic potential. In particular, in a murine model of non-small cell lung cancer, lack of adequate mitophagy was shown to inhibit tumor growth due to accumulation of defective mitochondria, converting aggressive adenocarcinomas into benign tumors [82].

Mitophagy as a quality control process is activated following mitochondrial fission (fragmentation), which not only increases the organelle number, but compartmentalizes mtDNA molecules, influencing the wild-type versus mutant molecule distribution and consequently mutation selection/purification. As demonstrated by isogenic induced pluripotent stem cells harboring different levels of the m.3243A > G mutation in *MT-TL1*, drops in mitochondrial membrane potential signal occurrence of mitochondrial dysfunction, stimulate fission and the removal of damaged mitochondria by mitophagy [83,84]. Depending on the cell's energy and metabolite requirements, fission is balanced with fusion (mitochondrial merging) (Fig. 17.6), maintaining the low mutant load status to prevent accumulation of pathogenic mtDNA mutations [85].

Another mechanism related to accumulation of mitochondrial damage in cancer regards the phenomenon of large mtDNA deletions. Some studies associate the common 4977-bp deletion with cancer [86]. Most likely, since mitochondrial replication is uncoupled from nuclear division, high rates of proliferation may induce preferential selection and amplification of deleted mtDNA

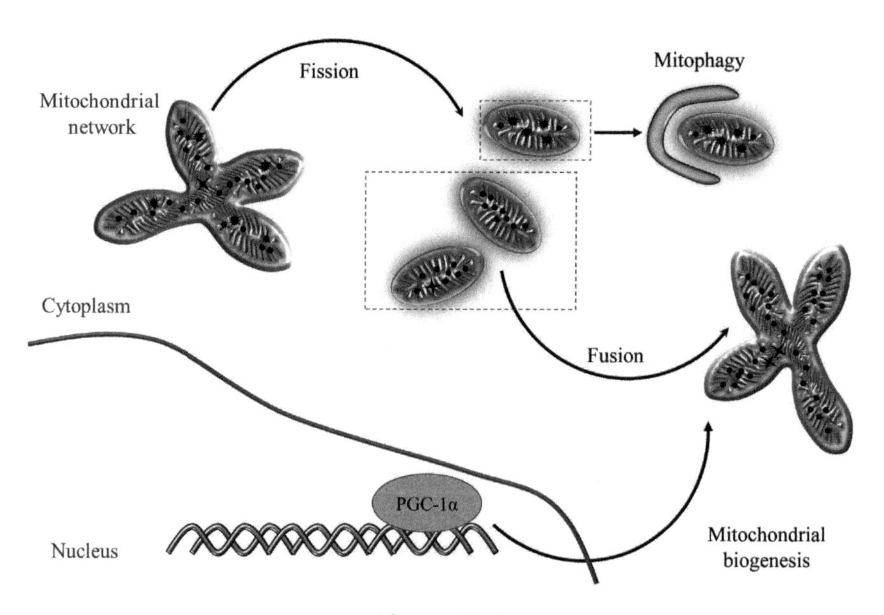

Figure 17.6

Mechanisms regulating purification of pathogenic mtDNA mutations. Mitophagy is involved in the removal of damaged mitochondria, a process triggered by organelle fragmentation (fission). Fusion of healthy mitochondria together with mitochondrial biogenesis regulated by PGC-1α ensures high mtDNA copy number required for maintenance of low mutant loads.

molecules, as the reduced size of the mtDNA harboring pathogenic big deletions might favor a faster rate of replication, conferring a selective advantage to the cell [86–88]. However, it is important to note that the common deletion is never found in homoplasmy, not even in benign tumors, strongly suggesting that such damage is disadvantageous for cancer.

17.3.2 Compensatory mechanisms to overcome mitochondrial dysfunction

The fact that solid tumors preferentially retain functional mitochondria is evident also from studies reporting the adaptive processes that are selected in cancer cells to compensate for an existing mitochondrial damage. Apart from the homeostasis between mitochondrial biogenesis and mitophagy discussed in the previous section, evidence is emerging on several additional mechanisms adopted by cancer cells to counterbalance for mtDNA mutation-related mitochondrial damage.

An intriguing discovery of inter-cell mtDNA exchange, so-called horizontal mtDNA transfer, has been proposed as an evolved physiological mechanism by which cells and tissues regulate their mtDNA heteroplasmic state and recover mitochondrial function [89,90], and is described in detail elsewhere in this book. Indeed, trading of whole organelles has been demonstrated to take place between cells of the tumor microenvironment and neoplastic cells, to support cancer progression and/or therapy resistance [91]. In particular, the horizontal transfer usually occurs between tumor and mesenchymal stem cells of the resident tissue, in order to rescue mitochondrial function in cancer cells with depleted or mutated mtDNA [92,93]. Interestingly, a more successful organelle exchange was observed in less differentiated cells [41]. The process is correlated to mitophagic activity in a way that the uptake of healthy mitochondria is required to ensure clearance of defective organelles [41,94]. However, it is important to note that the transfer may occur in presence of critically damaged mitochondria, but not in the case of cells harboring pathogenic mtDNA mutations, leaving open questions that warrant further investigations [95].

It is interesting to note that, apart from uptaking mtDNA molecules from surrounding stroma, cancer cells with severe mitochondrial defects were associated also with other microenvironment-mediated survival mechanisms, such as those guided by tumor associated macrophages [96]. Indeed, cancer metabolism is being recognized as one of the determining factors in immune-oncology [97], with functional mitochondrial metabolism recently identified as crucial in appropriate response of melanoma patients to immunotherapy [98], suggesting that mtDNA mutations may have their say in such contexts.

Another example of a compensatory mechanism activated in the presence of damaged mitochondria is the rewiring of glutamine metabolism, which is generally crucial in ensuring the metabolic plasticity required for cancer cell survival. To allow glutamine oxidation that promotes macromolecule biosynthesis within mitochondria, functional

OXPHOS is required in fast proliferating cells [99]. However, if mtDNA mutations affecting OXPHOS become fixed at the somatic level, cancer cells may metabolically adapt to compensate severe mitochondrial dysfunction and ensure tumor survival. For example, reductive carboxylation of glutamine or reverse TCA is activated in cancer cells suffering mitochondrial damage to support lipid biosynthesis [39]. This is a process by which glutamine-derived α-KG is reduced through the consumption of NADPH by isocitrate dehydrogenases 1 and 2 (IDH1 and IDH2, cytosolic and mitochondria isoforms, respectively) in the non-canonical reverse reaction to form citrate [39]. A large amount of this citrate pool is then carried into the cytosol to produce acetyl-CoA for fatty acids synthesis and oxaloacetate for anaplerotic pathways to guarantee the four-carbon intermediates needed to produce the remaining TCA metabolites and related macromolecular precursors such as aspartate.

17.3.3 MtDNA mutations in oncocytomas: an exception from the rule

Oncocytic tumors arise in endocrine and exocrine tissues, and are of epithelial origin, characterized by swollen cells with a large cytoplasm full of defective mitochondria [100]. Oncocytic cells or "oncocytes" on histology appear with small nuclei and a highly eosinophilic granular cytoplasm, routinely revealed through hematoxylin and eosin staining. On transmission electron microscopy, mitochondria in oncocytes present with low electron density, due to the loss of mitochondrial potential and aberrant mitochondrial cristae [101]. The most striking morphological feature of these neoplasms is the abnormal mitochondrial hyperplasia, which is most likely the result of a compensatory effect induced by the energetic crisis, possibly mediated by a retrograde mitochondria-nucleus signaling [102–104].

Oncocytomas are the only tumors in which severely pathogenic mutations are found in high frequency and at high mutant load, providing a curious example of pathogenic mtDNA mutations escaping purifying selection in cancer [15]. In particular, they often harbor highly damaging, frameshift or stop-gain mtDNA mutations in genes encoding CI subunits, even in homoplasmy, thus reaching a phenotypic effect of OXPHOS inhibition mostly due to respiratory CI damage [13,15]. This is seemingly in contrast with the currently accepted view that severe mitochondrial damage in cancer cells would lead to lethality. However, the paradox is only apparent, as oncocytic tumors with homoplasmic pathogenic mtDNA mutations are benign, such as in the case of renal neoplasia [105]. Indeed, by associating severe mtDNA damage with indolent phenotype, oncocytomas were one of the clues which triggered functional studies proving the anti-tumorigenic role of severe mtDNA mutations [16–18,103]. In particular, introduction of the oncocytoma-specific homoplasmic m.3571insC mutation in a context of a highly aggressive model of osteosarcoma, not only caused development of clearly oncocytic phenotype, but also significantly lowered tumorigenic potential [18], among other due to the lack of HIF-1 activation [49,106].

It is important to acknowledge that even though the oncocytic phenotype is generally associated with a low-proliferative index, one should distinguish the following definitions: (1) oncocytoma—benign lesion; (2) oncocytic tumor—any tumor with mitochondrial hyperproliferation, meaning it includes also malignant lesions (carcinomas). For instance, oncocytomas in the context of renal, parotid and pituitary gland are usually considered benign neoplasms, whereas thyroid tumors characterized by mitochondrial hyperplasia may display different degrees of aggressiveness [107]. This may be due to the nature and degree of mtDNA mutations occurring in these tumors. Indeed, somatic mtDNA mutations in thyroid oncocytic tumors are found in about 50% of patients. They are reported to be more often missense amino-acid substitutions, rather than frameshift or stop-gain mutations, and are mostly heteroplasmic, even though tumor heterogeneity makes it difficult to assess the heteroplasmic status with precision [105,108,109]. On the other hand, somatic mtDNA mutations in renal oncocytomas are more often severe and almost exclusively homoplasmic. Similarly, oncocytic lesions of the parotid gland harbor a higher mtDNA content in association with severe mtDNA mutations affecting CI [106,109].

Up to date, most oncocytic tumors have been reported as sporadic, carrying somatic mtDNA mutations at variable degrees of mutation load. Familial cases occur occasionally [110–112], sometimes with inherited disruptive mtDNA mutations shifting toward homoplasmy only in the oncocytoma tissue [7,113]. The mechanisms through which mtDNA mutations shift toward homoplasmy in oncocytomas are yet unknown. The most likely hypotheses regard either random drift due to relaxed negative selection allowed at certain stages of tumor progression [68], or potential damage in the mitophagic machinery [114,115], which eventually could lead to a short-circuit-like phenomenon of damaged mitochondria accumulation, preventing further progression to malignancy.

17.4 Clinical potential of cancer-associated mtDNA mutations

Currently, there are no available standardized clinical protocols exploiting the information of mtDNA mutations in cancer. Nonetheless, due to their frequency in many tumors, these variants have the potential of being used as markers in the diagnosis and clinical management of cancer patients, which is discussed in the following section.

17.4.1 MtDNA mutations and cancer treatment

17.4.1.1 Chemotherapy

MtDNA mutations have a dual role in chemoresistance. On one hand, they can arise due to the effects of chemotherapeutic drugs and, on the other, they have been shown to play a role in the acquisition of chemoresistance itself [116,117].

For example, a study including 20 leukemia patients found a higher mutation rate of heteroplasmic variants with amino-acid alterations in cases treated with a fludarabine/alkylator-based 6-month chemotherapy regimen compared to nontreated patients [118]. The complexities intrinsic to in vivo analyses do not allow determining if the elevated number of mutations is a direct consequence of chemotherapy-induced DNA damage or mediated by ROS generated due to the treatment, but it is possible that both scenarios occurred in vivo. Considering that untreated patients tended to have low rates of both hetero- and homoplasmic mutations, it is more likely that the high frequencies of heteroplasmic mutations observed in treated subjects are directly related to chemotherapy.

Moreover, the same study reported a higher mutation rate in non-responding patients, compared to the ones who responded to therapy, suggesting that apart from being a consequence of chemotherapy treatment, mtDNA mutations may play a role in acquiring chemoresistance. In this context, mtDNA mutations may be selected due to their advantage for overcoming chemotherapy-related selective pressures, for example, in terms of functional changes in the mitochondrial ETC leading to ROS increase. Indeed, Rho0 cells, depleted of their mtDNA, generated from normal intestinal epithelium showed a four- to five-fold increased resistance against cisplatin compared to their parental counterparts, implying mitochondrial damage may lead to chemoresistance [119]. Similarly, Singh et al. reported that Rho0 cells lacking mtDNA were resistant to adriamycin and photodynamic therapy, whereas the isogenic, wild-type cell line was sensitive [116]. Moreover, cybrids carrying mtDNA mutations found in pancreatic cancer were shown to grow more slowly than wild-type controls, and resulted resistant to 5-fluorouracil and cisplatin treatments [120]. All these results suggest that mitochondrial damage may be involved in the acquisition of resistance to cancer therapeutic agents [116]. However, even though there is plenty of literature pointing to a correlation between mitochondrial metabolism reprogramming and response to chemotherapics, the nature of this relationship is still controversial and not generalizable in all cancer contexts. For example, oxidative metabolism in cancer cells was associated both to chemoresistance and to enhanced chemosensitivity [121–125]. This is not surprising, considering the heterogeneity in the response to therapy among various tumor subtypes, individuals, and even within the same neoplasia. Due to such heterogeneity, it is difficult to determine to which extent mtDNA mutations may play a role in modifying chemotherapy response. Indeed, although mutations in mtDNA are common events in cancer cells, very few are specifically known for their role in chemoresistance, mainly due to the lack of functional studies. Certainly, mtDNA variants may be involved in metabolic reprogramming, as pathogenic mutations may promote rewiring of the bioenergetics profile in cancer cells through a mitochondria-to-nucleus signaling activated by dysfunctional mitochondria [123]. For example, a metabolic switch was associated with a *MT-CO1* mtDNA mutation and with the acquisition of resistance to PI3K/mTOR inhibitors in lung cancer cells [126].

Interestingly, the commonly used cisplatin accumulates in mitochondria causing an impairment of mtDNA [127] and induces a more widespread adduct formation in mtDNA than in the nuclear genome, resulting in de novo mutations and damage of mitochondrial function [128]. In particular, cisplatin adducts formation in head and neck squamous cell carcinoma cells was 300–500-fold higher in mtDNA compared to nuclear DNA [129]. Moreover, chemoresistance induced upon exposure of A549 non-small cell lung cancer cell line to cisplatin was associated with a hetero- to homoplasmic shift of a nonsynonymous mutation in the CI gene *MT-ND2* [130]. The shift was accompanied by a compensatory mechanism of increased mitochondrial biogenesis, likely mediated by the upregulation of PGC-1α and PGC-1β in the cisplatin-resistant cells, as the authors claim [130]. Similarly, in a study on gynecological cancer cell lines, cisplatin was shown to induce mtDNA mutations in *MT-ND5* and *MT-CO2* [131]. These platinum-induced deleterious mtDNA mutations cause a decrease in the mitochondrial respiratory function making cells less aggressive and less proliferative with the inability to migrate and invade. This may represent a selective advance for cancer cell to survive in a context of chemotherapeutic drugs that act on proliferating cells such as paclitaxel, a microtubules stabilizing agent [132]. In contrast, a study on small cell lung cancer cell lines found no differences in the mitochondrial genome of H446 and the isogenic resistant cell line but a difference in lactic acid and ROS generation, suggesting the involvement of mitochondrial dysfunction in the resistant phenotype [133].

In the last decade, occasional reports working on patient tissues and analyzing the association between chemotherapy and mtDNA mutations have become available. In a patient with residual serous ovarian cancer after chemotherapy, a novel missense m.10875T > C in *MT-ND4* was found in the chemoresistant post-chemo tissue associated with oncocytic phenotype. The novel mutation, randomly occurring or induced by carboplatin, may have determined an energetic defect facilitating resistance onset to paclitaxel until a threshold was reached for CI disruption, which then most likely caused a low-proliferative oncocytic phenotype. Paradoxically, the more quiescent, non-invasive profile was an advantage to resist to carboplatin and paclitaxel, which indeed affect actively dividing cells [134].

Taken together, all the studies described above point to a correlation between pathogenic mtDNA mutations and chemoresistance. However, there is evidence associating chemoresistance to maintenance of mitochondrial metabolism, in terms of functional OXPHOS. In this context, an intriguing insight comes from a recently emerging field of horizontal mitochondrial transfer [135], through which cancer cells acquire mtDNA from non-transformed cells within the tumor microenvironment [93], resulting in the boost or restoration of mitochondrial respiratory function [92,95], rescue from apoptotic cell death [136] and development of chemoresistance [137]. In particular, such transfer was reported to occur from endothelium to breast and ovarian cancer cells in association with chemoresistance to doxorubicin [137], from surrounding normal tissue to apoptotic pheochromocytoma cells stressed with UV light that were then able to survive [136], and in vivo in immunodeficient mouse xenograft models where mitochondrial exchange from bone marrow stromal cells to human leukemia initiating cells provided a survival advantage under chemotherapy treatment [138]. Moreover, in breast cancer, the horizontal transfer of mtDNA from extracellular vesicles caused exit from dormancy in therapy-induced cancer stem-like cells and led to resistance to hormonal therapy in OXPHOS-dependent masses [66]. Thus, in contrast to most of the earlier literature, which associates mtDNA mutations with chemoresistance, cancer cells acquiring healthy mitochondria via nanotube transfer seem to display lower sensitivity to several chemotherapeutics, in line with OXPHOS dependency described in the context of cancer stem cells and primary/acquired resistance to chemotherapy [139–141]. It is still unclear whether these effects are due to the acquired mtDNA only, or to the transfer of entire respiratory proteins from healthy mitochondria. Indeed, it is important to note that one study showed mitochondrial transfer was possible from human mesenchymal cells to human osteosarcoma 143B-derived Rho0 cells, but not if the recipient cells carried pathogenic mtDNA mutations, suggesting that cancer cells with severe mtDNA mutations are less able to uptake foreign mitochondria to rescue their phenotype [95]. This would mean that, rather than mtDNA, it is the organelles themselves that are responsible for metabolic rescue and development of chemoresistance.

In general, further and more detailed clinical studies are required to define the prognostic role of mtDNA mutations in the context of chemotherapy, which is most likely going to be specific for the drug, tumor type and mutation involved. Nonetheless, considering both in vitro and ex vivo data, it is plausible to envision that chemotherapy may induce mtDNA variants that in turn influence the response to therapy. In particular, the following scenarios may be hypothesized: (1) a mutation can arise due to the effect of chemotherapeutic drugs (direct DNA damage or secondary effect due to ROS induction) and positively selected during chemotherapy if it provides a selective advantage, (2) a very low heteroplasmy mutation originally present in a cancer cell may be selected during chemotherapy. However, it may not be excluded that mitochondrial damage in some contexts is disadvantageous for cancer cells that subdue chemotherapy, as suggested by the experiments on mitochondrial transfer. Moreover, mtDNA mutations may lead to an excessive ROS concentrations, amplifying the ROS-producing effect of chemotherapeutics themselves, leading to exhaustion of the cell's antioxidant capacity and apoptosis [142,143]. In line with the latter hypotheses, since "multidrug resistance" (MDR), driven by overexpression of ATP-binding cassette (ABC) proteins that extrude chemotherapeutic agents from the cells allowing them to survive, is an ATP-dependent process [144], it may be envisioned that mtDNA mutations affecting the energy content of a cell may sensitize cancer cells to chemotherapy.

17.4.1.2 Radiotherapy

Ionizing radiation (IR) is one of the most commonly used cancer treatment options resulting in cellular damage. Mitochondria were reported to be involved in downstream irradiation effects [145,146], but the exact role of mtDNA mutations is still unclear. For example, no evidence was found that radiotherapy is associated with higher mtDNA mutation rate by analyzing tissues of women who had previously been treated with radiotherapy to treat pediatric leukemia [147]. Moreover, no differences were reported in sensitivity to irradiation between osteosarcoma Rho0 and parental cell line [148]. On the contrary, as already discussed for the effects of chemotherapy on mtDNA mutations, there are indications that they may both be caused by radiotherapy and affect its outcome. In particular, an increased number of point mutations and deletions was observed in patients treated for cancer with radiotherapy [149], whereas a decrease in mtDNA content was reported in post-treatment head and neck cancer patients [150]. On one hand, mitochondrial dysfunction was associated with an increased radiosensitivity in mtDNA-depleted cells, as they displayed a regrowth delay of irradiated tumors in vivo, compared to wild-type controls [151]. In contrast, pancreatic cancer Rho0 cells and fibroblast Rho0 cells were shown to be more radioresistant than parental cell lines [152], meaning that the effect of mitochondrial damage on radiosensitivity may be dual, depending on the context.

An important factor influencing radiosensitivity is the intracellular oxygen levels. The levels of oxygenation within the same tumor are highly variable from one area to another and can change over time. Indeed, hypoxia significantly reduces the efficacy of the damage caused by IR, resulting in lower tumor response and decrease of overall survival [153–155]. In radiotherapy, the primary effect of radiation is the creation of ROS that irreversibly damage tumor cell DNA resulting in apoptosis and cell death [154]. The reason for this effect is that oxygen, being the most electron-affine molecule in the cell, reacts extremely rapidly with the lesion produced by IR, a radical in DNA, making the damage permanent. In the absence of oxygen, the radical damage can be restored to its undamaged form making IR not effective in killing hypoxic cells [156]. In the 1950s, Gray et al performed experiments showing that a three-fold higher dose of radiation was required to kill hypoxic cells compared to normoxic cells [153] therefore, due to the limited tolerance of normal tissues to radiation, it is generally not possible to use the required dose in the clinical practice. After radiation, hypoxic tumor cells may persist and then divide, resulting in tumor persistence and development of radiotherapy resistance [157]. In this context, pathogenic mtDNA mutations, such as those causing CI disruption and reduction in oxygen consumption [17,158], may confer to cancer cells a higher sensitivity to radiation therapy thanks to the elevated availability of intracellular oxygen. In conclusion, there is no clear evidence about how mtDNA mutations affect the response of cancer cells to radiotherapy, as mitochondrial dysfunction has been associated both with the acquisition of resistance and with a greater response to such treatment, leaving the clinical utility of mtDNA mutations in the context of radiotherapy still rather limited.

17.4.1.3 Interventions in cancer therapy based on mitochondrial functional status

The information gathered from the analyses of mtDNA mutations in cancer chemo- and radiotherapy may provide clues for development of adjuvant therapeutic strategies. For example, targeting glycolysis through inhibition of glucose uptake may be a potential strategy for tumors with OXPHOS dysfunction [159] or in chemoresistant cancers with disruptive mtDNA mutations that force the cells to depend on glycolysis. There are several glucose uptake blocking small molecules that may be exploited in this context, including fasentin, phloretin, STF-31, and WZB117 [160]. Similarly, glutamine metabolism intervening drugs such as glutaminase (GLS) inhibitors currently being tested in clinical trials [36,161,162] may be more effective in cancers with mtDNA damage. Finally, targeting OXPHOS may be envisaged as an adjuvant strategy in tumor types where mtDNA mutations have been associated with chemosensitivity or, in general, with reduction of tumor growth, indicating susceptibility to mitochondrial damage in such contexts. The most studied anti-cancer drug targeting mitochondria is metformin, a type II diabetes drug that is used to decrease gluconeogenesis. It inhibits ROS production and CI activity [163,164], activates adenosine monophosphate activated protein kinase (AMPK) resulting in mTORC1 inhibition with a consequent decrease in protein synthesis and cell proliferation [165], and reduces HIF-1 signaling [166].

Pharmacological impairment of OXPHOS by metformin was reported to increase the efficiency of radiotherapy and chemotherapy. These studies are in line with mitochondrial transfer related experiments, which suggest OXPHOS function promotes resistance. Similarly, in vitro studies point to a combination of metformin and radiotherapy, to increase radiosensitivity of cancer cells [167–170]. Numerous studies point to a combination of metformin and traditional chemotherapy as a preferred treatment option in clinics, to accelerate tumor regression [171] in different types of cancers such as colorectal, hepatic, gastric, ovarian, endometrial, pancreatic, breast, prostate, lung (for a review see Ref. [172]). However, the results are variable, depending on the drug combination and cancer type. Moreover, it is important to note that a great number of these studies reported that the positive effect of metformin is not due to mitochondrial CI inhibition but to other side effects of the combination with chemotherapeutic agents [172]. This evidence has led to several currently ongoing clinical trials that will be crucial for understanding whether the addition of metformin to standard chemotherapeutic treatments can improve progression free survival [172] and overall survival rate.

In addition to metformin, other mitochondria-targeting drugs including atovaquone, arsenic trioxide, ME344 and fenofibrate have shown to enhance the effect of radio- or chemotherapy [173]. Therefore, it seems that characterization of cancers metabolism by finding markers such as mtDNA mutations could improve cancer patient management by recognizing masses intrinsically sensitive to OXPHOS inhibition.

17.4.2 Cancer-specific mtDNA mutations as markers of tumor progression

In the event of simultaneously detected neoplastic lesions it is fundamental to assess whether they arise independently or as a result of metastatic dissemination, to ensure the correct patient management, prognosis, and therapeutic regimen choice [174].

Independent/synchronous lesions require implementation of therapeutic regimens aimed at contrasting both cancer types, avoiding the negative impact on the overall outcome due to increased toxicity or relevant pharmacological interplay [174]. In the case of clonal/metastatic masses, the cancer stage is more advanced than two synchronous tumors and a different therapeutic choice is indicated. A relevant part of simultaneously detected neoplastic lesions remains difficult to classify with confidence because of the tumor heterogeneity between primary lesion and metastasis or due to ambiguous histological features [175,176]. In these cases, the application of molecular techniques is recommended. Since mtDNA mutations occur in approximately 60% of solid cancers, they may be informative in understanding clonality. In particular, if a random tumor-specific mtDNA mutation is detected in two different neoplasms of the same patient, the masses are considered clonal, since two independent tumors arising in the same patient are highly unlikely to acquire the same somatic mtDNA genotype [175,177]. Therefore, mtDNA sequencing may be used as a tool to help the clinicians to distinguish clonal (metastatic) from synchronous masses (Fig. 17.7).

NGS technology is one of the ways to establish the tumor specificity and the heteroplasmy level of mtDNA mutations. It permits increase of the precision in using mtDNA mutations as clonality markers, since it allows detection of low-level heteroplasmies. Alternatively, fluorescent PCR (F-PCR), denaturing high performance liquid chromatography (dHPLC) or locked nucleic acid allele specific quantitative PCR may be used, which may detect heteroplasmy levels as low as 2% [113,178].

Several studies support a role for the high propensity of mtDNA to accumulate mutations in cancer for diagnostic purposes. For example, the relationship between primary oral squamous cell carcinoma and lymph node metastasis in a series of patients with synchronous and metachronous metastases was assessed using mtDNA, allowing to evaluate the genetic relationship among different tumor clones [179]. Similarly, the screening of the D-loop region was used to evaluate the clonal relationship between primary oral squamous cell carcinoma and secondary neoplasia appearing in the reconstructed skin graft after surgical resection revealed that the second tumor is a recurrence of the original lesion, rather than a second primary tumor [180]. Moreover, the finding of shared mutations between oral

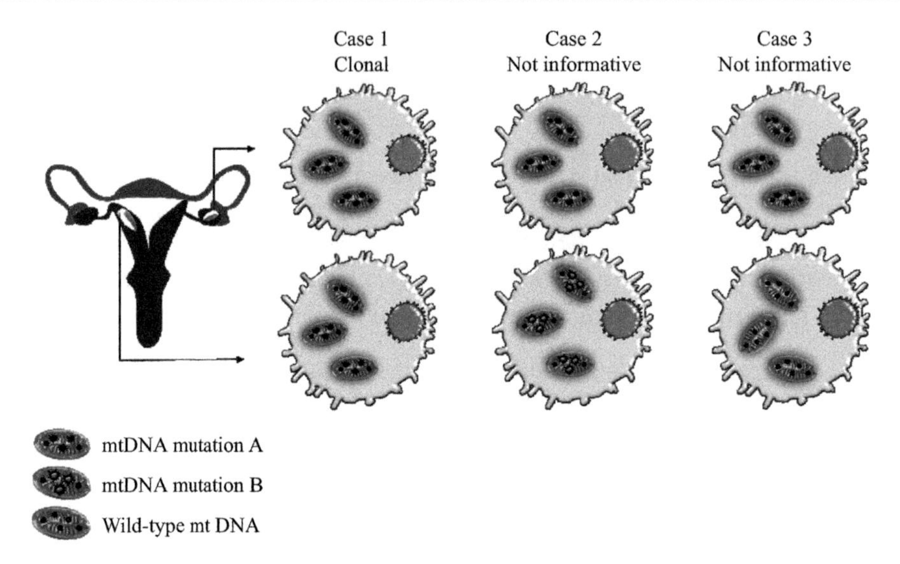

Figure 17.7
MtDNA mutations as markers of clonality in the diagnosis of ovarian and endometrial cancer. When the same private mtDNA somatic mutation is detected in both neoplasias of the same patient the tumors are defined as clonal (Case 1). On the other hand, if the two masses display two different mtDNA mutations (Case 2) or mtDNA mutation is detected only in one tumor (Case 3), mtDNA sequencing is not informative to infer diagnosis.

squamous cell carcinoma, recurrences and metastasis using mtDNA NGS indicated a clonal origin of cancer cells [177].

MtDNA screening was found to be particularly useful in the context of simultaneously detected gynecological lesions. A study on concurrent endometrial and ovarian carcinomas, where an optimized Sanger technique was used to sequence the entire mtDNA, revealed tumor-specific mutations in both endometrial and ovarian cancers, ruling out independence of the two neoplasms and indicating clonality in half of the analyzed cases [176]. MtDNA sequencing was also applied to recognize the monoclonal origin of the borderline ovarian tumors and the peritoneal lesions to trace a clonal spread, strongly supporting the hypothesis that the origin of implants may be monoclonal, from the primary tumor [181]. Moreover, in a single peculiar case of a woman with inguinal lymph node lesions of a well differentiated endometrial cancer, mtDNA sequencing was used to provide an additional diagnostic confirmation of the inguinal node metastasis [182].

It is important to acknowledge that mtDNA variants may not always distinguish two primaries from a metastatic disease, since during metastatic progression cancer cells may acquire or lose mtDNA mutations. In this context, mtDNA variants may arise subsequently to the initial clonal expansion, and in this case, when somatic mutations occur in only one of the two tumors, it is not possible to rule out a clonal origin of the two lesions [176]. Thus, mtDNA sequencing is not informative to detect the independent (non-clonal) origin of two masses when only one of the multiple neoplasias carries a mutation, but may be used to infer clonality when the same mtDNA mutation is present in both tumor masses (Fig. 17.7).

In order to assign unequivocal clonal origin of two lesions, one must ensure the identified mtDNA mutations are not of germline origin. Thus a precise quantification of the mtDNA mutation load and an efficient detection of low-level heteroplasmies are fundamental not only in tumor masses but also in the non-tumor matched control, such as blood, adjacent normal tissue or saliva. Absence of a low-level germline heteroplasmic variant in the non-tumor tissue confirms it is tumor-specific, and if the mutation is present in both tumors, a clonal relationship may be inferred. On the other hand, if low-heteroplasmic germline mtDNA mutations are found in non-tumor tissue, such mtDNA mutations become non-informative (Fig. 17.7).

Today it is recognized that therapy response and cancer patient survival depend not only on the tumor type, but may differ between individuals depending on many still unrecognized internal and external factors, leading the development of personalized medicine. Thus, correct description of the molecular signatures in tumors is crucial. In this context, there is evidence implying mtDNA analysis may aid in the assessment of tumor heterogeneity that often confounds correct patient management. In particular, multiplex PCR-based ultra-deep sequencing technology was used for the whole mtDNA analysis to demonstrate the monoclonal origin of histologically heterogeneous metastases of non-small cell lung cancers [183]. Despite the different histological subtypes or mixed-histology tumors, in most cases, primary tumor and multiple lesions bore common mtDNA mutations in each mass of an individual patient, indicating a single cell progeny and clonal relationship [183].

Thus, the analysis of mitochondrial propensity to accumulate mutations in cancer cells may be useful in understanding tumor heterogeneity, by defining clonal relationship of cells within a single tumor, or in cases of metastatic tumors with ambiguous histology.

17.5 Insights from next generation sequencing and bioinformatics approaches

17.5.1 Technical pitfalls and false discoveries of the past

Mitochondrial genetics studies in cancer have always been overshadowed by those on nuclear genetics, despite the high frequency of mtDNA mutations in tumors. Several technical issues contributed to neglect mitochondrial genomics in oncology, such as (1) the accuracy in the measurement of heteroplasmy levels; (2) sample contamination, which may result in the incorrect annotation of haplogroup-related variants as cancer-specific [184];

(3) the co-amplification of nuclear mitochondrial sequences (NUMTs) and the consequent difficulty in attributing a sequence change to nuclear or mtDNA [185]; and (4) the lack of analytical approaches allowing the correct annotation of pathogenic variants. Indeed, in past studies, the lack of well-curated databases and efficient tools to recognize the functionally important mitochondrial mutations among non-pathogenic variants caused haplogroup-specific changes to be described as pathogenic mtDNA mutations [184], mistakenly classifying many polymorphisms as cancer-associated mtDNA mutations [186,187].

17.5.2 Methodological recommendations for mtDNA mutation analysis in the advent of next generation sequencing in oncology

The attention for human mitochondrial genetic data is growing among both clinicians and researchers. Thanks to the recent advances in high-throughput sequencing techniques and the creation of public databases containing high number of sequences and related cancer patient metadata, an unprecedented amount of genetic information is today available, representing an opportunity for the understanding of the role of mtDNA variants in cancer.

Recent studies analyzing large datasets of cancer patients, such as The Cancer Genome Atlas (TCGA) or International Cancer Genome Consortium (ICGC), made accessible a great and constantly growing number of mitochondrial genomes from tumor tissues and cells. Most of the whole-exome capture kits do not include mtDNA genes [188]. However, because of the abundance of this DNA in human cells, a number of fragments can be retrieved from off-target sequences. Based on the type of protocol, the amount of these reads may often be sufficient for mitochondrial variant detection, allowing data extraction and analysis [189]. Indeed, this increasing amount of mitochondria targeted and off-target sequencing data pushed the development of bioinformatics pipelines that today are available to accurately annotate the mass sequencing-derived mtDNA variants, defining their pathogenic status and heteroplasmy levels.

Appropriate analysis of cancer samples involves: (1) distinguishing whether the variant is germline or somatic, (2) detection of the variant and ensuring it is not a technical false positive, (3) mutation annotation to exclude its polymorphic status, (4) connection to its potential function using prioritization to define pathogenicity status, and (5) mutant load evaluation (heteroplasmic fraction).

The use of paired tumor and normal tissue samples is essential for distinguishing somatic from germline variants. Furthermore, phylogenetic tools offer a valuable support for the recognition of polymorphisms [190], whereas online resources for mitochondrial data analysis, developed over the years to aid a comprehensive knowledge of the onset and development of mitochondria-related diseases, may be useful also in the context of cancer-related mtDNA mutation annotation.

The development of NGS techniques and dedicated bioinformatics pipelines has widely improved the detection of low-level heteroplasmic mtDNA variants. Although it is commonly accepted that they have a pathological effect only over a certain threshold, the significance of low-level heteroplasmy mutations and their possible role in cancer is still unclear. These may represent a predisposition for driving genetic variation in the mtDNA in certain contexts. Moreover, based on computational modeling, positive selection is not a required

Table 17.1: MtDNA mutations reported to affect tumor growth.

Mutation	Gene	Het/Hom status	AA change	Effect on tumor growth	References
m.3243A>G	MT-TL1	Hom	–	2	[17]
m.3460G>A	MT-ND1	Hom	A52T	0	[17]
m.3460G>A	MT-ND1	NA	A52T	1	[195]
m.3571insC	MT-ND1	Het	L89fs	0	[18]
m.3571insC	MT-ND1	Hom	L89fs	2	[18]
m.4605A>G	MT-ND2	NA	K46E	1	[196]
m.4776G>A	MT-ND2	NA	A103T	1	[47]
m.4831G>A	MT-ND2	NA	G121D	1	[196]
m.6124T>C	MT-COI	Het	M74T	1	[191]
m.6930G>A	MT-COI	Het/Hom	G343X	0	[197]
m.6930G>A	MT-COI	Hom	G343X	1	[198,199]
m.8344A>G	MT-TK	NA	–	NA	[200]
m.8363G>A	MT-TK	NA	–	2	[195]
m.8696T>C	MT-ATP6	NA	M57T	0	[120]
m.8993T>G	MT-ATP6	Hom	L156R	1	[201]
m.8993T>G	MT-ATP6	NA	L156R	1	[202]
m.8993T>G	MT-ATP6	Hom	L156R	1	[46]
m.9821insA	MT-TR	NA	–	1	[203]
m.10176G>A	MT-ND3	NA	G40S	0	[120]
m.10398G>A	MT-ND3	NA	A114T	1	[59]
m.10970T>C	MT-ND4	NA	W71R	0	[120]
m.11778G>A	MT-ND4	NA	R340H	1	[195]
m.12084C>T	MT-ND4	NA	S442F	1	[64]
m.12308A>G	MT-TL	NA	–	1	[57]
m.12417insA	MT-ND5	Het	N27fs	1	[16]
m.12417insA	MT-ND5	Het	N27fs	1	[197]
m.12417insA	MT-ND5	Hom	N27fs	2	[16]
m.13289G>A	MT-ND5	NA	G318D	1	[204]
m.13513G>A	MT-ND5	NA	D393N	NA	[200]
m.13885insC	MT-ND6	Hom	L517fs	1	[52]
m.13966A>G	MT-ND5	NA	T544A	1	[64]
m.13997G>A	MT-ND6	Hom	D554G	2	[52]
m.14484T>C	MT-ND6	NA	M64V	1	[195]
m.14787–14790del	MT-CYB	Hom	I14–15del	1	[198,199]
m.15642–15622del	MT-CYB	Het	L299fs	1	[205]

Het, Heteroplasmic; *Hom*, Homoplasmic; *AA*, Amino Acid. Effect on tumor growth: 0, neutral; 1, protumorigenic; 2, antitumorigenic; *NA*, not available.

process to reach expansion and fixation of low-frequency heteroplasmic mutations, indicating drift as a possible mechanism [68,191].

17.5.3 The influence of big data on what we know on mtDNA mutations in cancer

From studies on large databases, it was reported that 60% of solid cancers bear one or more mtDNA mutations [6,192]. The number of variants is heterogeneous in the different cancer types and their prevalence and abundance is in line with the pathogenicity; severe mutations in coding genes (nonsense and frameshift) are less common, whereas regulatory region substitutions are more common [24,192]. A great part of missense nucleotide changes was shown to be selectively neutral and often drifted toward homoplasmy over time. Conversely, damaging variants undergo negative selection and remain heteroplasmic [6]. Unexpectedly, somatic substitutions were reported to be predominately C>T and T>C transitions and did not follow the patterns associated with oxidative damage [24,192]. Rather, such pattern implies most of the mtDNA mutations in cancer occur due to replicative errors [6].

Nevertheless, the use of big dataset was not sufficient so far to fully understand the role of mtDNA mutations in tumor formation and metastasization, whereas it allowed the study of the correlation between nuclear and mitochondrial mutation profile to identified interactions in specific cancer types [193].

As discussed in Section 17.2, variants in the mtDNA of cancer cells may be involved in the adaptability of cell metabolism to the stress environment during tumor progression. In this context, the integration of different data types, such as sequence variants, gene expression and consequences on phenotype was suggested as the most promising research area to clarify the significance of mtDNA mutations [194].

Research perspectives

We may acknowledge today that the mitochondrial genome is no longer neglected. Development of new technologies and consortial efforts in meta-data analyses have resulted in emerging literature reporting novel insights on mtDNA mutations in cancer. Moreover, specific functional studies have pointed out to a double-edged role of mitochondrial genes, which depending on the type of mutations may exert opposite effects on tumor progression, coining a novel definition in oncology, the oncojanus. However, we are still far from understanding all the complexities behind mitochondrial biology, cancer heterogeneity and their crossroads. Further functional studies are required to clarify many of the context-dependent roles of mtDNA mutations in cancer. Here we tackled some of the exciting novel aspects, such as the horizontal mitochondrial transfer and relationship between cancer immunosuppression and metabolism. This and similar research will hopefully soon bring clarity as to what extent mtDNA mutations may find their potential use in the clinical setting, especially in the context of stratifying patients based on cancer metabolic phenotypes.

References

[1] Wallace DC. Mitochondrial diseases in man and mouse. Science 1999;283(5407):1482–8. Available from: https://doi.org/10.1126/science.283.5407.1482.

[2] Polyak K, Li Y, Zhu H, et al. Somatic mutations of the mitochondrial genome in human colorectal tumours. Nat Genet 1998;20(3):291–3. Available from: https://doi.org/10.1038/3108.

[3] Brandon M, Baldi P, Wallace DC. Mitochondrial mutations in cancer. Oncogene 2006;25(34):4647–62. Available from: https://doi.org/10.1038/sj.onc.1209607.

[4] Lu J, Sharma LK, Bai Y. Implications of mitochondrial DNA mutations and mitochondrial dysfunction in tumorigenesis. Cell Res 2009;19(7):802–15. Available from: https://doi.org/10.1038/cr.2009.69.

[5] Larman TC, DePalma SR, Hadjipanayis AG, et al. Spectrum of somatic mitochondrial mutations in five cancers. Proc Natl Acad Sci U S A 2012;109(35):14087–91. Available from: https://doi.org/10.1073/pnas.1211502109.

[6] Ju YS, Alexandrov LB, Gerstung M, et al. Origins and functional consequences of somatic mitochondrial DNA mutations in human cancer. Elife. 2014;3. Available from: https://doi.org/10.7554/eLife.02935.

[7] Gasparre G, Iommarini L, Porcelli AM, et al. An inherited mitochondrial DNA disruptive mutation shifts to homoplasmy in oncocytic tumor cells. Hum Mutat 2009;30(3):391–6. Available from: https://doi.org/10.1002/humu.20870.

[8] He Y, Wu J, Dressman DC, et al. Heteroplasmic mitochondrial DNA mutations in normal and tumour cells. Nature 2010;464(7288):610–14. Available from: https://doi.org/10.1038/nature08802.

[9] Tuppen HAL, Blakely EL, Turnbull DM, Taylor RW. Mitochondrial DNA mutations and human disease. Biochim Biophys Acta 2010;1797(2):113–28. Available from: https://doi.org/10.1016/j.bbabio.2009.09.005.

[10] Chatterjee A, Mambo E, Sidransky D. Mitochondrial DNA mutations in human cancer. Oncogene 2006;25(34):4663–74. Available from: https://doi.org/10.1038/sj.onc.1209604.

[11] Boesch P, Weber-Lotfi F, Ibrahim N, et al. DNA repair in organelles: pathways, organization, regulation, relevance in disease and aging. Biochim Biophys Acta 2011;1813(1):186–200. Available from: https://doi.org/10.1016/j.bbamcr.2010.10.002.

[12] Cline SD. Mitochondrial DNA damage and its consequences for mitochondrial gene expression. Biochim Biophys Acta 2012;1819(9-10):979–91. Available from: https://doi.org/10.1016/j.bbagrm.2012.06.002.

[13] Iommarini L, Calvaruso MA, Kurelac I, Gasparre G, Porcelli AM. Complex I impairment in mitochondrial diseases and cancer: parallel roads leading to different outcomes. Int J Biochem Cell Biol 2013;45(1):47–63. Available from: https://doi.org/10.1016/j.biocel.2012.05.016.

[14] Liu J, Wang L-D, Sun Y-B, et al. Deciphering the signature of selective constraints on cancerous mitochondrial genome. Mol Biol Evol 2012;29(4):1255–61. Available from: https://doi.org/10.1093/molbev/msr290.

[15] Pereira L, Soares P, Maximo V, Samuels DC. Somatic mitochondrial DNA mutations in cancer escape purifying selection and high pathogenicity mutations lead to the oncocytic phenotype: pathogenicity analysis of reported somatic mtDNA mutations in tumors. BMC Cancer 2012;12:53. Available from: https://doi.org/10.1186/1471-2407-12-53.

[16] Park JS, Sharma LK, Li H, et al. A heteroplasmic, not homoplasmic, mitochondrial DNA mutation promotes tumorigenesis via alteration in reactive oxygen species generation and apoptosis. Hum Mol Genet 2009;18(9):1578–89. Available from: https://doi.org/10.1093/hmg/ddp069.

[17] Iommarini L, Kurelac I, Capristo M, et al. Different mtDNA mutations modify tumor progression in dependence of the degree of respiratory complex I impairment. Hum Mol Genet 2014;23(6):1453–66. Available from: https://doi.org/10.1093/hmg/ddt533.

[18] Gasparre G, Kurelac I, Capristo M, et al. A mutation threshold distinguishes the antitumorigenic effects of the mitochondrial gene MTND1, an oncojanus function. Cancer Res 2011;71(19):6220–9. Available from: https://doi.org/10.1158/0008-5472.CAN-11-1042.

[19] Leone G, Abla H, Gasparre G, Porcelli AM, Iommarini L. The oncojanus paradigm of respiratory complex I. Genes (Basel) 2018;9(5). Available from: https://doi.org/10.3390/genes9050243.

[20] Warburg O, Posener K, Negelein E. üeber den Stoffwechsel der Tumoren. üeber den Stoffwechs der Tumoren 1924;319–44.

[21] Koppenol WH, Bounds PL, Dang CV. Otto Warburg's contributions to current concepts of cancer metabolism. Nat Rev Cancer 2011;11(5):325–37. Available from: https://doi.org/10.1038/nrc3038.

[22] Rigo P, Paulus P, Kaschten BJ, et al. Oncological applications of positron emission tomography with fluorine-18 fluorodeoxyglucose. Eur J Nucl Med 1996;23(12):1641–74. Available from: https://doi.org/10.1007/bf01249629.

[23] Warburg O. On respiratory impairment in cancer cells. Science 1956;124(3215):269–70.

[24] Gammage PA, Frezza C. Mitochondrial DNA: the overlooked oncogenome? BMC Biol 2019;17(1):53. Available from: https://doi.org/10.1186/s12915-019-0668-y.

[25] Kurelac I, Vidone M, Girolimetti G, Calabrese C, Gasparre G. Mitochondrial mutations in cancer progression: causative, bystanders, or modifiers of tumorigenesis? In: Mazurek S, Shoshan M, editors. Tumor cell metabolism. Vienna: Springer Vienna; 2015. p. 199–231. Available from: http://dx.doi.org/10.1007/978-3-7091-1824-5_10.

[26] Ward PS, Thompson CB. Metabolic reprogramming: a cancer hallmark even Warburg did not anticipate. Cancer Cell 2012;21(3):297–308. Available from: https://doi.org/10.1016/j.ccr.2012.02.014.

[27] Chen X, Qian Y, Wu S. The Warburg effect: evolving interpretations of an established concept. Free Radic Biol Med 2015;79:253–63. Available from: https://doi.org/10.1016/j.freeradbiomed.2014.08.027.

[28] DeBerardinis RJ, Chandel NS. Fundamentals of cancer metabolism. Sci Adv 2016;2(5):e1600200. Available from: https://doi.org/10.1126/sciadv.1600200.

[29] Martínez-Reyes I, Diebold LP, Kong H, et al. TCA cycle and mitochondrial membrane potential are necessary for diverse biological functions. Mol Cell 2016;61(2):199–209. Available from: https://doi.org/10.1016/j.molcel.2015.12.002.

[30] Vatrinet R, Iommarini L, Kurelac I, De Luise M, Gasparre G, Porcelli AM. Targeting respiratory complex I to prevent the Warburg effect. Int J Biochem Cell Biol 2015;63:41–5. Available from: https://doi.org/10.1016/j.biocel.2015.01.017.

[31] Nowell PC. The clonal evolution of tumor cell populations. Science 1976;194(4260):23–8. Available from: https://doi.org/10.1126/science.959840.

[32] Fouad YA, Aanei C. Revisiting the hallmarks of cancer. Am J Cancer Res 2017;7(5):1016–36.

[33] Smolková K, Plecitá-Hlavatá L, Bellance N, Benard G, Rossignol R, Ježek P. Waves of gene regulation suppress and then restore oxidative phosphorylation in cancer cells. Int J Biochem Cell Biol 2011;43 (7):950–68. Available from: https://doi.org/10.1016/j.biocel.2010.05.003.

[34] Lunt SY, Vander Heiden MG. Aerobic glycolysis: meeting the metabolic requirements of cell proliferation. Annu Rev Cell Dev Biol 2011;27:441–64. Available from: https://doi.org/10.1146/annurev-cellbio-092910-154237.

[35] Owen OE, Kalhan SC, Hanson RW. The key role of anaplerosis and cataplerosis for citric acid cycle function. J Biol Chem 2002;277(34):30409–12. Available from: https://doi.org/10.1074/jbc.R200006200.

[36] Hensley CT, Wasti AT, DeBerardinis RJ. Glutamine and cancer: cell biology, physiology, and clinical opportunities. J Clin Invest 2013;123(9):3678–84. Available from: https://doi.org/10.1172/JCI69600.

[37] Ogino S, Nosho K, Meyerhardt JA, et al. Cohort study of fatty acid synthase expression and patient survival in colon cancer. J Clin Oncol 2008;26(35):5713–20. Available from: https://doi.org/10.1200/JCO.2008.18.2675.

[38] Takahiro T, Shinichi K, Toshimitsu S. Expression of fatty acid synthase as a prognostic indicator in soft tissue sarcomas. Clin Cancer Res 2003;9(6):2204–12.

[39] Mullen AR, Wheaton WW, Jin ES, et al. Reductive carboxylation supports growth in tumour cells with defective mitochondria. Nature 2011;481(7381):385–8. Available from: https://doi.org/10.1038/nature10642.

[40] Cavalli LR, Varella-Garcia M, Liang BC. Diminished tumorigenic phenotype after depletion of mitochondrial DNA. Cell Growth Differ 1997;8(11):1189–98.

[41] Tan AS, Baty JW, Dong L-F, et al. Mitochondrial genome acquisition restores respiratory function and tumorigenic potential of cancer cells without mitochondrial DNA. Cell Metab 2015;21(1):81–94. Available from: https://doi.org/10.1016/j.cmet.2014.12.003.

[42] Semenza GL. Oxygen-dependent regulation of mitochondrial respiration by hypoxia-inducible factor 1. Biochem J 2007;405(1):1–9. Available from: https://doi.org/10.1042/BJ20070389.

[43] Ruas JL, Poellinger L. Hypoxia-dependent activation of HIF into a transcriptional regulator. Semin Cell Dev Biol 2005;16(4-5):514–22. Available from: https://doi.org/10.1016/j.semcdb.2005.04.001.

[44] Al Tameemi W, Dale TP, Al-Jumaily RMK, Forsyth NR. Hypoxia-modified cancer cell metabolism. Front Cell Dev Biol 2019;7:4. Available from: https://doi.org/10.3389/fcell.2019.00004.

[45] Luo W, Hu H, Chang R, et al. Pyruvate kinase M2 is a PHD3-stimulated coactivator for hypoxia-inducible factor 1. Cell 2011;145(5):732–44. Available from: https://doi.org/10.1016/j.cell.2011.03.054.

[46] Shidara Y, Yamagata K, Kanamori T, et al. Positive contribution of pathogenic mutations in the mitochondrial genome to the promotion of cancer by prevention from apoptosis. Cancer Res 2005;65(5):1655–63. Available from: https://doi.org/10.1158/0008-5472.CAN-04-2012.

[47] Sun W, Zhou S, Chang SS, McFate T, Verma A, Califano JA. Mitochondrial mutations contribute to HIF1alpha accumulation via increased reactive oxygen species and up-regulated pyruvate dehydrogenease kinase 2 in head and neck squamous cell carcinoma. Clin Cancer Res 2009;15(2):476–84. Available from: https://doi.org/10.1158/1078-0432.CCR-08-0930.

[48] Pagé EL, Robitaille GA, Pouysségur J, Richard DE. Induction of hypoxia-inducible factor-1alpha by transcriptional and translational mechanisms. J Biol Chem 2002;277(50):48403–9. Available from: https://doi.org/10.1074/jbc.M209114200.

[49] Calabrese C, Iommarini L, Kurelac I, et al. Respiratory complex I is essential to induce a Warburg profile in mitochondria-defective tumor cells. Cancer Metab 2013;1(1):11. Available from: https://doi.org/10.1186/2049-3002-1-11.

[50] De Luise M, Guarnieri V, Ceccarelli C, D'Agruma L, Porcelli AM, Gasparre G. A nonsense mitochondrial DNA mutation associates with dysfunction of HIF1α in a Von Hippel-Lindau renal oncocytoma. Oxid Med Cell Longev 2019;2019:8069583. Available from: https://doi.org/10.1155/2019/8069583.

[51] Yuan Y, Wang W, Li H, et al. Nonsense and missense mutation of mitochondrial ND6 gene promotes cell migration and invasion in human lung adenocarcinoma. BMC Cancer 2015;15:346. Available from: https://doi.org/10.1186/s12885-015-1349-z.

[52] Ishikawa K, Takenaga K, Akimoto M, et al. ROS-generating mitochondrial DNA mutations can regulate tumor cell metastasis. Science 2008;320(5876):661–4. Available from: https://doi.org/10.1126/science.1156906.

[53] Schafer ZT, Grassian AR, Song L, et al. Antioxidant and oncogene rescue of metabolic defects caused by loss of matrix attachment. Nature 2009;461(7260):109–13. Available from: https://doi.org/10.1038/nature08268.

[54] Jiang L, Shestov AA, Swain P, et al. Reductive carboxylation supports redox homeostasis during anchorage-independent growth. Nature 2016;532(7598):255–8. Available from: https://doi.org/10.1038/nature17393.

[55] Moloney JN, Cotter TG. ROS signalling in the biology of cancer. Semin Cell Dev Biol 2018;80:50–64. Available from: https://doi.org/10.1016/j.semcdb.2017.05.023.

[56] Sabharwal SS, Schumacker PT. Mitochondrial ROS in cancer: initiators, amplifiers or an Achilles' heel? Nat Rev Cancer 2014;14(11):709–21. Available from: https://doi.org/10.1038/nrc3803.

[57] Kulawiec M, Owens KM, Singh KK. Cancer cell mitochondria confer apoptosis resistance and promote metastasis. Cancer Biol Ther 2009;8(14):1378–85. Available from: https://doi.org/10.4161/cbt.8.14.8751.

[58] Qiao M, Sheng S, Pardee AB. Metastasis and AKT activation. Cell Cycle 2008;7(19):2991–6. Available from: https://doi.org/10.4161/cc.7.19.6784.

[59] Kulawiec M, Owens KM, Singh KK. mtDNA G10398A variant in African-American women with breast cancer provides resistance to apoptosis and promotes metastasis in mice. J Hum Genet 2009;54 (11):647–54. Available from: https://doi.org/10.1038/jhg.2009.89.

[60] Fischer ANM, Fuchs E, Mikula M, Huber H, Beug H, Mikulits W. PDGF essentially links TGF-beta signaling to nuclear beta-catenin accumulation in hepatocellular carcinoma progression. Oncogene 2007;26 (23):3395–405. Available from: https://doi.org/10.1038/sj.onc.1210121.

[61] Weber GF. Metabolism in cancer metastasis. Int J Cancer 2016;138(9):2061–6. Available from: https://doi.org/10.1002/ijc.29839.

[62] Aguilar E, Marin de Mas I, Zodda E, et al. Metabolic reprogramming and dependencies associated with epithelial cancer stem cells independent of the epithelial-mesenchymal transition program. Stem Cell 2016;34(5):1163–76. Available from: https://doi.org/10.1002/stem.2286.

[63] Fong MY, Zhou W, Liu L, et al. Breast-cancer-secreted miR-122 reprograms glucose metabolism in premetastatic niche to promote metastasis. Nat Cell Biol 2015;17(2):183–94. Available from: https://doi.org/10.1038/ncb3094.

[64] Imanishi H, Hattori K, Wada R, et al. Mitochondrial DNA mutations regulate metastasis of human breast cancer cells. PLoS One 2011;6(8):e23401. Available from: https://doi.org/10.1371/journal.pone.0023401.

[65] Yu L, Lu M, Jia D, et al. Modeling the genetic regulation of cancer metabolism: interplay between glycolysis and oxidative phosphorylation. Cancer Res 2017;77(7):1564–74. Available from: https://doi.org/10.1158/0008-5472.CAN-16-2074.

[66] Sansone P, Savini C, Kurelac I, et al. Packaging and transfer of mitochondrial DNA via exosomes regulate escape from dormancy in hormonal therapy-resistant breast cancer. Proc Natl Acad Sci U S A 2017;114(43):E9066–75. Available from: https://doi.org/10.1073/pnas.1704862114.

[67] Shpak M, Goldberg MM, Cowperthwaite MC. Rapid and convergent evolution in the glioblastoma multiforme genome. Genomics 2015;105(3):159–67. Available from: https://doi.org/10.1016/j.ygeno.2014.12.010.

[68] Coller HA, Khrapko K, Bodyak ND, Nekhaeva E, Herrero-Jimenez P, Thilly WG. High frequency of homoplasmic mitochondrial DNA mutations in human tumors can be explained without selection. Nat Genet 2001;28(2):147–50. Available from: https://doi.org/10.1038/88859.

[69] Yu M. Somatic mitochondrial DNA mutations in human cancers. Adv Clin Chem 2012;57:99–138.

[70] McMahon S, LaFramboise T. Mutational patterns in the breast cancer mitochondrial genome, with clinical correlates. Carcinogenesis 2014;35(5):1046–54. Available from: https://doi.org/10.1093/carcin/bgu012.

[71] Li X, Guo X, Li D, et al. Multi-regional sequencing reveals intratumor heterogeneity and positive selection of somatic mtDNA mutations in hepatocellular carcinoma and colorectal cancer. Int J Cancer 2018;143(5):1143–52. Available from: https://doi.org/10.1002/ijc.31395.

[72] Li D, Du X, Guo X, et al. Site-specific selection reveals selective constraints and functionality of tumor somatic mtDNA mutations. J Exp Clin Cancer Res 2017;36(1):168. Available from: https://doi.org/10.1186/s13046-017-0638-6.

[73] Grandhi S, Bosworth C, Maddox W, et al. Heteroplasmic shifts in tumor mitochondrial genomes reveal tissue-specific signals of relaxed and positive selection. Hum Mol Genet 2017;26(15):2912–22. Available from: https://doi.org/10.1093/hmg/ddx172.

[74] Volpe A, Novara G, Antonelli A, et al. Chromophobe renal cell carcinoma (RCC): oncological outcomes and prognostic factors in a large multicentre series. BJU Int 2012;110(1):76–83. Available from: https://doi.org/10.1111/j.1464-410X.2011.10690.x.

[75] Otten ABC, Smeets HJM. Evolutionary defined role of the mitochondrial DNA in fertility, disease and ageing. Hum Reprod Update 2015;21(5):671–89. Available from: https://doi.org/10.1093/humupd/dmv024.

[76] Mambo E, Gao X, Cohen Y, Guo Z, Talalay P, Sidransky D. Electrophile and oxidant damage of mitochondrial DNA leading to rapid evolution of homoplasmic mutations. Proc Natl Acad Sci U S A 2003;100(4):1838–43. Available from: https://doi.org/10.1073/pnas.0437910100.

[77] Lee H-C, Li S-H, Lin J-C, Wu C-C, Yeh D-C, Wei Y-H. Somatic mutations in the D-loop and decrease in the copy number of mitochondrial DNA in human hepatocellular carcinoma. Mutat Res 2004;547(1-2):71–8. Available from: https://doi.org/10.1016/j.mrfmmm.2003.12.011.

[78] Turner CJ, Granycome C, Hurst R, et al. Systematic segregation to mutant mitochondrial DNA and accompanying loss of mitochondrial DNA in human NT2 teratocarcinoma cybrids. Genetics 2005;170 (4):1879–85. Available from: https://doi.org/10.1534/genetics.105.043653.

[79] Ricketts CJ, De Cubas AA, Fan H, et al. The cancer genome atlas comprehensive molecular characterization of renal cell carcinoma. Cell Rep 2018;23(12):3698. Available from: https://doi.org/10.1016/j.celrep.2018.06.032.

[80] Reznik E, Miller ML, Şenbabaoğlu Y, et al. Mitochondrial DNA copy number variation across human cancers. Elife 2016;5. Available from: https://doi.org/10.7554/eLife.10769.

[81] Palikaras K, Tavernarakis N. Mitochondrial homeostasis: the interplay between mitophagy and mitochondrial biogenesis. Exp Gerontol 2014;56:182–8. Available from: https://doi.org/10.1016/j.exger.2014.01.021.

[82] Guo JY, Karsli-Uzunbas G, Mathew R, et al. Autophagy suppresses progression of K-ras-induced lung tumors to oncocytomas and maintains lipid homeostasis. Genes Dev 2013;27(13):1447–61. Available from: https://doi.org/10.1101/gad.219642.113.

[83] Twig G, Elorza A, Molina AJA, et al. Fission and selective fusion govern mitochondrial segregation and elimination by autophagy. EMBO J 2008;27(2):433–46. Available from: https://doi.org/10.1038/sj.emboj.7601963.

[84] Lin D-S, Huang Y-W, Ho C-S, et al. Oxidative insults and mitochondrial DNA mutation promote enhanced autophagy and mitophagy compromising cell viability in pluripotent cell model of mitochondrial disease. Cells 2019;8(1). Available from: https://doi.org/10.3390/cells8010065.

[85] Gilkerson RW, De Vries RLA, Lebot P, et al. Mitochondrial autophagy in cells with mtDNA mutations results from synergistic loss of transmembrane potential and mTORC1 inhibition. Hum Mol Genet 2012;21(5):978–90. Available from: https://doi.org/10.1093/hmg/ddr529.

[86] Yusoff AAM, Abdullah WSW, Khair SZNM, Radzak SMA. A comprehensive overview of mitochondrial DNA 4977-bp deletion in cancer studies. Oncol Rev 2019;13(1):409. Available from: https://doi.org/10.4081/oncol.2019.409.

[87] Diaz F, Bayona-Bafaluy MP, Rana M, Mora M, Hao H, Moraes CT. Human mitochondrial DNA with large deletions repopulates organelles faster than full-length genomes under relaxed copy number control. Nucleic Acids Res 2002;30(21):4626–33. Available from: https://doi.org/10.1093/nar/gkf602.

[88] Clark KA, Howe DK, Gafner K, et al. Selfish little circles: transmission bias and evolution of large deletion-bearing mitochondrial DNA in *Caenorhabditis briggsae* nematodes. PLoS One 2012;7(7):e41433. Available from: https://doi.org/10.1371/journal.pone.0041433.

[89] Viale A, Corti D, Draetta GF. Tumors and mitochondrial respiration: a neglected connection. Cancer Res 2015;75(18):3685–6. Available from: https://doi.org/10.1158/0008-5472.CAN-15-0491.

[90] Jayaprakash AD, Benson EK, Gone S, et al. Stable heteroplasmy at the single-cell level is facilitated by intercellular exchange of mtDNA. Nucleic Acids Res 2015;43(4):2177–87. Available from: https://doi.org/10.1093/nar/gkv052.

[91] Herst PM, Dawson RH, Berridge MV. Intercellular communication in tumor biology: a role for mitochondrial transfer. Front Oncol 2018;8:344. Available from: https://doi.org/10.3389/fonc.2018.00344.

[92] Spees JL, Olson SD, Whitney MJ, Prockop DJ. Mitochondrial transfer between cells can rescue aerobic respiration. Proc Natl Acad Sci U S A 2006;103(5):1283–8. Available from: https://doi.org/10.1073/pnas.0510511103.

[93] Dong L-F, Kovarova J, Bajzikova M, et al. Horizontal transfer of whole mitochondria restores tumorigenic potential in mitochondrial DNA-deficient cancer cells. Elife 2017;6. Available from: https://doi.org/10.7554/eLife.22187.

[94] Graef M, Nunnari J. Mitochondria regulate autophagy by conserved signalling pathways. EMBO J 2011;30(11):2101–14. Available from: https://doi.org/10.1038/emboj.2011.104.

[95] Cho YM, Kim JH, Kim M, et al. Mesenchymal stem cells transfer mitochondria to the cells with virtually no mitochondrial function but not with pathogenic mtDNA mutations. PLoS One 2012;7(3):e32778. Available from: https://doi.org/10.1371/journal.pone.0032778.

[96] Kurelac I, Iommarini L, Vatrinet R, et al. Inducing cancer indolence by targeting mitochondrial complex I is potentiated by blocking macrophage-mediated adaptive responses. Nat Commun 2019;10(1):903. Available from: https://doi.org/10.1038/s41467-019-08839-1.

[97] Schulze A, Yuneva M. The big picture: exploring the metabolic cross-talk in cancer. Dis Model Mech 2018;11(8). Available from: https://doi.org/10.1242/dmm.036673.

[98] Harel M, Ortenberg R, Varanasi SK, et al. Proteomics of melanoma response to immunotherapy reveals mitochondrial dependence. Cell 2019;179(1):236. Available from: https://doi.org/10.1016/j.cell.2019.08.012 250.e18.

[99] DeBerardinis RJ, Cheng T. Q's next: the diverse functions of glutamine in metabolism, cell biology and cancer. Oncogene 2010;29(3):313–24. Available from: https://doi.org/10.1038/onc.2009.358.

[100] De Luise M, Girolimetti G, Okere B, Porcelli AM, Kurelac I, Gasparre G. Molecular and metabolic features of oncocytomas: seeking the blueprints of indolent cancers. Biochim Biophys Acta Bioenerg 2017;1858(8):591–601. Available from: https://doi.org/10.1016/j.bbabio.2017.01.009.

[101] Eirin A, Lerman A, Lerman LO. The emerging role of mitochondrial targeting in kidney disease. Handb Exp Pharmacol 2017;240:229–50. Available from: https://doi.org/10.1007/164_2016_6.

[102] Müller-Höcker J, Schäfer S, Krebs S, et al. Oxyphil cell metaplasia in the parathyroids is characterized by somatic mitochondrial DNA mutations in NADH dehydrogenase genes and cytochrome c oxidase activity-impairing genes. Am J Pathol 2014;184(11):2922–35. Available from: https://doi.org/10.1016/j.ajpath.2014.07.015.

[103] Gasparre G, Romeo G, Rugolo M, Porcelli AM. Learning from oncocytic tumors: why choose inefficient mitochondria? Biochim Biophys Acta 2011;1807(6):633–42. Available from: https://doi.org/10.1016/j.bbabio.2010.08.006.

[104] Savagner F, Mirebeau D, Jacques C, et al. PGC-1-related coactivator and targets are upregulated in thyroid oncocytoma. Biochem Biophys Res Commun 2003;310(3):779–84. Available from: https://doi.org/10.1016/j.bbrc.2003.09.076.

[105] Gasparre G, Hervouet E, de Laplanche E, et al. Clonal expansion of mutated mitochondrial DNA is associated with tumor formation and complex I deficiency in the benign renal oncocytoma. Hum Mol Genet 2008;17(7):986–95. Available from: https://doi.org/10.1093/hmg/ddm371.

[106] Porcelli AM, Ghelli A, Ceccarelli C, et al. The genetic and metabolic signature of oncocytic transformation implicates HIF1alpha destabilization. Hum Mol Genet 2010;19(6):1019–32. Available from: https://doi.org/10.1093/hmg/ddp566.

[107] Tallini G. Oncocytic tumours. Virchows Arch 1998;433(1):5–12.

[108] Máximo V, Soares P, Lima J, Cameselle-Teijeiro J, Sobrinho-Simões M. Mitochondrial DNA somatic mutations (point mutations and large deletions) and mitochondrial DNA variants in human thyroid pathology: a study with emphasis on Hürthle cell tumors. Am J Pathol 2002;160(5):1857–65. Available from: https://doi.org/10.1016/S0002-9440(10)61132-7.

[109] Zimmermann FA, Mayr JA, Feichtinger R, et al. Respiratory chain complex I is a mitochondrial tumor suppressor of oncocytic tumors. Front Biosci (Elite Ed) 2011;3:315–25.

[110] Canzian F, Amati P, Harach HR, et al. A gene predisposing to familial thyroid tumors with cell oxyphilia maps to chromosome 19p13.2. Am J Hum Genet 1998;63(6):1743–8. Available from: https://doi.org/10.1086/302164.

[111] Schonewille H, Haak HL, Kerkhofs H, Gerrits WB. The effect of anticoagulants on the size of platelets in blood smears in the course of time. Clin Lab Haematol 1991;13(1):67–74.

[112] Weirich G, Glenn G, Junker K, et al. Familial renal oncocytoma: clinicopathological study of 5 families. J Urol 1998;160(2):335–40.

[113] Kurelac I, Salfi NC, Ceccarelli C, et al. Human papillomavirus infection and pathogenic mitochondrial DNA mutation in bilateral multinodular oncocytic hyperplasia of the parotid. Pathology 2014;46 (3):250–3. Available from: https://doi.org/10.1097/PAT.0000000000000079.
[114] Guo JY, White E. Autophagy is required for mitochondrial function, lipid metabolism, growth, and fate of KRAS(G12D)-driven lung tumors. Autophagy 2013;9(10):1636–8. Available from: https://doi.org/10.4161/auto.26123.
[115] Lee J, Ham S, Lee MH, et al. Dysregulation of Parkin-mediated mitophagy in thyroid Hurthle cell tumors. Carcinogenesis 2015;36(11):1407–18. Available from: https://doi.org/10.1093/carcin/bgv122.
[116] Singh KK, Russell J, Sigala B, Zhang Y, Williams J, Keshav KF, et al. Mitochondrial DNA determines the cellular response to cancer therapeutic agents. Oncogene 1999;18(48):6641–6. Available from: https://doi.org/10.1038/sj.onc.1203056.
[117] van Gisbergen MW, Voets AM, Starmans MHW, et al. How do changes in the mtDNA and mitochondrial dysfunction influence cancer and cancer therapy? Challenges, opportunities and models. Mutat Res Rev Mutat Res 2015;764:16–30. Available from: https://doi.org/10.1016/j.mrrev.2015.01.001.
[118] Carew JS, Zhou Y, Albitar M, Carew JD, Keating MJ, Huang P, et al. Mitochondrial DNA mutations in primary leukemia cells after chemotherapy: clinical significance and therapeutic implications. Leukemia 2003;17(8):1437–47. Available from: https://doi.org/10.1038/sj.leu.2403043.
[119] Qian W, Nishikawa M, Haque AM, et al. Mitochondrial density determines the cellular sensitivity to cisplatin-induced cell death. Am J Physiol, Cell Physiol 2005;289(6):C1466–75. Available from: https://doi.org/10.1152/ajpcell.00265.2005.
[120] Mizutani S, Miyato Y, Shidara Y, et al. Mutations in the mitochondrial genome confer resistance of cancer cells to anticancer drugs. Cancer Sci 2009;100(9):1680–7. Available from: https://doi.org/10.1111/j.1349-7006.2009.01238.x.
[121] Gentric G, Kieffer Y, Mieulet V, et al. PML-regulated mitochondrial metabolism enhances chemosensitivity in human ovarian cancers. Cell Metab 2019;29(1):156–73. Available from: https://doi.org/10.1016/j.cmet.2018.09.002 e10.
[122] Farge T, Saland E, de Toni F, et al. Chemotherapy-resistant human acute myeloid leukemia cells are not enriched for leukemic stem cells but require oxidative metabolism. Cancer Discov 2017;7(7):716–35. Available from: https://doi.org/10.1158/2159-8290.CD-16-0441.
[123] Guerra F, Arbini AA, Moro L. Mitochondria and cancer chemoresistance. Biochim Biophys Acta Bioenerg 2017;1858(8):686–99. Available from: https://doi.org/10.1016/j.bbabio.2017.01.012.
[124] Bokil A, Sancho P. Mitochondrial determinants of chemoresistance. Cancer Drug Resistance 2019;2:634–46. Available from: https://doi.org/10.20517/cdr.2019.46.
[125] Cruz-Bermúdez A, Laza-Briviesca R, Vicente-Blanco RJ, et al. Cisplatin resistance involves a metabolic reprogramming through ROS and PGC-1α in NSCLC which can be overcome by OXPHOS inhibition. Free Radic Biol Med 2019;135:167–81. Available from: https://doi.org/10.1016/j.freeradbiomed.2019.03.009.
[126] Koh KX, Tan GH, Hui Low SH, et al. Acquired resistance to PI3K/mTOR inhibition is associated with mitochondrial DNA mutation and glycolysis. Oncotarget 2017;8(66):110133–44. Available from: https://doi.org/10.18632/oncotarget.22655.
[127] Garrido N, Pérez-Martos A, Faro M, et al. Cisplatin-mediated impairment of mitochondrial DNA metabolism inversely correlates with glutathione levels. Biochem J 2008;414(1):93–102. Available from: https://doi.org/10.1042/BJ20071615.
[128] Olivero OA, Semino C, Kassim A, Lopez-Larraza DM, Poirier MC. Preferential binding of cisplatin to mitochondrial DNA of Chinese hamster ovary cells. Mutat Res 1995;346(4):221–30. Available from: https://doi.org/10.1016/0165-7992(95)90039-x.
[129] Yang Z, Schumaker LM, Egorin MJ, Zuhowski EG, Guo Z, Cullen KJ. Cisplatin preferentially binds mitochondrial DNA and voltage-dependent anion channel protein in the mitochondrial membrane of head and neck squamous cell carcinoma: possible role in apoptosis. Clin Cancer Res 2006;12 (19):5817–25. Available from: https://doi.org/10.1158/1078-0432.CCR-06-1037.

[130] Yao Z, Jones AWE, Fassone E, et al. PGC-1β mediates adaptive chemoresistance associated with mitochondrial DNA mutations. Oncogene 2013;32(20):2592–600. Available from: https://doi.org/10.1038/onc.2012.259.

[131] Catanzaro D, Gaude E, Orso G, et al. Inhibition of glucose-6-phosphate dehydrogenase sensitizes cisplatin-resistant cells to death. Oncotarget 2015;6(30). Available from: https://doi.org/10.18632/oncotarget.4945.

[132] Girolimetti G, Guerra F, Iommarini L, et al. Platinum-induced mitochondrial DNA mutations confer lower sensitivity to paclitaxel by impairing tubulin cytoskeletal organization. Hum Mol Genet 2017;26(15):2961–74. Available from: https://doi.org/10.1093/hmg/ddx186.

[133] Ma L, Wang R, Duan H, Nan Y, Wang Q, Jin F. Mitochondrial dysfunction rather than mtDNA sequence mutation is responsible for the multi-drug resistance of small cell lung cancer. Oncol Rep 2015;34(6):3238–46. Available from: https://doi.org/10.3892/or.2015.4315.

[134] Guerra F, Perrone AM, Kurelac I, et al. Mitochondrial DNA mutation in serous ovarian cancer: implications for mitochondria-coded genes in chemoresistance. J Clin Oncol 2012;30(36):e373–8. Available from: https://doi.org/10.1200/JCO.2012.43.5933.

[135] Berridge MV, Dong L, Neuzil J. Mitochondrial DNA in tumor initiation, progression, and metastasis: role of horizontal mtDNA transfer. Cancer Res 2015;75(16):3203–8. Available from: https://doi.org/10.1158/0008-5472.CAN-15-0859.

[136] Wang X, Gerdes H-H. Transfer of mitochondria via tunneling nanotubes rescues apoptotic PC12 cells. Cell Death Differ 2015;22(7):1181–91. Available from: https://doi.org/10.1038/cdd.2014.211.

[137] Pasquier J, Guerrouahen BS, Al Thawadi H, et al. Preferential transfer of mitochondria from endothelial to cancer cells through tunneling nanotubes modulates chemoresistance. J Transl Med 2013;11:94. Available from: https://doi.org/10.1186/1479-5876-11-94.

[138] Moschoi R, Imbert V, Nebout M, et al. Protective mitochondrial transfer from bone marrow stromal cells to acute myeloid leukemic cells during chemotherapy. Blood 2016;128(2):253–64. Available from: https://doi.org/10.1182/blood-2015-07-655860.

[139] Sica V, Bravo-San Pedro JM, Stoll G, Kroemer G. Oxidative phosphorylation as a potential therapeutic target for cancer therapy. Int J Cancer 2019;146(1):10–17. Available from: https://doi.org/10.1002/ijc.32616.

[140] Sancho P, Barneda D, Heeschen C. Hallmarks of cancer stem cell metabolism. Br J Cancer 2016;114(12):1305–12. Available from: https://doi.org/10.1038/bjc.2016.152.

[141] Kuntz EM, Baquero P, Michie AM, et al. Targeting mitochondrial oxidative phosphorylation eradicates therapy-resistant chronic myeloid leukemia stem cells. Nat Med 2017;23(10):1234–40. Available from: https://doi.org/10.1038/nm.4399.

[142] Kong Q, Beel JA, Lillehei KO. A threshold concept for cancer therapy. Med Hypotheses 2000;55(1):29–35. Available from: https://doi.org/10.1054/mehy.1999.0982.

[143] Zhou Y, Hileman EO, Plunkett W, Keating MJ, Huang P. Free radical stress in chronic lymphocytic leukemia cells and its role in cellular sensitivity to ROS-generating anticancer agents. Blood 2003;101(10):4098–104. Available from: https://doi.org/10.1182/blood-2002-08-2512.

[144] Altenberg GA. Structure of multidrug-resistance proteins of the ATP-binding cassette (ABC) superfamily. Curr Med Chem Anticancer Agents 2004;4(1):53–62. Available from: https://doi.org/10.2174/1568011043482160.

[145] Ferrari D, Stepczynska A, Los M, Wesselborg S, Schulze-Osthoff K. Differential regulation and ATP requirement for caspase-8 and caspase-3 activation during CD95- and anticancer drug-induced apoptosis. J Exp Med 1998;188(5):979–84. Available from: https://doi.org/10.1084/jem.188.5.979.

[146] Kroemer G, Galluzzi L, Brenner C. Mitochondrial membrane permeabilization in cell death. Physiol Rev 2007;87(1):99–163. Available from: https://doi.org/10.1152/physrev.00013.2006.

[147] Guo Y, Cai Q, Samuels DC, et al. The use of next generation sequencing technology to study the effect of radiation therapy on mitochondrial DNA mutation. Mutat Res 2012;744(2):154–60. Available from: https://doi.org/10.1016/j.mrgentox.2012.02.006.

[148] Yamazaki H, Yoshida K, Yoshioka Y, et al. Impact of mitochondrial DNA on hypoxic radiation sensitivity in human fibroblast cells and osteosarcoma cell lines. Oncol Rep 2008;19(6):1545–9.
[149] Wardell TM, Ferguson E, Chinnery PF, et al. Changes in the human mitochondrial genome after treatment of malignant disease. Mutat Res 2003;525(1-2):19–27. Available from: https://doi.org/10.1016/s0027-5107(02)00313-5.
[150] Jiang W-W, Rosenbaum E, Mambo E, et al. Decreased mitochondrial DNA content in posttreatment salivary rinses from head and neck cancer patients. Clin Cancer Res 2006;12(5):1564–9. Available from: https://doi.org/10.1158/1078-0432.CCR-05-1471.
[151] Bol V, Bol A, Bouzin C, et al. Reprogramming of tumor metabolism by targeting mitochondria improves tumor response to irradiation. Acta Oncol 2015;54(2):266–74. Available from: https://doi.org/10.3109/0284186X.2014.932006.
[152] Cloos CR, Daniels DH, Kalen A, et al. Mitochondrial DNA depletion induces radioresistance by suppressing G2 checkpoint activation in human pancreatic cancer cells. Radiat Res 2009;171(5):581–7. Available from: https://doi.org/10.1667/RR1395.1.
[153] Gray LH, Conger AD, Ebert M, Hornsey S, Scott OC. The concentration of oxygen dissolved in tissues at the time of irradiation as a factor in radiotherapy. Br J Radiol 1953;26(312):638–48. Available from: https://doi.org/10.1259/0007-1285-26-312-638.
[154] Rockwell S, Dobrucki IT, Kim EY, Marrison ST, Vu VT. Hypoxia and radiation therapy: past history, ongoing research, and future promise. Curr Mol Med 2009;9(4):442–58. Available from: https://doi.org/10.2174/156652409788167087.
[155] Crabtree HG, Cramer W. The action of radium on cancer cells. II. Some factors determining the susceptibility of cancer cells to radium. Proc R Soc B: Biol Sci 1933;113(782):238–50. Available from: https://doi.org/10.1098/rspb.1933.0044.
[156] Brown JM. The hypoxic cell: a target for selective cancer therapy—eighteenth Bruce F. Cain Memorial Award lecture. Cancer Res 1999;59(23):5863–70.
[157] Graham K, Unger E. Overcoming tumor hypoxia as a barrier to radiotherapy, chemotherapy and immunotherapy in cancer treatment. Int J Nanomed 2018;13:6049–58. Available from: https://doi.org/10.2147/IJN.S140462.
[158] Iommarini L, Porcelli AM, Gasparre G, Kurelac I. Non-canonical mechanisms regulating hypoxia-inducible factor 1 alpha in cancer. Front Oncol 2017;7:286. Available from: https://doi.org/10.3389/fonc.2017.00286.
[159] Pelicano H, Martin DS, Xu R-H, Huang P. Glycolysis inhibition for anticancer treatment. Oncogene 2006;25(34):4633–46. Available from: https://doi.org/10.1038/sj.onc.1209597.
[160] Akins NS, Nielson TC, Le HV. Inhibition of glycolysis and glutaminolysis: an emerging drug discovery approach to combat cancer. Curr Top Med Chem 2018;18(6):494–504. Available from: https://doi.org/10.2174/1568026618666180523111351.
[161] Meric-Bernstam F, Lee RJ, Carthon BC, et al. CB-839, a glutaminase inhibitor, in combination with cabozantinib in patients with clear cell and papillary metastatic renal cell cancer (mRCC): results of a phase I study. J Clin Oncol 2019;37(7 Suppl.):549. Available from: https://doi.org/10.1200/JCO.2019.37.7_suppl.549 -549.
[162] Katt WP, Cerione RA. Glutaminase regulation in cancer cells: a druggable chain of events. Drug Discovery Today 2014;19(4):450–7. Available from: https://doi.org/10.1016/j.drudis.2013.10.008.
[163] Owen MR, Doran E, Halestrap AP. Evidence that metformin exerts its anti-diabetic effects through inhibition of complex 1 of the mitochondrial respiratory chain. Biochem J 2000;348(Pt 3):607–14.
[164] Hou X, Song J, Li X-N, et al. Metformin reduces intracellular reactive oxygen species levels by upregulating expression of the antioxidant thioredoxin via the AMPK-FOXO3 pathway. Biochem Biophys Res Commun 2010;396(2):199–205. Available from: https://doi.org/10.1016/j.bbrc.2010.04.017.
[165] Li B, Chauvin C, De Paulis D, et al. Inhibition of complex I regulates the mitochondrial permeability transition through a phosphate-sensitive inhibitory site masked by cyclophilin D. Biochim Biophys Acta 2012;1817(9):1628–34. Available from: https://doi.org/10.1016/j.bbabio.2012.05.011.

[166] Kurelac I, Umesh Ganesh N, Iorio M, Porcelli AM, Gasparre G. The multifaceted effects of metformin on tumor microenvironment. Semin Cell Dev Biol 2019;98:90–7. Available from: https://doi.org/10.1016/j.semcdb.2019.05.010.
[167] Sanli T, Storozhuk Y, Linher-Melville K, et al. Ionizing radiation regulates the expression of AMP-activated protein kinase (AMPK) in epithelial cancer cells: modulation of cellular signals regulating cell cycle and survival. Radiother Oncol 2012;102(3):459–65. Available from: https://doi.org/10.1016/j.radonc.2011.11.014.
[168] Muaddi H, Chowdhury S, Vellanki R, Zamiara P, Koritzinsky M. Contributions of AMPK and p53 dependent signaling to radiation response in the presence of metformin. Radiother Oncol 2013;108(3):446–50. Available from: https://doi.org/10.1016/j.radonc.2013.06.014.
[169] Song CW, Lee H, Dings RPM, et al. Metformin kills and radiosensitizes cancer cells and preferentially kills cancer stem cells. Sci Rep 2012;2:362. Available from: https://doi.org/10.1038/srep00362.
[170] Storozhuk Y, Hopmans SN, Sanli T, et al. Metformin inhibits growth and enhances radiation response of non-small cell lung cancer (NSCLC) through ATM and AMPK. Br J Cancer 2013;108(10):2021–32. Available from: https://doi.org/10.1038/bjc.2013.187.
[171] Iliopoulos D, Hirsch HA, Struhl K. Metformin decreases the dose of chemotherapy for prolonging tumor remission in mouse xenografts involving multiple cancer cell types. Cancer Res 2011;71(9):3196–201. Available from: https://doi.org/10.1158/0008-5472.CAN-10-3471.
[172] Zhang H-H, Guo X-L. Combinational strategies of metformin and chemotherapy in cancers. Cancer Chemother Pharmacol 2016;78(1):13–26. Available from: https://doi.org/10.1007/s00280-016-3037-3.
[173] Ashton TM, McKenna WG, Kunz-Schughart LA, Higgins GS. Oxidative phosphorylation as an emerging target in cancer therapy. Clin Cancer Res 2018;24(11):2482–90. Available from: https://doi.org/10.1158/1078-0432.CCR-17-3070.
[174] Vogt A, Schmid S, Heinimann K, et al. Multiple primary tumours: challenges and approaches, a review. ESMO Open 2017;2(2):e000172. Available from: https://doi.org/10.1136/esmoopen-2017-000172.
[175] Perrone AM, Girolimetti G, Procaccini M, et al. Potential for mitochondrial DNA sequencing in the differential diagnosis of gynaecological malignancies. Int J Mol Sci 2018;19(7). Available from: https://doi.org/10.3390/ijms19072048.
[176] Guerra F, Girolimetti G, Perrone AM, et al. Mitochondrial DNA genotyping efficiently reveals clonality of synchronous endometrial and ovarian cancers. Mod Pathol 2014;27(10):1412–20. Available from: https://doi.org/10.1038/modpathol.2014.39.
[177] Kloss-Brandstätter A, Weissensteiner H, Erhart G, et al. Validation of next-generation sequencing of entire mitochondrial genomes and the diversity of mitochondrial DNA mutations in oral squamous cell carcinoma. PLoS One 2015;10(8):e0135643. Available from: https://doi.org/10.1371/journal.pone.0135643.
[178] Kurelac I, Lang M, Zuntini R, et al. Searching for a needle in the haystack: comparing six methods to evaluate heteroplasmy in difficult sequence context. Biotechnol Adv 2012;30(1):363–71. Available from: https://doi.org/10.1016/j.biotechadv.2011.06.001.
[179] Morandi L, Tarsitano A, Gissi D, et al. Clonality analysis in primary oral squamous cell carcinoma and related lymph-node metastasis revealed by TP53 and mitochondrial DNA next generation sequencing analysis. J Craniomaxillofac Surg 2015;43(2):208–13. Available from: https://doi.org/10.1016/j.jcms.2014.11.007.
[180] Foschini MP, Morandi L, Marchetti C, et al. Cancerization of cutaneous flap reconstruction for oral squamous cell carcinoma: report of three cases studied with the mtDNA D-loop sequence analysis. Histopathology 2011;58(3):361–7. Available from: https://doi.org/10.1111/j.1365-2559.2011.03754.x.
[181] Girolimetti G, De Iaco P, Procaccini M, et al. Mitochondrial DNA sequencing demonstrates clonality of peritoneal implants of borderline ovarian tumors. Mol Cancer 2017;16(1):47. Available from: https://doi.org/10.1186/s12943-017-0614-y.

[182] Perrone AM, Girolimetti G, Cima S, et al. Pathological and molecular diagnosis of bilateral inguinal lymph nodes metastases from low-grade endometrial adenocarcinoma: a case report with review of the literature. BMC Cancer 2018;18(1):7. Available from: https://doi.org/10.1186/s12885-017-3944-7.

[183] Amer W, Toth C, Vassella E, et al. Evolution analysis of heterogeneous non-small cell lung carcinoma by ultra-deep sequencing of the mitochondrial genome. Sci Rep 2017;7(1):11069. Available from: https://doi.org/10.1038/s41598-017-11345-3.

[184] Salas A, Yao Y-G, Macaulay V, Vega A, Carracedo A, Bandelt H-J. A critical reassessment of the role of mitochondria in tumorigenesis. PLoS Med 2005;2(11):e296. Available from: https://doi.org/10.1371/journal.pmed.0020296.

[185] Schon EA, DiMauro S, Hirano M. Human mitochondrial DNA: roles of inherited and somatic mutations. Nat Rev Genet 2012;13(12):878–90. Available from: https://doi.org/10.1038/nrg3275.

[186] Máximo V, Sobrinho-Simões M. Hürthle cell tumours of the thyroid. A review with emphasis on mitochondrial abnormalities with clinical relevance. Virchows Arch 2000;437(2):107–15. Available from: https://doi.org/10.1007/s004280000219.

[187] Setiawan VW, Chu L-H, John EM, et al. Mitochondrial DNA G10398A variant is not associated with breast cancer in African-American women. Cancer Genet Cytogenet 2008;181(1):16–19. Available from: https://doi.org/10.1016/j.cancergencyto.2007.10.019.

[188] Falk MJ, Pierce EA, Consugar M, et al. Mitochondrial disease genetic diagnostics: optimized whole-exome analysis for all MitoCarta nuclear genes and the mitochondrial genome. Discov Med 2012;14(79):389–99.

[189] Griffin HR, Pyle A, Blakely EL, et al. Accurate mitochondrial DNA sequencing using off-target reads provides a single test to identify pathogenic point mutations. Genet Med 2014;16(12):962–71. Available from: https://doi.org/10.1038/gim.2014.66.

[190] van Oven M, Kayser M. Updated comprehensive phylogenetic tree of global human mitochondrial DNA variation. Hum Mutat 2009;30(2):E386–94. Available from: https://doi.org/10.1002/humu.20921.

[191] Arnold RS, Sun Q, Sun CQ, et al. An inherited heteroplasmic mutation in mitochondrial gene COI in a patient with prostate cancer alters reactive oxygen, reactive nitrogen and proliferation. Biomed Res Int 2013;2013:239257. Available from: https://doi.org/10.1155/2013/239257.

[192] Stewart JB, Alaei-Mahabadi B, Sabarinathan R, et al. Simultaneous DNA and RNA mapping of somatic mitochondrial mutations across diverse human cancers. PLoS Genet 2015;11(6):e1005333. Available from: https://doi.org/10.1371/journal.pgen.1005333.

[193] Hopkins JF, Sabelnykova VY, Weischenfeldt J, et al. Mitochondrial mutations drive prostate cancer aggression. Nat Commun 2017;8(1):656. Available from: https://doi.org/10.1038/s41467-017-00377-y.

[194] Hertweck KL, Dasgupta S. The landscape of mtDNA modifications in cancer: a tale of two cities. Front Oncol 2017;7:262. Available from: https://doi.org/10.3389/fonc.2017.00262.

[195] Cruz-Bermúdez A, Vallejo CG, Vicente-Blanco RJ, et al. Enhanced tumorigenicity by mitochondrial DNA mild mutations. Oncotarget 2015;6(15):13628–43. Available from: https://doi.org/10.18632/oncotarget.3698.

[196] Zhou S, Kachhap S, Sun W, et al. Frequency and phenotypic implications of mitochondrial DNA mutations in human squamous cell cancers of the head and neck. Proc Natl Acad Sci U S A 2007;104(18):7540–5. Available from: https://doi.org/10.1073/pnas.0610818104.

[197] Sharma LK, Fang H, Liu J, Vartak R, Deng J, Bai Y. Mitochondrial respiratory complex I dysfunction promotes tumorigenesis through ROS alteration and AKT activation. Hum Mol Genet 2011;20(23):4605–16. Available from: https://doi.org/10.1093/hmg/ddr395.

[198] D'Aurelio M, Gajewski CD, Lin MT, et al. Heterologous mitochondrial DNA recombination in human cells. Hum Mol Genet 2004;13(24):3171–9. Available from: https://doi.org/10.1093/hmg/ddh326.

[199] Kenny TC, Hart P, Ragazzi M, et al. Selected mitochondrial DNA landscapes activate the SIRT3 axis of the UPRmt to promote metastasis. Oncogene 2017;36(31):4393–404. Available from: https://doi.org/10.1038/onc.2017.52.

[200] Hashimoto M, Bacman SR, Peralta S, et al. MitoTALEN: a general approach to reduce mutant mtDNA loads and restore oxidative phosphorylation function in mitochondrial diseases. Mol Ther 2015;23(10):1592–9. Available from: https://doi.org/10.1038/mt.2015.126.

[201] Petros JA, Baumann AK, Ruiz-Pesini E, et al. mtDNA mutations increase tumorigenicity in prostate cancer. Proc Natl Acad Sci U S A 2005;102(3):719–24. Available from: https://doi.org/10.1073/pnas.0408894102.

[202] Arnold RS, Sun CQ, Richards JC, et al. Mitochondrial DNA mutation stimulates prostate cancer growth in bone stromal environment. Prostate 2009;69(1):1–11. Available from: https://doi.org/10.1002/pros.20854.

[203] Jandova J, Shi M, Norman KG, Stricklin GP, Sligh JE. Somatic alterations in mitochondrial DNA produce changes in cell growth and metabolism supporting a tumorigenic phenotype. Biochim Biophys Acta 2012;1822(2):293–300. Available from: https://doi.org/10.1016/j.bbadis.2011.11.010.

[204] Dasgupta S, Soudry E, Mukhopadhyay N, et al. Mitochondrial DNA mutations in respiratory complex-I in never-smoker lung cancer patients contribute to lung cancer progression and associated with EGFR gene mutation. J Cell Physiol 2012;227(6):2451–60. Available from: https://doi.org/10.1002/jcp.22980.

[205] Dasgupta S, Hoque MO, Upadhyay S, Sidransky D. Mitochondrial cytochrome B gene mutation promotes tumor growth in bladder cancer. Cancer Res 2008;68(3):700–6. Available from: https://doi.org/10.1158/0008-5472.CAN-07-5532.

CHAPTER 18

MitoTALENs for mtDNA editing

Sandra R. Bacman and Carlos T. Moraes

Department of Neurology, University of Miami Miller School of Medicine, Miami, FL, United States

18.1 Introduction

The mitochondrial DNA (mtDNA) is present in thousands of copies in the different cell types. Most pathogenic mtDNA mutations coexist in variable amounts with wild-type mtDNA molecules (mtDNA heteroplasmy). Pathogenic mtDNA mutations include point mutations, either homo or heteroplasmic, and large-scale rearrangements [1,2].

The elimination of mutant mitochondrial genomes has been the goal in our lab for many years. The approach we have developed uses mitochondrial-targeted, site-specific nuclease, such as mitoTALEN, to reduce the mutant mtDNA to a level that is below the threshold for phenotypic expression (Fig. 18.1). Such change could eliminate the disease state. The phenotypic threshold value is commonly 60%–70% for mtDNA deletions, and around 80%–90% for mtDNA point mutations [3,4].

For the vast majority of patients with mitochondrial diseases, only supportive and symptomatic therapies are available. Currently, there is no treatment for the correction of the biochemical or genetic defect that causes the disease. Genome editing is a growing field that has advanced rapidly in the last 15 years, thanks to new tools and technologies. Therefore mtDNA editing is a very attractive tool to treat mitochondrial diseases.

We first used mitochondrial-targeted specific restriction endonucleases to cleave mtDNA. Due to the limited targeting sequence choices of these bacterial restriction endonucleases, we broaden the application by using transcription activator-like (TAL) effector nucleases (TALENs), to recognize specific pathogenic mutations in the mtDNA. In this chapter, we will describe these approaches, and the pros and cons of using mitoTALENs as tools to change mtDNA heteroplasmy.

18.2 The use of specific endonucleases to target mtDNA

Gene therapy as a tool for the treatment of mitochondrial diseases includes the use of engineered systems to correct or "edit" the genetic abnormality responsible for the disease. As

The Human Mitochondrial Genome.
DOI: https://doi.org/10.1016/B978-0-12-819656-4.00018-8

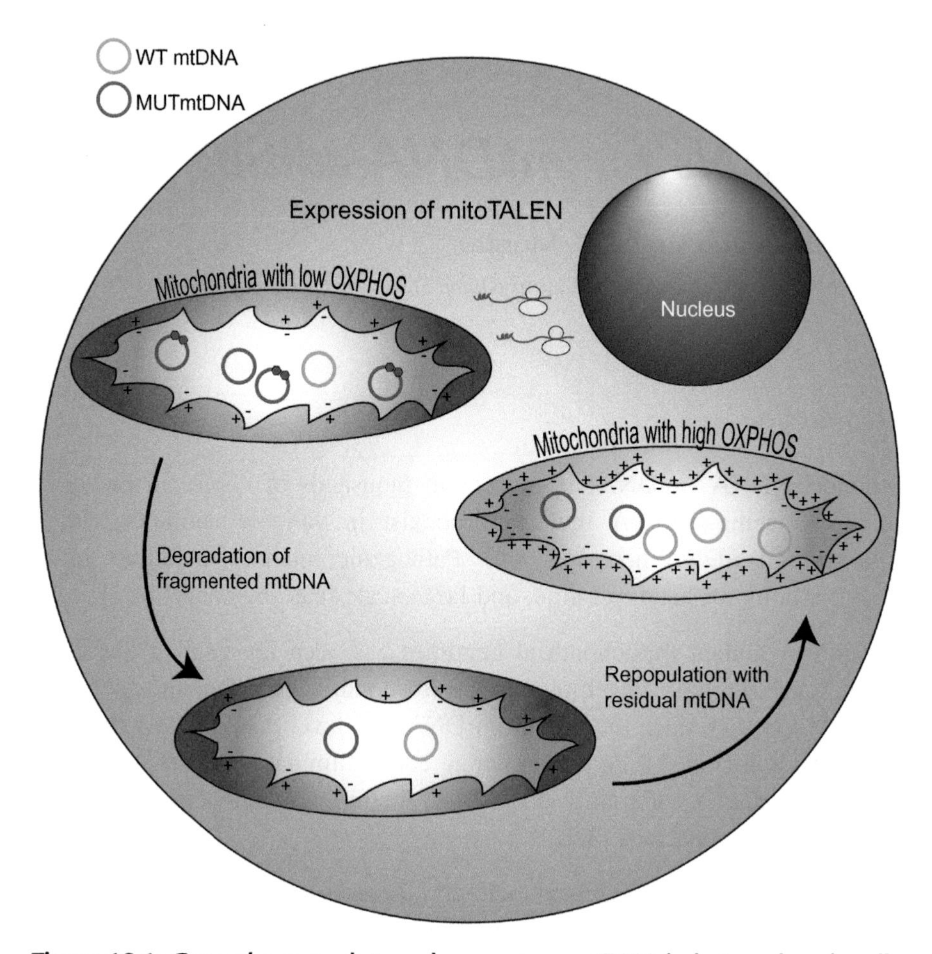

Figure 18.1: General approach to reduce mutant mtDNA in heteroplasmic cells. Mitochondrial-targeted specific nucleases, such as mitoTALENs are expressed from the nuclear–cytoplasmic compartment. A mitochondrial localization signal directs the monomers to the mitochondrial matrix, where they bind and cleave specifically mutant mtDNA. The DSBs generated in the mtDNA carrying the mutation lead to degradation. Repopulation with the residual mtDNA cells increases the levels of wild-type mtDNA, thereby improving OXPHOS capacity.

no technique is available to modify the mtDNA, we relied on the elimination of specific haplotypes in a heteroplasmic context. The first studies to change mtDNA heteroplasmy used mitochondrial-targeted restriction endonucleases to cleave mtDNA with the goal of changing heteroplasmy and avoid the pathogenic phenotype. Therefore the objective of mitochondrial-targeted endonucleases is to promote double-strand breaks (DSBs), which lead to degradation.

The original approach to change heteroplasmy in vitro was developed in our laboratory in the early 2000s. A restriction endonuclease (RE) was targeted to mitochondria and showed the ability to change heteroplasmy in a predicted way. The *Pst*I-RE could differentiate between mouse and rat mtDNA in xenomitochondrial cybrids harboring mouse (which contains two sites for *Pst*I) and rat (which contains none) mtDNA [5]. The mito*Pst*I construct was able to degrade the mouse mtDNA in the heteroplasmic rodent cell line and increase the rat mtDNA haplotype. These experiments were the first proof-of-principle that mitoRE are viable tools for genetic therapy of specific heteroplasmic mtDNA mutations [6]. This concept was soon after applied to a pathogenic mtDNA point mutation, a T > G transversion in the mitochondrial ATP6 gene (at mtDNA position m.8399) that creates a unique *Sma*I/*Xma*I site. This mutation has been associated with neuropathy, ataxia, and retinitis pigmentosa (NARP) with 60%–90% of m.8399T > G mutation [7] and maternally inherited Leigh syndrome (MILS) [8,9] with >90% mutant mtDNA [10]. After transfections with the *Xma*I targeted to mitochondria, high expression and elimination of mutant mtDNA in cells carrying the m.8993T > G mutation was observed [11].

To further understand this approach, we developed an inducible system using *Apa*LI-RE in a heteroplasmic mouse model that contains two mtDNA haplotypes (NZB/BALB, with one *Apa*LI site in the BALB and none in the NZB mtDNA) [12] (Fig. 18.2). Mito*Apa*LI was able to efficiently shift heteroplasmy increasing the NZB haplotype [13]. Our group also showed applicability in vivo, both in skeletal muscle and brain in NZB/BALB heteroplasmic mice using viral vectors (Adenovirus and AAV1). In both cases, increased levels of NZB mtDNA were obtained after injections of rAd[mito*ApaLI*] or rAAV1 in the right cerebral or rAd5[mito*ApaLI*] in skeletal muscle [13]. When mito*ApaLI* was targeted using a cardiotrophic viral vector AAV6, specific cardiac expression of rAAV6[mito*ApaLI*] induced a significant shift in mtDNA heteroplasmy in heart that persisted for 12 weeks after injection, demonstrating that a single injection can induce long-term heteroplasmy shift [14]. Delivery to newborns was also effective [15] using recombinant AAV9 carrying the mito*ApaLI*. Delivery via intraperitoneal (IP) or temporal vein (TV) injection in neonates [16] with a single systemic injection of rAAV9[mito*ApaLI*] induced shifts in mtDNA heteroplasmy in all skeletal muscles. These changes persisted for at least 6 months, in agreement with previous reports showing that robust transduction could be achieved in skeletal muscle after delivery of AAV9 vectors [17,18]. Because mito*ApaLI* targets a unique site in the mtDNA, we extended the studies using the mito*Sca*I-RE that recognizes multiple restriction sites in the mtDNA: five sites in the NZB mtDNA and three in BALB mtDNA (Fig. 18.2). Expression was obtained in both liver, after intravenous injection, and in skeletal muscle, after intramuscular injection of recombinant adenovirus (rAd5[mito*ApaLI*]). We observed modulation of heteroplasmy, with the haplotype that harbors fewer restriction sites (BALB mtDNA) being preferentially retained [19].

Because restriction endonucleases have the limitation of recognizing very short sequences in the mtDNA, the development of gene editing enzymes allowed for the recognition of longer DNA sequences. Therefore DNA editing platforms such as TALENs and zinc finger nucleases (ZFNs) broadened the spectrum of mutations in the mtDNA that can be targeted.

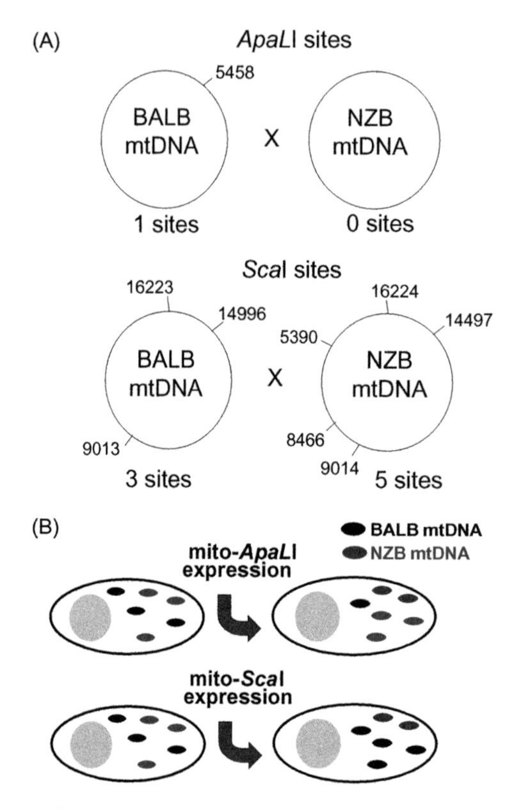

Figure 18.2: The use of specific endonucleases to target mtDNA.
(A) Diagram of the mtDNA of a heteroplasmic mice model (NZB/BALB) carrying two haplotypes of mtDNA and the respective localization of the restriction sites for *ApaLI* and *ScaI*. (B) After expression of mito*ApaLI* or mito*ScaI*, there is a shift in the mtDNA haplotype decreasing the mtDNA with less restriction sites for the respective nuclease.

18.3 The use of mitoTALENs to target mtDNA

TAL effectors are found in plant pathogenic bacteria, particularly members of the genus *Xanthomonas*. TAL effectors reprogram host cells by acting as transcriptional activators in the plant cell nucleus. They are injected into plant cells via the bacterial type III secretion system, imported into the plant cell nucleus, and targeted to effector-specific gene promoters [20]. TAL effector binding activates expression of downstream genes, which may contribute to bacterial colonization, symptom development, or pathogen dissemination [21]. They directly bind to DNA via a central domain of tandem repeats [22], nuclear localization signals (NLSs), and an acidic transcriptional activation domain (AD) [23]. The amino acid repeats of TAL effectors are generally composed of 34 amino acids. However, variants with 33 or 35 amino acids are also common, and the last repeat in the domain is truncated at 20 amino acids. Most TAL effectors have between 13 and 28 repeats [24]. TAL effectors recognize DNA in a modular fashion and target DNA using the amino acid repeats localized at positions 12 and 13, called the repeat-variable di-residue (RVD) [24]. Different RVDs preferentially associate with different nucleotides, with the four most common RVDs (HD, NG, NI, and NN) (Fig. 18.2A) promoting binding to each of the four nucleotides (C, T, A, and G, respectively). Thus the number of repeats is given by the number of RVDs plus the final truncated repeat [25,26]. TALENs are fusions of TAL effectors with nonspecific endonucleases such as *Fok*I that work as dimers [27] and cut within a 12- to 19-bp spacer sequence that separates each TALE binding site (Fig. 18.2B). Miller and colleagues devised an obligatory heterodimer *Fok*I that increases specificity [28]. TALENs can be designed to bind to essentially any DNA sequence [29] and can be used to create site-specific DNA DSBs (Fig. 18.3) [28,30,31].

When TALENs are targeted to the nucleus, the generated breaks activate the cell's DNA repair pathways, which can create specific DNA sequence modifications at or near the break site [24]. This can happen by one of two highly conserved processes, nonhomologous end joining (NHEJ), which often results in small insertions or deletions (INDELs), and homologous recombination (HR), which produce gene insertion or replacement [32–34]. The sequence of the repair template can be modified or amended to swap in specific mutations or additional sequences (referred to as DNA editing). When TALENs are targeted to the mitochondria, DSBs generated in the mtDNA [35] lead to mtDNA degradation and loss of the linear mtDNA molecules [36,37]. If a depletion of mtDNA copy number is sensed, mtDNA replication, which is carried out by the nuclear DNA encoded polymerase γ, ensues to normalize their levels [38]. MtDNA lacks a DSB repair mechanisms [39], consequently soon after the breaks, depletion of the mtDNA occurs. Therefore it is crucial to analyze mtDNA copy number after mitoTALEN expression to evaluate the recovery of mtDNA levels [40]. Besides mitoTALENs, specific DSBs can also be generated in the mtDNA by mitochondrial-targeted restriction endonucleases [19] and mitoZFNs [41–43].

18.4 Structure of mitoTALENs

This simple DNA recognition code and its modular nature makes TALENs an ideal platform for constructing custom-designed DNA nucleases [44,45] (Fig. 18.3A and B). The basic structure of the mitoTALENs constructs used to target different mutations in the mtDNA is depicted in Fig. 18.3C [46] and contains: (1) a specific TAL DNA-binding domain [34], which is relatively short and consists of 10–16 TAL repeats; (2) a mitochondrial localization signal (MLS) in the N′ terminus used to target proteins to the

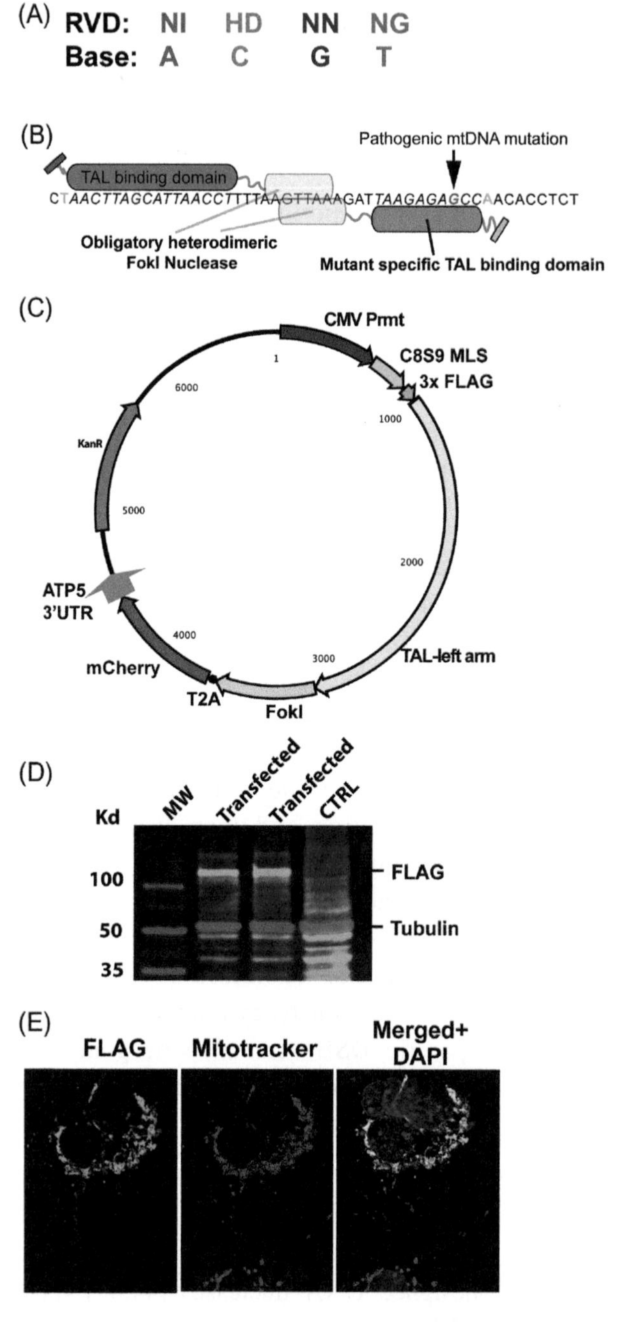

Figure 18.3: Construction and expression of plasmids carrying mitoTALEN monomers. (A) Within the TAL DNA-binding domain, different repeat-variable di-residues (RVDs) preferentially associate with specific nucleotides. The four most common RVDs (NI, HD, NN, and NG) can preferentially bind to each of the four nucleotides (A, C, G, and T), respectively. (B) Diagram

(Continued)

mitochondria; (3) an unique tag (Hemagglutinin [HA] or [Flag]) for immunological detection [47] (Fig. 18.3D and E); (4) eGFP or mCherry for sorting [48]; (5) a 3′UTR from ATP5B or SOD2, which are believed to help localize mRNA to ribosomes contacting mitochondria [49]; (6) a specific promoter (ubiquitous or organ-specific). We used the human cytomegalovirus (CMV) enhancer/promoter [50]; (7) a picornaviral 2A-like sequence (T2A′) between the mitoTALEN and the fluorescent marker to allow expression from the same promoter [51]; (8) removal of the stock nuclear localization signal; and (9) the use of an obligatory heterodimeric *Fok*I nuclease domains to reduce self-self off-site cleavage [31,52,53]. Some specifics about the mitoTALENs design are important when engineering to target a specific sequence, such as the presence of a T in the preceding portion of the targeted sequence (called T0). This naturally occurring N-terminus shows a predicted secondary structure similar to a repeat, though it does not contain a recognizable RVD [1,20,22]. When the design does not allow the use of T0 as the discriminant base (e.g., the mutation creates a new "T"), it can be challenging. The A > G transition, for example, can be difficult to discriminate as the conventional RVD for G binding is "NN," which is not a good discriminator between A and G [22,34]. However, in this case, in the antisense strand, the mutant gains a C, which can be recognized more specifically by an "HD" RVD. We took advantage of the T0 in many of our constructs [46], but for the m.8344A > G mutation, we chose to place the mutated G at position 3 of the antisense monomer, exploiting a "gain of C3" model [54].

18.5 MitoTALENs targeting mutations in cybrids

We designed mitoTALENs to specific mutations in the mtDNA in human cell lines. The designed inserts were cloned in a pVax backbone, a plasmid for mammalian transient expression. As explained above, two monomers must be designed for each mtDNA mutation. Table 18.1 shows different mitoTALENs to target mtDNA. We designed mitoTALENs against the mtDNA "common deletion" (m.8483_13459del4977)

◀ of the TALENs domains, their binding to a specific mtDNA sequence, and the heterodimeric *Fok*I nucleases. Note that one monomer binds specifically to the mutant strand, whereas the other monomer is common for mutant and wild-type mtDNA. (C) The basic structure of one mitoTALEN plasmid that contains: a human cytomegalovirus (CMV) promoter, a mitochondrial localization signal (MLS) in the N′ terminus, in this case C8S9; a unique tag (Flag) for immunological detection; a TAL DNA-binding domain, an obligatory heterodimeric *Fok*I nuclease domain; a picornaviral 2A-like sequence (T2A′) between the mitoTALEN and the fluorescent marker (mCherry); a 3′UTR untranslated region from a nuclear gene (ATP5B) and kanamycin for the bacteria resistant. (D) Western blot of a FLAG-tagged mitoTALEN after 24 h of transfection in Cos7 cells. The blot was also probed for tubulin. (E) Mitochondrial localization of the mitoTALEN monomer in Cos7 cells 24 h after transfection. The monomer carries a Flag tag (green) that colocalized with mitotracker (red).

Table 18.1: Different mitoTALENs targeted to mitochondria.

MitoTALEN	Model tested	Delivery	In vitro	In vivo	Results	Citation
"Common deletion" (m.8483_13459del4977) breakpoint (Δ5)	Osteosarcoma cybrids 70%–80% mutation	Transfection	Yes	No	Reduction mutant haplotype	[55,46]
m.14459G > A *MT-ND6* mutation	Osteosarcoma cybrids 90%–95% mutation	Transfection	Yes	No	Reduction mutant haplotype. Complex I activity recovery	[59,46]
m.8344A > G *tRNA*^{Lys} mutation	Osteosarcoma cybrids 55%–60% mutation	Transfection	Yes	No	Reduction mutant haplotype. OXPHOS recovery	[3,54]
m.13513G > A *ND5* mutation	Osteosarcoma cybrids 80%–85% mutation	Transfection	Yes	No	Reduction mutant haplotype. Complex I activity recovery	[93,54]
m.5024 C > T *tRNA*^{Ala} mutation	Mouse MEFs 60%–80% mutation	Transfection/ intramuscular and systemic injections	Yes	Yes	Reduction mutant haplotype, tRNA^{Ala} recovery	[65,64]
NZB mtDNA	NZB/BALB oocytes (70%–80% BALB mtDNA)	Transfection/ injection of RNA in oocytes	Yes	Yes	Reduction NZB haplotype. Prevent germline transmission	[13,71]
m.14459G > A LHOND	Fusion mouse oocytes carrying 80%–85% human mutation	Injection of RNA in oocytes	Yes	Yes	Reduction mutant haplotype	[59,71]
m.9176T > C NARP	Fusion mouse oocytes carrying > 95% human mutation	Injection of RNA in oocytes	Yes	Yes	Reduction mutant haplotype	[94,71]
m.3243A > G MELAS	Human IPSc > 80% mutation	Transfection	Yes	No	Reduction mutant haplotype. OXPHOS recovery	[95,69]
m.13513G > A MELAS	Human IPSc 60%–70% mutation	Transfection	Yes	No	Reduction mutant haplotype	[93,68]

breakpoint (Δ5-mitoTALEN), present in approximately 30% of all patients with mtDNA deletions [55] and also in normal aging tissues [46,56]. The cell line used was a cybrid previously characterized (BH10.9) [57] and also described elsewhere in this book [58]. In addition, mitoTALENs against the m.14459G > A mutation in MT-ND6, which causes Leber's hereditary optic neuropathy plus dystonia (LHON) [59], are described. Cybrid clones were derived by fusing dermal fibroblasts from a patient harboring the heteroplasmic missense mutation m.14459G > A in *MT-ND6* with an osteosarcoma cell line devoid of mtDNA (143B/206). MitoTALENs were also tested against the m.13513G > A ND5 mutation associated with MELAS/Leigh syndrome [54,60,61]. Likewise, heteroplasmic cybrids cells harboring the *ND5* m.13513G > A mutation were produced by fusing 143B/206 with patients-derived enucleated fibroblasts carrying the point mutation [54]. In the case of the m.8344A > G $tRNA^{Lys}$ mutation associated with myoclonic epilepsy with ragged red fibers (MERRF) [3,62], two monomers were designed to cleave the mutated region and tested in heteroplasmic cybrids [54,63].

Our lab also designed mitoTALENs to cleave the mouse mtDNA, at m.5024C > T $tRNA^{Ala}$ mutation. In this case, the $tRNA^{Ala}$ mutation is present in the mtDNA of a heteroplasmic mutant mouse line developed at the Max Plank Institute in the labs of Jim Stewart and Nils Larsson [64–66]. Mouse embryonic fibroblasts (MEFs) from these mice were immortalized with the E6-E7 gene from the human papilloma virus [67].

MitoTALENs were designed and tested for expression and mitochondrial localization (Fig. 18.2D and E). After expression of the mitoTALENs in cells harboring heteroplasmic mtDNA mutations (humans and mouse), we were able to shift heteroplasmy in the predicted direction, increasing the load of wild-type mtDNA. In the case of the mitoTALEN targeting the m.14459G > A mutation in the MT-ND6 gene, normal levels of complex I activity were obtained after mitoTALENs transfection of the cybrids carrying high levels of the m.14459G > A mutation that showed partial decreased complex I activity before the treatment [46]. Furthermore, heteroplasmic cells carrying the m.8344A > G $tRNA^{Lys}$ gene mutation improved their OXPHOS function when treated with mitoTALENs [54].

18.6 MitoTALENs in a heteroplasmic mouse model carrying a $tRNA^{Ala}$ mutation

A mouse model carrying a pathogenic heteroplasmic m.5024C > T mtDNA mutation in the $tRNA^{Ala}$ was used to attempt to alter mtDNA heteroplasmy in vivo. The m.5024C > T mutation was associated with $tRNA^{Ala}$ instability and a mild cardiac phenotype at older ages [65]. The mouse mutation resembles human mtDNA mutations in $tRNA^{Ala}$, which were associated with myopathies and OXPHOS deficiencies [66]. The mitoTALENs against the heteroplasmic m.5024C > T are the same as described above in MEFs carrying the $tRNA^{Ala}$ mutation. The basic structure of the mitoTALEN consisted of two adjacent monomers, one of which was specific for the $tRNA^{Ala}$ m.5024C > T mutation [64] inserted in an AAV2/9 backbone to produce viral particles. In this study, we were able to specifically shift heteroplasmy increasing the WT mtDNA after intramuscular injections. We injected $1.0–1.5 \times 10^{12}$ AAV9 particles of each monomer in the right tibialis anterior of 8–12-week-old mice and analyzed mice at 4, 6, 8, 12, and 24 weeks after injections. The shift in heteroplasmy persisted over 24 weeks as well as the recovery of the specific transcript ($tRNA^{Ala}$) to normal levels that were decreased in the targeted tissues of mice carrying more than 50% of mutant load [64]. We also injected 16–17-day-old mice retro-orbitally for systemic expression with $1.0-1.5 \times 10^{12}$ AAV9 particles of each monomer. Systemically injected mice also showed mitoTALENs expression in muscle and heart. Mutant mtDNA load was reduced in these tissues [64].

18.7 MitoTALENs and induced pluripotent stem cells

Induced pluripotent stem cells (iPSCs) derived from a diseased patient that harbored a high proportion of m.13513G > A and m.3243A > G mtDNA mutations and presented mitochondrial myopathy, encephalopathy, lactic acidosis, and stroke-like episodes (MELAS) were generated and used also as a proof of concept to shift mtDNA heteroplasmy. The m.13513G > A heteroplasmy level in MELAS-iPSCs was decreased in the short term by transduction of G13513A-mitoTALEN [68]. The mitoTALEN targeting the m.3243A > G was reported to eliminate the mutant mtDNA in iPSCs and rescue respiration and energy production [69]. Attempts by our lab to design mitoTALENs against the m.3243A > G MELAS mutation were not successful in decreasing mutant mtDNA loads (Claudia Pereira, unpublished observation). It is not clear why the differences between these studies, but a protein termed MTERF, which binds to the 3243 mtDNA region, may have a steric role, preventing TALEN binding [70].

18.8 Other uses of mitoTALENs

18.8.1 MitoTALENs in germline transmission

As a proof of concept, mitoTALENs were used in a heteroplasmic mouse model to reduce the fraction of a specific mtDNA haplotype in embryos. Also using mitoTALENs against the NARP m.9176T > C mutation and LHON m.14459G > A, human mutant mtDNA was reduced in manipulated murine oocytes carrying the human mtDNA mutations [71].

18.8.1.1 MitoTALENs to study mtDNA replication

Mitochondrially targeted TALEN against mtDNA regions involved in the formation of the "common deletion" (m.8483_13459del4977) showed that one of the sites had a critical role for the formation of the common deletion. These results suggested a unique replication-dependent repair pathway that triggers the formation of the mtDNA common deletion [72].

18.8.2 MitoTevTAL nuclease

Because the dimeric nature of mitoTALEN adds complexity to the system, efforts to develop monomeric-specific DNA editing enzymes were undertaken. A monomeric nuclease to target the m.8344A > G $tRNA^{Lys}$ gene mutation, mito-Tev-TALE, was successfully tested [73]. The construct is based on the GIY-YIG homing nuclease I-TevI, assembled as a monomeric chimera between I-TevI nuclease and a TAL DNA-binding domain [96,97]. We used a previously tested mitoTALEN against the MERRF mutation for the TAL DNA-binding domain [54]. This novel design is a smaller molecule for mitochondrial genome editing, which is more readily applicable to in vivo delivery using adeno-associated virus (AAV) vectors as only one monomer needs to be cloned in the viral vector.

18.9 Pros and cons of using mitoTALENs for gene therapy

18.9.1 Specificity and mtDNA depletion

MitoTALENs have proven to be highly specific. This characteristic might be due to the fact that two monomers are required, with the DSB mediated by the obligatory heterodimeric form of *Fok*I. We did not observe off-target DSBs in the mtDNA after injection of AAV [mitoTALENs] against the m.5024C > T $tRNA^{Ala}$ mutation [64]. MtDNA depletion was also not observed in in vivo studies, probably because of the partial nature of the DSB events and fast repopulation. A transient decrease in the mtDNA levels could be observed in cultured cells. When mitoTALENs were used against the m.14459G > A mutation in MT-ND6, a transient decrease in mtDNA copy number was observed, but these levels were restored as soon as 2 weeks after transfection of the heteroplasmic cybrids carrying high levels of the mutation [46]. Consequently, it is essential to carefully evaluate mtDNA copy number after targeting the mtDNA to prevent undesired side effect [40]. Because specific sequences of the mtDNA are targeted with the mitoTALEN, it is rare to find off-targeted DSBs in the mitochondria. However, ultra-deep mtDNA amplicon resequencing could be done to look for these rare events. In contrast to the nuclear genome, rare DSB in the mtDNA should not be a major concern because of the multiple copies and the fact that cleaved DNA fragments are rapidly degraded.

18.9.2 Off-target sequences in the nucleus

Subcellular localization studies showed that mitoTALENs localize specifically in the mitochondria [46]. Still, when targeting mtDNA with endonucleases, one potential side effect is the cleavage of off-target sequences in the nuclear DNA, which could lead to the formation of insertions or deletions (INDELs). MitoTALENs are designed to be used as dimers because of the obligatory heterodimeric *Fok*I nucleases, making mitoTALENs highly specific tools for gene editing. Although it is possible to make the mitoTALEN specific enough that it does not have a similar target in the nucleus, amplicon resequencing of potential nuclear DNA off-target sites can provide additional information [74].

18.9.3 Easy design of new recognition sites

The simple DNA recognition code and its modular nature make TALENs an ideal platform for constructing custom-designed DNA nucleases [44,45] and targeting nearly any mutation.

Using the appropriate mitochondrial targeting signal (MTS) is a key issue in the mitoTALENs design. Like signal peptides, MTSs are cleaved once mitochondrial matrix localization is complete [75]. The charge, length, and structure of the MTS are important for protein import into the mitochondria. We found that a hybrid Cox8Sub9 (Cox8 MTS from human plus ATPase subunit 9 MTS from *Neurospora crassa*) for one monomer, as well as SOD2 for the other, was suitable MTSs for our approach [14,64]. Furthermore, choosing the appropriate promoter when targeting different tissues may be important for high expression of mitoTALENs. When ubiquitous expression in all cell types is desired, the use of ubiquitous promoters such as the human elongation factor 1α-subunit (EF1α), immediate-early cytomegalovirus (CMV), chicken β-actin (CBA) and its derivative CAG, the β-glucuronidase (GUSB), or ubiquitin C (UBC) can be used [76–78]. When specific tissues are targeted, restricted expression can be achieved with tissue-specific promoters. In the case of skeletal muscle, creatine kinase (MCK) or desmin (1.7 kb) has been shown to be highly specificity with minimal undesired expression in the liver [79]. Because of the nature of mitochondrial diseases, which usually affect many organs and tissues, we decided to use an ubiquitous promoter in our experiments [14,64].

18.9.4 MitoTALEN gene size

MitoTALENs are large molecules that have to be inserted in the expression cassette of the viral vectors for the production of virus. A variety of different vectors and delivery techniques have been applied in gene therapy trials. Although nonviral approaches are becoming increasingly common (16.5% of trials), viral vectors remain by far the most popular and effective approach [80]. We have used adenovirus to transduce RE to the liver [19] and have tested different serotypes of the AAV such as AAV6 to target the heart [14] and also AAV9 for systemic delivery of REs and mitoTALENs in the skeletal muscle and the heart [16,64].

Recombinant AAVs do not integrate into the genome, and commonly stay as episomes, with just about 0.1% integrated into the host cell genome. AAV genomes decrease gradually in dividing cells and lead to decrease of a transgene expression level. For this reason, AAV is the best choice for transfection of slowly dividing cells [80]. There is a DNA packaging size constraint of the expression cassette in the AAV backbone (approximately 4.1–4.9 kb) [81,82]. MitoTALENs are bulky (more than 4 kb each monomer) so each monomer needs to be packaged into one AAV vector. Therefore two viral preparations were required to obtain the desired effect [64]. This might be a disadvantage because the final viral titer of each preparation will be diluted, which constitutes a limitation of using dimeric nucleases. Monomeric DNA editing enzymes might offer advantages as only one viral preparation would be needed [73].

18.9.5 The future of mitoTALENs as therapy

To date, only two DNA recognition platforms have been used to target mitochondria: ZFNs [41,83,84] and TALENs [25,46,54,64], although headways are being made on monomeric enzymes [73]. Clustered Regularly Interspaced Short Palindromic Repeats (CRISPR/Cas9) [85] were used in one report that showed that Cas9 can localize to mitochondria and edit mtDNA with sgRNAs targeting specific loci of the mitochondrial genome [86]. However, these results have yet to be reproduced by other labs and the difficulty to import RNA into mitochondrial is still debated [87–89]. Engineered meganucleases or homing endonucleases are a collection of naturally occurring enzymes that recognize and cleave long DNA sequences [90]. The monomeric nature of the meganucleases as well as chimeric I-TevI/TAL in mitoTevTal [73] makes them an attractive tool to be used as an effective method to edit mtDNA. As previously discussed, the monomeric nature and small size may confer great advantages for gene therapy [91].

Although there is no perfect system to target mtDNA, studies using mitoTALENs constituted a proof-of-principle that this editing tool when combined with the permanent growing collection of engineered virus types opens a promising path to treat heteroplasmic mtDNA diseases. The fact that a single injection was enough to promote a significant change in heteroplasmy makes this approach an attractive tool for gene therapy. The innate and acquired immunogenicity of AAV precludes effective serial administrations [92]. In our hands, a single injection was sufficient to change heteroplasmy and the effect lasted at least 24 weeks after injection [64]. Decreasing and maintaining heteroplasmy below a threshold for phenotypic expression is the ultimate goal to treat heteroplasmic mitochondrial diseases.

In the clinical context, the complete elimination of mutant mtDNA is not necessary as a moderate decrease in mutation load, which can be achieved with a single dose of mitoTALENs, maybe sufficient to produce long-lasting beneficial clinical outcomes. The risk of a rapid reduction in mtDNA copy numbers as well as a possible nuclear or mtDNA off-target side effects are remaining concerns that need to be carefully addressed when using DSB-causing enzymes for gene therapy.

Research perspectives

MitoTALENs are highly specific genetic tools to promote mtDNA DSB and change heteroplasmy in the predicted way. Their modular structure allows their design to target almost any mutation in the mtDNA. Although its dimeric design represents an advantage for specificity and to avoid off-targeted sequences, mitoTALENs require two recombinant AAV particles, making translation to human gene therapy treatment possible but costly. Monomeric versions might facilitate AAV-based delivery. The development of new AAV viral vectors will accompany the progress of the use of mitoTALENs to treat mitochondrial diseases. The fact that a single application might be sufficient to change heteroplasmy in a long-lasting manner makes this approach clinically attractive.

References

[1] Bacman SR, Williams SL, Pinto M, Moraes CT. The use of mitochondria-targeted endonucleases to manipulate mtDNA. Methods Enzymol 2014;547:373–97.

[2] Viscomi C, Bottani E, Zeviani M. Emerging concepts in the therapy of mitochondrial disease. Biochim Biophys Acta 2015;1847(6-7):544–57.

[3] Shoffner JM, Lott MT, Lezza AM, Seibel P, Ballinger SW, Wallace DC. Myoclonic epilepsy and ragged-red fiber disease (MERRF) is associated with a mitochondrial DNA tRNA(Lys) mutation. Cell 1990;61 (6):931–7.

[4] Shoffner JM, Kaufman A, Koontz D, Krawiecki N, Smith E, Topp M, et al. Oxidative phosphorylation diseases and cerebellar ataxia. Clin Neurosci 1995;3(1):43–53.

[5] Dey R, Barrientos A, Moraes CT. Functional constraints of nuclear-mitochondrial DNA interactions in xenomitochondrial rodent cell lines. J Biol Chem 2000;275(40):31520–7.

[6] Srivastava S, Moraes CT. Manipulating mitochondrial DNA heteroplasmy by a mitochondrially targeted restriction endonuclease. Hum Mol Genet 2001;10(26):3093–9.

[7] Tsao CY, Mendell JR, Bartholomew D. High mitochondrial DNA T8993G mutation (<90%) without typical features of Leigh's and NARP syndromes. J Child Neurol 2001;16(7):533–5.

[8] Holt IJ, Harding AE, Petty RK, Morgan-Hughes JA. A new mitochondrial disease associated with mitochondrial DNA heteroplasmy. Am J Hum Genet 1990;46(3):428–33.

[9] Tatuch Y, Pagon RA, Vlcek B, Roberts R, Korson M, Robinson BH. The 8993 mtDNA mutation: heteroplasmy and clinical presentation in three families. Eur J Hum Genet 1994;2(1):35–43.

[10] Tanaka M, Borgeld HJ, Zhang J, Muramatsu S, Gong JS, Yoneda M, et al. Gene therapy for mitochondrial disease by delivering restriction endonuclease SmaI into mitochondria. J Biomed Sci 2002;9(6 Pt 1):534–41.

[11] Alexeyev MF, Venediktova N, Pastukh V, Shokolenko I, Bonilla G, Wilson GL. Selective elimination of mutant mitochondrial genomes as therapeutic strategy for the treatment of NARP and MILS syndromes. Gene Ther 2008;15(7):516–23.

[12] Jenuth JP, Peterson AC, Shoubridge EA. Tissue-specific selection for different mtDNA genotypes in heteroplasmic mice. Nat Genet 1997;16(1):93–5.

[13] Bayona-Bafaluy MP, Blits B, Battersby BJ, Shoubridge EA, Moraes CT. Rapid directional shift of mitochondrial DNA heteroplasmy in animal tissues by a mitochondrially targeted restriction endonuclease. Proc Natl Acad Sci USA 2005;102(40):14392–7.

[14] Bacman SR, Williams SL, Garcia S, Moraes CT. Organ-specific shifts in mtDNA heteroplasmy following systemic delivery of a mitochondria-targeted restriction endonuclease. Gene Ther 2010;17(6):713–20.

[15] Uusimaa J, Remes AM, Rantala H, Vainionpaa L, Herva R, Vuopala K, et al. Childhood encephalopathies and myopathies: a prospective study in a defined population to assess the frequency of mitochondrial disorders. Pediatrics 2000;105(3 Pt 1):598–603.
[16] Bacman SR, Williams SL, Duan D, Moraes CT. Manipulation of mtDNA heteroplasmy in all striated muscles of newborn mice by AAV9-mediated delivery of a mitochondria-targeted restriction endonuclease. Gene Ther 2012;19(11):1101–6.
[17] Inagaki K, Fuess S, Storm TA, Gibson GA, McTiernan CF, Kay MA, et al. Robust systemic transduction with AAV9 vectors in mice: efficient global cardiac gene transfer superior to that of AAV8. Mol Ther 2006;14(1):45–53.
[18] Ghosh A, Yue Y, Long C, Bostick B, Duan D. Efficient whole-body transduction with trans-splicing adeno-associated viral vectors. Mol Ther 2007;15(6):1220.
[19] Bacman SR, Williams SL, Hernandez D, Moraes CT. Modulating mtDNA heteroplasmy by mitochondria-targeted restriction endonucleases in a 'differential multiple cleavage-site' model. Gene Ther 2007;14 (18):1309–18.
[20] Moscou MJ, Bogdanove AJ. A simple cipher governs DNA recognition by TAL effectors. Science 2009;326(5959):1501.
[21] Bogdanove AJ, Schornack S, Lahaye T. TAL effectors: finding plant genes for disease and defense. Curr Opin Plant Biol 2010;13(4):394–401.
[22] Boch J, Scholze H, Schornack S, Landgraf A, Hahn S, Kay S, et al. Breaking the code of DNA binding specificity of TAL-type III effectors. Science 2009;326(5959):1509–12.
[23] Schornack S, Meyer A, Romer P, Jordan T, Lahaye T. Gene-for-gene-mediated recognition of nuclear-targeted AvrBs3-like bacterial effector proteins. J Plant Physiol 2006;163(3):256–72.
[24] Boch J, Bonas U. Xanthomonas AvrBs3 family-type III effectors: discovery and function. Annu Rev Phytopathol 2010;48:419–36.
[25] Bogdanove AJ, Voytas DF. TAL effectors: customizable proteins for DNA targeting. Science 2011;333 (6051):1843–6.
[26] Zhang F, Cong L, Lodato S, Kosuri S, Church GM, Arlotta P. Efficient construction of sequence-specific TAL effectors for modulating mammalian transcription. Nat Biotechnol 2011;29(2):149–53.
[27] Bitinaite J, Wah DA, Aggarwal AK, Schildkraut I. FokI dimerization is required for DNA cleavage. Proc Natl Acad Sci USA 1998;95(18):10570–5.
[28] Miller JC, Tan S, Qiao G, Barlow KA, Wang J, Xia DF, et al. A TALE nuclease architecture for efficient genome editing. Nat Biotechnol 2011;29(2):143–8.
[29] Boch J. TALEs of genome targeting. Nat Biotechnol 2011;29(2):135–6.
[30] Christian M, Cermak T, Doyle EL, Schmidt C, Zhang F, Hummel A, et al. Targeting DNA double-strand breaks with TAL effector nucleases. Genetics 2010;186(2):757–61.
[31] Li T, Huang S, Jiang WZ, Wright D, Spalding MH, Weeks DP, et al. TAL nucleases (TALNs): hybrid proteins composed of TAL effectors and FokI DNA-cleavage domain. Nucleic Acids Res 2011;39(1):359–72.
[32] West SC. Molecular views of recombination proteins and their control. Nat Rev Mol Cell Biol 2003;4 (6):435–45.
[33] Urnov FD, Rebar EJ, Holmes MC, Zhang HS, Gregory PD. Genome editing with engineered zinc finger nucleases. Nat Rev Genet 2010;11(9):636–46.
[34] Cermak T, Doyle EL, Christian M, Wang L, Zhang Y, Schmidt C, et al. Efficient design and assembly of custom TALEN and other TAL effector-based constructs for DNA targeting. Nucleic Acids Res 2011;39 (12):e82.
[35] Moretton A, Morel F, Macao B, Lachaume P, Ishak L, Lefebvre M, et al. Selective mitochondrial DNA degradation following double-strand breaks. PLoS One 2017;12(4):e0176795.
[36] Nissanka N, Bacman SR, Plastini MJ, Moraes CT. The mitochondrial DNA polymerase gamma degrades linear DNA fragments precluding the formation of deletions. Nat Commun 2018;9(1):2491.
[37] Peeva V, Blei D, Trombly G, Corsi S, Szukszto MJ, Rebelo-Guiomar P, et al. Linear mitochondrial DNA is rapidly degraded by components of the replication machinery. Nat Commun 2018;9(1):1727.

[38] Copeland WC. The mitochondrial DNA polymerase in health and disease. Subcell Biochem 2010;50:211–22.

[39] Zinovkina LA. Mechanisms of mitochondrial DNA repair in mammals. Biochem (Mosc) 2018;83 (3):233–49.

[40] Rooney JP, Ryde IT, Sanders LH, Howlett EH, Colton MD, Germ KE, et al. PCR based determination of mitochondrial DNA copy number in multiple species. Methods Mol Biol 2015;1241:23–38.

[41] Gammage PA, Rorbach J, Vincent AI, Rebar EJ, Minczuk M. Mitochondrially targeted ZFNs for selective degradation of pathogenic mitochondrial genomes bearing large-scale deletions or point mutations. EMBO Mol Med 2014;6(4):458–66.

[42] Gammage PA, Viscomi C, Simard ML, Costa ASH, Gaude E, Powell CA, et al. Genome editing in mitochondria corrects a pathogenic mtDNA mutation in vivo. Nat Med 2018;24(11):1691–5.

[43] Minczuk M. Engineered zinc finger proteins for manipulation of the human mitochondrial genome. Methods Mol Biol 2010;649:257–70.

[44] Pan Y, Xiao L, Li AS, Zhang X, Sirois P, Zhang J, et al. Biological and biomedical applications of engineered nucleases. Mol Biotechnol 2013;55(1):54–62.

[45] Sung YH, Baek IJ, Kim DH, Jeon J, Lee J, Lee K, et al. Knockout mice created by TALEN-mediated gene targeting. Nat Biotechnol 2013;31(1):23–4.

[46] Bacman SR, Williams SL, Pinto M, Peralta S, Moraes CT. Specific elimination of mutant mitochondrial genomes in patient-derived cells by mitoTALENs. Nat Med 2013;19(9):1111–13.

[47] Terpe K. Overview of tag protein fusions: from molecular and biochemical fundamentals to commercial systems. Appl Microbiol Biotechnol 2003;60(5):523–33.

[48] Thorn K. Genetically encoded fluorescent tags. Mol Biol Cell 2017;28(7):848–57.

[49] Sylvestre J, Margeot A, Jacq C, Dujardin G, Corral-Debrinski M. The role of the 3′ untranslated region in mRNA sorting to the vicinity of mitochondria is conserved from yeast to human cells. Mol Biol Cell 2003;14(9):3848–56.

[50] Xia W, Bringmann P, McClary J, Jones PP, Manzana W, Zhu Y, et al. High levels of protein expression using different mammalian CMV promoters in several cell lines. Protein Expr Purif 2006;45 (1):115–24.

[51] Szymczak AL, Workman CJ, Wang Y, Vignali KM, Dilioglou S, Vanin EF, et al. Correction of multi-gene deficiency in vivo using a single 'self-cleaving' 2A peptide-based retroviral vector. Nat Biotechnol 2004;22(5):589–94.

[52] Doyon Y, Vo TD, Mendel MC, Greenberg SG, Wang J, Xia DF, et al. Enhancing zinc-finger-nuclease activity with improved obligate heterodimeric architectures. Nat Methods 2011;8(1):74–9.

[53] Miller JC, Holmes MC, Wang J, Guschin DY, Lee YL, Rupniewski I, et al. An improved zinc-finger nuclease architecture for highly specific genome editing. Nat Biotechnol 2007;25(7):778–85.

[54] Hashimoto M, Bacman SR, Peralta S, Falk MJ, Chomyn A, Chan DC, et al. MitoTALEN: a general approach to reduce mutant mtDNA loads and restore oxidative phosphorylation function in mitochondrial diseases. Mol Ther 2015;23(10):1592–9.

[55] Schon EA, Rizzuto R, Moraes CT, Nakase H, Zeviani M, DiMauro S. A direct repeat is a hotspot for large-scale deletion of human mitochondrial DNA. Science 1989;244(4902):346–9.

[56] Corral-Debrinski M, Horton T, Lott MT, Shoffner JM, Beal MF, Wallace DC. Mitochondrial DNA deletions in human brain: regional variability and increase with advanced age. Nat Genet 1992;2(4):324–9.

[57] Diaz F, Bayona-Bafaluy MP, Rana M, Mora M, Hao H, Moraes CT. Human mitochondrial DNA with large deletions repopulates organelles faster than full-length genomes under relaxed copy number control. Nucleic Acids Res 2002;30(21):4626–33.

[58] Bacman SR, Moraes CT. Transmitochondrial technology in animal cells. Methods Cell Biol 2007;80:503–24.

[59] Jun AS, Trounce IA, Brown MD, Shoffner JM, Wallace DC. Use of transmitochondrial cybrids to assign a complex I defect to the mitochondrial DNA-encoded NADH dehydrogenase subunit 6 gene mutation at

nucleotide pair 14459 that causes Leber hereditary optic neuropathy and dystonia. Mol Cell Biol 1996;16 (3):771–7.
[60] Chol M, Lebon S, Benit P, Chretien D, de Lonlay P, Goldenberg A, et al. The mitochondrial DNA G13513A MELAS mutation in the NADH dehydrogenase 5 gene is a frequent cause of Leigh-like syndrome with isolated complex I deficiency. J Med Genet 2003;40(3):188–91.
[61] Shanske S, Coku J, Lu J, Ganesh J, Krishna S, Tanji K, et al. The G13513A mutation in the ND5 gene of mitochondrial DNA as a common cause of MELAS or Leigh syndrome: evidence from 12 cases. Arch Neurol 2008;65(3):368–72.
[62] Berkovic SF, Shoubridge EA, Andermann F, Andermann E, Carpenter S, Karpati G. Clinical spectrum of mitochondrial DNA mutation at base pair 8344. Lancet 1991;338(8764):457.
[63] Masucci JP, Schon EA, King MP. Point mutations in the mitochondrial tRNA(Lys) gene: implications for pathogenesis and mechanism. Mol Cell Biochem 1997;174(1-2):215–19.
[64] Bacman SR, Kauppila JHK, Pereira CV, Nissanka N, Miranda M, Pinto M, et al. MitoTALEN reduces mutant mtDNA load and restores tRNA(Ala) levels in a mouse model of heteroplasmic mtDNA mutation. Nat Med 2018;24(11):1696–700.
[65] Kauppila JHK, Baines HL, Bratic A, Simard ML, Freyer C, Mourier A, et al. A phenotype-driven approach to generate mouse models with pathogenic mtDNA mutations causing mitochondrial disease. Cell Rep 2016;16(11):2980–90.
[66] Lehmann D, Schubert K, Joshi PR, Hardy SA, Tuppen HA, Baty K, et al. Pathogenic mitochondrial mt-tRNA(Ala) variants are uniquely associated with isolated myopathy. Eur J Hum Genet 2015;23 (12):1735–8.
[67] Lochmuller H, Johns T, Shoubridge EA. Expression of the E6 and E7 genes of human papillomavirus (HPV16) extends the life span of human myoblasts. Exp Cell Res 1999;248(1):186–93.
[68] Yahata N, Matsumoto Y, Omi M, Yamamoto N, Hata R. TALEN-mediated shift of mitochondrial DNA heteroplasmy in MELAS-iPSCs with m.13513G > A mutation. Sci Rep 2017;7(1):15557.
[69] Yang Y, Wu H, Kang X, Liang Y, Lan T, Li T, et al. Targeted elimination of mutant mitochondrial DNA in MELAS-iPSCs by mitoTALENs. Protein Cell 2018;9(3):283–97.
[70] Rebelo AP, Williams SL, Moraes CT. In vivo methylation of mtDNA reveals the dynamics of protein-mtDNA interactions. Nucleic Acids Res 2009;37(20):6701–15.
[71] Reddy P, Ocampo A, Suzuki K, Luo J, Bacman SR, Williams SL, et al. Selective elimination of mitochondrial mutations in the germline by genome editing. Cell 2015;161(3):459–69.
[72] Phillips AF, Millet AR, Tigano M, Dubois SM, Crimmins H, Babin L, et al. Single-molecule analysis of mtDNA replication uncovers the basis of the common deletion. Mol Cell 2017;65(3):527–38 e526.
[73] Pereira CV, Bacman SR, Arguello T, Zekonyte U, Williams SL, Edgell DR, et al. mitoTev-TALE: a monomeric DNA editing enzyme to reduce mutant mitochondrial DNA levels. EMBO Mol Med 2018;10(9).
[74] Gammage PA, Gaude E, Van Haute L, Rebelo-Guiomar P, Jackson CB, Rorbach J, et al. Near-complete elimination of mutant mtDNA by iterative or dynamic dose-controlled treatment with mtZFNs. Nucleic Acids Res 2016;44(16):7804–16.
[75] Brix J, Dietmeier K, Pfanner N. Differential recognition of preproteins by the purified cytosolic domains of the mitochondrial import receptors Tom20, Tom22, and Tom70. J Biol Chem 1997;272(33):20730–5.
[76] Husain T, Passini MA, Parente MK, Fraser NW, Wolfe JH. Long-term AAV vector gene and protein expression in mouse brain from a small pan-cellular promoter is similar to neural cell promoters. Gene Ther 2009;16(7):927–32.
[77] Qin JY, Zhang L, Clift KL, Hulur I, Xiang AP, Ren BZ, et al. Systematic comparison of constitutive promoters and the doxycycline-inducible promoter. PLoS One 2010;5(5):e10611.
[78] Norrman K, Fischer Y, Bonnamy B, Wolfhagen Sand F, Ravassard P, Semb H. Quantitative comparison of constitutive promoters in human ES cells. PLoS One 2010;5(8):e12413.
[79] Katwal AB, Konkalmatt PR, Piras BA, Hazarika S, Li SS, John Lye R, et al. Adeno-associated virus serotype 9 efficiently targets ischemic skeletal muscle following systemic delivery. Gene Ther 2013;20 (9):930–8.

[80] Lukashev AN, Zamyatnin AA. Viral vectors for gene therapy: current state and clinical perspectives. Biochem (Mosc) 2016;81(7):700–8.
[81] Dong JY, Fan PD, Frizzell RA. Quantitative analysis of the packaging capacity of recombinant adeno-associated virus. Hum Gene Ther 1996;7(17):2101–12.
[82] Kumar M, Keller B, Makalou N, Sutton RE. Systematic determination of the packaging limit of lentiviral vectors. Hum Gene Ther 2001;12(15):1893–905.
[83] Porteus MH, Carroll D. Gene targeting using zinc finger nucleases. Nat Biotechnol 2005;23(8):967–73.
[84] Gammage PA, Minczuk M. Enhanced manipulation of human mitochondrial DNA heteroplasmy in vitro using tunable mtZFN technology. Methods Mol Biol 2018;1867:43–56.
[85] Gaj T, Gersbach CA, Barbas CF. ZFN, TALEN, and CRISPR/Cas-based methods for genome engineering. Trends Biotechnol 2013;31(7):397–405.
[86] Jo A, Ham S, Lee GH, Lee YI, Kim S, Lee YS, et al. Efficient mitochondrial genome editing by CRISPR/Cas9. Biomed Res Int 2015;2015:305716.
[87] Gammage PA, Moraes CT, Minczuk M. Mitochondrial genome engineering: the revolution may not be CRISPR-ized. Trends Genet 2018;34(2):101–10.
[88] Loutre R, Heckel AM, Smirnova A, Entelis N, Tarassov I. Can mitochondrial DNA be CRISPRized: pro and contra. IUBMB Life 2018;70(12):1233–9.
[89] Fogleman S, Santana C, Bishop C, Miller A, Capco DG. CRISPR/Cas9 and mitochondrial gene replacement therapy: promising techniques and ethical considerations. Am J Stem Cell 2016;5(2):39–52.
[90] Boissel S, Jarjour J, Astrakhan A, Adey A, Gouble A, Duchateau P, et al. megaTALs: a rare-cleaving nuclease architecture for therapeutic genome engineering. Nucleic Acids Res 2014;42(4):2591–601.
[91] Silva G, Poirot L, Galetto R, Smith J, Montoya G, Duchateau P, et al. Meganucleases and other tools for targeted genome engineering: perspectives and challenges for gene therapy. Curr Gene Ther 2011;11 (1):11–27.
[92] Lotfinia M, Abdollahpour-Alitappeh M, Hatami B, Zali MR, Karimipoor M. Adeno-associated virus as a gene therapy vector: strategies to neutralize the neutralizing antibodies. Clin Exp Med 2019;19:289–98.
[93] Santorelli FM, Tanji K, Kulikova R, Shanske S, Vilarinho L, Hays AP, et al. Identification of a novel mutation in the mtDNA ND5 gene associated with MELAS. Biochem Biophys Res Commun 1997;238:326–8.
[94] Taylor RW, Turnbull DM. Mitochondrial DNA mutations in human disease. Nat Rev Genet 2005;6 (5):389–402.
[95] Goto Y, Nonaka I, Horai S. A mutation in the tRNA(Leu)(UUR) gene associated with the MELAS subgroup of mitochondrial encephalomyopathies. Nature 1990;348(6302):651–3.
[96] Kleinstiver BP, Wang L, Wolfs JM, Kolaczyk T, McDowell B, Wang X, et al. The I-TevI nuclease and linker domains contribute to the specificity of monomeric TALENs. G3 (Bethesda) 2014;4:1155–65.
[97] Beurdeley M, Bietz F, Li J, Thomas S, Stoddard T, Juillerat A, et al. Compact designer TALENs for efficient genome engineering. Nat Commun 2013;4:1762.

CHAPTER 19

Mitochondrially targeted zinc finger nucleases

Pedro Pinheiro[1], Payam A. Gammage[2,3] and Michal Minczuk[1]

[1]MRC Mitochondrial Biology Unit, University of Cambridge, Cambridge, United Kingdom, [2]CRUK Beatson Institute, Glasgow, United Kingdom, [3]Institute of Cancer Sciences, University of Glasgow, Glasgow, United Kingdom

19.1 Introduction

Human mitochondria have their own small, circular genome (mtDNA) that encodes essential subunits of the oxidative phosphorylation machinery. Defects of the mitochondrial genome cause a wide variety of genetic disorders with diverse clinical manifestations ranging from progressive muscle weakness to fatal infantile disease. This genome is also implicated in the mitochondrial theory of ageing, which proposes that progressive accumulation of somatic mutations in mtDNA leads to a decline of mitochondrial function and limits mammalian lifespan. However, the contribution of these mtDNA mutations to ageing is still under debate. Mammalian mtDNA mutations that impair mitochondrial function have also been proposed to contribute to age-associated disease, such as Parkinson's disease and cancer.

Genome editing methods, routinely used for modifying DNA in the cell nucleus, are currently not available for mammalian mtDNA, mainly because of the difficulty of delivering exogenous nucleic acids into mammalian mitochondria and the lack of efficient homologous DNA repair mechanisms operating in this organelle. The inability to modify mitochondria genetically is a serious impediment to basic research on mitochondrial biogenesis and limits development of animal models of mtDNA diseases. Accordingly, the only current option for editing mammalian mtDNA is site-specific endonuclease-mediated introduction of double-strand breaks at selected mtDNA sites. This approach has been explored to reduce the burden of mutant mtDNA molecules, while sparing wild-type mtDNA molecules. These approaches use mitochondrially targeted restriction enzymes or programmable nucleases that do not contain a RNA element, namely transcription activator-like effector nucleases (TALENs) and zinc finger nucleases (ZFNs). In this chapter, the development and

The Human Mitochondrial Genome.
DOI: https://doi.org/10.1016/B978-0-12-819656-4.00019-X

application of mitochondrially targeted ZFNs (mtZFNs) are discussed in the context of future adaptation of this technology for clinical and therapeutic use in humans.

19.2 Zinc finger domain—structure and interaction with DNA

Cys2His2 zinc finger domains are relatively small protein motifs of about 3 kDa, frequently present in proteins as repeated tandem arrays, capable of binding specific sequences of DNA or RNA by direct protein–nucleic acid interactions. Zinc fingers (ZFs) are the most common DNA-binding motifs encoded in eukaryotes, being components of a large subset of proteins with important functions in health and disease [1].

Cys2His2 zinc fingers were first discovered in 1985, while studying the transcription factor IIIA (TFIIIA) of *Xenopus laevis*. The TFIIIA protein structure was found to have repetitive modular sequence consisting of nine 30 amino acid units, which was stabilized by zinc (and not by other metals). The repetitive unit was then named zinc finger domain because it contained zinc (Zn) and gripped or grasped the DNA, resembling fingers [2,3]. More specifically, an individual ZF unit consists of approximately 30 amino acids in a compact ααββ fold internally stabilized by chelation of a single zinc ion. Coordination of the zinc atom is accomplished by two conserved cysteines (within β-strands) and two histidines (within α-helix) [4–6].

Each ZF recognizes three successive bases on one strand of the DNA. The primary contacts are made through specific hydrogen-bond interactions between the major groove of the DNA and amino acids at positions—1, 3, and 6 with respect to the start of α-helix [7]. Additionally, the amino acid at position 2 of the helix mediates an important secondary interaction with a base on the opposite strand of DNA, which is complementary to the target recognized by the amino acid at position 6 of the previous ZF [8,9]. This cross-strand interaction adds a link between neighboring ZFs converting them into an overlapping array and extends the recognition site of a 2-zinc finger protein (ZFP) from 6 to 7 bases (Fig. 19.1).

19.3 Designer zinc fingers

The relatively simple one-to-one mode of DNA recognition between individual amino acids of ZFs with individual DNA bases together with the ability of each motif to act independently made them an appealing platform for the design of custom DNA-binding proteins. Initial experiments have shown the ability to rationally modify the DNA binding sequence of zinc fingers by changing individual DNA-interacting amino acids [10–13]. Nevertheless, as a canonical recognition code was starting to arise, some exceptions were identified. For this reason, design and selection strategies to allow simultaneous analysis of vast ZF libraries, such as phage display, were adopted [14–19]. Phage display is based on the expression

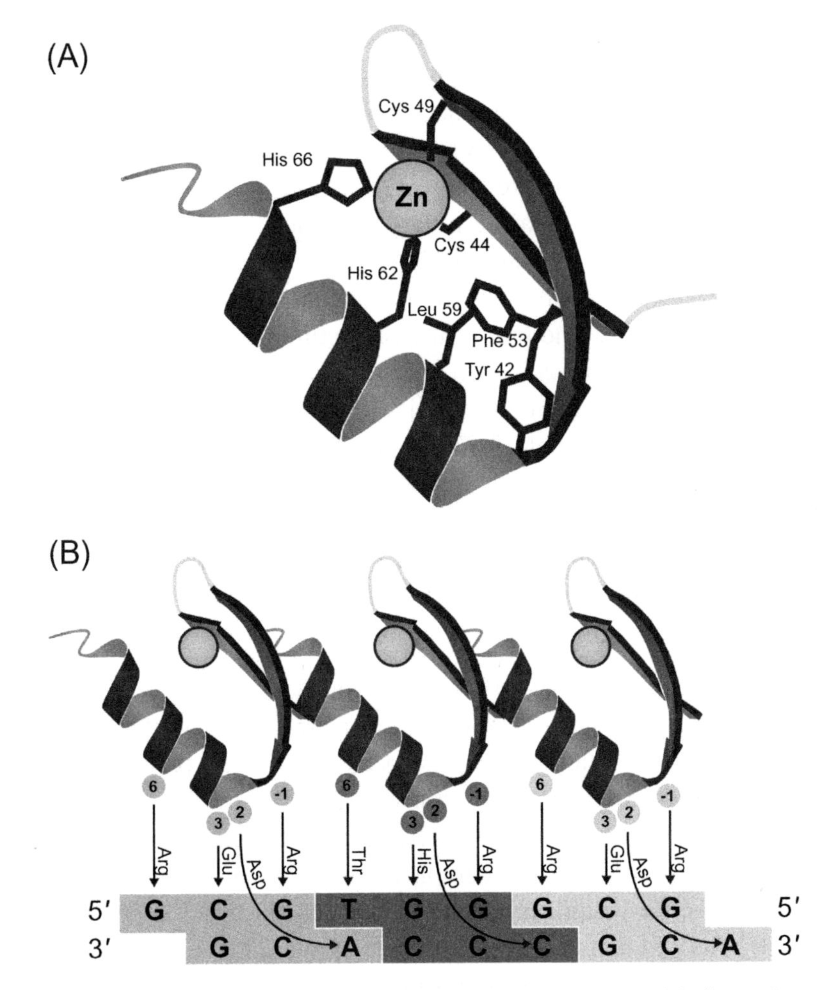

Figure 19.1: Structure of the Cys2His2 zinc finger fold and binding of DNA.
(A) Ribbon diagram of a Cys2His2 zinc finger, comprising an α-helix and two short β-strands. Side chains from invariant tyrosine (Tyr), phenylalanine (Phe), and leucine (Leu) residues form a hydrophobic core within which a single Zn^{2+} ion is coordinated by invariant cysteine (Cys) and His residues. Zn^{2+} is represented by a *yellow sphere* (*grey sphere in print version*). Structure from Ref. [6]. (B) Schematic of the Zif268 ZFP binding its cognate DNA target. Primary contacts are made by helical residues—1, 3, and 6, with cross-strand interactions mediated by residues at position 2. ZFPs bind DNA 5′−3′ in "reverse" orientation, from C-terminus to N-terminal. Source: *Images adapted from Klug A. The discovery of zinc fingers and their applications in gene regulation and genome manipulation. Annu Rev Biochem 2010;79:213−31.*

of protein domains on the tip of the capsid of bacteriophages [20,21]. Phages displaying the library of interest are affinity purified by binding to a DNA target of interest. The DNA sequence of candidate ZFs is then recovered by PCR from the purified phages. Using this method, ZF domains have been developed that recognize nearly all the 64 possible nucleotide triplets. This could be used for rational design of ZF arrays linked together capable of recognizing small stretches of DNA by mixing and matching individual specific ZFs. However, the cross-strand interaction of adjacent ZFs (see above and Fig. 19.1) could result in undesired loss of binding affinity. Therefore several strategies were employed to generate phage display libraries that would take into consideration the interaction between flanking ZFs [22–25]. One of these, the bipartite library approach, consists of two complementary libraries comprising variants of a 3-finger DNA-binding domain based on that of the mouse transcription factor Zif268. The first half-library contains randomizations in all the DNA-interacting positions of finger-1 and part of the contacting positions of finger-2, whereas the second library contains randomizations in the remaining part of finger-2 and all DNA-interacting positions of finger-3. This way, pairs of ZFs (1 + 2 and 2 + 3) are selected and then recombined in vitro to generate 3-finger proteins (1 + 2 + 3) [24,25]. Consequently, this strategy was exploited to create a large archive of ZFPs that recognize a vast number of 9 bp DNA sequences [26]. Also, with the accumulation of information concerning DNA binding by both natural and engineered ZFs, it was possible to postulate a general "recognition code." Zinc fingers created based solely on the recognition code were successfully applied in several organisms [27–30].

An array of three ZFs recognizes a 9-bp sequence, which would occur randomly many thousands of times in a large genome. Therefore longer arrays of zinc fingers are required to achieve higher binding specificity. For instance, a 6-finger protein recognizes 18-bp sequence, which can confer specificity within 68 billion bp of DNA (the human genome contains approximately 3 billion bp). However, designing 6-ZF proteins with conventional linkers can put excessive strain on the complex leading to an undesired small decrease in affinity. This is because the periodicity of the DNA does not match precisely the periodicity of packed ZFs [3]. The addition of an extra glycine or a glycine-serine-glycine to the canonical TGEKP linker between 2-finger modules makes the linker slightly longer and thus flexible [31]. Six-finger proteins generated by linking three 2-finger modules (3 × 2-fingers) can bind their targets with picomolar affinity. An alternative strategy is to assemble two 3-finger modules (2 × 3-fingers); however, the 3 × 2-finger architecture exhibits superior sensitivity to a mutation or insertion in the target sequence [3,31,32].

19.4 Chimeric zinc finger proteins—birth of zinc finger nuclease

The potential for ZFPs to recognize virtually any desired DNA sequence with high affinity and specificity made them an innovative platform with numerous possibilities to explore in research and medicine. In early experiments, ZFs were used as physical barrier for enzymes acting on DNA (e.g., inhibitory effect on RNA polymerase machinery) [33] or as competitors of other DNA binding proteins (e.g., competition with transcription factors in promoter binding sites) [34,35]. However, the competence of ZFs used in the absence of a functional domain is relatively weak. This has been improved by creating chimeric ZFPs containing effector domains, not only enhancing their efficiency but also opening the possibility of several other applications [36]. ZFs have been fused to transcription activators or repressor domains. For example, ZFs fused to the Krüppel-associated box (KRAB) repressor domain (KOX1) and targeted to the HIV promoter, repressed viral protein expression and attenuated viral infection [37]. In another study, linking zinc fingers with either VP16 or p65 transcriptional activation domains activates expression of the VEGF-A gene and induces the creation of new blood vessels in mouse ears [38,39]. While these chimeric ZFPs exert their function by interacting with other proteins, the use of effector domains acting directly on DNA was also explored to modify DNA in a site-specific manner [36]. Cytosine methylation can be accomplished in specific sequences by fusing a CpG-specific DNA methyltransferase. Targeted DNA methylation to a viral promoter was effective in reducing viral infection [40,41].

For many years, unsuccessful attempts have been made to create tailor-made endonucleases with desired DNA sequence specificity. A breakthrough in ZFP technology was the invention of ZFNs [42]. By combining ZFPs with the nonspecific C-terminal cleavage domain of type IIs endonuclease *Fok*I (R. FokI), the resulting chimeric protein can target and cleave virtually any required DNA sequence. Notably, fusion of the *Fok*I cleavage domain to a ZFP did not influence its sequence specificity nor alter significantly its binding affinity [43]. Later, it was shown that an efficient double-strand break requires two ZFNs binding two recognition sites in an inverted tail-to-tail orientation [44]. This requirement further increased ZFNs target specificity by necessitating longer target sites.

Interest in ZFNs broadened with the earlier discovery that double-strand breaks at specific loci radically potentiates homology-directed repair (HDR) [45,46]. Monogenic disorders are caused by the inheritance of defective alleles of a single gene; thus HDR-mediated gene correction triggered by double-strand breaks in the mutated allele using ZFNs, together with an extrachromosomal DNA donor, could restore the function of the gene. As such, ZFNs opened a new avenue for nuclear gene editing with a future view to correction therapies for several monogenic disorders. HDR-mediated gene correction by ZFNs was first utilized to generate *Drosophila melanogaster* strains with specific mutations, and in human cells to interrupt an exogenous GFP gene [47,48]. Later, correction of an endogenous human gene was reported for the first time using a ZFN pair designed against a X-linked severe combined immune deficiency (SCID) mutation in the IL2Rγ gene, with ZFN-driven HDR in patient-derived T-cells resulting in heterozygous and homozygous gene correction [49].

In addition to HDR, which requires a donor template, cells can also repair double-strand breaks by nonhomologous end joining (NHEJ). Through this imperfect repair mechanism, cells repair double-strand breaks without using a donor DNA template, often resulting in changes to the DNA sequence that can generate knock-out alleles. Therefore ZFNs can also be used to generate precise gene knockout in cells and organisms. Over the years, ZFNs have become important and powerful tools for precise genome editing, being used in a wide range of organisms to create gene disruptions, gene additions, and gene corrections [50].

19.5 Manipulation of the mammalian mitochondrial genome with mtZFNs

Manipulation of the mitochondrial genome within cells and organisms would facilitate exploration of normal mtDNA processes and enable development of therapies for currently incurable diseases arising from mutations or rearrangements of mtDNA. However, despite significant efforts, approaches to mitochondrial genetic transformation have failed so far [51]. The lack of an efficient mtDNA double-strand break repair mechanism [52], the existence of a highly efficient linear DNA degradation mechanism [53,54], and challenging delivery of exogenous DNA into mitochondria account for difficulty in successfully applying the same strategies used to manipulate nuclear DNA.

Mammalian cells possess several hundreds or thousands of copies of mtDNA, while disease-causing mutations and rearrangements often coexist with wild-type mtDNA, a condition called heteroplasmy. In the disease state, pathology is only present when the ratio of mutated mtDNA exceeds a certain threshold. For most mtDNA mutations and rearrangements, defective molecules should be present above 60% [55].

Mitochondrially targeted ZFNs, able to recognize and selectively degrade defective mtDNA in a heteroplasmic population, arose as a potential alternative to the direct correction of preexisting molecules. The success of this strategy is based on two biological processes, characteristic of mitochondria: (1) introduction of double-strand breaks into mtDNA leads to rapid degradation of these molecules that involves components of the mtDNA replication machinery [53,54] and (2) mtDNA copy number is maintained at cell-type specific steady-state levels. Therefore selective cleavage of mutant mtDNA by mtZFNs would lead to degradation of these molecules and stimulate copy number recovery using the remaining mtDNA pool of wild-type molecules, thus shifting the heteroplasmy ratio (Fig. 19.3).

19.5.1 The first step

In order to generate mtZFNs, it was necessary to accurately deliver functional ZFPs into the mitochondrial matrix. As mtDNA encodes only a small portion of the mitochondrial proteome, most nuclear-encoded mitochondrial proteins harbor a cleavable N-terminal mitochondrial targeting sequence (MTS), which guaranties their import into the mitochondrial matrix (Fig. 19.3).

However, the addition of a conventional MTS was not sufficient for exclusive trafficking of ZFPs to mitochondria, probably because ZFs have evolved primarily to operate in the cell nucleus. The way to overcome this was by adding a nuclear export signal (NES) in the C-terminal of ZFPs [56]. Thus incorporating both MTS and NES into ZFPs enabled their exclusive mitochondrial localization. Exclusion from the nucleus is also important to minimize possible off-targets in the nuclear DNA [35] (Fig. 19.3).

Once inside the mitochondrial matrix, ZFPs must fold correctly to bind mtDNA. In early experiments, we fused mtZFPs to DNA methylase effectors, instead of, for example, nucleases, enabling direct assays for sequence-specific DNA modification, CpG methylation in this case. When expressed in human cells, the mtZFP-methylase could methylate the mtDNA in a site-specific manner. Most importantly, it could discriminate between mtDNA sequences differing by one of nine base pairs, retaining the specificity reported for nuclear ZFPs [56]. This was the first evidence that mtZFPs could be used to recognize point mutations responsible for several mtDNA diseases.

Given the size of the human mitochondrial genome, the first generation of mtZFN was designed using single-chain (sc) ZFN architecture, in which a ZFP is linked to two *Fok*I catalytic domains within a single polypeptide. When single-chain mutation-specific (m.8993T > G) ZFNs were targeted to mitochondria in heteroplasmic cells harboring $\sim$85% mutant mtDNA, selective elimination of the m.8993T > G mtDNA was observed resulting in a twofold increase of wild-type mtDNA [57]. Although the mt-scZFN strategy was successful in targeting mtDNA point mutations, off-target cleavage due to constitutive dimerization of both *Fok*I domains in the same monomer was a concern. Furthermore, the mt-scZFN architecture could not be used to target mtDNA rearrangements, such as class I mtDNA deletions [58], which typically occur between and retain one of two direct repeats, rendering deleted molecules indistinguishable from wild-type mtDNA.

Second-generation mtZFNs were designed based on a conventional architecture of two ZFN monomers, each with a single *Fok*I catalytic domain, binding in a tail-to-tail orientation on complementary DNA strands. In this way, introduction of a DSB in the DNA requires binding of both mtZFNs monomers near each other to allow *Fok*I dimerization (Fig. 19.2). Further improvement in specificity was achieved by changing the wild-type *Fok*I domain for obligatory heterodimeric *Fok*I domains (ELD/KKR) [59]. These variants only cleave DNA when one *Fok*I-ELD dimerizes with one *Fok*I-KKR, reducing homodimer formation (ELD + ELD and KKR + KKR) activity by >40-fold [60]. Additional modifications to the architecture included rearrangement of the epitope tag and NES from C-terminal to N-terminal to avoid possible interference with *Fok*I domains [59] (Fig. 19.3).

The second generation of mtZFNs was used to target a large-scale mtDNA deletion of 4977 bp, the so-called common deletion (CD) [59]. This deletion spans a region between two 13-bp direct repeats (nucleotide positions 8470–8482 and 13,447–13,459). By using

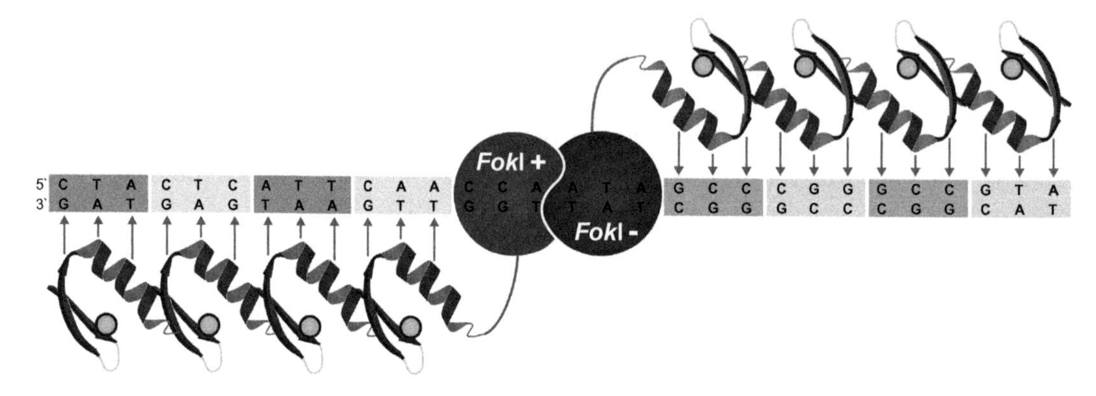

Figure 19.2: Schematic of zinc finger nuclease architecture.
ZFPs fused to the C-terminal domain of R. FokI are targeted to adjacent target sites on DNA, dimerizing within the spacer region when bound and inducing a DNA double-strand break. ZFP cross-strand interactions are not shown.

the conventional tail-to-tail mtZFN design, two different DNA-binding regions can be targeted, which will only be adjacent in deleted molecules. At the same time, wild-type molecules are spared because the binding sites will be separated by several kilobases, avoiding dimerization of *Fok*I domains. The CD-targeted mtZFNs successfully recognized and cut CD molecules, leading to substantial shifts in heteroplasmy toward wild-type mtDNA [59]. High levels of CD normally lead to low protein levels of all mtDNA-encoded genes due to insufficient levels of mRNAs and tRNAs that are encoded within the deletion, leading to defective OXPHOS. By eliminating CD mtDNA molecules using mtZFNs, cells previously harboring high heteroplasmic levels of this deletion demonstrate restored mitochondrial gene expression and OXPHOS function [59].

This new architecture of mtZFNs was also tested in heteroplasmic cells with high levels of the m.8993T > G mutation. A single transfection of these cells resulted in selective degradation of mutant molecules, which lead to a fivefold increase in wild-type mtDNA (outperforming the first-generation mtZFNs, which lead to a twofold increase) [56,59]. Furthermore, using each monomer independently did not exert any detectable effect on mtDNA, providing further evidence on specificity of mtZFN targeting. The second generation of mtZFNs is thus more effective and safer in targeting mtDNA mutations compared to first-generation mtZFNs. In addition, the new architecture expands the use of mtZFNs beyond point mutations, also being effective in other mtDNA rearrangements, such as mtDNA deletions.

A step forward in generating safer and more effective mtZFNs was finding the importance of controlled expression. Second-generation mtZFNs could produce directional shifts of mtDNA heteroplasmy; however, at high mtZFN expression levels, generalized (both

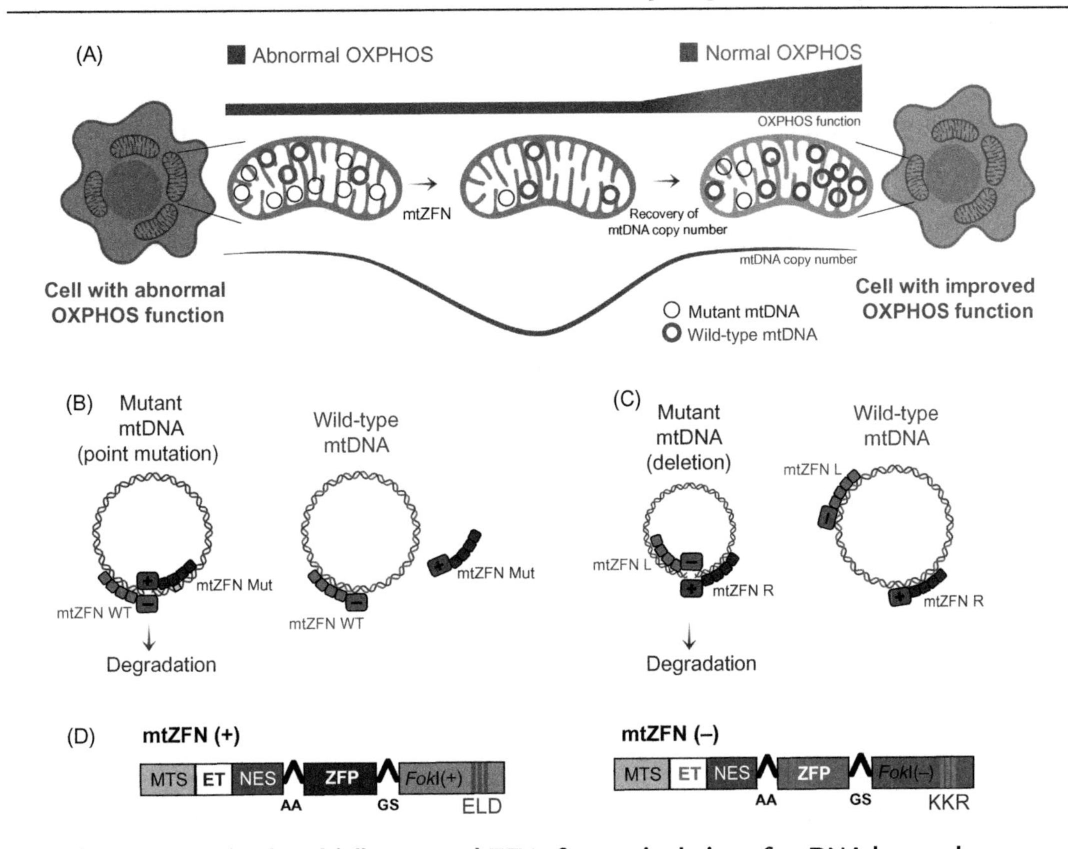

Figure 19.3: Mitochondrially targeted ZFNs for manipulation of mtDNA heteroplasmy. (A) The "antigenomic" approach to manipulate mtDNA heteroplasmy using mtZFNs. In heteroplasmic cells with perturbed OXPHOS due to a mutation in mtDNA, mtZFNs distinguish between the mutant (thin circles) than wild-type mtDNA molecules (thick circles). Cutting of the mutant mtDNA results in its degradation, while the wild-type molecules are spared. This leads to a reduction in the total mtDNA copy number. Upon the recovery of mtDNA copy number to the initial level, the cell is repopulated by residual, mainly wild-type, mtDNA molecules overcoming the pathogenicity threshold and restoring proper OXPHOS function. (B) Targeting mtDNA point mutations by mtZFNs. The mutant-specific mtZFN monomer (mtZFN Mut) binds to the mutation site and forms a dimer with the mtZFN monomer that recognizes mtDNA on the opposite strand (mtZFN WT). Dimerization of the *Fok*I domains leads to DNA double-strand breaks, with this linear mutant mtDNA being rapidly degraded. Because the mutation-specific mtZFN monomer does not bind to wild-type mtDNA, double-strand breaks are not introduced in this molecule, hence it is spared. (C) Targeting mtDNA large-scale deletions by mtZFNs. mtZFN monomers (mtZFN L and mtZFN R) dimerize at the deletion break-point; this results in DNA double-strand breaks, followed by degradation of mtDNA. mtZFN monomers do not dimerize on wild-type mtDNA as their binding sites are too far apart. (D) Schematic representation of the obligatory heterodimeric mtZFN constructs. MTS: mitochondrial targeting sequence; ET: epitope tag; NES: nuclear export signal; AA: two-alanine linker; ZFP: zinc finger peptide; GS: glycine-serine linker; *Fok*I (+) and *Fok*I(-): ELD and KKR modified *Fok*I domains. Source*: Images adapted from Gammage PA, Van Haute L, Minczuk M. Engineered mtZFNs for Manipulation of Human Mitochondrial DNA Heteroplasmy. Methods Mol Biol 2016(1351):145–62.*

wild-type and mutant) mtDNA depletion was also observed. This mtDNA depletion could also be observed when using mtZFNs nonspecific to endogenous mtDNA sequences expressed in mitochondria, further supporting the need for tight control of mtZFN expression to improve specific on-target nuclease activity. Initial experiments using m.8993T > G heteroplasmic cells showed that separating transfected cells into "low" and "high" expressions of mutation-specific mtZFNs produced different outcomes. Cells with "low" expression of mtZFNs had an increase in effectiveness of heteroplasmy shift compared with "high" expression cells. A greater mtDNA copy number depletion in "high" expression cells and an improved rate of copy number recovery in the "low" expressing population was also observed [61]. It was concluded that an optimized dose of mtZFNs can improve mtDNA heteroplasmy shifts and reduce deleterious effects of mtDNA copy number depletion via reducing off-target mtDNA nucleolysis.

To exert tight control of mtZFN expression and to provide a broad dynamic range of catalytic rate, we encoded engineered hammerhead ribozyme (HHR) into mtZFN mRNA. When placed upstream of the poly(A) signal of eukaryotic mRNAs, HHRs greatly reduce protein expression by autocatalysis of RNA. We used the 3′K19 version of an engineered HHR that additionally contains a tetracycline-responsive RNA aptamer. Binding of tetracycline rigidifies the RNA structure and inhibits catalysis by the HHR. Therefore supplementation of tetracycline inhibits the HHR in a dose-dependent manner and can control protein expression encoded by the mRNA in which it is contained [62,63]. Indeed, the addition of the HHR between the mtZFN STOP codon and poly(A) signal successfully enabled dynamic control of mtZFN expression. Using this modified strategy, mtZFNs were tested in m.8993T > G heteroplasmic cells. Here again, lower doses of mtZFNs proved to be beneficial in shifting mtDNA heteroplasmy with less pronounced copy number depletion, while higher doses produced unfavorable off-target effects and negligible shifts [61]. Thus a strict level of control over mtZFNs is necessary to avoid rapid and damaging effects of mtDNA depletion and a key step in achieving any positive therapeutic outcome.

Following heteroplasmy shifts triggered by mtZFNs in m.8993T > G cells, a functional rescue of mitochondrial respiration was achieved. In the same experiment, levels of key mitochondrial metabolites were recovered, suggesting a rescue of additional mitochondrial functions [61]. Remarkably, several sequential rounds of transfection of mtZFNs lead to a heteroplasmic shift from 25% to 90% of wild-type mtDNA [61,64]. This suggests that prolonged, controlled expression of mtZFNs could theoretically result in complete shifts toward the wild-type mtDNA.

19.5.2 In vivo *use of mtZFNs*

The *in vitro* experiments described above have highlighted the potential of mtZFN use as a therapeutic platform. Not only have mtZFNs been shown capable of selectively degrading mutant mtDNA, thus promoting a directional shift of heteroplasmy toward wild-type molecules, but importantly, the mtDNA heteroplasmic shifts were translated in biochemical and phenotypic rescues.

A crucial next step toward the therapeutic use of mtZFNs, as for most genetic therapies, is in vivo delivery and proof of efficacy. A current gold standard for transgene delivery in gene therapies are vectors based upon the recombinant adeno-associated virus (AAV). These vectors offer low immune response stimulation and low risk of genome integration, and it has been shown that viral genomes can persist for the entire life-span of lab animals in postmitotic or slowly dividing tissues [65,66].

Based on the encouraging outcome of experiments performed in cultured cells, mtZFNs were tested in a mouse model harboring a mutation in the mitochondrial $tRNA^{Ala}$ (mt-$tRNA^{Ala}$) gene at position m.5024C > T. The presence of m.5024C > T leads to lower steady-state levels of this tRNA and problems with intramitochondrial translation. A pair of second-generation mtZFNs were packaged individually into cardiotropic AAV9.45 capsids and systemically administered into mice at different doses (1×10^{12} viral genomes (vg)/mouse—low, 5×10^{12} vg/mouse—intermediate, and 1×10^{13} vg/mouse—high). At 65 days after injection, mtZFN-treated mice demonstrated specific elimination of the m.5024C > T mutant mtDNA in heart tissue, the efficacy of heteroplasmy shift being dependent on viral dose. The lowest dose did not result in heteroplasmy shifts, probably due to insufficient transduction of the heart. The highest dose, despite exhibiting significant shifts, also exhibited partial mtDNA copy number depletion. Importantly, the intermediate dose resulted in significant heteroplasmic shifts and did not lead to mtDNA copy number depletion, further supporting the importance of precise control of mtZFNs expression to achieve an optimal therapeutic effect. Notably, expression of mtZFNs for longer periods of time resulted in significant increases in heteroplasmy-shifting activity (unpublished data from our laboratory). This demonstrates that *in vivo* expression of mtZFNs over time was not only safe, but it can also have a beneficial, cumulative effect. Finally, the observed *in vivo* heteroplasmic shift of mtDNA molecules harboring m.5024C > T was realized in molecular and physiological rescue of disease phenotypes in heart tissue, assessed by stabilization of $tRNA^{Ala}$ levels and recovery of several mitochondrial metabolites [67]. The successful application of mtZFNs in selectively recognizing and degrading mutant mtDNA molecules *in vitro* and *in vivo* constitutes proof of concept that mtZFNs could be a promising therapeutic platform. Furthermore, the combination of mtZFNs with tissue-specific AAV serotypes may offer a potentially universal route to treatment for heteroplasmic mitochondrial diseases.

19.6 Concluding remarks

To date, successful strategies to manipulate mtDNA have focused on repopulating cells with healthy copies of mtDNA following selective elimination of disease-causing molecules. Early attempts made use of natural-occurring bacterial restriction endonucleases targeted to mitochondria (mtREs). These enzymes have been proven very effective in recognizing and specifically degrading mtDNA, both *in vitro* and *in vivo* [68,69]. However, REs are fundamentally unsuited to reprogramming of alternate DNA recognition sites and thus their applications. The ability to design mtZFNs to bind virtually any predetermined sequence overcomes this limitation. In addition, at optimized expression levels, mtZFNs have been shown to be as effective as REs in shifting heteroplasmy in cellular models [61]. Recent advancements in the ZFN technology that include attenuated DNA cleavage kinetics of *Fok*I (e.g., Q481A) and enhanced ZF specificity by removing on a nonspecific DNA contact (Arg-5) should lead to further improvements mtZFN specificity, enabling future introduction of double-strand breaks into mtDNA and heteroplasmy shifts with undetectable off-target activity [70].

Another class of engineered nucleases successfully applied to manipulate mtDNA heteroplasmy is mitoTALENs. Their DNA recognition code is significantly simpler than that of ZFPs, making mitoTALENs less laborious to design against a desired target site. A thymine (T) is generally required in the beginning of the target sequence, which could be advantageous to target N > T (or N > A) mutations. However, this feature restricts site selection. Also, the relatively large size of mitoTALENs could be a concern for in vivo delivery using AAVs, as these vectors have a limited packaging size of about 4.7 kb, generally precluding packaging both mitoTALEN monomers into a single AAV virion. On the contrary, the relatively small size of mtZFNs opens the possibility to package both monomers in the same viral genome, allowing reduction of viral doses. Nevertheless, mitoTALENs have been delivered by packaging into separate AAV virions and were, to a lesser extent, successful in correcting a mtDNA mutation in the mt-tRNA^{Ala} mouse model [71]. The CRISPR/Cas9 endonuclease has recently revolutionized nuclear DNA manipulation. The simplicity of using a single short RNA to guide Cas9 to a desired target site made this tool widely used by the scientific community. However, the use of CRISPR/Cas9 in mitochondria is limited due to challenges related to guide RNA import into mitochondria [72] and a limited number of targetable sites in mtDNA.

Research perspectives

Further applications of mtZFPs could be envisaged through use of different effector domains. In this chapter, we focused mainly on describing mtZFP fusion with the *Fok*I endonuclease domain. However, several other effector domains may potentially be repurposed to target mtDNA. Briefly described here, the fusion of mtZFPs with DNA methylase domains can be used for site-specific methylation of mtDNA. Another example is the fusion of mtZFPs with the recently developed base editor domains, theoretically allowing direct manipulation of single base pairs in the mtDNA.

While there are currently no treatments available for mitochondrial disease patients, the possibility of shifting heteroplasmy in mtDNA diseases using mtZFNs holds significant therapeutic potential. Given the specificity and effectiveness of mtZFNs in vivo, amelioration of clinical symptoms and/or halting of disease progression could be expected. Thus the successful translation of these tools into effective medicines could transform the prospects of many patients with mitochondrial disease. Accumulating data on the safe use of nuclear ZFNs in human clinical trials [73–76] augurs well for the use of mtZFNs, as the era of mitochondrial genetic medicine approaches.

Acknowledgments

We would like to thank the past and present members of the Mitochondrial Genetics Group at the MRC-MBU, University of Cambridge for stimulating discussions during the course of the work on the development of the mtZFN technology. The Medical Research Council, UK (MC_UU_00015/4), CRUK Beatson Institute, and The Champ Foundation are gratefully acknowledged for their support of our work. P. P. has been supported by the EU Horizon 2020 ITN "Regulation of Mitochondrial Gene Expression" **REMIX**-H2020-MSCA-**ITN**-2016.

References

[1] Cassandri M, et al. Zinc-finger proteins in health and disease. Cell Death Discov 2017;3:17071.

[2] Miller J, McLachlan AD, Klug A. Repetitive zinc-binding domains in the protein transcription factor IIIA from *Xenopus oocytes*. EMBO J 1985;4(6):1609–14.

[3] Klug A. The discovery of zinc fingers and their applications in gene regulation and genome manipulation. Annu Rev Biochem 2010;79:213–31.

[4] Berg JM. Proposed structure for the zinc-binding domains from transcription factor IIIA and related proteins. Proc Natl Acad Sci U S A 1988;85(1):99–102.

[5] Lee MS, et al. Three-dimensional solution structure of a single zinc finger DNA-binding domain. Science (New York, NY) 1989;245(4918):635–7.

[6] Neuhaus D, et al. Solution structures of two zinc-finger domains from SWI5 obtained using two-dimensional 1H nuclear magnetic resonance spectroscopy. A zinc-finger structure with a third strand of beta-sheet. J Mol Biol 1992;228(2):637–51.

[7] Pavletich NP, Pabo CO. Zinc finger–DNA recognition: crystal structure of a Zif268-DNA complex at 2.1 A. Science (New York, NY) 1991;252(5007):809–17.

[8] Fairall L, et al. The crystal structure of a two zinc-finger peptide reveals an extension to the rules for zinc-finger/DNA recognition. Nature 1993;366(6454):483–7.

[9] Elrod-Erickson M, et al. Zif268 protein–DNA complex refined at 1.6 A: a model system for understanding zinc finger–DNA interactions. Structure (London, England: 1993) 1996;4(10):1171–80.
[10] Desjarlais JR, Berg JM. Toward rules relating zinc finger protein sequences and DNA binding site preferences. Proc Natl Acad Sci U S A 1992;89(16):7345–9.
[11] Thukral SK, Morrison ML, Young ET. Mutations in the zinc fingers of ADR1 that change the specificity of DNA binding and transactivation. Mol Cell Biol 1992;12(6):2784–92.
[12] Shi Y, Berg JM. A direct comparison of the properties of natural and designed zinc-finger proteins. Chem Biol 1995;2(2):83–9.
[13] Desjarlais JR, Berg JM. Use of a zinc-finger consensus sequence framework and specificity rules to design specific DNA binding proteins. Proc Natl Acad Sci U S A 1993;90(6):2256–60.
[14] Choo Y, Klug A. Toward a code for the interactions of zinc fingers with DNA: selection of randomized fingers displayed on phage. Proc Natl Acad Sci U S A 1994;91(23):11163–7.
[15] Choo Y, Klug A. Selection of DNA binding sites for zinc fingers using rationally randomized DNA reveals coded interactions. Proc Natl Acad Sci U S A 1994;91(23):11168–72.
[16] Jamieson AC, Kim SH, Wells JA. In vitro selection of zinc fingers with altered DNA-binding specificity. Biochemistry 1994;33(19):5689–95.
[17] Jamieson AC, Wang H, Kim SH. A zinc finger directory for high-affinity DNA recognition. Proc Natl Acad Sci U S A 1996;93(23):12834–9.
[18] Rebar EJ, Pabo CO. Zinc finger phage: affinity selection of fingers with new DNA-binding specificities. Science (New York, NY) 1994;263(5147):671–3.
[19] Wu H, Yang WP, Barbas 3rd CF. Building zinc fingers by selection: toward a therapeutic application. Proc Natl Acad Sci U S A 1995;92(2):344–8.
[20] Smith GP. Filamentous fusion phage: novel expression vectors that display cloned antigens on the virion surface. Science (New York, NY) 1985;228(4705):1315–17.
[21] McCafferty J, et al. Phage antibodies: filamentous phage displaying antibody variable domains. Nature 1990;348(6301):552–4.
[22] Segal DJ, Barbas 3rd CF. Design of novel sequence-specific DNA-binding proteins. Curr Opin Chem Biol 2000;4(1):34–9.
[23] Segal DJ, et al. Toward controlling gene expression at will: selection and design of zinc finger domains recognizing each of the 5′-GNN-3′ DNA target sequences. Proc Natl Acad Sci U S A 1999;96(6):2758–63.
[24] Isalan M, Klug A, Choo Y. Comprehensive DNA recognition through concerted interactions from adjacent zinc fingers. Biochemistry 1998;37(35):12026–33.
[25] Isalan M, Klug A, Choo Y. A rapid, generally applicable method to engineer zinc fingers illustrated by targeting the HIV-1 promoter. Nat Biotechnol 2001;19(7):656–60.
[26] Jamieson AC, Miller JC, Pabo CO. Drug discovery with engineered zinc-finger proteins. Nat Rev Drug Discov 2003;2(5):361–8.
[27] Corbi N, et al. The artificial zinc finger coding gene 'Jazz' binds the utrophin promoter and activates transcription. Gene Therapy 2000;7(12):1076–83.
[28] Libri V, et al. The artificial zinc finger protein 'Blues' binds the enhancer of the fibroblast growth factor 4 and represses transcription. FEBS Lett 2004;560(1-3):75–80.
[29] Sera T. Inhibition of virus DNA replication by artificial zinc finger proteins. J Virol 2005;79(4):2614–19.
[30] Sera T, Uranga C. Rational design of artificial zinc-finger proteins using a nondegenerate recognition code table. Biochemistry 2002;41(22):7074–81.
[31] Kim JS, Pabo CO. Getting a handhold on DNA: design of poly-zinc finger proteins with femtomolar dissociation constants. Proc Natl Acad Sci U S A 1998;95(6):2812–17.
[32] Moore M, Klug A, Choo Y. Improved DNA binding specificity from polyzinc finger peptides by using strings of two-finger units. Proc Natl Acad Sci U S A 2001;98(4):1437–41.
[33] Choo Y, Sanchez-Garcia I, Klug A. In vivo repression by a site-specific DNA-binding protein designed against an oncogenic sequence. Nature 1994;372(6507):642–5.

[34] Bartsevich VV, Juliano RL. Regulation of the MDR1 gene by transcriptional repressors selected using peptide combinatorial libraries. Mol Pharmacol 2000;58(1):1–10.
[35] Papworth M, et al. Inhibition of herpes simplex virus 1 gene expression by designer zinc-finger transcription factors. Proc Natl Acad Sci U S A 2003;100(4):1621–6.
[36] Papworth M, Kolasinska P, Minczuk M. Designer zinc-finger proteins and their applications. Gene 2006;366(1):27–38.
[37] Reynolds L, et al. Repression of the HIV-1 5′ LTR promoter and inhibition of HIV-1 replication by using engineered zinc-finger transcription factors. Proc Natl Acad Sci U S A 2003;100(4):1615–20.
[38] Liu PQ, et al. Regulation of an endogenous locus using a panel of designed zinc finger proteins targeted to accessible chromatin regions. Activation of vascular endothelial growth factor A. J Biol Chem 2001;276(14):11323–34.
[39] Rebar EJ, et al. Induction of angiogenesis in a mouse model using engineered transcription factors. Nat Med 2002;8(12):1427–32.
[40] Xu GL, Bestor TH. Cytosine methylation targeted to pre-determined sequences. Nat Genet 1997;17 (4):376–8.
[41] Li F, et al. Chimeric DNA methyltransferases target DNA methylation to specific DNA sequences and repress expression of target genes. Nucleic Acids Res 2007;35(1):100–12.
[42] Kim YG, Cha J, Chandrasegaran S. Hybrid restriction enzymes: zinc finger fusions to Fok I cleavage domain. Proc Natl Acad Sci U S A 1996;93(3):1156–60.
[43] Smith J, Berg JM, Chandrasegaran S. A detailed study of the substrate specificity of a chimeric restriction enzyme. Nucleic Acids Res 1999;27(2):674–81.
[44] Smith J, et al. Requirements for double-strand cleavage by chimeric restriction enzymes with zinc finger DNA-recognition domains. Nucleic Acids Res 2000;28(17):3361–9.
[45] Jasin M. Genetic manipulation of genomes with rare-cutting endonucleases. Trends Genet: TIG 1996;12 (6):224–8.
[46] Rouet P, Smih F, Jasin M. Introduction of double-strand breaks into the genome of mouse cells by expression of a rare-cutting endonuclease. Mol Cell Biol 1994;14(12):8096–106.
[47] Bibikova M, et al. Enhancing gene targeting with designed zinc finger nucleases. Science (New York, NY) 2003;300(5620):764.
[48] Porteus MH, Baltimore D. Chimeric nucleases stimulate gene targeting in human cells. Science (New York, NY) 2003;300(5620):763.
[49] Urnov FD, et al. Highly efficient endogenous human gene correction using designed zinc-finger nucleases. Nature 2005;435(7042):646–51.
[50] Gaj T, Gersbach CA, Barbas 3rd CF. ZFN, TALEN, and CRISPR/Cas-based methods for genome engineering. Trends Biotechnol 2013;31(7):397–405.
[51] Patananan AN, et al. Modifying the mitochondrial genome. Cell Metab 2016;23(5):785–96.
[52] Alexeyev M, et al. The maintenance of mitochondrial DNA integrity—critical analysis and update. Cold Spring Harb Perspect Biol 2013;5(5):a012641.
[53] Nissanka N, et al. The mitochondrial DNA polymerase gamma degrades linear DNA fragments precluding the formation of deletions. Nat Commun 2018;9(1):2491.
[54] Peeva V, et al. Linear mitochondrial DNA is rapidly degraded by components of the replication machinery. Nat Commun 2018;9(1):1727.
[55] Gorman GS, et al. Mitochondrial diseases. Nat Rev Dis Prim 2016;2:16080.
[56] Minczuk M, et al. Sequence-specific modification of mitochondrial DNA using a chimeric zinc finger methylase. Proc Natl Acad Sci U S A 2006;103(52):19689–94.
[57] Minczuk M, et al. Development of a single-chain, quasi-dimeric zinc-finger nuclease for the selective degradation of mutated human mitochondrial DNA. Nucleic Acids Res 2008;36(12):3926–38.
[58] Nissanka N, Minczuk M, Moraes CT. Mechanisms of mitochondrial DNA deletion formation. Trends Genet: TIG 2019;35(3):235–44.

[59] Gammage PA, et al. Mitochondrially targeted ZFNs for selective degradation of pathogenic mitochondrial genomes bearing large-scale deletions or point mutations. EMBO Mol Med 2014;6(4):458–66.

[60] Miller JC, et al. An improved zinc-finger nuclease architecture for highly specific genome editing. Nat Biotechnol 2007;25(7):778–85.

[61] Gammage PA, et al. Near-complete elimination of mutant mtDNA by iterative or dynamic dose-controlled treatment with mtZFNs. Nucleic Acids Res 2016;44(16):7804–16.

[62] Berens C, Suess B. Riboswitch engineering—making the all-important second and third steps. Curr Opin Biotechnol 2015;31:10–15.

[63] Beilstein K, et al. Conditional control of mammalian gene expression by tetracycline-dependent hammerhead ribozymes. ACS Synth Biol 2015;4(5):526–34.

[64] Gaude E, et al. NADH shuttling couples cytosolic reductive carboxylation of glutamine with glycolysis in cells with mitochondrial dysfunction. Mol Cell 2018;69(4):581–93 e7.

[65] Mingozzi F, High KA. Therapeutic in vivo gene transfer for genetic disease using AAV: progress and challenges. Nat Rev Genet 2011;12(5):341–55.

[66] Wang D, Tai PWL, Gao G. Adeno-associated virus vector as a platform for gene therapy delivery. Nat Rev Drug Discov 2019;18(5):358–78.

[67] Gammage PA, et al. Genome editing in mitochondria corrects a pathogenic mtDNA mutation in vivo. Nat Med 2018;24(11):1691–5.

[68] Srivastava S, Moraes CT. Manipulating mitochondrial DNA heteroplasmy by a mitochondrially targeted restriction endonuclease. Hum Mol Genet 2001;10(26):3093–9.

[69] Tanaka M, et al. Gene therapy for mitochondrial disease by delivering restriction endonuclease SmaI into mitochondria. J Biomed Sci 2002;9(6 Pt 1):534–41.

[70] Miller JC, et al. Enhancing gene editing specificity by attenuating DNA cleavage kinetics. Nat Biotechnol 2019;37(8):945–52.

[71] Bacman SR, et al. MitoTALEN reduces mutant mtDNA load and restores tRNA(Ala) levels in a mouse model of heteroplasmic mtDNA mutation. Nat Med 2018;24(11):1696–700.

[72] Gammage PA, Moraes CT, Minczuk M. Mitochondrial genome engineering: the revolution may not be CRISPR-ized. Trends Genet: TIG 2018;34(2):101–10.

[73] *Ascending dose study of genome editing by the zinc finger nuclease (ZFN) therapeutic SB-913 in subjects with MPS II.* Available from: https://clinicaltrials.gov/show/NCT03041324.

[74] *Ascending dose study of genome editing by the zinc finger nuclease (ZFN) therapeutic SB-318 in subjects with MPS I.* Available from: https://clinicaltrials.gov/show/NCT02702115.

[75] Laoharawee K, et al. Dose-dependent prevention of metabolic and neurologic disease in murine MPS II by ZFN-mediated in vivo genome editing. Mol Therapy: J Am Soc Gene Therapy 2018;26(4):1127–36.

[76] Ou L, et al. ZFN-mediated in vivo genome editing corrects murine Hurler syndrome. Mol Therapy: J Am Soc Gene Therapy 2019;27(1):178–87.

CHAPTER 20

Mitochondrial movement between mammalian cells: an emerging physiological phenomenon

Michael V. Berridge[1], Patries M. Herst[1,2] and Carole Grasso[1]

[1]Malaghan Institute of Medical Research, Wellington, New Zealand, [2]Department of Radiation Therapy, University of Otago, Wellington, New Zealand

20.1 Introduction

Horizontal transfer of genetic material between individual cells is commonplace in the bacterial and archaeal domains of life. Eukaryotic life originated through one or several endosymbiotic events between a common archaeal ancestor and a proteobacterium [1,2], which gave the new aerobically respiring organism a strong competitive advantage over anaerobic unicellular organisms in an increasingly oxygenated environment [3]. While eubacteria and archaebacteria make up the large majority of biomass today, eukaryotic cells, with their newly acquired ability to exploit oxygen for increasingly diverse bioenergetic purposes, have adapted to live symbiotically in a microbial world. The human body harbors as many microbial cells as human cells, which are found mainly in the colon and on the skin [4]. In recent years, this new knowledge about our cellular origins and our relationship with microbes has challenged our concept of self and revolutionized our understanding of our place in the biosphere. Until recently, and with few exceptions, genes and genetic material in cells of higher eukaryotes were thought to be constrained within cells. During the last two decades, this view has been challenged with horizontal transfer of mitochondrial DNA (mtDNA) between cells having been demonstrated on many occasions phylogenetically, as well as in cell culture and physiologically. The discovery that maternally inherited mitochondria with their mtDNA can transfer between mammalian cells was unexpected given that horizontal transfer of nuclear genes has never been convincingly demonstrated. Thus it would appear that gene transfer between cells in the mammalian body is restricted to those within mitochondria. This chapter will review the literature of horizontal mtDNA transfer in mammals with particular emphasis on the potential translational benefits this process confers on recipient cells and on the organism as a whole.

The Human Mitochondrial Genome.
DOI: https://doi.org/10.1016/B978-0-12-819656-4.00020-6

20.2 Cell-to-cell transfer of mitochondria with mtDNA: a brief overview

The field of mitochondrial transfer between mammalian cells was foreshadowed by pioneering research in yeast involving the development of cells devoid of mtDNA [5–7], and by Clark and Shay [8] in 1982 who transferred isolated chloramphenicol-resistant mitochondria from mouse mammary cells into chloramphenicol-sensitive mouse adrenal cortical tumor cells. This transfer occurred at high frequencies through endocytosis in coculture experiments. Chloramphenicol resistance was conferred by a mutation in the 16S rRNA mtDNA gene (CAP^R), and transfer of antibiotic resistance as well as rhodamine 123 accumulation and recipient cell-specific inhibitor profiles marked mitochondrial transfer [8]. Rather than relying on endocytosis in coculture systems, King and Attardi [9] injected isolated chloramphenicol-resistant mitochondria from human fibroblasts into the cytoplasm of antibiotic sensitive human 143B osteosarcoma cells. The 143B cells had their mtDNA content depleted by treatment with low-dose ethidium bromide for 3–4 days before microinjection. Transformants were identified by antibiotic resistance, the presence of the CAP^R mutation and recipient-specific inhibitor profiles [9]. In a subsequent study, King and Attardi [10] completely depleted 143B cells of mtDNA (ρ^0 cells) by long-term culture with low-dose ethidium bromide. Respiration was restored after microinjection with chloramphenicol-resistant mitochondria isolated from HeLa cells. Antibiotic resistance and growth requirements for uridine and pyruvate were used as selective markers for respiration deficiency [11]. It would be another 15 years after the King and Attardi experiments, and 2 years after Rustom and colleagues showed that organelles transfer between human cells through "tunneling nanotubes" or TNTs [12], before Spees et al. demonstrated unequivocally that mitochondria and mtDNA transferred from human mesenchymal stem cells (MSCs) or fibroblasts to A549ρ^0 human lung adenocarcinoma cells, resulting in respiration recovery [13]. Mitochondrial transfer between human cells in coculture and between rodent cells has been reviewed by our group [14–19], and by others [20–22]. In most reports, mesenchymal stem/stromal cells, endothelial cells, or fibroblasts are the mitochondrial donors, and tumor cells without mtDNA, or with damaged mtDNA are the recipients. Transfer between similar cells has also been demonstrated [23]. Mitochondrially targeted fluorescent dyes have often been used to demonstrate mitochondrial transfer between cells, but the limitations of these methodologies have not always been considered [16]. Many of these studies are subject to dye leakage, are short-term in nature, and therefore preclude genetic analysis of mtDNA, restoration of respiration, and other cellular functions. Nevertheless, some studies have used cells expressing mitochondrially targeted fluorescent proteins and confocal microscopy approaches to show internalization of transferred mitochondria. In addition, some coculture studies have assessed respiratory function and/or used more rigorous genetic methodologies that allow verification of repopulation by donor mitochondria [13,24–28]. Recent reports of mitochondrial transfer between cells in coculture and in xenotransplantation studies have confirmed earlier results and demonstrated the broad physiological occurrence of intercellular mitochondria transfer between normal cells, in disease models and under pathological conditions. The ideas behind and practicalities of complementing or replacing respiration-incompetent mitochondria through transfer of healthy mitochondria will be explored further in the next section.

20.3 Translational benefits of mitochondrial transfer

Mitochondrial dysfunction lies at the core of neuromuscular and neurodegenerative mitochondriopathies [29–31]. However, loss of mitochondrial respiration has also been implicated in traumatic brain and spinal cord injury and stroke [32–35], ischemic heart disease [36] and in more global diseases such as diabetes [37], cardiovascular diseases [38,39], cancer [40–45], gastrointestinal disorders [46], skin disorders [47], and aging [48,49]. The underlying cause in most if not all of the mitochondriopathies is a nuclear or mtDNA mutation or epigenetic change. However, loss of mitochondrial respiratory function can also occur during life through damage to mtDNA or any of the components of the mitochondrial electron transport and oxidative phosphorylation systems, or the mitochondrial protein synthetic machinery. Ischemia-reperfusion injury is an example of such an injury, which occurs when arterial supply to an organ such as the heart, liver, kidneys, or brain is severely compromised. The ability to restore mitochondrial function would have a very significant impact on patient outcomes for all diseases with a mitochondrial component. Most research in the field has focused on interventions that affect a particular aspect of mitochondrial function, such as increasing antioxidants, or using mild uncouplers, alternative mitochondrial energy substrates, antiinflammatory agents, calcium channel antagonists, and ischemic preconditioning in the case of ischemic heart disease [32,50–53]. However, these therapies, alone or in combination, have either not been tried clinically or have not been very successful in clinical settings. A more holistic approach to countering mitochondrial defects would be to replace damaged mitochondria. The effect of respiratory deficiency on healthy cells and cancer cells as well as the beneficial and detrimental effects of replacing respiration-deficient mitochondria are summarized in Fig. 20.1. The next section reviews in vitro approaches and animal studies involving cell-to-cell mitochondrial transfer to ameliorate various disease conditions involving mitochondrial dysfunction.

20.3.1 Mitochondrial transfer between cells

The first indication that intercellular mitochondrial transfer occurs in nature in higher animals concerned phylogenetic analysis of mtDNA from a 6000 year-old canine transmissible venereal tumor. Following analysis of 37 tumors from 15 dogs, it was proposed that periodic transfer of mitochondria/mtDNA from host animals rescued tumor cells which had accumulated deleterious mtDNA mutations [54,55]. In another notable publication, Lei and Spradling showed that nurse germ cells in primordial cysts donate organelles, including mitochondria, to facilitate differentiation of mature mouse oocytes [56]. Many studies have

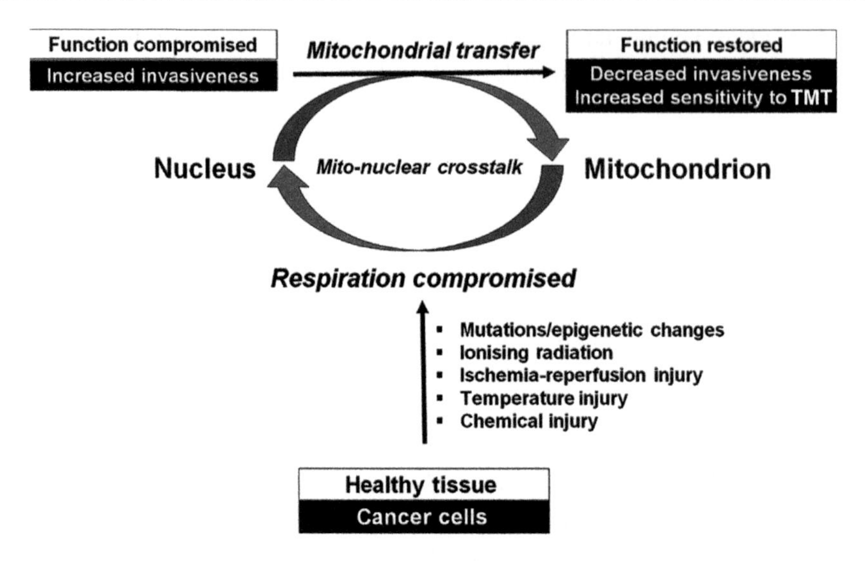

Figure 20.1: The benefits of mitochondrial transfer for recipient cells.
Mitochondrial respiration can be compromised by genetic/epigenetic changes or any type of injury that damages the mtDNA or components of the respiratory chain. Mito-nuclear crosstalk will compensate for a decrease in mitochondrial ATP production through mtDNA repair, mitochondrial biogenesis, building highly efficient mitochondrial networks, removal of damaged respiratory subunits, etc. Respiratory deficiency will affect cancer cells and normal healthy cells differently. Most cancer cells are able to shift from a mainly mitochondrial metabolism to extensively glycolytic metabolism, although this results in decreased proliferation and is accompanied by increased invasiveness. Mitochondrial transfer to respiration-compromised cancer cells will increase their respiratory capacity and proliferation rate; it will also make them less invasive and more sensitive to radiation therapy and certain forms of chemotherapy (TMT). Cancer cells without mtDNA (ρ^0 cells) are unable to proliferate in the absence of uridine in growth media or in mice because they are unable to produce pyrimidine precursors of nucleic acids. For these cells, mitochondrial transfer restores respiration and tumorigenicity. For healthy cells, respiratory insufficiency is always a life-threatening situation with few cellsable to survive on glycolysis for long. Transfer of healthy mitochondria will restore respiration, metabolic rates, and viability.

now shown mitochondrial transfer between cells, mostly to respiration-deficient tumor cells in vitro, with examples summarized in Table 20.1. Many of these studies involved restoration of respiration, loss of auxotrophy for uridine, and in some cases, increased or restored tumorigenic potential.

20.3.1.1 Mitochondrial transfer into tumor cells

Early studies showed that mitochondrial transfer from human MSCs to 143Bρ^0 osteosarcoma cells restored their ability to grow without uridine [25]. Low-level mitochondrial transfer that enabled the formation of small tumors was also observed following xenotransplantation of

Table 20.1: Examples of mitochondrial transfer in vitro and in vivo.

Donor cells	Recipient cells	Outcome of mitochondrial transfer	References
Mitochondrial transfer in vitro			
BM-MSC, fibroblasts (h)[a]	A549ρ^0 lung adenocarcinoma cells (h)	Repopulation of mitochondria and respiration rescue	[13]
BM-MSC (rat)	Lung endothelial cells (rat)	Apoptosis and capillary degeneration by donor cells	[57]
Adipose-derived MSC (h)	Cardiomyocytes (m)	Reprogramming toward a progenitor-like state	[58]
BM-MSC (m)	Lung epithelial (m and h)	Reduction in cell stress	[59]
iPSC and BM-MSC (h)	Bronchial epithelia (h)	Restored ATP after cigarette smoke-induced damaged	[26]
BM-MSC (h)	143Bρ^0 osteosarcoma cells (h)	Restoration of mitochondrial function	[25]
Wharton's jelly-derived MSC (h)	143Bρ^0 osteosarcoma cells (h)	Restoration of mitochondrial function	[27]
BM-MSC (h)	Neurons, astrocytes (rat)	N/A	[60]
BM-MSC (rat)	Cardiomyoblasts (rat)	Increased resistance to ischemia/reperfusion injury	[61]
Umbilical cord-derived MSC (h)	T cells (h)	Downregulated autophagy in systemic lupus erythematosus	[62]
iPSC MSC (h)	Corneal epithelial cells (r)	Protection from oxidative stress-induced mitochondrial damage	[63]
BM stromal cells (m)	AML cells (h)	Survival advantage following chemotherapy	[64]
BM-MSC (h)	Skin fibroblasts (h)	Rescue of fission morphology in an inherited mitochondrial disease	[65]
BM stromal cells (h)	AML blasts (h)	Increased mitochondrial respiration	[66]
BM-MSC (h)	Astrocytes (rat)	Restoration of bioenergetics following ischemic injury	[67]
iPSC mesenchymal stem cells (h)	Cardiomyocytes (m, h)	Rescue of anthracycline-induced injury	[68]
iPSC mesenchymal stem cells (h)	Bronchial epithelial (h)	Lower levels of apoptosis following hypoxic injury	[69]
BM stromal cells (h)	Multiple myeloma (h)	Evidence CD38 inhibits mitochondrial transfer	[70]
BM-MSC (rat)	Motor-neurons (rat)	Improved bioenergetic profile after oxygen-glucose deprivation	[71]

(*Continued*)

Table 20.1: (Continued)

Donor cells	Recipient cells	Outcome of mitochondrial transfer	References
Mitochondrial transfer in vitro			
BM-MSC (h)	T helper 17 (h)	Acquired an antiinflammatory phenotype	[72]
Endothelial progenitor cells (m)	Umbilical vein endothelial cells (h)	Rescue from adriamycin-associated neuropathy	[73]
Endothelial cells (h)	Breast cancer cells (h)	Chemoresistance	[74]
Cardiac fibroblasts (rat)	Cardiomyocytes (rat)	Attenuation of hypoxia/reoxygenation-induced apoptosis	[75]
Cardiomyocytes (rat)	Cardiofibroblasts (rat)	Calcium signal propagation between donor and recipient cells	[76]
Neurons (rat)	BM-MSC(h)	N/A	[60]
Neuronal cells (h)	Astrocytes (h)	Elevated mitochondrial membrane potential	[28]
Astrocytes (h)	Neurons (rat)	Mitigated inflammatory cell death after preconditioned hyperbaric oxygen therapy	[77]
Vascular smooth muscle cells (h)	BM mesenchymal stem cells (h)	Regulated cell proliferation	[78]
Germline cyst cells (m)	Germ cells (m)	Role in mammalian oocyte differentiation	[56]
T cell acute lymphoblastic leukemia (h)	BM-MSC(h)	Reduced ROS levels inducing chemoresistance	[79]
Mitochondrial transfer in vivo			
Stromal cells (c)	Transmissible venereal tumor cells (c)	Potential rescue of cell function	[80]
BM stromal cells (m)	Alveolar epithelia (m)	Increased ATP and protection against acute lung injury	[81]
BM-MSC (h)	Bronchial epithelial cells (m)	Rescue from cell death following airway injury	[59]
BM-MSC (h)	Alveolar macrophage (h)	Enhanced phagocytosis in acute respiratory damage model	[82]
iPSC and BM-MSC (h)	Bronchial epithelia (h)	Attenuated cigarette smoke-induced damaged	[26]
BM-MSC (h)	Neurons (rat)	Improved poststroke recovery	[60]
BM-MSC (h)	Alveolar macrophage (h)	Enhanced phagocytosis, improved bioenergetics	[82]

(Continued)

Table 20.1: (Continued)

Mitochondrial transfer in vivo			
iPSC-MSC (h)	Corneal epithelial cells (r)	Protection from oxidative stress-induced mitochondrial damage	[63]
BM-MSC (h)	Hepatocytes (m)	Enhanced OXPHOS in low-grade inflammation model	[65]
BM stromal cells (m)	AML cells (h)	Survival advantage following chemotherapy	[64]
BM stromal cells (m)	B16ρ^0 melanoma cells (m)	Restoration of mitochondrial function	[83]
BM stromal cells (m)	AML blasts (h)	AML disease progression	[66]
iPSC-MSC (h)	Bronchial epithelia (m)	Alleviated asthma inflammation	[69]
BM-MSC (rat)	Renal proximal tubular epithelial cells (m)	Structural and functional cell repair in diabetic nephropathy	[84]
iPSC-MSC (h)	Cardiomyocytes (m)	Rescue of anthracycline-induced cardiomyopathy	[68]
BM stromal cells (m)	Multiple myeloma (h)	Evidence CD38 inhibits mitochondrial transfer and increases survival	[70]
Optic nerve head retinal ganglia (m)	Astrocytes (m)	Mitochondrial degradation	[85]
Glioma cells (h)	Glioma cells (h)	Network communication	[86]
Stromal cells (m)	4T1ρ^0 breast cancer, B16ρ^0 melanoma (m)	Repopulation and restoration of mitochondrial function	[87]
Stromal cells (m)	143Bρ^0 osteosarcoma (h)	Repopulation and restoration of mitochondrial function	[88]

[a] *BM*, Bone marrow; *MSC*, mesenchymal stem cells; *BM-MSC*, bone marrow-derived MSC; *iPSC*, induced pluripotent stem cells; *AML*, acute myeloid leukemia; ρ^0, cells devoid of mtDNA; *m*, mouse; *r*, rabbit; *p*, pig; *c*, canine.

143Bρ^0 cells [88]. Subsequently, Moschoi et al. showed that primary and cultured human AML blasts acquired functional mitochondria from human and mouse bone marrow-derived (BM)-MSCs. Mitochondrial transfer increased survival of AML blasts in a mouse xenograft model which was enhanced following cytarabine chemotherapy [64]. Mitochondrial transfer in this study required cell-to-cell contact and involved endocytosis. In another similar study, Marlein et al. showed mitochondrial transfer from BM-MSCs to primary AML cells and cell lines. This transfer was dependent on NADPH oxidase 2 (NOX2), enhanced by reactive oxygen species (ROS) and inhibited by the NOX2 inhibitor, diphenyleneiodinium [66]. More recently, the same group showed that mitochondrial trafficking promoted bioenergetic plasticity in multiple myeloma (MM) cells in coculture and in xenografts in immunocompromised mice [70]. Transfer was achieved through TNTs and was dependent on CD38, a multifaceted cell surface enzyme that metabolizes NAD^+ and mediates extracellular nucleotide and intracellular calcium homeostasis, and plays a role in leukocyte diapedesis [89]. Daratumumab, an antibody against CD38, has shown some efficacy in a recent early stage II trial for heavily pretreated and refractory MM patients [90]. Mitochondrial abnormalities have also been noted in pancreatic ductal carcinoma models with cells exhibiting abnormally fragmented mitochondria [91]. Interestingly, reversing mitochondrial fragmentation using genetic approaches suppressed tumor growth and improved survival by inducing mitophagy in this preclinical model.

Intravital confocal imaging of human astrocytic brain tumors, including glioblastomas growing in immunocompromised mice revealed long membranous tubular protrusions referred to as tunneling nanotubes (TNTs: see Fig. 20.2) and tumor microtubes connecting groups of tumor cells in a multicellular syncytium [86,92]. Formation of these microtubular networks involved normal neuro-developmental mechanisms, including the growth-associated protein, GAP43. Microtubes were shown to transfer calcium, mitochondria, and even nuclei via connexin-43-containing gap junctions. The microtube network protected connected tumor

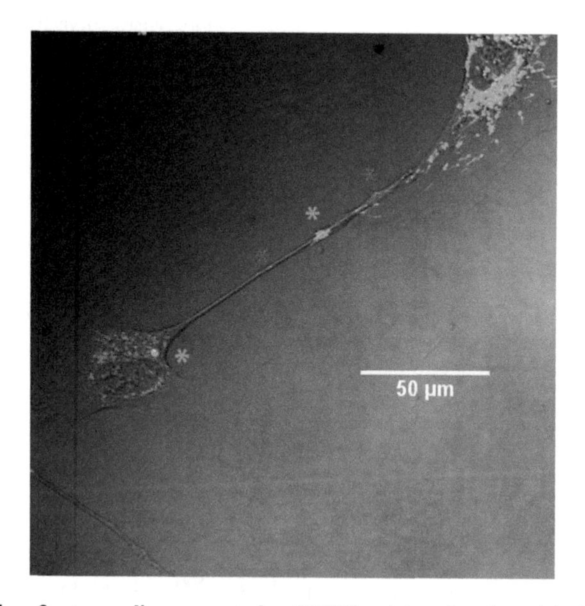

Figure 20.2: Example of a tunneling nanotube (TNT) with mitochondria between two human astrocytes.
Human SVG astrocytes stained with MitoTracker red or MitoTracker green were cocultured for 24 h and live cells were imaged using a fluorescent confocal microscope. A TNT extends between an astrocyte with red mitochondria and an astrocyte with green mitochondria. *: red and green mitochondria traveling inside the TNT.

cells from radiation therapy and chemotherapy and facilitated invasion and tumor progression. Targeting these networks could provide new opportunities to break treatment resistance in these highly aggressive brain tumors [93].

Our group became interested in mitochondrial transfer in 2008, following a decade in which we had generated or obtained more than a dozen human and mouse ρ^0 cell lines for characterizing plasma membrane electron transport [94]. All ρ^0 cells require uridine in their growth medium because respiration is required for dihydroorotate dehydrogenase (DHODH) activity. DHODH is essential for de novo pyrimidine biosynthesis and without respiration cells cannot generate the nucleic acids necessary for cell division [10,95]. Pyruvate supplementation was also used with ρ^0 cells because, in most cases, it promoted cell growth. Our interest was to determine whether or not ρ^0 cells with their complete reliance on glycolytic ATP production and uridine supplementation would be able to grow tumors, given that at that time, cancer stem cells were thought to be highly glycolytic. We found that both B16ρ^0 metastatic melanoma cells injected into syngeneic C57BL/6 mice and 4T1ρ^0 breast carcinoma cells injected orthotopically or subcutaneously into Balb/cJ mice were able to form tumors but only after a long lag period of 2–3 weeks [96]. Cell lines were derived from subcutaneous tumors and from lung metastasis, as well as from circulating tumor cells (see Fig. 20.3A). We initially thought that tumors developed because they had adapted to grow in vivo without mtDNA. However, to our surprise, we found that all derived cell lines contained host mtDNA as shown by the presence of mitochondrially encoded cytochrome b and mouse mtDNA polymorphisms present in both tumor models [87]. Tumor cell lines with host mtDNA were characterized by partial or complete recovery of respiration and were phenotypically and genotypically stable following transplantation. The purely glycolytic ρ^0 tumor models, with their complete inability to perform mitochondrial respiration, are extreme tumor models and were never designed to reflect actual physiological situations. Nevertheless, a number of anticancer drugs deplete mtDNA and push cells toward glycolytic metabolism. Therefore ρ^0 tumor models do have relevance in this field with mitochondrial transfer emerging as a potential mechanism of drug resistance [97,98]. In addition, these models generate information about the potential of tumor cells to grow and metastasize with compromised mitochondrial ATP production.

The question of whether or not mtDNA is transferred to tumor cells without mtDNA via intact mitochondria was addressed in subsequent experiments where C57BL/6N^{su9DsRed2} mice with nuclear-encoded, mitochondrially imported red fluorescent protein were injected subcutaneously with B16ρ^0 cells that had been stably transfected with nuclear blue fluorescent protein. Double-labeled cells were purified from the pretumor lesion at 11 days and following expansion in culture, formed tumors in C57BL/6 mice [83] (see Fig. 20.3B). To determine the timing of mitochondrial uptake by 4T1ρ^0 cells, we developed cell lines from injected tumor cells at various times following injection, and using digital droplet PCR, showed the presence of host mouse mtDNA polymorphisms as early as 5 days after cell inoculation [60]. To determine if tumor formation was dependent on mitochondrial ATP

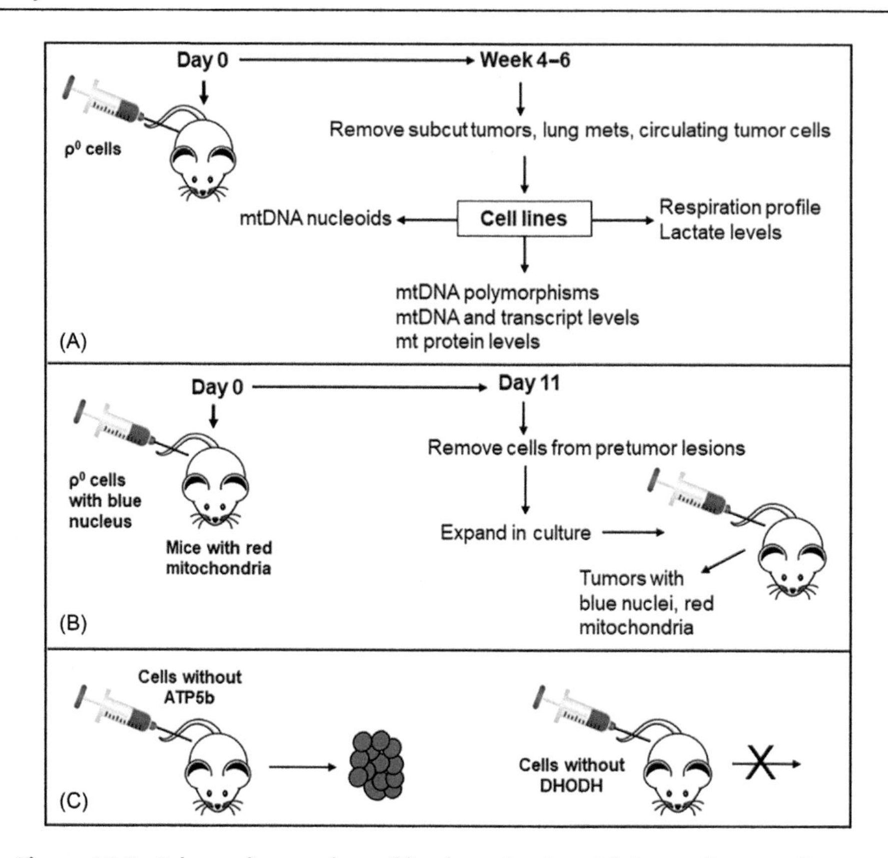

Figure 20.3: Schematic overview of in vivo mitochondrial transfer experiments. Experimental design for (A) generating cell lines from subcutaneous tumors, lung metastases, and circulating tumor cells; (B) creating tumors with blue nuclei and red mitochondria through mitochondrial transfer and (C) and the fate of tumor cells without ATP5b and tumor cells without DHODH when injected subcutaneously in mice.

production, we deleted the *Atp5b* gene essential for ATP synthase activity. Surprisingly, these cell lines maintained ATP levels and were able to grow tumors demonstrating that mitochondrial ATP is not essential for tumor formation in this model. In contrast, 4T1 cells without their *Dhodh* gene were unable to form tumors (see Fig. 20.3C). Similar results were obtained with B16 cells indicating common bioenergetic boundaries in these two quite disparate tumor models with different developmental origins, different genetic drivers and tumor suppressor status, and different mouse strain backgrounds [60].

20.3.1.2 Mitochondrial transfer into normal cells

A number of studies have shown the potential of transferring healthy mitochondria to respiration-deficient astrocytes or neural cells. In neurons, the mitochondrial GTPase-1 (Miro1) enhances transport of mitochondria toward the distal ends of neuronal processes

[99]. Miro1 overexpression enhanced mitochondrial transfer from human BM-MSCs via TNTs to oxygen-glucose starved rat astrocytes and neural pheochromocytoma PC12ρ^0 cells, a cellular model of brain ischemia [67]. Injection of MSCs overexpressing Miro1 also improved neurological outcomes following cerebral artery occlusion [67]. In another study, Li et al. [71] reported mitochondrial transfer from rat BM-MSCs to oxygen-glucose deprived spinal cord neurons via gap junctions, with reversal of neuronal apoptosis. Prophylactic hyperbaric oxygen treatment of rat neurons prior to TNF-α or LPS exposure was shown to reverse inflammatory responses and improve neuronal cell survival by mediating mitochondrial transfer [77]. Mitochondria transferred between human astrocytes and from human pluripotent stem cell-derived neural cells to astrocytes, and this process was facilitated by CD38/cADP-ribose, and Miro1 and Miro2. In contrast, neural cells with Alexander disease-associated GFAP mutations showed impaired astrocytic mitochondrial transfer [28]. These results are similar to previous seminal research where damaged or spent mitochondria from the optic nerve head and elsewhere in the brain were packaged into exosomes for recycling by surrounding astrocytes in a process termed "transmitophagy" or transcellular mitochondrial degradation [100]. Previously, it was thought that neuronal mitochondrial maintenance occurred by axonal transport to and from the cell body where mitophagy occurred via the lysosomal pathway [101]. Similar accumulations of degrading mitochondria were found in other areas of the brain and central nervous system (CNS) suggesting that local processes involving intercellular mitochondria transfer may contribute to mitochondrial maintenance (see also Fig. 20.2).

Mitochondrial transfer from BM-MSCs to lung epithelial cell lines was shown to involve cytoplasmic bridges or TNTs, which increased following rotenone or TNF-α treatment, and was shown to involve Connexin 43 [81] and Miro1 [59]. These authors also showed that Connexin 43 and Miro1 were involved in mitochondrial transfer from the BM-MSCs gavaged intranasally or intratracheally into bronchial epithelial cells in vivo following acute lung injury or induced allergic airway inflammation in mouse models. Connexin 43 was shown to be involved in mitochondrial transfer from human iPSC-MSCs to the bronchial epithelial cell line, BEAS-2B, treated with mitochondria-damaging $CoCl_2$, rescuing these cells from apoptosis [69]. The authors also showed involvement of Connexin 43 in mitochondrial transfer from iPSC-MSCs to lung epithelial cells in a mouse model of asthmatic inflammation. Transfer was associated with reduced T helper 2 cytokine production, suggesting a potential therapeutic strategy for targeting inflammation in asthma. With respect to T cells, Luz-Crawford et al. [72] reported mitochondrial transfer from human BM-MSCs to primary Th17 cells within 4 hours of coculture. Cell-to-cell contact impaired IL-17 production and conversion to regulatory T cells. Of note, mitochondrial transfer was impaired when synovial MSCs from patients with rheumatoid arthritis were used.

A recent report by Konari et al. [84] showed that mitochondrial transfer from BM-MSCs to renal proximal tubular epithelial cells in vitro reversed apoptosis and restored function, and that mitochondrial injection under the kidney capsule improved cellular morphology and reversed loss of basement membrane and brush border structural integrity. These results confirmed and extended an earlier report of mitochondrial transfer from human fetal MSCs to primary rat renal tubular cells [102]. Several in vitro studies have further shown that mitochondria can transfer from human MSCs [76,103,104] or from cardiac myofibroblasts [75] to damaged cardiomyocytes via TNTs, rescuing them from apoptosis.

With respect to MELAS patients, a subgroup of mitochondrial encephalomyopathies characterized by a specific leucine-tRNA mutation, Wharton's jelly-derived MSCs restored normal mitochondrial function to rotenone-stressed fibroblasts and the mutational burden to undetectable levels [105]. These Wharton's jelly-derived MSCs had previously been shown to transfer mitochondria to 143Bρ^0 cells, restoring aerobic respiration [27]. Newell et al. [65] showed that human MSCs alter mitochondrial dynamics and enhance respiratory function in skin fibroblasts of patients with mitochondrial diseases. Mitochondrial transfer has therapeutic potential across those mitochondriopathy patients with mtDNA damage.

20.3.1.3 Mitochondrial donation therapy to prevent mitochondrial diseases in offspring

The normal process of fertilization involves fusion of the spermatocyte nucleus with the oocyte nucleus inside the oocyte, leaving the oocytoplasm as the only source of mtDNA. Any paternal mitochondria from the spermatozoa tail are extensively eliminated by the oocyte through allogeneic (nonself) mitophagy [106]. This means that both mtDNA and associated mitochondriopathies are exclusively maternally inherited. Several techniques have been developed, detailed in a recent review [107], that circumvent this issue of maternal inheritance of mitochondrial diseases (see Fig. 20.4). These techniques all require an additional female donor with healthy mtDNA, which creates potential ethical issues around the embryo containing genetic material from three different biological parents: nuclear DNA from the mother and father, along with mtDNA from the donor egg [32,107–110]. Partial or complete ooplasmic transfer is the process of injecting a small portion or the entire volume of cytoplasm, which includes mRNA, protein, mitochondria, other cell organelles, and numerous other cell components, into an egg with insufficient or compromised cytoplasm [108]. The Muggleton-Harris group first successfully completed ooplasmic transfer in mice in 1982 [111], which was followed by a successful pregnancy after ooplasmic transfer in 1997 by the Cohen group [112]. Nuclear transfer involves removal of the nucleus and its genetic material from a donor egg with healthy mitochondria [109]. The nucleus from the mother's egg (containing unhealthy mitochondria) is removed and implanted into the donor egg, resulting in an egg with donor mitochondria and cytoplasm, and the mother's nucleus (with nuclear DNA). The egg is then fertilized with the father's sperm and again an embryo develops with three distinct genetic parents. Mitochondrial donation therapy is a modified version of in vitro fertilization, legalized in the United Kingdom with stringent conditions in early 2015, and by the Human Fertilisation and Embryology Authority

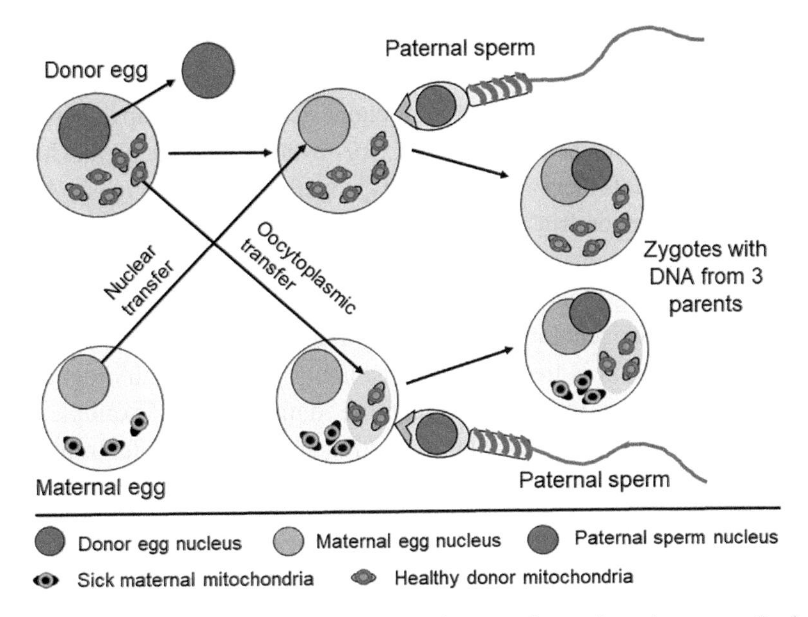

Figure 20.4: Schematic overview of oocytoplasmic transfer and nuclear transfer in assisted fertility.
In oocytoplasmic transfer a portion of the donor egg cytoplasm is transferred into the parental egg. In nuclear transfer the nucleus of the parental egg is transferred into the enucleated donor egg. After fertilization, in both cases, the zygote will have genetic material from three parents.

(HFEA) in the UK in December 2016 [109,113]. This means that fertility clinics can now apply for a license permitting the use of mitochondrial donation for prevention of serious mtDNA disease, with applications for each individual patient considered on a case-by-case basis. Similarly, restricted clinical trials of mitochondrial replacement techniques have been permitted in the USA since 2016 [109,113].

20.3.2 Transfer of isolated mitochondria

Caicedo et al. devised an artificial system of mitochondrial transfer they named "*MitoCeption*," whereby small numbers of mitochondria are isolated, counted, and transferred to recipient cells using two centrifugation steps of 1500 *g* for 15 minutes at 4°C [114,115]. Their approach of simulating the natural process of transferring small numbers of mitochondria and studying the effects of transfer on recipient cells in detail is in sharp contrast with the process of mitochondrial transplantation, whereby large numbers of isolated mitochondria are injected directly into ischemic parts of the heart and the brain as described in more detail below. MitoCeption is the ideal platform to investigate which cell types yield the "best" mitochondria for transfer or transplantation and once isolated, how they are best preserved [115]. A different approach was taken by Wu et al., who developed a photothermal nanoblade for the efficient transfer of mitochondria directly into the cytoplasm of mammalian cells [116]. Because of the very precise nature of this technique, the nanoblade could become a pathway to mitochondrial gene therapy, removing mutated mtDNA and replacing it with healthy mtDNA, for instance as part of in vitro fertilization. Both MitoCeption and photothermal nanoblades, together with previously used techniques of microinjection of mitochondria directly into cells, result in restoration of respiration in individual respiration-deficient recipient cells. However, when it comes to ischemia-reperfusion injury involving sheets of cells such as the myocardium, or parts of organs such as in stroke or in neurodegenerative disorders, a more wholesale approach is required. Mitochondrial transplantation is the process of isolating billions of autologous mitochondria and injecting these directly into organs that are damaged, for example, by ischemia-reperfusion injury. This approach is likely to give more immediate translatable benefits when dealing with heart failure and strokes. The next section will review animal and human studies that have used mitochondrial transplantation to ameliorate ischemic heart disease, neurodegenerative disorders and ischemic stroke as well as behavioral disorders. Representative studies are summarized in Table 20.2.

20.3.2.1 Ischemic heart disease

Compromised arterial supply is the leading cause of cardiac ischemia resulting in a decrease in mitochondrial ATP levels, which is sustained for up to 3 hours after restoration of the blood supply. Low ATP levels decrease myocardial cell viability, ultimately leading to heart failure [53,135–137]. The McCully group at Harvard Medical School and others have progressed the field of mitochondrial transplantation for cardioprotection from animal studies in rabbits [120], pigs [125], rats, and mice [127] to a case series of five pediatric patients [129]. A recent review by the group describes this progress and the protocols they have developed in detail [36]. In their animal studies, McCully and colleagues produced temporary ischemia by ligating the left anterior descending artery with a snare for 30 minutes, causing ischemia and loss of cell viability in 25%–30% of the ventricle, resulting in a 25%, loss of contractile force. This procedure mimics events occurring in human acute myocardial infarction caused by acute coronary artery obstruction followed by surgical intervention [118]. The authors obtained viable mitochondria from skeletal muscle of the same animal. After homogenizing and filtering, mitochondria were injected directly into the ischemic part of the heart just prior to reinfusion [118]. This procedure takes 20–30 minutes and produces mitochondria that appear healthy with normal membrane potential and oxygen consumption parameters. Mitochondria were found in the interstitial spaces between cardiomyocytes 10 minutes after injection. After 1 hour, approximately 40% of MitoTracker CMX-Ros stained mitochondria were found inside cardiomyocytes. In addition, hearts that received additional mitochondria had normal heart function 28 days after injection, whereas hearts that received vehicle control displayed continued myocardial dysfunction.

Table 20.2: Examples of mitochondrial transplantation in vitro and in vivo.

Donor cells	Recipient cells	Outcome of mitochondrial transfer	References
Mitochondrial transplantation in vitro			
Mammary cells (m)[a]	Adrenal cortical tumor cells (m)	Antibiotic resistance	[8]
Fibroblasts (h)	Osteosarcoma and fibrosarcoma (h)	Replacement of endogenous mtDNA after partial depletion	[9]
Kidney fibroblasts (hamster)	Sciatic nerve (rat)	Prevention of axonal degeneration	[117]
Cervical cancer cells (h)	143Bρ^0 osteosarcoma cells (h)	Mitochondrial repopulation of cells	[10]
Pectoralis major muscle cells (rat), cervical cancer (h)	Cardiomyocytes (r)	Increased ATP production and protection from ischemia-reperfusion injury	[118]
Liver carcinoma cells (h)	Neuroblastoma (h)	Increased ATP production	[119]
Endothelial progenitor cells (h)	Brain endothelial cells (h)	Supported energetics after oxygen-glucose deprivation	[120]
BM-MSC (h)	Breast adenocarcinoma (h)	Enhanced OXPHOS, favors cell proliferation and invasion	[114]
Umbilical cord-derived MSC (h)	Umbilical cord-derived MSC (h)	Enhanced cellular metabolic function	[121]
BM-MSC (rat)	Renal proximal tubular epithelial cells (rat)	Structural and functional cell repair in diabetic nephropathy	[84]
BM-MSC (h)	T helper 17 (h)	Reduction in IL-17 production	[72]
Osteosarcoma (h)	Pheochromocytoma (rat)	Resistance to neurotoxin-induced oxidative stress and apoptosis	[122]
Osteosarcoma (h)	Osteosarcoma MELAS A3243G mutation (h)	Improved mitochondrial function and cell viability	[123]
Osteosarcoma (h)	Breast cancer (h)	Apoptosis, decreased oxidative stress, chemotherapeutic sensitivity	[98]
Astrocytes (h)	Glioma cells (h)	Aerobic respiration rescue after serum- and glucose-free culture	[97]
Primary neurons (rat)	Hippocampal neurons (rat)	Promoted regeneration of injured cell, restored membrane potential	[124]
Mitochondrial transplantation in vivo			
Osteosarcoma (h), pheochromocytoma (rat)	Neurons (rat)	Mitochondrial function restoration, attenuation of Parkinson's disease neurotoxicity	[122]
Pectoralis major muscle cells (r) cervical cancer (h)	Cardiomyocytes (r)	Cardioprotection from ischemia-reperfusion injury	[118]

(*Continued*)

Table 20.2: (Continued)

Mitochondrial transplantation in vivo			
Pectoralis major muscle cells (p)	Cardiomyocytes (p)	Myocardial rescue from ischemia/ reperfusion injury	[125]
Pectoralis major muscle cells (rat)	Neurons (rat)	Reduced oxidative stress, apoptosis, attenuated astrogliosis, neurogenesis	[126]
Skeletal muscle cells (m)	Cardiomyocytes (m)	Recovery after ischemic injury	[123]
Gastrocnemius muscle cells (m)	Cardiomyocytes (m)	Prolonged cold ischemia time for transplanted hearts	[127]
Pheochromocytoma, soleus muscle cells (rat)	Spinal cord neurons (rat)	Maintained normal bioenergetics, but not long-term neuronal protection	[128]
Rectus abdominis muscle (h)	Cardiomyocytes (h)	Improvement in ventricular function after ischemia-reperfusion injury	[129]
Kidney fibroblasts (hamster)	Neurons (rat)	Neural protection against ischemic stress	[130]
Kidney fibroblasts (hamster)	Sciatic nerve (rat)	Prevention of axonal degeneration from crush injury	[117]
Cardiac fibroblasts (h)	Cardiomyocytes (r)	Cardioprotection from ischemia-reperfusion injury	[131]
Cardiomyocytes (r)	Cardiomyocytes (r)	Enhanced postischemic functional recovery and cellular viability	[120]
BM-MSC (h)	Alveolar macrophage (m)	Enhanced phagocytosis, improved bioenergetics	[132]
BM-MSC (rat)	Neurons (rat)	Improved locomotor function after spinal cord injury	[71]
Astrocytes (m)	Cortical neurons (m)	Amplified cell survival signals following focal cerebral ischemia	[133]
Astrocytes (h)	Glioma cells (h)	Inhibited glioma growth, enhanced radiation sensitivity	[97]
Hippocampal cells (m)	Astrocytes, microglia (m)	Ameliorated LPS-induced depressive-like behaviors	[134]
Liver carcinoma cells (h)	Various tissue (m)	Prevention of apoptosis and necrosis in a Parkinson's disease model	[119]

[a] *BM*, Bone marrow; *iPSC*, induced pluripotent stem cell; *AML*, acute myeloid leukemia; ρ^0, cells devoid of mtDNA; *m*, mouse; *r*, rabbit; *p*, pig; *HRPTE*, human renal proximal tubular epithelia.

Cowan et al. [131] injected autologous mitochondria into the coronary artery of rabbits. Approximately 25% of 18F-rhodamine 6G labeled mitochondria were associated with the cardiomyocytes 10 minutes after injection [131]. Both direct injection and injection into the arterial supply resulted in good cardiomyocyte protection. Mitochondrial transplantation significantly decreased infarct size and enhanced postischemic function recovery with improved contractility, increased ATP levels and an increase in cardioprotective cytokines that protect against apoptosis, facilitate cardiomyocyte remodeling, regeneration, and distribution. Mitochondrial transplantation did not affect heart rhythm or serial ECG, and there was no sign of fibrillation, conduction system defects, ventricular hypertrophy, valve dysfunction, fibrosis, or pericardial effusion at any time during the experiment, which lasted for 24 days. In addition, the autologous mitochondria did not generate inflammation; nor did they induce an immune response [118,131].

Pacak et al. [138] used specific inhibitors of internalization processes, that when delivered by direct injection, halted mitochondrial uptake via actin-mediated endocytosis as evidenced by complete blockage of mitochondrial uptake by cytochalasin D. The method of rapid extravasation of mitochondria into the myocardium after arterial injection is more complicated and differs from diapedesis involving cell adhesion proteins, which are lacking in mitochondria.

The McCully group has so far published only one study that evaluated the effect of mitochondrial transplantation on cardioprotection in humans. Emani and colleagues [129] reported on a case series of five pediatric patients (age 4 days to 2 years) with critical myocardial ischemia-reperfusion injury after heart surgery for various heart conditions. Mitochondria were isolated from the rectus abdominis and each patient received 10 injections of 10^7 mitochondria directly into the affected part of the heart. All patients saw a significant improvement in systolic function and ventricular function improved at 4–6 days. There were no complications associated with the procedure such as arrhythmia, intramyocardial hematoma, or scarring.

Moskowitzova et al. [127] applied mitochondrial transplantation to address a common problem encountered in heart transplantations. Hearts often have to travel considerable distances between donor and recipient and they are kept under cold ischemic conditions during that time. If hearts are kept at 4°C for longer than 4 hours the chance of a successful transplant decreases rapidly. In this study, mitochondria were injected into the coronary arteries before the hearts were taken out of donor mice and again after transplantation into the recipient mouse with the donor hearts having been kept at 4°C for 29 hours. The implanted hearts were evaluated after 24 hours and were found to have significantly higher beating scores and significantly lower levels of necrosis and inflammation than hearts that had not received additional mitochondria.

20.3.2.2 Neurodegenerative disorders and ischemic stroke

The role of mitochondrial dysfunction in neurodegenerative diseases such as Parkinson's disease, Alzheimer's disease, Huntington's disease, and multiple sclerosis, as well as in ischemia-based injuries such as stroke, traumatic brain injury, and spinal cord injury has been detailed in recent reviews [33,34,130]. Mitochondrial fission and fragmentation as well as the accumulation of damaged mitochondria and loss of synapses all lead to cognitive impairment in models of Alzheimer's and Parkinson's disease and in ischemic brain disease. Mitochondrial fission in ischemic stroke and traumatic brain and spinal injury is a double-edged sword; if fission leads to dysfunctional mitochondria, this will increase the size of the infarction and lead to a worse outcome. However, smaller mitochondria can be mobilized more easily to regenerate injured areas of the neurons, improving prognosis.

Compared with the field of cardioprotection, mitochondrial transplantation in the CNS for the purpose of neuroprotection is a more recent field with comparatively few animal studies to date. Several authors have reported an improvement in recovery from ischemic brain injuries through transplantation of mitochondria-containing microvesicles. Hayakawa et al. [139–141] injected MitoTracker Red CMX-Ros stained mitochondria in astrocyte microvesicles into the peri-infarct cortex 3 days after ischemic stroke in mice. Mitochondria-containing microvesicles were taken up by neurons within 24 hours with mitochondrially enhanced neurons displaying stronger survival signals as well as a reduced infarct site. Release of extracellular microvesicles containing mitochondria from astrocytes was mediated by calcium-dependent CD38 and cyclic ADP-ribose signaling. However, the question of whether mitochondria associated with damaged neurons were intracellular and functional remains contentious as most effects on respiration were relatively small and repopulation was not demonstrated [15].

Zhang et al. [126] infused MitoTracker Red CMX-Ros stained mitochondria isolated from autologous skeletal muscle during reperfusion in an ischemic rat stroke model induced by temporarily occluding the middle cerebral artery (MCAO). Intracerebroventricular injection led to wide distribution of mitochondria within the brain cortex and resulted in a decrease in size of the ischemic penumbra as well as an improvement in neurological and behavioral deficits. Huang et al. [130] found that transplanting exogenous hamster mitochondria via local intracerebral and intraarterial injection reduced brain lesions after ischemic cerebral stroke and restored motor function in MCAO-treated rats. Although mitochondrial transplantation has shown clear benefits in ischemic brain injury in animal models, this was not the case for ischemic spinal cord injury. When Gollihue et al. [128] injected turbo green fluorescent protein-labeled PC12-derived mitochondria into the L1/L2 ischemic area of rat spinal cords, they visualized exogenous mitochondria 24–48 hours after injection in several cell types but not in neurons and the additional mitochondria did not confer long-term functional neuroprotection. Very few exogenous mitochondria were present 7 days after injection and those that were present were mainly found inside macrophages and pericytes. After 5 weeks, locomotor scores were similar to those of untreated groups.

With respect to neurodegenerative disorders, a few studies have shown that mitochondrial transplantation seems to be beneficial in animal models of Parkinson's disease.

Chang et al. [122] visualized exogenous mitochondria up to 12 weeks after injection of cell-penetrating peptide (Pep-1)-conjugated mitochondria into 6-OHDA-induced Parkinson's disease rat brains. Mitochondrial transplantation rescued respiration, enhanced survival of dopaminergic neurons and sustained mitochondrial function as well as improving locomotor activity of the rats after 3 months. Shi et al. [119] showed that isolated mitochondria injected in the tail vein of mice were observed in brain, heart, liver, kidney, and muscle 2 hours after injection. The additional mitochondria normalized behavior and prevented PD progression in an MPTP-induced mouse model of Parkinson's disease.

20.3.2.3 Behavioral disorders

A potential role for mitochondrial dysfunction in the etiology of neuroimmune and neuropsychiatric disorders such as bipolar disorder, schizophrenia, depression, autism, and chronic fatigue syndrome has also recently been proposed. Although these mood disorders do not present with a consistent set of symptoms and underlying pathophysiology, they are all characterized by chronic oxidative stress and chronic systemic inflammation. These chronic conditions are closely aligned with and perpetuate and exacerbate mitochondrial dysfunction in neurons, with macrophages and microglia contributing to the proinflammatory environment (for a detailed review see [142]). A study by Wang et al. [134] showed that MitoTracker CMX-Ros stained mitochondria isolated from the hippocampus of mice decreased symptoms of depression when injected intravenously into LPS-treated mice. LPS causes depression in mice with symptoms similar to those of major depressive disorder in humans. The ameliorating effects of mitochondrial transplantation were of the same magnitude as those of the antidepressants, ketamine and fluoxetine. Mice performed better in the tail suspension test, forced swim test, sucrose preference test, and anxiety behavior test of time spent in open field. Mitochondrial transplantation also restored insufficiency in neurogenesis and decreased the level of neuroinflammation caused by LPS. Robicsek et al. investigated the effect of heterologous mitochondria on mitochondrial membrane potential, ATP levels, oxygen consumption and behavior in a poly-I:C rat model of schizophrenia in which pregnant Wistar rats are injected with poly-I:C [143]. The prelimbic cortex of adolescent offspring was injected with isolated JC-1-stained mitochondria isolated from healthy rat brains. During this adolescent period, rats are asymptomatic and atypical antipsychotic drugs have been shown to prevent emergence of brain and behavioral abnormalities in adult poly-I:C offspring. Heterologous mitochondria were used because all mitochondria of affected rat brains have deficits. Injection with healthy mitochondria prevented selective attention deficit in adulthood. Long-lasting improvement in attention span capacity, increased brain respiration activity, and oxygen consumption was seen. Of note, healthy adolescent rats injected with additional healthy mitochondria showed increased mitochondrial membrane potential and disruption of homeostasis [143].

20.3.2.4 Cancer sensitization to treatment

Sun et al. [97] implanted U87 glioblastoma multiforme xenografts in the flank of mice, and after establishing a significant level of hypoxia in the tumors, injected MitoTracker Red CMX-Ros stained mitochondria isolated from human astrocytes directly into the tumor. The additional mitochondria made the tumors more sensitive to radiation and restored respiration and oxygen levels required for radiation to be effective. The exogenous mitochondria seemed fully functional with normal levels of respiration, fission, and proliferation. Chang et al. [98] delivered mitochondria isolated from healthy individuals and dysfunctional mitochondria isolated from MERRF patients, to MCF-7 breast cancer cells. Only healthy mitochondria decreased MCF-7 viability, induced apoptosis, and increased sensitivity to doxorubicin and paclitaxel chemotherapy. Mitochondrial transplantation of either type of mitochondrion had no effect on the nontumorigenic MCF-10A cells.

20.4 Mechanisms of mitochondrial transfer

Several mechanisms of mitochondrial transfer have been reported in the literature. The first studies were based on endocytosis in coculture conditions [8] or microinjection into the cytoplasm of recipient cells [9]. Endocytosis has also been shown to be the mechanism of transfer behind MitoCeption, where small numbers of isolated mitochondria are centrifuged onto recipient cells [114], and mitochondrial transplantation whereby large numbers of mitochondria are injected into ischemic parts of the heart [36] or brain [33]. Another method of mitochondrial transfer that has received considerable attention is that of transfer through TNTs, first described by Rustom et al. [12], followed by many other authors [13,24,59,67,70,75,89,104,144–146]. Movement of mitochondria through Connexin-43 mediated gap junctions has also been described as a mechanism of mitochondrial transfer in some studies [69,81] as has the use of mitochondria isolated in microvesicles [81,140,147,148]. Several authors have described roles for CD38, [69,70,89] and Miro1 [59,67,104] in facilitating mitochondrial transfer. Our group has observed that cell fusion is also used in mitochondrial transfer from bone chip-derived MSCs and pericytes to astrocytes, glioma cells, breast cancer, and melanoma cells (manuscripts in preparation). Fig. 20.5 summarizes the different mechanism of mitochondrial transfer currently known.

20.5 Mito-nuclear crosstalk: potential consequences of mitochondrial transfer/transplantation

Two billion years of evolution have shaved down, through redundancy and nuclear transfer, the ancestral proteobacterial genome to a very small circular mitochondrial genome of 16,569 nucleotide pairs. mtDNA encodes only 13 polypeptides of the 85 polypeptides that make up the

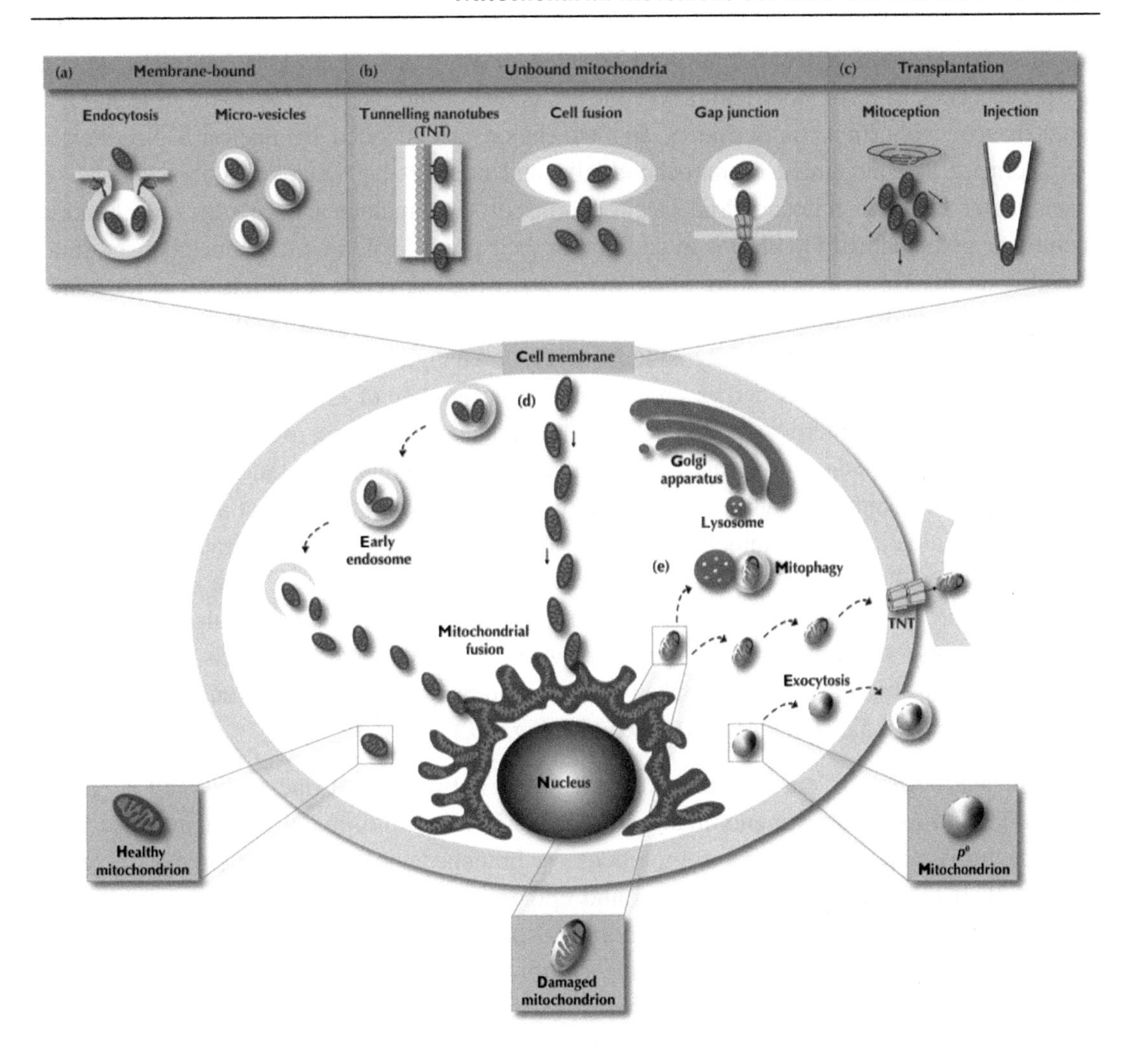

Figure 20.5: Methods of mitochondrial transfer and clearance of dysfunctional mitochondria. The top panel depicts (a) physiological routes of entry of membrane-associated mitochondria into cells through endocytosis and microvesicles fusing with the cell membrane and releasing mitochondria into the cytoplasm; (b) entry of mitochondria into cells via tunneling nanotubes that form between donor and recipient cells; cell–cell membrane fusion; Connexin 43 gap junctions that facilitate cell-to-cell attachment; (c) artificial methods of mitochondria transplantation into cells, using centrifugal force (MitoCeption), or direct injection into the cell; (d) membrane-bound mitochondrial entry into cells by endocytosis and release into the cytoplasm from early endosomes. Along with unbound mitochondria, mitochondria released from endosomes are actively transported to the mitochondrial network where fusion occurs. (e) Dysfunctional mitochondria or cells depleted of mtDNA (ρ^0 cells) either undergo mitophagy and recycling via lysosomal degradation, or are passed on to neighboring cells via TNTs, or cleared via exocytosis in microvesicles.

respirasome, 22 transfer RNAS, and the 12S and 16S ribosomal RNAs [149,150]. The remaining polypeptides of respiratory complexes I to V, as well as all the proteins involved in mtDNA maintenance, replication, transcription, and translation are encoded by nuclear DNA. Perfect stoichiometry and alignment of mitochondrially and nuclear-encoded respiratory subunits are essential for effective mitochondrial electron transport and maintenance of the mitochondrial membrane potential and OXPHOS. In addition to their pivotal role in energy metabolism, mitochondria are also biosynthetic hubs where many pathways converge, including those for nonessential amino acids, nucleotides, porphyrins, heme, glutathione, fatty acids, and cholesterol [151]. The ability to respond rapidly and effectively to changes in energy and nutrient demands is a requirement of all cells and is orchestrated by a number of complicated feedback loops often referred to as anterograde and retrograde signaling pathways, mitochondria-to-nucleus crosstalk, or "*mito-nuclear crosstalk*" [17]. Mitochondria continually update the nucleus about their bioenergetic status through a series of energy-linked metabolites that act as mito-stress signals; these include increased mitochondrial ROS levels, decreased ATP/AMP ratios and NAD^+/NADH ratios, and increased cytosolic Ca^{2+} levels. The nucleus reacts to these signals by activating one or several stress response signaling pathways to maintain cellular homeostasis. Nuclear response pathways include a shift between OXPHOS and glycolysis, mitochondrial repair, fission/fusion, biogenesis/mitophagy, the unfolding protein response, and the integrated stress response [17,152–154] (Fig. 20.6, see also Fig. 20.1).

Mito-nuclear crosstalk ensures the mitochondrial network will meet the cell's energy demands, with rapid adjustments made through fission and mitophagy if demands decrease, or biogenesis and fusion into comprehensive branched networks if demand increases [154,155]. If the cell, due to mutational and epigenetic changes or some form of injury, is unable to produce the required mitochondrial respiration, a robust membrane potential and OXPHOS, then cellular functions become compromised and cell viability may be threatened. Cancer cells are at an advantage here over normal body cells as they have the metabolic flexibility to quickly shift between OXPHOS and glycolysis to meet energy requirements [19]. Many highly glycolytic tumors shift from a fast growing, OXPHOS-based low invasive phenotype to a highly glycolytic slower growing but more invasive phenotype [156–158]. Normal body cells do not have this level of flexibility; although muscle cells can fall back on lactic acid fermentation to some extent, this process is not adequate in situations of arterial occlusion. Respiration, when compromised by ischemia or damage to respiratory subunits, becomes ineffective in that it no longer maintains a threshold mitochondrial membrane potential to support OXPHOS and generates a large amount of superoxide. For normal tissues, severe mitochondrial dysfunction that is not reversed by mito-nuclear crosstalk results in compromised organ function, or ultimately, organ failure.

As discussed in the previous sections, mitochondrial transfer and mitochondrial transplantation directly interfere with mito-nuclear crosstalk. The presence of fully functional exogenous mitochondria, likely from different cell types, in addition to dysfunctional endogenous

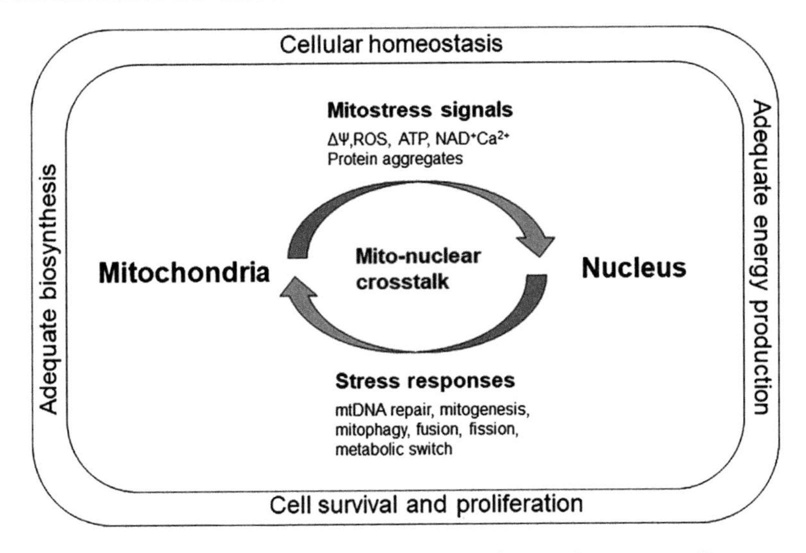

Figure 20.6: Simplified diagram of mito-nuclear crosstalk.
Ongoing communication between the nucleus and the mitochondria is essential to meet the needs for energy and biosynthesis required for cellular homeostasis to ensure cell survival and proliferation in response to changes in the microenvironment. Mito-nuclear communication is a two-way communication system also referred to as retrograde and antiretrograde signaling. The mitochondria update the nucleus on an ongoing basis about their bioenergetic status through a number of energy metabolites (mito-stress signals) such as the membrane potential over the inner mitochondrial membrane, the type and concentration of reactive oxygen species, ATP/AMP ratio, NAD^+/NADH ratio, Ca^{2+} concentration in the cytoplasm, presence of protein aggregates in the mitochondrial matrix, etc. The nucleus responds to these signals by activating the appropriate stress response pathways, including mtDNA repair, increasing the number of mitochondria as well as network complexity and switching energy generation from mitochondrial respiration to glycolysis.

mitochondria, poses an additional form of stress for the cell. In the short term, exogenous mitochondria will normalize ROS, ATP, NAD^+, and Ca^{2+} levels and dysfunctional mitochondria will be removed through mitophagy. However, long-term supplementation may interfere with other aspects of mitochondrial functioning, including potential "mismatched" biosynthetic pathways between the original nucleus and the remaining exogenous mitochondria. Recent work in a poly-I:C model of schizophrenia in rats suggests that injecting healthy mitochondria into healthy brains of adolescent rats has detrimental effects [143]. Some animal models suggest that the bioenergetic benefits from mitochondrial transplantation remain measurable for at least 3 months after the event in the case of improved locomotor activity of the rats [122]. These results suggest that mito-nuclear crosstalk can also be very plastic in normal cells and is able to accommodate exogenous mitochondria in the mixture to the ultimate benefit of the cell.

20.6 Concluding statement

There is no doubt that mitochondrial transfer between cells is a naturally occurring physiological process that involves membrane nanotube connections, microvesicles, endocytosis, direct cellular contact, partial or complete cytoplasmic fusion, and perhaps other poorly characterized mechanisms. In most instances this process is used to restore respiration to respiration-deficient recipient cells. However, dysfunctional mitochondria can also be shed by exocytosis or direct contact with an adjacent healthy cell for mitophagy. Mitochondria and mitochondrial fragments, either as free organelles or in exosomes, microvesicles or apoptotic bodies, have been shown in the circulation and in cerebrospinal fluid indicating that processes deep in our evolutionary past may have been retained and adapted for a variety of physiological purposes in the body.

Research perspectives

Mitochondrial transfer between cells and transplantation of healthy isolated mitochondria are new fields of endeavor that explore the benefits of replacing damaged mitochondria following trauma and tissue damage, and in diseases like cancer and following anticancer therapy. Mitochondrial transfer to cells with repopulating ability in vitro has the potential to be applied clinically to address mitochondrial damage as well as to provide novel pathways to autologous mitochondrial gene therapy. In these ways, mitochondrial transplantation has taken a step toward ameliorating life-threatening ischemia-reperfusion injuries of the heart and brain, and neurodegenerative, behavioral and neuromuscular diseases that involve progressive mitochondrial dysfunction. Exploiting intercellular mitochondrial transfer and mitochondrial transplantation for health benefits will be challenging given the complexity of mitochondrial genetics and biology and the very limited tools available for mitochondrial gene manipulation. The key importance of mitochondrial bioenergetics and cellular dynamics in health and in numerous diseases suggest that these new fields of endeavor will be fertile soil in the coming years with the potential to deliver substantial health benefits.

Acknowledgments

The authors received salary support from the Health Research Council of New Zealand, the Marsden Fund, the Cancer Society of New Zealand, the Malaghan Institute of Medical Research and the University of Otago, Wellington. We thank Remy Schneider for the image presented in Fig. 20.2.

References

[1] Margulis L. Origin of eukaryotic cells. New Haven, CT: Yale University Press; 1970.
[2] Martijn J, Vosseberg J, Guy L, Offre P, Ettema T. Deep mitochondrial origin outside the sampled alphaproteobacteria. Nature 2018;557 101–5.
[3] Lane N. The costs of breathing. Science 2011;334(6053) 184–5.

[4] Sender R, Fuchs S, Milo R. Revised estimates for the number of human and bacteria cells in the body. PLoS Biol 2016;14(8):1–14.
[5] Borst P, Grivell LA. The mitochondrial genome of yeasts. Cell [Internet] 1978;15(3) 705–23.
[6] Goldring ES, Grossman LI, Krupnick D, Cryer DR, Marmur J. The petite mutation in yeast. Loss of mitochondrial deoxyribonucleic acid during induction of petites with ethidium bromide. J Mol Biol 1970;52(2) 323–35.
[7] Nagley P, Linnane AW. Mutants of yeast. Biochem Biophys Res Commun 1970;39(5):989–96.
[8] Clark MA, Shay JW. Mitochondrial transformation of mammalian cells. Nature 1982;295(5850) 605–7.
[9] King M, Attardi G. Injection of mitochondria into human cells leads to rapid replacement of the endogenous mitochondrial DNA. Cell 1988;52:811–19.
[10] King MP, Attardi G. Human cells lacking mtDNA: repopulation with exogenous mitochondria by complementation. Science 1989;246(4929):500–3.
[11] Gregoire M, Morais R, Quilliam MA, Gravel D. On auxotrophy for pyrimidines of respiration-deficient chick embryo cells. Eur J Biochem 1984;142(1):49–55.
[12] Rustom A, Saffrich R, Markovic I, Walther P, Gerdes H-H. Nanotubular highways for intercellular organelle transport. Science 2004;303(5660) 1007–10.
[13] Spees J, Olson S, Whitney M, Prockop D. Mitochondrial transfer between cells can rescue aerobic respiration. Proc Natl Acad Sci U S A 2006;103(5) 1283–8.
[14] Berridge M, Grasso C, Neuzil J. Mitochondrial genome transfer to tumour cells breaks the rules and establishes a new precedent in cancer biology. Mol Cell Oncol 2015;5(5):e1023929.
[15] Berridge MV, Schneider RT, McConnell MJ. Mitochondrial transfer from astrocytes to neurons following ischemic insult: guilt by association? Cell Metab 2016;24(3) 376–8.
[16] Berridge M, Herst P, Rowe M, Schneider R, McConnell M. Mitochondrial transfer between cells: methodological constraints in cell culture and animal models. Anal Biochem 2018;552:75–80.
[17] Herst P, Rowe M, Carson G, Berridge M. Functional mitochondria in health and disease. Front Endocrinol (Lausanne) 2017;8:e296.
[18] Herst P, Dawson R, Berridge M. Intercellular communication in tumor biology: a role for mitochondrial transfer. Front Oncol 2018;8:e344.
[19] Herst PM, Grasso C, Berridge MV. Metabolic reprogramming of mitochondrial respiration in metastatic cancer. Cancer Metastasis Rev 2018;37(4) 643–53.
[20] Torralba D, Baixauli F, Sánchez-Madrid F. Mitochondria know no boundaries: mechanisms and functions of intercellular mitochondrial transfer. Front Cell Dev Biol 2016;4:1–11.
[21] Rodriguez A, Nakhle J, Griessinger E, Vignais M. Intercellular mitochondria trafficking highlighting the dual role of mesenchymal stem cells as both sensors and rescuers of tissue injury. Cell Cycle 2018;17(6):712–21.
[22] Griessinger E, Moschoi R, Biondani G, Peyron J. Mitochondrial transfer in the leukemia microenvironment. Trends Cancer 2017;3(12):828–39.
[23] Berridge MV, McConnell MJ, Grasso C, Bajzikova M, Kovarova J, Neuzil J. Horizontal transfer of mitochondria between mammalian cells: beyond co-culture approaches. Curr Opin Genet Dev [Internet] 2016;38:75–82.
[24] Spees J, Lee R, Gregory C. Mechanisms of mesenchymal stem/stromal cell function. Stem Cell Res Ther 2016;7(1):125.
[25] Cho Y, Kim J, Kim M, Park S, Koh S, Ahn H, et al. Mesenchymal stem cells transfer mitochondria to the cells with virtually no mitochondrial function but not with pathogenic mtDNA mutations. PLoS One 2012;7(3):0–7.
[26] Li X, Zhang Y, Yeung S, Liang Y, Liang X, Ding Y, et al. Mitochondrial transfer of induced pluripotent stem cell-derived mesenchymal stem cells to airway epithelial cells attenuates cigarette smoke-induced damage. Am J Respir Cell Mol Biol 2014;51(3):455–65.
[27] Lin H, Liou C, Chen S, Hsu T, Chuang J, Wang P, et al. Mitochondrial transfer from Wharton's jelly-derived mesenchymal stem cells to mitochondria-defective cells recaptures impaired mitochondrial function. Mitochondrion 2015;22:31–44.

[28] Gao L, Zhang Z, Lu J, Pei G. Mitochondria are dynamically transferring between human neural cells and Alexander disease-associated GFAP mutations impair the astrocytic transfer. Front Cell Neurosci [Internet] 2019;13(July):1–16.
[29] Swerdlow RH. The neurodegenerative mitochondriopathies. J Alzheimers Dis 2009;17(4):737–51.
[30] Hroudová J, Singh N, Fišar Z. Mitochondrial dysfunctions in neurodegenerative diseases: relevance to Alzheimer's disease. Biomed Res Int 2014;2014.
[31] Finsterer J. Cognitive dysfunction in mitochondrial disorders. Acta Neurol Scand 2012;126(1):1–11.
[32] Gollihue JL, Rabchevsky AG. Prospects for therapeutic mitochondrial transplantation. Mitochondrion 2017;35(859):70–9.
[33] Chang CY, Liang MZ, Chen L. Current progress of mitochondrial transplantation that promotes neuronal regeneration. Transl Neurodegener 2019;8(1):1–12.
[34] Gollihue JL, Patel SP, Rabchevsky AG. Mitochondrial transplantation strategies as potential therapeutics for central nervous system trauma. Neural Regen Res 2018;13(2):194–7.
[35] Roushandeh AM, Kuwahara Y, Roudkenar MH. Mitochondrial transplantation as a potential and novel master key for treatment of various incurable diseases. Cytotechnology [Internet] 2019;71(2):647–63.
[36] McCully JD, Cowan DB, Emani SM, del Nido PJ. Mitochondrial transplantation: from animal models to clinical use in humans. Mitochondrion 2017;34:127–34.
[37] Li R, Guan M-X. Human mitochondrial leucyl-tRNA synthetase corrects mitochondrial dysfunctions due to the tRNALeu(UUR) A3243G mutation, associated with mitochondrial encephalomyopathy, lactic acidosis, and stroke-like symptoms and diabetes. Mol Cell Biol 2010;30(9):2147–54.
[38] Ylikallio E, Suomalainen A. Mechanisms of mitochondrial diseases. Ann Med 2012;44(1):41–59.
[39] Finsterer J, Zarrouk-Mahjoub S. Mitochondrial vasculopathy. World J Cardiol 2016;8(5):333–9.
[40] Wallace D. Mitochondrial DNA in aging and disease. Nature 2016;535(7613):498–500.
[41] Coppotelli G, Ross J. Mitochondria in ageing and diseases: the super trouper of the cell. Int J Mol Sci 2016;17(5):711 (1–5).
[42] van Gisbergen MW, Voets AM, Starmans MHW, de Coo IFM, Yadak R, Hoffmann RF, et al. How do changes in the mtDNA and mitochondrial dysfunction influence cancer and cancer therapy? Challenges, opportunities and models. Mutat Res Rev Mutat Res 2015;764:16–30.
[43] Tuppen H, Blakely E, Turnbull D, Taylor R. Mitochondrial DNA mutations and human disease. Biochim Biophys Acta 2010;1797(2):113–28.
[44] Vyas S, Zaganjor E, Haigis M. Mitochondria and cancer. Cell 2016;166:555–66.
[45] Zong WX, Rabinowitz JD, White E. Mitochondria and cancer. Mol Cell 2016;61(5):667–76.
[46] Finsterer J, Frank M. Gastrointestinal manifestations of mitochondrial disorders: a systematic review. Ther Adv Gastroenterol 2017;10(1):142–54.
[47] Feichtinger RG, Sperl W, Bauer JW, Kofler B. Mitochondrial dysfunction: a neglected component of skin diseases. Exp Dermatol 2014;23(9):607–14.
[48] Larsson NG. Somatic mitochondrial DNA mutations in mammalian aging. Annu Rev Biochem 2010;79:683–706.
[49] Taylor SD, Ericson NG, Burton JN, Prolla TA, Silber JR, Shendure J, et al. Targeted enrichment and high-resolution digital profiling of mitochondrial DNA deletions in human brain. Aging Cell 2014;13(1):29–38.
[50] Chen Q, Moghaddas S, Hoppel CL, Lesnefsky EJ. Ischemic defects in the electron transport chain increase the production of reactive oxygen species from isolated rat heart mitochondria. Am J Physiol Cell Physiol 2008;294(2) C460–6.
[51] Orenes-Piñero E, Valdés M, Lip G, Marín F. A comprehensive insight of novel antioxidant therapies for atrial fibrillation management. Drug Metab Rev 2015;47(3):388–400.
[52] Hausenloy DJ, Barrabes JA, Bøtker HE, Davidson SM, Di Lisa F, Downey J, et al. Ischaemic conditioning and targeting reperfusion injury: a 30 year voyage of discovery. Basic Res Cardiol 2016;111(6).
[53] Fillmore N, Lopaschuk GD. Targeting mitochondrial oxidative metabolism as an approach to treat heart failure. Biochim Biophys Acta Mol Cell Res [Internet] 2013;1833(4) 857–65.

[54] Rebbeck CA, Thomas R, Breen M, Leroi AM, Burt A. Origins and evolution of a transmissible cancer. Evolution (N Y) 2009;63(9) 2340–9.
[55] Ganguly B, Das U, Das AK. Canine transmissible venereal tumour: a review. Vet Comp Oncol 2016;14 (1):1–12.
[56] Lei L, Spradling A. Mouse oocytes differentiate through organelle enrichment from sister cyst germ cells. Res Rep 2016;352(6281) 95–9.
[57] Otsu K, Das S, Houser SD, Quadri SK, Bhattacharya S, Bhattacharya J. Concentration-dependent inhibition of angiogenesis by mesenchymal stem cells. Blood 2009;113(18):4197–205.
[58] Acquistapace A, Bru T, Lesault P, Figeac F, Coudert A, O le C, et al. Human mesenchymal stem cells reprogram adult cardiomyocytes toward a progenitor-like state through partial cell fusion and mitochondria transfer. Stem Cell 2011;29(5) 812–24.
[59] Ahmad T, Mukherjee S, Pattnaik B, Kumar M, Singh S, Rehman R, et al. Miro1 regulates intercellular mitochondrial transport & enhances mesenchymal stem cell rescue efficacy. EMBO J 2014;33(9):994–1010.
[60] Babenko V, Silachev D, Zorova L, Pevzner I, Khutornenko A, Plotnikov E, et al. Improving the post-stroke therapeutic potency of mesenchymal multipotent stromal cells by cocultivation with cortical neurons: the role of crosstalk between cells. Stem Cell Transl Med 2015;4(9):1011–20.
[61] Han J, Kim B, Shin JY, Ryu S, Noh M, Woo J, et al. Iron oxide nanoparticle-mediated development of cellular gap junction crosstalk to improve mesenchymal stem cells' therapeutic efficacy for myocardial infarction. ACS Nano 2015;9(3):2805–19.
[62] Chen J, Wang Q, Feng X. Umbilical cord-derived mesenchymal stem cells suppress autophagy of T cells in patients with systemic lupus erythematosus via transfer of mitochondria. Stem Cell Int 2016;2016:4062789.
[63] Jiang D, Gao F, Zhang Y, Wong D, Li Q, Tse H, et al. Mitochondrial transfer of mesenchymal stem cells effectively protects corneal epithelial cells from mitochondrial damage. Cell Death Dis 2016;7(11):e2467.
[64] Moschoi R, Imbert V, Nebout M, Chiche J, Mary D, Prebet T, et al. Protective mitochondrial transfer from bone marrow stromal cells to acute myeloid leukemic cells during chemotherapy. Blood 2016;128:253–64.
[65] Newell C, Sabouny R, Hittel DS, Shutt TE, Khan A, Klein MS, et al. Mesenchymal stem cells shift mitochondrial dynamics and enhance oxidative phosphorylation in recipient cells. Front Physiol 2018;9:1–16.
[66] Marlein C, Zaitseva L, Piddock R, Robinson S, Edwards D, Shafat M, et al. NADPH oxidase-2 derived superoxide drives mitochondrial transfer from bone marrow stromal cells to leukemic blasts. Blood 2017;130(14):1649–60.
[67] Babenko VA, Silachev DN, Popkov VA, Zorova LD, Pevzner IB, Plotnikov EY, et al. Miro1 enhances mitochondria transfer from multipotent mesenchymal stem cells (MMSC) to neural cells and improves the efficacy of cell recovery. Molecules 2018;23(3):1–14.
[68] Zhang Y, Yu Z, Jiang D, Liang X, Liao S, Zhang Z, et al. iPSC-MSCs with high intrinsic MIRO1 and sensitivity to TNF-α yield efficacious mitochondrial transfer to rescue anthracycline-induced cardiomyopathy. Stem Cell Rep [Internet] 2016;7(4):749–63.
[69] Yao Y, Fan XL, Jiang D, Zhang Y, Li X, Xu ZB, et al. Connexin 43-mediated mitochondrial transfer of iPSC-MSCs alleviates asthma inflammation. Stem Cell Rep [Internet] 2018;11(5):1120–35.
[70] Marlein CR, Piddock RE, Mistry JJ, Zaitseva L, Hellmich C, Horton RH, et al. CD38-driven mitochondrial trafficking promotes bioenergetic plasticity in multiple myeloma. Cancer Res 2019;79(9):2285–97.
[71] Li H, Wang C, He T, Zhao T, Chen YY, Shen YL, et al. Mitochondrial transfer from bone marrow mesenchymal stem cells to motor neurons in spinal cord injury rats via gap junction. Theranostics 2019;9 (7):2017–35.
[72] Luz-Crawford P, Hernandez J, Djouad F, Luque-Campos N, Caicedo A, Carrère-Kremer S, et al. Mesenchymal stem cell repression of Th17 cells is triggered by mitochondrial transfer. Stem Cell Res Ther [Internet] 2019;10(1):232.
[73] Yasuda K, Park HC, Ratliff B, Addabbo F, Hatzopoulos AK, Chander P, et al. Adriamycin nephropathy: a failure of endothelial progenitor cell-induced repair. Am J Pathol [Internet] 2010;176(4):1685–95. Available from: <https://doi.org/10.2353/ajpath.2010.091071>.

[74] Pasquier J, Guerrouahen BS, Al Thawadi H, Ghiabi P, Maleki M, Abu-Kaoud N, et al. Preferential transfer of mitochondria from endothelial to cancer cells through tunneling nanotubes modulates chemoresistance. J Transl Med 2013;11(1):94.
[75] Shen J, Zhang JH, Xiao H, Wu JM, He KM, Lv ZZ, et al. Mitochondria are transported along microtubules in membrane nanotubes to rescue distressed cardiomyocytes from apoptosis article. Cell Death Dis [Internet] 2018;9(2).
[76] He K, Shi X, Zhang X, Dang S, Ma X, Liu F, et al. Long-distance intercellular connectivity between cardiomyocytes and cardiofibroblasts mediated by membrane nanotubes. Cardiovasc Res 2011;92(1):39–47.
[77] Lippert T, Borlongan CV. Prophylactic treatment of hyperbaric oxygen treatment mitigates inflammatory response via mitochondria transfer. CNS Neurosci Ther 2019;25(8):815–23.
[78] Vallabhaneni K, Haller H, Dumler I. Vascular smooth muscle cells initiate proliferation of mesenchymal stem cells by mitochondrial transfer via tunneling nanotubes. Stem Cell Dev 2012;21(17):3104–13.
[79] Wang J, Liu X, Qiu Y, Shi Y, Cai J, Wang B, et al. Cell adhesion-mediated mitochondria transfer contributes to mesenchymal stem cell-induced chemoresistance on T cell acute lymphoblastic leukemia cells. J Hematol Oncol 2018;11(1):11.
[80] Rebbeck C, Leroi A, Burt A. Mitochondrial capture by a transmissible cancer. Science 2011;331 (6015):303. Available from: https://doi.org/10.101126/science1197696.
[81] Islam M, Das S, Emin MT, Wei M, Sun L, Westphalen K, et al. Mitochondrial transfer from bone-marrow-derived stromal cells to pulmonary alveoli protects against acute lung injury. Nat Med 2012;18 (5):759–65.
[82] Jackson MV, Morrison TJ, Doherty DF, McAuley DF, Matthay MA, Kissenpfennig A, et al. Mitochondrial transfer via tunneling nanotubes (TNT) is an important mechanism by which mesenchymal stem cells enhance macrophage phagocytosis in the in vitro and in vivo models of ARDS. Stem Cell 2016;2210–23.
[83] Dong L, Kovarova J, Bajzikova M, Bezawork-Geleta A, Svec D, Endaya B, et al. Horizontal transfer of whole mitochondria restores tumorigenic potential in mitochondrial DNA-deficient cancer cells. Elife 2017;6:e22187.
[84] Konari N, Nagaishi K, Kikuchi S, Fujimiya M. Mitochondria transfer from mesenchymal stem cells structurally and functionally repairs renal proximal tubular epithelial cells in diabetic nephropathy in vivo. Sci Rep [Internet] 2019;9(1):1–14. Available from: <https://doi.org/10.1038/s41598-019-40163-y>.
[85] Davis C, Kim K-Y, Bushong E, Mills E, Boassa D, Shih T, et al. Transcellular degradation of axonal mitochondria. Proc Natl Acad Sci U S A 2014;111(26):9633–8.
[86] Osswald M, Jung E, Sahm F, Solecki G, Venkataramani V, Blaes J, et al. Brain tumour cells interconnect to a functional and resistant network. Nature 2015;528(7580):93–8.
[87] Tan A, Baty J, Dong L, Bezawork-Geleta A, Endaya B, Goodwin J, et al. Mitochondrial genome acquisition restores respiratory function and tumorigenic potential of cancer cells without mitochondrial DNA. Cell Metab 2015;21(1):81–94.
[88] Lee St. WTY, John JC. Mitochondrial DNA as an initiator of tumorigenesis. Cell Death Dis 2016;7 (3):3–4.
[89] Hogan KA, Chini CCS, Chini EN. The multi-faceted ecto-enzyme CD38: roles in immunomodulation, cancer, aging, and metabolic diseases. Front Immunol 2019;10:1–12.
[90] Lonial S, Weiss BM, Usmani SZ, Singhal S, Chari A, Bahlis NJ, et al. Daratumumab monotherapy in patients with treatment-refractory multiple myeloma (SIRIUS): an open-label, randomised, phase 2 trial. Lancet 2016;387(10027):1551–60.
[91] Yu M, Nguyen ND, Huang Y, Lin D, Fujimoto TN, Molkentine JM, et al. Mitochondrial fusion exploits a therapeutic vulnerability of pancreatic cancer. JCI Insight 2019;.
[92] Weil S, Osswald M, Solecki G, Grosch J, Jung E, Lemke D, et al. Tumor microtubes convey resistance to surgical lesions and chemotherapy in gliomas. Neuro Oncol 2017;19(10):1316–26.
[93] Osswald M, Solecki G, Wick W, Winkler F. A malignant cellular network in gliomas: potential clinical implications. Neuro Oncol 2016;18(4):479–85.

[94] Herst P, Berridge M. Plasma membrane electron transport: a new target for cancer drug development. Curr Mol Med 2006;6:895–904.
[95] Bajzikova M, Kovarova J, Coelho AR, Boukalova S, Oh S, Rohlenova K, et al. Reactivation of dihydroorotate dehydrogenase-driven pyrimidine biosynthesis restores tumor growth of respiration-deficient cancer cells. Cell Metab 2019;29:399–416.
[96] Berridge MV, Tan AS. Effects of mitochondrial gene deletion on tumorigenicity of metastatic melanoma: reassessing the Warburg effect. Rejuvenation Res 2010;13(2–3):139–41.
[97] Sun C, Liu X, Wang B, Wang Z, Liu Y, Di C, et al. Endocytosis-mediated mitochondrial transplantation: transferring normal human astrocytic mitochondria into glioma cells rescues aerobic respiration and enhances radiosensitivity. Theranostics 2019;9(12):3595–607.
[98] Chang JC, Chang HS, Wu YC, Cheng WL, Lin TT, Chang HJ, et al. Mitochondrial transplantation regulates antitumour activity, chemoresistance and mitochondrial dynamics in breast cancer. J Exp Clin Cancer Res 2019;38(1):1–16.
[99] MacAskill AF, Brickley K, Stephenson FA, Kittler JT. GTPase dependent recruitment of Grif-1 by Miro1 regulates mitochondrial trafficking in hippocampal neurons. Mol Cell Neurosci 2009;40(3):301–12.
[100] Davis CHO, Marsh-Armstrong N. Discovery and implications of transcellular mitophagy. Autophagy 2014;10(12):2383–4.
[101] Misgeld T, Schwarz TL. Mitostasis in neurons: maintaining mitochondria in an extended cellular architecture. Neuron [Internet] 2017;96(3):651–66.
[102] Plotnikov EY, Khryapenkova TG, Galkina SI, Sukhikh GT, Zorov DB. Cytoplasm and organelle transfer between mesenchymal multipotent stromal cells and renal tubular cells in co-culture. Exp Cell Res 2010;316(15):2447–55.
[103] Cselenyák A, Pankotai E, Horváth E, Kiss L, Lacza Z, Dayer M, et al. Mesenchymal stem cells rescue cardiomyoblasts from cell death in an in vitro ischemia model via direct cell-to-cell connections. BMC Cell Biol 2010;11(1):29.
[104] Zhang Y, Yu Z, Jiang D, Liang X, Liao S, Zhang Z, et al. iPSC-MSCs with high intrinsic MIRO1 and sensitivity to TNF-a yield efficacious mitochondrial transfer to rescue anthracycline-induced cardiomyopathy. Stem Cell Rep 2016;7(4):749–63.
[105] Lin TK, Chen S Der, Chuang YC, Lan MY, Chuang JH, Wang PW, et al. Mitochondrial transfer of Wharton's jelly mesenchymal stem cells eliminates mutation burden and rescues mitochondrial bioenergetics in rotenone-stressed MELAS fibroblasts. Oxid Med Cell Longev 2019;2019:9537504.
[106] Hajjar C, Sampuda KM, Boyd L. Dual roles for ubiquitination in the processing of sperm organelles after fertilization. BMC Dev Biol [Internet] 2014;14(1):1–11. Available from: BMC Developmental Biology.
[107] Cozzolino M, Marin D, Sisti G. New frontiers in IVF: mtDNA and autologous germline mitochondrial energy transfer. Reprod Biol Endocrinol 2019;17(1):1–11.
[108] Darbandi S, Darbandi M, Khorram Khorshid HR, Sadeghi MR, Agarwal A, Sengupta P, et al. Ooplasmic transfer in human oocytes: efficacy and concerns in assisted reproduction. Reprod Biol Endocrinol 2017;15(1):1–11.
[109] Craven L, Tang MX, Gorman GS, De Sutter P, Heindryckx B. Novel reproductive technologies to prevent mitochondrial disease. Hum Reprod Update 2017;23(5):501–19.
[110] Bredenoord AL, Pennings G, De Wert G. Ooplasmic and nuclear transfer to prevent mitochondrial DNA disorders: conceptual and normative issues. Hum Reprod Update 2008;14(6):669–78.
[111] Muggleton-Harris A, Whittingham D, Wilson L. Cytoplasmic control of preimplantation development in vitro in the mouse. Nature 1982;299(5882):321–2.
[112] Cohen J, Scott R, Schimmel T, Levron J, Willadsen S. Birth of infant after transfer of anucleate donor oocyte cytoplasm into recipient eggs. Lancet 1997;350(9072):186–7.
[113] Ishii T, Hibino Y. Mitochondrial manipulation in fertility clinics: regulation and responsibility. Reprod Biomed Soc Online [Internet] 2018;5:93–109.

[114] Caicedo A, Fritz V, Brondello J-M, Ayala M, Dennemont I, Abdellaoui N, et al. MitoCeption as a new tool to assess the effects of mesenchymal stem/stromal cell mitochondria on cancer cell metabolism and function. Sci Rep 2015;5(1):9073.

[115] Caicedo A, Aponte P, Cabrera F, Hidalgo C, Khoury M. Artificial mitochondria transfer: current challenges, advances, and future applications. Stem Cell Int 2017;2017.

[116] Wu TH, Sagullo E, Case D, Zheng X, Li Y, Hong JS, et al. Mitochondrial transfer by photothermal nanoblade restores metabolite profile in mammalian cells. Cell Metab [Internet] 2016;23 (5):921–9.

[117] Kuo C, Su H, Chang T, Chiang C, Sheu M, Cheng F, et al. Prevention of axonal degeneration by perineurium injection of mitochondria in a sciatic nerve crush injury model. Neurosurgery 2017;80 (3):475–88.

[118] Masuzawa A, Black KM, Pacak CA, Ericsson M, Barnett RJ, Drumm C, et al. Transplantation of autologously derived mitochondria protects the heart from ischemia-reperfusion injury. Am J Physiol Circ Physiol [Internet] 2013;304(7):H966–82.

[119] Shi X, Zhao M, Fu C, Fu A. Intravenous administration of mitochondria for treating experimental Parkinson's disease. Mitochondrion 2017;34:91–100.

[120] McCully JD, Cowan DB, Pacak CA, Toumpoulis IK, Dayalan H, Levitsky S. Injection of isolated mitochondria during early reperfusion for cardioprotection. Am J Physiol Circ Physiol 2009;296(1): H94–105.

[121] Kim MJ, Hwang JW, Yun CK, Lee Y, Choi YS. Delivery of exogenous mitochondria via centrifugation enhances cellular metabolic function. Sci Rep 2018;8(1):1–13.

[122] Chang J, Wu S, Liu K, Chen Y, Chuang C, Cheng F, et al. Allogeneic/xenogeneic transplantation of peptide-labeled mitochondria in Parkinson's disease: restoration of mitochondria functions and attenuation of 6-hydroxydopamine-induced neurotoxicity. Transl Res 2016;170:40–56.e3.

[123] Chang JC, Hoel F, Liu KH, Wei YH, Cheng FC, Kuo SJ, et al. Peptide-mediated delivery of donor mitochondria improves mitochondrial function and cell viability in human cybrid cells with the MELAS A3243G mutation. Sci Rep 2017;7(1):1–15.

[124] Chien L, Liang MZ, Chang CY, Wang C, Chen L. Mitochondrial therapy promotes regeneration of injured hippocampal neurons. Biochim Biophys Acta Mol Basis Dis [Internet] 2018;1864 (9):3001–12.

[125] Kaza AK, Wamala I, Friehs I, Kuebler JD, Rathod RH, Berra I, et al. Myocardial rescue with autologous mitochondrial transplantation in a porcine model of ischemia/reperfusion. J Thorac Cardiovasc Surg [Internet] 2017;153(4):934–43.

[126] Zhang Z, Ma Z, Yan C, Pu K, Wu M, Bai J, et al. Muscle-derived autologous mitochondrial transplantation: a novel strategy for treating cerebral ischemic injury. Behav Brain Res [Internet] 2019;356:322–31.

[127] Moskowitzova K, Shin B, Liu K, Ramirez-Barbieri G, Guariento A, Blitzer D, et al. Mitochondrial transplantation prolongs cold ischemia time in murine heart transplantation. J Hear Lung Transpl [Internet] 2019;38(1):92–9.

[128] Gollihue JL, Patel SP, Eldahan KC, Cox DH, Donahue RR, Taylor BK, et al. Effects of mitochondrial transplantation on bioenergetics, cellular incorporation, and functional recovery after spinal cord injury. J Neurotrauma 2018;35(15):1800–18.

[129] Emani SM, Piekarski BL, Harrild D, del Nido PJ, McCully JD. Autologous mitochondrial transplantation for dysfunction after ischemia-reperfusion injury. J Thorac Cardiovasc Surg 2017;154(1):286–9.

[130] Huang P-J, Kuo C-C, Lee H-C, Shen C-I, Cheng F-C, Wu S-F, et al. Transferring xenogenic mitochondria provides neural protection against ischemic stress in ischemic rat brains. Cell Transpl 2016;25 (5):913–27.

[131] Cowan DB, Yao R, Akurathi V, Snay ER, Thedsanamoorthy JK, Zurakowski D, et al. Intracoronary delivery of mitochondria to the ischemic heart for cardioprotection. PLoS One 2016;11(8):1–19.

[132] Jackson M, Morrison T, Doherty DF, McAuley D, Matthay M, Kissenpfennig A, et al. Mitochondrial transfer via tunneling nanotubes (TNT) is an important mechanism by which mesenchymal stem cells enhance macrophage phagocytosis in the in vitro and in vivo models of ARDS. Stem Cell 2016;2014:2210–23.
[133] Hayakawa K, Esposito E, Wang X, Terasaki Y, Liu Y, Xing C, et al. Transfer of mitochondria from astrocytes to neurons after stroke neurons can release damaged mitochondria and transfer them to astrocytes for disposal and recycling. Nat Publ Gr 2016;535(7613):551–5.
[134] Wang Y, Ni J, Gao C, Xie L, Zhai L, Cui G, et al. Mitochondrial transplantation attenuates lipopolysaccharide-induced depression-like behaviors. Prog Neuro-psychopharmacol Biol Psychiatry 2019;93:240–9.
[135] Kurian G, Berenshtein E, Kakhlon O, Cheviona M. Energy status determines the distinct biochemical and physiological behavior of interfibrillar and sub-sarcolemmal mitochondria. Biochem Biophys Res Commun 2012;428(3):376–82.
[136] Lesnefsky E, Hoppel C. Ischemia–reperfusion injury in the aged heart: role of mitochondria. Arch Biochem Biophys 2003;420(2):287–97.
[137] Lesnefsky EJ, Chen Q, Tandle B, Hoppel CL. Mitochondrial dysfunction and myocardial ischemia-reperfusion: implications for novel therapies. Annu Rev Pharmacol Toxicol 2017;57:535–65.
[138] Pacak CA, Preble JM, Kondo H, Seibel P, Levitsky S, del Nido PJ, et al. Actin-dependent mitochondrial internalization in cardiomyocytes: evidence for rescue of mitochondrial function. Biol Open 2015;4(5):622–6.
[139] Hayakawa K, Bruzzese M, Chou SHY, Ning MM, Ji X, Lo EH. Extracellular mitochondria for therapy and diagnosis in acute central nervous system injury. JAMA Neurol 2018;75(1):119–22.
[140] Hayakawa K, Esposito E, Wang X, Terasaki Y, Liu Y, Xing C, et al. Transfer of mitochondria from astrocytes to neurons after stroke. Nature 2017;535(7613):551–5.
[141] Hayakawa K, Chan SJ, Mandeville ET, Park JH, Bruzzese M, Montaner J, et al. Protective effects of endothelial progenitor cell-derived extracellular mitochondria in brain endothelium. Stem Cell 2018;36(9):1404–10.
[142] Morris G, Berk M. The many roads to mitochondrial dysfunction in neuroimmune and neuropsychiatric disorders. BMC Med 2015;13(1):1–24.
[143] Robicsek O, Ene HM, Karry R, Ytzhaki O, Asor E, McPhie D, et al. Isolated mitochondria transfer improves neuronal differentiation of schizophrenia-derived induced pluripotent stem cells and rescues deficits in a rat model of the disorder. Schizophr Bull 2018;44(2):432–42.
[144] Gurke S, Barroso JFV, Gerdes HH. The art of cellular communication: tunneling nanotubes bridge the divide. Histochem Cell Biol 2008;129(5):539–50.
[145] Wang Y, Cui J, Sun X, Zhang Y. Tunneling-nanotube development in astrocytes depends on p53 activation. Cell Death Differ 2011;18(4):732–42.
[146] Zhang Y. Tunneling-nanotube: a new way of cell-cell communication. Commun Integr Biol 2011;4(3):324–5.
[147] Phinney D, Di Giuseppe M, Njah J, Sala E, Shiva S, St Croix C, et al. Mesenchymal stem cells use extracellular vesicles to outsource mitophagy and shuttle microRNAs. Nat Commun 2015;6:8472.
[148] Falchi AM, Sogos V, Saba F, Piras M, Congiu T, Piludu M. Astrocytes shed large membrane vesicles that contain mitochondria, lipid droplets and ATP. Histochem Cell Biol 2013;139(2):221–31.
[149] Taanman J-W. The mitochondrial genome: structure, transcription, translation and replication. Biochim Biophys Acta Bioenerg 1999;1410:103–23.
[150] Anderson S, Bankier AT, Barrell BG, Bruijn MH, de Coulson AR, Drouin J, et al. Sequence and organization of the human mitochondrial genome. Nature 1981;290:457–65.
[151] Ahn C, Metallo C. Mitochondria as biosynthetic factories for cancer proliferation. Cancer Metab 2015;3(1):1–10.
[152] Arnould T, Michel S, Renard P. Mitochondria retrograde signaling and the UPR mt: where are we in mammals? Int J Mol Sci 2015;16(8):18224–51.

[153] Cagin U, Enriquez J. The complex crosstalk between mitochondria and the nucleus: what goes in between? Int J Biochem Cell Biol 2015;63:10–15.
[154] Ploumi C, Daskalaki I, Tavernarakis N. Mitochondrial biogenesis and clearance: a balancing act. FEBS J 2017;284(2):183–95.
[155] Carelli V, Maresca A, Caporali L, Trifunov S, Zanna C, Rugolo M. Mitochondria: biogenesis and mitophagy balance in segregation and clonal expansion of mitochondrial DNA mutations. Int J Biochem Cell Biol 2015;63:21–4.
[156] Danhier P, Bański P, Payen V, Grasso D, Ippolito L, Sonveaux P, et al. Cancer metabolism in space and time: beyond the Warburg effect. Biochim Biophys Acta Bioenerg 2017;1858(8):556–72.
[157] Deberardinis R, Chandel N. Fundamentals of cancer metabolism. Oncology 2016;2(5):e1600200.
[158] Vander HMG, Deberardinis RJ. Understanding the intersections between metabolism and cancer biology. Cell 2017;168(4):657–69.

Index

Note: Page numbers followed by "*f*," "*t*," and "*b*" refer to figures, tables, and boxes, respectively.

A

aa-RS. *See* Aminoacyl-tRNA synthetase (aa-RS)
AARS2 gene, 391–392
AAV. *See* Adeno-associated virus (AAV)
ABAT, 413*t*, 419–420
Aberrant nuclear DNA methylation, 72
Abf2p. *See* ARS-binding factor 2 protein (Abf2p)
Accessory subunit POL γB, 9
Activation domain (AD), 484–485
"Active elimination" model, 91–93
AD. *See* Activation domain (AD)
Ad hoc bioinformatic methods, 244
Ad hoc designed methods, 134–135
Adeno-associated virus (AAV), 435, 491, 509
  AAV1, 483
Adenosine monophosphate activated protein kinase (AMPK), 462
Adenovirus, 483
aDNA. *See* Ancient DNA (aDNA)
ADOA, 420
AdPEO. *See* Autosomal dominant PEO (AdPEO)
Adult-onset myopathy of *TK2*-related mitochondrial disease, 382
AFDIL. *See* Armed Forces DNA Identification Laboratory (AFDIL)
Affymetrix, 252–253
Affymetrix GeneChip Mitochondrial Resequencing Array, 252–253
Agilent, 325
Aging, 221–223
AGK, 413*t*, 419–420
AHS. *See* Alpers-Huttenlocher syndrome (AHS)
Allele search function in MITOMAP, 283
Allogeneic stem cell transplantations (alloSCTs), 431
Alpers-Huttenlocher syndrome (AHS), 379, 434
α-ketoglutarate (α-KG), 450–451
Alzheimer's disease, 531–532
Aminoacyl-tRNA synthetase (aa-RS), 389
Aminoacyl-tRNAs, 56–57
4-Aminobutyrate transaminase (ABAT), 382
AMPK. *See* Adenosine monophosphate activated protein kinase (AMPK)
Amplification refractory mutation system qPCR (ARMS-qPCR), 245, 249
Ancient DNA (aDNA), 279–280
Animal models, 311–320
  Caenorhabditis elegans, 311–313
  Drosophila melanogaster, 313–315
  mice, 315–320
  of mtDNA alterations, 320*t*
ANS. *See* Ataxia neuropathy spectrum (ANS)
ANT1, 419–420
Antimycin, 330–331
API. *See* Application Programming Interface (API)
Application Programming Interface (API), 280–281
Armed Forces DNA Identification Laboratory (AFDIL), 151
ARMS-qPCR. *See* Amplification refractory mutation system qPCR (ARMS-qPCR)
ARS-binding factor 2 protein (Abf2p), 17–18
Artificial recombinants, 151
Asian-specific haplogroup, 115
ATAD3 gene, 20–21, 421–422
ATAD3A. *See* ATP-ase family AAA domain containing 3A (ATAD3A)
Ataxia, 310–311
Ataxia neuropathy spectrum (ANS), 379–380
ATP production, 333–335
  data analysis of ATP synthesis assay, 336*b*
ATP-ase family AAA domain containing 3A (ATAD3A), 18, 413*t*, 421–422
ATP6 mutant, 45, 310–311, 314
Autosomal DNA markers, 145–146
Autosomal dominant PEO (AdPEO), 380
Autosomal STR marker profiles, 148–149
AwSomics Gene Explorer (AwSomicsGE), 292

B

B16ρ^0 metastatic melanoma cells, 523
BAM. *See* Binary Alignment Map (BAM)
Base excision repair (BER), 174–178, 176*f*, 229
  unfolded proteins, 178*f*
Behavioral disorders, 533

Behr syndrome, 420
Benign reversible COX deficiency, 364
BER. *See* Base excision repair (BER)
Beta-carotene, 140–141
β-glucuronidase (GUSB), 492
BglII-digested native mtDNA, 76–77
Biallelic pathogenic variants in TWNK, 380
Big Dye chemistry, 251–252
Bilateral striatal lesions, 310–311
Binary Alignment Map (BAM), 260
Biochemical threshold, 93–94
Bioinformatics strategies
  bioinformatic workflow of analysis of mtDNA, 261*f*
  for in silico analysis, 263*t*
  mitochondrial phylogenetic analysis, 266
  mitochondrial variant calling, 260–265
  reads mapping and genome assembly, 259–260
Bioinformatics systems, 278
BioMart, 290
Bisulfite sequencing, 73–75, 74*f*
  analysis of mtDNA, 76–78
bL17, 53
bL22, 53
bL23, 53
bL24, 53
bL29, 53
bL33, 53
BLAST algorithm, 132–133
BLAT analysis, 260
*Blimp*1, 95*b*
Blue Native PolyAcrylamide Gel Electrophoresis (BN-PAGE), 337–339, 338*f*
BM-MSCs. *See* Bone marrow-derived-MSCs (BM-MSCs)
BN-PAGE. *See* Blue Native PolyAcrylamide Gel Electrophoresis (BN-PAGE)
Bone marrow-derived-MSCs (BM-MSCs), 518–522
Bootlace model, 376–379
Bootlace/RITOLS model, 6
Bottleneck effect, 309–310
Bovine serum albumin (BSA), 150, 322–323
9-bp sequence, 502
Br-positive RNA foci, 60
BrdU. *See* 5-Bromo-2′-deoxyuridine (BrdU)
Broad-spectrum techniques for variants detection
  DHPLC, 250–251
  microarrays, 252–253
  Sanger sequencing, 251–252
  second-generation sequencing, 253–256
  SSCP, 250
  third-generation sequencing, 257
  in whole mtDNA genomes, 250–257
5-Bromo-2′-deoxyuridine (BrdU), 16*b*, 308–309
Bromouridine (BrU), 59
BSA. *See* Bovine serum albumin (BSA)

C

C-stretches, 156
C-terminal extensions (CTE), 55–56
*C*12orf62, mutations in, 397
C12orfF65 gene, mutations in, 396–397
C57BL/6N^{su9DsRed2} mice, 523–524
Caenorhabditis elegans. *See* Garden worm (*Caenorhabditis elegans*)
Cancer, 100–101
  cancer-associated mtDNA variants, 444
  cells, 536
  sensitization to treatment, 534
CAP^{R}. *See* 16S rRNA mtDNA gene (CAP^{R})
Carbonyl cyanide-ptrifluoromethoxyphenylhydrazone (FCCP), 330–331
CBA. *See* Chicken β-actin (CBA)
CD. *See* Common deletion (CD)
CD-targeted mtZFNs, 505–506
CDA. *See* Cytidine deaminase (CDA)
Cell proliferation assays, 308
Cell-penetrating peptide (Pep-1), 532–533
Cellular antioxidants, 335–337
Central nervous system (CNS), 382, 524–525
Cesium chloride (CsCl), 258
Chicken β-actin (CBA), 492
Childhood myocerebrohepatopathy spectrum, 379
Chimeric haplotypes, 151
Chinese hamster (*Cricetulus griseus*), 91
Chronic progressive external ophthalmoplegia (CPEO), 7–8
CI. *See* Complex I (CI)
CIA. *See* Cluster assembly complex (CIA)
Citrate synthase activity (CS activity), 325
CK. *See* Creatine kinase (CK)
Classic clark-type electrode methods, 328–330
CLIP. *See* Cross-linking and immunoprecipitation (CLIP)
Clonal expansion, 100–102, 101*f*
Clopper-Pearson method, 161
Cluster assembly complex (CIA), 180
Clustered Regularly Interspaced Short Palindromic Repeats (CRISPR/Cas9), 493
Clusters of region-specific types, 113
CMV. *See* Cytomegalovirus (CMV)
CNS. *See* Central nervous system (CNS)
CoA. *See* Coenzyme-A (CoA)
COA3. *See* COX assembly factor 3 (COA3)
Cockayne Syndrome (CS), 180
CODIS^{mt}. *See* FBI mtDNA Population Database
Coenzyme-A (CoA), 325
CoI mutants, 314
CoII mutant, 315
Common deletion (CD), 487–490, 505–506

Compartmentalization of gene expression, 58–61
mitochondrial DNA nucleoids, 59
mitochondrial RNA degradosome, 60–61
mitochondrial RNA granules, 59–60
Complex I (CI), 443–444
Complex II (CII), 323–324
Complex III (CIII), 310–311
Compromised arterial supply, 528
Computed tomography (CT), 366
Confidence intervals, 160
Connexin 43, 524–525
Conserved sequence blocks (CSB), 5
CSB2, 38, 43
Control region (CR), 5, 151
Coomassie Blue dye, 337–339
Counting method, 158–160
COX. *See* Cytochrome *c* oxidase (COX)
COX assembly factor 3 (COA3), 397
mutations in, 397
Cox8Sub9, 492
COXI gene, 97–98
CPEO. *See* Chronic progressive external ophthalmoplegia (CPEO)
CR. *See* Control region (CR)
Creatine kinase (CK), 366
Cricetulus griseus. *See* Chinese hamster (*Cricetulus griseus*)
CRISPR/Cas9 endonuclease, 510
CRISPR/Cas9. *See* Clustered Regularly Interspaced Short Palindromic Repeats (CRISPR/Cas9)
Cross-linking and immunoprecipitation (CLIP), 42*b*, 45
Crude mitochondria, 322–323
Cryoelectron microscopy (cryo-EM), 49–50, 52*b*
Cryoelectron tomography (cryo-ET), 58
CS. *See* Cockayne Syndrome (CS)
CS activity. *See* Citrate synthase activity (CS activity)
CSB. *See* Conserved sequence blocks (CSB)
CsCl. *See* Cesium chloride (CsCl)
CSF protein, 366
CT. *See* Computed tomography (CT)
Ct value. *See* Cycle threshold value (Ct value)
CTE. *See* C-terminal extensions (CTE)
Cybrids, 306–309
Cycle threshold value (Ct value), 248
Cys2His2 zinc finger domains, 500, 501*f*
Cytidine deaminase (CDA), 428
Cytochrome *c* oxidase (COX), 388
Cytochrome c oxidoreductase activity
CI + III activity, 324
CIII activity, 324
CIV activity, 324
Cytomegalovirus (CMV), 485–487, 492
Cytoplasmic bridges, 525
Cytoplasts, 308
Cytosine methylation in mtDNA, 72–76
Cytosolic de novo dNTP synthesis, 13–15

D

D-loop. *See* Displacement loop (D-loop)
D257A mouse model, 318–319
dAdo. *See* Deoxyadenosine (dAdo)
DAPI. *See* 4′,6-Diamidino-2-phenylindole (DAPI)
DARS2 gene, 391–392
DB. *See* Decylbenzoquinone (DB)
dC. *See* Deoxycytidine (dC)
DCFH-DA. *See* Dichlorodihydrofluorescein diacetate (DCFH-DA)
DCIP. *See* 2,6-Dichloroindophenol (DCIP)
dCK. *See* Deoxycytidine kinase (dCK)
dCtd. *See* Deoxycytidine (dC)
DDBJ, 279
ddNTPs. *See* 2′,3′-Dideoxynucleotides (ddNTPs)
DDX28 (DEAD-box helicase), 59–60
De novo somatic mtDNA mutations, 100–101, 101*f*
DEAD-box RNA helicases, 53–54
Decylbenzoquinone (DB), 323
Deletor mice, 319
Denaturing high-performance liquid chromatography (DHPLC), 250–251
Deoxyadenosine (dAdo), 418–419
Deoxycytidine (dC), 80–81, 418–419
Deoxycytidine kinase (dCK), 428
Deoxyguanosine kinase (dGK), 13, 382
Deoxyguanosine kinase (*DGUOK*) gene, 413*t*, 418–419
DGUOK-deficient hepatocyte-like cells, 423–424
5′-Deoxynucleotidases (5′-dNs), 15
Deoxyribonucleoside monophosphates (dNMPs), 13, 382
Deoxyribonucleoside triphosphates (dNTP), 12–13, 251–252, 252*f*, 381–382
defects of dNTP salvage pathway and nucleotide metabolism, 382–383
maintenance of dNTP pool, 381–382
Deoxyribonucleosides (dNs), 12–15
Deoxythymidine (dThd), 418–419
Designer zinc fingers, 500–502
Desmin, 492
dGK. *See* Deoxyguanosine kinase (dGK)
DHODH. *See* Dihydroorotate dehydrogenase (DHODH)
Dhodh gene, 523–524
DHPLC. *See* Denaturing high-performance liquid chromatography (DHPLC)
dHPLC. *See* High performance liquid chromatography (dHPLC)
Dialysis, 430
4′,6-Diamidino-2-phenylindole (DAPI), 15–16

Dichlorodihydrofluorescein diacetate (DCFH-DA), 335–337
2,6-Dichloroindophenol (DCIP), 323
2′,3′-Dideoxynucleotides (ddNTPs), 251–252
Digital PCR (dPCR), 245, 249–250
Digitonin, 337–339
3,3′-Dihexyloxacarbocyanine iodide (DiOC6(3)), 332–333
7,8-Dihydro-8-oxo-2′-deoxyguanosine (8-oxodG), 173–174, 174*f*
Dihydroorotate dehydrogenase (DHODH), 523
DiOC6(3). *See* 3,3′-Dihexyloxacarbocyanine iodide (DiOC6(3))
Dipstick assay, 328
Disease score threshold (DS_T), 288–289
Disease-tailored therapies, 427–435
Displacement loop (D-loop), 5, 35–36, 113–114, 151
  formation and transcription-replication switch, 39*f*
5,5′-Dithiobis(2-nitrobenzoic acid) (DTNB), 325
DNA
  DNA2, 12, 413*t*, 418
    mutations in, 381
  methylation, 72, 72*f*
  N-glycosylase enzymes, 175
  polymerases, 229, 230*f*
  7S DNA, 37
  Sanger sequencing, 251–252
  sequencing, 73–75
  structure and interaction with, 500
DNA ligase 3 (LIG3), 11, 36, 376–379
DNA-methyltransferases (DNMTs), 72
  DNAMT3A, 72
  DNMT1, 72
  MTS-GFP fusion proteins, 75
  DNMT3A, 75
  DNMT3B, 72, 75
DNM1L. *See* Dynamin-like 1 protein (*DNM1L*)
dnm2. *See* Dynamin 2 (dnm2)
dNMPs. *See* Deoxyribonucleoside monophosphates (dNMPs)
DNMTs. *See* DNA-methyltransferases (DNMTs)
dNs. *See* Deoxyribonucleosides (dNs)
5′-dNs. *See* 5′-Deoxynucleotidases (5′-dNs)
dNTP. *See* Deoxyribonucleoside triphosphates (dNTP)
Dominant optic atrophy (DOA), 420
Double-labeled cells, 523–524
Double-strand break repair (DSBR), 174, 182*f*
Double-strand breaks (DSBs), 136, 173–174, 180–184, 224, 481–482
Double-stranded DNA (dsDNA), 387
dPCR. *See* Digital PCR (dPCR)
Drosophila, 99
  mtSSB, 9–10
  POL γ synthesis, 9–10
  suboscura, 313–314
Drosophila melanogaster (Fruit fly), 90, 313–315, 503
  ATP6 mutant, 314
  cells, 99–100
  CoI mutants, 314
  CoII mutant, 315
  ND2 mutants, 314–315
DRP1. *See* Dynamin-related protein (DRP1)
drp1. *See* Dynamin-related/like protein 1 (drp1)
DS_T. *See* Disease score threshold (DS_T)
DSBR. *See* Double-strand break repair (DSBR)
DSBs. *See* Double-strand breaks (DSBs)
dsDNA. *See* Double-stranded DNA (dsDNA)
dThd. *See* Deoxythymidine (dThd)
DTNB. *See* 5,5′-Dithiobis(2-nitrobenzoic acid) (DTNB)
DTYMK, 413*t*, 419
Duplex DNA, 245–247
Dynamin 2 (dnm2), 420
Dynamin-like 1 protein (*DNM1L*), 413*t*, 420
Dynamin-related protein (DRP1), 21
Dynamin-related/like protein 1 (drp1), 420
Dysfunctional mitochondria clearance, 535*f*

E

ECAR. *See* Extracellular acidification rate (ECAR)
"Eccidio delle fosse Ardeatine" victims, 146
EdU. *See* 5-Ethynyl-2′-deoxyuridine (EdU)
EE-TP therapy. *See* Erythrocytes encapsulated thymidine phosphorylase therapy (EE-TP therapy)
EF-Ts, 396
EF1α. *See* Elongation factor 1α-subunit (EF1α)
EFO. *See* Experimental Factor Ontology (EFO)
EI. *See* Exercise intolerance (EI)
ElaC ribonuclease Z 2 (ELAC2), 385–386
ELAC2 enzyme, 46–47
ELAC2. *See* ElaC ribonuclease Z 2 (ELAC2)
Electron microscopy (EM), 367
Electron transport chain (ETC), 444
ELISA. *See* Enzyme-linked immunosorbent assay (ELISA)
Elongation factor 1α-subunit (EF1α), 492
Elongation of mtDNA replication, 8–10
  mitochondrial DNA replisome, 10
  mtSSB, 9–10
  POL γ (mitochondrial DNA polymerase), 8–9
  Twinkle, 9
EM. *See* Electron microscopy (EM)
Embryonic stem cells (ESCs), 76–77, 316

EMPOP. *See* European DNA Profiling Group mitochondrial DNA population database project (EMPOP)
EMT. *See* Epithelial to mesenchymal transition (EMT)
ENA, 279
Endocytosis, 516–517, 534
Endonuclease, 11
Endonuclease G, 90–91
Endoplasmic reticulum (ER)-mitochondrial junctions, 20–21
Endoplasmic reticulum-mitochondria encounter structure complex (ERMES complex), 20–21
Ensembl, 290
ENTs. *See* Equilibrative nucleoside transporters (ENTs)
Enzyme replacement therapy, 430
  source for, 434
Enzyme-linked immunosorbent assay (ELISA), 79–80
Epithelial to mesenchymal transition (EMT), 451–452
Epstein–Barr virus, 306
Equilibrative nucleoside transporters (ENTs), 13–15
ERMES complex. *See* Endoplasmic reticulum-mitochondria encounter structure complex (ERMES complex)
Errors detection, 162
Erythrocytes encapsulated thymidine phosphorylase therapy (EE-TP therapy), 432–434
Escherichia coli (*E. coli*), 73–75, 176–177, 184, 206–207
  K-12, 78
ESCs. *See* Embryonic stem cells (ESCs)
ETC. *See* Electron transport chain (ETC)
5-Ethynyl-2′-deoxyuridine (EdU), 16*b*
Eukaryotic life, 515
European DNA Profiling Group mitochondrial DNA population database project (EMPOP), 161–162
Exercise intolerance (EI), 356
Exo-Sap. *See* Exonuclease I and shrimp alkaline phosphatase (Exo-Sap)
Exomiser, 292
Exonuclease G (EXOG), 11
Exonuclease I and shrimp alkaline phosphatase (Exo-Sap), 165
Experimental Factor Ontology (EFO), 290
Extracellular acidification rate (ECAR), 330–331
Extracellular nucleotide, 518–522
Extraction methods, 150
Eye-visible dyes, 245–247

F

F-PCR. *See* Fluorescent PCR (F-PCR)
Fas-activated serine/threonine kinase (FASTK), 46, 49, 60–61
  FASTK3, 49–50
FASTKD2, mutations in, 393
FASTQ, 253–255, 260
FBI mtDNA Population Database, 162
FBXL4, 413*t*, 421
FCCP. *See* Carbonyl cyanide-ptrifluoromethoxyphenylhydrazone (FCCP)
FDA. *See* Food and Drug Administration (FDA)
FEN1. *See* Flap-structure specific endonuclease 1 (FEN1)
FIBER-FISH technique, 136
First peopling of Americas, 120–121
FISH technique. *See* Fluorescence in situ hybridization technique (FISH technique)
FLAIR. *See* Fluid-attenuated inversion recovery (FLAIR)
Flap-structure specific endonuclease 1 (FEN1), 11, 376–379, 412
Fluid-attenuated inversion recovery (FLAIR), 366
Fluorescence in situ hybridization technique (FISH technique), 136
Fluorescent dye, 245–247
Fluorescent PCR (F-PCR), 245–247, 463
3′ Fluorescently labeled oligos, 245–247
Fluoxetine, 533
55S-fMettRNA ^{Met}-mtIF2 complex, 55–56
*Fok*I, 491
Food and Drug Administration (FDA), 429
Forensics
  alignment, 153–154
  DNA profiling, 149
  genetics, 145
  guidelines and recommendations, 163–164
  mtDNA
    population databases used in, 161–163
    typing, in historical forensic identification, 146–149
  notation for, 154–155
Fruit fly. *See Drosophila melanogaster* (Fruit fly)
Functional studies on mtDNA mutations
  animal models, 311–320
  methods for assessment of functional defects induced by mtDNA alterations, 321–339
    ATP production, 333–335
    Blue Native PolyAcrylamide Gel Electrophoresis, 337–339
    mitochondrial membrane potential determination, 332–333
    OXPHOS complexes activity, 322–328
    oxygen consumption, 328–331
    reactive oxygen species measurement, 335–337
  models for mtDNA mutations study, 306–311
    cybrids, 306–309
    human primary cell lines, 306
    patient-specific induced pluripotent stem cells, 309–310

Functional studies on mtDNA mutations (*Continued*)
yeast, 310–311

G

G-quadruplexes (G4s), 4–5, 38, 60–61
G-rich sequence binding factor 1 (GRSF1), 59–60
G13513A-mitoTALEN, 490
G3P. *See* Glycerol-3-phosphate (G3P)
G4s. *See* G-quadruplexes (G4s)
Garden worm (*Caenorhabditis elegans*), 90, 311–313
GARS. *See* Glycyl-ARS (GARS)
GenBank, 279
GenCode track, 140–141
Gene editing, 339
Gene therapy, 435, 481–482
Genetic testing, 367–369
Genome assembly, 259–260
Genome editing, 481, 499–500
Genomes of the Netherlands (GoNL), 98–99
Genomic revolution, 265
"Germline Genetic Bottleneck" theory, 95–97, 96*f*
Germline segregation of mtDNA mutations and genetic bottleneck, 94–97
Germline transmission, mitoTALEN in, 490
GFM1, mutations in, 396
GFM2, mutations in, 397
GFPs. *See* Green fluorescent proteins (GFPs)
Global genome nucleotide excision repair (GG-NER), 178–179
GLUT1, 450
Glutaminase inhibitors (GLS inhibitors), 462
Glutamine metabolism, 455–456
Glutathione peroxidase (GPX4), 231
Glycerol-3-phosphate (G3P), 329
Glycolysis, 452, 536
Glycolysis Stress Test, 325
Glycolytic Rate Assay, 325
Glycolytic ρ^0 tumor models, 523
Glycyl-ARS (GARS), 389
GMP. *See* Good manufacturing practice (GMP)
Gomori trichrome, 367
GoNL. *See* Genomes of the Netherlands (GoNL)
Good manufacturing practice (GMP), 429
GPX4. *See* Glutathione peroxidase (GPX4)
Green fluorescent proteins (GFPs), 73
GRSF1. *See* G-rich sequence binding factor 1 (GRSF1)
GTP
GTP-bound IF2mt, 56
hydrolysis, 56–57
GTP-bound mitochondrial elongation factor Tu (mtEFTu: GTP), 394
GTPBP3. *See* Mitochondrial GTP-binding protein 3 (GTPBP3)
Guanosine triphosphatases (GTPases), 53–54
GUSB. *See* β-glucuronidase (GUSB)

H

H-strand promoter (HSP), 5, 35–36
HSP1, 40–41
HSP2, 40–41
H-strand replication (O_H), 376–379
H-strand synthesis (O_H), 5–6
H3f1, 115–117
H3f2, 115–117
HAM. *See* Hearing loss-ataxia-myoclonus (HAM)
Hammerhead ribozyme (HHR), 508
HaploGrep, 294
Haplogroups, 111, 277–278
analysis, 266
assignment, 162
in MITOMAP, 282–283
first peopling of Americas, 120–121
nomenclature of human mtDNA, 114–115
"Out of Africa Exit", 118–120
PCR in mtDNA world, 113–114
peopling of island in Mediterranean Sea, 121–123
RFLP studies, 112–113
survey of entire mitogenomes, 115–118
Harding syndrome, 357–358
HCT116 mtDNA regions, 73–75
HDR. *See* Homology-directed repair (HDR)
Hearing loss-ataxia-myoclonus (HAM), 362
Heavy strands (H strands), 71, 376
HeLa cells, 15–16
Hematopoietic stem cell transplantation (HSCT), 431–432
Hemodialysis, 430
Heredity and segregation of mtDNA, 87
characteristics of human nuclear and mitochondrial genomes, 88*t*
general principles of mtDNA segregation, 87–90
germline segregation of mtDNA mutations and genetic bottleneck, 94–97
mtDNA mutations, 93–94
paternal leakage during mtDNA inheritance, 92–93
purifying selection against mtDNA mutations in germline, 97–100
somatic mtDNA mutations and clonal expansion, 100–102
uniparental maternal inheritance of mitochondrial DNA, 90–92
Hermaphrodite worms, 311–313
Heteroplasmy, 3, 87–88, 93–94, 155–157, 243
bioinformatics strategies to detect, 259–266
heteroplasmic mtDNA mutations, 94, 365
heteroplasmic mutations, 97
HFEA. *See* Human Fertilisation and Embryology Authority (HFEA)

HHR. *See* Hammerhead ribozyme (HHR)
High mobility group (HMG), 17–18
High performance liquid chromatography (dHPLC), 463
High scoring pairs (HSPs), 132–133
High throughput sequencing, 134–135
High-quality mtDNA sequence databases, 161
High-resolution respirometry, 330
High-resolution RFLP analysis, 114
High-throughput next-generation sequencing, 375–376
HLA. *See* Human leukocyte antigen (HLA)
5hmC. *See* 5-Hydroxymethylcytosine (5hmC)
HMG. *See* High mobility group (HMG)
HmtDB, 284–286
  API, 286
  of HmtDB query criteria, 286*t*
  information with each HmtDB genome, 285*t*
HmtNote, 294
HmtPhenome, 289–290
  databases, 279
HmtVar, 287–289
  databases, 279
  query criteria implemented in, 288*t*
  tiers of pathogenicity of nonsynonymous and tRNA mtDNA variants, 288–289
  variants pathogenicity assessment, 288–289
Holliday-junction resolvases, 43
Homologous interaction, 296–297
Homologous recombination (HR), 181, 485
Homology-directed repair (HDR), 503
Homoplasmy, 87–88, 93–94, 243
Horizontal transfer of genetic material, 515
HPO. *See* Human Phenotype Ontology (HPO)
HR. *See* Homologous recombination (HR)
HSCT. *See* Hematopoietic stem cell transplantation (HSCT)
*HSD*17*B*10 gene, 386
HSP. *See* H-strand promoter (HSP)
HSPs. *See* High scoring pairs (HSPs)
hSuv3p, 387
Human DNA Pol γ, 177
Human female germline development, 95*b*
Human Fertilisation and Embryology Authority (HFEA), 526–527
Human leukocyte antigen (HLA), 431
Human mitochondrial DNA (Human mtDNA)
  genomes and variants
    Human MitoCompendium, 284–290
    MITOMAP, 279–283
    primary databases, 279
  MSeqDR Consortium, 291–292
  nuclear encoded mitochondrial gene databases, 295–297
  repair
    BER, 175–178
    of bulky lesions, 178–180, 179*f*
    DSBs, 180–184
    mismatch repair, 184–185
    translesion synthesis, 185–186
  specialized human mitochondrial databases, 293
  tools for variant annotations, 293–295
    HaploGrep, 294
    HmtNote, 294
    MitImpact3D, 295
    PON-mt-tRNA, 295
  variability, 277–278
Human mitochondrial proteome, 295
Human MitoCompendium, 279
  HmtDB, 284–286
  HmtPhenome, 289–290
  HmtVar, 287–289
Human Phenotype Ontology (HPO), 290
Human primary cell lines, 306
Human TP-deficient cell, 427–428
Humanin, 140–141
Huntington's disease, 531–532
HUPRA. *See* Hyperuricemia, Pulmonary hypertension, Renal failure in infancy, and Alkalosis (HUPRA)
HV1. *See* Hypervariable region I (HV1)
HV2. *See* Hypervariable region II (HV2)
HV3. *See* Hypervariable region III (HV3)
HVSI. *See* Hypervariable segment I (HVSI)
Hybridization, 247, 258
Hydrogen peroxide (H_2O_2), 173
Hydrolytic activity of ATP synthase, 325
Hydrophobic membrane proteins, 53, 58
Hydroxyl radicals ($^{\bullet}OH$), 173
3′-Hydroxyl (3′-OH), 251–252
5-Hydroxymethylcytosine (5hmC), 73
Hyperuricemia, Pulmonary hypertension, Renal failure in infancy, and Alkalosis (HUPRA), 391–392
Hypervariable region I (HV1), 151–152
Hypervariable region II (HV2), 151–152
Hypervariable region III (HV3), 151
Hypervariable segment I (HVSI), 113–114
Hypoxic stress, 448*f*, 450–451

I

ICGC. *See* International Cancer Genome Consortium (ICGC)
ICT1, 394
IF1. *See* Initiation factor 1 (IF1)
IGA. *See* In-Gel Activity (IGA)
Illumina, 252–253
  Miseq, 256
  sequencing, 253–255

IMM. *See* Inner mitochondrial membrane (IMM)
Immunocapture-based assays, 325–328
IMPI. *See* Integrated Mitochondrial Protein Index (IMPI)
IMS. *See* Intermembrane space (IMS)
In silico human NumtS detection based on reference genomes, 132–134
In silico modeling, 88
In vitro MGME1, 11
In vitro NumtS identification, 135–136
In vivo models, 305–306
In-Gel Activity (IGA), 337–339
IND. *See* Investigational New Drug (IND)
INDELs. *See* Insertions or deletions (INDELs)
Induced pluripotent stem cells (iPSCs), 235–236, 305–306, 309–310, 423–424, 490
  mitoTALEN, 490
  technology, 339
Infantile-onset *TK2*-related disease, 382
Initiation factor 1 (IF1), 54
Initiation factor 2 (IF2), 54
Initiation factor 3 (IF3), 54
Inner mitochondrial membrane (IMM), 381–382
  insertion of mtDNA-encoded
    OXPHOS proteins, 397–398
Inorganic phosphate (Pi), 325
INSDC. *See* International Nucleotide Sequence Database Collaboration (INSDC)
Insertions or deletions (INDELs), 485, 491–492
Integrated Mitochondrial Protein Index (IMPI), 296
Intercellular mitochondrial transfer, 538
Intermembrane space (IMS), 387
InterMine, 296
Internal positive control (IPC), 150–151
International Cancer Genome Consortium (ICGC), 466
International Nucleotide Sequence Database Collaboration (INSDC), 279
International Society for Forensic Genetics (ISFG), 155, 163, 163*t*
International Union of Pure and Applied Chemistry code (IUPAC code), 154
Interspecies genomic variability, 131
Intraspecies genomic variability, 131
Intracellular calcium homeostasis, 518–522
Intracerebroventricular injection, 532
Intraperitoneal injection (IP injection), 483
Investigational New Drug (IND), 429
Ion semiconductor technologies, 256
Ion Torrent sequencing, 253–255
Ionizing radiation (IR), 461
IP injection. *See* Intraperitoneal injection (IP injection)
IPC. *See* Internal positive control (IPC)
iPSCs. *See* Induced pluripotent stem cells (iPSCs)
IR. *See* Ionizing radiation (IR)
Ischemic heart disease, 528–531
Ischemic stroke, 531–533
ISFG. *See* International Society for Forensic Genetics (ISFG)
Isocitrate dehydrogenases 1 (IDH1), 455–456
Isocitrate dehydrogenases 2 (IDH2), 455–456
Isolated PEO, 363
IUPAC code. *See* International Union of Pure and Applied Chemistry code (IUPAC code)

J

Jelly-derived MSCs, 526

K

Kearns-Sayre syndrome (KSS), 356
Ketamine, 533
Kilo base pairs (kbp), 4–5
Kinases, 53–54
KRAB repressor domain (KOX1), 502–503
Krebs cycle, 447–448
Krüppel-associated box (KRAB), 502–503
KSS. *See* Kearns-Sayre syndrome (KSS)

L

L-strand promoter (LSP), 5, 35–36, 43–44
L-strand replication (O_L), 376–379
L3 mtDNAs, 118–119
L3 nodal haplotype, 120
Lactate dehydrogenase (LDH), 325
LARS. *See* Lysyl-ARS (LARS)
Last Hot-Cycle PCR, 245–247
Lauryl maltoside, 337–339
LC/MS. *See* Liquid chromatography/mass spectrometry (LC/MS)
LDH. *See* Lactate dehydrogenase (LDH)
Leber-dystonia, 357
Leber's Hereditary Optic Neuropathy (LHON), 199, 244–245, 309–310, 353–355, 357–358, 487–489
Leigh syndrome (LS), 310–311, 358–359, 366, 388
Leigh Syndrome French-Canadian type (LSFC type), 388
Length heteroplasmy (LHP), 155, 157*f*
Lentiviruses, 435
Leucine-rich PPR cassette (LRPPRC), 46, 48–49, 388
  LRPPRC–SLIRP, 48–49
  mutations in, 388
LHON. *See* Leber's Hereditary Optic Neuropathy (LHON)
LHP. *See* Length heteroplasmy (LHP)
LIG3. *See* DNA ligase 3 (LIG3)
Ligation, 11
Light strands (L strands), 71, 154, 376
Likelihood ratios (LRs), 161

Liquid chromatography/mass spectrometry (LC/MS), 79–80
experimental flow of mtDNA preparation for, 81*f*
5mC in mtDNA by nucleoside, 79–81
Liver transplant, 434
lncCYTB gene, 45
lncND5 gene, 45
lncND6 gene, 45
Locus Specific DataBase (LSDB), 292
Lon protease, 18
"Long-patch" BER (LP-BER), 175
Long-range PCR method (LR-PCR method), 245, 247, 258–259
LP-BER. *See* "Long-patch" BER (LP-BER)
LR-PCR method. *See* Long-range PCR method (LR-PCR method)
LRPPRC. *See* Leucine-rich PPR cassette (LRPPRC)
LRs. *See* Likelihood ratios (LRs)
LS. *See* Leigh syndrome (LS)
LSDB. *See* Locus Specific DataBase (LSDB)
LSFC type. *See* Leigh Syndrome French-Canadian type (LSFC type)
LSP. *See* L-strand promoter (LSP)
Lyase activity, 177
Lysyl-ARS (LARS), 389

M

m.5024C > T cells, 509–510
m.8993T > G cells, 508
Macromolecule and energy biosynthesis, 449*f*
Magnetic resonance imaging (MRI), 366
Mammalian mitochondrial genome manipulation with mtZFNs, 504–510, 507*f*
mRNAs, 52–53
Mammalian mtDNA, 71
epigenetic features of, 81–82
mutations, 499
Mammalian target of rapamycin pathway (mTOR pathway), 424–427
Mammalian TFAM, 384
Mass spectrometry (MS), 52*b*, 59
Massive parallel sequencing (MPS), 153
of full mitochondrial genome, 153
Maternally inherited diabetes and deafness (MIDD), 358
Maternally inherited Leigh syndrome (MILS), 358–359, 482–483
5mC. *See* 5-Methylcytosine (5mC)
MCAO. *See* Middle cerebral artery (MCAO)
MCHS. *See* Myocerebrohepatopathy spectrum (MCHS)
MCK. *See* Muscle creatine kinase (MCK)
McrBC endonuclease, mtDNA methylation estimation with, 78–79, 79*f*
meDIP. *See* Methylated DNA immunoprecipitation (meDIP)
Mediterranean Sea, peopling of island in, 121–123
MEFs. *See* Mouse embryonic fibroblasts (MEFs)
Mega NUMTs, 364
MELAS. *See* Mitochondrial encephalomyopathy with lactic acidosis and stroke-like episodes (MELAS)
MELAS/MIDD. *See* Mitochondrial Encephalomyopathy Lactic acidosis and stroke-like episodes/Maternal Inherited Diabetes and Deafness (MELAS/MIDD)
MEMSA. *See* Myoclonic epilepsy myopathy sensory ataxia (MEMSA)
Mentha, 295
MERRF. *See* Myoclonic epilepsy with ragged red fibers (MERRF)
Mesenchymal stem cells (MSCs), 516–517
Metabolic
adaptation, 447–450, 448*f*, 467*t*
decompensation, 359
labeling, 52*b*
Metastatic
niches, 452
progression, 451–452
5-Methyl-dCTP, 82
Methylated DNA immunoprecipitation (meDIP), 73–75
5-Methylcytosine (5mC), 72–73, 76
in mtDNA by nucleoside LC/MS, 79–81
5-Methyldeoxycytidine (m^5dC), 80–81
Methylene blue, 245–247
5′-Methylguanosine cap, 49–50
5-Methyltetrahydrofolate (5MTHF), 366
METTL15 protein, 392
MFN2 heterozygous variant, 413*t*, 420–421
MGME1. *See* Mitochondrial genome maintenance exonuclease 1 (MGME1)
MHRA. *See* Rules and Guidance for Pharmaceutical Manufacturers 2007 (MHRA)
Mice, 315–320
Mgme1, 319–320
mt-Co1, 316
mt-Nd6, 316–317
mt-tA ($tRNA^{Ala}$), 317–318
mt-tK ($tRNA^{Lys}$), 317
mtDNA deletions, 315–316
PolgA, 318–319
Twnk, 319
MICOS. *See* Mitochondrial contact site and cristae organizing system (MICOS)
Microarrays, 244, 252–253, 254*f*
Microhomology-mediated end joining (MMEJ), 181–182
Microrespirometry on multiwell plate, 330–331
MicroRNA hypothesis, 140–141
Microtubes, 522–523
MIDD. *See* Maternally inherited diabetes and deafness (MIDD)

Middle cerebral artery (MCAO), 532
MILS. *See* Maternally inherited Leigh syndrome (MILS)
MIM protein. *See* Mitochondrial inner membrane protein (MIM protein)
MIRAS. *See* Mitochondrial recessive ataxia syndrome (MIRAS)
Miro1. *See* Mitochondrial GTPase-1 (Miro1)
Miro2, 524–525
Mismatch repair (MMR), 184–185, 229
Mitchell's chemiosmotic theory, 332–333
MitImpact3D, 295
Mito-nuclear crosstalk, 534–537, 537*f*
mito-PstI. *See* Mitochondrial-targeted PstI (mito-PstI)
Mito-stress signals, 537*f*
mito-*XhoI*. *See* Mitochondrial-targeted *XhoI* (mito-*XhoI*)
Mitocarta, 295–296
MitoCarta2.0, 295–296
MitoCeption, 527–528, 534
MitoChip array, 252–253
Mitochondria(l), 35, 335–337, 524–525
  abnormalities, 518–522
  ATP levels, 528
  base calling analysis, 253
  biogenesis, 42–43, 207, 422–424, 453–454
  defects
    in mitochondrial dynamics, 420–421
    in mitochondrial nucleotides pool balance, 418–420
  degradosome, 387
  donation therapy, 526–527
    to prevent mitochondrial diseases in offspring, 526–527
  dysfunction, 517
    compensatory mechanisms to overcome, 455–456
  fission, 420, 531–532
  fraction, 80–81
  fragmentation, 99, 531–532
  fusion, 420
  genome, 151, 375, 534–536
    editing, 491
  IF2, 56
  membrane potential
    determination, 332–333
    TRMR-based assay for mitochondrial membrane potential, 334*b*
  mitochondria-containing microvesicles, 532
  mitochondrially targeted fluorescent dyes, 516–517
  mRNAs, 45
  phosphorylation system, 329–330
  phylogenetic analysis, 266
  pseudogenes, 131–132
  R-loop, 7
  reactive oxygen species, 102
  ribosomes, 35–36
  RNA degradosome, 60–61
  RNA-binding proteins and RNA biology, 45–50
    mitochondrial RNA chaperones and mRNA stability, 48–49
    mitochondrial RNA maturation, 47–48
    mitochondrial RNA processing, 46–47
    mitochondrial RNA translation activators, 49–50
  sRNAs, 45
  transcription, 40, 45
    defects, 384–385
    mechanism, 383–384
  transplantation, 527–528, 531
    in vitro and in vivo, 529*t*
  variant calling, 260–265
Mitochondrial aminoacyl-tRNA synthetases (mttRNA aa-RS), 391–392
  mutations in, 391–392
Mitochondrial contact site and cristae organizing system (MICOS), 20–21
Mitochondrial DNA (mtDNA), 3, 17, 35, 71–72, 111, 145, 200, 243, 277–278, 305–306, 353–355, 375, 411, 443, 481, 499, 515
  amplicons, 114
  approach to reducing mutant mtDNA in heteroplasmic cells, 482*f*
  bioinformatics strategies to detect mitochondrial variants and heteroplasmy, 259–266
  bisulfite sequencing analysis, 76–78
  and cancer treatment
    chemotherapy, 457–460
    interventions in cancer therapy, 462–463
    radiotherapy, 461–462
  cell-to-cell transfer of mitochondria with, 516–517
  challenges in mitochondrial variant studies, 257–259
    NumtS contamination, 259
  characteristics, 3
  clearance of toxic metabolites, 430
  compartmentalization of gene expression, 58–61
  control region, 37–39
    mitochondrial displacement loop, 37–38
    switch between replication and transcription, 38–39
  coordination of mitochondrial DNA replication and transcription, 36–39
  cytosine methylation in, 72–76
  defects
    in mitochondrial dynamics, 420–421
    in mitochondrial nucleotides pool balance, 418–420
    in mtDNA replisome, 412–418
    in nucleoid proteins, 421–422
  deletions, 224–225, 315–316, 319
    and aging, 224–226
    expansion of, during aging, 224–225
    origin of, 224
    role in aging, 225–226

syndrome, 12–13
disease-tailored therapies, 427–435
EE-TP therapy, 432–434
enzyme replacement, 430
epigenetic features of mammalian mtDNA, 81–82
experimental therapies, 422
gene expression, 377*f*
gene therapy, 435
genetics from perspective of aging, 223–224
haplogroup nomenclature of human mtDNA, 114–115
hematopoietic stem cell transplantation, 431–432
heteroplasmy, 204, 365, 481
and segregation, 89*f*
human mtDNA variants detection, 243–244
insights from next generation sequencing and bioinformatics approaches
influence of big data, 468
methodological recommendations, 466–468
technical pitfalls and false discoveries of past, 465–466
interpretation of mtDNA results, 157–161
sequence comparison, 157–158, 159*t*
statistical evaluation, 158–161
isolation, 258–259
lineage of Tsar Nicholas II, 147*f*
liver transplant, 434
maintenance, 21, 411
as markers
of clonality, 464*f*
of tumor progression, 463–465
5mC in mtDNA by nucleoside LC/MS, 79–81
methylation estimation with McrBC endonuclease, 78–79, 79*f*
mitochondrial dNTP supply, 12–15, 14*f*
mitochondrial nucleoids, 15–21, 19*f*
composition, 17–18
localization, 20–21
nucleoid-associated proteins, 17–18
segregation, 21
topology, 18–20
mitochondrial RNA-binding proteins and RNA biology, 45–50
mitochondrial transcriptome, 44–45
mitochondrial translation, 50–58
mitoTALEN use to target, 484–485
mtDNA-Server tool, 260–262
mtDNAmanager, 162–163
mutations, 93–94
criteria to designate primary mtDNA mutation, 200
and hypoxic stress, 450–451
and metabolic adaptation, 447–450, 448*f*, 467*t*
and metastatic progression, 451–452
compensatory mechanisms to overcome mitochondrial dysfunction, 455–456
effects of respiratory CI mtDNA mutations on tumor progression, 445*f*
fate of severely pathogenic mtDNA mutations, 452–457
functional effects, 447–452
germline segregation of mtDNA mutations and genetic bottleneck, 94–97
in oncocytomas, 456–457
landscape, 443–446
molecular mechanisms behind selection and purification, 453–455
oncojanus gene, 446*f*
pathogenic mtDNA mutations, 454*f*
purifying selection against mtDNA mutations in germline, 97–100
nucleoids, 59
NumtS in mtDNA sequencing and disease, 140
old and new mitochondrial theories of aging, 221–223, 222*f*, 226*f*
organization of human mitochondrial genome, 4–5
PCR in mtDNA world, 113–114
pharmacological approaches
targeting mitochondrial biogenesis, 422–424
targeting mTOR pathway, 424–427
therapies tested in vitro and/or in vivo, 425*t*
phylogenetic trees, 114
platelet infusion, 431
point mutations, 227–233
during aging, 227
origin of mtDNA mutations, 231–232
oxidative stress *vs.* replication errors, 227
point mutations, 232–233
polymerase gamma, 90–91
population databases used in forensics, 161–163
quantification by real-time PCR, 150–151
replication, 5–12, 6*f*, 36, 485, 491
defects of mtDNA replication, 379–381
elongation of mtDNA replication, 8–10
fork proteins, 8*f*
mechanisms, 5–6, 376–379
mitoTALEN to study, 490
priming, 7–8
proteins in mtDNA replication, 12
termination of mtDNA replication, 10–12
sequencing in forensic practice, 149–153
somatic mtDNA mutations and aging, 233–235

Mitochondrial DNA (mtDNA) (*Continued*)
tissue-specific consequences of, 234–235
stem cells, 235–236
techniques for detecting mitochondrial variants, 244–257
broad-spectrum techniques for variants detection, 250–257
PCR–based methods and mtDNA rearrangements, 245–250
therapeutic approaches, 424*f*
transcription, 40–44, 41*f*
elongation, 43
initiation, 41–43
termination, 43–44
typing, in historical forensic identification, 146–149
uniparental maternal inheritance, 90–92
use of specific endonucleases to target, 481–483, 484*f*
Mitochondrial Disease Sequence Data Resource Consortium (MSeqDR Consortium), 291–292
MvTool, 292
Mitochondrial diseases. *See* Neuromuscular disorders
Mitochondrial elongation factor G1 (mtEFG1), 57, 394
Mitochondrial elongation factor G2 (mtEFG2), 57–58, 394
Mitochondrial elongation factor Tu (mtEF-Tu), 56–57
Mitochondrial elongation factor-Ts (mtEF-Ts), 56–57, 394
Mitochondrial Encephalomyopathy Lactic acidosis and stroke-like episodes/Maternal Inherited Diabetes and Deafness (MELAS/MIDD), 426
Mitochondrial encephalomyopathy with lactic acidosis and stroke-like episodes (MELAS), 71, 309–310, 359–360, 363, 490
"Mitochondrial Eve", 113, 119–120
Mitochondrial genome maintenance exonuclease 1 (MGME1), 11, 36, 412, 413*t*, 418
mutations in, 381
Mitochondrial GTP-binding protein 3 (GTPBP3), 390
Mitochondrial GTPase-1 (Miro1), 524–525
Mitochondrial initiation factor 2 (mtIF2), 54
Mitochondrial initiation factor 3 (mtIF3), 54, 393–394
Mitochondrial inner membrane protein (MIM protein), 18
Mitochondrial localization signal (MLS), 485–487
Mitochondrial methionyl-tRNA formyltransferase (MTFMT), 54, 395
mutations in, 395
Mitochondrial movement between mammalian cells
cell-to-cell transfer of mitochondria with mtDNA, 516–517
mito-nuclear crosstalk, 534–537, 537*f*
mitochondrial transfer mechanisms, 534, 535*f*
translational benefits of mitochondrial transfer, 517–534
mitochondrial transfer between cells, 517–527, 518*f*
transfer of isolated mitochondria, 527–534
Mitochondrial Myopathy, Lactic acidosis, and Sideroblastic Anemia (MLASA), 389–392
Mitochondrial myopathy and cardiomyopathy (MMC), 360
Mitochondrial neurogastrointestinal encephalomyopathy (MNGIE), 418–419, 430–431
Mitochondrial recessive ataxia syndrome (MIRAS), 379–380
Mitochondrial ribosome release factor (mtRRF), 394
Mitochondrial RNA granules (MRGs), 20, 59–60
Mitochondrial RNA polymerase (mtRNAP), 412
Mitochondrial RNA polymerase (POLRMT), 7, 36, 38, 40–44, 376–379, 383–384
Mitochondrial RNase P protein 3 (MRPP3), 46–47
Mitochondrial RNase P proteins I–III (MRPP1, 2, and 3), 386
Mitochondrial single-stranded DNA-binding protein (mtSSB), 6, 8–10, 18, 20–21, 59–60, 376–379
Mitochondrial targeting signal (MTS), 73, 492, 504
Mitochondrial transcription elongation factor (TEFM), 7, 36, 40, 43
Mitochondrial transcription factor A (TFAM), 17–18, 36, 38, 40–43, 59, 383–384, 413*t*, 421
mutations in, 385
Mitochondrial transcription factor B2 (TFB2M), 38, 40–42, 383–384
mutations in, 385
Mitochondrial transfer
between cells, 517–527, 518*f*
of isolated mitochondria, 527–534
behavioral disorders, 533
cancer sensitization to treatment, 534
ischemic heart disease, 528–531
neurodegenerative disorders and ischemic stroke, 531–533
mechanisms, 534, 535*f*
mitochondrial transfer/transplantation consequences, 534–537
into normal cells, 524–526

mitochondrial donation therapy, 526–527
oocytoplasmic transfer and nuclear transfer, 527*f*
translational benefits of, 517–534
into tumor cells, 518–524
in vitro and in vivo, 519*t*
in vivo mitochondrial transfer experiments, 524*f*
Mitochondrial translation, 50–58
cotranslational membrane insertion of newly synthesized polypeptides, 58
elongation, 56–57
experimental approaches for study of, 52*b*
initiation, 54–56
machinery, 51–54
mitoribosome biogenesis, 53–54
mitoribosome structure, 51–53
mechanism, 393–394
mitochondrial protein synthesis, 54–58
in mitoribosomes, 55*f*
termination and mitoribosome recycling, 57–58
Mitochondrial translation optimization 1 (MTO1), 390
mutations in, 390
Mitochondrial translation release factor A (mtRF1a), 57, 394
Mitochondrial tRNA Informatics Predictor (MitoTIP), 283
Mitochondrial-targeted PstI (mito-PstI), 315–316
Mitochondrial-targeted *XhoI* (mito-*XhoI*), 314
Mitochondrially targeted zinc finger nucleases (mtZFNs), 499–500
chimeric ZFPs, 502–504
designer zinc fingers, 500–502
mammalian mitochondrial genome manipulation with, 504–510, 507*f*
in vivo use, 509–510
Mitochondriopathies, 517
MitoDel pipeline, 260
Mitofusins, 21, 99
MFN1, 21, 420
MFN2, 21, 420
Mitogenomes, survey of, 115–118
MitoInteractome, 295
MITOMAP, 279–283
allele search function in, 283
haplogroup assignment in, 282–283
MitoTIP, 283
variant status in, 281–282
MITOMASTER tool, 279–280
analyzing mtDNA variability using, 283
MitoMiner, 296
Mitophagy, 453–454
MitoProteome, 295–297
Mitoribosome proteins (MRPs), 51
mutations in, 394–395
Mitoribosomes, 35–36, 53, 58
mitochondrial translation in, 55*f*
recycling, 57–58
structure, 51–53
MitoStress Test kit, 330–331
MitoTALEN, 481
construction and expression of plasmids, 486*f*
future as therapy, 493
gene size, 492–493
in germline transmission, 490
in heteroplasmic mouse model, 489–490
and iPSCs, 490
MitoTevTAL nuclease, 491
pros and cons for gene therapy
easy design of new recognition sites, 492
off-target sequences in nucleus, 491–492
specificity and mtDNA depletion, 491
structure, 485–487
to study mtDNA replication, 490
targeting mutations in cybrids, 487–489
use to target mtDNA, 484–485
MitoTIP. *See* Mitochondrial tRNA Informatics Predictor (MitoTIP)
MitoTracker CMX-Ros, 528
MitoTracker Red CMX-Ros, 532, 534
MitVarProt algorithm, 284–285
mL45, 53, 58
MLASA. *See* Mitochondrial Myopathy, Lactic acidosis, and Sideroblastic Anemia (MLASA)
MLS. *See* Mitochondrial localization signal (MLS)
MM cells. *See* Multiple myeloma cells (MM cells)
MMC. *See* Mitochondrial myopathy and cardiomyopathy (MMC)
MMEJ. *See* Microhomology-mediated end joining (MMEJ)
MMR. *See* Mismatch repair (MMR)
MNGIE. *See* Mitochondrial neurogastrointestinal encephalomyopathy (MNGIE)
Molecular genetics of mitochondrial DNA SLSDs and point mutations, 364–365
Molecular techniques, 308
Mouse embryonic fibroblasts (MEFs), 489
Mouse mtDNA, 76–77
MPS. *See* Massive parallel sequencing (MPS)
*Mpv*17
knockout mouse model, 381–382, 413*t*, 419–420, 435
mutations in, 383
MRGs. *See* Mitochondrial RNA granules (MRGs)
MRI. *See* Magnetic resonance imaging (MRI)
*MRM*1 mutation, 392
*MRM*2, mutations in, 392–393
*MRM*3 mutation, 392
*MRPL*3 protein, 394–395
*MRPL*12 protein, 394–395
*MRPL*44 protein, 394–395
MRPP3. *See* Mitochondrial RNase P protein 3 (MRPP3)
MRPs. *See* Mitoribosome proteins (MRPs)
*MRPS*2 protein, 394–395

MRPS7 protein, 394–395
MRPS14 protein, 394–395
MRPS16 protein, 394–395
MRPS22 protein, 394–395
MRPS23 protein, 394–395
MRPS25 protein, 394–395
MRPS28 protein, 394–395
MRPS34 protein, 394–395
MRPS39 protein, 394–395
MS. *See* Mass spectrometry (MS)
mS27 (mitoribosome SSU protein), 46
mS39 (mitoribosome SSU protein), 46, 49–50, 56
MSCs. *See* Mesenchymal stem cells (MSCs)
MSeq-OpenCGA, 292
MSeqDR Consortium. *See* Mitochondrial Disease Sequence Data Resource Consortium (MSeqDR Consortium)
MSTO1, 413*t*, 421
5MTHF. *See* 5-Methyltetrahydrofolate (5MTHF)
MT-ATP6 transcripts, 385
MT-CO1 transcripts, 45, 316, 385
MT-CO2 transcripts, 459
MT-CYB transcripts, 385, 448–450
mt-mRNA maturation and turnover, 387–388
  defects, 388
MT-ND4L transcripts, 45, 385
MT-ND5 transcripts, 385, 448–450, 459
MT-ND6 transcript, 47, 316–317, 387, 451
mt-rRNA
  defects of mt-rRNA maturation, modification, and stability, 392–393
  maturation, 392
MT-TI gene, 360
mt-tK (tRNA^{Lys}), 317
MT-TL1 gene, 360, 453
mt-tRNA maturation, 389
  defects, 389–392
MTATP8 mRNAs, 45
MTCO3 mRNA, 49–50, 56
mtDNA. *See* Mitochondrial DNA (mtDNA)
mtDNA mutations
  clinical and biochemical correlates, 201
  genetic drift, 206–207
  mitochondrial DNA
    abnormalities, 195–200
    maintenance disorders, 208–209
    network and implications for heteroplasmy, 210–213
    primary mitochondrial DNA mutants, 195–200
    selection, 207
  mitochondrial genetic rules, 201–202
  ribonucleotide incorporation, 209–210
  selection and counterselection of deleterious mtDNA variants, 202–206, 212*f*
    metabolic configuration and nutrient availability, 205–206
    phenotypic selection, 202
    propagation of dysfunctional mitochondria, 202–203
    selfish mechanisms, 203–205, 205*f*
  stable heteroplasmy, 208
mtEF-Tu. *See* Mitochondrial elongation factor Tu (mtEF-Tu)
mtEFG1. *See* Mitochondrial elongation factor G1 (mtEFG1)
mtEFTu:GTP. *See* GTP-bound mitochondrial elongation factor Tu (mtEFTu:GTP)
MTERF, 489–490
  MTERF1, 38, 40–41, 43–44, 384
MTFMT. *See* Mitochondrial methionyl-tRNA formyltransferase (MTFMT)
mtIF2. *See* Mitochondrial initiation factor 2 (mtIF2)
MTND1, 45
MTND6 mRNA, 49, 60–61
MTO1. *See* Mitochondrial translation optimization 1 (MTO1)
MToolBox pipeline, 260–262
mTOR pathway. *See* Mammalian target of rapamycin pathway (mTOR pathway)
MTPAP, mutations in, 388
mtRF1a. *See* Mitochondrial translation release factor A (mtRF1a)
mtRNAP. *See* Mitochondrial RNA polymerase (mtRNAP)
mtRRF. *See* Mitochondrial ribosome release factor (mtRRF)
mtRRF1, 57–58
MTS. *See* Mitochondrial targeting signal (MTS)
mtSSB. *See* Mitochondrial single-stranded DNA-binding protein (mtSSB)
mttRNA aa-RS. *See* Mitochondrial aminoacyl-tRNA synthetases (mttRNA aa-RS)
mtZFNs. *See* Mitochondrially targeted zinc finger nucleases (mtZFNs)
Muller's Ratchet process, 97
Multiple Displacement Amplification (MDA). *See* Rolling circle amplification
Multiple myeloma cells (MM cells), 518–522
Multiple sclerosis, 531–532
Multiple tRNA pathogenic mutations, 310–311
Mus musculus, 93
Mus spretus, 93
Muscle creatine kinase (MCK), 492
Muscle homogenate, 367
Mutalyzer nomenclatures, 292
Mutations
  in *C12orf65*, 396–397
  in *COA3* and *C12orf62*, 397
  in *DNA2*, 381
  in *FASTKD2*, 393
  in *GFM1*, 396
  in *GFM2*, 397
  in LRPPRC, 388

in *MGME*1, 381
in mitochondrial aminoacyl-tRNA synthetases, 391–392
in mitoribosomal proteins, 394–395
in *MPV*17, 383
in *MRM*2, 392–393
in MTFMT, 395
in *MTO*1 and *GTPBP*3, 390
in MTPAP, 388
in *POLG*, 379–380
in *PUS*1, 389–390
in RMND1, 395
in RNase P complex, 386–387
in RNase Z, 387
in *RRM2B*, 383
in *TACO*1, 397
in *TFAM*, 385
in *TFB2M*, 385
in *TK*2, 382
in TRIT1, 391
in TRMU, 390–391
in *TRNT*1, 389
in TSFM, 396
in *TWNK*, 380
MutPred, 288–289
MvTool, 292
6MWT. *See* Six minute walking test (6MWT)
Myocerebrohepatopathy spectrum (MCHS), 434
Myoclonic epilepsy myopathy sensory ataxia (MEMSA), 379–380
Myoclonic epilepsy with ragged red fibers (MERRF), 353–355, 361
Mytilus, 90

N

N,N,N′,N′-tetramethyl-p-phenylenediamine (TMPD), 329
N-terminal extensions (NTE), 55–56
NAD, 422
NADH, 323–324
NADPH, 455–456
NADPH oxidase 2 (NOX2), 518–522
Nanga Parbat mystery, 148
NARP. *See* Neuropathy, ataxia, and retinitis pigmentosa (NARP)
Native American mtDNA, 114–115, 116*f*
NCR. *See* Noncoding region (NCR)
*ND*2 mutants, 314–315
*ND*4 mutants, 45
ND6 gene, 97–98
Ndfus$^{4-/-}$knockout mouse model (KO mouse model), 426
nDNA. *See* Nuclear DNA (nDNA)
NDPK. *See* Nucleotide diphosphate kinase (NDPK)
NDPs. *See* Ribonucleoside diphosphates (NDPs)
Neanderthal mtDNA, 119–120
Neomycin, 308–309
NER. *See* Nucleotide excision repair (NER)
Nernst equation, 330–331
NES. *See* Nuclear export signal (NES)
Neurodegenerative disorders, 531–533
Neuroimaging, 366
Neuromuscular disorders, 305–306
Neuropathy, 310–311
Neuropathy, ataxia, and retinitis pigmentosa (NARP), 310–311, 361–362, 482–483
Neuroprotection, 532
Neurospora crassa, 90
Next-generation sequencing (NGS), 42*b*, 134–135, 153, 244, 375–376, 443
software, 297
technology, 463
N–formylmethionine-tRNA^{Met} (fMet-tRNA^{Met}), 54
NGS. *See* Next-generation sequencing (NGS)
NHEJ. *See* Nonhomologous end joining (NHEJ)
Nicotinamide riboside (NR), 422–424
NLSs. *See* Nuclear localization signals (NLSs)
NMPK. *See* Nucleotide monophosphate kinase (NMPK)
NMPKs. *See* Nucleoside monophosphate kinases (NMPKs)
Noncoding region (NCR), 5, 37, 73, 376–379
Nonhomologous end joining (NHEJ), 181, 197, 485, 503–504
machinery, 136, 137*f*
Nonsyndromic sensorineural hearing loss, 362
Nonsynonymous mutations, 97–98
NOX2. *See* NADPH oxidase 2 (NOX2)
NR. *See* Nicotinamide riboside (NR)
NTE. *See* N-terminal extensions (NTE)
Nuclear DNA (nDNA), 87, 150–151, 244, 277–278
methylation, 73
Nuclear encoded mitochondrial genes databases, 295–297
Nuclear export signal (NES), 505
Nuclear genetic disorders of mtDNA gene expression defects of
dNTP salvage pathway and nucleotide metabolism, 382–383
maturation of pre mt-RNA, 386–387
mt-mRNA maturation and turnover, 388
mt-rRNA maturation, modification, and stability, 392–393
mtDNA replication, 379–381
translation elongation, 396
translation initiation, 395
translation termination and mitoribosome recycling, 396–397
translational activation and coupling, 397
IMM insertion of mtDNA-encoded OXPHOS proteins, 397–398

Nuclear genetic disorders of mtDNA gene expression (*Continued*)
maintenance of dNTP pool, 381–382
mitochondrial transcription defects, 384–385
mechanism, 383–384
mitochondrial translation mechanism, 393–394
mt-mRNA maturation and turnover, 387–388
mt-tRNA maturation, 389, 392
mtDNA replication mechanisms, 376–379
mutations
in mitochondrial aminoacyl-tRNA synthetases, 391–392
in mitoribosomal proteins, 394–395
nuclear-encoded mitochondrial disease genes, 378*f*
transcript processing, 385–386
Nuclear localization signals (NLSs), 484–485
Nuclear mitochondrial sequences (NumtS), 131, 255–257
annotation, 140–141
contamination, 259
detection, 132–136
sample-specific NumtS, 134–135
in silico human NumtS detection, 132–134
UCSC NumtS on mitochondrion track, 134*f*
in vitro NumtS identification, 135–136
discovery, 131–132
in mtDNA sequencing and disease, 140
numtogenesis, 136–137
origin and evolution, 138*f*
sequence difference between NumtS and mtDNA counterpart, 139*f*
variability and polymorphisms, 137–139
Nuclear mitochondrial sequences (NUMTs), 465–466
Nuclear-encoded POL γ, 8
Nucleoid proteins, defect in, 421–422
Nucleoids, 16*b*
Nucleos(t)ide supplementation therapies, 427–430
Nucleoside diphosphate kinases (NDPKs), 13
Nucleoside monophosphate kinases (NMPKs), 13
Nucleotide diphosphate kinase (NDPK), 382
Nucleotide excision repair (NER), 174
Nucleotide monophosphate kinase (NMPK), 382
Numtogenesis, 136–137
NumtS. *See* Nuclear mitochondrial sequences (NumtS)
NUMTs. *See* Nuclear mitochondrial sequences (NUMTs)
NZB/BALB, 483
NZB/BALBc embryos, 94

O

OCR. *See* Oxygen consumption rate (OCR)
Off-target sequences in nucleus, 491–492
Oligomycin, 330–331
Oligonucleotides, 252–253
OMIM. *See* Online Mendelian Inheritance in Man (OMIM)
Oncocytes. *See* Oncocytic cells
Oncocytic cells, 456
Online Mendelian Inheritance in Man (OMIM), 199, 290
ONT. *See* Oxford Nanopore Technologies (ONT)
Oocytes, 95*b*
OPA1. *See* Optic atrophy 1 (OPA1)
Optic atrophy 1 (OPA1), 21, 413*t*, 420
Oroboros O2K instrument, 330
Orphanet, 290
143Bρ^0 Osteosarcoma cells, 518–522
"Out of Africa Exit", 118–120
OXA1L, 397–398
machinery, 53
mutations in, 398
Oxford Nanopore Technologies (ONT), 257
Oxidative damage, 227–228, 228*f*
mtDNA susceptible to, 229–231
Oxidative phosphorylation (OXPHOS), 4–5, 35–36, 71, 88, 195, 305–306, 375, 444–446, 448–450, 452, 455–456, 536
function, 367–369, 489
pharmacological impairment of, 462
system, 321–322
8-Oxo-guanosine, 43
8-oxodG. *See* 7,8-Dihydro-8-oxo-2′-deoxyguanosine (8-oxodG)
8-Oxoguanine (8-oxoG), 227–228
OXPHOS. *See* Oxidative phosphorylation (OXPHOS)
Oxygen consumption
classic clark-type electrode methods, 328–330
data analysis of oxygen consumption rate using MitoStress test, 332*b*
high-resolution respirometry, 330
microrespirometry on multiwell plate, 330–331
Oxygen consumption rate (OCR), 330–332

P

PacBio technologies, 257
PALM. *See* Photoactivated localization microscopy (PALM)
PANTHER, 288–289
PAP. *See* Poly(A) polymerase (PAP)
Parallel Analysis of RNA ends (PARE), 42*b*
PARKIN/MUL1-dependent pathway, 91–92

Parkinson's disease, 100–101, 531–532
"Passive dilution" model, 91–93
Paternal leakage during mtDNA inheritance, 92–93
Paternal mitochondrial elimination (PME), 90
Paternal mtDNA
elimination models, 91, 92*f*
inheritance, 90
Pathogenic mtDNA heteroplasmies, 97–98
Patient-specific induced pluripotent stem cells, 309–310
PCR. *See* Polymerase chain reaction (PCR)
Pearson marrow pancreas syndrome, 362–363
Pearson syndrome (PS), 309–310
PEIMAP. *See* Protein experimental interactome MAP (PEIMAP)
Pentatricopeptide repeat domain (PPR domain), 40, 46
PEO. *See* Progressive external ophthalmoplegia (PEO)
PEO/PEO plus. *See* Progressive external ophthalmoplegia/ progressive external ophthalmoplegia plus (PEO/PEO plus)
Peopling of island in Mediterranean Sea, 121–123
Peopling of Sardinia, 121–122
Pep-1. *See* Cell-penetrating peptide (Pep-1)
Peptidyl transferase center (PTC), 48, 392
Peritoneal dialysis, 430
Peroxisome proliferator-activated receptor alpha (PPAR-α), 422
Perrault syndrome, 380, 412
Personal NumtS, 136
PET. *See* Positron emission tomography (PET)
Petite frequency integration 1 (PIF1), 12
PGC1α. *See* Proliferator-activated receptor gamma coactivator 1-alpha (PGC1α)
PGCs. *See* Primordial germ cells (PGCs)
Phage display, 500–502
PHD hydroxylate. *See* Prolyl hydroxylase hydroxylate (PHD hydroxylate)
PhD-SNP, 288–289
Phi29 polymerase, 259
Photoactivated localization microscopy (PALM), 16*b*
Photothermal nanoblades, 527–528
PHP. *See* Point heteroplasmy (PHP)
Phylogeny, 284
Phylogeography approach, 115
PhyloTree, 114, 162, 279, 294
Phylotree human phylogeny, 266
Phylotree-based Haplogrep2 engine, 283
Physarum polycephalum. *See* Slime mold (*Physarum polycephalum*)
PI3K/mTOR inhibitors, 458
PicoGreen, 15–16
PIF1. *See* Petite frequency integration 1 (PIF1)
PILEUP files, 265
PK. *See* Pyruvate kinase (PK)
PKM2 gene, 450
Platelet infusion, 431
Pluripotent stem cells, 411
PME. *See* Paternal mitochondrial elimination (PME)
PNPase. *See* Polynucleotide phosphorylase (PNPase)
PNPT1 gene, 387–388
Point estimation for probabilistic approaches, 160–161
Point heteroplasmy (PHP), 155
Point mutations
clinical syndromes of mitochondrial DNA-related diseases associated with, 353–364, 354*t*
diagnostic approach to mitochondrial DNA-related diseases associated with, 365–369
human mtDNA molecule annotated with confirmed pathogenic, 355*f*
molecular genetics, 364–365
Pol γ. *See* Polymerase γ (Pol γ)
POLG. *See* Polymerase gamma (POLG)
POLG-related disorders, 412
POLG2, 412, 413*t*
PolgA, 318–319, 376–379
PolgA exo$^-$ mtDNA mutator mouse model, 97–98, 101
POLRMT. *See* Mitochondrial RNA polymerase (POLRMT)
Poly-I:C model of schizophrenia, 536–537
Poly(A) polymerase (PAP), 387
Polyadenylation, 47
Polymerase chain reaction (PCR), 111–112, 245, 308, 367–369
amplification, 151–152
in mtDNA world, 113–114
PCR–based methods and mtDNA rearrangements detection, 245–250
PCR RFLP, 245–247
pyrosequencing, 248
quantitative polymerase chain reaction, 248–249
single molecule–based detection techniques, 249–250
Southern blotting and long-range PCR, 247
RFLP, 245–247
Polymerase gamma (POLG), 229, 413*t*, 434
mutations in, 379–380
Polymerase γ (Pol γ), 7–9, 36, 318, 376–379
Polynucleotide phosphorylase (PNPase), 387
Polypeptides, 35–36
Polyphen-2, 288–289
Population databases, 161
Positron emission tomography (PET), 444–446

Post-PCR, 113
PPAR-α. *See* Peroxisome proliferator-activated receptor alpha (PPAR-α)
PPIs. *See* Protein–protein interactions (PPIs)
PPR domain. *See* Pentatricopeptide repeat domain (PPR domain)
PPR domain-containing proteins 1 and 2 (PTCD1 and 2), 46–47
Pre mt-RNA, defects of maturation of, 386–387
  mutations in RNase P complex, 386–387
  mutations in RNase Z, 387
"Pre-L3" mitogenomes, 118–119
Pre-PCR, 113
"Preconditioning of the niche", 452
Primary mitochondrial DNA mutants, 195–200
  mitochondrial DNA point mutants, 199–200
  overlaps between nuclear defects in mtDNA maintenance system and, 210
  pathological mtDNA, 196*f*
  rearrangements, 195–199
Primer removal process, 10–11
Priming, 7–8
Primordial germ cells (PGCs), 95–97, 95*b*, 411
PrimPol (DNA polymerase/primase), 12
Probatio diabolica, 82
Processivity factor. *See* Accessory subunit POL γB
Progressing solid tumors, 452–457
Progressive external ophthalmoplegia (PEO), 12–13, 318, 356
Progressive external ophthalmoplegia/progressive external ophthalmoplegia plus (PEO/PEO plus), 363
Proliferator-activated receptor gamma coactivator 1-alpha (PGC1α), 422
Prolyl hydroxylase hydroxylate (PHD hydroxylate), 450
PRORP gene, 386–387
Protein experimental interactome MAP (PEIMAP), 296–297
Protein structural interactome MAP (PSIMAP), 296–297
Protein synthesis, 50
  mitochondrial, 54–58
Protein-coding genes, 4–5
Protein–protein interactions (PPIs), 296–297
Proteins in mtDNA replication, 12
PS. *See* Pearson syndrome (PS)
Pseudouridylate synthase 1 (PUS1), 389–390
  mutations in, 389–390
Pseudouridylation, 48
PSICQUIC, 295–297
PSIMAP. *See* Protein structural interactome MAP (PSIMAP)
PTC. *See* Peptidyl transferase center (PTC)
PTCD1 and 2. *See* PPR domain-containing proteins 1 and 2 (PTCD1 and 2)
PTCD3 gene, 394–395
PUS1. *See* Pseudouridylate synthase 1 (PUS1)
Pyrosequencing, 245, 248
Pyruvate kinase (PK), 325

Q

Quality filtering, 253
Quantitative polymerase chain reaction (qPCR), 150–151, 248–249, 308
Quantitative real-time PCR
Quasimedian network analysis, 162
Quick-Mitome platform, 292

R

R. FokI. *See* Type IIs endonuclease *Fok*I (R. FokI)
Radiosensitivity, 461–462
Ragged red fibers (RRFs), 202–203, 315–316
"Random drift" of heteroplasmy levels, 88
Rapid screening assay for mtDNA type, 152–153
Rarity of mtDNA profile, 161
Rat liver, 15–16
RBPs. *See* RNA-binding proteins (RBPs)
RCR. *See* Respiratory control ratio (RCR)
rCRS. *See* Revised Cambridge Reference Sequenc (rCRS)
RE. *See* Restriction endonuclease (RE)
Reactive biogenesis, 102
Reactive oxygen species (ROS), 173, 206, 221–223, 311–313, 385, 443, 518–522
  measurement, 335–337
Real-time PCR, mtDNA quantification by, 150–151
Real-time polymerase chain reaction. *See* Quantitative polymerase chain reaction (qPCR)
Real-time quantitative PCR, 245
Recessive *DNA2* variants, 381
Recessive nuclear marker, 308
Recognition code, 500–502
Reconstructed Sapiens Reference Sequence (RSRS), 278
Redox sensitive green fluorescent protein (RoGFP), 335–337
Relaxed replication, 88, 90
Repeat-variable di-residue (RVD), 484–485
Respirasome, 534–536
Respiratory chain biochemistry, 367
Respiratory complexes activity
  immunocapture-based assays, 325–328
  spectrophotometric methods, 322–325
Respiratory control ratio (RCR), 329–330
Restriction endonuclease (RE), 482–483
Restriction fragment length polymorphism (RFLP) studies, 112–113, 245
RETROspective study, 429
Reversible infantile mitochondrial myopathy, 364

Revised Cambridge Reference Sequenc (rCRS), 132–133, 252–253, 278
REXO2. *See* RNA exonuclease 2 (REXO2)
Rh123. *See* Rhodamine 123 (Rh123)
Rho0 cells, 306–309
Rhodamine 123 (Rh123), 332–333
Ribonuclease H1 (RNase H1), 36, 376–379
Ribonuclease P (RNaseP), 385–386
  mutations in RNase P complex, 386–387
  RNase H1, 7–8
  RNase P, 59
Ribonucleic acid (RNA), 3
  degradosome, 49
  footprinting, 42*b*
  modification enzymes, 53–54
  primers, 376–379
Ribonucleoside diphosphates (NDPs), 12–13
Ribonucleotide reductase (RNR), 13, 381–382
Ribonucleotide reductase regulatory TP53 inducible subunit M2B (*RRM2B*), 413*t*, 418–419
  mutations in, 383
2′-*O*-Ribose methyltransferase MRM1, 392–393
2′-*O*-Ribosemethylation, 48
Ribosomal RNAs (rRNAs), 4–5, 376
12S Ribosomal RNAs (12*S*-rRNAs), 40–41, 48, 52–53, 534–536
16S Ribosomal RNAs (16*S*-rRNAs), 48, 534–536
Ribosome profiling, 52*b*
RMND1, mutations in, 395
RNA. *See* Ribonucleic acid (RNA)
RNA exonuclease 2 (REXO2), 49, 387
RNA-binding proteins (RBPs), 45
RNA-seq-based approaches for study of mitochondrial transcriptome, 42*b*
RNase H1. *See* Ribonuclease H1 (RNase H1)
RNase Z, mutations in, 387
RNASEH1, 413*t*, 418
RNaseP. *See* Ribonuclease P (RNaseP)
RNaseP enzyme, 46–47
RNaseZ enzyme, 46–47
RNR. *See* Ribonucleotide reductase (RNR)
Roche/454, 253–255
RoGFP. *See* Redox sensitive green fluorescent protein (RoGFP)
Rolling circle amplification, 259
ROS. *See* Reactive oxygen species (ROS)
Rotenone, 330–331
RRFs. *See* Ragged red fibers (RRFs)
RRM2B. *See* Ribonucleotide reductase regulatory TP53 inducible subunit M2B (*RRM2B*)
16S rRNA mtDNA gene (CAP^R), 516–517
rRNAs. *See* Ribosomal RNAs (rRNAs)
RSRS. *See* Reconstructed Sapiens Reference Sequence (RSRS)
Rules and Guidance for Pharmaceutical Manufacturers 2007 (MHRA), 433–434
RVD. *See* Repeat-variable di-residue (RVD)

S

Saccharomyces cerevisiae, 17–18, 203–204, 310
SAM. *See* Sequence Alignment Map (SAM)
SAM domain and HD domain-containing protein 1 (SAMHD1), 15
SAMHD1. *See* SAM domain and HD domain-containing protein 1 (SAMHD1)
SANDO. *See* Sensory ataxia, neuropathy, dysarthria, and ophthalmoplegia (SANDO)
Sanger sequencing technique, 152, 243–244, 251–252, 252*f*
Sardinian-specific haplogroup (SSH), 115–117, 122
Sardinians, 121–122
SARS2 gene, 391–392
sc ZFN architecture. *See* Single-chain ZFN architecture (sc ZFN architecture)
SCAE. *See* Spinocerebellar ataxia with epilepsy (SCAE)
SCID. *See* Severe combined immune deficiency (SCID)
Scientific Working Group on DNA Methods (SWGDAM), 155, 157–158, 163
SDH. *See* Succinate dehydrogenase (SDH)
Sea Urchin nuclear genomic regions, 131–132
Seahorse XF Analyzer, 330–331
Seckel syndrome, 381
Second-generation mtZFNs, 505
Second-generation sequencing, 253–256
Segregation of mtDNA, general principles, 87–90
Selfish drive, 99–100
Semiautomatic electrophoretic system, 250
Sensorineural hearing loss (SNHL), 356, 362
Sensory ataxia, neuropathy, dysarthria, and ophthalmoplegia (SANDO), 379–380
Separation of synthetized molecules, 11–12
SeqScape program, 153–154
Sequence Alignment Map (SAM), 161–162, 260
Sequence-specific oligonucleotide probes (SSO probes), 146
Sequencer program, 153–154
Sequencing of mtDNA in forensic practice, 149–153
  extraction, 150
  MPS of full mitochondrial genome, 153
  rapid screening assay for mtDNA type, 152–153
  targeted region and PCR amplification, 151–152

Sequencing technology, 278
Severe combined immune deficiency (SCID), 503
Shine/Dalgarno sequences, 49–50
Short tandem repeat system (STR system), 145
"Short-patch" BER (SP-BER), 175
Sideroblastic anemia, B-cell immunodeficiency, fever, and developmental delay (SIFD), 389
SILAC. *See* Stable isotope labeling with amino acids in cell culture (SILAC)
Silver staining dyes, 245–247
Single large-scale deletions (SLSDs), 353–355, 362–363
  clinical syndromes of mitochondrial DNA-related diseases, 353–364, 354*t*
    exercise intolerance, 356
    Kearns-Sayre syndrome, 356
    Leber hereditary optic neuropathy, 357–358
    Leber-dystonia, 357
    MELAS, 359–360
    MERRF, 361
    MIDD, 358
    MILS, 358–359
    MMC, 360
    NARP, 361–362
    nonsyndromic sensorineural hearing loss, 362
    Pearson marrow pancreas syndrome, 362–363
    PEO/PEO plus, 363
    reversible infantile mitochondrial myopathy, 364
  diagnostic approach to mitochondrial DNA-related diseases associated with, 365–369
    electron microscopy, 367
    genetic testing, 367–369
    histochemical staining of skeletal muscle sections, 368*f*
    laboratory tests, 366
    neuroimaging, 366
    respiratory chain biochemistry, 367
    skeletal muscle histochemistry, 367
  management of mitochondrial diseases
    emerging therapies, 369–370
    reproductive options, 370
    supportive therapies, 369
    vitamins and cofactors, 369
  molecular genetics, 364–365
Single molecule–based detection techniques, 249–250
Single-chain ZFN architecture (sc ZFN architecture), 505
Single-molecule PCR (smPCR), 245, 247, 249–250
Single-molecule real-time sequencing technology (SMRT technology), 257
Single-strand annealing (SSA), 181
Single-strand break (SSB), 174
Single-strand conformation polymorphism (SSCP), 250
Single-stranded mtDNA (ssmtDNA), 418
Sirtuin 1 (Sirt1), 422
SiteVar algorithm, 284–285
Six minute walking test (6MWT), 429
Skeletal muscle histochemistry, 367
SLC25A33 gene, 13–15
SLC25A36 gene, 13–15
SLC25A4 gene, 413*t*
Slime mold (*Physarum polycephalum*), 15–16
SLIRP. *See* Stem-loop interacting RNA-binding protein (SLIRP)
SLSDs. *See* Single large-scale deletions (SLSDs)
SmaI/XmaI site, 482–483
smPCR. *See* Single-molecule PCR (smPCR)
SMRT technology. *See* Single-molecule real-time sequencing technology (SMRT technology)
SNHL. *See* Sensorineural hearing loss (SNHL)
SNP&GO, 288–289
SOD2. *See* Superoxide dismutase (SOD2)
Solexa, 253–255
Somatic mtDNA mutations, 100–102, 457
Southern blot genetic techniques, 367–369
Southern blotting PCR, 247
*SOX*17 gene, 95*b*
SP-BER. *See* "Short-patch" BER (SP-BER)
Species-specific NumtS, 132
Specificity of mitoTALEN, 491
Spectrophotometric methods, 322–325
  citrate synthase activity, 325
  cytochrome c oxidoreductase activity
    CIII activity, 324
    CIV activity, 324
  data analysis, 327*b*
  experimental setting of spectrophotometric assays, 326*t*
  hydrolytic activity of ATP synthase, 325
  NADH, 323–324
  sample preparation, 322–323
  succinate, 323–324
  ubiquinol, 324
"SPECTRUM", 422–423
Sperm mtDNA, 92
Spinocerebellar ataxia with epilepsy (SCAE), 379–380
SSA. *See* Single-strand annealing (SSA)
SSB. *See* Single-strand break (SSB)
*SSBP*1 gene, 413*t*, 418
SSCP. *See* Single-strand conformation polymorphism (SSCP)
ssDNA replication intermediates, 6
SSH. *See* Sardinian-specific haplogroup (SSH)
ssmtDNA. *See* Single-stranded mtDNA (ssmtDNA)
SSO probes. *See* Sequence-specific oligonucleotide probes (SSO probes)
Stable isotope labeling with amino acids in cell culture (SILAC), 52*b*

Statistical evaluation, 158–161
Staurosporine (STS), 140–141
STED microscopy. *See* Stimulation emission depletion microscopy (STED microscopy)
Stem cells, 235–236
Stem-loop interacting RNA-binding protein (SLIRP), 388
Stimulation emission depletion microscopy (STED microscopy), 16*b*
STR system. *See* Short tandem repeat system (STR system)
Strand displacement model, 376–379
  of mtDNA replication, 5
STS. *See* Staurosporine (STS)
Succinate, 323–324
Succinate dehydrogenase (SDH), 367
Succinyl CoA ligase (SUCL), 382
*SUCLA*2, 413*t*, 419–420
*SUCLG*1, 413*t*, 419–420
Superoxide anions (O_2^-), 173
Superoxide dismutase (SOD2), 231
SWGDAM. *See* Scientific Working Group on DNA Methods (SWGDAM)
Synonymous mutations, 97–98
Synthetic mtDNA, 76–77

T

T7 RNA polymerase (T7 RNAP), 383–384
*TACO*1, 49–50
  mutations in, 397
TAL. *See* Transcription activator-like (TAL)
TAL effectors, 484–485
TALENs. *See* Transcription activator-like effector nucleases (TALENs)
Taq polymerase, 113
Taqman probes, 150–151
TAS. *See* Termination–associated sequence (TAS)
TCA cycle. *See* Tricarboxylic acid cycle (TCA cycle)
TCGA. *See* The Cancer Genome Atlas (TCGA)
TCR. *See* Transcription coupled repair (TCR)
TEFM. *See* Mitochondrial transcription elongation factor (TEFM)
Temporal vein injection (TV injection), 483
Termination of mtDNA replication, 10–12
  ligation, 11
  primer removal, 10–11
  separation of synthetized
    molecules, 11–12
Termination–associated sequence (TAS), 5, 37–38
Tetramethylrhodamine ethyl ester (TMRE ester), 332–333
Tetramethylrhodamine methyl ester (TMRM), 330
TFAM. *See* Mitochondrial transcription factor A (TFAM)
TFB1M, 41–42
TFB2M. *See* Mitochondrial transcription factor B2 (TFB2M)
TFIIIA. *See* Transcription factor IIIA (TFIIIA)
The Cancer Genome Atlas (TCGA), 466
Thenoyltrifluoroacetone (TTFA), 328
Third generation sequencing techniques (3rd generation sequencing techniques), 244, 257
Thymidine kinase (TK), 308–309
  TK1, 13, 382, 418–419, 428
Thymidine phosphorylase (TP), 15, 418–419
Thymidylate synthase (TS), 381–382
Thymine, 432–433
TIM44, 58
Tissue-specific disorder, 420–421, 434
*TK*2 gene, 413*t*
  mutations in, 382
TMPD. *See* N,N,N′,N′-tetramethyl-p-phenylenediamine (TMPD)
TMRE ester.
  See Tetramethylrhodamine ethyl ester (TMRE ester)
TMRM.
  See Tetramethylrhodamine methyl ester (TMRM)
TNTs. *See* Tunneling nanotubes (TNTs)
TOP3A gene, 11–12
TOP3α. *See* Topoisomerase 3 alpha (TOP3α)
Topoisomerase
  TOP1MT, 12
  TOP1mt, 36
  TOP2β, 12
  TOP3A, 36
  TOP3α, 11–12
Topoisomerase 3 alpha (TOP3α), 376–379
Toxic metabolites, clearance of, 430
TP. *See* Thymidine phosphorylase (TP)
Transcript processing, 385–386
Transcription activator-like (TAL), 481
Transcription activator-like effector nucleases (TALENs), 369–370, 481, 483, 485, 499–500
Transcription coupled repair (TCR), 178–179
Transcription factor IIIA (TFIIIA), 500
Transcriptional regulators, 38
Transcytoplasmic hybrid cells, 305–306
Transfer RNAs (tRNAs), 4–5, 45, 376
  punctuation model, 47, 385–386
  $tRNA^{Ala}$ mutation, 490
  $tRNA^{Leu}$ gene, 40–41, 44
  $tRNA^{Phe}$ gene, 40–41
  $tRNA^{Val}$ gene, 40–41, 51
22 Transfer RNAS, 534–536
Transmitochondrial mouse model, 316–317
Transmitophagy, 524–525
Tricarboxylic acid cycle (TCA cycle), 447–448
TRIT1. *See* tRNA iso-pentenyltransferase (TRIT1)
TRMR-based assay for mitochondrial membrane potential, 334*b*

TRMT10C gene, 386
TRMT61B gene, 48, 392
tRNA 5-methylaminomethyl-2-thiouridylate methyltransferase (TRMU), 390–391
  mutations in, 390–391
tRNA iso-pentenyltransferase (TRIT1), 391
  mutations in, 391
tRNA nucleotidyltransferase 1 enzyme (TRNT1), 48, 389
  mutations in, 389
tRNAs. *See* Transfer RNAs (tRNAs)
TRNT1. *See* tRNA nucleotidyltransferase 1 enzyme (TRNT1)
TS. *See* Thymidylate synthase (TS)
TSFM, mutations in, 396
TTFA. *See* Thenoyltrifluoroacetone (TTFA)
Tunneling nanotubes (TNTs), 516–517, 522–523, 522*f*, 525
TV injection. *See* Temporal vein injection (TV injection)
Twinkle gene (*TWNK*), 8–9, 18, 20–21, 60, 319, 376–379, 412, 413*t*
  mutations in, 380
TWNK. *See* Twinkle gene (*TWNK*)
TYMP gene, 413*t*, 418–419
*Tymp/Upp*1 double knockout mouse model, 427–428
Type IIs endonuclease *Fok*I (R. FokI), 503

U

UBC. *See* Ubiquitin C (UBC)
Ubiquinol, 324
Ubiquinone oxidoreductase activity
  CI activity, 323
  CII activity, 323
Ubiquitin C (UBC), 492
UNG1. *See* Uracil N-glycosylase (UNG1)
Uniparental maternal inheritance of mitochondrial DNA, 90–92
5′-Untranslated region (5′-UTR), 49–50
Uracil, 432–433
Uracil N-glycosylase (UNG1), 177
5′-UTR. *See* 5′-Untranslated region (5′-UTR)

V

Vegetative segregation, 88, 90

W

Warburg effect, 447*f*
Warburg's hypothesis, 444–446, 447*f*
Whole Exome Sequencing (WES), 253–256
Whole Genome Sequencing (WGS), 134–135, 253–256, 375–376
Wild-type genomes, 102
Wild-type mtDNA, 245–247

X

X mtDNAs, 120
Xanthomonas, 484–485
Xenopus laevis, 15–16, 500
Xenopus oocytes, 59

Y

17 Y-STR loci, 148–149
Yarrowia lipolytica, 311
*YARS*2 gene, 391–392
Yeast, 310–311

Z

ZF domain. *See* Zinc finger domain (ZF domain)
ZFNs. *See* Zinc finger nucleases (ZFNs)
ZFP. *See* 2-Zinc finger protein (ZFP)
Zif268, 500–502
Zinc (Zn), 500
Zinc finger domain (ZF domain), 500
Zinc finger nucleases (ZFNs), 369–370, 499–500, 502–504, 506*f*
2-Zinc finger protein (ZFP), 500
  chimeric, 502–504